Combinatorial Games

Traditional game theory has been successful at developing strategy in games of incomplete information: when one player knows something that the other does not. But it has little to say about games of complete information, for example Tic-Tac-Toe, solitaire, and hex. This is the subject of Combinatorial Game Theory. Most board games are a challenge for mathematics: to analyze a position one has to examine the available options, and then the further options available after selecting any option, and so on. This leads to combinatorial chaos, where brute force study is impractical.

In this comprehensive volume, József Beck shows readers how to escape from the combinatorial chaos via the fake probabilistic method, a game-theoretic adaptation of the probabilistic method in combinatorics. Using this, the author is able to determine the exact results about infinite classes of many games, leading to the discovery of some striking new duality principles.

J ó Z S E F B E C K is a Professor in the Mathematics Department of Rutgers University. He has received the Fulkerson Prize for research in Discrete Mathematics and has written around 100 research publications. He is the co-author, with W. L. Chen, of the pioneering monograph *Irregularities of Distribution*.

ENCYCLOPEDIA OF MATHEMATICS AND ITS APPLICATIONS

Combinatorial Games

Tic-Tac-Toe Theory

JÓZSEF BECK

Rutgers University

CAMBRIDGE
UNIVERSITY PRESS

CAMBRIDGE UNIVERSITY PRESS
Cambridge, New York, Melbourne, Madrid, Cape Town, Singapore, São Paulo, Delhi

Cambridge University Press
The Edinburgh Building, Cambridge CB2 8RU, UK

Published in the United States of America by Cambridge University Press, New York

www.cambridge.org
Information on this title: www.cambridge.org/9780521461009

© Cambridge University Press 2008

First published 2008

A catalogue record for this publication is available from the British Library

ISBN 978-0-521-46100-9 hardback

Transferred to digital printing 2009

Dedicated to
my mother who taught me how to play Nine Men's Morris ("Mill")

Contents

Preface

There is an old story about the inventor of Chess, which goes something like this. When the King learned the new game, he quickly fell in love with it, and invited the inventor to his palace. "I love your game," said the King, "and to express my appreciation, I decided to grant your wish." "Oh, thank you, Your Majesty," began the inventor, "I am a humble man with a modest wish: just put one piece of rice on the first little square of the chess board, 2 pieces of rice on the second square, 4 pieces on the third square, 8 pieces on the fourth square, and so on; you double in each step." "Oh, sure," said the King, and immediately called for his servants, who started to bring in rice from the huge storage room of the palace. It didn't take too long, however, to realize that the rice in the palace was not enough; in fact, as the court mathematician pointed out, even the rice produced by the whole world in the last thousand years wouldn't be enough to fulfill the inventor's wish ($2^{64} - 1$ pieces of rice). Then the King became so angry that he gave the order to execute the inventor. This is how the King discovered Combinatorial Chaos.

Of course, there is a less violent way to discover Combinatorial Chaos. Any attempt to analyze unsolved games like Chess, Go, Checkers, grown-up versions of Tic-Tac-Toe, Hex, etc., lead to the same conclusion: we get quickly lost in millions and millions of cases, and feel shipwrecked in the middle of the ocean.

To be fair, the hopelessness of Combinatorial Chaos has a positive side: it keeps the games alive for competition.

Is it really hopeless to escape from Combinatorial Chaos? The reader is surely wondering: "How about Game Theory?" "Can Game Theory help here?" Traditional Game Theory focuses on games of *incomplete information* (like Poker where neither player can see the opponent's cards) and says very little about Combinatorial Games such as Chess, Go, etc. Here the term Combinatorial Game means a 2-player zero-sum game of skill (no chance moves) with *complete information*, and the payoff function has 3 values only: win, draw, and loss.

The "very little" that Traditional Game Theory can say is the following piece of advice: try a backtracking algorithm on the game-tree. Unfortunately, backtracking leads to mindless exponential-time computations and doesn't give any insight; this is better than nothing, but not much. Consequently, computers provide remarkably little help here; for example, we can easily simulate a *random play* on a computer, but it is impossible to simulate an *optimal play* (due to the enormous complexity of the computations). We simply have no data available for these games; no data to extrapolate, no data to search for patterns.

The 3-dimensional $5 \times 5 \times 5$ version of Tic-Tac-Toe, for instance, has about 3^{125} positions (each one of the 5^3 cells has 3 options: either marked by the first player, or marked by the second player, or unmarked), and backtracking on a graph of 3^{125} vertices ("position graph") takes at least 3^{125} steps, which is roughly the third power of the "chaos" the chess-loving King was facing above. No wonder the $5 \times 5 \times 5 = 5^3$ Tic-Tac-Toe is unsolved!

It is even more shocking that we know only two(!) explicit winning strategies in the whole class of $n \times n \times \cdots \times n = n^d$ Tic-Tac-Toe games: the 3^3 version (which has an easy winning strategy) and the 4^3 version (which has an extremely complicated winning strategy).

If traditional Game Theory doesn't help, and the computer doesn't really help either, then what can we do? The objective of this book is exactly to show an escape from Combinatorial Chaos, to win a battle in a hopeless war. This "victory" on the class of Tic-Tac-Toe-like games is demonstrated. Tic-Tac-Toe itself is for children (a very simple game really), but there are many grown-up versions, such as the $4 \times 4 \times 4 = 4^3$ game, and, in general, the $n \times n \times \cdots \times n = n^d$ hypercube versions, which are anything but simple. Besides hypercube Tic-Tac-Toe, we study Clique Games, Arithmetic Progression Games, and many more games motivated by Ramsey Theory. These "Tic-Tac-Toe-like games" form a very interesting sub-class of Combinatorial Games: these are games for which the standard algebraic methods fail to work. The main result of the book is that for some infinite families of natural "Tic-Tac-Toe-like games with (at least) 2-dimensional goals" we know the exact value of the phase transition between "Weak Win" and "Strong Draw." We call these thresholds Clique Achievement Numbers, Lattice Achievement Numbers, and in the Reverse Games, Clique Avoidance Numbers and Lattice Avoidance Numbers. These are game-theoretic analogues of the Ramsey Numbers and Van der Waerden Numbers. Unlike the Ramsey Theory thresholds, which are hopeless in the sense that the best-known upper and lower bounds are very far from each other, here we can find the exact values of the game numbers. For precise statements see Sections 6, 8, 9, and 12.

To prove these exact results we develop a "fake probabilistic method" (we don't do case studies!); the name *Tic-Tac-Toe theory* in the title of the book actually refers to this "fake probabilistic method." The "fake probabilistic method" has two steps: (1) randomization and (2) derandomization. *Randomization* is a game-theoretic adaptation of the so-called Probabilistic Method ("Erdős Theory"); *derandomiza-tion* means to apply potential functions ("resource count"). The Probabilistic Method (usually) gives existence only; the potential technique, on the other hand, supplies explicit strategies. What is more, many of our explicit winning and drawing strategies are very efficient combinatorial algorithms (in fact, the most efficient ones that we know).

The "fake probabilistic method" is not the first theory of Combinatorial Games. There is already a well-known and successful theory: the *addition theory* of "Nim-like compound games." It is an algebraic theory designed to handle complicated games which are, or eventually turn out to be, compounds of several very simple games. "Nim-like compound games" is the subject of the first volume of the remarkable *Winning Ways for your Mathematical Plays* written by *Berlekamp, Conway*, and *Guy* (published in 1982). Volume 1 was called *Theory*, and volume 2 had the more prosaic name of *Case Studies*. As stated by the authors: "there are lots of games for which the theories we have now developed are useful, and even more for which they are not." The family of Tic-Tac-Toe-like games – briefly discussed in Chapter 22 of the *Winning Ways* (vol. 2) – definitely belongs to this latter class. By largely extending Chapter 22, and systematically using the "fake probabilistic method" – which is completely missing(!) from the *Winning Ways* – in this book an attempt is made to upgrade the Case Studies to a Quantitative Theory.

The algebraic and probabilistic approaches represent two entirely different view-points, which apparently complement each other. In contrast to the *local* viewpoint of the addition theory, the "fake probabilistic method" is a *global* theory for games which do *not* decompose into simple sub-games, and remain as single coherent entities throughout play. A given position P is evaluated by a score-system which has some natural probabilistic interpretation such as the "loss probability in the ran-domized game starting from position P." Optimizing the score-system is how we cut short the exhaustive search, and construct efficient ("polynomial time") strategies.

The "fake probabilistic method" works best for large values of the parameters – a consequence of the underlying "laws of large numbers." The "addition theory," on the other hand, works best for little games.

The pioneering papers of the subject are:

1. *Regularity and Positional Games*, by A. W. Hales and R. I. Jewett from 1963;
2. *On a Combinatorial Game*, by P. Erdős and J. Selfridge from 1973;
3. *Biased Positional Games* by V. Chvátal and P. Erdős from 1978; and, as a guiding motivation,
4. the Erdős–Lovász 2-Coloring Theorem from 1975.

The first discovered fundamental connections such as "strategy stealing and Ramsey Theory" and "pairing strategy and Matching Theory", and introduced our basic game class ("positional games"). The last three papers (Erdős with different co-authors) initiated and motivated the "games, randomization, derandomization" viewpoint, the core idea of the book. What is developed here is a far-reaching extension of these ideas – it took 25 years hard labor to work out the details. The majority of the results are published here for the first time.

Being an enthusiastic teacher myself, I tried to write the book in a lecture series format that I would like to use myself in the classroom. Each section is basically an independent lecture; most of them can be covered in the usual 80-minute time frame.

Beside the Theory the book contains dozens of challenging Exercises. The reader is advised to find the solutions to the exercises all by him/herself.

The notation is standard. For example, $c, c_0, c_1, c_2, \ldots$ denote, as usual, positive absolute constants (that I could but do not care to determine); "$a_n = o(1)$" and "$a_n = O(1)$" mean that $a_n \to 0$ and $|a_n| < c$ as $n \to \infty$; and, similarly, "$f(n) = o(g(n))$" and "$f(n) = O(g(n))$" mean that $f(n)/g(n) \to 0$ and $|f(n)/g(n)| < c$ as $n \to \infty$. Also $\log x$, $\log_2 x$, and $\log_3 x$ stand for, respectively, the natural logarithm, the base 2 logarithm, and the base 3 logarithm of x.

There are two informal sections: *A summary of the book in a nutshell* at the beginning, and *An informal introduction to Game Theory* at the end of the book in Appendix D. Both are easy reading; we highly recommend the reader to start the book with these two sections.

Last but not least, I would like to thank the Harold H. Martin Chair at Rutgers University and the National Science Foundation for the research grants supporting my work.

A summary of the book in a nutshell

Mathematics is spectacularly successful at making generalizations: the more than 2000-year old arithmetic and geometry were developed into the monumental fields of calculus, modern algebra, topology, algebraic geometry, and so on. On the other hand, mathematics could say remarkably little about nontraditional complex systems. A good example is the notorious "$3n+1$ problem." If n is even, take $n/2$, if n is odd, take $(3n+1)/2$; show that, starting from an arbitrary positive integer n and applying the two rules repeatedly, eventually we end up with the periodic sequence $1,2,1,2,1,2,\dots$. The problem was raised in the 1930s, and after 70 years of diligent research it is still completely hopeless!

Next consider some games. Tic-Tac-Toe is an easy game, so let's switch to the 3-space. The $3 \times 3 \times 3$ Tic-Tac-Toe is a trivial first player win, the $4 \times 4 \times 4$ Tic-Tac-Toe is a very difficult first player win (computer-assisted proof by O. Patashnik in the late 1970s), and the $5 \times 5 \times 5$ Tic-Tac-Toe is a hopeless open problem (it is conjectured to be a draw game). Note that there is a general recipe to analyze games: perform backtracking on the game-tree (or position graph). For the $5 \times 5 \times 5$ Tic-Tac-Toe this requires about 3^{125} steps, which is totally intractable.

We face the same "combinatorial chaos" with the game of Hex. Hex was invented in the early 1940s by Piet Hein (Denmark), since when it has become very popular, especially among mathematicians. The board is a rhombus of hexagons of size $n \times n$; the two players, White (who starts) and Black, take two pairs of opposite sides of the board. The two players alternately put their pieces on unoccupied hexagons (White has white pieces and Black has black pieces). White (Black) wins if his pieces connect his opposite sides of the board.

In the late 1940s John Nash (*A Beautiful Mind*) proved, by a pioneering application of the Strategy Stealing Argument, that Hex is a first player win. The notorious open problem is to find an *explicit* winning strategy. It remains open for every $n \geq 8$. Note that the standard size of Hex is $n = 11$, which has about 3^{121} different positions.

1

What is common in the $3n+1$ problem, the $5 \times 5 \times 5$ Tic-Tac-Toe, and Hex? They all have extremely simple rules, which unexpectedly lead to chaos: exhibiting unpredictable behavior, without any clear order, without any pattern. These three problems form a good sample, representing a large part (perhaps even the majority) of the applied world problems. Mathematics gave up on these kinds of problems, sending them to the dump called "combinatorial chaos." Is there an escape from the combinatorial chaos?

It is safe to say that understanding/handling combinatorial chaos is one of the main problems of modern mathematics. However, the two game classes (n^d Tic-Tac-Toe and $n \times n$ Hex) represent a bigger challenge, they are even more hopeless, than the $3n+1$ problem. For the $3n+1$ problem we can at least carry out computer experimentation; for example, it is known that the conjecture is true for every $n \le 10^{16}$ (a huge data bank is available): we can search the millions of solved cases for hidden patterns; we can try to extrapolate (which, unfortunately, has not led us anywhere yet).

For the game classes, on the other hand, only a half-dozen cases are solved. Computers do not help: it is easy to simulate a *random play*, but it is impossible to simulate an *optimal play* – this hopelessness leaves the games alive for competition. We simply have no data available; it is impossible to search for patterns if there are no data. (For example, we know only two(!) explicit winning strategies in the whole class of $n \times n \times \cdots \times n = n^d$ Tic-Tac-Toe games: the 3^3 version, which has an easy winning strategy, and the 4^3 version, which has an extremely complicated winning strategy.) These Combinatorial Games represent a humiliating challenge for mathematics!

Note that the subject of Game Theory was created by the Hungarian–American mathematician John von Neumann in a pioneering paper from 1928 and in the well-known book *Theory of Games and Economic Behavior* jointly written with the economist Oscar Morgenstern in 1944. By the way, the main motivation of von Neumann was to understand the role of bluffing in Poker. (von Neumann didn't care, or at least had nothing to say, about combinatorial chaos; the von Neumann–Morgenstern book completely avoids the subject!) Poker is a card game of incomplete information: the game is interesting because neither player knows the opponent's cards. In 1928 von Neumann proved his famous minimax theorem, stating that in games of incomplete information either player has an optimal strategy. This optimal strategy is typically a randomized ("mixed") strategy (to make up for the lack of information).

Traditional Game Theory doesn't say much about games of complete information like Chess, Go, Checkers, and grown-up versions of Tic-Tac-Toe; this is the subject of Combinatorial Game Theory. So far Combinatorial Game Theory has developed in two directions:

(I) the theory of "Nim-like games," which means games that fall apart into simple subgames in the course of a play, and

(II) the theory of "Tic-Tac-Toe-like games," which is about games that do not fall apart, but remain a coherent entity during the course of a play.

Direction (I) is discussed in the first volume of the well-known book *Winning Ways* by Berlekamp, Conway, and Guy from 1982. Direction (II) is discussed in this book.

As I said before, the main challenge of Combinatorial Game Theory is to handle combinatorial chaos. To analyze a position in a game (say, in Chess), it is important to examine the options, and all the options of the options, and all the options of the options of the options, and so on. This explains the exponential nature of the game tree, and any intensive case study is clearly impractical even for very simple games, like the $5 \times 5 \times 5$ Tic-Tac-Toe. There are dozens of similar games, where there is a clearcut natural conjecture about which player has a winning strategy, but the proof is hopelessly out of reach (for example, 5-in-a-row in the plane, the status of "Snaky" in Animal Tic-Tac-Toe, Kaplansky's 4-in-a-line game, Hex in a board of size at least 8×8, and so on, see Section 4).

Direction (I), "Nim-like games," basically avoids the challenge of chaos by restricting itself to games with simple components, where an "addition theory" can work. Direction (II) is a desperate attempt to handle combinatorial chaos.

The first challenge of direction (II) is to pinpoint the reasons why these games are hopeless. Chess, Tic-Tac-Toe and its variants, Hex, and the rest are all "Who-does-it-first?" games (which player gives the first checkmate, who gets the first 3-in-a-row, etc.). "Who-does-it-first?" reflects competition, a key ingredient of game playing, but it is not the most fundamental question. The most fundamental question is "What are the achievable configurations, achievable, but not necessarily first?" and the complementary question "What are the impossible configurations?" Drawing the line between "doable" and "impossible" (doable, but not necessarily first!) is the primary task of direction (II). First we have to clearly understand "what is doable"; "what is doable first" is a secondary question. "Doing-it-first" is the ordinary win concept; it is reasonable, therefore, to call "doing it, but not necessarily first" a Weak Win. If a player fails to achieve a Weak Win, we say the opponent forced (at least) a Strong Draw.

The first idea is to switch from ordinary win to Weak Win; the second idea of direction (II) is to carefully define its subject: "generalized Tic-Tac-Toe." Why "generalized Tic-Tac-Toe"? "Tic-Tac-Toe-like games" are the simplest case in the sense that they are static games. Unlike Chess, Go, and Checkers, where the players repeatedly relocate or even remove pieces from the board ("dynamic games"), in Tic-Tac-Toe and Hex the players make permanent marks on the board, and relocating or removing a mark is illegal. (Chess is particularly complicated. There are 6 types of pieces: King, Queen, Bishop, Knight, Rook, Pawn, and each one has its own set of rules of "how to move the piece." The instructions of playing Tic-Tac-Toe is just a couple of lines, but the "instructions of playing Chess" is several pages long.) The "relative" simplicity of games such as "Tic-Tac-Toe" makes them ideal candidates for a mathematical theory.

What does "generalized Tic-Tac-Toe" mean? Nobody knows what "generalized Chess" or "generalized Go" are supposed to mean, but (almost) everybody would agree on what "generalized Tic-Tac-Toe" should mean. In Tic-Tac-Toe the "board" is a $3 \times 3 = 9$ element set, and there are 8 "winning triplets." Similarly, "generalized Tic-Tac-Toe" can be played on an arbitrary finite hypergraph, where the hyperedges are called "winning sets," the union set is the "board," the players alternately occupy elements of the "board." Ordinary win means that a player can occupy a whole "winning set" first; Weak Win simply means to occupy a whole winning set, but not necessarily first.

How can direction (II) deal with combinatorial chaos? The exhaustive search through the exponentially large game-tree takes an *enormous* amount of time (usually more than the age of the universe). A desperate(!) attempt to make up for the lack of time is to study the *random walk* on the game-tree; that is, to study the *randomized game* where both players play randomly.

The extremely surprising message of direction (II) is that the probabilistic analysis of the randomized game can often be *converted* into optimal Weak Win and Strong Draw strategies via potential arguments. It is basically a game-theoretic adaptation of the so-called Probabilistic Method in Combinatorics ("Erdős Theory"); this is why we refer to it as a "fake probabilistic method."

The fake probabilistic method is considered a mathematical paradox. It is a "paradox" because Game Theory is about *perfect* players, and it is shocking that a play between *random generators* ("dumb players") has anything to do with a play between perfect players! "Poker and randomness" is a natural combination: mixed strategy (i.e. random choice among deterministic strategies) is necessary to make up for the lack of complete information. On the other hand, "Tic-Tac-Toe and randomness" sounds like a mismatch. To explain the connection between "Tic-Tac-Toe" and "randomness" requires a longer analysis.

First note that the connection is not trivial in the sense that an optimal strategy is never a "random play." In fact, a "random play" usually leads to a quick, catastrophic defeat. It is a simple general fact that for games of "complete information" the optimal strategies are always deterministic ("pure"). The fake probabilistic method is employed to *find* an explicit deterministic optimal strategy. This is where the connection is: the fake probabilistic method is *motivated* by traditional Probability Theory, but eventually it is *derandomized* by *potential arguments*. In other words, we eventually get rid of Probability Theory completely, but the intermediate "probabilistic step" is an absolutely crucial, inevitable part of the understanding process.

The fake probabilistic method consists of the following main chapters:

  (i) game-theoretic first moment,
 (ii) game-theoretic second and higher moments,
(iii) game-theoretic independence.

By using the fake probabilistic method, we can find the *exact* solution of infinitely many natural "Ramseyish" games, thought to be completely hopeless before, like some Clique Games, 2-dimensional van der Waerden games, and some "subspace" versions of multi-dimensional Tic-Tac-Toe (the goal sets are at least "2-dimensional").

As said before, nobody knows how to win a "who-does-it-first game." We have much more luck with Weak Win where "doing it first" is ignored. A Weak Win Game, or simply a Weak Game, is played on an arbitrary finite hypergraph, the two players are called Maker and Breaker (alternative names are Builder and Blocker). To achieve an ordinary win a player has to "build and block" at the same time. In a Weak Game these two jobs are separated, which makes the analysis somewhat *easier*, but not *easy*. For example, the notoriously difficult Hex is clearly equivalent to a Weak Game, but it doesn't help to find an explicit first player's winning strategy.

What we have been discussing so far was the achievement version. The Reverse Game (meaning the avoidance version) is equally interesting, or perhaps even more interesting.

The general definition of the *Reverse Weak Game* goes as follows. As usual, it is played on an arbitrary finite hypergraph. One player is a kind of "anti-builder": he wants to avoid occupying a whole winning set – we call him Avoider. The other player is a kind of "anti-blocker": he wants to force the reluctant Avoider to build a winning set – "anti-blocker" is officially called Forcer.

Why "Ramseyish" games? Well, Ramsey Theory gives some partial information about ordinary win. We have a chance, therefore, to compare what we know about ordinary win with that of Weak Win.

The first step in the fake probabilistic method is to describe the majority play, and then, in the second step, try to find a connection between the majority play and the optimal play (the surprising part is that it works!).

The best way to illustrate this is to study the Weak and Reverse Weak versions of the (K_n, K_q) Clique Game: the players alternately take new edges of the complete graph K_n; Maker's goal is to occupy a large clique K_q; Breaker wants to stop Maker. In the Reverse Game, Forcer wants to force the reluctant Avoider to occupy a K_q.

If $q = q(n)$ is "very small" in terms of n, then Maker (or Forcer) can easily win. On the other hand, if $q = q(n)$ is "not so small" in terms of n, then Breaker (or Avoider) can easily win. Where is the game-theoretic breaking point? We call the breaking point the Clique Achievement (Avoidance) Number.

For "small" ns no one knows the answer, but for "large" ns we know the exact value of the breaking point! Indeed, assume that n is sufficiently large like $n \geq 2^{10^{10}}$. If we take the lower integral part

$$q = \lfloor 2\log_2 n - 2\log_2 \log_2 n + 2\log_2 e - 3 \rfloor$$

(base 2 logarithm), then Maker (or Forcer) wins. On the other hand, if we take the upper integral part

$$q = \lceil 2\log_2 n - 2\log_2 \log_2 n + 2\log_2 e - 3 \rceil,$$

then Breaker (or Avoider) wins.

For example, if $n = 2^{10^{10}}$, then

$$2\log_2 n - 2\log_2 \log_2 n + 2\log_2 e - 3 =$$

$$= 2 \cdot 10^{10} - 66.4385 + 2.8854 - 3 = 19,999,999,933.446,$$

and so the largest clique size that Maker can build (Forcer can force Avoider to build) is $19,999,999,933$.

This level of accuracy is even more striking because for smaller values of n we do not know the Clique Achievement Number. For example, if $n = 20$, then it can be either 4 or 5 or 6 (which one?); if $n = 100$, then it can be either 5 or 6 or 7 or 8 or 9 (which one?); if $n = 2^{100}$, then it can be either 99 or 100 or 101 or ... or 188 (which one?), that is there are 90 possible candidates. (Even less is known about the small Avoidance Numbers.) We will (probably!) never know the exact values of these game numbers for $n = 20$, or for $n = 100$, or for $n = 2^{100}$, but we know the exact value for a monster number such as $n = 2^{10^{10}}$. This is truly surprising! This is the complete opposite of the usual induction way of discovering patterns from the small cases (the method of direction (I)).

The explanation for this unusual phenomenon comes from our technique: the fake probabilistic method. Probability Theory is a collections of Laws of Large Numbers. Converting the probabilistic arguments into a potential strategy leads to certain "error terms"; these "error terms" become negligible compared to the "main term" if the board is large.

It is also very surprising that the Weak Clique Game and the *Reverse* Weak Clique Game have *exactly* the same breaking point: Clique Achievement Number = Clique Avoidance Number. This contradicts common sense. We would expect that an eager Maker in the "straight" game has a good chance to build a larger clique than a reluctant Avoider in the Reverse version, but this "natural" expectation turns out to be wrong. We cannot give any *a priori* reason why the two breaking points coincide. All that can be said is that the highly technical proof of the "straight" case (around 30 pages) can be easily adapted (like *maximum* is replaced by *minimum*) to yield the same breaking point for the Reverse Game, but this is hardly the answer that we are looking for.

What is the mysterious expression $2\log_2 n - 2\log_2\log_2 n + 2\log_2 e - 3$? An expert of the theory of Random Graphs immediately recognizes that $2\log_2 n - 2\log_2\log_2 n + 2\log_2 e - 3$ is exactly 2 less than the Clique Number of the symmetric Random Graph $\mathbf{R}(K_n, 1/2)$ ($1/2$ is the edge probability).

A combination of the first and second moment methods (standard Probability Theory) shows that the Clique Number $\omega(\mathbf{R}(K_n, 1/2))$ of the Random Graph has a very strong concentration. Typically it is concentrated on a *single* integer with probability $\to 1$ as $n \to \infty$ (and even in the worst case there are at most two values). Indeed, the expected number of q-cliques in $\mathbf{R}(K_n, 1/2)$ equals

$$f(q) = f_n(q) = \binom{n}{q}2^{-\binom{q}{2}}.$$

The function $f(q)$ drops under 1 around $q \approx 2\log_2 n$. The real solution of the equation $f(q) = 1$ is

$$q = 2\log_2 n - 2\log_2\log_2 n + 2\log_2 e - 1 + o(1), \tag{1}$$

which is exactly 2 more than the game-theoretic breaking point

$$q = 2\log_2 n - 2\log_2\log_2 n + 2\log_2 e - 3 + o(1) \tag{2}$$

mentioned above.

To build a clique K_q of size (1) by Maker (or Avoider in the Reverse Game) on the board K_n is the majority outcome. The majority play outcome differs from the optimal play outcome by a mere additive constant 2.

The strong concentration of the Clique Number of the Random Graph is not that terribly surprising as it seems at first sight. Indeed, $f(q)$ is a very rapidly changing function

$$\frac{f(q)}{f(q+1)} = \frac{q+1}{n-q}2^q = n^{1+o(1)}$$

if $q \approx 2\log_2 n$. On an intuitive level, it is explained by the obvious fact that if q switches to $q+1$, then $\binom{q}{2}$ switches to $\binom{q+1}{2} = \binom{q}{2} + q$, which is a large "square-root size" increase.

Is there a "reasonable" variant of the Clique Game for which the breaking point is exactly (1), i.e. the Clique Number of the Random Graph? The answer is "yes," and the game is a "Picker–Chooser game." To motivate the "Picker–Chooser game," note that the alternating Tic-Tac-Toe-like play splits the board into two equal (or almost equal) parts. But there are many other ways to divide the board into two equal parts. The "I-cut-you'll-choose way" (motivated by how a couple shares a single piece of cake after dinner) goes as follows: in each move, Picker picks two previously unselected points of the board, Chooser chooses one of them, and the other one goes back to Picker. In the *Picker–Chooser* game Picker is the builder (i.e. he wants to occupy a whole winning set) and Chooser is the blocker (i.e. his goal is to mark every winning set).

When Chooser is the builder and Picker is the blocker, we call it the *Chooser–Picker* game.

The proof of the theorem that the "majority clique number" (1) is the exact value of the breaking point for the (K_n, K_q) Picker–Chooser Clique Game (where of course the "points" are the edges of K_n) is based on the concepts of:

(a) game-theoretic first moment; and
(b) game-theoretic second moment.

The proof is far from trivial, but not so terribly difficult either (because Picker has so much control of the game). It is a perfect stepping stone before conquering the much more challenging Weak and Reverse Weak, and also the Chooser–Picker versions. The last three Clique Games all have the *same* breaking point, namely (2). What is (2)?

Well, (2) is the real solution of the equation

$$\binom{n}{q} 2^{-\binom{q}{2}} = f(q) = \frac{\binom{n}{2}}{2\binom{q}{2}}. \tag{3}$$

The intuitive meaning of (3) is that the overwhelming majority of the edges of the random graph are covered by exactly one copy of K_q. In other words, the Random Graph may have a large number of copies of K_q, but they are well-spread (uncrowded); in fact, there is room enough to be typically pairwise edge-disjoint. This suggests the following *intuition*. Assume that we are at a "last stage" of playing a Clique Game where Maker (playing the Weak Game) has a large number of "almost complete" K_qs: "almost complete" in the sense that, (a) in each "almost complete" K_q all but *two edges* are occupied by Maker, (b) all of these edge-pairs are unoccupied yet, and (c) these extremely dangerous K_qs are pairwise edge-disjoint. If (a)–(b)–(c) hold, then Breaker can still escape from losing: he can block these disjoint unoccupied edge-pairs by a simple Pairing Strategy! It is exactly the Pairing Strategy that distinguishes the Picker–Chooser game from the rest of the bunch. Indeed, in each of the Weak, Reverse Weak, and Chooser–Picker games, "blocker" can easily *win* the Disjoint Game (meaning the trivial game where the winning sets are disjoint and contain at least two elements each) by employing a Pairing Strategy. In sharp contrast, in the Picker–Chooser version Chooser always loses a "sufficiently large" Disjoint Game (more precisely, if there are at least 2^n disjoint n-element winning sets, then Picker wins the Picker–Chooser game).

This is the best intuitive explanation that we know to understand breaking point (2). This intuition requests the "Random Graph heuristic," i.e., to (artificially!) introduce a random structure in order to understand a deterministic game of complete information.

But the connection is much deeper than that. To *prove* that (2) is the exact value of the game-theoretic breaking point, one requires a fake probabilistic method. The main steps of the proof are:

(i) game-theoretic first moment,
(ii) game-theoretic higher moments (involving "self-improving potentials"), and
(iii) game-theoretic independence.

Developing (i)–(iii) is a long and difficult task. The word "fake" in the fake probabilistic method refers to the fact that, when an optimal strategy is actually defined, the "probabilistic part" completely disappears. It is a metamorphosis: as a caterpillar turns into a butterfly, the probabilistic arguments are similarly converted into (deterministic) potential arguments.

Note that potential arguments are widely used in puzzles ("one-player games"). A well-known example is Conway's *Solitaire Army* puzzle: arrange men behind a line and then by playing "jump and remove", horizontally or vertically, move a man as far across the line as possible. Conway's beautiful "golden ratio" proof, a striking potential argument, shows that it is *impossible* to send a man forward 5 (4 is possible). Conway's result is from the early 1960s. (It is worthwhile to mention the new result that if "to jump a man diagonally" is permitted, then 5 is replaced by 9; in other words, it is impossible to send a man forward 9, but 8 is possible. The proof is similar, but the details are substantially more complicated.)

It is quite natural to use potential arguments to describe *impossible configurations* (as Conway did). It is more surprising that potential arguments are equally useful to describe *achievable configurations* (i.e. Maker's Weak Win) as well. But the biggest surprise of all is that the Maker's Building Criterions and the Breaker's Blocking Criterions often coincide, yielding *exact* solutions of several seemingly hopeless Ramseyish games. There is, however, a fundamental difference: Conway's argument works for small values such as 5, but the fake probabilistic method gives sharp results only for "large values" of the parameters (we refer to this mysterious phenomenon as a "game-theoretic law of large numbers").

These exact solutions all depend on the concept of "game-theoretic independence" – another striking connection with Probability Theory. What is game-theoretic independence? There is a trivial and a non-trivial interpretation of game-theoretic independence.

The "trivial" (but still very useful) interpretation is about *disjoint* games. Consider a set of hypergraphs with the property that, in each one, Breaker (as the second player) has a strategy to block (mark) every winning set. If the hypergraphs are pairwise disjoint (in the strong sense that the "boards" are disjoint), then, of course, Breaker can block the union hypergraph as well. Disjointness guarantees that in any component either player can play independently from the rest of the components. For example, the concept of the pairing strategy is based on this simple observation.

In the "non-trivial" interpretation, the initial game does *not* fall apart into disjoint components. Instead Breaker can *force* that eventually, in a much later stage of the play, the family of unblocked (yet) hyperedges does fall apart into much smaller (disjoint) components. This is how Breaker can eventually finish the job of blocking the whole initial hypergraph, namely "blocking componentwise" in the "small" components.

A convincing probabilistic intuition behind the non-trivial version is the well-known Local Lemma (or Lovász Local Lemma). The Local Lemma is a remarkable probabilistic sieve argument to prove the *existence* of certain very complicated structures that we are unable to construct directly.

A typical application of the Local Lemma goes as follows:

Erdős–Lovász 2-Coloring Theorem (1975). *Let* $\mathcal{F} = \{A_1, A_2, A_3, \ldots\}$ *be an n-uniform hypergraph. Suppose that each A_i intersects at most 2^{n-3} other $A_j \in \mathcal{F}$ ("local size"). Then there is a 2-coloring of the "board" $V = \bigcup_i A_i$ such that no $A_i \in \mathcal{F}$ is monochromatic.*

The conclusion (almost!) means that there exists a *drawing terminal position* (we have cheated a little bit: in a drawing terminal position, the two color classes have equal size). The very surprising message of the Erdős–Lovász 2-Coloring Theorem is that the "global size" of hypergraph $\mathcal{F}$ is irrelevant (it can even be infinite!), only the "local size" matters.

Of course, the existence of a single (or even several) drawing terminal position does *not* guarantee the existence of a *drawing strategy*. But perhaps it is still true that under the Erdős–Lovász condition (or under some similar but slightly weaker local condition), Breaker (or Avoider, or Picker) has a blocking strategy, i.e. he can block every winning set in the Weak (or Reverse Weak, or Chooser–Picker) game on $\mathcal{F}$. We refer to this "blocking draw" as a Strong Draw.

This is a wonderful problem; we call it the Neighborhood Conjecture. Unfortunately, the conjecture is still open in general, in spite of all efforts trying to prove it during the last 25 years.

We know, however, several partial results, which lead to interesting applications. A very important special case, when the conjecture is "nearly proved," is the class of Almost Disjoint hypergraphs: where any two hyperedges have at most one common point. This is certainly the case for "lines," the winning sets of the n^d Tic-Tac-Toe.

What do we know about the multidimensional n^d Tic-Tac-Toe? We know that it is a draw game even if the dimension d is as large as $d = c_1 n^2 / \log n$, i.e. nearly quadratic in terms of (the winning size) n. What is more, the draw is a Strong Draw: the second player can mark every winning line (if they play till the whole board is occupied). Note that this bound is nearly best possible: if $d > c_2 n^2$, then the second player *cannot* force a Strong Draw.

How is it that for the Clique Game we know the *exact* value of the breaking point, but for the multidimensional Tic-Tac-Toe we could not even find the asymptotic truth (due to the extra factor of $\log n$ in the denominator)? The answer is somewhat technical. The winning lines in the multidimensional n^d Tic-Tac-Toe form an extremely *irregular* hypergraph: the maximum degree is *much* larger than the average degree. This is why one cannot apply the Blocking Criterions directly to the "n^d hypergraph." First we have to employ a Truncation Procedure to bring the maximum degree close to the average degree, and the price that we pay for this degree reduction is the loss of a factor of $\log n$.

However, if we consider the n^d *Torus* Tic-Tac-Toe, then the corresponding hypergraph becomes perfectly uniform (the torus is a *group*). For example, every point of the n^d *Torus* Tic-Tac-Toe has $(3^d - 1)/2$ winning lines passing through it. This uniformity explains why for the n^d *Torus* Tic-Tac-Toe we can prove asymptotically sharp thresholds.

A "winning line" in the n^d Tic-Tac-Toe is a set of n points on a straight line forming an n-term Arithmetic Progression. This motivates the "Arithmetic Progression Game": the board is the interval $1, 2, \ldots, N$, and the goal is to build an n-term Arithmetic Progression. The corresponding hypergraph is "nearly regular"; this is why we can prove asymptotically sharp results.

Let us return to the n^d Torus Tic-Tac-Toe. If the "winning line" is replaced by "winning plane" (or "winning subspace of dimension ≥ 2" in general), then we can go far beyond "asymptotically sharp": we can even determine the *exact* value of the game-theoretic threshold, as in the Clique Game. For example, a "winning plane" is an $n \times n$ lattice in the n^d Torus. This is another rapidly changing 2-dimensional configuration: if n switches to $n+1$, then $n \times n$ switches to $(n+1) \times (n+1)$, which is again a "square-root size" increase just as in the case of the cliques. This formal similarity to the Clique Game (both have "2-dimensional goals") explains why there is a chance to find the *exact* value of the game-theoretic breaking point (the actual proofs are rather different).

It is very difficult to visualize the d-dimensional torus if d is large; here is an easier version: a game with 2-dimensional goal sets played on the plane.

Two-dimensional Arithmetic Progression Game. A natural way to obtain a 2-dimensional arithmetic progression (AP) is to take the Cartesian product. The Cartesian product of two q-term APs with the same gap is a $q \times q$ Aligned Square Lattice.

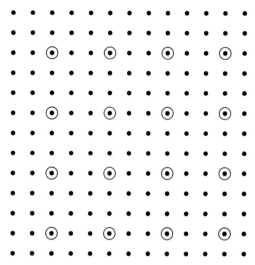

Figure 1 4×4 Aligned Square Lattice on a 13×13 board

Let $(N \times N, q \times q$ Square Lattice) denote the game where the board is the $N \times N$ chessboard, and the winning sets are the $q \times q$ Aligned Square Lattices (see Figure 1 for $N = 13$, $q = 4$, and for a particular 4×4 winning set). Again we know the exact value of the game-theoretic breaking point: if

$$q = \left\lfloor \sqrt{\log_2 N + o(1)} \right\rfloor,$$

then Maker can always build a $q \times q$ Aligned Square Lattice, and this is the best that Maker can achieve. Breaker can always prevent Maker from building a $(q + 1) \times (q + 1)$ Aligned Square Lattice. Again the error term $o(1)$ becomes negligible if N is large. For example, $N = 2^{10^{40} + 10^{20}}$ is large enough, and then

$$\sqrt{\log_2 N} = \sqrt{10^{40} + 10^{20}} = 10^{20} + \frac{1}{2} + O(10^{-20}),$$

so $\sqrt{\log_2 N}$ is not too close to an integer (in fact, it is almost exactly in the middle), which guarantees that $q = 10^{20}$ is the largest Aligned Square Lattice size that Maker can build.

Similarly, $q = 10^{20}$ is the largest Aligned Square Lattice size that Forcer can force Avoider to build.

Here is an interesting detour: consider (say) the biased (2:1) avoidance version where Avoider takes 2 points and Forcer takes 1 point of the $N \times N$ board per move. Then again we know the exact value of the game-theoretic breaking point: if

$$q = \left\lfloor \sqrt{\log_{\frac{3}{2}} N + o(1)} \right\rfloor,$$

then Forcer can always force Avoider to build a $q \times q$ Aligned Square Lattice, and this is the best that Forcer can achieve. Avoider can always avoid building a $(q+1) \times (q+1)$ Aligned Square Lattice. Notice that the base of the logarithm changed from 2 to 3/2.

How about the biased (2:1) achievement version where Maker takes 2 points and Breaker takes 1 point of the $N \times N$ board per move? Then we know the following lower bound: if

$$q = \left\lfloor \sqrt{\log_{\frac{3}{2}} N + 2\log_2 N + o(1)} \right\rfloor,$$

then Maker can always build a $q \times q$ Aligned Square Lattice. We conjecture (but cannot prove) that this is the best that topdog Maker can achieve (i.e. Breaker can always prevent Maker from building a $(q+1) \times (q+1)$ Aligned Square Lattice). Notice that in the biased (2:1) game (eager) Maker can build a substantially larger Aligned Square Lattice than (reluctant) Avoider; the ratio of the corresponding qs is at least as large as

$$\frac{\sqrt{\frac{1}{\log(3/2)} + \frac{2}{\log 2}}}{\sqrt{\frac{1}{\log(3/2)}}} = 1.473.$$

This makes the equality Achievement Number = Avoidance Number in the fair (1:1) games even more surprising.

We can prove the exact formulas only for large board size, such as K_N with $N \geq 2^{10^{10}}$ (Clique Game) and the $N \times N$ grid with $N \geq 2^{10^{40}}$ (Square Lattice Game), but we are convinced that the exact formulas give the truth even for small board sizes like 100 and 1000.

We summarize the meaning of "game-theoretic independence" in the (1:1) game as follows. It is about games such as Tic-Tac-Toe for which the local size is much smaller than the global size. Even if the game starts out as a coherent entity, either player can force it to develop into smaller, local size composites. A sort of intuitive explanation behind it is the Erdős–Lovász 2-Coloring Theorem, which itself is a sophisticated application of statistical independence. Game-theoretic independence is about how to sequentialize statistical independence.

Here we stop the informal discussion, and begin the formal treatment. It is going to be a long journey.

Part A

Weak Win and Strong Draw

Games belong to the oldest experiences of mankind, well before the appearance of any kind of serious mathematics. ("Serious mathematics" is in fact very young: Euclid's *Elements* is less than three-thousand years old.) The playing of games has long been a natural instinct of all humans, and is why the *solving* of games is a natural instinct of mathematicians. Recreational mathematics is a vast collection of all kinds of clever observations ("pre-theorems") about games and puzzles, the perfect empirical background for a mathematical theory. It is well-known that games of chance played an absolutely crucial role in the early development of Probability Theory. Similarly, Graph Theory grew out of puzzles (i.e. 1-player games) such as the famous Königsberg bridge problem, solved by Euler ("Euler trail"), or Hamilton's roundtrip puzzle on the graph of the dodecahedron ("Hamilton cycle problem"). Unlike these two very successful theories, we still do not have a really satisfying quantitative theory of games of pure skill with complete information, or as they are usually called nowadays: Combinatorial Games. Using technical terms, Combinatorial Games are 2-player zero-sum games, mostly finite, with complete information and no chance moves, and the payoff function has three values $\pm 1, 0$ as the first player wins or loses the play, or it ends in a draw.

Combinatorial Game Theory attempts to answer the questions of "who wins," "how to win," and "how long does it take to win." Naturally "win" means "forced win," i.e. a "winning strategy."

Note that Graph Theory and Combinatorial Game Theory face the very same challenge: combinatorial chaos. Given a general graph G, the most natural questions are: what is the chromatic number of G? What is the length of the longest path in G? In particular, does G contain a Hamiltonian path, or a Hamiltonian cycle? What is the size of the largest complete subgraph of G? All that Graph Theory can say is "try out everything," i.e. the brute force approach, which leads to combinatorial chaos.

Similarly, to find a winning strategy in a general game (of complete information) all we can do is backtracking of the enormous game-tree, or position-graph, which also leads to combinatorial chaos.

How do we escape from the combinatorial chaos? In particular, when and how can a player win in a game such as Tic-Tac-Toe? And, of course, what are the "Tic-Tac-Toe like games"? This is the subject of Part A. We start slowly: in Chapter I we discuss many concrete examples and prove a few simple (but important!) theorems. In Chapter II we formulate the main results, and prove a few more simple theorems. The hard proofs come later in Parts C and D.

Chapter I
Win vs. Weak Win

Chess, Tic-Tac-Toe, and Hex are among the most well-known games of complete information with no chance move. What is common in these apparently very different games? In either game the player that wins is the one who achieves a "winning configuration" first. A "winning configuration" in Tic-Tac-Toe is a "3-in-a-row," in Hex it is a "connecting chain of hexagons," and in Chess it is a "capture of the opponent's King" (called a checkmate).

The objective of other well-known games of complete information like Checkers and Go is more complicated. In Checkers the goal is to be the first player either to capture all of the opponent's pieces (checkers) or to build a position where the opponent cannot make a move. The capture of a single piece (jumping over) is a "mini-win configuration," and, similarly, an arrangement where the opponent cannot make a move is a "winning configuration."

In Go the goal is to capture as many stones of the opponent as possible ("capturing" means to "surround a set of opponent's stones by a connected set").

These games are clearly very different, but the basic question is always the same: "Which player can achieve a winning configuration **first**?".

The bad news is that no one knows *how* to achieve a winning configuration first, except by exhaustive case study. There is no general theorem whatsoever answering the question of *how*. The well-known strategy stealing argument gives a partial answer to *when*, but doesn't say a word about *how*. (Note that "doing it first" means competition, a key characteristic of game playing.)

For example, the $4 \times 4 \times 4 = 4^3$ Tic-Tac-Toe is a first player's win, but the winning strategy is extremely complicated: it is the size of a phone-book (computer-assisted task due to O. Patashnik). The $5 \times 5 \times 5 = 5^3$ version is expected to be a draw, but no one can prove it.

In principle, we could search *all* strategies, but it is absurdly long: the total number of strategies is a double exponential function of the board-size. The exhaustive search through all positions (backtracking the "game tree," or the "position graph"), which is officially called Backward Labeling, is more efficient, but still requires exponential time ("hard").

Doing it first is hopeless, but if we ignore "first," then an even more fundamental question arises: "What can a player achieve by his own moves against an adversary?" "Which configurations are achievable (but not necessarily first)?" Or the equivalent complementary question: "What are the impossible configurations?"

To see where our general concepts (to be defined in Section 5) come from, in Sections 1–4 we first inspect some particular games.

1

Illustration: every finite point set in the plane is a Weak Winner

1. Building a congruent copy of a given point set. The first two sections of Chapter I discuss an amusing game. The objective is to demonstrate the power of the potential technique – the basic method of the book – with a simple example. Also it gives us the opportunity for an early introduction to some useful Potential Criterions (see Theorems 1.2–1.4).

To motivate our concrete game, we start with a trivial observation: every 2-coloring of the vertices of an equilateral triangle of side length 1 yields a side where both endpoints have the same color (and have distance 1).

This was trivial, but how about 3 colors instead of 2? The triangle doesn't work, we need a more sophisticated geometric graph: the so-called "7-point Moser-graph" in the plane – which has 11 edges, each of length 1 – will do the job.

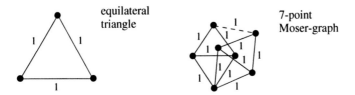

The Moser-graph has the combinatorial property that every 3-coloring of the vertices yields an edge where both endpoints have the same color (and have distance 1).

How about 4 colors instead of 3? Does there exist a geometric graph in the plane such that every edge has length 1, and every 4-coloring of the vertices of this graph yields an edge where both endpoints have the same color? This innocent-looking question was unsolved for more than 50 years, and became a famous problem under the name of *the chromatic number of the plane*. Note that the answer to the question is negative for 7 colors; nothing is known about 4, 5, and 6 colors.

An interesting branch of Ramsey Theory, called Euclidean Ramsey Theory, studies the following more general problem: consider a finite set of points X in some Euclidean space $\mathbf{R}^d$. We would like to decide whether or not for every partition of $X = A_1 \cup A_2 \cup \cdots \cup A_r$ into r subsets, it is always true that some A_i contains a congruent copy of some given point set S (the "goal set"). The partition $X = A_1 \cup A_2 \cup \cdots \cup A_r$ is often called an r-coloring of X, where $A_1, \ldots, A_r$ are the color classes. For example, if X is the 7-point Moser-graph in the plane, the "goal set" S consists of two points a unit distance apart, and $r = 3$. Then the answer to the question above is "yes."

Unfortunately Euclidean Ramsey Theory is very under-developed: it has many interesting conjectures, but hardly any general result (see e.g. Chapter 11 in the *Handbook of Combinatorics*). Here we study a game-theoretic version, and prove a very general result in a surprisingly elementary way.

The game-theoretic version goes as follows: there are two players, called Maker and Breaker, who alternately select new points from some Euclidean space $\mathbf{R}^d$. Maker marks his points red and Breaker marks his points blue. Maker's goal is to build a congruent copy of some given point set S, Breaker's goal is simply to stop Maker (Breaker doesn't want to build).

The board of the game is infinite, in fact uncountable, so how to define the *length* of the game? It is reasonable to assume that the length of the game is $\leq \omega$, where ω denotes, as usual, the first infinite ordinal number. In other words, if Maker cannot win in a finite number of moves, then the players take turns for every natural number, and the play declares that Breaker wins (a draw is impossible). We call this game the "S-building game in $\mathbf{R}^d$."

Example 1: Let the goal set $S = S_3$ be a 3-term arithmetic progression (A.P.) where the gap is 1, and let $d = 1$ (we play on the line). Can Maker win? The answer is an easy "no." Indeed, divide the infinite line into disjoint pairs of points at distance 1 apart – by using this pairing strategy (if a player takes one member of the pair, the opponent takes the other one) Breaker can prevent Maker from building a congruent copy of S_3. This example already convinces us that the 1-dimensional case is not very interesting: Maker can build hardly anything. In sharp contrast, the 2-dimensional case will turn out to be very "rich": Maker can build a congruent copy of *any* given finite plane set, even if he is the underdog!

But before proving this surprisingly general theorem, let us see more concrete examples. Of course, in the plane Maker can easily build a congruent copy of the 3-term A.P. S_3 in 3 moves, and also the 4-term A.P. S_4 (the gap is 1) in 5 moves. Indeed, the trick is to start with an equilateral triangle, which has 3 ways to be extended to a "virgin configuration."

How about the 5-term and 6-term A.P.s S_5 and S_6 (the gap is always 1)?

Illustration 21

S_5: •——•——•——•——• S_6: •——•——•——•——•——•

We challenge the reader to spend some time with these concrete goal sets S_5 and S_6 before reading the proof of the general theorem below.

Example 2: Let the goal set be the 4 vertices of the "unit square" $S = S_4$

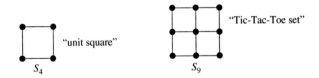

A simple pairing strategy shows that Maker cannot build a "unit square" S_4 on the infinite grid $\mathbb{Z}^2$, but he can easily do it on the whole plain. The trick is to get a trap

We challenge the reader to show that Maker can always build a congruent copy of the "unit square" S_4 in the plane in his 6th move (or before).

Example 3: Let us switch from the 2×2 S_4 to the 3×3 goal set $S = S_9$. We call S_9 the "Tic-Tac-Toe set." Let $d = 2$; can Maker win? If "yes," how long does it take to win?

Example 4: Example 2 was about the "unit square"; how about the regular pentagon $S = S_5$ or the regular hexagon $S = S_6$?

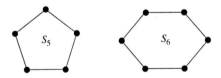

Let $d = 2$; can Maker win? If "yes," how long does it take to win? We challenge the reader to answer these questions before reading the rest of the section.

2. A positive result. The objective is to prove:

Theorem 1.1 *Let S be an arbitrary finite point set in the plane, and consider the following S-building game: 2 players, called Maker and Breaker, alternately pick new points from the plane, each picks 1 point per move; Maker's goal is to build a congruent copy of S in a finite number of moves, and Breaker's goal is to stop Maker.*

   Given an arbitrary finite point set S in the plane, Maker always has a winning strategy in the S-building game in the plane.

Proof. We label the points of S as follows: $S = \{P_0, P_1, \ldots, P_k\}$, i.e. S is a $(k+1)$-element set. We pick P_0 (an arbitrary element of S) and consider the k vectors $v_j = \vec{P_0 P_j}$, $j = 1, 2, \ldots, k$ starting from P_0 and ending at P_j. The k planar vectors $v_1, v_2, \ldots, v_k$ may or may not be linearly independent over the *rationals*.   □

Remark. It is important to emphasize that the field of *rational numbers* is being discussed and not *real numbers*. For example, if S is the regular pentagon in the unit circle with $P_0 = 1$ ("complex plane"), then the 4 vectors v_1, v_2, v_3, v_4 are linearly independent over the rational numbers (because the cyclotomic field $\mathbf{Q}(e^{2\pi i/5})$ with $i = \sqrt{-1}$ a 4-dimensional vector space over $\mathbf{Q}$), but, of course, the same set of 4 vectors are *not* linearly independent over the real numbers (since the dimension of the plane is 2). In other words, the "rational dimension" is 4, but the "real dimension" is 2.

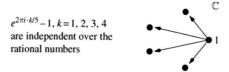

$e^{2\pi i \cdot k/5} - 1, k = 1, 2, 3, 4$
are independent over the
rational numbers

   On the other hand, for the 9-element "Tic-Tac-Toe set" $S = S_9$ (see Example 3), the 8 vectors $v_1, v_2, \ldots, v_8$ have the form $kv_1 + lv_2$, $k \in \{0, 1, 2\}$, $l \in \{0, 1, 2\}$, $k + l \geq 1$, implying that the rational and the real dimensions coincide: either one is 2.

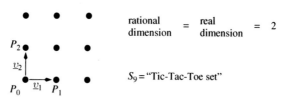

rational dimension = real dimension = 2

S_9 = "Tic-Tac-Toe set"

   For an arbitrary point set S with $|S| = k+1$, let $m = m(S)$ denote the maximum number of vectors among $v_j = \vec{P_0 P_j}$, $j = 1, 2, \ldots, k$, which are linearly independent over the *rational* numbers. Note that m may have any value in the interval $1 \leq m \leq k$.

   For notational convenience assume that v_j, $j = 1, 2, \ldots, m$ are linearly independent over the *rational numbers*; then, of course, the rest can be expressed as

$$v_{m+i} = \sum_{j=1}^{m} \alpha_{m+i}^{(j)} v_j, \quad i = 1, 2, \ldots, k - m, \tag{1.1}$$

where the coefficients $\alpha_{m+i}^{(j)}$ are all rational numbers.

Illustration 23

The basic idea of the proof is to involve a very large number of *rotated* copies of the given goal set S. The lack of rotation makes the 1-dimensional game disappointingly restrictive, and the possibility of using rotation makes the 2-dimensional game rich.

We are going to define a large number of angles

$$0 < \theta_1 < \theta_2 < \theta_3 < \cdots < \theta_i < \cdots < \theta_r < 2\pi, \tag{1.2}$$

where $r = r(S)$ is an integral parameter to be specified later (r will depend on goal set S only).

Let Rot_θ denote the operation "rotated by angle θ"; for example, $Rot_{\theta_i} v_j$ denotes the rotated copy of vector v_j, rotated by angle θ_i. For notational convenience write $v_{j,i} = Rot_{\theta_i} v_j$, including the case $\theta_0 = 0$, that is $v_{j,0} = v_j$.

The existence of the desired angles θ_i in (1.2) is guaranteed by:

Lemma 1: *For every integer $r \geq 1$, we can find r real-valued angles $0 = \theta_0 < \theta_1 < \cdots < \theta_i < \cdots < \theta_r < 2\pi$ such that the $m(r+1)$ vectors $v_{j,i}$ ($1 \leq j \leq m$, $0 \leq i \leq r$) are linearly independent over the rational numbers.*

Proof. We use the well-known fact that the set of rational numbers is countable, but the set of real numbers is uncountable. We proceed by induction on r; we start with $r = 1$. Assume that there exist rational numbers $a_{j,i}$ such that

$$\sum_{i=0}^{1} \sum_{j=1}^{m} a_{j,i} v_{j,i} = 0, \tag{1.3}$$

where in the right-hand side of (1.3) 0 stands for the "zero vector"; then $\sum_{j=1}^{m} a_{j,1} v_{j,1} = -\sum_{j=1}^{m} a_{j,0} v_{j,0}$, or, equivalently, $Rot_{\theta_1} u = w$, where both vectors u and w belong to the set

$$Y = \left\{ \sum_{j=1}^{m} a_j v_j : \text{ every } a_j \text{ is rational and } \sum_{j=1}^{m} a_j^2 \neq 0 \right\}. \tag{1.4}$$

Since set Y defined in (1.4) is countable, there is only a countable set of solutions for the equation $Rot_\theta u = w$, $u \in Y$, $w \in Y$ in variable θ.

Choosing a real number $\theta = \theta_1$ in $0 < \theta_1 < 2\pi$, which is *not* a solution, we conclude that

$$\sum_{i=0}^{1} \sum_{j=1}^{m} a_{j,i} v_{j,i} = 0, \quad \text{where } \sum_{i=0}^{1} \sum_{j=1}^{m} a_{j,i}^2 \neq 0,$$

can never happen, proving Lemma 1 for $r = 1$.

The general case goes very similarly. Assume that we already constructed $r - 1$ (≥ 1) angles $0 = \theta_0 < \theta_1 < \cdots < \theta_i < \cdots < \theta(r-1) < 2\pi$ such that the mr vectors

$v_{j,i}$ ($1 \le j \le m$, $0 \le i \le r-1$) are linearly independent over the rationals. Assume that there exist rational coefficients $a_{j,i}$ such that

$$\sum_{i=0}^{r}\sum_{j=1}^{m} a_{j,i}v_{j,i} = 0, \tag{1.5}$$

where in the right-hand side of (1.5) 0 stands for the "zero vector"; then $\sum_{j=1}^{m} a_{j,r}v_{j,r} = -\sum_{i=0}^{r-1}\sum_{j=1}^{m} a_{j,i}v_{j,i}$, or, equivalently, $Rot_{\theta_1} u = w$, where $u \in Y$ (see (1.4)) and w belongs to the set

$$Z = \left\{ \sum_{i=0}^{r-1}\sum_{j=1}^{m} b_{j,i}v_{j,i} : \text{ every } b_{j,i} \text{ is rational and } \sum_{i=0}^{r-1}\sum_{j=1}^{m} b_{j,i}^2 \neq 0 \right\}.$$

Since both Y and Z are countable, there is only a countable set of solutions for the equation $Rot_\theta u = w$, $u \in Y$, $w \in Z$ in variable θ.

By choosing a real number $\theta = \theta_r$ in $\theta_{r-1} < \theta_r < 2\pi$, which is *not* a solution, Lemma 1 follows. $\square$

The heart of the proof is the following "very regular, lattice-like construction" (a finite point set in the plane)

$$X = X(r; D; N) = \left\{ \sum_{i=0}^{r}\sum_{j=1}^{m} \frac{d_{j,i}}{D}v_{j,i} : \text{ every } d_{j,i} \text{ is an integer with } |d_{j,i}| \le N \right\}, \tag{1.6}$$

where both new integral parameters D and N will be specified later (together with $r = r(S)$).

Notice that the "very regular, lattice-like plane set" (1.6) is the projection of an $m(r+1)$-dimensional $(2N+1) \times \cdots \times (2N+1) = (2N+1)^{m(r+1)}$ hypercube to the plane.

The key property of point set $X = X(r; D; N)$ is that it is very "rich" in congruent copies of goal set S. $\square$

Lemma 2: *Point set X – defined in (1.6) – has the following two properties:*

(a) *the cardinality $|X|$ of set X is exactly $(2N+1)^{m(r+1)}$;*
(b) *set X contains at least $(2N+1-C_0)^{m(r+1)} \cdot (r+1)$ distinct congruent copies of goal set S, where $C_0 = C_0(S)$ is an absolute constant depending only on goal set S, but entirely independent of the parameters (r, D, and N) of the proof.*

Proof. By Lemma 1 the $m(r+1)$ vectors $v_{j,i}$ ($1 \le j \le m$, $0 \le i \le r$) are linearly independent over the rationals. So different vector sums in (1.6) determine different points in the plane, which immediately proves part (a).

To prove part (b), fix an arbitrary integer i_0 in $0 \le i_0 \le r$, and estimate from below the number of translated copies of $Rot_{\theta_{i_0}} S$ in set X (i.e. angle θ_{i_0} is fixed). Select an arbitrary point

Illustration 25

$$Q_0 = \sum_{i=0}^{r} \sum_{j=1}^{m} \frac{d_{j,i}}{D} v_{j,i} \in X, \tag{1.7}$$

and consider the k points $Q_l = Q_0 + v_{l,i_0}$, where $l = 1, 2, \ldots, k$. The $(k+1)$-element set $\{Q_0, Q_1, Q_2, \ldots, Q_k\}$ is certainly a translated copy of $Rot_{\theta_{i_0}} S$, but when can we guarantee that set $\{Q_0, Q_1, Q_2, \ldots, Q_k\}$ is *inside* X? Visualizing plane set X as a "hypercube," the answer becomes very simple: the set $\{Q_0, Q_1, Q_2, \ldots, Q_k\}$ is inside X if point Q_0 is "far from the border of the hypercube." The following elementary calculations make this vague statement more precise.

We divide the k points $Q_1, Q_2, \ldots, Q_k$ into two parts. First, consider an arbitrary Q_l with $1 \le l \le m$: by definition

$$Q_l = \sum_{i=0}^{r} \sum_{j=1}^{m} \frac{d_{j,i} + \delta(l, i_0; j, i) D}{D} v_{j,i}, \tag{1.8}$$

where $\delta(l, i_0; j, i) = 1$ if $(l, i_0) = (j, i)$ and $\delta(l, i_0; j, i) = 0$ if $(l, i_0) \ne (j, i)$.

If $m + 1 \le l \le k$, then by (1.1)

$$v_l = \sum_{j=1}^{m} \alpha_l^{(j)} v_j = \frac{1}{D} \sum_{j=1}^{m} C_l^{(j)} v_j, \tag{1.9}$$

where D is the least common denominator of all rational coefficients $\alpha_l^{(j)}$, and, of course, all $C_l^{(j)}$ are integers. So if $m + 1 \le l \le k$, then by (1.9)

$$Q_l = \sum_{i=0}^{r} \sum_{j=1}^{m} \frac{d_{j,i} + \delta(i_0; i) C_l^{(j)}}{D} v_{j,i}, \tag{1.10}$$

where $\delta(i_0; i) = 1$ if $i_0 = i$ and $\delta(i_0; i) = 0$ if $i_0) \ne i$. Let

$$C^* = \max_{1 \le j \le m} \max_{m+1 \le l \le k} |C_l^{(j)}|, \quad \text{and} \quad C^{**} = \max\{C^*, D\}. \tag{1.11}$$

Now if $|d_{j,i_0}| \le N - C^{**}$ holds for every $j = 1, 2, \ldots, m$ (meaning "the point Q_0 is far from the border of hypercube X"), then by (1.6)–(1.11) the set $\{Q_0, Q_1, Q_2, \ldots, Q_k\}$ is *inside* X. We recall that $\{Q_0, Q_1, Q_2, \ldots, Q_k\}$ is a translated copy of $Rot_{\theta_{i_0}} S$; therefore, if the inequality $|d_{j,i}| \le N - C^{**}$ holds for every $j = 1, 2, \ldots, m$ and $i = 0, 1, \ldots, r$, then by (1.6) the point $Q_0 \in X$ (defined by (1.7)) is contained in at least $(r+1)$ distinct copies of goal set S (namely, in a translated copy of $Rot_{\theta_i} S$ with $i = 0, 1, \ldots, r$).

Let $\#[S \subset X]$ denote the total number of congruent copies of goal set S; the previous argument gives the lower bound

$$\#[S \subset X] \ge (2(N - C^{**}) + 1)^{m(r+1)} \cdot (r+1) = (2N + 1 - C_0)^{m(r+1)} \cdot (r+1), \tag{1.12}$$

where $C_0 = 2C^{**}$, completing the proof of Lemma 2. $\qquad\square$

The fact that "set X is rich in congruent copies of goal set S" is expressed in quantitative terms as follows (see Lemma 2 and (1.12))

$$\frac{\#[S \subset X]}{|X|} \geq \left(1 - \frac{C_0}{2N+1}\right)^{m(r+1)} \cdot (r+1). \tag{1.13}$$

In (1.13) $C_0 = 2C^{**}$ and $m \,(\leq k = |S| - 1)$ are "fixed" (i.e. they depend on goal set S only), but at this stage parameters r and N are completely arbitrary. It is crucial to see that parameters r and N can be specified in such a way that

$$\frac{\#[S \subset X]}{|X|} \geq \frac{r+1}{2} \geq \text{"arbitrarily large."} \tag{1.14}$$

Indeed, for every "arbitrarily large" value of r there is a sufficiently large value of N such that

$$\left(1 - \frac{C_0}{2N+1}\right)^{m(r+1)} \geq \frac{1}{2},$$

and then (1.13) implies (1.14).

We emphasize that inequality (1.14) is the key quantitative property of our point set X (see (1.6)).

After these preparations we are now ready to explain how Maker is able to build a congruent copy of the given goal set S. The reader is probably expecting a quick greedy algorithm, but what we are going to do here is in fact a slow indirect procedure:

(i) Maker will stay strictly inside (the huge!) set X;
(ii) Maker will always choose his next point by optimizing an appropriate *potential function* (we define it below);
(iii) whenever set X is completely exhausted, Maker will end up with a congruent copy of goal set S.

Steps (i)–(iii) describe Maker's indirect building strategy. Of course, Breaker doesn't know about Maker's plan to stay inside the set X (Breaker doesn't know about the set X at all!), and Maker doesn't know in advance whether Breaker's next move will be inside or outside of X – but these are all irrelevant, Maker will own a congruent copy of S anyway.

The main question remains: "What kind of potential function does Maker use?" Maker will use a natural Power-of-Two Scoring System. As far as we know, the first appearance of the Power-of-Two Scoring System is in a short but important paper of Erdős and Selfridge [1973], see Theorem 1.4 below. We will return to the "potential technique" and the Erdős–Selfridge Theorem in Section 10.

3. Potential criterions. It is convenient to introduce the hypergraph $\mathcal{F}$ of "winning sets": a $(k+1)$-element subset $A \subset X$ of set X (defined in (1.6)) is a hyperedge

Illustration 27

$A \in \mathcal{F}$ if and only if A is a congruent copy of goal set S. Hypergraph $\mathcal{F}$ is $(k+1)$-uniform with size $|\mathcal{F}| = \#[S \subset X]$; we refer to X as the "board," and call $\mathcal{F}$ the "family of winning sets." We will apply the following very general hypergraph result; it plays the role of our "Lemma 3," but for later applications we prefer to call it a "theorem."

Theorem 1.2 *Let $(V, \mathcal{F})$ be a finite hypergraph: V is an arbitrary finite set, and $\mathcal{F}$ is an arbitrary family of subsets of V. The Maker–Breaker Game on $(V, \mathcal{F})$ is defined as follows: the two players, called Maker and Breaker, alternately occupy previously unoccupied elements of "board" V (the elements are called "points"); Maker's goal is to occupy a whole "winning set" $A \in \mathcal{F}$, Breaker's goal is to stop Maker. If $\mathcal{F}$ is n-uniform and*

$$\frac{|\mathcal{F}|}{|V|} > 2^{n-3} \cdot \Delta_2(\mathcal{F}),$$

where $\Delta_2(\mathcal{F})$ is the Max Pair-Degree, then Maker, as the first player, has a winning strategy in the Maker–Breaker Game on $(V, \mathcal{F})$.

The Max Pair-Degree is defined as follows: assume that, fixing any 2 distinct points of board V, there are $\leq \Delta_2(\mathcal{F})$ winning sets $A \in \mathcal{F}$ containing both points, and equality occurs for some point pair. Then we call $\Delta_2(\mathcal{F})$ the Max Pair-Degree of $\mathcal{F}$.

In particular, for **Almost Disjoint** *hypergraphs, where any two hyperedges have at most one common point (like a family of "lines"), the condition simplifies to $|\mathcal{F}| > 2^{n-3}|V|$.*

Remark. If $\mathcal{F}$ is n-uniform, then $\frac{|\mathcal{F}|}{|V|}$ is $\frac{1}{n}$ times the *Average Degree*. Indeed, this equality follows from the easy identity $n|\mathcal{F}| = \text{AverDeg}(\mathcal{F})|V|$.

The hypothesis of Theorem 1.2 is a simple Density Condition: in a "dense" hypergraph Maker can always occupy a whole winning set.

First we explain how Theorem 1.2 completes the proof of Theorem 1.1, and discuss the proof of Theorem 1.2 later. Of course, we apply Theorem 1.2 with $V = X$, where X is defined in (1.6), and $\mathcal{F}$ is the $(k+1)$-uniform hypergraph such that a $(k+1)$-element subset $A \subset X$ of set X (see (1.6)) is a hyperedge $A \in \mathcal{F}$ if and only if A is a congruent copy of goal set S. There is, however, an almost trivial formal difficulty in the application of Theorem 1.2 that we have to point out: in Theorem 1.2 Breaker always replies in set X, but in the "S-building game" Breaker may or may not reply in set X (Breaker has the whole plane to choose from). We can easily overcome this formal difficulty by introducing "fake moves": whenever Breaker's move is outside of set X, Maker chooses an arbitrary unoccupied point in X and declares this fake move to be "Breaker's move." If later Breaker actually occupies this fake point (i.e. the "fake move" becomes a "real move"), then Maker chooses another unoccupied point and declares this fake move

"Breaker's move," and so on. By using the simple trick of "fake moves," there is no difficulty whatsoever in applying Theorem 1.2 to the "S-building game in the plane." As we said before, we choose $V = X$ (see (1.6)) and

$$\mathcal{F} = \{A \subset X : \ A \text{ is a congruent copy of } S\}.$$

Clearly $n = |S| = k + 1$, but what is the Max Pair-Degree $\Delta_2(\mathcal{F})$? The exact value is a difficult question, but we don't need that – the following trivial upper bound suffices

$$\Delta_2(\mathcal{F}) \leq \binom{|S|}{2} = \binom{k+1}{2}. \tag{1.15}$$

Theorem 1.2 applies if (see Lemma 2)

$$\frac{|\mathcal{F}|}{|V|} = \frac{\#[S \subset X]}{|X|} \geq \left(1 - \frac{C_0}{2N+1}\right)^{m(r+1)} \cdot (r+1) > 2^{n-3} \cdot \Delta_2(\mathcal{F}). \tag{1.16}$$

(1.16) is satisfied if (see (1.15))

$$\left(1 - \frac{C_0}{2N+1}\right)^{m(r+1)} \cdot (r+1) > 2^{n-3} \cdot \Delta_2(\mathcal{F}) \geq \binom{k+1}{2} 2^{k-2}. \tag{1.17}$$

Here C_0 and $m(\leq k = |S| - 1)$ are fixed (in the sense that they depend only on set S), but parameters r and N are completely free. Now let $r = (k+1)^2 2^{k-2}$, then inequality (1.17) holds if N is sufficiently large. Applying Theorem 1.2 (see (1.16)–(1.17)) the proof of Theorem 1.1 is complete. □

It remains, of course, to prove Theorem 1.2.

Proof of Theorem 1.2. Assume we are in the middle of a play where Maker (the first player) already occupies $x_1, x_2, \ldots, x_i$, and Breaker (the second player) occupies $y_1, y_2, \ldots, y_i$. The question is how to choose Maker's next point x_{i+1}. Those winning sets that contain at least one $y_j (j \leq i)$ are "useless" for Maker; we call them "dead sets." The winning sets which are not "dead" (yet) are called "survivors." The "survivors" have a chance to be completely occupied by Maker. What is the total "winning chance" of the position? We evaluate the given position by the following "opportunity function" (measuring the *opportunity* of winning): $T_i = \sum_{s \in S_i} 2^{n-u_s}$, where u_s is the number of unoccupied points of the "survivor" A_s ($s \in S_i =$ "index-set of the survivors," and index i indicates that we are at the stage of choosing the $(i+1)$st point x_{i+1} of Maker. Note that the "opportunity" can be much greater than 1 (i.e. it is *not* a probability), but it is always non-negative.

A natural choice for x_{i+1} is to maximize the "winning chance" T_{i+1} at the next turn. Let x_{i+1} and y_{i+1} denote the next moves of the 2 players. What is their effect on T_{i+1}? Well, first x_{i+1} doubles the "chances" for each "survivor" $A_s \ni x_{i+1}$, i.e. we have to add the sum $\sum_{s \in S_i : \ x_{i+1} \in A_s} 2^{n-u_s}$ to $\ni T_i$.

Illustration 29

On the other hand, y_{i+1} "kills" all the "survivors" $A_s \ni y_{i+1}$, which means we have to subtract the sum

$$\sum_{s \in S_i:\ y_{i+1} \in A_s} 2^{n-u_s}$$

from T_i.

Warning: we have to make a correction to those "survivors" A_s that contain both x_{i+1} and y_{i+1}. These "survivors" A_s were "doubled" first and "killed" second. So what we have to subtract from T_i is not

$$\sum_{s \in S_i:\ \{x_{i+1}, y_{i+1}\} \subset A_s} 2^{n-u_s}$$

but the twice as large

$$\sum_{s \in S_i:\ \{x_{i+1}, y_{i+1}\} \subset A_s} 2^{n-u_s+1}.$$

It follows that

$$T_{i+1} = T_i + \sum_{s \in S_i:\ x_{i+1} \in A_s} 2^{n-u_s} - \sum_{s \in S_i:\ y_{i+1} \in A_s} 2^{n-u_s} - \sum_{s \in S_i:\ \{x_{i+1}, y_{i+1}\} \subset A_s} 2^{n-u_s}.$$

Now the natural choice for x_{i+1} is the unoccupied z for which $\sum_{s \in S_i:\ z \in A_s} 2^{n-u_s}$ attains its maximum. Then clearly

$$T_{i+1} \geq T_i - \sum_{s \in S_i:\ \{x_{i+1}, y_{i+1}\} \subset A_s} 2^{n-u_s}.$$

We trivally have

$$\sum_{s \in S_i:\ \{x_{i+1}, y_{i+1}\} \subset A_s} 2^{n-u_s} \leq \Delta_2 \cdot 2^{n-2}.$$

Indeed, there are at most Δ_2 winning sets A_s containing the given 2 points $\{x_{i+1}, y_{i+1}\}$, and $2^{n-u_s} \leq 2^{n-2}$, since x_{i+1} and y_{i+1} were definitely unoccupied points at the previous turn.

Therefore

$$T_{i+1} \geq T_i - \Delta_2 2^{n-2}. \tag{1.18}$$

What happens at the end? Let ℓ denote the number of turns, i.e. the ℓth turn is the last one. Clearly $\ell = |V|/2$. Inequality $T_\ell = T_{last} > 0$ means that at the end Breaker could not "kill" (block) all the winning sets. In other words, $T_\ell = T_{last} > 0$ means that Maker was indeed able to occupy a whole winning set.

So all what we have to check is that $T_\ell = T_{last} > 0$. But this is trivial; indeed, $T_{start} = T_0 = |\mathcal{F}|$, so we have

$$T_{last} \geq |\mathcal{F}| - \frac{|V|}{2} \Delta_2 2^{n-2}. \tag{1.19}$$

It follows that, if $|\mathcal{F}| > 2^{n-3}|V|\Delta_2$, then $T_{last} > 0$, implying that at the end of the play Maker was able to completely occupy a winning set. $\qquad \square$

Under the condition of Theorem 1.2, Maker can occupy a whole winning set $A \in \mathcal{F}$, but how long does it take for Maker to do this? The minimum number of moves needed against a perfect opponent is called the Move Number.

A straightforward adaptation of the proof technique of Theorem 1.2 gives a simple but very interesting lower bound on the Move Number.

Move Number. Assume that $\mathcal{F}$ is an n-uniform hypergraph with Max Pair-Degree $\Delta_2(\mathcal{F})$; how long does it take for Maker to occupy a whole $A \in \mathcal{F}$? The following definition is crucial: a hyperedge $A \in \mathcal{F}$ becomes *visible* if Maker has at least 2 marks in it. Note that "2" is critical in the sense that 2 points determine at most $\Delta_2(\mathcal{F})$ hyperedges. We follow the previous proof applied for the *visible sets*. Assume that we are in the middle of a play: $x_1, x_2, \ldots, x_i$ denote the points of Maker and $y_1, y_2, \ldots, y_{i-1}$ denote the points of Breaker up to this stage (we consider the "worse case" where Breaker is the second player). The "danger function" is defined as

$$D_i = \sum_{A \in \mathcal{F}:\ **} 2^{|A \cap \{x_1, x_2, \ldots, x_i\}|}$$

where ** means the double requirement $A \cap \{y_1, y_2, \ldots, y_{i-1}\} = \emptyset$ and $|A \cap \{x_1, x_2, \ldots, x_i\}| \geq 2$; Breaker chooses that new point $y = y_i$ for which the sum

$$D_i(y) = \sum_{y \in A \in \mathcal{F}:\ **} 2^{|A \cap \{x_1, x_2, \ldots, x_i\}|}$$

attains its maximum. What is the effect of the consecutive moves y_i and x_{i+1}? We clearly have

$$D_{i+1} = D_i - D_i(y_i) + D_i(x_{i+1}) - \sum_{\{y_i, x_{i+1}\} \subset A \in \mathcal{F}:\ **} 2^{|A \cap \{x_1, x_2, \ldots, x_i\}|} + \sum_{A \in \mathcal{F}:\ ***} 2^2,$$

where *** means the triple requirement $x_{i+1} \in A$, $|A \cap \{x_1, x_2, \ldots, x_i\}| = 1$, and $A \cap \{y_1, y_2, \ldots, y_i\} = \emptyset$. Since 2 points determine at most $\Delta_2(\mathcal{F})$ hyperedges, we obtain the simple inequality

$$D_{i+1} \leq D_i - D_i(y_i) + D_i(x_{i+1}) + \sum_{A \in \mathcal{F}:\ ***} 2^2 \leq D_i + 4i \cdot \Delta_2(\mathcal{F}).$$

Now assume that Maker can occupy a whole winning set for the first time at his Mth move (M is the Move Number); then

$$2^n \leq D_M \leq D_{M-1} + 4\Delta_2(\mathcal{F})(M-1) \leq D_{M-2} + 4\Delta_2(\mathcal{F})(M-1+M-2) \leq \cdots$$

$$\leq D_1 + 4\Delta_2(\mathcal{F})(1 + 2 + \ldots + (M-1)) = D_1 + 2\Delta_2(\mathcal{F})M(M-1).$$

Notice that $D_1 = 0$ (for at the beginning Maker does not have 2 points yet), so $2^n \leq D_M \leq D_1 + 2\Delta_2(\mathcal{F})M(M-1) = 2\Delta_2(\mathcal{F})M(M-1)$, implying the following: *exponential* lower bound.

Illustration 31

Theorem 1.3 *If the underlying hypergraph is n-uniform and the Max Pair-Degree is* $\Delta_2(\mathcal{F})$, *then playing the Maker–Breaker game on hypergraph* $\mathcal{F}$, *it takes at least*

$$\frac{2^{(n-1)/2}}{\sqrt{\Delta_2(\mathcal{F})}}$$

moves for Maker (the first player) to occupy a whole winning set.

What it means is that there is no quick win in an Almost Disjoint, or nearly Almost Disjoint, hypergraph. Even if the first player has a winning strategy, the second player can postpone it for an exponentially long time! Exponential time practically means "it takes forever."

The potential proof technique of Theorems 1.2–1.3 shows a striking similarity with the proof of the well-known Erdős–Selfridge Theorem. This is not an accident: the Erdős–Selfridge Theorem was the pioneering result, and Theorems 1.2–1.3 were discovered after reading the Erdős–Selfridge paper [1973], see Beck [1981a] and [1981b].

The Erdős–Selfridge Theorem is a draw criterion, and proceeds as follows.

Theorem 1.4 *("Erdős–Selfridge Theorem") Let* $\mathcal{F}$ *be an n-uniform hypergraph, and assume that* $|\mathcal{F}| + MaxDeg(\mathcal{F}) < 2^n$, *where* $MaxDeg(\mathcal{F})$ *denotes the maximum degree of hypergraph* $\mathcal{F}$. *Then playing the Maker–Breaker game on* $\mathcal{F}$, *Breaker (the second player) can put his mark in every* $A \in \mathcal{F}$.

Remark. If the second player can block every winning set, then the first player can also block every winning set (hint: either use the general "strategy stealing" argument, or simply repeat the whole proof); consequently, the "generalized Tic-Tac-Toe game" on $\mathcal{F}$ is a draw.

Exercise 1.1 *Prove the Erdős–Selfridge Theorem.*

We will return to Theorem 1.4 in Section 10, where we include a proof, and show several generalizations and applications.

For many more quick applications of Theorems 1.2–1.4, see Chapter III (Sections 13–15). The reader can jump ahead and read these self-contained applications right now.

What we do in this book is a "hypergraph theory with a game-theoretic motivation." Almost Disjoint hypergraphs play a central role in the theory. The main task is to develop sophisticated "reinforcements" of the simple hypergraph criterions Theorems 1.2–1.4. For example, see Theorem 24.2 (which is an "advanced" building criterion) and Theorems 34.1, 37.5, and 40.1 (the three "ugly" blocking criterions).

2

Analyzing the proof of Theorem 1.1

1. How long does it take to build? Let us return to Example 1: in view of Theorem 1.1, Maker can always build a congruent copy of the 5-term A.P. S_5 in the plane

but how long does it take to build S_5? Analyzing the proof of Theorem 1.1 in this very simple special case, we have

$$k = 4, \quad m = 1, \quad D = 1, \quad C^* = C^{**} = 4, \quad C_0 = 2C^{**} = 8, \quad \Delta_2 = 4,$$

and the key inequality (1.17) is

$$\left(1 - \frac{8}{2N+1}\right)^{r+1} \cdot (r+1) > 2^{4-2} \cdot \Delta_2 = 2^2 \cdot 4. \tag{2.1}$$

By choosing (say) $r = 43$ and $N = 154$, inequality (2.1) holds. It follows that, if Maker restricts himself to set X (see (1.6)), then, when X is exhausted, Maker will certainly own a congruent copy of the 5-term A.P. S_5. Since $|X| = (2N+1)^{m(r+1)}$ (see Lemma 2 (a)), this gives about $309^{44} \approx 10^{110}$ moves even for a very simple goal set like S_5. This bound is rather disappointing!

How about the 6-term A.P. S_6?

The proof of Theorem 1.1 gives

$$k = 5, \quad m = 1, \quad D = 1, \quad C^* = C^{**} = 5, \quad C_0 = 2C^{**} = 10, \quad \Delta_2 = 5,$$

and the key inequality (1.17) is

$$\left(1 - \frac{10}{2N+1}\right)^{r+1} \cdot (r+1) > 2^{5-2} \cdot \Delta_2 = 2^3 \cdot 5. \tag{2.2}$$

By choosing (say) $r = 108$ and $N = 491$, inequality (2.2) holds. Since $|X| = (2N+1)^{m(r+1)}$ (see Lemma 2 (a)), the argument gives about $983^{109} \approx 10^{327}$ moves for goal set S_6.

Next we switch to Example 3, and consider the 3×3 "Tic-Tac-Toe set" S_9; how long does it take to build a congruent copy of S_9?

The proof of Theorem 1.1 gives

$$k = 8, \ m = 2, \ D = 1, \ C^* = C^{**} = 2, \ C_0 = 2C^{**} = 4, \ \Delta_2 = 12,$$

and the key inequality (1.17) is

$$\left(1 - \frac{4}{2N+1}\right)^{2(r+1)} \cdot (r+1) > 2^{8-2} \cdot \Delta_2 = 2^6 \cdot 12,$$

which is satisfied with $r = 2087$ and $N = 10450$, so $|X| = (2N+1)^{m(r+1)} = 20901^{4176} \approx 10^{18,041}$ moves suffice to build a congruent copy of goal set S_9.

Example 4 is about the regular pentagon and the regular hexagon; the corresponding "rational dimensions" turn out to be $m = 4$ and $m = 2$, respectively. Why 4 and 2? The best is to understand the general case: the regular n-gon for arbitrary $n \geq 3$. Consider the regular n-gon in the unit circle of the complex plane where $P_0 = 1$

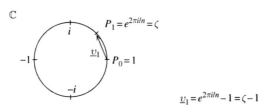

$$P_1 = e^{2\pi i/n} = \zeta$$
$$v_1 = e^{2\pi i/n} - 1 = \zeta - 1$$

By using the notation of the proof of Theorem 1.1, we have $v_j = \zeta^j - 1$, $1 \leq j \leq n-1$, where $\zeta = e^{2\pi i/n}$ (and, of course, $i = \sqrt{-1}$). For later application note that

$$\sum_{j=1}^{n-1} v_j = \sum_{j=1}^{n-1} (\zeta^j - 1) = -1 - (n-1) = -n. \tag{2.3}$$

Note that the algebraic number field $\mathbf{Q}(\zeta) = \mathbf{Q}(e^{2\pi i/n})$ is called the cyclotomic field of the nth roots of unity. It is known from algebraic number theory that $\mathbf{Q}(\zeta)$ is a $\phi(n)$-dimensional vector space over the rationals $\mathbf{Q}$, where $\phi(n)$ is Euler's function: $\phi(n)$ is the number of integers t in $1 \le t \le n$ which are relatively prime to n. We have the elementary product formula

$$\phi(n) = n \prod_{p \mid n:\ p=prime} \left(1 - \frac{1}{p}\right);$$

for example, $\phi(5) = 4$ and $\phi(6) = 2$.

Since $0 = \zeta^n - 1 = (\zeta - 1)(1 + \zeta + \zeta^2 + \ldots + \zeta^{n-1})$, we have $\zeta^{n-1} = -(1 + \zeta + \zeta^2 + \ldots + \zeta^{n-2})$, so the cyclotomic field $\mathbf{Q}(\zeta)$ is generated by $1, \zeta, \zeta^2, \ldots, \zeta^{n-2}$. The $n-1$ generators $1, \zeta, \zeta^2, \ldots, \zeta^{n-2}$ can be expressed in terms of the $n-1$ vectors $v_1, v_2, \ldots, v_{n-1}$; indeed, by (2.3) we have

$$1 = -\frac{v_1 + v_2 + \cdots + v_{n-1}}{n}, \quad \zeta^j = v_j - \frac{v_1 + v_2 + \cdots + v_{n-1}}{n}, \quad 1 \le j \le n-2.$$

Therefore, in the special case of the regular n-gon, the "rational dimension" of vector set $\{v_1, \ldots, v_{n-1}\}$, denoted by m in the proof of Theorem 1.1, is exactly the Euler's function $m = \phi(n)$.

For the regular pentagon (see Example 4) the proof of Theorem 1.1 gives

$$k = 4, \quad m = \phi(5) = 4, \quad D = C^* = C^{**} = 1, \quad C_0 = 2C^{**} = 2, \quad \Delta_2 = 5,$$

and the key inequality (1.17) is

$$\left(1 - \frac{2}{2N+1}\right)^{4(r+1)} \cdot (r+1) > 2^{4-2} \cdot \Delta_2 = 2^2 \cdot 5,$$

which is satisfied with $r = 54$ and $N = 110$, so $|X| = (2N+1)^{m(r+1)} = 221^{220} \approx 10^{516}$ moves suffice to build a congruent copy of a regular pentagon.

Finally consider the regular hexagon: the proof of Theorem 1.1 gives

$$k = 5, \quad m = \phi(6) = 2, \quad D = 1, \quad C^* = C^{**} = 2, \quad C_0 = 2C^{**} = 4, \quad \Delta_2 = 6,$$

and the key inequality (1.17) is

$$\left(1 - \frac{4}{2N+1}\right)^{2(r+1)} \cdot (r+1) > 2^{5-2} \cdot \Delta_2 = 2^3 \cdot 6,$$

which is satisfied with $r = 130$ and $N = 393$, so $|X| = (2N+1)^{m(r+1)} = 787^{262} \approx 10^{759}$ moves suffice to build a congruent copy of a regular hexagon.

The wonderful thing about Theorem 1.1 is that it is strikingly general. Yet there is an obvious handicap: these upper bounds to the Move Number are all ridiculously large. We are convinced that Maker can build each one of the concrete goal sets listed in Examples 1–4 in (say) less than 1000 moves, but do not have the slightest idea how to prove it. The problem is that any kind of brute force case study becomes hopelessly complicated.

By playing on the plane, Maker can build a congruent copy of a given 3-term A.P. in 3 moves, a 4-term A.P. in 5 moves, an equilateral triangle in 3 moves, a square in 6 moves. These are economical, or nearly economical, strategies. How about some more complicated goal sets such as a regular 30-gon or a regular 40-gon ("polygons")? Let $\mathcal{F}$ denote the family of all congruent copies of a given regular 30-gon in the plane. The Max Pair-Degree of hypergraph $\mathcal{F}$ is 2 (a regular polygon uniquely defines a circle, and for the family of all circles of fixed radius the Max Pair-Degree is obviously 2). Applying Theorem 1.3 we obtain that Maker needs at least $2^{15-1} = 2^{14} > 16\,000$ moves to build a given regular 30-gon. For the regular 40-gon the same argument gives at least $2^{20-1} = 2^{19} > 524\,000$ moves. Pretty big numbers!

Note that for an arbitrary n-element goal set S in the plane the corresponding Max Pair-Degree is trivially less than $\binom{n}{2} < n^2/2$ (a non-trivial bound comes from estimating the maximum repetition of the same distance; this is Erdős's famous Unit Distance Problem). Thus Theorem 1.3 gives the general lower bound $\geq \frac{1}{n}2^{n/2}$ for the Move Number. This is exponentially large, meaning that for large goal sets the "building process" is extremely slow; anything but economical!

The trivial upper bound $\leq \binom{n}{2}$ for the Max Pair-Degree (see above) can be improved to $\leq 4n^{4/3}$; this is the current record in the Unit Distance Problem (in Combinatorial Geometry), see Székely's elegant paper [1997]. Replacing the trivial bound $\binom{n}{2}$ with the hard bound $4n^{4/3}$ makes only a slight ("logarithmic") improvement in the given exponential lower bound for the Move Number.

2. Effective vs. ineffective. Let us return to Theorem 1.1. One weakness is the very poor upper bound for the Move Number (such as $\leq 10^{18041}$ moves for the 9-element "Tic-Tac-Toe set" S_9 in Example 3), but an even more fundamental weakness is the lack of an upper bound depending only on $|S|$ (the size of the given goal set S). The appearance of constant C_0 in key inequality (1.17) makes the upper bound "ineffective"!

What does "ineffective" mean here? What is "wrong" with constant C_0? The obvious problem is that a rational number $\alpha = C/D$ is "finite but not bounded", i.e. the numerator C and the denominator D can be arbitrarily large. Indeed, in view of (1.1), the rational coefficients $\alpha_l^{(j)}$ in $v_l = \sum_{j=1}^m \alpha_l^{(j)} v_j$, $l = m+1, m+2, \ldots, k$, may have arbitrarily large common denominator D and arbitrarily large numerators: $\alpha_l^{(j)} = C_l^{(j)}/D$, which implies that

$$C^* = \max_{1 \leq j \leq m} \max_{m+1 \leq l \leq k} |C_l^{(j)}| \quad \text{and} \quad C^{**} = \max\{C^*, D\},$$

are both "finite but not bounded in terms of S." Since $C_0 = 2C^{**}$ and C_0 show up in the key inequality (1.17), the original proof of Theorem 1.1 does not give any hint of how to bound the Move Number in terms of the single parameter $|S|$.

We have to *modify* the proof of Theorem 1.1 to obtain the following *effective* version.

Theorem 2.1 *Consider the "S-building game in the plane" introduced in Theorem 1.1: Maker can always build a congruent copy of any given finite point set S in at most*

$$2^{2^{|S|^2}} \quad \text{moves if } |S| \geq 10.$$

Question: Can we replace the doubly exponential upper bound in Theorem 2.1 by a plain exponential bound? Notice that "plain exponential" is necessary.

For the sake of completeness we include a **proof of Theorem 2.1.** A reader in a rush is advised to skip the technical proof below, and to jump ahead to Theorem 2.2.

As said before, we are going to modify the original proof of Theorem 1.1, but the beginning of the proof remains the same. Consider the $k = |S| - 1$ vectors $v_1, \ldots, v_k$, and again assume that among these k vectors exactly the first m (with some $1 \leq m \leq k$) are linearly independent over the rationals; so $v_1, \ldots, v_m$ are linearly independent over the rationals, and the rest can be written in the form

$$v_l = \sum_{j=1}^{m} \alpha_l^{(j)} v_j, \quad l = m+1, m+2, \ldots, k$$

with rational coefficients

$$\alpha_l^{(j)} = \frac{A_1(j, l)}{B_1(j, l)};$$

here $A_1(j, l)$ and $B_1(j, l)$ are relatively prime integers. Write

$$C_l = \max_{1 \leq j \leq m} \{|A_1(j, l)|, |B_1(j, l)|\}, \quad l = m+1, m+2, \ldots, k. \tag{2.4}$$

For notational convenience we can assume

$$C_{m+1} \geq C_{m+2} \geq \cdots \geq C_k \tag{2.5}$$

(if (2.5) is violated, then simply rearrange the lower indexes!), and take the largest one $M_1 = C_{m+1}$.

Next we basically repeat the previous step with the one larger set $\{v_1, v_2, \ldots, v_m, v_{m+1}\}$ instead of $\{v_1, v_2, \ldots, v_m\}$: for every $l = m+2, m+3, \ldots, k$ consider the solutions of the linear equation

$$v_l = \sum_{j=1}^{m+1} \beta_l^{(j)} v_j, \tag{2.6}$$

with *rational* coefficients

$$\beta_l^{(j)} = \frac{A_2(j, l)}{B_2(j, l)};$$

here $A_2(j, l)$ and $B_2(j, l)$ are relatively prime integers. Since the vector set $\{v_1, v_2, \ldots, v_{m+1}\}$ is not linearly independent any more, system (2.6) may have several solutions; one solution comes from the previous step $\beta_l^{(j)} = \alpha_l^{(j)}$ $(j = 1, \ldots, m)$, $\beta_l^{(m+1)} = 0$ for every $l = m+2, m+3, \ldots, k$. Since (2.6) may have many solutions, we optimize: for every $l = m+2, m+3, \ldots, k$ we choose that particular solution of (2.6) for which

$$C'_l = \max_{1 \le j \le m+1} \{|A_2(j, l)|, |B_2(j, l)|\}$$

attains its *minimum*. For notational convenience we can assume

$$C'_{m+2} \ge C'_{m+3} \ge \cdots \ge C'_k \tag{2.7}$$

(if (2.7) is violated, then simply rearrange the lower indexes!), and take the largest one $M_2 = C'_{m+2}$. Then, of course, $M_1 \ge M_2$.

Again we repeat the previous step with the one longer set $\{v_1, v_2, \ldots, v_{m+1}, v_{m+2}\}$ instead of $\{v_1, v_2, \ldots, v_{m+1}\}$: for every $l = m+3, m+4, \ldots, k$ consider the solutions of the linear equation

$$v_l = \sum_{j=1}^{m+1} \gamma_l^{(j)} v_j \tag{2.8}$$

with *rational* coefficients

$$\gamma_l^{(j)} = \frac{A_3(j, l)}{B_3(j, l)};$$

here $A_3(j, l)$ and $B_3(j, l)$ are relatively prime integers. Since the vector set $\{v_1, \ldots, v_{m+2}\}$ is not linearly independent, system (2.8) may have several solutions; one solution comes from the previous step $\gamma_l^{(j)} = \beta_l^{(j)}$ $(j = 1, \ldots, m+1)$, $\gamma_l^{(m+2)} = 0$ for every $l = m+3, m+4, \ldots, k$. Since (2.8) may have many solutions, we optimize: for every $l = m+3, m+4, \ldots, k$ we choose that particular solution for which

$$C''_l = \max_{1 \le j \le m+2} \{|A_3(j, l)|, |B_3(j, l)|\}$$

attains its *minimum*. For notational convenience, we can assume

$$C''_{m+3} \ge C''_{m+3} \ge \cdots \ge C''_k \tag{2.9}$$

(if (2.9) is violated, then simply rearrange the lower indexes!), and take the largest one $M_3 = C''_{m+3}$. Then, of course, $M_1 \ge M_2 \ge M_3$.

By repeating this argument, we obtain a decreasing sequence

$$M_1 \ge M_2 \ge M_3 \ge \ldots \ge M_{k-m}. \tag{2.10}$$

Notice that sequence (2.10) depends only on the given point set S; the elements of the sequence can be *arbitrarily large* (since a rational number is "finite but unbounded").

We are going to define a decreasing sequence of *constants*

$$\Omega_1 > \Omega_2 > \Omega_3 > \ldots > \Omega_{k-m} \tag{2.11}$$

depending only on the *size* $|S| = k+1$ of S (the explicit form of (2.11) is "ugly," see (2.19) and (2.20) later), and compare the "arbitrary" sequence (2.10) with the "constant" sequence (2.11).

Assume that there is an index $\nu \geq 1$ such that

$$M_\nu > \Omega_\nu \text{ but } M_{\nu+1} \leq \Omega_{\nu+1}. \tag{2.12}$$

We modify our "board set" X (see (1.6)): let

$$\widetilde{X} = \widetilde{X}(r, D; N) = \left\{ \sum_{i=0}^{r} \sum_{j=1}^{m+\nu} \frac{d_{j,i}}{D} v_{j,i} : \text{ every } d_{j,i} \text{ is an integer with } |d_{j,i}| \leq N \right\}; \tag{2.13}$$

here again r, D, and N are unspecified integral parameters.

Notice that (2.13) is the projection of an $(m+\nu)(r+1)$-dimensional $(2N+1) \times \cdots \times (2N+1) = (2N+1)^{(m+\nu)(r+1)}$ hypercube; the value of $\nu(\geq 1)$ is defined by (2.12). The meaning of (2.13) is that, although the "extra" vectors $v_{m+1}, \ldots, v_{m+\nu}$ are not rationally independent of $v_1, \ldots, v_m$, we still handle the $m + \nu$ vectors $v_1, \ldots, v_m, v_{m+1}, \ldots, v_{m+\nu}$ like an independent vector set (because any dependence among them requires rationals with too large numerator/denominator).

We define parameters N and D in (2.13) as follows

$$N = \Omega_\nu \text{ and } D = \prod_{j=1}^{m+\nu} \prod_{l=m+\nu+1}^{k} B_{\nu+1}(j, l), \tag{2.14}$$

i.e. D is the "product of the denominators" showing up in the $(\nu+1)$st step of the iterated procedure above (see (2.5)–(2.10)). By (2.12) and (2.14)

$$D \leq (\Omega_{\nu+1})^{k^2}. \tag{2.15}$$

By repeating the proof of Lemma 2 in Section 1, we get the following "effective" analogue.

Lemma 1: *Point set $\widetilde{X}$ – defined in (2.13) – has the following 2 properties:*

(a) the cardinality $|\widetilde{X}|$ of set $\widetilde{X}$ is exactly $(2N+1)^{(m+\nu)(r+1)}$;
(b) set $\widetilde{X}$ contains at least $(2N+1-\widetilde{C})^{(m+\nu)(r+1)} \cdot (r+1)$ distinct congruent copies of goal set S, where

$$\widetilde{C} = 2(\Omega_{\nu+1})^{k^2}$$

is an "effective" constant.

We apply Theorem 1.2 to the new "board" $V = \widetilde{X}$ (see (2.13)) with the simple trick of "fake moves," and, of course

$$\mathcal{F} = \{A \subset \widetilde{X} : A \text{ is a congruent copy of } S\};$$

clearly $n = |S| = k + 1$.

Theorem 1.2 applies if (see Lemma 1)

$$\frac{|\mathcal{F}|}{|V|} = \frac{\#[S \subset \widetilde{X}]}{|\widetilde{X}|} \geq \left(1 - \frac{\widetilde{C}}{2N+1}\right)^{(m+\nu)(r+1)} \cdot (r+1) > 2^{n-3} \cdot \Delta_2(\mathcal{F}). \tag{2.16}$$

(2.16) is satisfied if

$$\left(1 - \frac{\tilde{C}}{2N+1}\right)^{(m+v)(r+1)} \cdot (r+1) > 2^{n-3} \cdot \Delta_2(\mathcal{F}) \geq \binom{k+1}{2} 2^{k-2}. \tag{2.17}$$

By choosing $r = (k+1)^2 2^k$ and

$$N \geq (k+1)^3 2^k (\Omega_{v+1})^{k^2}, \tag{2.18}$$

inequality (2.17) holds.

Now it is clear how to define the constants in (2.12). We proceed backward; we start with the last one Ω_{k-m}: let

$$\Omega_{k-m} = (k+1)^3 2^k, \tag{2.19}$$

and define the backward recurrence relation

$$\Omega_v = (k+1)^3 2^k (\Omega_{v+1})^{k^2}. \tag{2.20}$$

Formulas (2.19)–(2.20) are clearly motivated by (2.14) and (2.18).

Now we are ready to complete the proof of Theorem 2.1: Theorem 1.2 implies that, staying in $\tilde{X}$ as long as possible, at the end Maker will own a congruent copy of goal set S; this gives the following upper bound for the Move Number (see (2.14))

$$\leq |\tilde{X}| = (2N+1)^{(m+v)(r+1)} = (2\Omega_v + 1)^{(m+v)(r+1)} \leq (2\Omega_v + 1)^{(k+1)^3 2^k}. \tag{2.21}$$

By (2.20) the first (i.e. largest) member of the constant sequence in (2.12) is less than

$$\Omega_1 \leq \left(\left(\left(2^{2k}\right)^{2^2}\right)^{2k^2} \cdots\right)^{2k^2} = 2^{2k \cdot 2k^2 \cdot 2k^2 \cdots 2k^2} \leq 2^{(2k^2)^{k+1}},$$

so, by (2.20), the "Move Number" is less than

$$\left(2^{(2k^2)^{k+1}}\right)^{(k+1)^3 2^k} < 2^{2^{(k+1)^2}} \quad \text{if} \quad |S| = k+1 \geq 10. \tag{2.22}$$

Finally, consider the last case when there is *no* index $v \geq 1$ such that $M_v > \Omega_v$ but $M_{v+1} \leq \Omega_{v+1}$, i.e. when (2.2) fails. Then

$$\Omega_1 \geq M_1, \Omega_2 \geq M_2, \ldots, \Omega_{k-m} \geq M_{k-m},$$

so the original proof of Theorem 1.1 already gives the "effective" upper bound (2.22). This completes the proof of Theorem 2.1. □

3. The biased version. Last question: What happens if Maker is the "underdog"? More precisely, what happens in the "biased" case where Maker takes 1 point per move but Breaker takes several, say, $b \ (\geq 2)$ points per move? For example, let (say) $b = 100$. Can Maker still build a congruent copy of any given finite point set S in the plane? The answer is "yes"; all what we need to do is to replace the

"fair" (1:1) type building criterion Theorem 1.2 with the following "biased" $(p:q)$
version (see Beck [1982]):

Theorem 2.2 ("biased building") *If*

$$\sum_{A \in \mathcal{F}} \left(\frac{p+q}{p} \right)^{-|A|} > p^2 \cdot q^2 \cdot (p+q)^{-3} \cdot \Delta_2(\mathcal{F}) \cdot |V(\mathcal{F})|,$$

*where $\Delta_2(\mathcal{F})$ is the Max Pair-Degree of hypergraph $\mathcal{F}$, and $V(\mathcal{F})$ is the board,
then the first player can occupy a whole winning set $A \in \mathcal{F}$ in the biased $(p:q)$
play on $\mathcal{F}$ (the first player takes p new points and the second player takes q new
points per move).*

Applying Theorem 2.2 instead of Theorem 1.2 we immediately obtain the following
"biased" version of Theorem 2.1.

Theorem 2.3 *Let S be an arbitrary finite set of points in the Euclidean plane, let
$b \geq 1$ be an arbitrary integer, and consider the $(1:b)$ version of the S-building
game where Maker is the underdog: Maker and Breaker alternately pick new points
in the plane, Maker picks one point per move, Breaker picks $b \geq 1$ point(s) per
move; Maker's goal is to build a congruent copy of S in a finite number of moves,
and Breaker's goal is to stop Maker. For every finite S and $b \geq 1$, Maker can build
a congruent copy of S in less than*

$$(b+1)^{(b+1)|S|^2} \quad \text{moves if } |S| \geq 10.$$

Not surprisingly, the proof of the "biased criterion" Theorem 2.2 is very similar
to that of the "fair" Theorem 1.2. Assume we are in the middle of a $(p:q)$ play,
Maker (the first player) already occupies

$$X(i) = \{x_1^{(1)}, \ldots, x_1^{(p)}, x_2^{(1)}, \ldots, x_2^{(p)}, \ldots, x_i^{(1)}, \ldots, x_i^{(p)}\}$$

and Breaker (the second player) occupies

$$Y(i) = \{y_1^{(1)}, \ldots, y_1^{(q)}, y_2^{(1)}, \ldots, y_2^{(q)}, \ldots, y_i^{(1)}, \ldots, y_i^{(q)}\};$$

at this stage of the play the "weight" $w_i(A)$ of an $A \in \mathcal{F}$ is either 0 or an integral
power of $\frac{p+q}{p}$. More precisely, either (1) $w_i(A) = 0$ if $A \cap Y(i) \neq \emptyset$ (meaning that
$A \in \mathcal{F}$ is a "dead set"), or (2)

$$w_i(A) = \left(\frac{p+q}{p} \right)^{|A \cap X(i)|} \quad \text{if } A \cap Y(i) = \emptyset$$

(meaning that $A \in \mathcal{F}$ is a "survivor"). Maker evaluates his chances to win by using
the following "opportunity function"

$$T_i = \sum_{A \in \mathcal{F}} w_i(A). \tag{2.23}$$

We stop here and challenge the reader to complete the proof of Theorem 2.2. Just in case, we include the whole proof at the end of Section 20.

It is very important to see what motivates "opportunity function" (2.23). The motivation is "probabilistic," and goes as follows. Assume we are right after the ith turn of the play with sets $X(i)$ and $Y(i)$ defined above, and consider a Random 2-Coloring of the unoccupied points of board V with odds $(p : q)$, meaning that, we color a point red (Maker's color) with probability $\frac{p}{p+q}$ and color a point blue (Breaker's color) with probability $\frac{q}{p+q}$, the points are colored independently of each other. Then the renormalized "opportunity function" (see (2.23))

$$\left(\frac{p}{p+q}\right)^n T_i = \left(\frac{p}{p+q}\right)^n \sum_{A \in \mathcal{F}} w_i(A)$$

is exactly the *expected number of completely red winning sets* (red means: Maker's own). In other words, a probabilistic argument motivates the choice of our potential function! We refer to this proof technique as a "fake probabilistic method." We will return to the probabilistic motivation later in great detail.

Theorem 2.2 is the biased version of Theorem 1.2; the biased version of Theorem 1.4 will be discussed, with several interesting applications, in Section 20 (see Theorem 20.1); a biased version of Theorem 1.3 will be applied in Section 15 (see Lemma 2 there).

3

Examples: Tic-Tac-Toe games

1. Weak Winners and Winners. The game that we have been studying in Sections 1–2 (the "S-building game in the plane," where S is a given finite point set) was a Maker–Breaker game. One player – called Maker – wanted to build a goal set (namely, a congruent copy of S), and the other player – called Breaker – simply wanted to stop Maker. Tic-Tac-Toe and its variants are very different: they are *not* Maker–Breaker games, they are games where both players want to build, and the player declared the winner is the player who occupies a whole goal set *first*. The main question is "Who does it first?" instead of "Can Maker do it or not?".

The awful truth is that we know almost nothing about "Who does it first?" games. For example, consider the "Who does it first?" version of the S-building game in Section 1 – we restrict ourselves to the fair (1:1) version and assume that Maker is the first player. We know that Maker can always build a congruent copy of the 2×2 goal set S_4

$S_4 = 2 \times 2$ $S_9 = 3 \times 3$

in ≤ 6 moves. What is more, the same case study shows that Maker can always build a congruent copy of S_4 *first* (again in ≤ 6 moves). But how about the 3×3 goal set S_9 in Example 3? Can Maker build S_9 *first*? If "yes," how long does it take to build S_9 *first*?

If Maker can do it, but not necessarily first, we call it a *Weak Winner;* if Maker can do it first, we call it a *Winner*.

So the previous question goes as follows: "Is S_9 a Winner?" Is the regular pentagon a Winner? Is the regular hexagon a Winner? we don't know!

The message of Section 1 is that "every finite point set in the plane is a Weak Winner" – explaining the title of the section. Is it true that every finite point set in the

42

plane is a Winner? The answer is "No." For example, the following irregular pentagon is *not* a Winner; the key property is that angle α is an irrational multiple of π

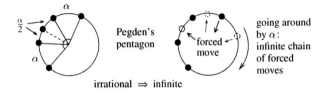

irrational $\Rightarrow$ infinite

This example is due to Wesley Pegden (Ph.D. student at Rutgers). Pegden's pentagon consists of 4 consecutive points with common gap α plus the point in the middle. The idea of the proof is that the second player can always threaten the first player with an *infinite* sequence of forced moves, namely with irrational rotations of angle α along a fixed circle, before the first player can complete his own copy of the goal pentagon. More precisely, at an early stage of the play, the second player can achieve that on some circle he owns 3 points from some consecutive 4 with gap α, and he can take the 4th point in his next move. Then of course the first player is forced to take the middle point. After that the second player takes the 5th point with gap α, which again forces the first player to take the middle point of the last four; after that the second player takes the 6th point with gap α, which again forces the first player to take the middle point of the last four; and so on. We challenge the reader to clarify this intuition, and to give a precise proof that Pegden's irrational pentagon is not a Winner. The proof requires some case study(!) which we skip here, see Pegden [2005].

The underlying idea of Pegden's irrational pentagon construction is illustrated on the following oversimplified "abstract" hypergraph game. Consider a binary tree of 3 levels; the players take vertices

the "winning sets" are the 4 full-length branches (3-sets) of the binary tree. This is a simple first player win game; however, adding infinitely many disjoint 2-element "extra" winning sets to the hypergraph enables the second player to postpone his inevitable loss by infinitely many moves!

infinitely many
pairs

In other words, by adding the extra 2-sets, the first player cannot force a finite win anymore.

Pegden's clever construction is one example (a pentagon); are there infinitely many examples of "Weak Winner≠Winner" with arbitrarily large size (number of points)? In general, we can ask:

Open Problem 3.1 *Is there a finite procedure to decide whether or not a given finite point set S in the plane is a Winner? In other words, is there a way to characterize those finite point sets S in the plane for which Maker, as the first player, can always build a congruent copy of S in the plane* **first** *(i.e. before the opponent could complete his own copy of S)?*

We think Open Problem 3.1 is totally hopeless even for medium size point sets S, especially that "building is exponentially slow." What we are referring to here is Theorem 1.3, which gives the exponential lower bound $\geq \frac{1}{n} 2^{n/2}$ for the Move Number of an arbitrary S with $|S| = n$.

2. Tic-Tac-Toe games on the plane. The S-building game was an artificial example, constructed mainly for illustration purposes. It is time now to talk about a "real" game: Tic-Tac-Toe and its closest variants. We begin with Tic-Tac-Toe itself, arguably the simplest, oldest, and most popular board game in the world.

Why is Tic-Tac-Toe so popular? Well, there are many reasons, but we think the best explanation is that it is the simplest example to demonstrate the difference between "Weak Win" and ordinary "win"! This will be explained below.

Let us emphasize again: we can say very little about ordinary win in general, nothing other than exhaustive search; the main subject of the book is "Weak Win."

Now let's return to Tic-Tac-Toe. The rules of Tic-Tac-Toe (called Noughts-and-Crosses in the UK) are well-known; we quote Dudeney: "Every child knows how to play this game; whichever player first gets three in a line, wins with the exulting cry:

> *Tit, tat, toe, My last go;*
> *Three jolly butcher boys All in a row.*"

The **Tic-Tac-Toe** board is a big square which is partitioned into $3 \times 3 = 9$ congruent small squares. The first player starts by putting an X in one of the 9 small squares, the second player puts an O into any other small square, and then they alternate X and O in the remaining empty squares until one player wins by getting three of his own squares in a line (horizontally, vertically, or diagonally). If neither player gets three in a line, the play ends in a draw. Figure 3.1 below shows the board and the eight winning triplets of Tic-Tac-Toe.

Figure 3.2 below shows a "typical" play: $X_1, \ldots, X_5$ denote the moves of the first player, and $O_1, \ldots, O_4$ denote the moves of the second player in this order. This particular play ends in a draw.

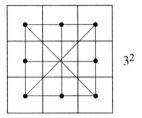

3^2

Figure 3.1

$$O_3 \quad X_2 \quad O_1$$
$$X_4 \quad X_1 \quad O_4$$
$$X_5 \quad O_2 \quad X_3$$

Figure 3.2

X_2	O_1	
X_3	X_1	?
?		O_2

Figure 3.3

Figure 3.3 above shows another play in which the second player's opening move O_1 was a "mistake": the first player gets a "winning trap" and wins in his 4th move.

Every child "knows" that Tic-Tac-Toe is a *draw game*, i.e. either player can force a draw. We mathematicians have higher standards: we want a *proof*. Unfortunately, there is nothing to be proud of about the proof that we are going to give below. The proof is an ugly case study (anyone knows an elegant proof?).

The first half of the statement is easy though: the first player can always force a draw by an easy pairing strategy. Indeed, the first player opens with the center, which blocks 4 winning lines, and the remaining 4 lines are easily blocked by the following pairing

$$\begin{bmatrix} | & - & - \\ | & X & | \\ - & - & | \end{bmatrix}$$

This means the first player can always put his mark in every winning set, no matter what the opponent is doing. By contrast, the second player *cannot* put his mark in every winning set. In other words, the second player cannot prevent the first player from occupying a winning set. The reader is probably wondering, "Wait a minute, this seems to contradict the fact that Tic-Tac-Toe is a draw!" But, of course, there is no real contradiction here: the first player can occupy a whole winning set, but he cannot occupy it first if the opponent plays rationally. Indeed, let the first player's opening move be the center

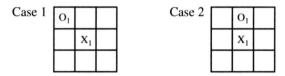

Then the second player has two options: either he takes a corner, or a side. In either case the first player occupies X_2, X_3, X_4, and completes a winning triplet. Of course, this way the opponent's winning triplet (O_1, O_2, O_3 if the second player plays rationally) comes first; notice that here we *changed the rule* and assumed that the players do *not* quit playing even after some winning set is completely occupied by either player, they just keep playing till the whole board is completed. We refer to this as the Full Play Convention.

Case 1

O_1		
	X_1	
X_2	X_3	X_4

Case 2

O_1		
	X_1	
X_2	X_4	X_3

Occupying a whole winning set, but not necessarily first, is what we call a *Weak Win*. We have just learned that in Tic-Tac-Toe the first player can achieve a Weak Win (assuming the Full Play Convention!).

The complement of Weak Win is called a *Strong Draw*. Tic-Tac-Toe is a draw game (we prove this fact below) but not a Strong Draw. We sometimes refer to this property – draw but not a Strong Draw – as a "Delicate Draw."

Tic-Tac-Toe is a "3-in-a-row" game on a 3×3 board. A straightforward 2-dimensional generalization is the "n-in-a-row" game on an $n \times n$ board; we call it the $n \times n$ Tic-Tac-Toe, or simply the n^2 game. The n^2 game has $2n + 2$ winning sets: n horizontals, n verticals, and 2 diagonals, each one of size n. The n^2 games are rather dull: the 2^2 game is a trivial first player win, and the rest of the n^2 games ($n \geq 3$) are all simple draw games – see Theorems 3.1–3.3 below. We begin with:

Theorem 3.1 *Ordinary 3^2 Tic-Tac-Toe is a draw but not a Strong Draw.*

The only **proof** that we know is a not too long but still unpleasant(!) case study. However, it seems ridiculous to write a whole book about games such as Tic-Tac-Toe, and not to solve Tic-Tac-Toe itself. To emphasize: by including this case study an exception was made; the book is *not* about case studies.

We have already proved that the first player can force a draw in the 3^2 game; in fact, by a pairing strategy. It remains to give the second player's drawing strategy. Of course, we are not interested in how poorly the second player can play; all that we care about is the second player's optimal play. Therefore, when we describe a second player's drawing strategy, a substantial reduction in the size of the case study comes from the following two assumptions:

(i) each player completes a winning triplet if he/she can;
(ii) each player prevents the opponent from doing so in his/her next move.

It is either player's best interest to follow rules (i) and (ii). A second source of reduction comes from using the *symmetries* of the board. We label the 9 little squares in the following natural way

1	2	3
4	5	6
7	8	9

The center ("5") is the "strongest point": it is the only cell contained by 4 winning lines. The second player's drawing strategy has 3 parts according, as the opponent's opening move is in the center (threatening four winning triplets), or in a corner (threatening 3 winning triplets), or on a side (threatening two winning triplets). If the first player starts in the center, then the second player's best response is in the corner, say in 1. The second player can force a draw in the $5 \rightarrow 1$ end game. This part of second player's drawing strategy is the following

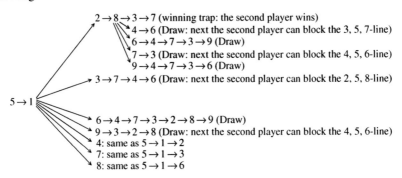

$2 \rightarrow 8 \rightarrow 3 \rightarrow 7$ (winning trap: the second player wins)
$4 \rightarrow 6$ (Draw: next the second player can block the 3, 5, 7-line)
$6 \rightarrow 4 \rightarrow 7 \rightarrow 3 \rightarrow 9$ (Draw)
$7 \rightarrow 3$ (Draw: next the second player can block the 4, 5, 6-line)
$9 \rightarrow 4 \rightarrow 7 \rightarrow 3 \rightarrow 6$ (Draw)
$3 \rightarrow 7 \rightarrow 4 \rightarrow 6$ (Draw: next the second player can block the 2, 5, 8-line)

$5 \rightarrow 1$

$6 \rightarrow 4 \rightarrow 7 \rightarrow 3 \rightarrow 2 \rightarrow 8 \rightarrow 9$ (Draw)
$9 \rightarrow 3 \rightarrow 2 \rightarrow 8$ (Draw: next the second player can block the 4, 5, 6-line)
4: same as $5 \rightarrow 1 \rightarrow 2$
7: same as $5 \rightarrow 1 \rightarrow 3$
8: same as $5 \rightarrow 1 \rightarrow 6$

If the first player does not start in the center, then, not surprisingly, the second player's best response is in the center. If the first player starts in the corner, say in 1, then the second player's drawing strategy is the following

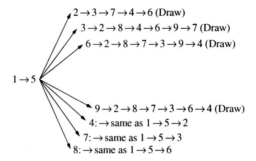

$1 \to 5$
$2 \to 3 \to 7 \to 4 \to 6$ (Draw)
$3 \to 2 \to 8 \to 4 \to 6 \to 9 \to 7$ (Draw)
$6 \to 2 \to 8 \to 7 \to 3 \to 9 \to 4$ (Draw)
$9 \to 2 \to 8 \to 7 \to 3 \to 6 \to 4$ (Draw)
$4: \to$ same as $1 \to 5 \to 2$
$7: \to$ same as $1 \to 5 \to 3$
$8: \to$ same as $1 \to 5 \to 6$

Finally, if the first player starts on the side, say in 2, then the second player's drawing strategy is the following

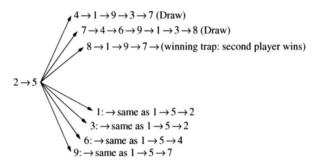

$2 \to 5$
$4 \to 1 \to 9 \to 3 \to 7$ (Draw)
$7 \to 4 \to 6 \to 9 \to 1 \to 3 \to 8$ (Draw)
$8 \to 1 \to 9 \to 7 \to$ (winning trap: second player wins)
$1: \to$ same as $1 \to 5 \to 2$
$3: \to$ same as $1 \to 5 \to 2$
$6: \to$ same as $1 \to 5 \to 4$
$9: \to$ same as $1 \to 5 \to 7$

This completes the case study, and Theorem 3.1 follows. $\qquad\square$

The 4^2 Tic-Tac-Toe is a "less interesting" draw: neither player can force a Weak Win (assuming the Full Play Convention). First we show that the first player can put his mark in every winning set (4-in-a-row) of the 4^2 board, i.e. he can force a Strong Draw. In fact, this Strong Draw is a Pairing Strategy Draw: first player's opening move is in the middle

$$\begin{bmatrix} * & - & | & - \\ | & X_1 & / & | \\ | & / & - & - \\ - & - & | & | \end{bmatrix}$$

X_1 in the middle blocks 3 winning lines, 7 winning lines remain unblocked, and we have $4^2 - 1 = 15$ cells to block them. An explicit pairing strategy is given on the picture above (the first player doesn't even need the asterisk-marked cell in the upper-left corner).

Next we show how the second player can put his mark in every winning set ("Strong Draw"). This case is more difficult since the second player *cannot* have a Pairing Strategy Draw. Indeed, the cell/line ratio is *less* than 2 (there are $4^2 = 16$

cells and $4+4+2 = 10$ winning lines), so it is impossible that every winning line owns a "private pair of cells."

Nevertheless the second player can block by using a combination of 3 different pairing strategies! Indeed, apart from symmetry there are three possible opening moves of the first player, marked by X. The second player's reply is 0, and for the rest of the play the second player employs the pairing strategy direction marked in the corresponding picture (for every move of the first player in a marked square the second player takes the similarly marked square in the direction indicated by the mark).

$$
\begin{bmatrix}
X & - & | & - \\
| & O & / & | \\
| & / & - & - \\
- & - & | & |
\end{bmatrix}
\qquad
\begin{bmatrix}
| & X & - & - \\
| & O & / & | \\
- & / & | & - \\
- & - & | & |
\end{bmatrix}
\qquad
\begin{bmatrix}
\backslash & | & - & - \\
| & X & O & | \\
- & - & \backslash & | \\
| & | & - & -
\end{bmatrix}
$$

This elegant argument is due to *David Galvin*.

This shows that the 4^2 game comes very close to having a Pairing Strategy Draw: *after* the first player's opening move the second player can always employ a draw-forcing pairing strategy.

Theorem 3.2 *The 4^2-game is a Strong Draw, but not a Pairing Strategy Draw (because the second player cannot force a draw by a single pairing strategy).*

The n^2 Tic-Tac-Toe with $n \geq 5$ is particularly simple: either player can force a draw by a pairing strategy; in fact, both players may use the same pairing.

Case $n = 5$

In the 5^2-game either player can force a draw by employing the following "pairing"

$$
\begin{bmatrix}
11 & 1 & 8 & 1 & 12 \\
6 & 2 & 2 & 9 & 10 \\
3 & 7 & * & 9 & 3 \\
6 & 7 & 4 & 4 & 10 \\
12 & 5 & 8 & 5 & 11
\end{bmatrix}
$$

Note that every winning line has its own *pair* ("private pair"): the first row has two 1s, the second row has two 2s,..., the fifth row has two 5s, the first column has two 6s,..., the fifth column has two 10s, the diagonal of slope -1 has two 11s, and finally, the other diagonal of slope 1 has two 12s. Either player can occupy at least 1 point from each winning line: indeed, whenever the first (second) player occupies a numbered cell, the opponent takes the other cell of the same number. (In the first player's strategy, the opening move is the asterisk-marked center; in the second player's strategy, if the first player takes the center, then the second player may take any other cell.)

By using this *pairing strategy* either player can block every winning set, implying that 5^2 is a draw game.

A more suggestive way to indicate the same pairing strategy for the 5^2-game is shown below: all we have to do is to make sure that for every move of the opponent in a marked square we take the similarly marked square in the direction indicated by the mark.

$$\begin{bmatrix} \backslash & - & | & - & / \\ | & - & - & | & | \\ - & | & * & | & - \\ | & | & - & - & | \\ / & - & | & - & \backslash \end{bmatrix}$$

Case $n = 6$

The 6^2-game is another Pairing Strategy Draw. An explicit pairing goes as follows

$$\begin{bmatrix} 13 & 1 & 9 & 10 & 1 & 14 \\ 7 & * & 2 & 2 & * & 12 \\ 3 & 8 & * & * & 11 & 3 \\ 4 & 8 & * & * & 11 & 4 \\ 7 & * & 5 & 5 & * & 12 \\ 14 & 6 & 9 & 10 & 6 & 13 \end{bmatrix}$$

Whenever the first (second) player occupies a numbered cell, the opponent takes the other cell of the same number. We do not even need the eight asterisk-marked cells in the 2 diagonals.

Case $n \geq 7$

By using the special cases $n = 5$ and 6, the reader can easily solve all n^2-games with $n \geq 7$.

Exercise 3.1 *Let $n \geq 5$ be an integer. Show that if the n^2-game has a Pairing Strategy Draw, then the $(n+2)^2$-game also has a Pairing Strategy Draw.*

Theorem 3.3 *The $n \times n$ Tic-Tac-Toe is a Pairing Strategy Draw for every $n \geq 5$.*

Ordinary 3^2 Tic-Tac-Toe turns out to be the most interesting member of the (dull) family of n^2-games. This is where the "phase transition" happens: 2^2 is a trivial first player win (the first player always wins in his second move), 3^2 is a draw, and the rest – the n^2-games with $n \geq 4$ – are all drawn, too.

This gives a complete understanding of the 2-dimensional $n \times n$ Tic-Tac-Toe.

3. Tic-Tac-Toe in higher dimensions. Unfortunately, we know much less about the 3-dimensional $n \times n \times n = n^3$ Tic-Tac-Toe. The $2 \times 2 \times 2 = 2^3$-game (just like the 2^2-game) is a trivial win. Every play has the same outcome: first player win.

The 3^3 Tic-Tac-Toe is a less trivial but still easy win. Every play has only two possible outcomes: either a (1) first player win, or a (2) second player win. That is, no draw play is possible. Thus the first player can force a win – this follows from a well-known general argument called "Strategy Stealing." We return to "Strategy Stealing" later in Section 5 (We recommend the reader to finish the proof him/herself).

In the 3^3-game there are 49 winning lines (3-in-a-row): 4 *space-diagonals* (joining opposite corners of the cube), 18 *plane-diagonals* (in fact 12 of them are on the 6 faces joining opposite corners of some face, and 6 more plane-diagonals inside the cube), and 27 axis-parallel lines (parallel with one of the 3 coordinate axes). We illustrate 4 particular winning lines in the left-hand side of the figure below.

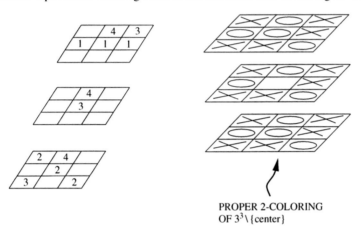

PROPER 2-COLORING
OF $3^3 \setminus \{center\}$

Exercise 3.2 *Find an explicit first player winning strategy in the 3^3-game.*

Exercise 3.3

(a) *Show that no draw is possible in the 3^3-game: every play must have a winner.*
(b) *Show that every 2-coloring of the cells in the 3^3-board yields a monochromatic winning line.*

Exercise 3.4 *Consider the Reverse version of the 3^3-game: the player who gets 3-in-a-row first is the loser. Which player has a winning strategy?*

The right-hand side of the figure above demonstrates an interesting property of the 3^3-game. We already know that Drawing Position cannot exist (see Exercise 3.3 (b)), *but* if we remove the center and the 13 winning lines going through the center, then the "truncated" 3^3-hypergraph of $3^3 - 1 = 26$ points and $49 - 13 = 36$ winning lines has a *Proper 2-coloring*, i.e. there is no monochromatic winning line.

It is interesting to note that the sizes of the two color classes differ by 2, so this is *not* a Drawing Terminal Position. In fact, a Drawing Terminal Position (i.e. Proper *Halving* 2-Coloring) cannot even exist!

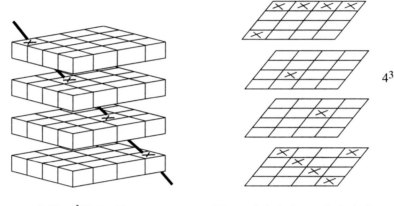

Qubic = 4^3 Tic-Tac-Toe   Three typical winning sets in $4 \times 4 \times 4$ game

The first variant of Tic-Tac-Toe, which is truly exciting to play, is the 3-dimensional $4 \times 4 \times 4 = 4^3$ game: it was marketed by Parker Brothers as *Qubic*; henceforth we often refer to it as such. A remarkable property of Qubic is that it has a Drawing Terminal Position, but the first player can nevertheless force a win.

In Qubic there are 4 *space-diagonals* (joining opposite corners of the cube), 24 *plane-diagonals* (in fact 12 of them are on the 6 faces joining opposite corners of some face, and 12 more plane-diagonals inside the cube), and 48 axis-parallel lines (parallel with one of the 3 coordinate axes); altogether 76 4-in-a-row. As said before, Qubic has a Drawing Terminal Position: we can put 32 Xs and 32 Os on the board such that every 4-in-a-row contains both marks – see the figure below.

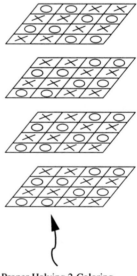

Proper Halving 2-Coloring
of 4^3 Tic-Tac-Toe

Qubic is a first player win just like the 3^3 game, but there is a big difference: the winning strategy in Qubic is extremely complicated! The first explicit winning strategy was found by Oren Patashnik in 1977, and it was a celebrated victory for Computer Science (and Artificial Intelligence). The solution involved a most intricate human–computer interaction; for the details we refer the reader to Patashnik's fascinating survey paper [1980]. Patashnik's solution employes hundreds of long sequences of *forced moves*. A sequence of *forced moves* means that the second player must continually block first player's 3-in-a-line until at some move the first player has a *winning trap*: the second player must simultaneously block two such 3-in-a-line, this is clearly impossible, so the first player wins. Patashnik's solution contains a "dictionary" of 2 929 "strategic moves." The first player forces a win as follows:

(1) if he can make 4-in-a-row with this move, he does it;
(2) he blocks the opponent's 3-in-a-line if he must;
(3) he looks for a sequence of forced moves, and employs it if he finds one;
(4) otherwise he consults Patashnik's dictionary.

After this brief discussion of the 4^3 game ("Qubic"), we switch to the general case, called *hypercube Tic-Tac-Toe*, which is formally defined as follows.

n^d **hypercube Tic-Tac-Toe** or simply the n^d **game**. The board V of the n^d game is the d-dimensional hypercube of size $n \times \cdots \times n = n^d$, that is, the set of d-tuples

$$V = \left\{ \mathbf{a} = (a_1, a_2, \ldots, a_d) \in \mathbf{Z}^d : 1 \le a_j \le n \text{ for each } 1 \le j \le d \right\}.$$

The winning sets of the n^d-game are the n-in-a-line sets, i.e. the n-element sequences

$$\left(\mathbf{a}^{(1)}, \mathbf{a}^{(2)}, \ldots, \mathbf{a}^{(n)} \right)$$

of the board V such that, for each j, the sequence $a_j^{(1)}, a_j^{(2)}, \ldots, a_j^{(n)}$ composed of the jth coordinates is either $1, 2, 3, \ldots, n$ ("increasing"), or $n, n-1, n-2, \ldots, 1$ ("decreasing"), or a *constant*. The two players alternately put their marks (X and O) in the previously unmarked cells (i.e. unit cubes) of the d-dimensional solid hypercube n^d of side n. Each player marks one cell per move. The winner is the player to occupy a whole winning set *first*, i.e. to have n of his marks in an n-in-a-line *first*. In other words, the winning sets are exactly the n-in-a-line in the n^d hypercube; here, of course, each elementary "cell" is identified with its own center. If neither player gets n-in-a-line, the play is a draw. The special case $n = 3, d = 2$ gives ordinary Tic-Tac-Toe. Note that in higher dimensions most of the n-in-a-line are some kind of diagonal.

The winning sets in the n^d game are "lines," or "winning lines." The number of winning lines in the 3^2 and 4^3 games are 8 and 76. In the general case we have an elegant short formula for the number of "winning lines," see Theorem 3.4 (a) below. In the rest of the book we often call the cells "points" (identifying a cell with its own center).

Theorem 3.4

(a) *The total number of winning lines in the n^d-game is $\left((n+2)^d - n^d\right)/2$.*

(b) *If n is odd, there are at most $(3^d - 1)/2$ winning lines through any point, and this is attained only at the center of the board. In other words, the maximum degree of the n^d-hypergraph is $(3^d - 1)/2$.*

(c) *If n is even ("when the board does not have a center"), the maximum degree drops to $2^d - 1$, and equality occurs if there is a common $c \in \{1, \ldots, n\}$ such that every coordinate c_j equals either c or $n+1-c$ ($j = 1, 2, \ldots, d$).*

Proof. To prove (a) note that for each $j \in \{1, 2, \ldots, d\}$, the sequence $a_j^{(1)}, a_j^{(2)}, \ldots, a_j^{(n)}$ composed of the jth coordinates of the points on a winning line is either strictly *increasing* from 1 to n, or strictly *decreasing* from n to 1, or a *constant* $c = c_j \in \{1, 2, \ldots, n\}$. Since for each coordinate we have $(n+2)$ possibilities $\{1, 2, \ldots, n, increasing, decreasing\}$, this gives $(n+2)^d$, but we have to subtract n^d because at least one coordinate must change. Finally, we have to divide by 2, since every line has two orientations.

An alternative geometric/intuitive way of getting the formula $\left((n+2)^d - n^d\right)/2$ goes as follows. Imagine the board n^d is surrounded by an additional layer of cells, one cell thick. This new object is a cube

$$(n+2) \times (n+2) \times \cdots \times (n+2) = (n+2)^d.$$

It is easy to see that every winning line of the n^d-board extends to a uniquely determined *pair* of cells in the new surface layer. So the total number of lines is $\left((n+2)^d - n^d\right)/2$.

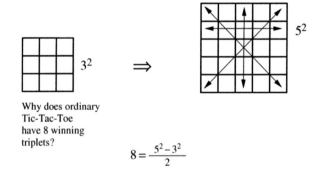

Why does ordinary Tic-Tac-Toe have 8 winning triplets?

$$8 = \frac{5^2 - 3^2}{2}$$

Next we prove (b): let n be odd. Given a point $\mathbf{c} = (c_1, c_2, \ldots, c_d) \in n^d$, for each $j \in \{1, 2, \ldots, d\}$ there are three options: the jth coordinates of the points on an *oriented* line containing $\mathbf{c}$:

(1) either increase from 1 to n,

(2) or decrease from n to 1,

(3) or remain constant c_j.

Since every line has two orientations, and it is impossible that all coordinates remain constant, the maximum degree is $\leq (3^d - 1)/2$, and we have equality for the center (only).

This suggests that the center of the board is probably the best opening move (n is odd).

Finally, assume that n is even. Let $\mathbf{c} = (c_1, c_2, \ldots, c_d) \in n^d$ be a point, and consider the family of those n-in-a-line which contain $\mathbf{c}$. Fixing a proper subset index-set $I \subset \{1, 2, \ldots, d\}$, there is at most *one* n-in-a-line in this family for which the jth coordinates of the points on the line remain constant c_j for each $j \in I$, and increase or decrease for each $j \notin I$. So the maximum degree is $\leq \sum_{i=0}^{d-1} \binom{d}{i} = 2^d - 1$, and equality occurs if for some fixed $c \in \{1, \ldots, n\}$ every coordinate c_j equals c or $n + 1 - c$ ($j = 1, 2, \ldots, d$). $\qquad\qquad\square$

4. Where is the phase transition? We know that the n^2 games are rather dull (with the possible exception of ordinary 3^2 Tic-Tac-Toe itself); the 3^3 game is too easy, the 4^3 game is very interesting and difficult, but it is completely solved; how about the next one, the 5^3 game? Is it true that 5^3 is a draw game? How about the 5^4 game? Is it true that 5^4 is a first player win? Unfortunately these are hopeless questions.

Open Problem 3.2 *Is it true that 5^3 Tic-Tac-Toe is a draw game? Is it true that 5^4 Tic-Tac-Toe is a first player win?*

Very little is known about the n^d games with $d \geq 3$, especially about winning. We know that the first player can achieve a 4-in-a-row first in the 3-space (4^3 Tic-Tac-Toe); how about achieving a 5-in-a-row? In other words, the first player wants a winning strategy in some 5^d Tic-Tac-Toe. Let d_0 denote the smallest dimension d when the first player has a forced win in the 5^d game; how small is d_0? (A famous result in Ramsey Theory, called the Hales–Jewett Theorem, see Section 7, guarantees that d_0 is finite.) The second question in Open Problem 3.2 suggests that $d_0 = 4$, but what can we prove? Can we prove that $d_0 \leq 1000$? No, we cannot. Can we prove that $d_0 \leq 1000^{1000}$? No, we cannot. Can we prove that $d_0 \leq 1000^{1000^{1000}}$? No, we cannot prove that either. Even if we iterate this 1000 times, we still cannot prove that this "1000-tower" is an upper bound on d_0. Unfortunately, the best-known upper bound on d_0 is embarrassingly poor. For more about d_0, see Section 7.

Another major problem is the following. We know an explicit dimension d_0 such that in the 5^{d_0} Tic-Tac-Toe the first player has a winning strategy: (1) it is bad enough that the smallest d_0 we know is enormous, but (2) it is even worse that the proof does not give the slightest hint how the winning strategy actually looks (!), see Theorems 5.1 and 6.1 later.

Next we mention two conjectures about hypercube Tic-Tac-Toe (published in Patashnik [1980]), which represent a very interesting but failed(!) attempt to describe the "phase transition" from draw to win in simple terms. The first one, called "modification of Gammill's conjecture" by Patashnik [1980], predicted that:

Conjecture A ("Gammill") *The n^d game is a draw if and only if there are more points than winning lines.*

For example, the 3^2 and 4^3 games both support this conjecture. Indeed, the 3^2 game is a draw and number-of-lines $= 8 < 9 =$ number-of-points; on the other hand, the 4^3 game is a first player win and number-of-lines $= 76 > 64 =$ number-of-points.

In the 5^3 game, which is believed to be a draw, there are $(7^3 - 5^3)/2 = 109$ lines and $5^3 = 125$ points. On the other hand, in the 5^4 game, which is believed to be a first player win, there are $(7^4 - 5^4)/2 = 938$ lines and $5^4 = 625$ points.

A modification of Citrenbaum's conjecture (see Patashnik [1980]) predicted that:

Conjecture B ("Citrenbaum") *If $d > n$, then the first player has a winning strategy in the n^d game.*

Of course, we have to be very critical about conjectures like these two: it is difficult to make any reasonable prediction based on such a small number of solved cases. And indeed, both Conjectures A and B turned out to be false; in Section 34, we prove that both have infinitely many counterexamples.

Unfortunately, our method doesn't work in lower dimensions: an explicit relatively low-dimensional counter-example to Conjecture A that we could come up with is the 144^{80}-game (it has more lines than points), and an explicit counter-example to Conjecture B is the 214^{215}-game which is a draw. These are pretty large dimensions; we have no idea what's going on in low dimensions.

The failure of the at-first-sight-reasonable Conjectures A and B illustrates the difficulty of coming up with a "simple" conjecture about the "phase transition" from draw to win for hypercube Tic-Tac-Toe games. We don't feel confident enough to formulate a conjecture ourselves. We challenge the reader to come up with something that makes sense. Of course, to formulate a conjecture is one thing (usually the "easy part"), and to prove it is a totally different thing (the "hard part").

Before discussing more games, let me stop here, and use the opportunity to emphasize the traditional *viewpoint* of Game Theory. First of all, we assume that the reader is familiar with the concept of **strategy** (the basic concept of Game Theory!). Of course, strategy is such a natural/intuitive "common sense" notion that we are tempted to skip the formal definition; but just in case, if there is any doubt, the reader can always consult Appendix C for a formal treatment. Now the traditional viewpoint: Game Theory is about **optimal strategies**, which is shortly expressed in the vague term: "the players play rationally." We certainly share the traditional viewpoint: we always assume that either player knows an optimal strategy, even if finding one requires "superhuman powers" such as performing a case study of size (say) $10^{1000!}$.

A pithy way to emphasize the traditional viewpoint is to name the two players after some *gods*. Let us take, for example, the most famous war in Greek Mythology: the *Trojan War* between the Greeks and Troy, see *The Iliad* of Homer. Motivated by the Trojan War we may call the first player *Xena* (or *Xenia*) and the second player *Apollo*. Of course, Xena uses mark X and Apollo uses mark O. Xena is an epithet of Pallas Athena, meaning "hospitable." Xena (alias Pallas Athena), goddess of wisdom, sided with the Greeks, and Apollo, god of arts and learning, sided with Troy.

It is most natural to expect a god/goddess to know his/her optimal strategy; carrying out a case study of size (say) $10^{1000!}$ shouldn't be a problem for them (but it is a BIG problem for us humans!).

The only reason why we don't follow the advice, and don't use a name-pair such as Xena/Apollo (or something similar), is to avoid the awkwardness of he/she, him/her, and his/her (the gender problem of the English language).

We conclude Section 3 with an entertaining observation, and a picture.

Consider the following number game: two players alternately select integers from 1 to 9 and no number may be used twice. A player wins by getting three numbers the sum of which is 15; the first to do that is declared to be the winner. Who wins?

Well, this is a mathematical joke! The number game is just Tic-Tac-Toe in disguise! Indeed, there are exactly 8 solutions (a, b, c) of the equation $a + b + c = 15$, $1 \leq a < b < c \leq 9$, and the 8 solutions are represented by the 8 winning lines of Tic-Tac-Toe (3 horizontals, 3 verticals, and 2 diagonals). Therefore, this number game is a draw.

2	7	6
9	5	1
4	3	8

Finally, a figure illustrating the hopeless Open Problem 3.2.

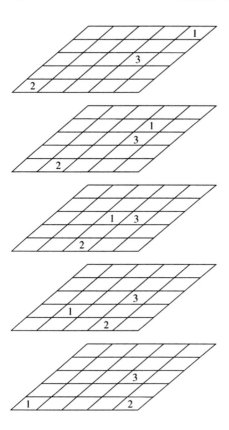

A hopeless problem:
5^3 Tic-Tac-Toe
Is this a draw game?

$5^3 = 125$ points ("cells")

$6^3 - 5^3 = 91$ combinatorial lines like (1) and (3), and 18 geometric lines which are not combinatorial lines like (2)

(1) xxx $x = 1, 2, 3, 4, 5$

(2) $x1x'$ $x' = 6 - x$

(3) $43x$

4

More examples: Tic-Tac-Toe like games

Tic-Tac-Toe itself is a simple game, but some natural changes in the rules quickly lead to very difficult or even hopelessly difficult games. We have already mentioned the 3-dimensional $4 \times 4 \times 4$ version ("Qubic"), which was solved by a huge computer-assisted case study (it is a first player win). The next case, the $5 \times 5 \times 5$ version, is expected to be a draw, but there is no hope of proving it (brute force is intractable). A perhaps more promising direction is to go back to the plane, and study 2-dimensional variants of Tic-Tac-Toe. We will discuss several 2-dimensional variants: (1) unrestricted n-in-a-row, (2) Harary's Animal Tic-Tac-Toe, (3) Kaplansky's n-in-a-line, (4) Hex, and (5) Gale's Bridge-it game. They are all "who does it first" games.

1. Unrestricted n-in-a-row. The 5^2 Tic-Tac-Toe, that is, the "5-in-a-row on a 5×5 board" is a very easy draw game, but if the 5×5 board is extended to the whole plane, we get a very interesting and still unsolved game called *unrestricted* 5-in-a-row. *Unrestricted* means that the game is played on an *infinite* chessboard, infinite in every direction. In the *unrestricted* 5-in-a-row game the players alternately occupy little squares of an infinite chessboard; the first player marks his squares by X, and the second player marks his squares by O. The person who first gets 5 consecutive marks of his own in a row horizontally, or vertically, or diagonally (of slope 1 or -1) is the winner; if no one succeeds, the play ends in a draw. *Unrestricted* n-in-a-row differs in only one aspect: the winning size is n instead of 5.

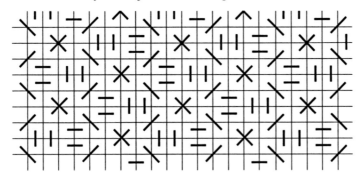

Similarly to the "*S*-building game" in Section 1, unrestricted *n*-in-a-row is a *semi-infinite* game: the board is infinite but the winning sets are all finite. Since the board is infinite, we have to define the *length* of the game. We assume that the two players take turns until either of them wins in a finite number of moves, or until they have taken their *n*th turns for every natural number *n*. In other words, the length of a play is at most ω, where ω denotes (as usual) the first countable (infinite) ordinal number.

Every n^d Tic-Tac-Toe is *determined*, meaning that:

(i) either the first player has a winning strategy,
(ii) or the second player has a winning strategy,
(iii) or either player can force a draw (called "draw game"); see Appendix C.

This simple fact is a special case of a general theorem about *finite Combinatorial Games*, usually called "Zermelo's Theorem," see Appendix C.

Unrestricted *n*-in-a-row is not finite, but the standard proof easily extends to semi-infinite games as follows. If player P (first or second) has no winning strategy, then the opponent, called player Q, can always make a next move so that player P still has no winning strategy, and this is exactly how player Q can force a draw. The point is that the winning sets are finite, so, if a player wins, he wins in a finite number of moves; this is why player P cannot win and the opponent–player Q–forces a draw.

Note that alternative (ii) *cannot* occur for (the finite) n^d Tic-Tac-Toe or for (the semi-infinite) *unrestricted n*-in-a-row; in other words, the second player cannot have a winning strategy. This follows from the well-known "Strategy Stealing" argument, see Appendix C. Later we will settle this issue under much more general circumstances.

Thus we have two alternatives only: (i) either the first player has a winning strategy, (ii) or the game is a draw.

Exercise 4.1 *Prove that unrestricted 4-in-a-row is a first player win.*

Open Problem 4.1 *Is it true that unrestricted 5-in-a-row is a first player win?*

Open Problem 4.2 *Is it true that unrestricted n-in-a-row is a draw for every $n \geq 6$?*

In Open Problem 4.2 the real question is the two cases $n = 6$ and $n = 7$; for $n \geq 8$ we know that the game is a draw. We give a proof at the end of Section 10. The figure above illustrates the weaker result that $n = 9$ is a draw (why?).

2. Harary's Animal Tic-Tac-Toe. Unrestricted *n*-in-a-row is a very interesting 2-dimensional variant of Tic-Tac-Toe, which is wide open for $n = 5, 6, 7$. In the 1970s, Frank Harary introduced another 2-dimensional variant that he called *Animal Tic-Tac-Toe*. The first step in Animal Tic-Tac-Toe is to choose an arbitrary polyomino, or, using the old terminology, cell animal, and declare it to be the objective of the game. Note that a cell animal is by definition edge connected (or using chess terminology: rookwise connected in the sense that a rook can visit every cell of the set). For example, the "diagonal 3-in-a-row," a winning set in ordinary Tic-Tac-Toe, is *not* a cell animal.

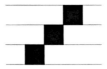

The two players play on an $n \times n$ board (of course, the board is assumed to be large enough to contain at least one congruent copy of the goal animal), and as in ordinary Tic-Tac-Toe, they alternately mark the cells with X and O. Each player tries to mark the cells to build a congruent copy of the goal animal. The player who builds a congruent copy of the goal animal with his own mark first is the winner; otherwise the play ends in a draw. For example, if the goal animal is the 3-cell

called "Tic," the first player wins on every $n \times n$ board with $n \geq 4$ in 3 moves (the fastest way!) – We leave the easy proof to the reader.

Next consider "El," "Knobby," and "Elly" (the funny names were coined by Harary and his colleagues)

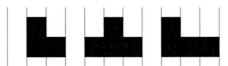

The first player can build a congruent copy of "El" on every $n \times n$ board with $n \geq 3$ in 3 moves; a congruent copy of "Knobby" on every $n \times n$ board with $n \geq 5$ in 4 moves; and a congruent copy of "Elly" on every $n \times n$ board with $n \geq 4$ in 4 moves. Harary calls "Tic," "El," "Knobby," and "Elly" *Economical Winners* for the obvious reason that the first player can build each one without having to take any cell that is not part of the winning goal animal.

There are exactly 5 4-cell animals: besides "Knobby" and "Elly" we have

"Fatty" is a *Loser* (i.e. not a Winner); indeed, either player can prevent the opponent from building a congruent copy of "Fatty" by applying the following simple domino tiling (pairing strategy)

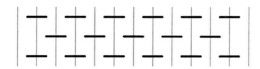

This pairing strategy proof is due to Andreas R. Blass; the terms *Winner*, *Economical Winner*, and *Loser* were coined by Harary.

"Skinny" and "Tippy" are both *Winners*, but neither one is an Economical Winner. Indeed, a mid-size case study proves that the first player can build a congruent copy of "Skinny" first on every $n \times n$ board with $n \geq 7$ in at most 8 moves, and he can also build a congruent copy of "Tippy" first on every $n \times n$ board with $n \geq 3$ in at most 5 moves.

We have a complete understanding of the 4-cell animals; let's move next to the 5-cell animals. There are exactly 12 5-cell animals: 9 are Losers and 3 are Winners (neither one is an Economical Winner). Haray calls a Loser that contains no Loser of lower order ("order" is the number of cells) a Basic Loser. "Fatty" is the smallest Basic Loser; among the 9 5-cell Losers 8 are Basic Losers

not a Basic Loser

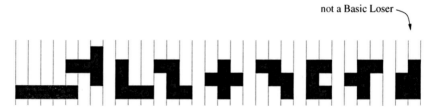

Exercise 4.2 *Find a domino tiling pairing strategy for each one of the 8 5-cell Basic Losers.*

There are exactly 35 6-cell animals, all but 4 contain one of the 9 Basic Losers of order ≤ 5; the 4 exceptions are

Snaky

The first 3 turned out to be Losers (consequently Basic Losers).

Exercise 4.3 *Find a domino tiling pairing strategy for each one of the first 3 6-cell Basic Losers (not "Snaky").*

The status of "Snaky" is undecided–this is the big open problem of Animal Tic-Tac-Toe.

Open Problem 4.3 *Is it true that "Snaky" is a Winner? In particular, is it true that "Snaky" is a Winner on every $n \times n$ board with $n \geq 15$, and the first player can always build a congruent copy of "Snaky" first in at most 13 moves?*

This conjecture, formulated by Harary 20 years ago, remains unsolved.

If "Snaky" is a Winner, there are 12 Basic Losers of order ≤ 6: one of order 4, eight of order 5, and three of order 6. The 5 domino tilings (pairing strategy) provide the proof:

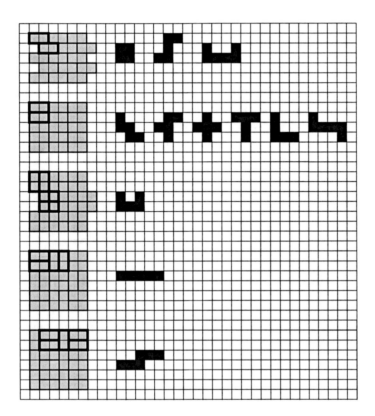

Here comes the surprise: there are 107 7-cell animals, each containing one of the 12 Basic Losers of order ≤ 6 (we don't need "Snaky"!), so even if "Snaky" remains unsolved, Haray and his colleagues could still prove the following elegant result: *there are no Winners among the cell animals of order ≥ 7.* The proof is elementary: domino tiling pairing strategy and exhaustive search suffice.

In other words, every Animal Tic-Tac-Toe is "easy": either a Winner in a few moves, or a Loser by a simple pairing strategy (domino tiling), with the exception of "Snaky"! For more about Animal Tic-Tac-Toe, see Chapter 37 in Gardner [2001].

Brute force complexity. Let me return to the conjecture about "Snaky": "Snaky" is conjectured to be a Winner on every $n \times n$ board with $n \geq 15$, and the first player is expected to be able to build a copy of "Snaky" in at most 13 moves. Why is this innocent-looking conjecture still open? Consider the special case $n = 15$; the "brute force complexity" of this special case of the conjecture is about

$$\sum_{m=0}^{25} \binom{225}{m} \binom{m}{m/2} \approx 10^{44}.$$

What we mean by the "brute force complexity" is the running time of the *backward labeling* algorithm, the only known general method to solve an arbitrary finite *Combinatorial Game* (for the concepts of *Combinatorial Game* and *backward labeling* we refer the reader to Appendix C). We give a nutshell illustration of *backward labeling* as follows. A position is called even or odd depending on the parity of the total number of marks made by the two players together; for example, the starting position ("empty board") is even. In an even (or odd) position, it is the first (or second) player who makes the next move. The sum $\sum_{m=0}^{25} \binom{225}{m} \binom{m}{m/2}$ above gives the total number of positions on a 15×15 board in which the first player made ≤ 13 moves and the second player made ≤ 12 moves; among these positions we label those odd positions which were just won by the first player (i.e. the first player owns a whole winning set but the second player doesn't own a whole winning set yet): the label reads "I-wins." The *backward labeling* algorithm means to apply the following two rules:

Rule 1: if an even position P is a predecessor of a "I-wins" position, then P is also a "I-wins" position;

Rule 2: if an odd position P has the property that all of its options ("successors") are "I-wins" positions, then P is also a "I-wins" position; otherwise P is not labeled.

The brute force way to prove the above-mentioned conjecture about "Snaky" is to show that, by repeated application of Rules 1–2, the starting position becomes labeled with "I-wins." The brute force way to decide whether this is the case or not takes about $\sum_{m=0}^{25} \binom{225}{m} \binom{m}{m/2} \approx 10^{44}$ steps, which is beyond hope. This is the "combinatorial chaos" that prevents us from proving (or disproving) the "Snaky conjecture" (Open Problem 4.3), and the same applies for Open Problem 3.2.

3. Kaplansky's n-in-a-line. We already saw two different 2-dimensional variants of Tic-Tac-Toe (the unrestricted *n*-in-a-row and Harary's Animal Tic-Tac-Toe); a third, no less interesting, variant was invented by Kaplansky, and goes as follows. Two players move alternately by marking unmarked integer lattice points in the plane; for example, the first player may color his points red and the second player may color his points blue. If, during a play, there ever occurs a configuration where some straight line contains *n* points of one color and no points anywhere on the line of the other color, then whoever does it first is declared the winner; otherwise the play ends in a draw. The length of the play is $\leq \omega$ exactly like in the unrestricted *n*-in-a-row or the *S*-building game in Section 1. We refer to this game as Kaplansky's n-in-a-line. The novelty of this variant is that the whole line has to be "opponent-free."

The cases $n = 1, 2, 3$ are all trivial (why?), but the case $n = 4$ is already unsolved.

Open Problem 4.4 *Is it true that Kaplansky's 4-in-a-line is a draw game? Is it true that Kaplansky's n-in-a-line is a draw game for every $n \geq 4$?*

This conjecture is due to Kleitman and Rothschild [1972]; it is about 35-years-old now. There is no progress whatsoever. We will return to this game in Section 15, where we discuss a weaker version.

The next game is well known among mathematicians.

4. Hex. The popular game of **Hex** was invented by Piet Hein in the early 1940s, and it is still "unsolved." The board of Hex is a rhombus of hexagons of size $n \times n$ (the standard size is $n = 11$). Why hexagons (why not squares or triangles)? The hexagons have the unique property that, if two hexagons share a corner point, then they must share a whole edge. In other words, in a hexagon lattice *vertex-connected* and *edge-connected* mean the same thing.

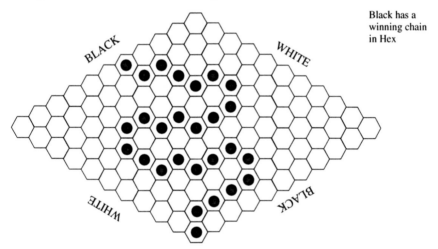

Black has a winning chain in Hex

The two players, White (the first player) and Black (the second player) take the two pairs of opposite sides of the board. The players alternately put their pieces on unoccupied hexagons (White has white pieces, and Black has black pieces). White wins if his pieces connect his opposite sides of the board, and Black wins if his pieces connect the other pair. Observe that Hex is *not* a positional game: the winning sets for White and Black are (mostly) *different*. In fact the only common winning sets are the chains connecting diagonally opposite corners (a small minority among all winning sets).

In the late 1940s John Nash proved, by a pioneering application of the Strategy Stealing argument (see Section 5), that Hex is a first player win.

Exercise 4.4 *By using the strategy stealing argument show that the first player has a winning strategy in Hex.*

It is a famous open problem to find an *explicit* winning strategy (unsolved for $n \geq 8$).

Open Problem 4.5 *Find an explicit first player ("White") winning strategy in $n \times n$ Hex for every $n \geq 8$. In particular, find one for the standard size $n = 11$.*

5. Bridge-it. Bridge-it is a variant of Hex. It was invented by D. Gale in the 1960s. The board of the game is a pair of interlaced n by $n+1$ lattices (a white one and a black one). White (the first player) joins to adjacent (horizontal or vertical) spots of the white lattice, and Black (the second player) joins to adjacent (horizontal or vertical) spots of the black lattice. There is a restriction: no two moves may cross. White wins if he forms a white chain connecting a left-most spot to a right-most spot; Black wins if he forms a black chain connecting a top-most spot to a bottom-most spot.

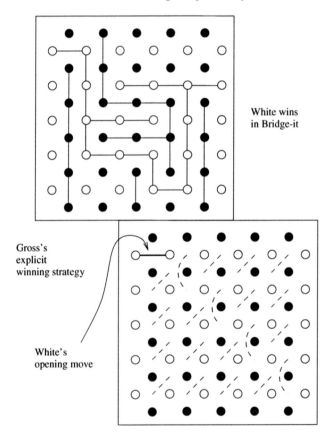

Exercise 4.5 *By using the strategy stealing argument show that the first player has a winning strategy in Bridge-it.*

To find an *explicit* first player winning strategy in Hex remains a famous unsolved problem; however, for Bridge-it we know several *explicit* winning strategies! The first one was found by Oliver Gross, which is vaguely indicated on the second picture above (it will be clarified below).

Before studying Gross's explicit first player winning strategy, it is very instructive to solve a simpler game first that we call **2-Colored Dots-and-Cycles**. Two players start from a rectangular array of dots and take turns to join two horizontally or vertically adjacent dots; the first player uses red pen and the second player uses blue pen. That player who completes a cycle of his own color first is declared the winner.

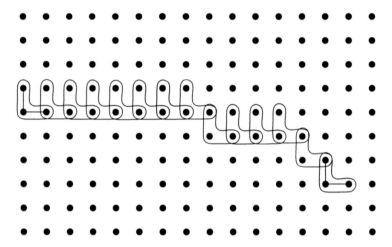

Exercise 4.6 *Prove that 2-Colored Dots-and-Cycles is a draw game. What is more, it is a Strong Draw game: either player can prevent the opponent from occupying a monochromatic cycle (by using a pairing strategy).*

Now we are ready to explain Gross's explicit first player's winning strategy (see the picture above). It is a shockingly simple Pairing Strategy which guarantees a first player win as follows: White (first player) opens with the move on the picture above, thereafter whenever Black's play crosses the end of a dotted line, White plays by crossing the other end of the same line.

Exercise 4.7 *Prove that, by using Gross's "dotted line" pairing strategy, the first player can force a win in Bridge-it.*

A completely different solution of Bridge-it was given later by Alfred Lehman [1964]. In fact, Lehman solved a large class of games, the *Multigraph Connectivity Games* (it is also called "Shannon's Switching Game").

6. Multigraph Connectivity Game. This game is defined as follows: two players, Inker (basically "Maker") and Eraser (basically "Breaker"), play a game on a *multigraph* $\mathbf{G} = (V, E)$; *multigraph* means a graph where two vertices may be joined by two or more edges. The players take turns, with Eraser going first, and each of them in his turn claims some previously unclaimed edge of G. Inker's aim is to claim all the edges of some spanning tree of **G**; Eraser's aim is simply to prevent Inker from achieving his goal.

Each edge of the multigraph represents a *permissible* connection between the vertices at its ends; begin the game with the edges drawn in *pencil*. Inker at his move may establish one of these connections permanently (ink over a penciled edge) and attempts to form a *spanning tree* (i.e. a tree containing every vertex). The opponent (Eraser) may permanently prevent a possible connection (erase a penciled edge) and his goal is to "cut" Inker's graph forever. When can Inker win?

Lehman [1964] completely answered this question by the following simple/elegant criterion.

Theorem 4.1 *In the Multigraph Connectivity Game played on a multigraph* **G**, *Inker, as the second player, can win if and only if* **G** *contains two edge-disjoint spanning trees.*

Proof. Both parts are surprisingly simple. The "only if" part goes as follows. If Inker, as the second player, has a win in the game, then by the Strategy Stealing argument, there must be two edge-disjoint spanning trees. Indeed, an extra move is no disadvantage, so both players can play Inker's strategy. If they do this, two disjoint spanning trees will be established.

Next the "if" part: whenever Eraser's move cuts one of the two trees into two parts, say, A and B, Inker's reply is an edge in the other tree joining a vertex of A to one of B. Identifying the two endpoints, we obtain a multigraph with one less vertex, which again contains two edge-disjoint spanning trees, and keep doing this.   □

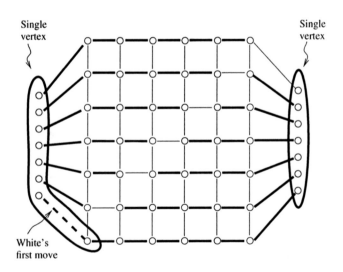

The key fact is that Bridge-it is a particular case of the Multigraph Connectivity Game. Indeed, we can assume that both players play on White's lattice. White (the first player in Bridge-it) starts with an edge (say) in the lower-left corner. Identifying the endpoints, we get a multigraph which contains two edge-disjoint spanning trees: the "horizontal" tree (marked with heavy lines) and the complementary "vertical" tree, see the figure above. From now on Black becomes Eraser and White becomes Inker. White (i.e. Inker) wins Bridge-it by using the above-mentioned "joining a vertex of piece *A* to a vertex of piece *B*" strategy.

This strategy is not a pairing strategy, but there is a strong resemblance: we need a *pair* of spanning trees (see the two edge-disjoint spanning trees on the figure above). If the opponent damages one tree, we use the other one to fix the broken tree.

Lehman's Theorem inspired some important work in Matching Theory. Tutte [1961] and Nash-Williams [1961] proved independently of each other that the non-existence of two disjoint spanning trees of a multigraph *G* is equivalent to the existence of a set *S* of edges whose deletion splits *G* into at least $(|S|+3)/2$ components. When such an edge-set exists, Eraser wins simply by claiming edges from *S* so long as there remain any. In fact the theorem of Tutte and Nash-Williams is more general. It asserts that *G* has *k* pairwise disjoint spanning trees if and only if there is no set *S* of edges whose deletion splits *G* into more than $(|S|+1)/k$ components. This theorem has then been generalized by Edmonds [1965]. Edmonds's Theorem is accompanied by an efficient algorithm which, in the special case of multigraphs, terminates by constructing either the set *S* or the *k* pairwise disjoint spanning trees. It follows that the inefficient "only if" part in the proof of Theorem 4.1 ("Strategy Stealing") can be replaced by an efficient polynomial strategy.

Next we discuss a Reverse generalization of Lehman's Theorem where the goal becomes the anti-goal.

7. Multigraph Connectivity Game: a Biased Avoidance version. Very recently Hefetz, Krivelevich, and Szabó [2007] made the surprising/elegant observation that Theorem 4.1 ("Lehman's Theorem") can be extended to the $(1 : b)$ Avoidance play as follows. Let $b \geq 1$ be a fixed integer. In the $(1 : b)$ Avoidance version the board remains the same *multigraph* $\mathbf{G} = (V, E)$, and the two players remain the same: Inker and Eraser. The players take turns, with Eraser going first; Inker claims one previously unclaimed edge of $\mathbf{G}$ and Eraser claims *b* previously unclaimed edge(s) of $\mathbf{G}$. Eraser's goal is to force Inker to claim all the edges of some spanning tree of $\mathbf{G}$; Inker's aim is simply to avoid building a spanning tree.

Each edge of the multigraph represents a *permissible* connection between the vertices at its ends; begin the game with the edges drawn in *pencil*. Inker at his move may establish one of these connections permanently (ink over a penciled edge) and attempts to form a *spanning tree* (i.e. a tree containing every vertex). The opponent (Eraser) may permanently prevent *b* possible connection(s) (erase penciled edges). When can Eraser win, i.e. when can he force the reluctant Inker to build a spanning tree?

Hefetz, Krivelevich, and Szabó [2007] discovered the following simple criterion, a kind of generalization of Lehman's Theorem (the special case of (1:1) play was independently discovered by Brian Cornell, my former Ph.D. student at Rutgers).

Theorem 4.2 *In the Multigraph Connectivity Game played on a multigraph* **G**, *if* **G** *contains* $b+1$ *edge-disjoint spanning trees, then topdog Eraser can always win the* (1:b) *Avoidance game. In other words, Eraser can always force underdog Inker to occupy a whole spanning tree.*

Proof. Eraser's strategy is to "prevent cycles." Let $T_1, T_2, \ldots, T_{b+1}$ be $b+1$ pairwise disjoint spanning trees in multigraph **G**. For simplicity assume that **G** equals the union of these $b+1$ spanning trees, and Eraser is the second player. Let e_1 denote Inker's first move, and assume that $e_1 \in T_j$. For every $1 \le i \le b+1$ with $i \ne j$ consider the spanning tree T_i: adding the extra edge e_1 to T_i there is a uniquely determined cycle C_i in $T_i \cup \{e_1\}$ containing e_1. For every $1 \le i \le b+1$ with $i \ne j$ Eraser picks an edge f_i from cycle C_i which is different from e_1, this is Eraser's first move. Eraser defines the new spanning trees: $T'_j = T_j$, and $T'_i = T_i \cup \{e_1\} \setminus \{f_i\}$ for every $1 \le i \le b+1$ with $i \ne j$. Identifying the two endpoints of e_1 we obtain a multigraph with one less vertex, which again contains $b+1$ pairwise disjoint spanning trees as before. Thus Eraser has no difficulty repeating the first step.

We leave the general case – when **G** is strictly larger than the union of $b+1$ spanning trees, and Eraser is the first or second player – to the reader as an exercise.   □

Here is a natural question: can Theorem 4.2 be generalized to the (1:b) Achievement game, where underdog Inker is eager to build a spanning tree? The answer is "no"; we will discuss the proof in Section 20.

We conclude Section 4 with another reverse game; in fact, with a Reverse Clique Game.

8. Sim and other Clique Games on graphs. A well-known puzzle states that in a party of 6 people there is always a group of 3 who either all know each other or are all strangers to each other (this is the simplest special case of Ramsey's well-known theorem). This fact motivates the very entertaining game of *Sim*. The board of Sim is K_6, a complete graph on 6 vertices.

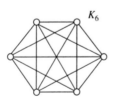

Red ———
Blue – – – – –

Blue does not
want to take
the *ab*-edge

There are 2 players: Red and Blue. At a turn a player colors a previously uncolored edge with his own color. Sim is a *Reverse Game*: that player *loses* who builds a monochromatic K_3 first; otherwise the play ends in a draw. In other words, in Sim either player's goal is to force the opponent to complete a monochromatic triangle *first*. The game is named after Gustavus J. Simmons, see [1969].

The beauty of this game lies in its simplicity and the fact that a draw is impossible (this follows from the above-mentioned puzzle).

In *Sim* the second player has a winning strategy, i.e. alternately coloring the edges of the complete graph K_6 red and blue, the second player can force the opponent to build a monochromatic triangle first. This result was proved by an exhaustive computer analysis. We don't know any simple winning strategy; for a relatively simple one we refer the reader to O'Brian [1978–79].

Sim is a **Reverse Clique Game** ("clique" is a nickname for "complete graph"); we denote it by $(K_6, K_3, -)$. The notation is clear: K_6 is the board, "$-$" stands for "Reverse," and K_3 is the "anti-goal." How about the "normal" version? The "normal" version, the (K_6, K_3) **Clique Game**, is just too easy: the first player can always have a triangle of his own first in his 4th move (or before). We leave the trivial details to the reader.

An advanced version of the initial puzzle goes as follows: in a party of 18 people there is always a group of 4 who either all know each other or are all strangers to each other. This implies that if the board is K_{18} and the goal (or anti-goal) is K_4, then again a draw is impossible. There are two game versions: the normal (K_{18}, K_4) Clique Game and the Reverse $(K_{18}, K_4, -)$. Both are good games, although the difficulty of identifying tetrahedrons (K_4) makes them somewhat hard to play. The normal (K_{18}, K_4) Clique Game is known to be a first player win (why? see Theorem 6.1 later), but we don't know how he wins. The Reverse version is a complete mystery; we don't know anything.

Open Problem 4.6

(a) *Find an explicit first player winning strategy in the (K_{18}, K_4) Clique Game.*
(b) *Which player has a winning strategy in the Reverse Clique Game $(K_{18}, K_4, -)$? If you know who wins, find an explicit winning strategy.*

The Clique Games are clearly related to the Ramsey Theorem, and the n^d hypercube Tic-Tac-Toe is clearly related to the Hales–Jewett Theorem – two basic results in Ramsey Theory. In the next few sections we explore the connection between Games and Ramsey Theory.

5

Games on hypergraphs, and the combinatorial chaos

1. Positional Games. In Sections 1–4 we discussed several classes of concrete games:

(1) S-building game on the plane;
(2) n^d Tic-Tac-Toe;
(3) unrestricted n-in-a-row;
(4) Harary's Animal Tic-Tac-Toe;
(5) Kaplansky's n-in-a-line;
(6) Hex on an $n \times n$ board;
(7) Bridge-it on $n \times (n+1)$ boards;
(8) Multigraph Connectivity Game;
(9) Sim and other Clique Games.

We could find a satisfying solution only for each of the two Maker–Breaker games, classes (1) and (8), but the "who does it first" classes, namely classes (2)–(6) and (9), seem to be hopelessly difficult, with the lucky exception of Bridge-it (class (7)) and the small game of Sim. The humiliating Open Problems 3.1–3.2 and 4.1–4.6 do not give too much reason for optimism. Why are "who does it first" games so difficult? Why do they lead to combinatorial chaos? Before addressing these exciting questions, we need to introduce some very general concepts, the basic concepts of the book.

Is there a common way to generalize the seemingly very different classes (1)–(9)? A common feature is that there is always an underlying set V that we may call the "board," and there is always a family $\mathcal{F}$ of "winning sets" (a family of subsets of V).

Classes (2) and (4) are special cases of the following very general concept.

Positional Games. Let $(V, \mathcal{F})$ be an arbitrary *finite hypergraph*. A "finite hypergraph" means that V is an arbitrary finite set, called the **board** of the game, and $\mathcal{F}$ is an arbitrary family of subsets of V, called the family of **winning sets**. The two players, the first player and the second player, alternately occupy previously unoccupied elements ("points") of board V. That player wins who occupies all the elements of some winning set $A \in \mathcal{F}$ first; otherwise the play ends in a draw.

V : board

$\mathcal{F}$: family of winning sets

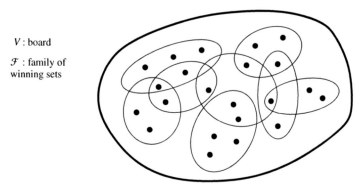

Sometimes we just give the family $\mathcal{F}$ of winning sets, then the board V is the union $\bigcup_{A \in \mathcal{F}} A$ of all winning sets. Note that the board V is also called the "vertex set," the elements of V are also called "points" or "vertices," and the winning sets $A \in \mathcal{F}$ are also called "hyperedges."

Every Positional Game is *determined*, which means that either:

(a) the first player has a winning strategy, or
(b) the second player has a winning strategy, or
(c) both of them have a drawing strategy.

Remarks. Alternatives **(a), (b), (c)** are what we call the three **outcomes of a game**. Of course, every single *play* has three possible outcomes: either the first player wins, or the second player wins, or the play ends in a draw, but the outcome of a particular play has nothing to do with the *outcome of the game*. For example, the second player can easily lose in ordinary Tic-Tac-Toe (e.g. if the first player opens in the center, and the second player replies on the side), even if Tic-Tac-Toe is a *draw game* (i.e. the game itself belongs to class **(c)**).

The proof of the fact that Positional Games are all determined is very simple: it is a straightforward application of De Morgan's law. Indeed, there are only three alternatives: either:

(a) the first player $= \mathbf{I}$ has a winning strategy: $\exists x_1 \forall y_1 \exists x_2 \forall y_2 \cdots$ such that $\mathbf{I}$ wins; or
(b) the second player $= \mathbf{II}$ has a winning strategy: $\forall x_1 \exists y_1 \forall x_2 \exists y_2 \cdots$ such that $\mathbf{II}$ wins;
(c′) or the negation of (a)$\vee$(b)

$$\neg\Big((\exists x_1 \forall y_1 \exists x_2 \forall y_2 \cdots \ \mathbf{I} \ wins) \vee (\forall x_1 \exists y_1 \forall x_2 \exists y_2 \cdots \ \mathbf{II} \ wins)\Big),$$

which, by De Morgan's law, is equivalent to

$$(\forall x_1 \exists y_1 \forall x_2 \exists y_2 \cdots \text{ I } loses \text{ } or \text{ } draw) \wedge (\exists x_1 \forall y_1 \exists x_2 \forall y_2 \cdots \text{ II } loses \text{ } or \text{ } draw).$$

So the third alternative is that both players have a drawing strategy, which is exactly case **(c)**.

This explains why *strategy* is the primary concept of Game Theory; in fact Game Theory is often called the *Theory of Strategies*.

Every game can be visualized as a "tree of all possible plays," called the *game-tree*. Backtracking of the game-tree gives a *constructive* proof for the previous *existential* argument. It provides an explicit winning (or drawing) strategy. The bad news is that backtracking is usually impractical. From a complexity viewpoint it is better to work with the (usually smaller) position graph; then "backtracking" is called "backward labeling of the position graph." Unfortunately, "backward labeling" is still impractical: it takes "exponential time" (see Appendix C).

(It is worth noting that *infinite* Positional Games are not necessarily determined. This is a famous paradox, see Theorems C.4 and C.5 in Appendix C.)

2. Strategy Stealing: a remarkable existence argument. It is well known that whoever plays first in a Positional Game can force at least a draw. In other words, for Positional Games case **(b)** cannot occur. This seems very "plausible" because Positional Games are *symmetric* (i.e. both players want the same thing: to occupy a whole winning set first), and the first player has the *advantage* of the first move which "breaks" the symmetry. We will return to this "plausible proof" later.

Theorem 5.1 *("Strategy Stealing") Let $(V, \mathcal{F})$ be an arbitrary finite hypergraph. Then playing the Positional Game on $(V, \mathcal{F})$, the first player can force at least a draw, i.e. a draw or possibly a win.*

Remark. Theorem 5.1 seems to be *folklore*. "Strategy stealing" was definitely used by Nash in the late 1940s (in his "existential" solution of Hex), but the first publication of Theorem 5.1 is probably in Hales and Jewett [1963].

Proof. Assume that the second player (**II**) has a winning strategy, *STR*, and we want to obtain a contradiction. The idea is to see what happens if the first player (**I**) steals and uses *STR*. A winning strategy for a player is a list of instructions telling the player that if the opponent does this, then he does that, so if the player follows the instructions, he will always win. Now **I** can use **II**'s winning strategy *STR* to win as follows. **I** takes an arbitrary first move, and *pretends* to be the second player (he ignores his first move). After each move of **II**, **I**, as a fake second player, reads the instruction in *STR* before taking action. If **I** is told to take a move that is still available, he takes it. If this move was taken by him before as his ignored "arbitrary" first move, then he takes another "arbitrary move." The crucial point here is that an extra move, namely the last "arbitrary move," only *benefits* **I** in a positional game. □

This was a non-constructive proof: we did *not* construct the claimed drawing strategy. In a non-constructive proof it is particularly important to be very precise, so a true algebraist reader – requesting the highest level of precision – is justly wondering: "Where is the formal definition of the concept of *strategy?*" This criticism is especially well-founded so that we referred to Game Theory above as the *Theory of Strategies*.

Well, we have to admit: this approach was informal. We felt the concept of *strategy* was so natural/intuitive that it could be taken it for granted, something like "common sense." Those who agree on this may go on reading the rest of the book without interruption. Those who don't agree and feel it was "cheating" (they cannot be blamed; they have a point), should consult Appendix C *first* (before reading the rest of the book).

Let's return to "strategy stealing"; the fact that it does *not* supply an explicit strategy is a major weakness of the argument. It is an unlimited source of hopeless open problems. Here is a striking example. Open Problem 4.6 can be extended as follows: we know from Theorem 5.1 that the (K_n, K_4) Clique Game is a first player win for every $n \geq 18$. Is there a uniform upper bound for the Move Number? More precisely

Open Problem 4.6 (c) *Is there an absolute constant $C_4 < \infty$ such that the first player can always win in less than C_4 moves in every (K_n, K_4) Clique Game with $n \geq 18$?*

Is there an absolute constant $C_5 < \infty$ such that the first player can always win in less than C_5 moves in every (K_n, K_5) Clique Game with $n \geq 49$?

These questions were asked more than 20 years ago, and it is unlikely that a solution for K_5 (or K_6 or K_7) will be found in our lifetimes.

Computational Complexity in a nutshell. Strategy stealing doesn't say a word about how to find the existing strategy. What can we do then? As a last resort, we can always try *exhaustive search*, which means backtracking of the game-tree (or the position-graph). What is the complexity of the *exhaustive search*? To answer the question, consider a Positional Game, i.e. a finite hypergraph. If the board V is N-element, then the total number of positions is obviously $O(3^N)$. Indeed, in any particular instant of a play each point on the board can have 3 options: either marked by the first player, or marked by the second player, or unmarked yet. It follows that the *size* of the position-graph, i.e. the number of edges, is $O(N \cdot 3^N)$.

The size of the game-tree is clearly $O(N!)$. Indeed, a *play* is a permutation of the board. Note that for large N, $O(N \cdot 3^N)$ is substantially less than $O(N!)$. This saving, $O(N \cdot 3^N)$ instead of $O(N!)$, is the reason why it is more efficient to work with the position-graph instead of the game-tree.

The exhaustive search of the position-graph, often called "backward labeling," describes the winner and provides an explicit winning or drawing strategy in linear time – linear in terms of the position-graph. For Positional Games the running time is $O(N \cdot 3^N)$, where N is the size of the board.

At first sight this seems to be a satisfying answer. Well, not exactly. An obvious problem is that the exhaustive search is hardly more than mindless computation, lacking any kind of "understanding." The practical problem is computational complexity: A 3^N–step algorithm is intractable; to perform 3^N operations is far beyond the capacity of the fastest computer even for a relatively small board-size such as (say) $N = 100$. This means that, unless we find some substantial shortcut, the exhaustive search of a positional game with board-size ≥ 100 leads to completely hopeless combinatorial chaos. For example, the 5^3 Tic-Tac-Toe game has 125 cells ("Open Problem 3.1") and K_{18} has 153 edges ("Open Problem 4.6"); in both cases the board-size is well over 100.

This concludes our brief discussion of the "complexity of the exhaustive search." For a more detailed (and more precise!) treatment of the subject we refer the sceptical reader to Appendix C.

3. Reverse and semi-infinite games. Let us return to the "plausible proof" of Theorem 5.1: in a Positional Game the two players have exactly the same goal ("to occupy a whole winning set first"), but the first player has the "first-move-advantage," which breaks the symmetry in favor of him, and guarantees at least a first player's drawing strategy. This sounds pretty convincing.

Here comes the surprise: the argument is faulty! Indeed, if we repeat the argument for the *Reverse* Positional Game (where that player *loses* who occupies a whole winning set first, see e.g. Sim in Section 4), then it leads to a *false* conclusion, namely to the conclusion that "the second player cannot lose." Indeed, we can argue that in a Reverse Positional Game, the two players have exactly the same goal, but the first player has the "first-move-disadvantage," which breaks the symmetry in favor of the opponent, implying that the second player cannot lose. But the conclusion is *false*: it is *not* true that the second player cannot lose a Reverse Positional Game. There are infinitely many Reverse Positional Games in which the first player has a winning strategy, e.g. the Reverse 3^3 Tic-Tac-Toe.

Theorem 5.2 ("Reverse n^d Tic-Tac-Toe") *Consider the* **Reverse** n^d *game: the only difference in the rule is that the player who occupies a whole n-in-a-line first is the loser. If n is odd (i.e. the geometric center of the board is a "cell"), the first player has an* **explicit** *drawing strategy. If n is even (i.e. the geometric center of the board is not a "cell"), the second player has an* **explicit** *drawing strategy.*

The following shockingly simple **proof** is due to Golomb and Hales [2002]. If n is *odd*, then first player's opening move is the center C, and whenever the second player claims a point (i.e. a cell) P, the first player chooses the *reflection* P' of P with respect to center C (i.e. P, C, P' are on the same line and the PC distance equals the CP' distance). Assume the first player colors his points red, and the second player colors his points blue. We show that the first player cannot lose. Indeed, assume that the first player loses, and L is the *first* n-in-a-line owned by a player during the course of a play (i.e. L is a red line). Observe that L cannot contain center C. Indeed, every completed n-in-a-line containing the center has precisely $(n+1)/2$ red and $(n-1)/2$ blue points. If L doesn't contain the center, then its reflection L' is a complete blue line, and since L' was completed *before* L, we get a contradiction.

On the other hand, if n is *even*, then the second player can use the "reflection strategy," i.e. choosing P' if first player's last move is P, and achieve at least a draw. □

Note that the 3^3 board does *not* have a drawing terminal position ("easy case study"), so first player's explicit drawing strategy in the Reverse 3^3 game is automatically "upgraded" to a winning strategy.

In general, for every n (≥ 3) there is a finite threshold $d_0 = d_0(n)$ such that the n^d board does *not* have a Drawing Terminal Position if $d \geq d_0$. This follows from a famous result from Ramsey Theory ("Hales–Jewett Theorem"). So, if $d \geq d_0$, then first player's (resp. second player's) explicit drawing strategy if n is odd (resp. even) in Theorem 5.2 is automatically upgraded to a *winning* strategy. This proves that, for *Reverse* Positional Games each one of the 3 possible outcomes – see (a), (b), and (c) above – can *really* occur.

The concept of Positional Games covers class (2) ("n^d Tic-Tac-Toe") and class (4) ("Harary's Animal Tic-Tac-Toe"). In order to include class (3) ("unrestricted n-in-a-row") we just need a slight generalization: the concept of *semi-infinite* Positional Games. The minor difference is that the board V is infinite, but the winning sets $A \in \mathcal{F}$ are all finite. The length of a semi-infinite Positional Games is defined to be the usual $\leq \omega$.

Semi-infinite Positional Games and Reverse Positional Games remain determined: (a) either the first player has a winning strategy, (b) or the second player has a winning strategy), (c) or either player can force a draw. We can basically repeat the proof of the finite case as follows. If player P (first or second) has no winning strategy, then the opponent, called player Q, can always make a next move so that player P still has no winning strategy, and this is exactly how player Q can force a draw. The point is that the winning sets are finite, so if a player wins, he wins in a finite number of moves; this is why player P cannot win and the opponent, player Q, forces a draw.

Similarly, for semi-infinite Positional Games one alternative is excluded: the second player cannot have a winning strategy. Again the same proof works.

In the rest of the section we focus on Positional Games.

We are ready to address the main question of the section: Why are "who does it first" games so difficult? Consider, for example, Open Problem 4.1: *Is it true that unrestricted 5-in-a-row is a first player win?* This remains open in spite of the fact that for the 19×19 board there *is* an explicit first player's winning strategy! (It is a computer-assisted proof, a huge case study. The 19×19 board comes up naturally as the crosspoints of the Go board.)

The game 5-in-a-row on a 19×19 board *has* an explicit first player win strategy, but, as far as we know, no one can extend it to the whole plane. At least we don't know any rigorous proof. How is it possible? Well, this is the "curse of the Extra Set Paradox"; in fact, the "curse of the Induced Extra Set Paradox." The Extra Set Paradox says that:

4. Winning in Positional Games is not monotone. Let us start with a simple first player win hypergraph; for example, consider a binary tree of (say) 3 levels: the winning sets are the 4 full-length branches (3-sets, the players take vertices) of the binary tree. Adding n disjoint 2-element "extra sets" to the hypergraph enables the second player to postpone his inevitable loss by n moves (of course n can be infinite like $n = \omega$).

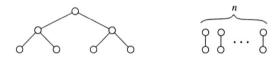

By the way, the second player can easily *postpone* his inevitable loss by adding pairwise disjoint 2-element sets to *any* first player win hypergraph.

In the previous example the second player can postpone his loss by n moves, where n can be arbitrarily large, but eventually the second player will lose the play. The *Extra Set Paradox* is more sophisticated: it says that we can construct a finite hypergraph (i.e. a Positional Game) which is a first player win, but adding an extra winning set turns it into a draw game. What it means is that *winning is not monotone!*

The simplest example we know is the following: the hypergraph on picture 2 consists of the 8 full branches (4-sets, the players take vertices) of the binary tree plus a 3-element Extra Set. The 8 4-sets form an economical winner for the first player; adding the 3-element Extra Set turns it into a draw game.

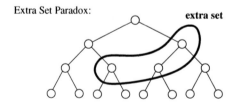

Extra Set Paradox:

This simple/elegant construction is due to R. Schroeppel.

In Schroeppel's constructions neither hypergraph was uniform. In 2002, two former Ph.D. students K. Kruczek and E. Sundberg were able to demonstrate the *Uniform* Extra Set Paradox: they constructed a hypergraph of 10 4-sets which is a draw, but deleting the $\{a, b, c, d\}$ set, the remaining hypergraph is a first player win, see figure below.

Uniform Extra Set Paradox:

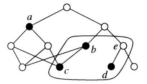

$2^3 = 8$ full branches plus
$\{b, c, d, e\}$ is a first player
win; adding **extra set**
$\{a, b, c, d\}$ to the hypergraph
turns it into a draw

Later we learned a simpler 3-uniform example from the "colossal book" of Martin Gardner [2001].

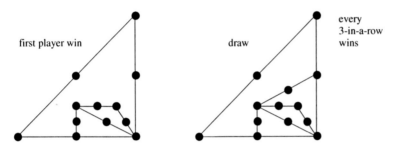

An even smaller 3-uniform example was found very recently by Sujith Vijay (Ph.D. student at Rutgers); we will discuss this example later.

Yet another kind of example is the *Induced* Extra Set Paradox: There is a hypergraph such that the Positional Game on it is a draw, but an *induced* sub-hypergraph is a first player win. The picture below shows a hypergraph on 7 points such that the Positional Game is a draw, but an induced sub-hypergraph on 5 points is a first player win. The board consists of $1, 2, \ldots, 7$, the winning sets are $\{1, 2, 3\}$, $\{1, 3, 4\}$, $\{1, 4, 5\}$, $\{1, 2, 5\}$, $\{6, 7\}$, $\{4, 5, 6\}$, and $\{4, 5, 7\}$. The induced sub-hypergraph has the sub-board $1, 2, 3, 4, 5$, and the 4 winning sets contained by this sub-board, see the figure below.

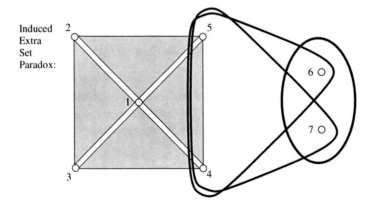

The first player wins the Positional Game on the sub-hypergraph induced by the sub-board 1, 2, 3, 4, 5 by taking first 1. On the whole hypergraph, however, the first player cannot win. Indeed, if his first move is 1, then the second player subsequently takes 6, 4, 2, forcing the first player to take 7, 5, and it is a draw. If the first player's first two moves are (say) 6, 1, then the second player subsequently takes 7, 4, 2, forcing a draw. Finally, if the first player's first 3 moves are (say) 6, 4, 1, then the second player subsequently takes 7, 5, 3, forcing a draw. This construction is due to Fred Galvin; it is about 10 years old.

Galvin's construction was not uniform; how about the *Uniform Induced Extra Set Paradox*? Does there exist a uniform hypergraph such that the Positional Game on it is a draw, but an induced sub-hypergraph is a first player win? Very recently, Ph.D. student Sujith Vijay was able to construct such an example; in fact, a 3-uniform example.

We begin the discussion with a simpler construction: Sujith Vijay was able to find the smallest possible example of the Uniform Extra Set Paradox: a 3-uniform hypergraph on 7 points with 7 3-sets. Consider the hypergraph $H = (V, E)$, where the vertex-set $V = \{1, 2, \ldots, 7\}$ and the edge-set $E = \{(1, 2, 3), (1, 2, 4), (1, 2, 5), (1, 3, 4), (1, 5, 6), (3, 5, 7)\}$. Let P_1 denote the first player and P_2 denote the second player.

$P_1 = $ first player $P_2 = $ second player

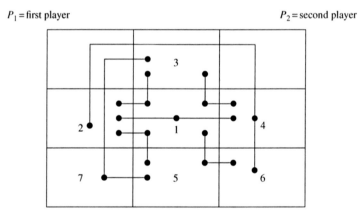

P_1 can force a win on this hypergraph as follows:

Move 1: P_1 picks 1. P_2 is forced to pick 2, for otherwise P_1 will pick 2 and eventually wins, since at most two 3-sets can be blocked in 2 moves.

Move 2: P_1 picks 3. P_2 is forced to pick 4 (immediate threat!).

Move 3: P_1 picks 5. P_2 is forced to pick 6 (immediate threat!).

Move 4: P_1 picks 7 and wins.

Now add the extra set: $E' = E \cup \{(2, 4, 6)\}$. We claim that the game played on $H' = (V, E')$ is a draw (with optimal play).

In order to have any chance of winning, P_1's opening move has to be 1. P_2 responds by picking 2.

If P_1 picks 3 (respectively 5) in Move 2, P_2 picks 4 (respectively 6). P_1 is then forced to pick 6 (respectively 4) and P_2 picks 5 (respectively 3), forcing a draw.

If P_1 picks 4 (respectively 6) in Move 2, P_2 picks 3 (respectively 5), forcing a draw.

Picking 7 in Move 2 makes a draw easier for P_2.

Thus P_2 can always force a draw on H'.

Furthermore, it is clear that any such 3-uniform example requires at least 7 vertices, since an extra edge cannot prevent a 3-move win.

Now we are ready to discuss the *Uniform Induced Extra Set Paradox*. Consider the hypergraph $H'' = (V'', E'')$ where the vertex-set $V'' = \{1, 2, \ldots, 9\}$ and the edge-set E'' consist of the 7 3-sets $\{(1, 2, 3), (1, 2, 4), (1, 2, 5), (1, 3, 4), (1, 5, 6), (3, 5, 7)\}$ of E above, plus the two new 3-sets $(2, 4, 8)$ and $(2, 6, 9)$. Again P_1 denotes the first player and P_2 denotes the second player.

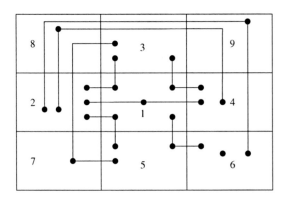

As we have already seen, P_1 can force a win on the sub-hypergraph induced by $V = \{1, 2, \ldots, 7\}$, i.e. on $H = (V, E)$. We claim that the game on the entire hypergraph H'' is a draw (with optimal play).

In order to have any chance of winning, P_1's opening move has to be 1. P_2 responds by picking 2.

If P_1 picks 3 (respectively 5) in Move 2, P_2 picks 4 (respectively 6). P_1 is then forced to pick 8 (respectively 9) and P_2 picks 5 (respectively 3), forcing a draw.

If P_1 picks 4 (respectively 6) in Move 2, P_2 picks 3 (respectively 5), forcing a draw.

Picking 7, 8, or 9 in Move 2 makes a draw easier for P_2.

Thus P_2 can always force a draw on H''.

Now something different: let me recall Open Problem 4.1: *Is unrestricted 5-in-a-row a first player win?* This is difficult because of the Extra Set Paradox. A related question is:

Open Problem 5.1 *Consider the unrestricted 5-in-a-row; can the first player always win in a bounded number of moves, say, in less than 1000 moves?*

Open Problem 4.1 and 5.1 are two different questions! It is possible (but not very likely) that the answer to Open Problem 4.1 is a "yes," and the answer to Open Problem 5.1 is a "no."

The Extra Set Paradox shows that (finite) Positional Games can have some rather surprising properties. Semi-infinite Positional Games provide even more surprises. The following two constructions are due to Ajtai–Csirmaz–Nagy [1979].

Finite but not time-bounded. The first example is a semi-infinite Positional Game in which the first player can always win in a finite number of moves, but there is no finite n such that he can always win in less than n moves. The hypergraph has 8 4-element sets and infinitely many 3-element sets:

The 8 4-element sets are the full-length branches of binary tree B_{15} of 15 vertices. To describe the 3-element sets, consider trees T_{2n+1}, $n = 1, 2, 3, \ldots$ as shown on the figure. Tree T_{2n+1} has $2n+1$ vertices which are labeled by $0, 1, 1', 2, 2', \ldots, n, n'$. The 3-sets are $\{0, 1, 1'\}$, $\{1, 2, 2'\}$, $\{2, 3, 3'\}$, $\cdots$, $\{n-1, n, n'\}$ as $n = 1, 2, 3, \ldots$. The first player can win by occupying a 4-set from B_{15}. Indeed, his first move is the root of binary tree B_{15}. If the second player stays in B_{15}, then the first player occupies a 4-set in his 4th move. If the second player makes a move in a T_{2n+1} the first time, then first player's next move is still in B_{15}. When the second player makes his move in the same T_{2n+1} the second time, the first player is forced to stay in this T_{2n+1} for at most n more moves, before going back to B_{15} to complete a 4-set. So the first player can always occupy a 4-set, but the second player may postpone his defeat for n moves by threatening in T_{2n+1}. In other words, the first player wins in a finite number of moves but doesn't have a *time-bounded* winning strategy.

Finite but not space-bounded. In the next example the first player wins but doesn't have a *space-bounded* winning strategy: there is no finite subset of the board such that in every play the first player can win by occupying elements of this finite subset only.

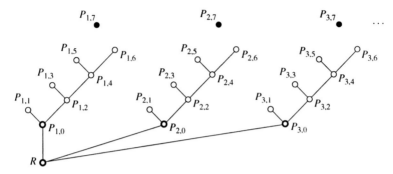

Figure 5.1

Let the board be $X = \{R\} \bigcup \{P_{i,j} : 1 \leq i < \omega, 0 \leq j \leq 7\}$, see Figure 5.1 below. Let $X_n = \{R\} \bigcup \{P_{i,j} : 1 \leq i \leq n, 0 \leq j \leq 7\}$, $n = 1, 2, 3, \ldots$ be finite initial segments of the infinite board.

The winning sets are

$$\{R, P_{i,0}, P_{i,1}\}, \ \{R, P_{i,0}, P_{i,2}, P_{i,3}\}, \ \{R, P_{i,0}, P_{i,2}, P_{i,4}, P_{i,5}\}, \ \{R, P_{i,0}, P_{i,2}, P_{i,4}, P_{i,6}\}$$

for all $1 \leq i < \omega$, and $\{P_{i,1}, P_{i,3}, P_{i,5}, P_{k,7}\}$ for all $1 \leq k < \omega$. The first player has a time-bounded winning strategy: he can win in his 5th move. He starts with occupying root R. The second player's first move is an element of sub-board X_n. Then the first player threatens and wins by occupying $P_{n+1,0}, P_{n+1,2}, P_{n+1,4}, P_{n+1,6}$ in succession. On the other hand, the first player cannot win by restricting himself to any finite sub-board X_n. Indeed, if the first player has to stay in X_n, then the second player can avoid losing by picking first either R or $P_{n,7}$. If the second player occupies root R, then he controls the play. If he occupies $P_{n,7}$, then he can prevent the first player's only real threat: $P_{i,0}, P_{i,2}, P_{i,4}, P_{i,6}$ in succession for some $i \leq n$ (since $\{P_{i,1}, P_{i,3}, P_{i,5}, P_{n,7}\}$ is a winning set). So the first player has a time-bounded winning strategy, but not a space-bounded winning strategy.

The concept of finite and semi-infinite Positional Games covers game classes (2)–(4), but how about class (6) ("Hex") and class (7) ("Bridge-it")? Hex and Bridge-it are not Positional Games: the winning sets for the two players are not the same; in Bridge-it the two players don't even share the same board: White moves in the white lattice and Black moves in the black lattice.

5. Weak Win: Maker–Breaker games. A natural way to include Hex and Bridge-it is the concept of the *Maker–Breaker game*, which was introduced in Theorem 1.2. We recall the definition: on a finite hypergraph $(V, \mathcal{F})$ we can play the "symmetric" Positional Game and also the "one-sided" Maker–Breaker game, where the only difference is in the goals:

(1) Maker's goal is to occupy a whole winning set $A \in \mathcal{F}$, but not necessarily first, and

(2) Breaker's goal is simply to stop Maker (Breaker does not want to occupy any winning set). The player who achieves his goal is declared the winner – so a draw is impossible by definition. Of course, there are two versions: Maker can be the first or second player.

There is a trivial implication: if the first player can force a win in the Positional Game on $(V, \mathcal{F})$, then the same play gives him, as Maker, a win in the Maker–Breaker game on $(V, \mathcal{F})$. The converse is not true: ordinary Tic-Tac-Toe is a simple counter-example.

We often refer to Maker's win as a Weak Win (like Weak Winner in Section 1). Weak Win is easier than ordinary win in a Positional Game. While playing the Positional Game on a hypergraph, both players have their own threats, and either of them, fending off the other's, may build his own winning set. Therefore, a play is a delicate balancing between threats and counter-threats and can be of very intricate structure even if the hypergraph itself is simple.

The Maker–Breaker version is usually somewhat simpler. Maker doesn't have to waste valuable moves fending off his opponent's threats. Maker can simply concentrate on his own goal of *building*, and Breaker can concentrate on *blocking* the opponent (unlike the positional game in which either player has to *build and block* at the same time). Doing one job at a time is definitely simpler.

For "ordinary win" even the most innocent-looking questions are wide open, such as the following "plausible conjecture."

Open Problem 5.2 *Is it true that, if the n^d Tic-Tac-Toe is a first player win, then the n^D game, where $D > d$, is also a win?*

This is again the "curse of the Extra Set Paradox." In sharp contrast with ordinary win, there is no Extra Set Paradox for Weak Win! The Weak Win version of Open Problem 5.2 is trivial: Maker simply uses a Weak Win strategy within a d-dimensional sub-cube of the n^D cube.

The twin brother of Open Problem 5.2 is the following:

Open Problem 5.3 *Is it true that, if the n^d game is a draw, then the $(n+1)^d$ game is also a draw?*

Note that Golomb and Hales [2002] proved a weaker result, that we formulate as an exercise.

Exercise 5.1 *If the n^d game is a draw, then the $(n+2)^d$ game is also a draw.*

The reader is probably wondering: "Why is it easier to go from n to $n+2$ than from n to $n+1$?" A good reason is that the n^d board is the "interior" of the $(n+2)^d$

board. The lack of any similar simple (geometric) relation between the n^d and $(n+1)^d$ games raises the possibility of occasional *non-monotonicity* here. That is, it may be, for some n and d, that the n^d game is a draw, the $(n+1)^d$ game is a first player win, and the $(n+2)^d$ game is a draw again.

We believe that in the class of n^d Tic-Tac-Toe *non-monotonicity* never happens. We are convinced (but cannot prove!) that for each given value of d, there is a *critical value* $n_0(d)$ of n below which the first player can always force a win, while at or above this critical value the n^d game is a draw.

In contrast, *monotonicity* is totally trivial for Weak Win.

Weak Win is not easy! As said before, Weak Win is obviously easier than ordinary win, because Maker doesn't have to occupy a winning set first (this is why *non-monotonicity* never happens), but "easier" does not mean "easy." Absolutely not! For example, the notoriously difficult game of *Hex* is equivalent to a Maker–Breaker game, but this fact doesn't help to find an explicit winning strategy. We prove the equivalence: let *WeakHex* denote the Maker–Breaker game in which the board is the $n \times n$ Hex board, Maker=White, Breaker=Black, and the winning sets are the connecting chains of White. We claim that Hex and *WeakHex* are equivalent. To show the equivalence, first notice that in Hex (and also in *WeakHex*) a draw is impossible. Indeed, in order to prevent the opponent from making a connecting chain, we must build a "river" separating the opponent's sides, and the "river" itself must contain a chain connecting the *other* pair of opposite sides. (This "topological" fact seems plausible, but the precise proof is not completely trivial, see Gale [1979].) This means that Breaker's goal in *WeakHex* (i.e. "blocking") is identical to Black's goal in Hex (i.e. "building first"). Here "identical" means that if Breaker has a winning strategy in *WeakHex*, then the same strategy works for Black as a winning strategy in Hex, and vice versa. Since a draw play is impossible in either game, Hex and *WeakHex* are equivalent.

The concept of Maker–Breaker game clearly covers class (1) ("S-building game in the plane") and class (8) ("Multigraph Connectivity Game"); it has just been explained why it covers class (6) ("Hex"), and, of course, the same argument applies for class (7) ("Bridge-it"), which is covered by class (8) anyway. The concept of finite and semi-infinite Positional Games covers classes (2)–(3)–(4); but how about class (5): Kaplansky's n-in-a-line? This is a **Shutout Game**; we can play a *Shutout Game* on an arbitrary hypergraph $(V, \mathcal{F})$, finite or infinite. First we choose a goal integer $n(\geq 1)$; an n-Shutout Game on $(V, \mathcal{F})$ is similar to the Positional Game in the sense that that the players alternate, but the goal is different. Instead of "complete occupation" the players want an "n-shutout;" i.e. if during a play the first time there is a winning set $A \in \mathcal{F}$ such that one of the players, we call him player P, owns n elements of A and the other player owns no element of A, then player P is declared the winner. If V is infinite, the length of the play is $\leq \omega$. If an ω-play does not have a winner, the play is declared a draw.

Kaplansky's n-in-a-line is an n-Shutout Game with hypergraph $V =$"Euclidean plane" and $\mathcal{F} =$"the family of all straight lines in the plane."

If $\mathcal{F}$ happens to be an n-uniform hypergraph, then the concepts of n-Shutout Game and Positional Game become identical.

6. Why does ordinary win lead to combinatorial chaos? We start with a short answer: exponentially long play and non-monotonicity! Next comes the long, detailed answer: notice that the three concepts of Positional Game, Maker–Breaker Game, and n-Shutout Game are all defined for every finite hypergraph. Let $\mathcal{A}$ denote the family of all finite hypergraphs; the concept of Positional Game splits $\mathcal{A}$ into two natural classes. Indeed, for every finite hypergraph $\mathcal{F} \in \mathcal{A}$ there are two options: the Positional Game played on $\mathcal{F}$ is a (1) first player win, or a (2) draw game, that is

$$\mathcal{A} = \mathcal{A}_{win} \cup \mathcal{A}_{draw}.$$

Let $\mathcal{A}[\geq ThreeChrom]$ denote the sub-family of all finite hypergraphs with chromatic number ≥ 3 (for the definition of the well-known concept of *chromatic number* see the last paragraph before Theorem 6.1 in Section 6). If $\mathcal{F} \in \mathcal{A}[\geq ThreeChrom]$, then playing the Positional Game on $\mathcal{F}$ there is no drawing terminal position, implying that the first player (at least) drawing strategy in Theorem 5.1 is automatically upgraded to a winning strategy. Formally

$$\mathcal{A}[\geq ThreeChrom] \subset \mathcal{A}_{win}.$$

A fundamental difference between the two classes is that the class $\mathcal{A}[\geq ThreeChrom]$ is *monotone increasing* and the other class $\mathcal{A}_{win}$ is *non-monotonic*. *Monotone increasing* simply means that, if $\mathcal{F} \in \mathcal{A}[\geq ThreeChrom]$ and $\mathcal{F} \subset \mathcal{G}$, then $\mathcal{G} \in \mathcal{A}[\geq ThreeChrom]$. This property is clearly violated for $\mathcal{A}_{win}$: the curse of the Extra Set Paradox.

Class $\mathcal{A}[\geq ThreeChrom]$ is the subject of Ramsey Theory, a higly difficult/respected chapter of Combinatorics. The *non-monotonicity* of class $\mathcal{A}_{win}$ indicates that understanding ordinary win is far more difficult than Ramsey Theory (which is already difficult enough).

Another good reason why ordinary win is so hard is the size of the Move Number. Theorem 1.3 gives an exponential lower bound for the Move Number of several games discussed so far:

(1) n^d hypercube Tic-Tac-Toe;
(2) Kaplansky's n-in-a-line game;
(3) S-building game in the plane (see Sections 1–2).

In games (1)–(2) the family of winning sets is Almost Disjoint, and Theorem 1.3 gives the lower bound $\geq 2^{(n-1)/2}$ for the Move Number. Game class (3) defines a nearly Almost Disjoint hypergraph, and Theorem 1.3 gives the slightly weaker but still exponential lower bound $\geq \frac{1}{n}2^{n/2}$ for the Move Number, where $|S| = n$. Exponential building time is a far cry from economical winners!

Another example is the class of:

(4) Clique Games.

Class (4) differs from (1)–(3) above in the sense that Theorem 1.3 doesn't work, but using Theorem 1.4 (instead of Theorem 1.3) and a simple "embedding trick," we can easily show that it takes at least $2^{q/2}$ moves to build a K_q. The details go as follows: the first $2^{q/2}$ moves give $2 \cdot 2^{q/2}$ edges with at most $4 \cdot 2^{q/2}$ endpoints. It follows that the first $2^{q/2}$ moves in every play of the Clique Game can be embedded into a play in the canonical board K_N with vertex set $\{1, 2, \ldots, N\}$, where $N = 4 \cdot 2^{q/2}$. The point is that the Clique Game (K_N, K_q) with $N = 4 \cdot 2^{q/2}$ is a draw; this easily follows from Theorem 1.4. Indeed, the condition of Theorem 1.4 applies if

$$\binom{N}{q} < 2^{\binom{q}{2}-1} \quad \text{where} \quad N = 4 \cdot 2^{q/2}. \tag{5.1}$$

We leave it to the reader to verify that inequality (5.1) holds for all $q \geq 20$, proving that building a K_q takes at least $2^{q/2}$ moves.

Note that the mere fact of "exponentially long play" does not necessarily mean "combinatorial chaos." Consider, for example, a biased (2:1) play on a family of 2^{n-2} pairwise disjoint n-sets. Maker is the first player and the "topdog." Maker takes 2 points per move; Breaker is the "underdog": he takes 1 point per move. Maker can easily occupy a whole given n-set (how?), but it takes him at least 2^{n-2} moves to do so (if Breaker plays rationally). This is an exponentially long building strategy for an n-set, but no one would call it difficult. It is a "halving" strategy which has a transparent self-similar nested structure. The transparency of the winning strategy comes from "disjointness."

What makes a hypergraph in $\mathcal{A}_{win}$ such a strong candidate for "combinatorial chaos" is that for the majority of Positional Games, such as classes (1)–(4) above, winning takes at least exponentially many moves, and, at the same time, we face *non-monotonicity* (of course, the hypergraphs in $\mathcal{A}_{win}$ are completely different from the "disjoint hypergraph" example above!). Exponential time implies that the number of relevant positions is (at least) doubly exponential, and non-monotonicity indicates a lack of order, unpredictable behavior. The analysis of a Positional Game on a hypergraph from class $\mathcal{A}_{win}$ exhibits a "doubly exponential disorder," or using the vague/popular term, "combinatorial chaos."

Of course, this was not a rigorous proof, just a naive attempt to understand why ordinary win is so hard. (Another good reason is that every finite game of complete information – including Chess and Go – can be simulated by a Positional Game; see Appendix C at the end of the book.)

Now let's switch from ordinary win to Weak Win: to achieve a Weak Win (i.e. to build but not necessarily first) still takes an exponentially long time (at least for game classes (1)–(4)), but Weak Win *is* a monotonic property, which makes it so much easier to handle. And, indeed, in the next chapter we will be able to describe the *exact* value of the phase transition from Weak Win to Strong Draw for several classes of very interesting games (Clique Games, 2-dimensional arithmetic progression games, etc.). These exact solutions are the main results of the book.

The 2-class decomposition

$$\mathcal{A} = \mathcal{A}_{win} \cup \mathcal{A}_{draw}$$

of all finite hypergraphs can be extended into a natural 6-class decomposition in the following way.

7. Classification of All Finite Hypergraphs. Let $\mathcal{F}$ be an arbitrary finite hypergraph, and consider the Positional Game played on $\mathcal{F}$ (the board V is the union set); hypergraph $\mathcal{F}$ belongs to one of the following 6 disjoint classes.

(0) **Class 0 ("Trivial win"):** It contains those hypergraphs $\mathcal{F}$ for which every play is a first player win.

   This is a dull class; we can easily characterize the whole class as follows. Let n be the minimum hyperedge size in $\mathcal{F}$ and let V be the board; then $|V| \geq 2n - 1$ and every n-element subset of V must be a hyperedge in $\mathcal{F}$. The reader is challenged to prove this.

   The next class is much more interesting.

(1) **Class 1 ("Draw is impossible: forced win"):** In this class every play has a winner; in other words, a Draw can never occur.

   Every Positional Game in Class 1 is a first player win. Indeed, first player's (at least) drawing strategy – see Theorem 5.1 – is automatically upgraded to a winning strategy (this simple observation is our Theorem 6.1 later). Of course, we have no clue how first player actually wins.

(2) **Class 2 ("Forced win but Drawing Position exists: delicate win"):** It contains those hypergraphs $\mathcal{F}$ which have a Drawing Position, but the first player can nevertheless force a win.

(3) **Class 3 ("Delicate Draw"):** It contains those hypergraphs $\mathcal{F}$ for which the Positional Game is a Draw but the first player can still force a Weak Win (the Full Play Convention applies!).

(4) **Class 4 ("Strong Draw"):** It contains those hypergraphs $\mathcal{F}$ for which the second player has a Strong Draw, but there is no Pairing Strategy Draw.

(5) **Class 5 ("Pairing Strategy Draw"):** It contains those hypergraphs $\mathcal{F}$ for which the second player has a Pairing Strategy Draw (the simplest kind of draw).

Note that each class contains an example from n^d Tic-Tac-Toe. Indeed, each 2^d Tic-Tac-Toe belongs to the trivial Class 0; 3^3 Tic-Tac-Toe belongs to Class 1, and, in general, for every $n \geq 3$ there is a finite threshold $d_0 = d_0(n)$ such that, if $d \geq d_0$, n^d Tic-Tac-Toe belongs to Class 1 ("Hales–Jewett Theorem"). Both Classes 0 and 1 contain infinitely many n^d Tic-Tac-Toe games, but the 4^3 is the only Tic-Tac-Toe game in Class 2 that we know. Similarly, the 3^2 is the only Tic-Tac-Toe game in Class 3 that we know. The 4^2 game belongs to Class 4; then comes a big "gap" in our knowledge: the next game we know to be in Class 4 is the 16-dimensional 44^{16} game. By Theorem 3.4 (a) the Point/Line ratio in the n^d Tic-Tac-Toe is

$$\frac{n^d}{((n+2)^d - n^d)/2} = \frac{2}{\left(1 + \frac{2}{n}\right)^d - 1},$$

which in the special case $n = 44$, $d = 16$ equals

$$\frac{2}{\left(1 + \frac{2}{44}\right)^{16} - 1} = \frac{2}{1.0365},$$

a fraction less than 2, proving that 44^{16} Tic-Tac-Toe cannot have a Pairing Strategy Draw.

The fact that the 44^{16} game is a Strong Draw is more complicated (see Part D). Besides 44^{16} Tic-Tac-Toe there are infinitely many "high-dimensional" n^d games in Class 4; we prove it in Part D. We don't know any n^d game in Class 4 between dimensions 5 and 15.

Finally Class 5: it contains all n^2 games with $n \geq 5$ (see Theorem 3.3).

Open Problem 5.4 *Is it true that each hypergraph class contains infinitely many n^d games? The unknown cases are Class 2 and Class 3.*

What we can prove is that Class 2 and Class 3 *together* are infinite. For example, each n^d Tic-Tac-Toe with $d = n^3$ and n sufficiently large belongs to either Class 2 or Class 3, but we cannot decide which one; for the details, see Section 12.

Pairing Strategy Draw (see Class 5) is well understood by Matching Theory, and Class 1 is basically Ramsey Theory. We cannot distinguish Class 2 from Class 3, but we know a lot about Class 4: Class 4 is a central issue of the book.

We conclude Section 5 with a possible:

Common generalization of Positional and Maker–Breaker Games. On the same finite hypergraph $(V, \mathcal{F})$ we can play the Positional Game ("generalized Tic-Tac-Toe") and the Maker–Breaker game. A common generalization is the concept of **Two-Hypergraph Game** $(V, \mathcal{F}, \mathcal{G})$: Let $\mathcal{F}$ an $\mathcal{G}$ be two hypergraphs on the same board V; the first player wins if he can occupy an $A \in \mathcal{F}$ before the second player occupyies a $B \in \mathcal{G}$; the second player wins if he can occupy a $B \in \mathcal{G}$ before the first player occupyies an $A \in \mathcal{F}$; otherwise the play ends in a draw.

If $\mathcal{F}$ and $\mathcal{G}$ are different, then it is called an **Asymmetric Two-Hypergraph Game**. Hex is clearly an Asymmetric Two-Hypergraph Game. The symmetric case $\mathcal{F} = \mathcal{G}$ gives back the class of "Positional Games," and the other special case $\mathcal{G} = \text{Transv}(\mathcal{F})$ gives the class of "Maker–Breaker Games." We owe the reader the definition of the *transversal hypergraph*: For an arbitrary finite hypergraph $\mathcal{F}$ write

$$\text{Transv}(\mathcal{F}) = \{S \subset V(\mathcal{F}) : S \cap A \neq \emptyset \quad \text{for all } A \in \mathcal{F}\};$$

$\text{Transv}(\mathcal{F})$ is called the *transversal hypergraph* of $\mathcal{F}$.

In the Reverse version of the Two-Hypergraph Game $(V, \mathcal{F}, \mathcal{G})$ the player who loses is the player who occupies a whole winning set from his own hypergraph first; otherwise the play ends in a draw.

Note that the class of Two-Hypergraph Games is universal(!); every finite game of complete information can be simulated by a Two-Hypergraph Game, see the part of "Simulation" in Appendix C.

The class of Two-Hypergraph Games, including the Reverse version, and the biased versions, is a very large class that covers most of the games discussed in this book, but not all. Some games will go beyond this framework; for example, the (1) "Picker–Chooser" and "Chooser–Picker" games, and the (2) "Shutout Games."

A last remark about Theorem 5.1: in view of Theorem 5.1 the second player cannot have a winning strategy in a Positional Game. If he cannot win, then what is the best that the second player can still hope for? This exciting question will be addressed at the end of Section 12, see "The second player can always avoid a humiliating defeat!" and "Second player's Moral Victory." Theorem 12.7 is the "moral-victory" result. It has the most difficult proof in the book; this is why we have to postpone the long proof to the end of the book, see Section 45.

Chapter II

The main result: exact solutions for infinite classes of games

Winning in "who does it first" games seems to be hopeless. We know nothing other than exhaustive search, which leads to combinatorial chaos. Weak Win is doable by the potential technique, see e.g. Sections 1–2. The potential technique is very flexible, but it gives terribly weak upper bounds for the "Move Number" (such as $\leq 10^{500}$ moves for the regular pentagon or $\leq 10^{18,000}$ moves for the 9-element 3×3 Tic-Tac-Toe set S_9, see Example 3 in Section 1). It seems that the potential technique provides very good qualitative but ridiculous quantitative results.

The surprising good news is that, under some special circumstances – namely for "2-dimensional goal sets in degree-regular hypergraphs" – the potential technique is capable of giving excellent quantitative results, even *exact solutions!* This includes many natural positional games, like Cliques Games and sub-lattice games; for them we can determine the exact value of the "phase transition" from Strong Draw to Weak Win. In fact, we can determine these "game numbers" for infinite classes of games.

The bad news is that the proofs are difficult, and they work only for large values of the parameters (when the "error terms" become negligible compared to the "main term").

As a byproduct of the *exact solutions* we obtain the unexpected equality

$$\text{Achievement Number} = \text{Avoidance Number}$$

which holds for our "Ramseyish" games, but fails, in general, for arbitrary hypergraphs.

Another exciting byproduct of the exact solutions is "second player's moral victory," see the end of Section 12.

The Weak Win thresholds in our "Ramseyish" games and the corresponding Ramsey Theory thresholds turn out to be (typically) very different.

6

Ramsey Theory and Clique Games

1. Achievement Games and Avoidance Games. At the end of Section 5 a classification of finite hypergraphs (Classes 0–5) was introduced. Everything known to the author about Class 1 ("Draw is impossible: forced win") comes from Ramsey Theory.

The graph version of Ramsey's well-known theorem says that, for any q, there is a (least) finite threshold $R(q)$ such that, for any 2-coloring of the edges of the complete graph K_N with $N = R(q)$ vertices, there is always a monochromatic copy of K_q.

$R(q)$ is called the *Ramsey Number*. For example, $R(3) = 6$ (the goal is a "triangle") and $R(4) = 18$ (the goal is a "tetrahedron").

Unfortunately the exact value of the next Ramsey Number $R(5)$ is unknown, but we know the close bounds $43 \leq R(5) \leq 49$; for $R(6)$ we know much less: $102 \leq R(6) \leq 165$, and for larger values of q it's getting much, much worse. For example, the current record for K_{10} is $798 \leq R(10) \leq 12\,677$.

The close connection between Ramsey Theory and games is clearly illustrated by the entertaining "Ramseyish" game of Sim, introduced at the end of Section 4. Sim, denoted by $(K_6, K_3, -)$, is in fact a Reverse Positional Game, a Reverse Clique Game. The notation is clear: K_6 is the board, "$-$" stands for "Reverse," and K_3 is the "anti-goal." The "normal" version (K_6, K_3) is far too easy: the first player can always have a triangle of his own first in his 4th move (or before).

If the board is K_6 and the goal is K_3, then a draw is impossible. Similarly, if the board is K_{18} and the goal is K_4, then again a draw is impossible (18 is the Ramsey Number for K_4). There are two versions: the normal (K_{18}, K_4) Clique Game and the Reverse $(K_{18}, K_4, -)$. In the normal (K_{18}, K_4) Clique Game, first player's (at least) drawing strategy in Theorem 5.1 is upgraded to a winning strategy. Can you find an *explicit* one? This was formulated in Open Problem 4.6. How about replacing the goal K_4 with K_5 or K_6?

Open Problem 6.1

(a) *Find an explicit first player's winning strategy in the* (K_{49}, K_5) *Clique Game.*

(b) *Find an explicit first player's winning strategy in the* (K_{165}, K_6) *Clique Game.*

Both seem to be hopeless.

Given an arbitrary finite hypergraph, we can, of course, play the Positional Game (defined in Section 5), but also we can play the **Reverse Positional Game**, which differs from the ordinary Positional Game in one respect only: in the *Reverse Positional Game* that player *loses* who occupies a whole winning set first (otherwise the play ends in a draw). We already (briefly) mentioned this concept before in Theorem 5.2.

The Reverse Game is a complete mystery. Who wins? How do you win? What happens if the anti-goal remains the same, but the board is increasing?

Open Problem 6.2 *Which player has a winning strategy in the Reverse Clique Game* $(K_{49}, K_5, -)$? *How about the* $(K_N, K_5, -)$ *game with* $N \geq 49$, *where* $N \to \infty$?

How about the Reverse Clique Game $(K_{165}, K_6, -)$? *How about the* $(K_N, K_6, -)$ *game with* $N \geq 165$, *where* $N \to \infty$? *In each case find an explicit winning strategy.*

Since ordinary win ("doing it first") looks hopeless, we are forced to change the subject and study Weak Win, i.e. the Maker–Breaker version. We have 4 different Clique Games:

(1) the "normal" (K_N, K_q) Clique Game;

(2) the Reverse Clique Game $(K_N, K_q; -)$;

(3) the Maker–Breaker Clique Game $[K_N, K_q]$, and its

(4) Reverse version $[K_N, K_q, -]$.

Games (1)–(2) are about ordinary win and games (3)–(4) are about Weak Win. We distinguish the two win concepts by using "(...)" and "[...]."

For each one of these 4 Clique Games, the board is K_N, the players alternately take new edges, and the goal (or anti-goal) is a copy of K_q.

In the Reverse version of the Maker–Breaker Clique Game $[K_N, K_q]$, denoted by $[K_N, K_q, -]$ in (4), the two players are called **Avoider** ("Anti-Maker") and **Forcer** ("Anti-Breaker"). Avoider loses if at the end of the play he owns a K_q. In other words, Forcer's goal is to force **Avoider** to occupy a whole K_q. If Forcer achieves his goal, he wins; if he fails to achieve his goal, he loses.

In general, for an arbitrary hypergraph, the Reverse of the Maker–Breaker game is called the **Avoider–Forcer Game**. **Avoider** ("Anti-Maker") and **Forcer** ("Anti-Breaker") alternate the usual way, and **Forcer** wins if he can force **Avoider** to occupy a whole winning set; otherwise, of course, **Avoider** wins. So a draw is impossible by definition.

If $q = q(N)$ is "small" in terms of N, then Maker (resp. Forcer) wins; if $q = q(N)$ is "large" in terms of N, then Breaker (resp. Avoider) wins. Where is the game-theoretic breaking point for Weak Win?

This may seem to be just another hopeless problem, but very surprisingly (at least for large N) we know the *exact value* of the breaking point! Indeed, consider the "lower integral part"

$$q = q(N) = \lfloor 2\log_2 N - 2\log_2 \log_2 N + 2\log_2 e - 3 \rfloor, \tag{6.1}$$

where $\log_2$ stands for the base 2 logarithm; then Maker (resp. Forcer) has a winning strategy in the Maker–Breaker Clique Game $[K_N, K_q]$ (resp. the Avoider–Forcer Clique Game $[K_N, K_q, -]$).

On the other hand, if we take the "upper integral part"

$$q = q(N) = \lceil 2\log_2 N - 2\log_2 \log_2 N + 2\log_2 e - 3 \rceil, \tag{6.2}$$

then Breaker (resp. Avoider) has a winning strategy.

(6.1) and (6.2) perfectly complement each other! This is wonderful, but we have to admit, for the sake of simplicity, we "cheated" a little bit at two points:

(1) the method works only for large N, in the range of $N \geq 2^{10^{10}}$; and
(2) if the logarithmic expression $f(N) = 2\log_2 N - 2\log_2 \log_2 N + 2\log_2 e - 3$ is "very close" to an integer, then that single integer value of q is "undecided": we don't know who wins the $[K_N, K_q]$ (resp. $[K_N, K_q, -]$) game. But for the overwhelming majority of Ns the function $f(N)$ is not too close to an integer, meaning that we know exactly who wins.

For example, let $N = 2^{10^{10}}$ (i.e. N is large enough); then $2\log_2 N = 2 \cdot 10^{10}$, $2\log_2 \log_2 N = 66.4385$, and $2\log_2 e = 2.8854$, so

$$2\log_2 N - 2\log_2 \log_2 N + 2\log_2 e - 3 =$$
$$= 2 \cdot 10^{10} - 66.4385 + 2.8854 - 3 = 19,999,999,933.446.$$

Since the fractional part .446 is not too close to an integer, Maker can build a copy of K_{q_0} with $q_0 = 19,999,999,933$. On the other hand, Breaker can prevent Maker from building a one larger clique K_{q_0+1}.

Similarly, Forcer can force the reluctant Avoider to build a copy of K_{q_0} with the same $q_0 = 19,999,999,933$, but K_{q_0+1} is "impossible" in the sense that Avoider can avoid doing it.

We find it very surprising that the "straight" Maker–Breaker and the "reverse" Avoider–Forcer Clique Games have the same breaking point. We feel this contradicts common sense. Indeed, it is very tempting to expect that the eager Maker can always outperform the reluctant Avoider, a little bit as economists explain the superiority of the capitalistic economy over the communist system. With harsh

anti-communist over-simplification we can argue that the capitalist system is successful because the people are eager to work (like Maker does), motivated by the
higher salaries; on the other hand, the communist system fails because the people
are reluctant to work (like Avoider does) for the very low salaries, and the only
motivation, the only reason, why the people still keep working is the fear of the
police ("Forcer").

Well, this "natural" expectation (eager Maker outperforms reluctant Avoider)
turned out to be plain wrong: eager Maker and reluctant Avoider end up with
exactly the same clique size!

We are sure the reader is wondering about the mysterious function

$$f(N) = 2\log_2 N - 2\log_2 \log_2 N + 2\log_2 e - 3.$$

What is this $f(N)$? An expert in the theory of Random Graphs must know that
$2\log_2 N - 2\log_2 \log_2 N + 2\log_2 e - 1$ is the "Clique Number" of the symmetric Random Graph $\mathbf{R}(K_N, 1/2)$ ($1/2$ is the "edge probability"). In other words, $f(N)$ is 2
less than the Clique Number of the Random Graph. Recall that the *Clique Number*
$\omega(G)$ of a graph G is the number of vertices in the largest complete subgraph of G.

How to make the somewhat vague statement "$f(N)$ is 2 less than the Clique
Number of the Random Graph" more precise? First notice that the expected number
of q-cliques in $\mathbf{R}(K_N, 1/2)$ equals

$$E(q) = E_N(q) = \binom{N}{q} 2^{-\binom{q}{2}}.$$

The function $E(q)$ drops under 1 around $q = (2 + o(1))\log_2 N$. The "real solution"
of the equation $E(q) = 1$ is

$$g(N) = 2\log_2 N - 2\log_2 \log_2 N + 2\log_2 e - 1 + o(1), \tag{6.3}$$

which is exactly 2 more than the $f(N)$ in (6.1)–(6.2).

Elementary Probability theory – a combination of the first and second moment
methods – shows that the Clique Number $\omega(\mathbf{R}(K_N, 1/2))$ of the Random Graph has
an extremely strong concentration. Typically it is concentrated on a single integer,
namely on $\lfloor g(N) \rfloor$ (with probability tending to 1 as $N \to \infty$); and even in the worst
case (which is rare) there are at most two values: $\lfloor g(N) \rfloor$ and $\lceil g(N) \rceil$.

Recovering from the first shock, one has to realize that the strong concentration
of the Clique Number of the Random Graph is not that terribly surprising after all.
Indeed, $E(q)$ is a *rapidly* changing function

$$\frac{E(q)}{E(q+1)} = \frac{q+1}{N-q} 2^q = N^{1+o(1)}$$

if $q = (2 + o(1))\log_2 N$. On an intuitive level, it is explained by the trivial fact that,
when q switches to $q+1$, the goal size $\binom{q}{2}$ switches to $\binom{q+1}{2} = \binom{q}{2} + q$, which is a
"square-root size" increase.

We call $\lfloor g(N) \rfloor$ (see (6.3)) the Majority-Play Clique Number of board K_N. What it refers to is the statistics of all plays with "dumb Maker" and "dumb Breaker" (resp. "dumb Avoider" and "dumb Forcer"). If the "rational" players are replaced by two "dumb" random generators, then for the overwhelming majority of all plays on board K_N, at the end of the play the largest clique in dumb Maker's graph (resp. dumb Avoider's graph) is K_q with $q = \lfloor g(N) \rfloor$.

With two "rational" players – which is the basic assumption of Game Theory – the largest clique that Maker can build (resp. Forcer can force Avoider to build) is the 2-less clique K_{q-2}. We refer to this $q - 2 = \lfloor f(N) \rfloor$ as the Clique Achievement Number on board K_N in the Maker–Breaker Game, and the Clique Avoidance Number in the Avoider–Forcer Game.

Therefore, we can write

$$\text{Clique Achievement Number} = \text{Clique Avoidance Number}$$

$$= \text{Majority Clique Number} - 2.$$

So far the discussion has been a little bit informal; it is time now to switch to a rigorous treatment. First we return to (6.3), and carry out the calculation.

A routine calculation. We have to show that the real solution of the equation

$$\binom{N}{q} = 2^{\binom{q}{2}} \tag{6.4}$$

is

$$q = 2\log_2 N - 2\log_2 \log_2 N + 2\log_2 e - 1 + o(1). \tag{6.5}$$

The deduction of (6.5) from (6.4) is completely routine to people working in areas such as the *Probabilistic Method* or *Analytic Number Theory*, but it may cause some headache to others. As an illustration we work out the details of this particular calculation. First take qth root of (6.4), and apply Stirling's formula $q! \approx (q/e)^q$, which gives the following equivalent form of (6.5)

$$\frac{e \cdot N}{q} \le 2^{(q-1)/2},$$

or equivalently, $e \cdot \sqrt{2} \cdot N \le q \cdot 2^{q/2}$. Taking binary logarithm of both sides, we get

$$q \ge 2\log_2 N - 2\log_2 q + 2\log_2 e - 1 + o(1). \tag{6.6}$$

Let q_0 be the smallest value of q satisfying (6.6); then trivially $q_0 = (2 + o(1))\log_2 N$, and taking binary logarithm, $\log_2 q_0 = 1 + o(1) + \log_2 \log_2 N$. Substituting this back to (6.6), we get (6.5).

The rest of the book is full of *routine calculations* such as this, and in most cases we skip the boring details (we hope the reader agrees with this).

2. Game-theoretic thresholds. Next we give a precise definition of natural concepts such as Win Number, Weak Win Number, Achievement Number, and their Reverse versions.

The Win Number is about ordinary win. Consider the "normal" (K_N, K_q) Clique Game; the *Win Number* $\mathbf{w}(K_q)$ denotes the threshold (least integer) such that the first player has a winning strategy in the (K_N, K_q) Clique Game for all $N \geq \mathbf{w}(K_q)$.

The particular goal graph K_q can be replaced by an arbitrary finite graph G: let (K_N, G) denote the Positional Game where the board is K_N, the players take edges, and the winning sets are the isomorphic copies of G in K_N. For every finite graph G, let *Win Number* $\mathbf{w}(G)$ denote the threshold (least integer) such that the first player has a winning strategy in the (K_N, G) game for all $N \geq \mathbf{w}(G)$.

The *Reverse Win Number* $\mathbf{w}(K_q; -)$ is the least integer such that for all $N \geq \mathbf{w}(K_q; -)$ one of the players has a winning strategy in the Reverse Clique Game $(K_N, K_q; -)$.

Similarly, for any finite graph G, one can define the *Reverse Win Number* $\mathbf{w}(G; -)$.

For an arbitrary finite graph G, let $R(G)$ denote the generalized Ramsey Number: $R(G)$ is the least N such that any 2-coloring of the edges of K_N yields a monochromatic copy of G.

Observe that $\mathbf{w}(G) \leq R(G)$, and similarly $\mathbf{w}(G; -) \leq R(G)$ (why? see Theorem 6.1 below).

Open Problem 6.3

(i) Is it true that $\mathbf{w}(K_q) < R(q)$ for all sufficiently large values of q ? Is it true that

$$\frac{\mathbf{w}(K_q)}{R(q)} \longrightarrow 0 \quad as \quad q \to \infty?$$

(ii) Is it true that $\mathbf{w}(K_q; -) < R(q)$ for all sufficiently large values of q ? Is it true that

$$\frac{\mathbf{w}(K_q; -)}{R(q)} \longrightarrow 0 \quad as \quad q \to \infty?$$

Note that for $G = K_3$ we have the *strict inequality*

$$\mathbf{w}(K_3) = 5 < 6 = \mathbf{w}(K_3; -) = R(3),$$

on the other hand, for the 4-cycle C_4 we have *equality*

$$\mathbf{w}(C_4) = \mathbf{w}(C_4; -) = R(C_4) = 6.$$

The following old problem goes back to Harary [1982].

Open Problem 6.4 *Is it true that $\mathbf{w}(K_4) < \mathbf{w}(K_4; -) < 18 = R(4)$?*

It is humiliating how little we know about Win Numbers, Reverse Win Numbers, and their relation to the Ramsey Numbers.

A quantitative version of Ramsey's graph theorem is the old Erdős–Szekeres Theorem from 1935 (still basically the best; no real progress in the last 70 years!). The Erdős–Szekeres Theorem states that, given any 2-coloring of the edges of the complete graph K_N with $N \geq \binom{2q-2}{q-1}$ vertices, there is always a monochromatic copy of K_q.

Consider the "straight" (K_N, K_q) Clique Game: the board is K_N and the goal is to have a copy of K_q first. If

$$N \geq \binom{2q-2}{q-1} = (1 + o(1)) \frac{4^{q-1}}{\sqrt{\pi q}},$$

then draw is impossible, so the existing (at least) drawing strategy of the first player (see Theorem 5.1) is automatically upgraded to a winning strategy! Unfortunately this argument ("Strategy Stealing") doesn't say a word about what first player's winning strategy actually looks like. We can generalize Open Problem 6.1 as follows:

Open Problem 6.5 *Consider the (K_N, K_q) Clique Game, and assume that the Erdős–Szekeres bound applies: $N \geq \binom{2q-2}{q-1}$. Find an explicit first player's winning strategy.*

Let us return to Weak Win: another reason why Weak Win is simpler than ordinary win is that "strategy stealing" can sometimes be replaced by an *explicit* strategy, see the twin theorems Theorems 6.1 and 6.2 below. The first one is about ordinary win and the second one is about Weak Win. In the latter we have an explicit strategy.

A drawing terminal position in a positional game $(V, \mathcal{F})$ gives a *halving 2-coloring* of the board V such that no winning set $A \in \mathcal{F}$ is monochromatic; We call it a *Proper Halving 2-Coloring* of hypergraph $(V, \mathcal{F})$. Of course, *halving 2-coloring* means to have $\lceil |V|/2 \rceil$ of one color and $\lfloor |V|/2 \rfloor$ of the other color.

A slightly more general concept is when we allow *arbitrary* 2-colorings, not just halving 2-colorings. This leads to the chromatic number.

The *chromatic number* $\chi(\mathcal{F})$ of hypergraph $\mathcal{F}$ is the least integer $r \geq 2$ such that the elements of the board V can be colored with r colors yielding no monochromatic $A \in \mathcal{F}$. *Ramsey Theory* is exactly the theory of hypergraphs with chromatic number at least 3 (see Graham, Rothschild, and Spencer [1980]).

Theorem 6.1 ("Win by Ramsey Theory") *Suppose that the board V is finite, and the family $\mathcal{F}$ of winning sets has the property that there is no Proper Halving 2-Coloring; this happens, for example, if $\mathcal{F}$ has chromatic number at least three. Then the first player has a winning strategy in the Positional Game on $(V, \mathcal{F})$.*

We already used this several times before: if a draw is impossible, then first player's (at least) drawing strategy in Theorem 5.1 is automatically upgraded to a winning strategy.

Theorem 6.1 describes a subclass of Positional Games with the remarkable property that we can easily determine the winner without being able to say how we win. (In most applications, we will probably never find an explicit winning strategy!)

Why Ramseyish games? Theorem 6.1 is the reason why we focus on "Ramseyish" games. Ramsey Theory gives some partial information about ordinary win. We have a chance, therefore, to compare what we know about ordinary win with that of Weak Win.

Theorem 6.1 is a "soft" existential criterion about ordinary win. Since the main objective of Game Theory is to find an explicit winning or drawing strategy, we have to conclude that ordinary win is far more complex than Ramsey Theory!

For example, it is hugely disappointing that we know only two(!) explicit winning strategies in the whole class of $n \times n \times \cdots \times n = n^d$ Tic-Tac-Toe games (the 3^3 version, which has an easy winning strategy, and the 4^3 version, which has an extremely complicated winning strategy).

In sharp contrast, the Maker–Breaker game exhibits an explicit version of Theorem 6.1: Weak Win is guaranteed by a simple copycat pairing strategy.

Theorem 6.2 ("Weak Win by Ramsey Theory") *Let* $(V, \mathcal{F})$ *be a finite hypergraph of chromatic number* ≥ 3, *and let* $(V', \mathcal{F}')$ *be a point-disjoint copy of* $(V, \mathcal{F})$. *Assume* Y *contains* $V \cup V'$ *and* $\mathcal{G}$ *contains* $\mathcal{F} \cup \mathcal{F}'$. *Then Maker has an* **explicit** *Weak Win strategy playing on* $(Y, \mathcal{G})$.

Theorem 6.2 seems to be *folklore* among Ramsey theorists. An interesting infinite version is published in Baumgartner [1973]; perhaps this is the first publication of the "copycat strategy" below.

Proof. Let $f : V \to V'$ be the isomorphism between $(V, \mathcal{F})$ and $(V', \mathcal{F}')$. We show that Maker can force a Weak Win by using the following copycat pairing strategy. If the opponent's last move was $x \in V$ or $x' \in V'$, then Maker's next move is $f(x) \in V'$ or $f^{-1}(x') \in V$ (unless it was already occupied by Maker before; then Maker's next move is arbitrary). Since the chromatic number of $(V, \mathcal{F})$ is at least three, one of the two players will completely occupy a winning set. If this player is Maker, we are done. If the opponent occupies some $A \in \mathcal{F}$, then Maker occupies $f(A) \in \mathcal{F}'$, and we are done again. □

For example, if $N \geq 2\binom{2q-2}{q-1}$, then Theorem 6.2 applies to the Maker–Breaker Clique Game $[K_N, K_q]$: the condition guarantees that the board K_N contains two disjoint copies of K_m with $m = \binom{2q-2}{q-1}$ ("Erdős–Szekeres threshold"), and the copycat pairing in Theorem 6.2 supplies an *explicit* Weak Win strategy for either player.

The *Weak Win* and *Reverse Weak Win Numbers* are defined in the following natural way. Let $\mathbf{ww}(K_q)$ denote the least threshold such that for every $N \geq \mathbf{ww}(K_q)$

the first player (as Maker) can force a Weak Win in the $[K_N, K_q]$ Clique Game ("**ww**" stands for "Weak Win").

Similarly, let $\mathbf{ww}(K_q; -)$ denote the least threshold such that for every $N \geq \mathbf{ww}(K_q; -)$ one of the players can force the other one to occupy a copy of K_q in the Reverse Clique Game $[K_N, K_q; -]$.

Trivially

$$\mathbf{ww}(K_q) \leq \mathbf{w}(K_q) \leq R(q),$$

and the same for the Reverse version. It is easily seen that $\mathbf{ww}(K_3) = \mathbf{w}(K_3) = 5 < 6 = R(3)$.

Open Problem 6.6

(a) *What is the relation between the Weak Win and Reverse Weak Win Numbers* $\mathbf{ww}(K_q)$ *and* $\mathbf{ww}(K_q; -)$*? Is it true that* $\mathbf{ww}(K_q) \leq \mathbf{ww}(K_q; -)$ *holds for every q?*

(b) *Is it true that* $\mathbf{ww}(\mathbf{K_q}) < \mathbf{w}(K_q)$ *for all sufficiently large values of q? Is it true that*

$$\frac{\mathbf{ww}(K_q)}{\mathbf{w}(K_q)} \longrightarrow 0 \text{ as } q \to \infty?$$

(c) *Is it true that* $\mathbf{ww}(K_q; -) < \mathbf{w}(K_q; -)$ *for all sufficiently large values of q? Is it true that*

$$\frac{\mathbf{ww}(K_q; -)}{\mathbf{w}(K_q; -)} \longrightarrow 0 \text{ as } q \to \infty?$$

(d) *Is it true that*

$$\frac{\mathbf{ww}(K_q)}{R(q)} \longrightarrow 0 \text{ and } \frac{\mathbf{ww}(K_q; -)}{R(q)} \longrightarrow 0 \text{ as } q \to \infty?$$

Theorem 6.2 combined with the Erdős–Szekeres bound gives the upper bound $\mathbf{ww}(K_q) < 4^q$; that is, if $q = \frac{1}{2}\log_2 N$, then the first player (Maker) can occupy a copy of K_q.

Besides Theorem 6.2 ("copycat criterion") we have another Weak Win Criterion: the "potential criterion" Theorem 1.2. Let us apply Theorem 1.2. It yields the following: if $q < const \cdot \sqrt{\log N}$, then the first player can occupy a copy of K_q. Indeed, Theorem 1.2 implies a Weak Win if

$$\binom{N}{q} > 2^{\binom{q}{2}-3} \binom{N}{2} \Delta_2.$$

For this particular family of winning sets the Max Pair-Degree $\Delta_2 \leq \binom{N}{q-3}$. Indeed, two distinct edges determine at least 3 different vertices. Therefore, we have to check

$$\binom{N}{q} > 2^{\binom{q}{2}-3} \binom{N}{2} \binom{N}{q-3},$$

which yields $N \geq 2^{q^2/2}$, or, in terms of N, $q < \sqrt{2 \log_2 N}$. Unfortunately this is a very disappointing quantitative result! It is asymptotically much weaker than the "Ramsey Theory bound" $q = \frac{1}{2} \log_2 N$.

3. Separating the Weak Win Numbers from the higher Ramsey Numbers in the Clique Games. Theorem 1.2 worked very poorly for ordinary graphs, but it still gives a very interesting result for p-graphs with $p \geq 4$. We show that the trivial implication

$$\text{draw play is impossible} \Rightarrow \text{winning strategy} \Rightarrow \text{Weak Win strategy}$$

fails to have a converse: the converse is *totally* false!

First we define the Clique Game for p-graphs where $p \geq 3$ – this corresponds to *higher Ramsey Numbers*, see Appendix A. This means a straightforward generalization of the complete graph, where the board is a complete p-uniform hypergraph instead of K_N (and, of course, the players claim p-sets instead of edges).

For every natural number N write $[N] = \{1, 2, \ldots, N\}$. If S is a set, let $\binom{S}{p}$ denote the family of all p-element subsets of S. Then $\binom{[N]}{2}$ can be interpreted as a complete graph with N vertices, i.e. $\binom{[N]}{2} = K_N$.

Let $2 \leq p < q < N$. I define the (K_N^p, K_q^p) Clique Game as follows: the board of this Positional Game is $K_N^p = \binom{[N]}{p}$, and the family of winning sets consists of all possible copies of K_q^p in K_N^p; i.e. all possible $\binom{S}{k}$, where $S \in \binom{[N]}{q}$. The family of winning sets is a $\binom{q}{p}$-uniform hypergraph of size $\binom{N}{q}$.

The general form of the Ramsey Theorem states that for every $p \geq 2$ and for every $q > p$, there is a least finite threshold number $R_p(q)$ such that the family of winning sets of the (K_N^p, K_q^p) Clique Game has a chromatic number of at least 3 if $N \geq R_p(q)$. If $N \geq R_p(q)$, then by Theorem 6.1 the first player has an ordinary win in the (K_N^p, K_q^p) Clique Game (but we don't know what the winning strategy looks like), and if $N \geq 2R_p(q)$, then by Theorem 6.2 the first player has an explicit "copycat" Weak Win Strategy in the (K_N^p, K_q^p) Clique Game.

What do we know about the size of the higher Ramsey Numbers $R_p(q)$? We collect the relevant results in Appendix A. Let $tower_x(k)$ denote the k-fold iteration of the exponential function: $tower_x(1) = 2^x$ and for $k \geq 2$, $tower_x(k) = 2^{tower_x(k-1)}$. So $tower_x(2) = 2^{2^x}$, $tower_x(3) = 2^{2^{2^x}}$, and so on; we call the parameter k in $tower_x(k)$ the *height*.

For graphs (i.e. $p = 2$), by the Erdős–Szekeres Theorem and by Erdős's well-known lower bound

$$2^{q/2} < R_2(q) < 4^q.$$

For p-graphs with $p \geq 3$

$$2^{q^2/6} < R_3(q) < 2^{2^{4q}}, \tag{6.7}$$

$$2^{2^{q^2/24}} < R_4(q) < 2^{2^{2^{4q}}}, \tag{6.8}$$

and in general

$$tower_{4^{3-p}q^2/6}(p-2) < R_p(q) < tower_{4q}(p-1). \tag{6.9}$$

The last two bounds are due to Erdős, Hajnal, and Rado, see Appendix A.

First let $p = 3$. If $q = c_1 \cdot \log \log N$, then by (6.7) the first player has an ordinary Win in the (K_N^3, K_q^3) Clique Game (but the winning strategy is not known); on the other hand, by Theorem 6.2 under the same condition $q = c_1 \cdot \log \log N$ Maker has an explicit (copycat) Weak Win strategy. Similarly, if $q = c_2 \cdot \log \log \log N$, then by (6.8) the first player has an ordinary Win in the (K_N^4, K_q^4) Clique Game (strategy is unknown), and, under the same condition, Maker has an explicit Weak Win strategy, and so on.

What happens if we replace Theorem 6.2 with Theorem 1.2? The potential Weak Win Criterion (Theorem 1.2) applies to the (K_N^p, K_q^p) Clique Game when

$$\binom{N}{q} > 2^{\binom{q}{p}-3} \binom{N}{p} \Delta_2.$$

For this particular family of winning sets the Max Pair-Degree Δ_2 satisfies the obvious inequality $\Delta_2 \le \binom{N}{q-p-1}$; indeed, two distinct p-sets cover at least $p+1$ points.

This leads to the inequality

$$\binom{N}{q} > 2^{\binom{q}{p}-3} \binom{N}{p} \binom{N}{q-p-1},$$

which means $N \ge 2^{q^p/p!}$, or in terms of N, $q \le (p! \log_2 N)^{1/p}$. Therefore, if

$$q = c_p (\log N)^{1/p},$$

then Maker has a Weak Win in the (K_N^p, K_q^p) Clique Game. This gives:

Theorem 6.3 *Consider the (K_N^p, K_q^p) Clique Game for p-graphs with $p \ge 4$; then the Weak Win Number*

$$\mathbf{ww}(K_q^p) \le (p! \log_2 N)^{1/p}, \tag{6.10}$$

implying that

$$\frac{\mathbf{ww}(K_q^p)}{R_p(q)} \longrightarrow 0 \text{ as } p \ge 4 \text{ is fixed and } q \to \infty. \tag{6.11}$$

That is, the Weak Win Ramsey Criterion (Theorem 6.2) definitely fails to give the true order of magnitude of the breaking point for Weak Win, and the same holds for the Reverse version.

Proof. By (6.9) the Ramsey Theory threshold $R_p(q)$ is greater than the tower function $tower_{q^2/6}(p-2)$ of height $p-2$, i.e. the height is linearly increasing with p. On the other hand, Maker can force a Weak Win around $N = 2^{q^p/p!}$, which has a *constant height* independent of p. It is easily seen that $tower_{q^2/6}(p-2)$ is asymptotically *much larger* than $N = 2^{q^p/p!}$ if $p \geq 4$; in fact, they have completely different asymptotic behaviors. $\qquad\square$

(6.11) solves Open Problem 6.6 (d) for p-graphs with $p \geq 4$; the cases $p = 2$ (ordinary graph, the original question) and $p = 3$ remain open.

Theorem 6.3 is a good illustration of a phenomenon that we call "Weak Win beyond Ramsey Theory" (we will show many more examples later in Section 14). A weakness of Theorem 6.3 is that (6.10) fails to give the true order of magnitude. The good news is that we will be able to determine the true order of magnitude(!), see Theorems 6.4 (b)–(c) below. Note that Theorem 1.2 is not good enough; we have to develop a new, more powerful potential technique in Chapter V.

First let's go back to ordinary graphs. We will prove the following asymptotic formula

$$\mathbf{ww}(K_q) = \frac{\sqrt{2}}{e} q 2^{q/2} (1+o(1)). \tag{6.12}$$

Notice that (6.12) is a precise form of the somewhat vague (6.1)–(6.2). Indeed, by elementary calculations

$$N = \frac{\sqrt{2}}{e} q 2^{q/2} (1+o(1)) \iff q = 2\log_2 N - 2\log_2 \log_2 N + 2\log_2 e - 3 + o(1). \tag{6.13}$$

We have the same asymptotic for the Reverse Weak Win Number

$$\mathbf{ww}(K_q; -) = \frac{\sqrt{2}}{e} q 2^{q/2} (1+o(1)); \tag{6.14}$$

i.e. the Weak Win and the Reverse Weak Win Numbers are asymptotically equal. Can "asymptotically equal" be upgraded to "equal"? Are they equal for every single q? Are they equal for all but a finite number of qs? Are they equal for infinitely many qs?

Open Problem 6.7 *Is it true that* $\mathbf{ww}(K_q) = \mathbf{ww}(K_q; -)$ *for every q? Is it true that* $\mathbf{ww}(K_q) = \mathbf{ww}(K_q; -)$ *for all but a finite number of qs? Is it true that they are equal for infinitely many qs?*

(6.12) is even more impressive if we express q in terms of N; i.e. if we take the *inverse function*. Let K_N be the board, and consider the largest value of q such that the first player ("Maker") can build a copy of K_q. In view of (6.13), we can reformulate (6.12) in terms of the *inverse* of the Weak Win Number

$$\mathbf{ww}^{-1}(K_N) = \lfloor 2\log_2 N - 2\log_2 \log_2 N + 2\log_2 e - 3 + o(1) \rfloor, \tag{6.15}$$

where the *inverse* of the Weak Win Number $\mathbf{ww}(K_q)$ is formally defined in the following natural way

$$\mathbf{ww}^{-1}(K_N) = q \text{ if } \mathbf{ww}(K_q) \leq N < \mathbf{ww}(K_{q+1}). \tag{6.16}$$

We call $\mathbf{ww}^{-1}(K_N)$ the **Achievement Number** for Cliques on board K_N, and prefer to use the alternative notation

$$\mathbf{A}(K_N; \text{clique}) = \mathbf{ww}^{-1}(K_N) \tag{6.17}$$

(**A** is for "Achievement"). In view of this, (6.12) can be restated in a yet another way

$$\mathbf{A}(K_N; \text{clique}) = \lfloor 2\log_2 N - 2\log_2\log_2 N + 2\log_2 e - 3 + o(1)\rfloor \tag{6.18}$$

We can define the *inverse* of the Reverse Weak Win Number in a similar way

$$\mathbf{ww}^{-1}(K_N; -) = q \text{ if } \mathbf{ww}(K_q; -) \leq n < \mathbf{ww}(K_{q+1}; -). \tag{6.19}$$

We call $\mathbf{ww}^{-1}(K_N; -)$ the **Avoidance Number** for Cliques on board K_N, and prefer to use the alternative notation

$$\mathbf{A}(K_N; \text{clique}; -) = \mathbf{ww}^{-1}(K_N; -) \tag{6.20}$$

("**A**" combined with "−" means "Avoidance").

In view of Theorem 6.1 we have the general lower bound

$$\text{Clique Achievement Number} \geq \text{inverse of the Ramsey Number.} \tag{6.21}$$

4. The First Main Result of the book. The advantage of the new notation is that the asymptotic equality (6.12) can be restated in the form of an ordinary equality.

Theorem 6.4

(a) *For ordinary graphs the Clique Achievement Number*

$$\mathbf{A}(K_N; \text{clique}) = \lfloor 2\log_2 N - 2\log_2\log_2 N + 2\log_2 e - 3 + o(1)\rfloor,$$

and similarly, the Avoidance Number

$$\mathbf{A}(K_N; \text{clique}; -) = \lfloor 2\log_2 N - 2\log_2\log_2 N + 2\log_2 e - 3 + o(1)\rfloor,$$

implying the equality $\mathbf{A}(K_N; \text{clique}) = \mathbf{A}(K_N; \text{clique}; -)$ *for the overwhelming majority of Ns.*

(b) *For 3-graphs*

$$\mathbf{A}(K_N^3; \text{clique}) = \lfloor \sqrt{6\log_2 N} + o(1)\rfloor,$$

and similarly

$$\mathbf{A}(K_N^3; \text{clique}; -) = \lfloor \sqrt{6\log_2 N} + o(1)\rfloor.$$

(c) *For 4-graphs*

$$\mathbf{A}(K_N^4; \text{clique}) = \lfloor (24\log_2 N)^{1/3} + 2/3 + o(1)\rfloor,$$

and similarly

$$\mathbf{A}(K_N^4; \text{clique}; -) = \lfloor (24\log_2 N)^{1/3} + 2/3 + o(1) \rfloor.$$

In general, for arbitrary p-graphs with $p \geq 3$ (p is fixed and N tends to infinity)

$$\mathbf{A}(K_N^p; \text{clique}) = \lfloor (p!\log_2 N)^{1/(p-1)} + p/2 - p/(p-1) + o(1) \rfloor,$$

and similarly

$$\mathbf{A}(K_N^p; \text{clique}; -) = \lfloor (p!\log_2 N)^{1/(p-1)} + p/2 - p/(p-1) + o(1) \rfloor.$$

Theorem 6.4 is the first *exact solution*; it is one of the main results of the book. The proof is long and difficult, see Sections 24, 25, and 38. The eager reader may jump ahead and start reading Section 24.

Deleting the additive term $-p/(p-1)$ in the last formula above we obtain the Majority-Play Clique Number for the complete p-graph.

An analysis of the proof shows that the "$o(1)$" in Theorem 6.4 (a) becomes negligible in the range $N \geq 2^{10^{10}}$. We are convinced that the "$o(1)$" in Theorem 6.4 (a) is uniformly ≤ 3 for all N, including the small Ns. (Our choice of constant "3" was accidental; maybe the uniform error is ≤ 2 or perhaps even ≤ 1.) Can the reader prove this?

7

Arithmetic progressions

1. Van der Waerden's Theorem. The well-known motto of Ramsey Theory goes as follows: *Every "irregular" structure, if it is large enough, contains a "regular" substructure of some given size.* In Section 6, we discussed the connection between Ramsey's well-known theorem (proved in 1929) and some Clique Games. Ramsey Theory was named after Ramsey, but the most influential result of Ramsey Theory is van der Waerden's Theorem on *arithmetic progressions*. It is interesting to know that van der Waerden's Theorem was proved in 1927, 2 years before Ramsey's work. (It is a very sad fact that Ramsey died at a very young age of 26, shortly after his combinatorial result was published; van der Waerden, on the other hand, moved on to Algebra, and never returned to Combinatorics again.)

Van der Waerden's famous combinatorial theorem, which solved a decade-old conjecture of Schur, goes as follows:

Theorem 7.1 *(B. L. van der Waerden [1927]) For all positive integers n and k, there exists an integer W such that, if the set of integers $\{1, 2, \ldots, W\}$ is k-colored, then there exists a monochromatic n-term arithmetic progression.*

Let $W(n, k)$ be the least such integer; we call it the van der Waerden threshold. The *size* of the van der Waerden threshold turned out to be a central problem in Combinatorics.

Note that van der Waerden's Theorem was originally classified as a result in number theory–see for example the wonderful book of Khintchine titled *Three Pearls of Number Theory*–and it was only in the last few decades that van der Waerden's theorem, with its several generalizations (such as the Hales–Jewett theorem in 1963 and the Szemerédi theorem in 1974), became a cornerstone of Combinatorics.

But what is the connection of van der Waerden's theorem with our main topic: Tic-Tac-Toe games? An obvious connection is that every "winning set" in Tic-Tac-Toe (or in any multidimensional version) is an arithmetic progression on a straight

106

line. But there is a much deeper reason: the beautiful Hales–Jewett Theorem, which is, roughly speaking, the "combinatorial content" of the van der Waerden's proof.

The original van der Waerden's proof was based on the idea of studying *iterated arithmetic progressions*; i.e. progressions of progressions of progressions of ... progressions of arithmetic progressions. This is exactly the *combinatorial structure of the family of n-in-a-line's in the d-dimensional* $n \times n \times \cdots \times n = n^d$ *hypercube.* This observation is precisely formulated in the Hales–Jewett Theorem (see later in this section).

We feel that the reader must know at least the intuition behind the original proof of van der Waerden, so we included a brief outline. Watch out for the enormous constants showing up in the argument!

2. An outline of the original "double induction" proof of van der Waerden. The basic idea is strikingly simple: the proof is a repeated application of the pigeonhole principle. First we study the simplest non-trivial case $W(3, 2)$: we show that given any 2-coloring (say, red and blue) of the integers $\{1, 2, \ldots, 325\}$ there is a monochromatic 3-term arithmetic progression. (Of course, 325 is an "accidental" number; the exact value of the threshold is actually known: it is the much smaller value of $W(3, 2) = 9$.) The proof is explained by the following picture

$$a \ldots a \ldots b \leftarrow d \rightarrow a \ldots a \ldots b \leftarrow d \rightarrow ? \ldots ? \ldots ?$$
$$\mathbf{a} \ldots a \ldots b \leftarrow d \rightarrow a \ldots \mathbf{a} \ldots b \leftarrow d \rightarrow ? \ldots ? \ldots \mathbf{a}$$
$$a \ldots a \ldots \mathbf{b} \leftarrow d \rightarrow a \ldots a \ldots \mathbf{b} \leftarrow d \rightarrow ? \ldots ? \ldots \mathbf{b}$$

What this picture means is the following. It is easy to see that any block of 5 consecutive integers contains a 3-term arithmetic progression of the color code $a \ldots a \ldots a$ or $a \ldots a \ldots b$. The first case is a monochromatic 3-term arithmetic progression (A.P.), and we are done. The second case is a 3-term A.P. where the first two terms have the same color and the third term has the other color. The pigeonhole principle implies that the *same ab-triplet* shows up twice

$$a \ldots a \ldots b \qquad a \ldots a \ldots b$$

Indeed, divide the $\{1, \ldots, 325\}$ interval into 65 blocks of length 5. Since each block has 5 numbers, and we have 2 colors, there are $2^5 = 32$ ways to 2-color a 5-block. By the pigeonhole principle, among the first 33 blocks there are two which are colored in exactly the same way. Assume that the distance between these two identically colored 5-blocks is d, and consider the third 5-block such that the blocks form a 3-term A.P.

$$a \ldots a \ldots b \quad \leftarrow d \rightarrow \quad a \ldots a \ldots b \quad \leftarrow d \rightarrow \quad ? \ldots ? \ldots ?$$

There are two possibilities. If the last ? has color a, then $\mathbf{a}$ forms a monochromatic 3-term A.P.

$$\mathbf{a} \ldots a \ldots b \quad \leftarrow d \rightarrow \quad a \ldots \mathbf{a} \ldots b \quad \leftarrow d \rightarrow \quad ? \ldots ? \ldots \mathbf{a}$$

If the last ? has color b, then **b** forms a monochromatic 3-term A.P.

$$a\ldots a\ldots \mathbf{b} \quad \leftarrow d \rightarrow \quad a\ldots a\ldots \mathbf{b} \quad \leftarrow d \rightarrow \quad ?\ldots?\ldots \mathbf{b}$$

This completes the proof of the case of 2 colors. Note that $325 = 5(2 \cdot 2^5 + 1)$. If we break up the line

$$a\ldots a\ldots b \quad \leftarrow d \rightarrow \quad a\ldots a\ldots b \quad \leftarrow d \rightarrow \quad ?\ldots?\ldots?$$

as follows

$$?\ldots?\ldots?$$

$$a\ldots a\ldots b$$

$$a\ldots a\ldots b$$

then the argument above resembles to a play of Tic-Tac-Toe in which someone wins. Notice that this is a Tic-Tac-Toe with *seven* winning triplets instead of the usual eight

$(1,3)$	$(2,3)$	$(3,3)$
$(1,2)$	$(2,2)$	$(3,2)$
$(1,1)$	$(2,1)$	$(3,1)$

where the diagonal $\{(1,3),(2,2),(3,1)\}$ does not show up in the argument.

Next consider the case of 3 colors, and again we want a monochromatic 3-term A.P. Repeating the previous argument, we obtain the ab-configuration as before

$$a\ldots a\ldots b \quad a\ldots a\ldots b \quad ?\ldots?\ldots?$$

This time we are not done yet, since the last ? can have the third color

$$a\ldots a\ldots b \quad a\ldots a\ldots b \quad ?\ldots?\ldots c$$

However, the pigeonhole principle implies that the same abc-block shows up twice

$$a\ldots a\ldots b \quad a\ldots a\ldots b \quad ?\ldots?\ldots c \quad a\ldots a\ldots b \quad a\ldots a\ldots b \quad ?\ldots?\ldots c$$

Consider the third block such that the 3 blocks form a 3-term A.P.

$$a..a..b \quad a..a..b \quad ?..?..c \quad a..a..b \quad a..a..b \quad ?..?..c \quad ?..?..? \quad ?..?..? \quad ?..?..?$$

This time there are 3 possibilities. If the last ? has color a, then **a** forms a monochromatic 3-term A.P.

$$\mathbf{a}..a..b \quad a..a..b \quad ?..?..c \quad a..a..b \quad a..\mathbf{a}..b \quad ?..?..? \quad ?..?..? \quad ?..?..? \quad ?..?..\mathbf{a}$$

If the last ? has color b, then **b** forms a monochromatic 3-term A.P.

$$a..a..\mathbf{b} \quad a..a..b \quad ?..?..c \quad a..a..b \quad a..a..\mathbf{b} \quad ?..?..c \quad ?..?..? \quad ?..?..? \quad ?..?..\mathbf{b}$$

And, finally, if the last ? has color c, then **c** forms a monochromatic 3-term A.P.

$a..a..b \quad a..a..b \quad ?..?..\mathbf{c} \qquad a..a..b \quad a..a..b \quad ?..?..\mathbf{c} \qquad ?..?..? \quad ?..?..? \quad ?..?..\mathbf{c}$

This is how we can force a monochromatic 3-term A.P. if there are 3 colors. The argument gives the upper bound

$$W(3,3) \leq 7(2 \cdot 3^7 + 1)(2 \cdot 3^{7(2 \cdot 3^7 + 1)} + 1) < 3^{20,000}.$$

The case of 3 colors resembles to a play of Tic-Tac-Toe on a 3-dimensional $3 \times 3 \times 3 = 3^3$ board" (instead of the usual $3 \times 3 = 3^2$ board) with three players in which someone will win.

If there are 4 colors, then we get the same *abc*-configuration as before, but the last ? can have the 4th color:

$a..a..b \quad a..a..b \quad ?..?..c \qquad a..a..b \quad a..a..b \quad ?..?..c \qquad ?..?..? \quad ?..?..? \quad ?..?..d$

Again by the pigeonhole principle, the following configuration will definitely show up

$a.a.b \quad a.a.b \quad ?.?.c \qquad a.a.b \quad a.a.b \quad ?.?.c \qquad ?.?.? \quad ?.?.? \quad ?.?.d$

$\qquad a.a.b \quad a.a.b \quad ?.?.c \qquad a.a.b \quad a.a.b \quad ?.?.c \qquad ?.?.? \quad ?.?.? \quad ?.?.d$

$\qquad\qquad ?.?.? \quad ?.?.? \quad ?.?.? \qquad ?.?.? \quad ?.?.? \quad ?.?.? \qquad ?.?.? \quad ?.?.? \quad ?.?.?$

What it means is that there are two identical *abcd*-blocks, separated from each other, and ? ... ? stands for the third block such that the 3 blocks form a 3-term A.P.

There are 4 possibilities. If the last ? has color a, then **a** forms a monochromatic 3-term A.P.; if the last ? has color b, then **b** forms a monochromatic 3-term A.P.; if the last ? has color c, then **c** forms a monochromatic 3-term A.P.; and, finally, if the last ? has color d, then **d** forms a monochromatic 3-term A.P. (we replace the last ? by •)

$\mathbf{a}.a.b \quad a.a.b \quad ?.?.\mathbf{c} \qquad a.a.b \quad a.a.b \quad ?.?.\mathbf{c} \qquad ?.?.? \quad ?.?.? \quad ?.?.\mathbf{d}$

$\qquad a.a.b \quad a.a.b \quad ?.?.c \qquad a.a.b \quad a.\mathbf{a}.b \quad ?.?.\mathbf{c} \qquad ?.?.? \quad ?.?.? \quad ?.?.\mathbf{d}$

$\qquad\qquad ?.?.? \quad ?.?.? \quad ?.?.? \qquad ?.?.? \quad ?.?.? \quad ?.?.? \qquad ?.?.? \quad ?.?.? \quad ?.?.\bullet$

This is how we can force a monochromatic 3-term A.P. if there are 4 colors. The case of 4 colors resembles to a play of Tic-Tac-Toe on a 4-dimensional $3 \times 3 \times 3 \times 3 = 3^4$ board with four players in which someone will win.

Repeating this argument we get a finite bound for arbitrary number of colors: $W(3,k) < \infty$; the bad news is that the upper bound for $W(3,k)$ is basically a tower function of height k. But *how to get a monochromatic 4-term A.P.?* Consider the simplest case of two colors. We recall the (very clumsy) upper bound $W(3,2) \leq 325$. It follows that 2-coloring any block of 500 consecutive integers, there is always a configuration $a \ldots a \ldots a \ldots a$ or $a \ldots a \ldots a \ldots b$. In the first case we are done. In the second case, we can force the existence of a 3-term A.P. of identical ab-blocks

$$a \ldots a \ldots a \ldots b \qquad a \ldots a \ldots a \ldots b \qquad a \ldots a \ldots a \ldots b$$

Indeed, any 500-block has 2^{500} possible 2-colorings, so if we take $W(3, 2^{500})$ consecutive 500-blocks, then we get a 3-term A.P. of identical 500-blocks. (Unfortunately the argument above gives an extremely poor upper bound: $W(3, 2^{500})$ is less than a tower of height 2^{500}–a truly ridiculous bound!) Consider the 4th block such that the 4 blocks form a 4-term A.P.

$$a \ldots a \ldots a \ldots b \qquad a \ldots a \ldots a \ldots b \qquad a \ldots a \ldots a \ldots b \qquad ? \ldots ? \ldots ? \ldots ?$$

Now there are two cases. If the last ? has color a, then **a** forms a monochromatic 4-term A.P.

$$\mathbf{a} \ldots a \ldots a \ldots b \qquad a \ldots \mathbf{a} \ldots a \ldots b \qquad a \ldots a \ldots \mathbf{a} \ldots b \qquad ? \ldots ? \ldots ? \ldots \mathbf{a}$$

If the last ? has color b, then **b** forms a monochromatic 4-term A.P.

$$a \ldots a \ldots a \ldots \mathbf{b} \qquad a \ldots a \ldots a \ldots \mathbf{b} \qquad a \ldots a \ldots a \ldots \mathbf{b} \qquad ? \ldots ? \ldots ? \ldots \mathbf{b}$$

This is how we can force a monochromatic 4-term A.P. if there are two colors. The case of 2 colors resembles to a play of Tic-Tac-Toe on a 2-dimensional $4 \times 4 = 4^2$ board" (2 players) in which someone wins.

Studying these special cases, it is easy to see how the *double induction* proof of van der Waerden's Theorem goes in the general case. This completes the outline of the proof.

We challenge the reader to finish van der Waerden's proof.

3. Hales–Jewett Theorem. Van der Waerden's double induction proof was adapted by Hales and Jewett [1963] to find monochromatic n-in-a-line's in an arbitrary k-coloring of the d-dimensional $n \times n \times \cdots \times n = n^d$ hypercube (provided d is sufficiently large). The Hales–Jewett Theorem has a wonderful application to the hypercube Tic-Tac-Toe: it implies that the d-dimensional n^d Tic-Tac-Toe is a first player's win if the dimension d is large enough in terms of the winning size n. This is a deep qualitative result; unfortunately, the quantitative aspects are truly dreadful!

Actually the Hales–Jewett proof gives more: it guarantees the existence of a monochromatic *combinatorial line*. A *combinatorial line* is basically a "1-parameter set"; to explain what it means, let $[n] = \{1, 2, \ldots, n\}$. An *x-string* is a finite word $a_1 a_2 a_3 \cdots a_d$ of the symbols $a_i \in [n] \cup \{x\}$, where at least one symbol a_i is x. An x-string is denoted by $\mathbf{w}(x)$. For every integer $i \in [n]$ and x-string $\mathbf{w}(x)$, let $\mathbf{w}(x; i)$ denote the string obtained from $\mathbf{w}(x)$ by replacing each x by i. A *combinatorial line* is a set of n strings $\{\mathbf{w}(x; i) : i \in [n]\}$, where $\mathbf{w}(x)$ is an x-string.

Every combinatorial line is a geometric line, i.e. *n*-in-a-line, but the converse is not true. Before showing a counter-example note that a *geometric line* can be described as an *xx'-string* $a_1 a_2 a_3 \cdots a_d$ of the symbols $a_i \in [n] \cup \{x\} \cup \{x'\}$, where at least one symbol a_i is *x* or *x'*. An *xx'-string* is denoted by $\mathbf{w}(xx')$. For every integer $i \in [n]$ and *xx'*-string $\mathbf{w}(xx')$, let $\mathbf{w}(xx'; i)$ denote the string obtained from $\mathbf{w}(xx')$ by replacing each *x* by *i* and each *x'* by $(n+1-i)$. A *directed geometric line* is a *sequence* $\mathbf{w}(xx'; 1), \mathbf{w}(xx'; 2), \mathbf{w}(xx'; 3), \ldots, \mathbf{w}(xx'; n)$ of *n* strings, where $\mathbf{w}(xx')$ is an *xx'*-string. Note that every geometric line has two orientations.

As we said before, it is *not* true that every geometric line is a combinatorial line. What is more, it is clear from the definition that there are substantially more geometric lines than combinatorial lines: in the n^d game there are $((n+2)^d - n^d)/2$ geometric lines and $(n+1)^d - n^d$ combinatorial lines. Note that the maximum degree of the family of combinatorial lines is $2^d - 1$, and the maximum is attained in the points of the "main diagonal" $(j, j, \ldots, j)$, where *j* runs from 1 to *n*.

For example, in ordinary Tic-Tac-Toe

(1, 3)	(2, 3)	(3, 3)
(1, 2)	(2, 2)	(3, 2)
(1, 1)	(2, 1)	(3, 1)

the "main diagonal" $\{(1, 1), (2, 2), (3, 3)\}$ is a combinatorial line defined by the *x*-string *xx*, $\{(1, 1), (2, 1), (3, 1)\}$ is another combinatorial line defined by the *x*-string *x*1, but the "other diagonal"

$$\{(1, 3), (2, 2), (3, 1)\}$$

is a geometric line defined by the *xx'*-string *xx'*. The "other diagonal" is the only geometric line of the 3^2 game which is *not* a combinatorial line.

The Hales–Jewett threshold $HJ(n, k)$ is the smallest integer *d* such that in each *k*-coloring of $[n]^d = n^d$ there is a monochromatic *geometric* line. The modified Hales–Jewett threshold $HJ^c(n, k)$ is the smallest integer *d* such that in each *k*-coloring of $[n]^d = n^d$ there is a monochromatic *combinatorial* line ("c" stands for "combinatorial"). Trivially

$$HJ(n, k) \leq HJ^c(n, k).$$

In the case of "two colors" $(k = 2)$, we write: $HJ(n) = HJ(n, 2)$ and $HJ^c(n) = HJ^c(n, 2)$; trivially $HJ(n) \leq HJ^c(n)$.

In 1963 Hales and Jewett made the crucial observation that van der Waerden's (double-induction) proof can be adapted to the n^d board, and proved that $HJ^c(n, k) < \infty$ for all positive integers *n* and *k*. This, of course, implies $HJ(n, k) < \infty$ for all positive integers *n* and *k*.

4. Shelah's new bound. How large is $HJ(n) = HJ(n, 2)$? This is a famous open problem. Unfortunately, in spite of all efforts, our present knowledge on the Hales–Jewett threshold number $HJ(n)$ is still rather disappointing. The best-known upper bound on $HJ(n)$ was proved by Shelah [1988]. It is a primitive recursive function (the *supertower* function), which is much-much better than the original van der Waerden–Hales–Jewett bound. The original "double-induction" argument gave the totally ridiculous Ackermann function; the much better Shelah's bound is still far too large for "layman combinatorics."

For a precise discussion, we have to introduce the so-called Grzegorczyk hierarchy of primitive recursive functions. In fact, we define the *representative* function for each class. (For a more detailed treatment of primitive recursive functions we refer the reader to any monograph of *Mathematical Logic*.)

Let $g_1(n) = 2n$, and for $i > 1$, let $g_i(n) = g_{i-1}\big(\ldots g_{i-1}(g_{i-1}(1)\ldots)\big)$, where g_{i-1} is taken n times. An equivalent definition is $g_i(n+1) = g_{i-1}\big(g_i(n)\big)$. For example, $g_2(n) = 2^n$ is the exponential function and

$$g_3(n) = 2^{2^{2^{\cdot^{\cdot^{\cdot^{2}}}}}}$$

is the "tower function" of height n. The next function $g_4(n+1) = g_3\big(g_4(n)\big)$ is what we call the "Shelah's supertower function" because this is exactly what shows up in Shelah's proof. Note that $g_k(x)$ is the *representative* function of the $(k+1)$st Grzegorczyk class.

The original van der Waerden–Hales–Jewett proof proceeded by a *double induction* on n ("length") and k ("number of colors"), and yielded an extremely large upper bound for $HJ^c(n, k)$. Actually, the original argument gave the same upper bound $U(n, k)$ for both $HJ^c(n, k)$ and $W(n, k)$ ("van der Waerden threshold"). We define $U(n, k)$ as follows. For $n = 3$: $U(3, 2) = 1000$ and for $k \geq 2$, $U(3, k+1) = (k+1)^{U(3,k)}$. For $n = 4$: $U(4, 2) = U\big(3, 2^{U(3,2)}\big)$ and for $k \geq 2$

$$U(4, k+1) = U\big(3, (k+1)^{U(4,k)}\big).$$

In general, for $n \geq 4$: let

$$U(n, 2) = U\big(n-1, 2^{U(n-1,2)}\big)$$

and for $k \geq 2$

$$U(n, k+1) = U\big(n-1, (k+1)^{U(n,k)}\big).$$

It is easy to see that for every $n \geq 3$ and $k \geq 2$, $U(n, k) > g_n(k)$. It easily follows that the function $U(x, 2)$ (i.e. the case of two colors) eventually *majorizes* $g_n(x)$ for every n (we recall that $g_n(x)$ is the representative function of the $(n+1)$th Grzegorczyk class). It follows that $U(x, 2)$ is *not* primitive recursive. In fact, $U(x, 2)$ behaves like the well-known Ackermann function $A(x) = g_x(x)$, the classical example of a recursive but not primitive recursive function. In plain language, the original Ackermann function upper bound was ENOOOOORMOUSLY LARGE BEYOND IMAGINATION!!!

In 1988 Shelah proved the following much better upper bound.

Shelah's primitive recursive upper bound: *For every $n \geq 1$ and $k \geq 1$*

$$HJ^c(n, k) \leq \frac{1}{(n+1)k} g_4(n+k+2).$$

That is, given any k-coloring of the hypercube $[n]^d = n^d$ where the dimension $d \geq \frac{1}{(n+1)k} g_4(n+k+2)$, there is always a monochromatic **combinatorial** *line.*

Consider $HJ(n) = HJ(n, 2)$ and $HJ^c(n) = HJ^c(n, 2)$. An easy case study shows that $HJ(3) = HJ^c(3) = 3$, but the numerical value of $HJ(4)$ remains a complete mystery. We know that it is ≥ 5 (see Golomb and Hales [2002]), and also that it is finite, but no one can prove a "reasonable" upper bound such as $HJ(4) \leq 1000$ or even a much weaker bound such as $HJ(4) \leq 10^{1000}$. Shelah's proof gives the explicit upper bound

$$HJ(4) \leq HJ^c(4) \leq g_3(24) = 2^{2^{2^{\cdot^{\cdot^{\cdot^2}}}}},$$

where the "height" of the tower is 24. This upper bound is still *absurdly large*. It is rather disappointing that Ramsey Theory is unable to provide a "reasonable" upper bound even for the first "non-trivial" value $HJ(4)$ of the Hales and Jewett function $HJ(n)$.

In general, it is an open problem to decide whether or not $HJ(n)$ is less than the "plain" tower function $g_3(n)$; perhaps $HJ(n)$ is simply exponential.

It seems to be highly unlikely that the game-theoretic "phase transition" between win and draw for the n^d game is anywhere close to the Hales–Jewett number $HJ(n)$, but no method is known for handling this problem.

We have already introduced the *Win Number* for Clique Games, now we introduce it for the n^d game. Let $\mathbf{w}(n\text{–line})$ denote the least threshold such that for every $d \geq \mathbf{w}(n\text{–line})$ the n^d game is a first player win ("w" stands for "win"). Theorem 6.1 yields the inequality $\mathbf{w}(n\text{–line}) \leq HJ(n)$. By Patashnik's work we know that $\mathbf{w}(4\text{–line}) = 3$, so luckily we don't really need to know the value of the difficult threshold $HJ(4)$. On the other hand, we are out of luck with $\mathbf{w}(5\text{–line})$, which remains a complete mystery. The upper bound $\mathbf{w}(5\text{–line}) \leq HJ(5)$ is "useless" in the sense that Shelah's proof gives a totally ridiculous upper bound for $HJ(5)$.

Open Problem 7.1 *Is it true that $\mathbf{w}(n\text{–line}) < HJ(n)$ for all sufficiently large values of n? Is it true that*

$$\frac{\mathbf{w}(n\text{–line})}{HJ(n)} \longrightarrow 0 \quad \text{as} \quad n \to \infty?$$

The winning sets of an n^d Tic-Tac-Toe game are n-term arithmetic progressions in the space. This motivates the "Arithmetic Progression Game": this is a Positional Game in which the board is the set of the first N integers $[N] = \{1, 2, 3, \ldots, N\}$, and the winning sets are the n-term arithmetic progressions in $[N]$. We call this "Arithmetic Progression Game" the (N, n) **van der Waerden Game**. An obvious motivation for the name is the van der Waerden's Theorem: for every n there is a (least) threshold $W(n) = W(n, 2)$ such that given any 2-coloring of $[N]$ with $N = W(n)$ there is always a monochromatic n-term arithmetic progression. $W(n)$ is called the *van der Waerden Number*. If $N \geq W(n)$, then Theorem 6.1 applies, and yields that the first player has a winning strategy in the (N, n) **van der Waerden Game**. Actually, in view of Theorem 6.1, $W(n)$ can be replaced by its *halving* version $W_{1/2}(n)$. $W_{1/2}(n)$ is defined as the least integer N such that each *halving* 2-coloring of the interval $[N]$ yields a monochromatic n-term arithmetic progression. Trivially, $W_{1/2}(n) \leq W(n)$; is there an n with strict inequality? Unfortunately, we don't know.

Similarly, let $HJ_{1/2}(n)$ denote the least integer d such that in each *halving* 2-coloring of n^d there is a monochromatic n-in-a-line (i.e. *geometric* line). We call $HJ_{1/2}(n)$ the *halving* version of the Hales–Jewett number. By definition

$$HJ_{1/2}(n) \leq HJ(n).$$

Is there an n^d game for which *strict inequality* holds? We don't know the answer to this question, but we *do* know an "*almost n^d game*" for which strict inequality holds: it is the "$3^3 \setminus \{center\}$ game" in which the center of the 3^3 cube is removed, and also the 13 3-in-a-line's going through the center are removed. The "$3^3 \setminus \{center\}$ game" – a truncated version of the 3^3 game – has $3^3 - 1 = 26$ points and $(5^3 - 3^3)/2 - 13 = 49 - 13 = 36$ winning triplets.

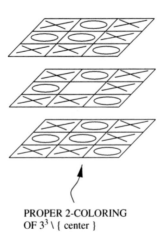

It is impossible to find a Proper Halving 2-Coloring of $3^3 \setminus \{ \text{center} \}$

PROPER 2-COLORING
OF $3^3 \setminus \{ \text{center} \}$

The "$3^3 \setminus \{center\}$ game" has chromatic number two, but every proper 2-coloring has the type (12,14) meaning that one color class has 12 points and the other one has 14 points; *Proper Halving* 2-Coloring, therefore, does not exist. (By the way, this implies, in view of Theorem 6.1, that the "$3^3 \setminus \{center\}$ game" is a first player win.)

The "$3^3 \setminus \{center\}$ game" was a kind of "natural game" example. In the family of *all* hypergraphs it is easy to find examples distinguishing *proper 2-coloring* from *Proper Halving 2-Coloring* in a much more dramatic fashion. We don't even need hypergraphs, it suffices to consider graphs: consider the complete bipartite graph $K_{a,b}$ (i.e. let A and B be disjoint sets where A is a-element and B is b-element, and take the ab point-pairs such that one point is from A and the other one is from B).

Graph $K_{a,b}$ has chromatic number two, and the *only* proper 2-coloring of the $(a+b)$-element point-set is the (A, B)-coloring. If $a = 1$ and b is "large," then the proper 2-coloring is very far from a *halving* 2-coloring.

The case $a = 1$ is the "star." The "star" easily generalizes to n-uniform hypergraphs as follows. Modify a complete $(n - 1)$-uniform hypergraph by adding the same new vertex v_0 to every hyperedge. Clearly the resulting n-uniform hypergraph has a proper 2-coloring (make v_0 red and everything else blue), but the color class of v_0 cannot have size $\geq n$ in a proper 2-coloring. (This construction is due to Wesley Pegden.)

Of course, we can also define the *halving* version of $HJ^c(n)$: $HJ^c_{1/2}(n)$ is the smallest integer d such that in each *halving* 2-coloring of $[n]^d = n^d$ there is a monochromatic *combinatorial* line. By definition $HJ^c_{1/2}(n) \leq HJ^c(n)$.

After this short detour on halving versions let's return to the van der Waerden number; what do we know about $W(n) = W(n, 2)$? First note that

$$W(n, k) \leq n^{HJ(n,k)}.$$

Indeed, we can embed the d-dimensional cube $[n]^d$ into the interval $\{0, 1, 2, \ldots, n^d - 1\}$ by the following natural 1-to-1 mapping: given any string $\mathbf{w} = a_1 a_2 \cdots a_d \in [n]^d$, let

$$f(\mathbf{w}) = (a_1 - 1) + (a_2 - 1)n + (a_3 - 1)n^2 + \ldots + (a_d - 1)n^{d-1}. \tag{7.1}$$

Observe that f maps any n-in-a-line ("geometric line") into an n-term arithmetic progression. It follows that

$$W(n, k) \leq n^{HJ(n,k)}.$$

Shelah's Theorem above immediately gives the following primitive recursive upper bound for the van der Waerden threshold: $W(n, k) \leq g_4(n+k+3)$ for all $n \geq 3$ and $k \geq 2$. This is again the supertower function.

5. Gowers's bound on W(n). The supertower upper bound was enormously improved by a recent breakthrough of Gowers: he pushed $W(n)$ down *well* below the "plain" tower function $g_3(n)$ by using analytic techniques instead of combinatorics.

In fact, Gowers [2001] proved much more: he proved a quantitative *Szemerédi theorem*, i.e. a quantitative *density* version of the van der Waerden's Theorem. (Szemerédi's theorem, which was proved only in 1974, is generally regarded a very deep result; its known proofs are much more difficult than that of van der Waerden's Theorem.) To formulate Gowers's bound we use the arrow-notation $a \uparrow b$ for a^b, with the obvious convention that $a \uparrow b \uparrow c$ stands for $a \uparrow (b \uparrow c) = a^{b^c}$. The relevant "two-color" special case (which has game-theoretic consequences) of Gowers's more general theorem goes as follows.

Gowers's analytic upper bound: *Let*

$$N \geq 2 \uparrow 2 \uparrow 2 \uparrow 2 \uparrow 2 \uparrow (n+9) = 2^{2^{2^{2^{2^{2^{n+9}}}}}}, \tag{7.2}$$

and let S be an arbitrary subset of $\{1, 2, \ldots, N\}$ of size $\geq N/2$. Then S contains an n-term arithmetic progression.

In the general case $N/2$ is replaced by εN with arbitrary $\varepsilon > 0$, and then $1/\varepsilon$ shows up in the tower expression.

Of course, (7.2) implies that $W(n) = W(n, 2) \leq 2 \uparrow 2 \uparrow 2 \uparrow 2 \uparrow 2 \uparrow (n+9)$. This bound is a huge improvement to Shelah's supertower function, but, unfortunately, this is still far too large for "layman combinatorics" (the best-known lower bound is plain exponential, see Section 11). Note that Gowers's paper is extremely complicated: it is 128 pages long and uses deep analytic techniques. Shelah's proof, on the other hand, is relatively short and uses only elementary combinatorics (see Appendix B).

Gowers's Density Theorem implies that, if

$$N \geq 2 \uparrow 2 \uparrow 2 \uparrow 2 \uparrow 2 \uparrow (n+9),$$

then the first player has a winning strategy in the (N, n) van der Waerden Game. This is an application of Theorem 6.1, so we have no idea what first player's winning strategy actually looks like.

Open Problem 7.2 *Consider the (N, n) van der Waerden Game where $N \geq W(n)$; for example, let*

$$N \geq 2 \uparrow 2 \uparrow 2 \uparrow 2 \uparrow 2 \uparrow (n+9).$$

*Find an **explicit** first player's winning strategy.*

It seems to be highly unlikely that the Hales–Jewett number $HJ(n)$ is anywhere close to Shelah's supertower function. Similarly, it seems highly unlikely that the van der Waerden number $W(n)$ is anywhere close to Gowers's 5-times iterated exponential function. Finally, it seems highly unlikely that the "phase transition" between win and draw for the van der Waerden game is anywhere close to the van der

Waerden number $W(n)$. The corresponding *Win Number* is defined as follows. Let $\mathbf{w}(n\text{–term A.P.})$ denote the least threshold such that for every $N \geq \mathbf{w}(n\text{–term A.P.})$ the (N, n) van der Waerden game is a first player win.

Theorem 6.1 implies $\mathbf{w}(n\text{–term A.P.}) \leq W(n)$. It is easily seen

$$\mathbf{w}(3\text{–term A.P.}) = 5 < 9 = W(3).$$

We don't know the exact value of $\mathbf{w}(4\text{–term A.P.})$ but we know that it is less than $W(4) = 35$, and similarly, $\mathbf{w}(5\text{–term A.P.})$ is unknown but it is definitely less than $W(5) = 178$.

Open Problem 7.3 *Is it true that* $\mathbf{w}(n\text{–term A.P.}) < W(n)$ *for all sufficiently large values of n? Is it true that*

$$\frac{\mathbf{w}(n\text{–term A.P.})}{W(n)} \longrightarrow 0 \text{ as } n \to \infty?$$

We conclude Section 7 with a picture.

How to visualize a 4-dimensional game: 3^4 **Tic-Tac-Toe**

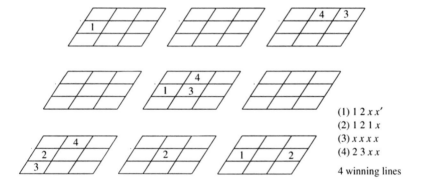

(1) 1 2 x x'
(2) 1 2 1 x
(3) x x x x
(4) 2 3 x x

4 winning lines

8

Two-dimensional arithmetic progressions

1. Weak Win: switching to 2-dimensional goal sets. Open Problems 7.1–7.3 are very depressing. There is no hope of solving them, ever. To gain some self-confidence, we do what we did in Section 6: we switch from ordinary win to Weak Win.

Consider the n^d game with $d \geq HJ(n) + 1$, where $HJ(n)$ is the Hales–Jewett number. The condition guarantees that the board contains two disjoint copies of $n^{HJ(n)}$. The copycat strategy of Theorem 6.2 supplies an *explicit* Weak Win strategy for either player. Theorem 6.2 solves the Weak Win version of Open Problem 7.2, except for the "boundary case" $d = HJ(n)$.

Next consider the (N, n) van der Waerden Game, and assume that

$$N \geq 2 \uparrow 2 \uparrow 2 \uparrow 2 \uparrow 2 \uparrow (n+9)$$

("Gowers's upper bound" for $W(n)$; the $\uparrow$-notation was defined at the end of Section 7). If the first player just wants a *Weak Win* (i.e. to occupy an n-term arithmetic progression, but not necessarily first), then he does *not* need to follow any particular strategy, simply "showing up" is enough. Indeed, at the end of a play he will certainly occupy half of $[N]$ ($N/2$ integers), and by Gowers's Theorem (which is a density theorem, i.e. a quantitative version of Szemerédi's Theorem), he *must* have an n-term arithmetic progression no matter how he plays.

By contrast, to achieve a Weak Win in the n^d game is definitely harder, simply "showing up" is *not* enough. The first player must do something special, but this "special" is not too difficult; a "copycat" pairing strategy does the trick (see Theorem 6.2).

Weak Win motivates the introduction of the *Weak* versions of the *Win Numbers*. Note that we already introduced this concept for clique Games in Section 6. Let **ww**(n–line) denote the least threshold such that for every $d \geq$ **ww**(n–line) the first player can force a Weak Win in the n^d Tic-Tac-Toe game ("**ww**" stands for "Weak Win").

The study of ordinary Tic-Tac-Toe yields **ww**(3–line) $= 2 < 3 =$ **w**(3–line). Patashnik's well-known computer-assisted study of the 4^3 game yields **ww**(4–line) $=$ **w**(4–line) $= 3$ (see Patashnik [1980]).

Open Problem 8.1

(a) *Is it true that* $\mathbf{ww}(n\text{–line}) < \mathbf{w}(n\text{–line})$ *for all sufficiently large values of n? Is it true that*

$$\frac{\mathbf{ww}(n\text{–line})}{\mathbf{w}(n\text{–line})} \longrightarrow 0 \text{ as } n \to \infty ?$$

(b) *Is it true that*

$$\frac{\mathbf{ww}(n\text{–line})}{HJ(n)} \longrightarrow 0 \text{ as } n \to \infty ?$$

Note that Open Problem 8.1 (b) has a positive solution; see Section 12.

Next consider the van der Waerden game. Let $\mathbf{ww}(n\text{–term A.P.})$ denote the least threshold such that for every $N \geq \mathbf{ww}(n\text{–term A.P.})$ first player can force a Weak Win in the (N, n) van der Waerden game.

It is easily seen that

$$\mathbf{ww}(3\text{–term A.P.}) = \mathbf{w}(3\text{–term A.P.}) = 5 < 9 = W(3).$$

Open Problem 8.2

(a) *Is it true that* $\mathbf{ww}(n\text{–term A.P.}) < \mathbf{w}(n\text{–term A.P.})$ *for all sufficiently large values of n? Is it true that*

$$\frac{\mathbf{ww}(n\text{–term A.P.})}{\mathbf{w}(n\text{–term A.P.})} \longrightarrow 0 \text{ as } n \to \infty ?$$

(b) *Is it true that*

$$\frac{\mathbf{ww}(n\text{–term A.P.})}{W(n)} \longrightarrow 0 \text{ as } n \to \infty ?$$

We summarize the "trivial inequalities"

$$\mathbf{ww}(n\text{–line}) \leq \mathbf{w}(n\text{–line}) \leq HJ_{1/2}(n) \leq HJ(n),$$

$$\mathbf{ww}(\text{comb. } n\text{–line}) \leq \mathbf{w}(\text{comb. } n\text{–line}) \leq HJ^c_{1/2}(n) \leq HJ^c(n),$$

and

$$\mathbf{ww}(n\text{–term A.P.}) \leq \mathbf{w}(n\text{–term A.P.}) \leq W_{1/2}(n) \leq W(n).$$

The Achievement and Avoidance Numbers for Clique Games were defined in Section 6, and we stated the exact values in Theorem 6.4. The Achievement and Avoidance Numbers for A.P.s ("Arithmetic Progression") on the board $[N] = \{1, 2, 3, \ldots, N\}$ are defined in the usual way as the inverse of the Weak Win and Reverse Weak Win Numbers

$$\mathbf{A}([N]; \text{A.P.}) = n \text{ if } \mathbf{ww}(n\text{–term A.P.}) \leq N < \mathbf{ww}((n+1)\text{–term A.P.}),$$

$$\mathbf{A}([N]; \text{A.P.}; -) = n \text{ if } \mathbf{ww}(n\text{–term A.P.}; -) \leq N < \mathbf{ww}((n+1)\text{–term A.P.}; -).$$

What can we prove about these Achievement and Avoidance Numbers?

Theorem 8.1

$$A([N]; \text{A.P.}) = (1 + o(1)) \log_2 N, \tag{8.1}$$

$$A([N]; \text{A.P.}; -) = (1 + o(1)) \log_2 N. \tag{8.2}$$

Unfortunately (8.1)–(8.2) is not nearly as striking as (6.1)–(6.2): Theorem 8.1 is just an *asymptotic* result (by the way, we guess that the phase transition from Weak Win to Strong Draw happens at $\log_2 N + O(1)$, i.e. the uncertainty is $O(1)$ instead of $o(\log N)$, see formula (9.2) later). Theorem 8.1 is an *asymptotic* result and Theorem 6.4 is an *exact* result – what a big difference! Where does the difference come from? By comparing the goal-sets of the Van der Waerden Game with the goal-sets of the Clique Game, it is easy to see the difference: K_q is a *rapidly* changing "2-dimensional" (or we may call it "quadratic") configuration, switching q to $q+1$ the size $\binom{q+1}{2} = \binom{q}{2} + q$ makes a "square-root size increase." The n-term AP, on the other hand, is a *slowly* changing linear ("1-dimensional") configuration. This is where the difference is: K_q is a *quadratic* goal and the n-term AP is a *linear* goal.

Two-dimensional Arithmetic Progressions. A natural way to obtain a 2-dimensional version of an n-term A.P. is to take the Cartesian product (for a coherent notation we switch from n to q).

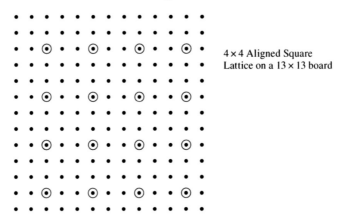

4 × 4 Aligned Square
Lattice on a 13 × 13 board

The Cartesian product of two q-term A.P.s with the same gap is a $q \times q$ (aligned) **Square Lattice**; the Cartesian product of two q-term A.P.s with not necessarily the same gap is a $q \times q$ (aligned) **rectangular lattice**.

Let $(N \times N, q \times q$ Square Lattice) denote the positional game where the board is the $N \times N$ chessboard, and the winning sets are the $q \times q$ (aligned) Square Lattices. Roughly speaking, $(N \times N, q \times q$ Square Lattice) is the "Cartesian square" of the (N, q) van der Waerden game. The total number of winning sets is (k denotes the "gap")

$$\sum_{1 \le k \le (N-1)/(q-1)} (N - k(q-1))^2 = \frac{N^3}{3(q-1)} + O(N^2). \qquad (8.3)$$

Similarly, let $(N \times N, q \times q$ rectangle lattice) denote the positional game where the board is the $N \times N$ chessboard, and the winning sets are the $q \times q$ (aligned) rectangle lattices. The $(N \times N, q \times q$ rectangle lattice) game is another type of "Cartesian product" of the (N, q) van der Waerden game.

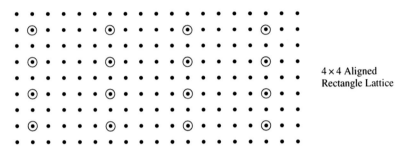

4 × 4 Aligned
Rectangle Lattice

The total number of winning sets is (j and k denote the "gaps")

$$\sum_{1 \le j \le (N-1)/(q-1)} \sum_{1 \le k \le (N-1)/(q-1)} (N - j(q-1))(N - k(q-1)) = \frac{N^4}{4(q-1)^2} + O(N^3).$$

$$(8.4)$$

Let $\mathbf{A}(N \times N;$ Square Lattice) denote the Achievement Number: $q_0 = \mathbf{A}(N \times N;$ Square Lattice) is the largest value of q such that Maker (as the first player) has a Weak Win in the $(N \times N, q \times q$ Square Lattice) game. We shall prove the exact result

$$\mathbf{A}(N \times N; \text{Square Lattice}) = \left\lfloor \sqrt{\log_2 N} + o(1) \right\rfloor,$$

and the same for the Avoidance Number

$$\mathbf{A}(N \times N; \text{Square Lattice}; -) = \left\lfloor \sqrt{\log_2 N} + o(1) \right\rfloor.$$

What happens if the (aligned) Square Lattice is replaced by the (aligned) rectangle lattice? The only difference is an extra factor of $\sqrt{2}$

$$\mathbf{A}(N \times N; \text{rectangle lattice}) = \left\lfloor \sqrt{2 \log_2 N} + o(1) \right\rfloor,$$

$$\mathbf{A}(N \times N; \text{rectangle lattice}; -) = \left\lfloor \sqrt{2 \log_2 N} + o(1) \right\rfloor.$$

What happens if the Aligned Square Lattices are replaced by the **tilted** Square Lattices?

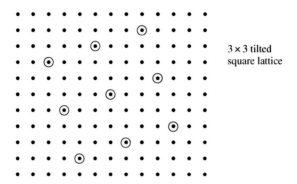

3 × 3 tilted
square lattice

The Achievement Numbers are

$$\mathbf{A}(N \times N; \text{tilted Square Lattice}) = \left\lfloor \sqrt{2\log_2 N} + o(1) \right\rfloor,$$

and the same for the Avoidance Numbers.

The tilted Square Lattice has 3 kinds of generalizations: (a) the tilted rectangle lattice, (b) the rhombus lattice, and, finally, (c) the parallelogram lattice.

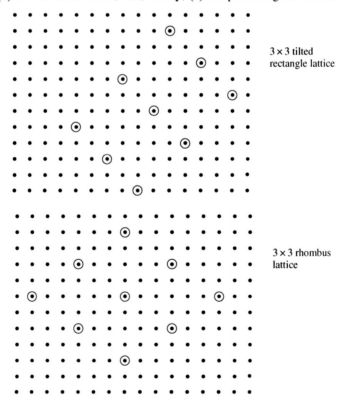

3 × 3 tilted
rectangle lattice

3 × 3 rhombus
lattice

Somewhat surprisingly (a) and (b) have the same threshold as the tilted Square Lattice.

$$\mathbf{A}(N \times N; \text{tilted rectangle lattice}) = \left\lfloor \sqrt{2 \log_2 N} + o(1) \right\rfloor,$$

$$\mathbf{A}(N \times N; \text{rhombus lattice}) = \left\lfloor \sqrt{2 \log_2 N} + o(1) \right\rfloor,$$

and the same for the Avoidance Numbers.

The tilted rectangle lattice and the rhombus lattice are special cases of the class of parallelogram lattices.

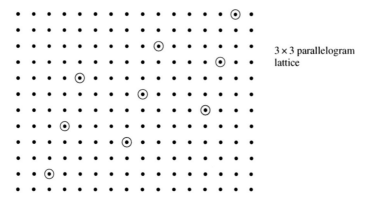

3 × 3 parallelogram lattice

The corresponding Achievement Number is twice as large as that of the Aligned Square Lattice and the same for the Avoidance Number.

$$\mathbf{A}(N \times N; \text{parallelogram lattice}) = \left\lfloor \sqrt{4 \log_2 N} + o(1) \right\rfloor = \left\lfloor 2\sqrt{\log_2 N} + o(1) \right\rfloor,$$

Area-one lattices: It means the class of *parallelogram lattices where the fundamental parallelogram has area one.*

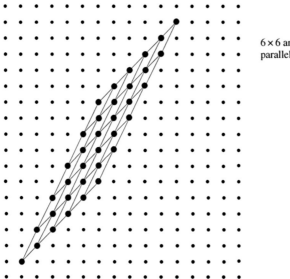

6 × 6 area-one parallelogram lattice

An infinite *area-one lattice* is equivalent to $\mathbf{Z}^2$: it lists every integral lattice point. If the integral vectors $\mathbf{v} = (a, b)$ and $\mathbf{w} = (c, d)$ satisfy $ad - bc = \pm 1$, and $\mathbf{u}$ is also integral, then $\{\mathbf{u} + k\mathbf{v} + l\mathbf{w} : (k, l) \in \mathbf{Z}^2\}$ is an infinite or *unrestricted* area-one lattice in $\mathbf{Z}^2$, and every unrestricted area-one lattice in $\mathbf{Z}^2$ can be obtained this way. A $q \times q$ ("restricted") version means

$$\{\mathbf{u} + k\mathbf{v} + l\mathbf{w} : (k, l) \in \mathbf{Z}^2, 0 \le k \le q-1, 0 \le l \le q-1\}.$$

Let $(N \times N, q \times q$ area-one lattice) denote the positional game where the board is the $N \times N$ chessboard, and the winning sets are the $q \times q$ area-one lattices.

Let $\mathbf{A}(N \times N;$ area-one lattice) denote the Achievement Number: $q_0 = \mathbf{A}(N \times N;$ area-one lattice) is the largest value of q such that Maker (as the first player) has a Weak Win in the $(N \times N, q \times q$ area-one lattice) game. The Avoidance Number $\mathbf{A}(N \times N;$ area-one lattice; $-)$ is defined in the usual way. We have

$$\mathbf{A}(N \times N; \text{area-one lattice}) = \left\lfloor \sqrt{2 \log_2 N} + o(1) \right\rfloor,$$

and, of course, the same for the Avoidance Number

$$\mathbf{A}(N \times N; \text{area-one lattice}; -) = \left\lfloor \sqrt{2 \log_2 N} + o(1) \right\rfloor.$$

Note that instead of "area one" we can have "area A" with any fixed integer A in the range $O(N^2/\log N)$. Needless to say, this result is "beyond Ramsey Theory"; in Ramsey Theory we cannot specify the area.

We started with the "squares," and proceeding in small steps we arrived at the "parallelograms," still a close relative. The far side of the spectrum is when we get rid of geometry completely, and take the Cartesian product of two *arbitrary* q-element subsets of $[N]: B \times C$ where $B \subset [N]$, $C \subset [N]$, $|B| = |C| = q$. This leads to the case of **Complete Bipartite Graphs**: the board is the complete bipartite graph $K_{N,N}$, and the winning sets are the copies of $K_{q,q}$; this is the Bipartite version of the (K_N, K_q) Clique Game. We call $K_{q,q}$ a *symmetric bipartite clique.*

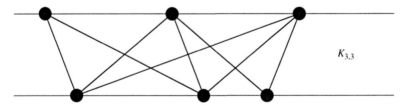

$K_{3,3}$

Not surprisingly, the corresponding Achievement Number is the same as in Theorem 6.4 (a)

$A(K_{N,N}; \text{sym. bipartite clique}) = \lfloor 2 \log_2 N - 2 \log_2 \log_2 N + 2\log_2 e - 3 + o(1) \rfloor$,

and, of course, the same for the Avoidance Number.

2. The Second Main Result of the book. Let me summarize the "lattice results" in a single statement.

Theorem 8.2 *Consider the $N \times N$ board; the following Achievement Numbers are known*

(a) $A(N \times N; \text{Square Lattice}) = \left\lfloor \sqrt{\log_2 N} + o(1) \right\rfloor$,

(b) $A(N \times N; \text{rectangle lattice}) = \left\lfloor \sqrt{2 \log_2 N} + o(1) \right\rfloor$,

(c) $A(N \times N; \text{tilted Square Lattice}) = \left\lfloor \sqrt{2 \log_2 N} + o(1) \right\rfloor$,

(d) $A(N \times N; \text{tilted rectangle lattice}) = \left\lfloor \sqrt{2 \log_2 N} + o(1) \right\rfloor$,

(e) $A(N \times N; \text{rhombus lattice}) = \left\lfloor \sqrt{2 \log_2 N} + o(1) \right\rfloor$,

(f) $A(N \times N; \text{parallelogram lattice}) = \left\lfloor 2\sqrt{\log_2 N} + o(1) \right\rfloor$,

(g) $A(N \times N; \text{area one lattice}) = \left\lfloor \sqrt{2 \log_2 N} + o(1) \right\rfloor$,

(h) $A(K_{N,N}; \text{sym. bipartite clique}) = \lfloor 2 \log_2 N - 2 \log_2 \log_2 N + 2\log_2 e - 3 + o(1) \rfloor$,

and the same for the corresponding Avoidance Number.

This book is basically about 2 exact solutions: Theorem 6.4 and Theorem 8.2. They are the main results. To see the proof of Theorem 8.2 the eager reader can jump ahead to Section 23, and after that to Part D.

3. Generalizations. To solve these particular games we are going to develop some very general hypergraph techniques, which have many more interesting applications. We cannot help discussing here an amusing generalization of Theorem 8.2: we extend Theorem 8.2 from the "shape of square" to any other "lattice polygon" (alternative name: "lattice animal"). What we mean by this is the following: Theorem 8.2 was about $q \times q$ sub-lattices of the $N \times N$ board, a $q \times q$ Aligned Square Lattice ("Case (a)" above) is a *homothetic* copy of the $(q-1)$th member of the "quadratic sequence"

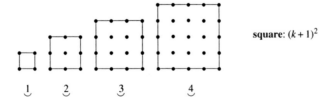

square: $(k+1)^2$

1   2   3   4

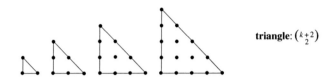

triangle: $\binom{k+2}{2}$

Instead of starting with the "unit square," we could also start with a "triangle," or a "pentagon," or a "hexagon," or an "octagon." In general, we could start with an arbitrary *lattice polygon* (where each vertex is a lattice point) such as the "arrow," the "duck," the "fish," or the "dog" (see the 2 figures below).

Let S be an arbitrary lattice polygon (for example the "fish"); let $S(1) = S$. How do we define the "quadratic sequence" $S(2)$, $S(3)$, ..., $S(k)$, ...? The figures above are rather self-explanatory: let E be an arbitrary *edge* of the boundary of S (*edge* means that the endpoints of E are consecutive lattice points on a straight line). Lattice polygon $S(k)$ arises from S by a magnification of k such that every edge of S is divided into k equal parts by $k-1$ consecutive lattice points on a straight line. For example, the second figure below shows the "fish of order 3."

What is the number of lattice points in the magnified lattice polygon $S(k)$ (the boundary is always included)? By using Pick's well-known theorem, we can easily express the answer in terms of the area A of $S = S(1)$ and the number B of lattice points on the boundary of S. Indeed, the area of $S(k)$ is $k^2 A$, and the number of lattice points on the boundary of $S(k)$ is kB, so by Pick's Theorem the number $I(k)$ of lattice points inside $S(k)$ satisfies the equation $k^2 A = I(k) - 1 + kB/2$. Therefore, the number of lattice points in $S(k)$ ("boundary plus inside") equals the quadratic polynomial $Ak^2 + 1 + Bk/2$. We call $Ak^2 + 1 + Bk/2$ the *quadratic polynomial* of lattice polygon S.

A note about the lattice-point counting formula $Ak^2 + 1 + Bk/2$: the coefficients A and $B/2$ are "half-integers," so it is not obvious why the polynomial $Ak^2 + 1 + Bk/2$ is in fact *integer valued*. The equivalent form (via Pick's Theorem) $Ak^2 + 1 + Bk/2 = B\binom{k+1}{2} + (I-1)k^2 + 1$ takes care of this "formal problem," and clearly demonstrates that the polynomial is integer valued (here is I is the number of lattice points inside S).

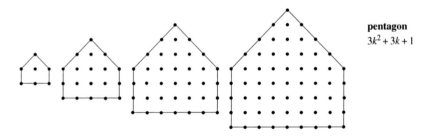

pentagon
$3k^2 + 3k + 1$

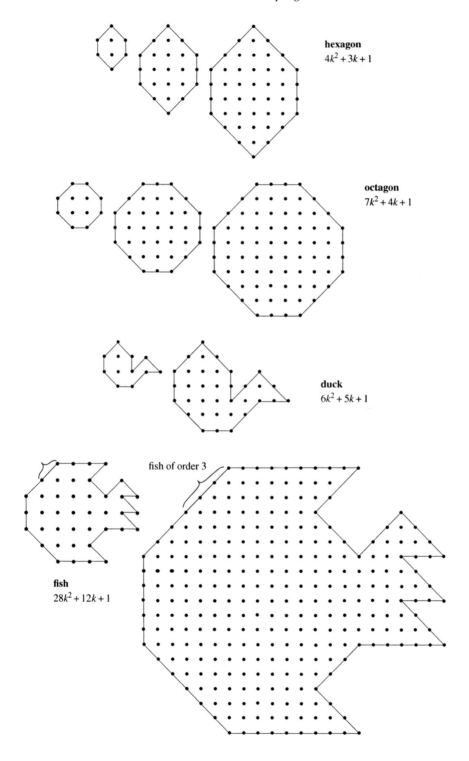

hexagon
$4k^2 + 3k + 1$

octagon
$7k^2 + 4k + 1$

duck
$6k^2 + 5k + 1$

fish of order 3

fish
$28k^2 + 12k + 1$

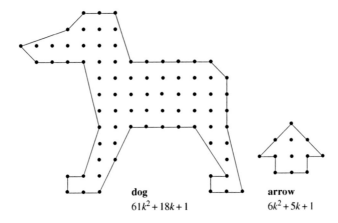

dog   **arrow**
$61k^2 + 18k + 1$ $6k^2 + 5k + 1$

For example, the "triangle," the "pentagon," the "hexagon," the "octagon," the "arrow," the "duck," the "fish," and the "dog" introduced above have *quadratic polynomials* respectively $\binom{k+2}{2}$, $3k^2 + 3k + 1$, $4k^2 + 3k + 1$, $7k^2 + 4k + 1$, $6k^2 + 5k + 1$, $6k^2 + 5k + 1$, $28k^2 + 12k + 1$, and $61k^2 + 18k + 1$. Different shapes may have the same quadratic polynomial, see e.g. the "arrow" and the "duck" (or notice that each "empty triangle" has the same polynomial $\binom{k+2}{2}$).

What happens in Theorem 8.2 (a) if the "square" is replaced by an arbitrary lattice polygon S with area A and "boundary" B? What is the largest "order" k such that by playing on an $N \times N$ board the first player can still achieve a *homothetic* copy of $S(k)$? The largest achievable order $k = k(S; N)$ equals

$$\left\lfloor \frac{1}{\sqrt{A}}\sqrt{\log_2 N} - \frac{B}{4A} + o(1) \right\rfloor, \tag{8.5}$$

which is the perfect generalization of Theorem 8.2 (a).

The similar generalization of Theorem 8.2 (c) ("tilted square") goes as follows. What is the largest "order" k such that playing on an $N \times N$ board the first player can still achieve a *similar* copy of $S(k)$? The largest achievable order $k = k(S; N)$ equals

$$\left\lfloor \frac{1}{\sqrt{A}}\sqrt{2\log_2 N} - \frac{B}{4A} + o(1) \right\rfloor, \tag{8.6}$$

which is the perfect generalization of Theorem 8.2 (c).

A similar generalization of Theorem 6.4 (a) and Theorem 8.2 (h) goes as follows. Let G be an arbitrary finite graph with $V = V(G)$ vertices and $E = E(G)$ edges. A **magnification** $G(q)$ of order q means a graph where each vertex of G is replaced by a q-element set and each edge is replaced by a $q \times q$ complete bipartite graph. $G(q)$ has $q \cdot V$ vertices and $q^2 \cdot E$ edges. Playing the usual (1:1) game on K_N, what is the largest value of q such that Maker can always build an isomorphic copy of $G(q)$? Here comes the exact answer: let $H \subseteq G$ be the subgraph of G with maximum average degree (of course, $H = G$ is allowed), let $A = A(H)$ denote the number of vertices of H and let $B = B(H)$ denote the number of edges of H, then

$$q = \left\lfloor \frac{A}{B} \left(\log_2 N - \log_2 \log_2 N - \log_2 \left(\frac{A}{B} \right) + \log_2 e \right) - \frac{2}{A} + o(1) \right\rfloor. \qquad (8.7)$$

If the "holes" in $G(q)$ are filled up with K_qs, then the analogue of (8.7) is

$$q = \left\lfloor \frac{2A}{A+2B} \left(\log_2 N - \log_2 \log_2 N - \log_2 \left(\frac{2A}{A+2B} \right) + \log_2 e \right) + 1 - 2\frac{A+2B}{A^2} \right\rfloor. \qquad (8.8)$$

Finally, let's go back to ordinary win. We want to define the analogue of the Achievement Number. If the inverse of the Weak Win Number is called the Achievement Number, then what do we call the inverse of the Win Number? We suggest the name Over-Achievement Number. For example, in the van der Waerden Game

$$\mathbf{OA}([N]; \text{A.P.}) = n \quad \text{if} \quad \mathbf{w}(n\text{-term A.P.}) \leq N < \mathbf{w}((n+1)\text{-term AP}).$$

Unfortunately, we know very little about the Over-Achievement Numbers. For example, in the van der Waerden Game the following upper and lower bounds are known

$$(1 + o(1)) \log_2 N \geq \mathbf{OA}([N]; \text{A.P.}) \geq \log_2 \log_2 \log_2 \log_2 \log_2 N.$$

The huge gap is rather disappointing.

Unfortunately, the situation is even *much worse* when the goal sets are 2-dimensional arithmetic progressions. Then the Over-Achievement Number is a total mystery: nobody has the slightest idea of what the true order might be; no one dares to come up with any kind of conjecture. This is just one more reason why we should cherish the exact solutions of the Achievement (Avoidance) Numbers in Theorem 8.2.

Of course, we can define the Over-Achievement Number for the Clique Games as well. For a play on the (ordinary) complete graph K_N let

$$\mathbf{OA}(K_N; \text{clique}) = q \quad \text{if} \quad \mathbf{w}(K_q) \leq N < \mathbf{w}(K_{q+1}),$$

and for a play on the complete p-graph K_N^p with $p \geq 3$ let

$$\mathbf{OA}(K_N^p; \text{clique}) = q \quad \text{if} \quad \mathbf{w}(K_q^p) \leq N < \mathbf{w}(K_{q+1}^p).$$

We know the following upper and lower bounds

$$2\log_2 N \geq \mathbf{OA}(K_N; \text{clique}) \geq \frac{1}{2}\log_2 N,$$

$$\sqrt{6\log_2 N} \geq \mathbf{OA}(K_N^3; \text{clique}) \geq c_1 \log\log N,$$

$$(24\log_2 N)^{1/3} \geq \mathbf{OA}(K_N^4; \text{clique}) \geq c_2 \log\log\log N,$$

$$(120\log_2 N)^{1/4} \geq \mathbf{OA}(K_N^5; \text{clique}) \geq c_3 \log\log\log\log N,$$

and so on. The upper bounds come from Theorem 1.4 (Erdős–Selfridge Theorem) and the lower bounds come from Ramsey Theory (see (6.7)–(6.9)). Again there is a striking contrast between these huge gaps and the exact values in Theorem 6.4.

We know little about the Ramsey Theory thresholds, we know little about the Over-Achievement Numbers, but we know a lot about the Achievement and Avoidance Numbers.

9

Explaining the exact solutions: a Meta-Conjecture

1. What is going on here? Theorems 6.4 and 8.2 described the exact values of infinitely many Achievement and Avoidance Numbers. We are sure the reader is wondering: "What are these *exact values*?" "Where did they come from?" The answer to these questions is surprisingly simple.

Simple Answer: In each one of the "exact solution games" (i.e. games with *quadratic goals*) the "phase transition" from Weak Win to Strong Draw happens when the **winning set size equals the binary logarithm of the Set/Point ratio of the hypergraph**, formally, $\log_2(|\mathcal{F}|/|V|)$.

For example, in the (K_N, K_q) Clique Game the *Set/Point* ratio is $\binom{N}{q}\binom{N}{2}^{-1}$, and the equation

$$\binom{q}{2} = \log_2\left(\binom{N}{q}\binom{N}{2}^{-1}\right)$$

has the real solution

$$q = q(N) = 2\log_2 N - 2\log_2\log_2 N + 2\log_2 e - 3 + o(1),$$

which is exactly (6.1)–(6.2) (the calculations are similar to (6.4)–(6.6)).

In the Aligned Square Lattice Game on an $N \times N$ board, by (8.3) the equation

$$q^2 = \log_2\left(\frac{N^3}{3(q-1)} \cdot N^{-2}\right)$$

has the real solution

$$q = q(N) = \sqrt{\log_2 N} + o(1),$$

which is exactly Theorem 8.2 (a).

In the Aligned Rectangle Lattice Game on an $N \times N$ board, by (8.4) the equation

$$q^2 = \log_2\left(\frac{N^4}{4(q-1)^2} \cdot N^{-2}\right)$$

has the real solution

$$q = q(N) = \sqrt{2\log_2 N} + o(1),$$

which is exactly Theorem 8.2 (b).

We challenge the reader to double-check my **Simple Answer** above by carrying out the analogous calculations for the rest of the "exact solution games" (such as Theorems 6.4 (b)–(c) and 8.2 (c)–(g)).

An alternative way to formulate the **Simple Answer** above is to consider the sum

$$\frac{1}{|V|}\left(\sum_{A\in\mathcal{F}} 2^{-|A|}\right) \tag{9.1}$$

associated with hypergraph $(V, \mathcal{F})$, and to look at (9.1) as a "gauge" for Weak Win: if sum (9.1) is "large" (meaning "much larger than one"), then the positional game is a Weak Win, and if sum (9.1) is "small" (meaning "much smaller than one"), then the positional game is a Strong Draw.

Perhaps sum (9.1) is too crude, and a more delicate sum such as

$$\frac{1}{|V|}\left(\sum_{A\in\mathcal{F}} |A|\cdot 2^{-|A|}\right) \tag{9.2}$$

reflects the "phase transition" from Weak Win to Strong Draw somewhat better.

For example, consider the following n-uniform hypergraph: the board V is a $(2n-1)$-element set $\{w, x_1, x_2, \cdots, x_{n-1}, y_1, y_2, \cdots, y_{n-1}\}$. The family of winning sets consists of all possible n-element subsets A of V with the following two properties: (1) $w \in A$, (2) A contains exactly 1 point from each pair $\{x_i, y_i\}$, $i = 1, 2, \ldots, n-1$. The number of winning sets is 2^{n-1}, and the first player can occupy a winning set in the fastest possible way in n turns. Notice that for this hypergraph sum (9.2) is the "right" gauge to separate Weak Win from Strong Draw.

Notice that (9.2) is exactly the quantitative form of the intuition explained after formula (3) in our informal introduction ("A summary of the book in a nutshell").

For the "exact solution hypergraphs" with "quadratic" goal sets, sums (9.1) and (9.2) are basically the same: the goal size is either $n = \binom{q}{2}$ or $n = q^2$, and if q switches to $q+1$, then n undergoes a "square-root size increase." The effect of this "square-root size increase" in sum (9.1) is much larger than the effect of the extra factor $|A|$ that distinguishes (9.2) from (9.1).

We should be able to separate (9.2) from (9.1) when the goal sets are "linear." Natural examples are the n-term arithmetic progressions in the (N, n) Van der Waerden Game, and the n-in-a-lines's in the n^d hypercube Tic-Tac-Toe. In the (N, n) Van der Waerden Game (9.2) is (basically) equivalent to

$$\frac{\frac{N^2}{n}n2^{-n}}{N} = O(1) \Longleftrightarrow n = n(N) = \log_2 N + O(1).$$

Is it true that in the (N, n) Van der Waerden Game the phase transition from Weak Win to Strong Draw happens at $n = n(N) = \log_2 N + O(1)$?

In the n^d hypercube Tic-Tac-Toe, (9.2) is (basically) equivalent to

$$\frac{\frac{(n+2)^d - n^d}{2} n 2^{-n}}{n^d} = O(1) \Longleftrightarrow d = d(n) = \frac{(\log 2)n^2 - n \log n}{2} + O(n).$$

Is it true that in the n^d hypercube Tic-Tac-Toe the phase transition from Weak Win to Strong Draw happens at $d = d(n) = ((\log 2)n^2 - n \log n)/2 + O(n)$?

We believe both questions have a positive answer, but we don't have a clue how to prove such delicate bounds (we can prove weaker results). The second conjecture about the n^d Tic-Tac-Toe is particularly risky, because the family of n-in-a-line's in the n^d hypercube is extremely degree irregular: the maximum degree is much larger than the average degree (for a degree reduction, see Theorem 12.2; see also Theorem 12.5).

Next consider the general case of *arbitrary* hypergraphs. Is it true that for every uniform hypergraph *the "phase transition" from Weak Win to Strong Draw happens when the winning set size equals the binary logarithm of the Set/Point ratio of the hypergraph?* Is sum (9.1) (or sum (9.2)) the right "gauge" to separate Weak Win from Strong Draw? A general result like that would settle the whole issue once and for all, but of course we are not that lucky: the answer to the general question is an easy "no." This means that the "exact solution hypergraphs" must have some special properties.

What are the "special properties" of the "exact solution hypergraphs"? This is a hard question that requires a longer discussion. We start the discussion by showing first a large class of hypergraphs for which the **Simple Answer** above is false in the strongest possible sense. This is the class of **Strictly Even Hypergraphs**.

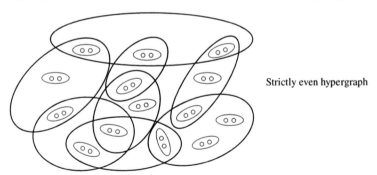

Strictly even hypergraph

The board V of a **Strictly Even Hypergraph** is an even-size set, say a $(2M)$-element set, representing the inhabitants of a little town: M married couples, M husbands, and M wives. The citizens of this little town have a habit of forming clubs, small and large. The same citizen may have membership in many different clubs at the same time, but there is a rule which is strictly enforced: if a citizen is the member of a club, then his/her spouse is automatically a member, too. Each club represents a hyperedge of a Strictly Even Hypergraph (and vice versa). In technical terms, a Strictly Even Hypergraph has an underlying "pairing," and if a hyperedge intersects a "pair," then the hyperedge must contain the whole "pair."

Note that in a Strictly even Hypergraph, every winning set has even size, and, in general, the intersection of an arbitrary family of winning sets has even size too (explaining the name "strictly even").

The Positional Game played on an arbitrary Strictly Even Hypergraph is trivial: the first player cannot occupy a whole winning set. In fact, the first player cannot even achieve a "lead" by one! Indeed, by using a Pairing Strategy, the second player can take the exact half from each winning set (preventing any "lead" of the opponent).

The hypergraph of the (K_N, K_q) Clique Game is very different from a Strictly Even Hypergraph. One of the many peculiarities of a Strictly Even Hypergraph is that its Maximum Pair-Degree equals the Maximum Degree; for the hypergraph of the (K_N, K_q) Clique Game, on the other hand, the Maximum Pair-Degree is much smaller than the Maximum Degree: the hypergraph of the (K_N, K_q) Clique Game is very "homogeneous."

Meta-Conjecture Let $\mathcal{F}$ be an n-uniform hypergraph, and let V denote the union set.

(a) Assume that $\mathcal{F}$ is "homogeneous," which vaguely means the "complete opposite of Strictly Even Hypergraphs." Is it true that if $n < \log_2(|\mathcal{F}|/|V|)+??$, then the first player can force a Weak Win?

(b) Assume that $\mathcal{F}$ is "reasonable." Is it true that if $n > \log_2(|\mathcal{F}|/|V|)+???$, then the second player can force a Strong Draw?

(c) How about the Avoider–Forcer version? Is it true that, under the condition of (a), if $n < \log_2(|\mathcal{F}|/|V|)+??$, then Forcer can force Avoider to occupy a whole winning set?

(d) Is it true that, under the condition of (b), if $n > \log_2(|\mathcal{F}|/|V|)+???$, then Avoider can avoid occupying a whole winning set?

The Meta-Conjecture states that the "phase transition" happens at the same time for both the Maker–Breaker and the Avoider–Forcer games. In fact, there is a third game with exactly the same phase transition (the concept is due to W. Pegden [2005]). We call it the:

Improper–Proper Game: Mr. Improper and Ms. Proper take turns occupying new points of a finite hypergraph; Mr. Improper colors his points red and Ms. Proper colors her points blue. Ms. Proper wins if and only if at the end of the play they produce a proper 2-coloring of the hypergraph (i.e. no hyperedge is monochromatic). Mr. Improper wins if at the end they produce an improper 2-coloring of the hypergraph (there is a monochromatic hyperedge), so draw is impossible by definition.

Notice that this new game has a one-sided connection with both the Maker–Breaker and the Avoider–Forcer Games: a Maker's winning strategy is automatically a Mr. Improper's winning strategy (he can force a red hyperedge), and, similarly, a Forcer's winning strategy is automatically a Mr. Improper's winning strategy (he can force a blue hyperedge).

There is also a fourth game with exactly the same phase transition: the Chooser–Picker Game; for the definition see Section 22.

Let's return now to the Meta-Conjecture: it is a very vague statement, and the reader is justly irritated by its clumsyness. We didn't define what "homogeneous" and "complete opposite of Strictly Even Hypergraphs" mean in (a); we didn't define what "reasonable" means in (b); and, finally, we didn't specify the meaning of the question marks "??" and "???".

How can we clarify the annoyingly vague Meta-Conjecture? Let us start with (a): the "Weak Win criterion." Is there a natural class of hypergraphs which is a complete opposite of Strictly Even Hypergraphs? The class of Almost Disjoint hypergraphs is a good candidate: in an Almost Disjoint hypergraph the Max Pair-Degree is 1, i.e. as small as possible; in a Strictly Even Hypergraph, on the other hand, the Maximum Pair-Degree equals the Maximum Degree, i.e. as large as possible. This suggests that Meta-Conjecture (a) has a good chance of being true for Almost Disjoint hypergraphs. And, indeed, Theorem 1.2 is exactly the result what we are looking for.

Therefore, in the special case of Almost Disjoint hypergraphs, Theorem 1.2 yields the following clarification of the Meta-Conjecture (a):

(a) Assume that $\mathcal{F}$ is an Almost Disjoint n-uniform hypergraph. If $n < \log_2(|\mathcal{F}|/|V|) + 3$, then the first player can force a Weak Win.

If the hypergraph is "far" from being Almost Disjoint, e.g. the Clique Game hypergraph, then the clarification of Meta-Conjecture (a) is much more difficult: this is the main subject of Chapter V.

After this more-or-less successful clarification of Meta-Conjecture (a), we try to clarify Meta-Conjecture (b): the "Strong Draw criterion." We conjecture that the vague requirement to be "reasonable" in (b) simply means to be (nearly) **degree-regular**, where degree-regular means that the *Maximum Degree* is (nearly) equal to the *Average Degree*. Note that if $\mathcal{F}$ is an n-uniform hypergraph and V is the board, then

$$n \cdot |\mathcal{F}| = \text{AverDeg}(\mathcal{F}) \cdot |V|,$$

implying that the Average Degree equals n-times $|\mathcal{F}|/|V|$, i.e. n-times the Set/Point ratio. This is how we get back to the Set/Point ratio, the key parameter of the Meta-Conjecture.

My vague conjecture that *reasonable* actually means *degree-regular* has the following precise form called the Neighborhood Conjecture (see below). First we need to introduce the concept of the *Maximum Neighborhood Size* of a hypergraph. If $\mathcal{F}$ is a hypergraph and $A \in \mathcal{F}$ is a hyperedge, then the $\mathcal{F}$-neighborhood of A is $\mathcal{F}_A = \{B \in \mathcal{F} : B \cap A \neq \emptyset\}$, i.e. the set of elements of $\mathcal{F}$ which intersect A, including A itself. Now the Maximum Neighborhood Size of $\mathcal{F}$ is the maximum of $|\mathcal{F}_A| = |\{B \in \mathcal{F} : B \cap A \neq \emptyset\}|$, where A runs over all elements of $\mathcal{F}$.

The Maximum Neighborhood Size is very closely related to the Maximum Degree. Indeed, if $\mathcal{F}$ is n-uniform, its Maximum Degree is D, and its Maximum Neighborhood Size is S, then $D + 1 \leq S \leq n(D-1) + 1$.

Open Problem 9.1 ("Neighborhood Conjecture") *(a) Assume that $\mathcal{F}$ is an n-uniform hypergraph, and its Maximum Neighborhood Size is less than 2^{n-1}. Is it true that by playing on $\mathcal{F}$ the second player has a Strong Draw?*

Notice that a positive solution would imply a far-reaching generalization of the Erdős–Selfridge Theorem.

Maybe the sharp upper bound $< 2^{n-1}$ is not quite right, and an "accidental" counterexample disproves it. The weaker version (b) below would be equally interesting.

Open Problem 9.1

(b) *If (a) is too difficult (or false), then how about if the upper bound on the Maximum Neighborhood Size is replaced by an upper bound $2^{n-c}/n$ on the Maximum Degree, where c is a sufficiently large positive constant?*

(c) *If (b) is still too difficult, then how about a polynomially weaker version where the upper bound on the Maximum Degree is replaced by $n^{-c} \cdot 2^n$, where $c > 1$ is a positive absolute constant?*

(d) *If (c) is still too difficult, then how about an exponentially weaker version where the upper bound on the Maximum Degree is replaced by c^n, where $2 > c > 1$ is an absolute constant?*

(e) *How about if we make the extra assumption that the hypergraph is Almost Disjoint (which holds for the n^d Tic-Tac-Toe anyway)?*

(f) *How about if we just want a Proper Halving 2-Coloring (i.e. Drawing Terminal Position)?*

The Neighborhood Conjecture, which is an elegant clarification of one-half of the Meta-Conjecture, is a central issue of the book. A good motivation (but not a proof!) for the Neighborhood Conjecture is the "Probabilistic Method," in particular the Erdős–Lovász 2-Coloring Theorem, see Section 11.

Note that part (f) is clearly the easiest problem, and there is a partial result: the answer is positive if the Maximum Degree is $\leq (3/2 - o(1))^n$, see the very end of Section 11. So the real question in (f) is whether or not 3/2 can be replaced by 2.

The board size is (nearly) irrelevant! A good illustration of the Neighborhood Conjecture is the following extension of Theorem 8.2. First, we fix a lattice type from (a) to (g) (the *complete bipartite graph* is excluded!); second, we extend the board from $N \times N$ to a much larger $M \times M$ board, *but* we keep the size of the winning lattices unchanged: the winning lattices are the $q \times q$ lattices in the $M \times M$ board that have diameter, say, $\leq 2N$. An inspection of the proof of Theorem 8.2 (a)–(g) shows that the new board size $M = M(N)$ can be as large as N^2, or N^3, or N^4, or N^{100}, or even superpolynomial like $N^{(\log N)^{c_0}}$ with a small constant $c_0 = 10^{-4}$; if the winning set diameter remains $\leq 2N$, then the Lattice Achievement and Avoidance Numbers remain unchanged. We refer to this phenomenon as the **Irrelevance of the Board Size**. We return to this issue at the end of Section 44.

Of course, Theorem 6.4 (Clique Game) also has a similar extension. Let M be much larger than N, and let K_M denote the clique where the vertex set is the set of consecutive integers $\{1, 2, \ldots, M\}$. A subclique $K_q \subset K_M$ is a winning set if and only if the vertex set $\{i_1, i_2, \ldots, i_q\}$ has the Diameter Property that $\max |i_j - i_l| \leq N$ for any two vertices i_j, i_l of K_q. An inspection of the proof of Theorem 6.4 (a) shows that the new board size parameter $M = M(N)$ can be as large as N^2, or N^3, or N^4, or N^{100}, or even superpolynomial like $N^{(\log N)^{c_0}}$ with a small positive absolute constant c_0; if the winning set diameter remains $\leq N$, then the Clique Achievement and Avoidance Numbers remain unchanged. This is another example of the **Irrelevance of the Board Size**.

Let us return to Open Problem 9.1. Unfortunately it cannot be solved in general, but we can prove some important partial results. This is how we settle the Strong Draw parts of the *exact solutions*; this is the subject of Part D of the book.

Most success is achieved in the Almost Disjoint case. If hypergraph $\mathcal{F}$ is (nearly) uniform, (nearly) degree-regular, (nearly) Almost Disjoint, and the global size $|\mathcal{F}|$ is not extremely large, then the Meta-Conjecture holds. More precisely, we have the following:

Theorem for Almost Disjoint Hypergraphs: *Assume that hypergraph $\mathcal{F}$ is n-uniform and Almost Disjoint; let V denote the union set.*

(a) If the Average Degree is sufficiently large

$$\mathrm{AverDeg}(\mathcal{F}) = \frac{n|\mathcal{F}|}{|V|} > n \cdot 2^{n-3},$$

then the first player can occupy a whole $A \in \mathcal{F}$ ("Weak Win").

(b) If the global size $|\mathcal{F}|$ and the Max Degree satisfy the upper bounds

$$|\mathcal{F}| < 2^{n^{1.1}} \quad \text{and} \quad \mathrm{MaxDegree}(\mathcal{F}) < 2^{n - 4n^{2/5}},$$

then the second player can put his mark in every $A \in \mathcal{F}$ ("Strong Draw").

If the hypergraph is (nearly) degree-regular, then the Average Degree and the Max Degree are (nearly) equal, see (a) and (b). Observe that part (a) is just a restating of the simple Theorem 1.2; on the other hand, part (b) is the very difficult "third ugly theorem," which will be proved in Chapter VIII.

Breaking the square-root barrier. If we have *quadratic* goal sets of size $n = n(q) = q^2$ or $\binom{q}{2}$, then switching from q to $(q+1)$ means a (roughly) $\sqrt{n}$ increase in the size. This explains why "breaking the square-root barrier," i.e. proving the error term $o(\sqrt{n})$ in the exponent of 2, is so crucial. Notice that the error term $4n^{2/5}$ in (b) above is clearly $o(\sqrt{n})$; this is a good indication of why we have a chance to find the exact solution of infinitely many games, and also why we need large n.

Actually this is an over-simplification, because neither of our main games (Clique Games or Lattice Games) gives an Almost Disjoint hypergraph. The two Square Lattice Games, aligned and tilted, come very close to Almost Disjointness: in both cases the Max Pair-Degree $\Delta_2 = \Delta_2(\mathcal{F})$ is "negligible" (a superlogarithmic function of the board parameter N). More precisely:

(1) the Aligned Square Lattice Game on $N \times N$: $\Delta_2 \le \binom{q^2}{2} = O((\log N)^2)$;
(2) the Tilted Square Lattice Game on $N \times N$: $\Delta_2 \le \binom{q^2}{2} = O((\log N)^2)$.

Classes (1)–(2) represent the "easy" lattice games. How about the rest of the lattice games, such as the aligned rectangle lattice or the parallelogram lattice games? The most general class is the parallelogram lattice game, and even this one shows some resemblance to Almost Disjointness: any 3 non-collinear points in the $N \times N$ board determine $\le \binom{q^2}{3} = O((\log N)^3)$ $q \times q$ parallelogram lattices (i.e. a kind of triplet-degree is negligible).

So why can we find the exact solution for the Lattice Games? Well, a short/intuitive explanation is that (1) we have the "Almost Disjoint Theorem," and (2) the corresponding hypergraph for each Lattice Game is "nearly" Almost Disjoint.

Unfortunately, the *real* explanation – the detailed proof – is very long; this is why we made this extra effort to give some intuition.

Now we more-or-less "understand" the Lattice Games, at least on an intuitive level. (Notice that the same intuition doesn't apply to the Clique Games: the clique hypergraph is a far cry from Almost Disjointness.)

Summary of the main result: If $(V, \mathcal{F})$ is a "homogeneous" hypergraph, which includes the "uniform," or at least "nearly uniform," and also the "nearly degree-regular," then a sum such as

$$\frac{1}{|V|} \left(\sum_{A \in \mathcal{F}} 2^{-|A|} \right)$$

(see (9.1)) associated with hypergraph $(V, \mathcal{F})$ is the right gauge to describe Weak Win: if the sum is larger than one, the Positional Game is expected to be a Weak Win; if the sum is smaller than one, the Positional Game is expected to be a Strong Draw.

The same applies for the Reverse ("Avoider–Forcer") game: if the sum is larger than one, then Forcer can force Avoider to occupy an whole hyperedge of $\mathcal{F}$, and if the sum is smaller than 1, then Avoider can avoid accupying any hyperedge.

This is certainly true for the class of "degree regular Ramsey-type games with quadratic goals" (the main subject of the book). It is an exciting research project to extend these sporadic results to larger classes of positional games; for more about this, see Section 46.

The Meta-Conjecture is about "local randomness." A different way to put it is the so-called "Phantom Decomposition Hypothesis"; this new viewpoint will be discussed at the end of Section 19.

2. Extensions of the main result. There are two natural ways to generalize the concept of the Positional Game: one way is the (1) *discrepancy version*, where Maker wants (say) 90% of some hyperedge instead of 100%; another way is the (2) *biased version* like the (1 : 2) play, where (say) Maker claims 1 point per move and Breaker claims 2 points per move.

Neither generalization is a perfect success; the *discrepancy version* generalizes more smoothly; the *biased version*, unexpectedly, leads to some tormenting(!) technical difficulties. We will discuss the details in Chapter VI; here just a summary is given.

In the α-Discrepancy Game, where Maker wants an α-part from some $A \in \mathcal{F}$ $(1 > \alpha > 1/2$ is a given constant, such as $\alpha = .9$, meaning that "90% majority suffices to win"), it is plausible to replace sum (9.1) with

$$\frac{1}{|V|} \left(\sum_{A \in \mathcal{F}} \left(\sum_{k=\alpha|A|}^{|A|} \binom{|A|}{k} \right) 2^{-|A|} \right) \approx \frac{1}{|V|} \left(\sum_{A \in \mathcal{F}} 2^{-(1-H(\alpha))|A|} \right). \tag{9.3}$$

The Shannon's entropy $H(\alpha) = -\alpha \log_2 \alpha - (1 - \alpha) \log_2(1 - \alpha)$ comes from applying Stirling's formula to the binomial coefficients.

The first surprise is that we cannot solve the α-Discrepancy problem for the Clique Game, but we succeed for all Lattice Games. To understand why, the reader should consult Sections 28–29.

Next consider the biased version. In the $(m:b)$ play Maker takes m new points and Breaker takes b new points per move. The obvious analogue of (9.1) is the sum

$$\frac{1}{|V|} \left(\sum_{A \in \mathcal{F}} \left(\frac{m+b}{m} \right)^{-|A|} \right). \tag{9.4}$$

Here is a short list of "successes" and "failures" in the biased case.

(1) In the (2:1) Avoidance version of all Lattice Games (Forcer is the underdog), we have the exact solution, and it is given by formula (9.4).

(2) In the (2:1) Achievement version of the Aligned Square Lattice Game (Breaker is the underdog), formula (9.4) fails to give the truth.

(3) In the (1:2) Achievement version of the Aligned Square Lattice Game (Maker is the underdog), we have the exact solution, and it is given by formula (9.4).

(4) In the (1:2) Achievement version of the Aligned Rectangle Lattice Game (Maker is the underdog), we don't know the exact solution.

(5) In the (1:2) Chooser–Picker version (in each turn Picker picks 2 new points from the board and offers them to Chooser, Chooser chooses one of them and the remaining two go back to Picker; Chooser is the "builder"), we come very close to the exact solution for all Clique and Lattice Games, and these solutions are given by formula (9.4).

(6) In the (2:2) and (2:1) Achievement versions, we don't know the exact solution for any class of games, but in many cases we can prove the "building part" (the "blocking part" remains open).

Fact (2) on the list above indicates that the biased version of the Meta-Conjecture has to be more complicated than formula (9.4) (at least when Maker is the top-dog). At the end of Section 30 a detailed discussion is given of what we believe to be the correct form of the **Biased Meta-Conjecture**. In a nutshell, our conjecture says that the threshold $n = \log_2(|\mathcal{F}|/|V|)$ in the Meta-Conjecture has to be replaced by

$$n = \log_{\frac{m+b}{m}}(|\mathcal{F}|/|V|) + \log_{\frac{m}{m-b}}|V|, \qquad (9.5)$$

when Maker is the topdog and plays the $(m : b)$ achievement version on $(V, \mathcal{F})$ (i.e. $m > b \geq 1$). Threshold (9.5) is motivated by the "Random Play plus Cheap Building" intuition.

List (1)–(6) above clearly demonstrates that the biased case is *work in progress*. There are many non-trivial partial results (see Sections 30–33), but we are very far from a complete solution. The reader is challenged to participate in this exciting research project.

Two sporadic results are mentioned as a sample. For every α with $1/2 < \alpha \leq 1$, let $\mathbf{A}(N \times N;$ Square Lattice; $\alpha)$ denote the largest value of q such that Maker can always occupy $\geq \alpha$ part of some $q \times q$ Aligned Square Lattice in the $N \times N$ board. We call it the α-*Discrepancy Achievement Number*. We can similarly define the α-Discrepancy Achievement Number for the rest of the Lattice Games, and also for the Avoidance version.

Theorem 9.1 *Consider the $N \times N$ board; this is what we known about the α-Discrepancy Achievement Numbers of the Lattice Games:*

(a) $\left\lceil \sqrt{\frac{\log_2 N}{1-H(\alpha)}} + o(1) \right\rceil \geq A(N \times N; \text{ Square Lattice}; \alpha) \geq \left\lfloor \sqrt{\frac{\log_2 N}{1-H(\alpha)}} - c_0(\alpha) - o(1) \right\rfloor$

(b) $\left\lceil \sqrt{\frac{2\log_2 N}{1-H(\alpha)}} + o(1) \right\rceil \geq A(N \times N; \text{ rectangle lattice}; \alpha) \geq \left\lfloor \sqrt{\frac{2\log_2 N}{1-H(\alpha)}} - c_0(\alpha) - o(1) \right\rfloor$

(c) $\left\lceil \sqrt{\frac{2\log_2 N}{1-H(\alpha)}} + o(1) \right\rceil \geq A(N \times N; \text{ tilted square latt.}; \alpha) \geq \left\lfloor \sqrt{\frac{2\log_2 N}{1-H(\alpha)}} - c_0(\alpha) - o(1) \right\rfloor$

(d) $\left\lceil \sqrt{\frac{2\log_2 N}{1-H(\alpha)}} + o(1) \right\rceil \geq A(N \times N; \text{ tilt. rect. lattice}; \alpha) \geq \left\lfloor \sqrt{\frac{2\log_2 N}{1-H(\alpha)}} - c_0(\alpha) - o(1) \right\rfloor$

(e) $\left\lceil \sqrt{\frac{2\log_2 N}{1-H(\alpha)}} + o(1) \right\rceil \geq A(N \times N; \text{ rhombus lattice}; \alpha) \geq \left\lfloor \sqrt{\frac{2\log_2 N}{1-H(\alpha)}} - c_0(\alpha) - o(1) \right\rfloor$

(f) $\left\lceil 2\sqrt{\frac{\log_2 N}{1-H(\alpha)}} + o(1) \right\rceil \geq A(N \times N; \text{ parall. lattice}; \alpha) \geq \left\lfloor 2\sqrt{\frac{\log_2 N}{1-H(\alpha)}} - c_0(\alpha) - o(1) \right\rfloor$

(g) $\left\lceil \sqrt{\frac{2\log_2 N}{1-H(\alpha)}} + o(1) \right\rceil \geq A(N \times N; area-one lattice; \alpha) \geq \left\lfloor \sqrt{\frac{2\log_2 N}{1-H(\alpha)}} - c_0(\alpha) - o(1) \right\rfloor$

and the same for the corresponding Avoidance Number. Here the function $H(\alpha) = -\alpha \log_2 \alpha - (1-\alpha) \log_2(1-\alpha)$ *is the Shannon's entropy, and*

$$c_0(\alpha) = \sqrt{\frac{\log 2 \cdot \alpha(1-\alpha)}{2}} \log_2\left(\frac{\alpha}{1-\alpha}\right)$$

is a constant (depending only on α*) which tends to 0 as* $\alpha \to 1$.

If α is close to 1, then the additive constant $c_0(\alpha)$ is small, so the upper and lower bounds coincide. Thus for the majority of the board size N we know the exact value of the α-Discrepancy Achievement and Avoidance Numbers; well, at least for the Lattice Games (when α is close to 1).

Next consider the $(a:1)$ Avoidance Game where $a \geq 2$; this means Avoider takes a points and Forcer takes 1 point per move. Let $A(N \times N; \text{ Square Lattice}; a:1; -)$ denote the largest value of q such that Forcer can always force Avoider to occupy a whole $q \times q$ Aligned Square Lattice in the $N \times N$ board. This is called the *Avoidance Number of the biased* $(a:1)$ game where Forcer is the underdog.

Theorem 9.2 *Consider the* $N \times N$ *board; let* $a \geq 2$ *and consider the* $(a:1)$ *Avoidance Game where Forcer is the underdog. We know the biased Avoidance Numbers:*

(a) $A(N \times N; \text{ Square Lattice}; a:1; -) = \left\lfloor \sqrt{\frac{\log N}{\log(1+\frac{1}{a})}} + o(1) \right\rfloor$,

(b) $A(N \times N; \text{ rectangle lattice}; a:1; -) = \left\lfloor \sqrt{\frac{2\log N}{\log(1+\frac{1}{a})}} + o(1) \right\rfloor$,

(c) $A(N \times N; \text{ tilted Square Lattice}; a:1; -) = \left\lfloor \sqrt{\frac{2\log N}{\log(1+\frac{1}{a})}} + o(1) \right\rfloor$, *(d)* $\mathbf{A}(N \times N;\ \text{tilted rectangle lattice};\ a:1;-) = \left\lfloor \sqrt{\frac{2\log N}{\log(1+\frac{1}{a})}} + o(1) \right\rfloor,$

(e) $\mathbf{A}(N \times N;\ \text{rhombus lattice};\ a:1;-) = \left\lfloor \sqrt{\frac{2\log N}{\log(1+\frac{1}{a})}} + o(1) \right\rfloor,$

(f) $\mathbf{A}(N \times N;\ \text{parallelogram lattice};\ a:1;-) = \left\lfloor 2\sqrt{\frac{\log N}{\log(1+\frac{1}{a})}} + o(1) \right\rfloor,$

(g) $\mathbf{A}(N \times N;\ \textit{area-one lattice};\ a:1;-) = \left\lfloor \sqrt{\frac{2\log N}{\log(1+\frac{1}{a})}} + o(1) \right\rfloor.$

For the proofs of (1)–(6) and Theorems 9.1–9.2, the reader is referred to Chapter VI.

3. Some counter-examples. Let us return to the Meta-Conjecture one more time: a surprising byproduct is the unexpected coincidence of the Achievement and Avoidance Numbers, at least for "homogeneous" hypergraphs in the usual (1:1) play. This is interesting because, for *arbitrary* hypergraphs, we can distinguish the Maker–Breaker game from the Avoider–Forcer version. A hypergraph example where Maker can over-perform Avoider is the following 3-uniform hypergraph with five 3-sets on a 6-element board:

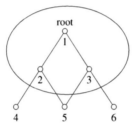

winning sets: $\{1, 2, 3\}$,
$\{1, 2, 4\}, \{1, 2, 5\}$
$\{1, 3, 5\}, \{1, 3, 6\}$

Here Maker (the first player) has an easy winning strategy. By contrast, in the Reverse version, Forcer does *not* have a winning strategy. Indeed, if Forcer is the second player, then Avoider wins by avoiding the "root." If Forcer is the first player, then Avoider can still win – the easy case study is left to the reader.

This simple construction can be easily "amplified" by taking the Cartesian product. Let us take $2n$ copies of the 3-uniform hypergraph, and define a $3n$-uniform hypergraph of $\binom{2n}{n}5^n$ winning sets as follows: a winning set consists of n 3-sets, where the 3-sets are from n distinct copies of the 3-uniform hypergraph.

1 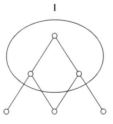 $\cdots$ 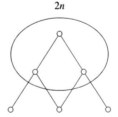 $2n$

$2n$ copies

It is clear that Maker can occupy a whole $3n$-element winning set (no matter he is the first or second player).

In the Reverse Game, however, Avoider can avoid taking more than $2n$ elements from any winning set (no matter whether he is the first or second player). This means Forcer cannot force 66.7% of what Maker can achieve; a big quantitative difference between "achievement" and "avoidance."

How about the other direction? Can Forcer force *more* than what Maker can achieve? For a long time we couldn't find any example for this, and started to believe that there is a *one-sided* implication here which is formally expressed as

$$\text{Forcer's win} \Rightarrow \text{Maker's win.}$$

We pretty much believed in the intuition that "the *reluctant* Avoider can never have more than the *eager* Maker," which has the following precise form:

> *Let $\mathcal{F}$ be an arbitrary finite hypergraph, and assume that in the Avoider–Forcer game on $\mathcal{F}$ Forcer has a winning strategy (i.e. Forcer can force the reluctant Avoider to occupy a whole winning set). Is it true that by playing the Maker–Breaker game on the same hypergraph Maker has a winning strategy (i.e. the eager Maker himself can occupy a whole winning set)?*

The guess, "yes," turned out to be wrong. Two students, Narin Dickerson (Princeton) and Brian Cornell (Rutgers), independently of each other, came up with two different counter-examples. Dickerson's example is remarkably simple, and goes as follows: consider a cycle of length 5, the players take vertices, and the winning sets are the five 3-consecutives. It is easy to see that Breaker can prevent Maker from occupying 3 consecutive vertices on the 5-cycle; on the other hand, Forcer, as the second player, can force Avoider to occupy 3 consecutive vertices on the 5-cycle.

Again the construction can be "amplified" by taking the Cartesian product. Let us take $2n$ disjoint copies of Dickerson's 3-uniform hypergraph, and define a $3n$-uniform hypergraph of $\binom{2n}{n}5^n$ winning sets as follows: a winning set consists of n 3-sets, where the 3-sets are from n distinct copies of the 3-uniform hypergraph.

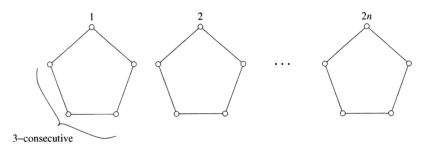

2n copies of Dickerson's construction

The point is that in the Reverse Game, Avoider can be forced to be the "first" in at least n copies, and so he can be forced to occupy a whole $3n$-element winning set (no matter whether Avoider is the first or second player).

In the normal version, however, Maker can be prevented from occupying a winning triplet from any copy, so Maker can be prevented from taking more than $2n$ elements from any $3n$-element winning set (no matter whether Maker is the first or second player). This means Forcer can be forced to take a whole winning set, but Maker cannot even occupy 66.67% of a winning set; a big quantitative ("factor 3/2") difference between "achievement" and "avoidance" in favor of "avoidance."

Cornell's construction is a *natural* example: it is an arithmetic progression game; the board is the set of the first 12 integers $[12] = \{1, 2, \ldots, 12\}$ and the winning sets are the 4-term arithmetic progressions in $[12]$. A rather complicated case study shows that Maker cannot occupy a 4-term arithmetic progression (no matter whether he is the first or second player); in the Reverse game; however, Avoider, as the second player, can be forced to occupy a 4-term arithmetic progression. We challenge the reader to double-check this statement. Needless to say, Cornell's construction can be also "amplified" by taking the Cartesian product: it leads to the slightly weaker "factor 4/3" (instead of 3/2).

This is the short list of hypergraphs for which we know that the Maker–Breaker and Avoider–Forcer versions differ from each other substantially. Are there other examples? Can the "factor 3/2" be pushed up to (say) 100?

It is important to see that the Maker–Breaker and Avoider–Forcer games exhibit very different behavior. The first one is a "hot" game with a possibility for quick win by Maker; the second one is a "cold" game where Avoider may lose in the last move. A good illustration is the following "binary tree hypergraph": the full-length branches of a binary tree with n levels form an n-uniform family of 2^{n-1} winning sets (the players occupy vertices of the binary tree). Playing on this hypergraph Maker (the first player) can occupy a full branch in n moves (as quick as possible: Economical Winner). In the Reverse Game Forcer (the second player) wins (by an easy Pairing Strategy), but Avoider can postpone his loss till the last move (by taking the "root"), i.e. Forcer's win is as *slow* as possible.

The same thing happens for the "triangle" game on a large clique: the first player can own a K_3 in his 4th move (or before); the Reverse "triangle" game, on the other hand, is very slow, see the next exercise.

Exercise 9.2 *Consider Sim on a very large board: the board is a complete graph K_N on N vertices, the players alternately claim a previously unoccupied edge per move, and that player loses who gets a triangle K_3 of his own first.*

(a) *Show that either player can avoid losing in less than $\frac{1}{4}\binom{N}{2}$ turns (i.e. when less than half of the edges are taken).*
(b) *Every play must have a loser in $\left(\frac{2}{5} + o(1)\right)\binom{N}{2}$ turns.*

Here is another question: We know that Sim, the $(K_6, K_3, -)$ Clique Game, is a second player win. Is it true that the $(K_N, K_3, -)$ Clique Game remains a second player win for every $N \geq 6$?

The many sophisticated counter-examples above make it even more interesting that the Achievement and Avoidance Numbers for Cliques and Sublattices are *equal*. More precisely, we can prove that they are equal for the overwhelming majority of the vertex size N in K_N. In fact, much more is true: the equality of the Achievement and Avoidance Numbers is a *typical* property. Typical in the sense that K_N can be replaced by a typical graph G_N on N vertices, the complete 3-uniform hypergraph K_N^3 can be replaced by a typical 3-graph $G_N^3 \subset K_N^3$, the $N \times N$ grid can be replaced by a typical subset; playing on these new boards the equality still holds! At the beginning of Section 46 we will return to this far-reaching and exciting generalization.

It is difficult to know how to explain the (typical) equality of the Achievement and Avoidance Numbers. All that can be said is that the highly technical and long proof of the Achievement case can be easily adapted, like *maximum* is replaced by *minimum*, to yield exactly the same "boundary" for the Avoidance game. There should be a better, shortcut explanation for this coincidence that we happen to overlook! Can the reader help out here?

10

Potentials and the Erdős–Selfridge Theorem

1. Perfect play vs. random play. Let us go back to Section 6 and consider the Clique Game on board K_N; what is the largest clique that Maker can build? The answer is K_{q_1} with

$$q_1 = q_1(N) = \lfloor 2\log_2 N - 2\log_2 \log_2 N + 2\log_2 e - 3 + o(1) \rfloor, \qquad (10.1)$$

where $\log_2$ stands for the base 2 logarithm. In other words, a "perfect" Maker's best achievement is K_{q_1} with (10.1). ("Perfect" Maker means God, or at least a world champion chess player like Kasparov).

How about a "random" Maker, who chooses among his options in a random way by uniform distribution? More precisely, what happens if *both* players play "randomly," i.e. if a "random" Maker plays against a "random" Breaker? Then the answer is a K_{q_2} with

$$q_2 = q_2(N) = \lfloor 2\log_2 N - 2\log_2 \log_2 N + 2\log_2 e - 1 \rfloor. \qquad (10.2)$$

The two thresholds are strikingly close to each other: $q_1 = q_2 - 2$, i.e. the Clique Achievement Number is 2 less than the Majority-Play Clique Number.

Note that the Majority-Play Number is a much more accessible, much "easier" concept than the Achievement Number. Indeed, we can easily simulate a *random play* on a computer, but it is impossible to simulate an *optimal play* (because of the enormous computational complexity).

We interrupt the discussion with a short detour. There are two more natural combinations that are worth mentioning: (1) a perfect Maker plays against a "random" Breaker, and (2) a "random" Maker plays against a perfect Breaker. What is the largest clique that Maker can achieve? Case (1) is trivial: Maker just keeps building one giant clique, and "random" Breaker will not notice it in the first $O(N^{4/3})$ moves; i.e. Maker can build a polynomial(!) clique of $N^{2/3}$ vertices.

Case (2) is not trivial, and we simply don't know what is going on. Perhaps a "random" Maker can build a clique of size close to (10.1), but we don't know how to prove it. A "random strategy" may sound simple, but the best that we can expect

146

from a "random strategy" is for it to perform well with probability close to 1, and this is not nearly as satisfying as a deterministic optimal strategy that performs well all the time. In this book, we always supply an explicit potential strategy.

For an interesting application of "random strategy" see Section 49: Bednarska–Luczak Theorem.

Even allowing for strong reservations about "random strategies," it is still quite interesting to ask the following:

Open Problem 10.1 *Is it true that the "Maker's building" results in the book, proved by using explicit potentials, can be also achieved by a Random Strategy (of course, this means the weaker sense that the strategy works with probability tending to 1 as the board size tends to infinity)?*

Let us leave the Clique Games, and switch to the lattice games of Section 8. Theorem 8.2 describes the Achievement Numbers, but how about the corresponding Majority-Play Numbers? Are they still close to each other? The answer is "no." The *negligible* additive constant 2 in the Clique Game becomes a *substantial* multiplicative constant factor (larger than 1) for the lattice games! Indeed, the "lattices" are quadratic goal sets, so the Majority-Play Number is the solution of a simple equation: "the expected number of the monochromatic goal lattices in a Random 2-Coloring of the $N \times N$ board equals 1." For different types of lattices, we get different equations; each one is simple (because the expected value is linear). For example, if the goal is a $q \times q$ Aligned Square Lattice, the Achievement Number is (see Theorem 8.2 (a))

$$q_1 = q_1(N) = \left\lfloor \sqrt{\log_2 N} + o(1) \right\rfloor. \tag{10.3}$$

The corresponding Majority-Play Number is the solution of the equation

$$\frac{N^3}{3(q-1)} = 2^{q^2}$$

in $q = q(N)$ (see (8.3)), which gives

$$q = q_2 = q_2(N) = \left\lfloor \sqrt{3\log_2 N} + o(1) \right\rfloor. \tag{10.4}$$

They **are not** too close: the q_2/q_1 ratio is $\sqrt{3}$. Next consider the case where the goal is a $q \times q$ aligned rectangle lattice. The Achievement Number is (see Theorem 8.2 (b))

$$q_1 = q_1(N) = \left\lfloor \sqrt{2\log_2 N} + o(1) \right\rfloor. \tag{10.5}$$

The corresponding Majority-Play Number is the solution of the equation

$$\frac{N^4}{4(q-1)^2} = 2^{q^2}$$

in $q = q(N)$ (see (8.4)), which gives

$$q = q_2 = q_2(N) = \left\lfloor 2\sqrt{\log_2 N} + o(1) \right\rfloor. \tag{10.6}$$

This time the ratio is $q_2/q_1 = \sqrt{2}$, still larger than 1.

In these examples the Achievement Number is always less than the Majority-Play Number: slightly less in the Clique Game and subtantially less in the Lattice Games.

What the Erdős–Selfridge Theorem (see Theorem 1.4) really says is the following: the Achievement Number is *always* less (or equal) than the Majority-Play Number, and this is a very general inequality, it holds for every finite hypergraph.

As far as we know the Erdős–Selfridge Theorem (published in 1973) was the first potential criterion for 2-player games. Of course, potentials were widely used well before 1973, in both physics and mathematics. First we say a few words about potentials in general, and discuss the Erdős–Selfridge Theorem later.

The origins of the *potential technique* goes back to physics. Consider for example a pendulum, a favorite example of undergraduate Newtonian Mechanics. When a pendulum is at the top of its swing, it has a certain potential energy, and it will attain a certain speed by the time it reaches the bottom. Unless it receives extra energy, it cannot attain more than this speed, and it cannot swing higher than its starting point.

An even more elementary (highschool-level) example is shown in the figure below: slightly pushing the ball at the start, it will go up the first hill (we ignore friction), it will go up the second hill, but how about the third hill? Can the ball go up the third hill?

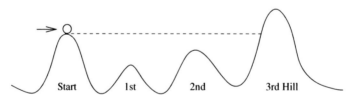

The answer is an obvious "no": the third hill is higher than the start, and all that matters here is the "height" (we ignore friction and air resistance). Please, don't laugh about this example: this childishly simple argument with the figure is a perfect illustration for the core idea of this book!

Now we leave physics and move to mathematics; first we study puzzles (1-player games). The potential technique is very successful in different versions of *Peg Solitaire*. A striking example is Conway's famous solution of the *Solitaire Army*.

2. Positions with limited potential: Solitaire Army. The common feature of the *Solitaire* puzzles is that each one is played with a board and men (pegs), the board contains a number of holes each of which can hold 1 man. Each move consists of a jump by 1 man over 1 or more other men, the men jumped over being removed

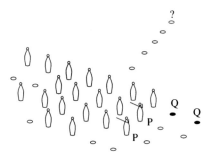

Figure 10.1

from the board. Each move therefore reduces the number of men on the board (for more, see Beasley [1992], an excellent little book).

The *Solitaire Army* is played on the infinite plane and the holes are in the lattice points (see pp. 715–717 in the *Winning Ways*). The permitted move is to jump a man horizontally or vertically but not diagonally. Let us draw a horizontal line across the infinite board and start with all men behind this line. Assume this line is the horizontal axis, so all men are in the lower half-plane. How many men do we need to send one man forward 1, 2, 3, 4, or 5 holes into the upper half-plane?

Obviously 2 men are needed to send a man forward 1 hole, and 4 men are needed to send a man forward 2 holes. Eight men are enough to send a man forward 3 holes. Twenty men are enough to send a man forward 4 holes, see Figure 10.1.

But the really surprising result is the case of 5 holes: it is *impossible* to send a man forward 5 holes into the upper half-plane. This striking result was discovered by Conway in 1961.

The idea behind Conway's resource count is the following. We assign a weight to each hole subject, with the condition that if $H1$, $H2$, $H3$ are *any* 3 consecutive holes in a row or in a column, and $w(H1)$, $w(H2)$, $w(H3)$ are the corresponding weights, then $w(H1) + w(H2) \geq w(H3)$. We can evaluate a position by the sum of the weights of those holes that are occupied by men – this sum is called the *value* of the position,

The meaning of inequality $w(H1) + w(H2) \geq w(H3)$ is very simple. The effect of a move where a man in $H1$ jumps over another man in $H2$ and arrives at $H3$ is that we replace men with weights $w(H1)$ and $w(H2)$ by a man with weight $w(H3)$. Since $w(H1) + w(H2) \geq w(H3)$, this change cannot be an increase in the value of the new position.

Inequality $w(H1) + w(H2) \geq w(H3)$ guarantees that **no play is possible from an initial position to a target position if the target position has a higher value**.

Let w be a positive number satisfying $w + w^2 = 1$: w equals the golden section $\frac{\sqrt{5}-1}{2}$. Now Conway's resource counting goes as follows. Assume that one succeeded in sending a man 5 holes forward into the upper half-plane by starting from a configuration of a finite number of men in the lower half-plane. Write 1 where the man stands 5 holes forward into the upper half-plane, and extend it in the following way:

$$1$$
$$w$$
$$w^2$$
$$w^3$$
$$w^4$$
$$\cdots\ w^9\ w^8\ w^7\ w^6\ w^5\ w^6\ w^7\ w^8\ w^9\ \cdots$$
$$\cdots\ w^{10}\ w^9\ w^8\ w^7\ w^6\ w^7\ w^8\ w^9\ w^{10}\ \cdots$$
$$\cdots\ w^{11}\ w^{10}\ w^9\ w^8\ w^7\ w^8\ w^9\ w^{10}\ w^{11}\ \cdots$$

The value of the top line of the lower half-plane is

$$w^5+2w^6+2w^7+2w^8+\cdots = w^5+2\frac{w^6}{1-w}=w^5+2\frac{w^6}{w^2}=w^5+2w^4=w^3+w^4=w^2.$$

So the value of the whole lower half-plane is

$$w^2(1+w+w^2+w^3+\cdots)=w^2\frac{1}{1-w}=w^2\frac{1}{w^2}=1,$$

which is exactly the value of the target position. So no finite number of men in the lower half-plane will suffice to send a man forward 5 holes into the upper half-plane.

We can even show that 8 men are in fact *needed* to send a man forward 3 holes, and, similarly, 20 men are *needed* to send a man forward 4 holes (i.e. the doubling pattern breaks for 4 holes – the first indication of the big surprise in the case of "5 holes forward"). Indeed, the fact that 8 men are necessary can be seen from the resource count of Figure 10.2 below, for the target position has value 21 and the

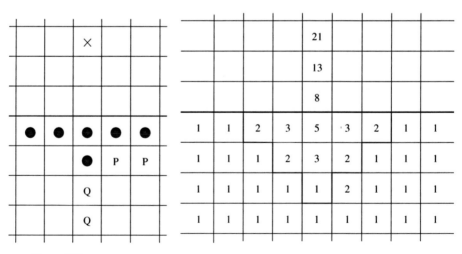

Figure 10.2

highest value that can be achieved with only 7 men below the line is 20. The 2 possible solutions with 8 men are shown, PP and QQ being alternatives.

To send a man forward 4 holes requires 20 men (*not* 16). The two possible solutions are shown in Figure 10.1, PP and QQ being alternatives. The proof that 20 men are *necessary* is more complicated, and goes as follows.

The resource count on the left side of Figure 10.3 shows that an 18-men solution is impossible and that a 19-men solution, if one exists, must occupy the 16 holes marked A in the middle of Figure 10.3 and an additional 3 of the holes marked B,C,D,E,F,G,H,I,J. The resource count of the right side of Figure 10.3 now shows that hole B must be occupied, for we need a position with a value of at least 55, the 16 holes marked A contribute only 48, and no 3 holes from C, D,E,F,G,H,I,J can contribute the remaining 7. The same argument shows that hole C must also be occupied, so a 19-men-solution must contain the 16 holes marked by A, and the 2 holes marked by B and C. Finally, a sophisticated "parity-check" shows that there is no way of placing a 19th man to get a solution (see Chapter 4 in Beasley [1992]).

Exercise 10.3 *Prove that "to send a man forward 4 holes" does require at least 20 men.*

Exercise 10.4 *We generalize Solitaire Army in such a way that "to jump a man diagonally" is permitted. Show that it is impossible to send a man forward 9 holes.*

The Solitaire Army puzzle is a wonderful illustration of a problem that, by using a potential technique, can be solved *exactly*. The subject of this book is 2-player games (*not* puzzles), but the main objective is the same: we try to find exact solutions.

3. The Erdős–Selfridge Theorem: Theorem 1.4. The pioneering application of the potential technique for 2-player games is a theorem by Erdős and Selfridge from 1973. This simple Strong Draw criterion in a 2-page paper made a huge impact on

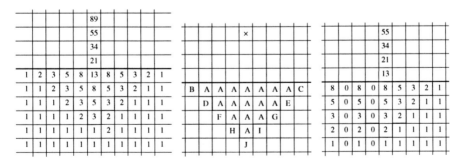

Figure 10.3

the subject; it completely changed the outlook: it shifted the emphasis from Ramsey Theory and Matching Theory to the Probabilistic Method (the meaning of this will become quite clear soon). In sharp contrast with the pairing strategy (and other local approaches), the Erdős–Selfridge Theorem is a *global* criterion. We have already formulated it at the end of Section 1 as Theorem 1.4; recall the statement.

Theorem 1.4 *("Erdős–Selfridge Theorem") Let $\mathcal{F}$ be an n-uniform hypergraph, and assume that $|\mathcal{F}| + MaxDeg(\mathcal{F}) < 2^n$, where $MaxDeg(\mathcal{F})$ denotes the maximum degree of hypergraph $\mathcal{F}$. Then playing the positional game on $\mathcal{F}$ the second player can force a Strong Draw.*

Remark. If the second player can force a Strong Draw in a Positional Game, then the first player can also force a Strong Draw (why?).

Proof. Let $\mathcal{F} = \{A_1, A_2, \ldots, A_M\}$. Assume we are at the stage of the play where the first player already occupies $x_1, x_2, \ldots, x_i$, and the second player occupies $y_1, y_2, \ldots, y_{i-1}$. The question is how to choose second player's next point y_i. Those winning sets which contain at least one $y_j (j \leq i-1)$ are "harmless" – we call them "dead sets." The winning sets which are not "dead" are called "survivors." The "survivors" have a chance to be completely occupied by the first player at the end of the play, so they each represent some "danger." What is the total "danger" of the *whole* position? We evaluate the given position by the following expression, called "danger-function": $D_i = \sum_{s \in S} 2^{-u_s}$, where u_s is the number of unoccupied elements of the "survivor" A_s ($s \in S_i =$ "index-set of the survivors"), and index i indicates that we are at the stage of choosing the ith point y_i of the second player. A natural choice for y_i is to minimize the "danger" D_{i+1} at the next stage. How to do that? The simple linear structure of the danger-function D_i gives an easy answer to this question. Let y_i and x_{i+1} denote the next two moves. What is the effect of these two points on D_i? How do we get D_{i+1} from D_i? Well, y_i "kills" all the "survivors" $A_s \ni y_i$, which means we have to subtract the sum

$$\sum_{s \in S_i:\ y_i \in A_s} 2^{-u_s}$$

from D_i. On the other hand, x_{i+1} doubles the "danger" of each "survivor" $A_s \ni x_{i+1}$; that is, we have to add the sum $\sum_{s \in S_i:\ x_{i+1} \in A_s} 2^{-u_s}$ back to D_i. *Warning:* if some "survivor" A_s contains both y_i and x_{i+1}, then we do not have to give the corresponding term 2^{-u_s} back because that A_s was previously "killed" by y_i.

The natural choice for y_i is the unoccupied z for which $\sum_{s \in S_i:\ z \in A_s} 2^{-u_s}$ attains its maximum. Then what we subtract is at least as large as what we add back

$$D_{i+1} \le D_i - \sum_{s \in S_i:\ y_i \in A_s} 2^{-u_s} + \sum_{s \in S_i:\ x_{i+1} \in A_s} 2^{-u_s}$$

$$\le D_i - \sum_{s \in S_i:\ y_i \in A_s} 2^{-u_s} + \sum_{s \in S_i:\ y_i \in A_s} 2^{-u_s} = D_i.$$

In other words, the second player can force the *decreasing property* $D_1 \ge D_2 \ge \cdots \ge D_{last}$ of the danger-function.

The second player's ultimate goal is to prevent the first player from completely occupying some $A_j \in \mathcal{F}$, i.e. to avoid $u_j = 0$. If $u_j = 0$ for some j, then $D_{last} \ge 2^{-u_j} = 1$. By hypothesis

$$D_{start} = D_1 = \sum_{A:\ x_1 \in A \in \mathcal{F}} 2^{-n+1} + \sum_{A:\ x_1 \notin A \in \mathcal{F}} 2^{-n} \le (|\mathcal{F}| + \mathrm{MaxDeg}(\mathcal{F}))2^{-n} < 1,$$

so by the *decreasing property* of the danger function, $D_{last} < 1$. This completes the proof of the Erdős–Selfridge Theorem. $\square$

Remarks.

(1) If $\mathcal{F}$ is n-uniform, then multiplying the *danger* 2^{-u_s} of a *survivor* by 2^n, the renormalized danger becomes 2^{n-u_s}. The exponent, $n - u_s$, is the number of the first player's marks in a *survivor* (i.e. second player-free) set (u_s denotes the number of unoccupied points). This means the following *Power-of-Two Scoring System*. A winning set containing an O (second player's mark) scores 0, a blank winning set scores 1, a set with a single X (first player's mark) and no O scores 2, a set with two Xs and no O scores 4, a set with three Xs and no O scores 8, and so on (i.e. the "values" are integers rather than small fractions). Occupying a whole n-element winning set, scores 2^n, i.e. due to the renormalization, the "target value" becomes 2^n (instead of 1).

It is just a matter of taste which scoring system is prefered: the first one, where the "scores" were negative powers of 2 and the "target value" was 1, or the second one, where the "scores" were positive powers of 2 and the "target value" was 2^n.

(2) The most frequently applied special case of the Erdős–Selfridge Theorem is the following: *If $\mathcal{F}$ is n-uniform and $|\mathcal{F}| < 2^n$ or $< 2^{n-1}$, then playing on $(V, \mathcal{F})$ the first or second player can force a Strong Draw.*

(3) The proofs of Theorems 1.2–1.3 at the end of Section 1 are just an adaptation of the Erdős–Selfridge proof technique.

(4) To have a better understanding of what is going on here, it is worth to study the *randomized game* where both players are "random generators." The calculation somewhat simplifies if we study the *Random 2-Coloring* instead: the points of

the board are colored (say) red and blue independently of each other with probability $p = 1/2$. (This model is a little bit different from considering the halving 2-colorings only: the case which corresponds to the randomized game.) Indeed, in the Random 2-Coloring model the *expected number* of monochromatic winning sets is clearly $2^{-n+1}|\mathcal{F}|$, which is less than 1 (by the hypothesis of the Erdős–Selfridge Theorem; the case of the second player). So there must *exist* a terminal position with no monochromatic winning set: a drawing terminal position (this is Theorem 11.3 from the next section).

Now the real meaning of the Erdős–Selfridge Theorem becomes clear: it "upgrades" the existing drawing terminal position to a Drawing Strategy. The Erdős–Selfridge proof is a "derandomization," in fact a pioneering application of the method of conditional probabilities, see Alon–Spencer [1992].

(5) As we mentioned already at the beginning of this section, the Erdős–Selfridge Theorem gives the "majority outcome" in the *randomized game* where both players play randomly. We refer to the "majority outcome" as the *Majority-Play Number*; note that the *Majority-Play Number* usually differs from the Achievement and Avoidance Numbers.

(6) Theorem 1.4 is tight: the full-length branches of a binary tree with n levels form an n-uniform family of 2^{n-1} winning sets such that the first player can occupy a full branch in n moves (the players take vertices of the tree).

(7) The proof of Theorem 1.4 can be easily extended to *Shutout Games*. We recall the concept: we can play a Shutout Game on an arbitrary hypergraph $(V, \mathcal{F})$, finite or infinite. First we choose a goal integer n (≥ 1); an n-Shutout Game on $(V, \mathcal{F})$ is similar to the Positional Game in the sense that the players alternate, but the goal is different: instead of "complete occupation" the players want an "n-shutout," i.e. if during a play the first time there is a winning set $A \in \mathcal{F}$ such that one of the players, we call him player P, owns n elements of A and the other player owns no element of A, then player P is declared the winner. If V is infinite, the length of the play is $\leq \omega$; if an ω-play does not have a winner, the play is declared a draw. By repeating the proof of Theorem 1.4, we can easily prove the following shutout result: Let $\mathcal{F}$ be a hypergraph with $|\mathcal{F}| + MaxDeg(\mathcal{F}) < 2^n$ and $\max_{A \in \mathcal{F}} |A| \geq n$, then, playing on $\mathcal{F}$, the second player can prevent the first player from achieving an n-shutout in any $A \in \mathcal{F}$.

(8) The Reverse version of Theorem 1.4 goes as follows: Let $\mathcal{F}$ be an n-uniform hypergraph with $|\mathcal{F}| + MaxDeg(\mathcal{F}) < 2^n$, then, playing on $\mathcal{F}$, Avoider, as the first player, can avoid occupying a whole $A \in \mathcal{F}$.

The proof of "Reverse Theorem 1.4" goes exactly the same way as the original one, except that Avoider chooses a point of *minimum* value. The duality between *maximum* and *minimum* explains the striking equality

Achievement Number = Avoidance Number

for our "Ramseyish" games with quadratic goals. We will return to this interesting issue at the end of the book in Section 47.

Note that "Reverse Theorem 1.4" is also tight. Indeed, consider the hypergraph for which the board V is the $(2n - 1)$-element set $V = \{w, x_1, y_1, x_2, y_2, \ldots, x_{n-1}, y_{n-1}\}$, and the winning sets are all possible n-element subsets A of V satisfying the following two properties: (1) $w \in A$, (2) A contains exactly one element from each pair $\{x_i, y_i\}$, $i = 1, 2, \ldots, n-1$. The number of winning sets is exactly 2^{n-1}, and Forcer can force Avoider (the *first* player) to occupy a whole winning set. Forcer's strategy is very simple: if Avoider takes an element from a pair $\{x_i, y_i\}$, then Forcer takes the other one.

Notice that this example is another extremal system for the original (Maker–Breaker) Erdős–Selfridge Theorem: playing on this hypergraph, Maker, as the first player, can always win in n moves.

4. Applications. By using Theorem 1.4 it is easy to give an alternative solution for the 4^2 game (which cannot have a Pairing Strategy Draw, see Theorem 3.2), and also the 3-dimensional $8^3 = 8 \times 8 \times 8$ game, without any case study!

Theorem 10.1 *In both of the 4^2 and 8^3 Tic-Tac-Toe games the second player can force a Strong Draw.*

Proof. In the 4^2 game there are 10 winning lines, and the maximum degree is 3. Since $3 + 10 < 2^4 = 16$, Theorem 1.4 applies, and we are done.

In the 3-dimensional 8^3 game there are $(10^3 - 8^3)/2 = 244$ winning lines, and the maximum degree is $2^3 - 1 = 7$ (why?). Since $244 + 7 < 2^8 = 256$, Theorem 1.4 applies, and we are done again. $\qquad\square$

In the 4^2 game the Point/Line ratio is less than 2, implying that the 4^2 game is a draw but *not* a Pairing Strategy Draw. In the 8^3 game, however, there are more than twice as many points ("cells") as winning lines (indeed, $8^3 = 512 > 2 \cdot (10^3 - 8^3)/2 = 488$), so there is a chance to find a draw-forcing pairing strategy. And indeed there is one: a symmetric pairing (strategy draw), due to S. Golomb, is described on pp. 677–678 of Berlekamp, Conway, and Guy [1982] (volume two) or in Golomb and Hales [2002].

Unfortunately, we don't know any similar elegant/short proof for ordinary 3^2 Tic-Tac-Toe; the dull case study proof in Section 3 is the only proof we know. The situation is much worse for the 3-dimensional n^3 games with $n = 5, 6, 7$: they are all conjectured to be draw games, but no proof is known (exhaustive search is beyond hope).

Next comes another quick application. What happens if we generalize the notion of *cell animal* ("polyomino") in Section 4? We can relax the requirements by which the cells are connected: we can define *pseudo-animals* as edge or corner connected, or using chess terminology: "kingwise" connected.

For example, the diagonal 3-in-a-row in ordinary Tic-Tac-Toe is a *pseudo-animal*

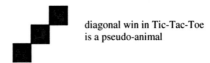

diagonal win in Tic-Tac-Toe
is a pseudo-animal

Pseudo-animal Tic-Tac-Toe. There is 1, 1-cell pseudo-animal, there are 2, 2-cell pseudo-animals; 5, 3-cell pseudo-animals; and 22, 4-cell pseudo-animals (the analogous numbers for ordinary animals are 1,1,2,5). The pseudo-animal kingdom is much bigger than the animal kingdom.

What is the largest pseudo-animal Winner? This is a very difficult question (we couldn't even solve the analogous problem for ordinary animals: the status of "Snaky" is undecided yet, see Section 4), but by using the Erdős–Selfridge Theorem it is easy to give at least some (weak) upper bound on the order of the largest pseudo-animal Winner.

Theorem 10.2 *There are only a finite number of pseudo-animal Winners, and the largest order is* ≤ 72.

Remark. The upper bound ≤ 72 is obviously very weak; the truth is probably less than 10.

Proof. We partition the plane into infinitely many pairwise disjoint *subboards* of size 73×73. Every pseudo-animal of order 73 is contained in a 73×73 "underlying square." This "underlying square" is divided by the disjoint *subboards* into 4 parts; the "largest part" of the pseudo-animal must have order ≥ 19. Either player's strategy is to play on the disjoint subboards independently, and in each subboard use the Erdős–Selfridge Theorem to block all possible "largest parts" of the pseudo-animal on the subboard.

The total number of these "largest parts" in a fixed subboard is $\leq 4 \cdot 73^2 \cdot 8$; indeed, $4 \cdot 73^2$ comes from the number of possibilities for the position of the lower left corner of the "underlying square", and "8" is the number of symmetries of the square. The Erdős–Selfridge Theorem applies if $4 \cdot 73^2 \cdot 8 < 2^{19-1}$, which is really true. Theorem 10.2 follows. $\qquad \square$

So far we discussed 2 types of blocking: (1) the global Erdős–Selfridge Theorem, and (2) the local Pairing Strategy. Next we compare the power of these two very different methods on a simple example: we give 6 different proofs of the amusing fact that the *Unrestricted n*-in-a-row in the plane is a draw game if *n* is sufficiently large (different proofs give different values of *n*).

Unrestricted *n*-in-a-row game on the plane. We recall from Section 4 that *Unrestricted* means the game is played on an *infinite* chessboard (infinite in every direction). In the *Unrestricted 5-in-a-row game* the players alternately occupy little squares of an infinite chessboard, the first player marks his squares by X, and the second player marks his squares by O. That player wins who first gets 5 consecutive marks of his own in a row horizontally, or vertically, or diagonally (of slope 1 or −1). *Unrestricted n-in-a-row* differs in only one aspect: the winning size is *n* instead of 5.

Unrestricted *n*-in-a-row on the plane is a *semi-infinite* game (in fact, a semi-infinite positional game): the board is infinite but the winning sets are all finite. Since the board is infinite, we have to define the *length* of the game: we assume the length of a play is at most ω, where ω denotes the first countable ordinal number. Semi-infinite positional games are all *determined*.

The Unrestricted 4-in-a-row game is an easy first player win. The Unrestricted 5-in-a-row game is conjectured to be a first player win, too, but I don't know any rigorous proof.

Does there exist a (finite) *n* such that the Unrestricted *n*-in-a-row is a Draw game? The answer is "yes."

Theorem 10.3 *The unrestricted n-in-a-row on the plane is a draw game for all sufficiently large n.*

We give 6 proofs: 2 proofs use the Erdős–Selfridge Theorem, 2 proofs use Pairing Strategy, and 2 more use a different decomposition technique.

First Proof: Unrestricted 40-in-a-row is a Strong Draw. It is a straightforward application of the Erdős–Selfridge Theorem. We divide the plane into $n \times n$ squares, where *n* will be specified later. The second player plays in the $n \times n$ squares independently of each other: when the first player makes a move in an $n \times n$ squares, the second player responds in the same large square. Every *n*-in-a-row on the plane intersects some $n \times n$ square in a block of length $\geq n/3$.

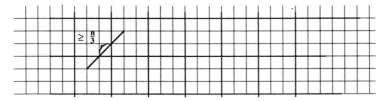

The Erdős–Selfridge Theorem applies if $4n^2 < 2^{\lceil n/3 \rceil - 1}$ (note that factor 4 comes from the 4 winning directions); the inequality holds for all $n \geq 40$, which completes the proof.   □

Second Proof: Unrestricted 13-in-a-row is a Strong Draw. This is a more sophisticated application of the Erdős–Selfridge Theorem. We employ the non-uniform version: If

$$\sum_{A \in \mathcal{F}} 2^{-|A|} + \max_{x \in V} \sum_{A \in \mathcal{F}: \, x \in A} 2^{-|A|} < 1,$$

then the second player can block every winning set $A \in \mathcal{F}$.

This time we divide the plane into 9×9 squares. Again the second player plays in the 9×9 squares independently of each other: when the first player makes a move in a large square, the second player responds in the same large square. Every 13-in-a-row on the plane intersects one of the 9×9 squares in one of the following "winning sets":

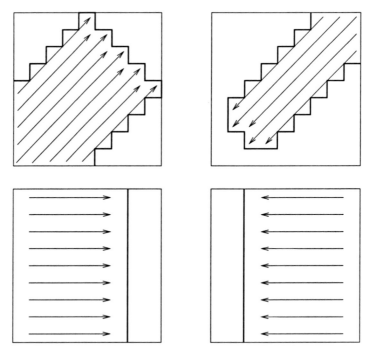

and also for the vertical and the "other diagonal" (of slope -1) directions.

There are 44 7-sets, 12 6-sets, and 8 5-sets, so

$$\sum_{A \in \mathcal{F}} 2^{-|A|} = \frac{44}{2^7} + \frac{12}{2^6} + \frac{8}{2^5} = \frac{25}{32}.$$

On the other hand, we trivially have

$$\max_{x \in V} \sum_{A \in \mathcal{F}:\, x \in A} 2^{-|A|} \le \frac{2}{2^5} + \frac{4}{2^7} = \frac{3}{32}.$$

Since $\dfrac{25+3}{32} = \dfrac{28}{32} < 1$, we are done. $\qquad\square$

Third Proof: Unrestricted 12-in-a-row is a Draw. Similarly to the first two solutions we divide the plane into infinitely many non-interacting games, but this time we apply a Pairing Strategy for the component games. Each component game will be "one-dimensional." The decomposition goes as follows. Extend the following 4×4 direction marking periodically over the whole plane

$$n = 4: \quad \begin{bmatrix} - & / & / & \backslash \\ | & | & / & - \\ \backslash & - & | & | \\ \backslash & / & - & \backslash \end{bmatrix}$$

For each move of the first player, the second player replies by taking a similarly marked square in the direction of the mark, by using a straightforward Pairing Strategy. The longest possible n-in-a-row occupied by the first player looks like this (if it is horizontal):

$$\cdots\ -\ \cdots\ -\ \cdots$$

which is an 11-in-a-row. This proves that 12-in-a-row is a draw. $\qquad\square$

Fourth Proof: Unrestricted 9-in-a-row is a Strong Draw. We can improve on the previous solution by employing the following 8×8 matrix instead of the 4×4 (the rest of the proof goes similarly)

$$n = 8: \quad \begin{bmatrix} \backslash & \backslash & - & - & | & | & / & / \\ - & - & \backslash & \backslash & / & / & | & | \\ / & / & | & | & - & - & \backslash & \backslash \\ \backslash & \backslash & | & | & - & - & / & / \\ | & | & \backslash & \backslash & / & / & - & - \\ | & | & / & / & \backslash & \backslash & - & - \\ / & / & - & - & | & | & \backslash & \backslash \end{bmatrix}$$

What this 8 by 8 matrix represents is a direction-marking of the $4 \cdot 8 = 32$ "torus-lines" of the 8×8 torus. The direction-marks $-$, $|$, $\backslash$, and $/$ mean (respectively) "horizontal," "vertical," "diagonal of slope -1," and "diagonal of slope 1." Each one of the 32 torus-lines contains 2 marks of its own direction. The periodic extension of the 8 by 8 matrix over the whole plane gives a Pairing Strategy Draw for the unrestricted 9-in-a-row game. Either player responses to the opponent's last move by taking the *nearest* similarly marked square in the direction indicated by the mark in the opponent's last move square.

This solution is taken from the *Winning Ways* (see p. 677 in vol. 2). What it says is that the "8^2 torus Tic-Tac-Toe game" is a Pairing Strategy Draw (the winning sets are the 32 8-element full-length lines on the torus). □

Fifth Proof: Unrestricted 9-in-a-row is a Strong Draw. Tile the plane with H-shaped heptominos ("seven-squares"). The second player plays on these heptominos independently of each other. In each heptomino second player's goal is to block a 3-in-a-row in either diagonal, or the horizontal, or the right vertical, see the figure below.

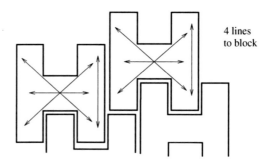

4 lines to block

It is easy to see that the second player can achieve his "blocking goal" in every heptomino, which implies that the first player cannot occupy a 9-in-a-row. This elegant solution is due to Pollak and Shannon (1954). □

Sixth Proof: Unrestricted 8-in-a-row is a Strong Draw. This is the best-known result, the current record. It is due to a group of Dutch mathematicians, see *American Mathematic Monthly (1980)*: 575–576.

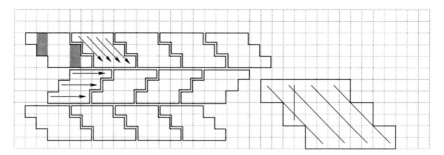

The idea is the same as in the previous solution, but here one tiles the plane with congruent zig-zag shaped regions of area 12. Of course, a more complicated shape leads to a more complicated case study. Again the second player plays on each 12-square tile independently: on each tile his goal is to prevent the first player from getting 4-in-a-row horizontally, or 3-in-a-row diagonally, or two in a shaded column vertically. The forbidden lines are indicated on the figure below.

If the second player can achieve his goal on each tile, then the first player cannot get more than 7-in-a-row diagonally, 6-in-a-row horizontally, or 6-in-a-row vertically, which is more than what we have to prove.

Our 12-square tile is equivalent to a 3×4 rectangle, where the 3 rows, the 4 columns, and the 2 indicated diagonal pairs are forbidden.

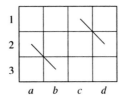

It is easier to explain second player's strategy on the 3×4 rectangle, and it goes as follows: If the first player starts:

(1) in column a, then the second player replies $b3$;
(2) in column b, then the second player replies $a2$;
(3) in the right-hand side of the rectangle, then the second player replies symmetrically, i.e. by $c1$ or $d2$.

Any move which is not a direct threat is answered by taking a point of the remaining diagonal pair. This leads to a position which, up to isomorphism, equals one of the following two cases:

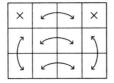 or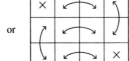

where columns b and c contain one mark of the second player (X indicates the first player). The Pairing Strategy indicated on the picture suffices to stop the first player. (Note that the 2 diagonal pairs have been taken care of, and the indicated Pairing guarantees that the first player cannot occupy a whole row or column.) □

5. Extremal systems of the Erdős–Selfridge Theorem. The simplest form of the Erdős–Selfridge Theorem goes as follows: if hypergraph $\mathcal{F}$ is n-uniform and has fewer than 2^{n-1} winning sets, then Breaker (as the second player) can always win the Maker–Breaker game on $\mathcal{F}$.

An n-uniform hypergraph $\mathcal{F}$ is called an *Extremal System* if $|\mathcal{F}| = 2^{n-1}$ and it is a Maker's win (Maker is the first player). The Erdős–Selfridge Theorem is tight in the sense that there exist Extremal Systems. We have already given two examples. In the original Erdős–Selfridge construction the board V is the $(2n-1)$-element set $V = \{w, x_1, y_1, x_2, y_2, \ldots, x_{n-1}, y_{n-1}\}$, and the winning sets are all possible n-element subsets A of V satisfying the following two properties: (a) $w \in A$, (b) A

contains exactly one element from each pair $\{x_i, y_i\}$, $i = 1, 2, \ldots, n-1$. The second example is my binary tree construction: the full-length branches of a binary tree with n levels form an n-uniform family of 2^{n-1} winning sets such that Maker, as the first player, can occupy a full branch in n moves (the players take vertices of the tree).

About 15 years ago we asked the following two questions: (1) Can we describe all possible Extremal Systems? Both examples above were Economical Winners, so it is natural to ask: (2) Is there an n-uniform Extremal System where Maker (as the first player) needs *more* than n turns to win?

We can report major progress for the second question. In 2003 my former Ph.D. students K. Kruczek and E. Sundberg could construct an n-uniform Extremal System where Maker (as the first player) needs at least $2n$ turns to win (the result is part of their Ph.D. thesis). Later A.J. Sanders (Cambridge, UK) came up with a startling improvement: there is an n-uniform Extremal System (for every $n \geq 5$) where Maker (as the first player) needs *exponential time*; namely at least 2^{n-4} turns to win! In fact, the maximum is between 2^{n-4} and 2^{n-1}. For the details of the construction we refer the interested reader to Sanders [2004].

Unfortunately, the first question remains wide open. At first sight this may seem rather surprising since the Erdős–Selfridge Theorem has such an easy, short/transparent proof. On the other hand, Sanders's startling construction is a warning: the easy-looking first question is actually very hard.

6. If a positional game played on a (finite) hypergraph is a draw game, then by definition either player can force a draw. If either player uses his draw-forcing strategy, the play ends in a drawing position, which is a Proper Halving 2-Coloring.

We know that the converse is not true: the existence of a Proper Halving 2-Coloring does *not* imply that the positional game is a draw game; see for example the 4^3 Tic-Tac-Toe ("Qubic"). Of course, there are many more hypergraph examples demonstrating the failure of the converse. Both of the Extremal Systems mentioned above are good examples: (1) In the original Erdős–Selfridge construction, where the board is the $(2n-1)$-element set $\{w, x_1, y_1, x_2, y_2, \ldots, x_{n-1}, y_{n-1}\}$, it suffices to color a pair $\{x_i, y_i\}$ red and another pair $\{x_j, y_j\}$ blue; then, independently of the coloring of the rest of the board, we obtain a Proper 2-Coloring. (2) In the binary tree construction, it suffices to color the root red and the two neighbors of the root blue; then, independently of the coloring of the rest of the board, we obtain a Proper 2-Coloring.

Even if the converse is not true, it is still a very good idea to approach the extremely complex concept of the Drawing Strategy from the angle of the much more accessible concept of Proper 2-Coloring (accessible via the Probabilistic Method). In other words, when can we "upgrade" a Proper 2-Coloring (existing via the Probabilistic Method) to a Drawing Strategy? This is the question that we are going to study in the second half of Section 11 (see Theorems 11.3 and 11.4) and in Section 12.

11

Local vs. Global

The Erdős–Selfridge theorem is a *global* blocking criterion. The (hypothetical) Neighborhood Conjecture – see Open Problem 9.1 – is a far-reaching *local* generalization of the Erdős–Selfridge Theorem: the global condition is reduced to a (much weaker) local condition. What supports the Neighborhood Conjecture? We can give two good reasons.

1. Pairing strategy. The Neighborhood Conjecture states, in a quantitative form, that when "playing the positional game on a hypergraph the local size is what really matters, the global size is completely irrelevant," and there is indeed a criterion with the same message, though much weaker than the Neighborhood Conjecture (see Theorem 11.2 below). This weaker result is about the Pairing Strategy Draw. The Pairing Strategy Draw is the simplest possible way to force a Strong Draw.

It is fair to say that the pairing strategy is the most common technique in the whole of Game Theory. We have already used the pairing strategy several times in the book. For example, the copycat strategy of Theorem 6.2 and Gross's explicit winning strategy in Bridge-it (see Section 4) are both *pairing strategies.*

Pairing strategy is a *local* strategy: it means a decomposition of the board (or some part of the board) into disjoint pairs, and when your opponent takes one member from a pair, you take the other one. It is applied when the opponent cannot achieve the objective (win or draw) without choosing both points of at least 1 pair.

We recall that a hypergraph is called *Almost Disjoint* if any two hyperedges have at most 1 point in common.

If the family of winning sets is *Almost Disjoint,* then the question "can pairing strategy work here" becomes a standard *perfect matching* problem. Indeed, in an Almost Disjoint family two distinct winning sets cannot share the same pair of points. So the pairing strategy works if and only if we can find a family of disjoint 2-element representatives of the hypergraph of winning sets. But to find a family of disjoint 2-element representatives of a given hypergraph is a well-characterized, completely solved problem in Matching Theory ("Bigamy version of Hall's Marriage Theorem").

In general, if the hypergraph of winning sets is *not* necessarily Almost Disjoint, then the existence of a family of disjoint 2-element representatives is a *sufficient but not necessary* condition for the existence of a pairing strategy. The following two criterions were (probably) first published in Hales and Jewett [1963] (and independently rediscovered later in several other papers).

Theorem 11.1 *("Pairing Strategy Draw") Consider the Positional Game on* $(V, \mathcal{F})$, *and assume that for every subfamily* $\mathcal{G} \subseteq \mathcal{F}$

$$\left| \bigcup_{A \in \mathcal{G}} A \right| \geq 2|\mathcal{G}|.$$

Then either player can force a Pairing Strategy Draw.

Theorem 11.2 *("Degree Criterion for Pairing Strategy Draw") Let* $\mathcal{F}$ *be an* n-*uniform hypergraph, i.e.* $|A| = n$ *for every* $A \in \mathcal{F}$. *Further assume that the Maximum Degree is at most* $n/2$, *every* $x \in V$ *is contained in at most* $n/2$ *elements of* $\mathcal{F}$. *Then playing the Positional Game on* $\mathcal{F}$, *either player can force a Pairing Strategy Draw.*

Notice that the two Pairing Strategy Criterions (Theorems 11.1–11.2) are very general. They are *local* conditions in the sense that they don't give any restriction on the global size of hypergraph $\mathcal{F}$. Both hold for an infinite board as well.

Proof of Theorem 11.1. The well-known Hall's Theorem ("Marriage Theorem") applies here. We can find disjoint 2-element representatives: $h(A) \subset A$ for all $A \in \mathcal{F}$ with $|h(A)| = 2$, and $h(A_1) \cap h(A_2) = \emptyset$ whenever A_1 and A_2 are different winning sets from $\mathcal{F}$. There is, however, a little technical twist involved here: we have to apply the Marriage Theorem to the "double" of $\mathcal{F}$, i.e. every $A \in \mathcal{F}$ is taken in two copies. In other words, one applies the Bigamy Corollary of the Marriage Theorem: "every man needs two wives."   □

Remark. It is important to know that Matching Theory provides several efficient ("polynomial") algorithms to actually *find* a family of disjoint 2-element representatives. For example, the Augmenting Path Algorithm has running time $O(N^{5/2})$, where N is the size of the board.

Proof of Theorem 11.2. For an arbitrary subfamily $\mathcal{G} \subseteq \mathcal{F}$ a standard double-counting argument gives

$$n|\mathcal{G}| \leq \sum_{A \in \mathcal{G}} |A| = \sum_{x \in \bigcup_{A \in \mathcal{G}} A} \sum_{A \in \mathcal{G}:\ x \in A} 1 \leq \left| \bigcup_{A \in \mathcal{G}} A \right| \cdot \frac{n}{2},$$

which means that Theorem 11.1 applies here and completes the proof.   □

Theorem 11.2 is a linear local criterion and the hypothetical Neighborhood Conjecture is an exponential local criterion. Of course there is a major difference between "linear" and "exponential," but still the existence of any kind of local criterion is a good reason to believe in the exponential Neighborhood Conjecture.

2. The Probabilistic Method and the Local Lemma. A second good reason to believe in the Neighborhood Conjecture is the Erdős–Lovász 2-Coloring Theorem (see below), which was the original application of the famous *Local Lemma* (or "Lovász Local Lemma"), a cornerstone of the Probabilistic Method. The Local Lemma is an advanced result; it is better to start with the basic result.

The following old theorem of Erdős can be justly considered the starting point of a long line of research, which culminated in the so-called "Probabilistic Method in Combinatorics" (see Erdős [1947], [1961], [1963]).

Theorem 11.3 *("Erdős 1947") Let $\mathcal{F}$ be an n-uniform hypergraph, and assume that $|\mathcal{F}| < 2^{n-1}$. Then:*

(a) there is a Proper 2-Coloring; and what is somewhat more,
(b) there also is a Proper Halving 2-Coloring (i.e. Drawing Terminal Position).

Both (a) and (b) can be proved by a simple "counting argument." The proof of (a) goes as follows. Let $N = |V|$ denote the size of the union set ("board") V of hypergraph $\mathcal{F}$. A simple counting argument shows that under the condition $|\mathcal{F}| < 2^{n-1}$ there exists a Proper 2-Coloring. Indeed, there are 2^N 2-colorings of board V, and for every single winning set $A \in \mathcal{F}$ there exists 2^{N-n+1} "bad" 2-colorings which are monochromatic on A. By hypothesis $2^N - |\mathcal{F}|2^{N-n+1} > 0$, which implies throwing out all "bad" 2-colorings, there must remain at least one Proper 2-Coloring (i.e. no $A \in \mathcal{F}$ is monochromatic).

To prove Theorem 11.3 (b) we have to find a Drawing Terminal Position (i.e. a 2-coloring of the board by "colors" X and O such that the 2 color classes have the same size, and each winning set contains both marks). For notational simplicity assume that N is even. The idea is exactly the same as that of (a), except that we restrict ourselves to the $\binom{N}{N/2}$ Halving 2-Colorings instead of the 2^N (arbitrary) 2-colorings. The analogue of $2^N - |\mathcal{F}|2^{N-n+1} > 0$ is the following requirement: $\binom{N}{N/2} - 2|\mathcal{F}|\binom{N-n}{N/2} > 0$. This holds because

$$\frac{\binom{N-n}{N/2}}{\binom{N}{N/2}} = \frac{N/2}{N} \frac{(N/2)-1}{N-1} \frac{(N/2)-2}{N-2} \cdots \frac{(N/2)-n+1}{N-n+1} \le 2^{-n},$$

and (b) follows. $\qquad\square$

The previous argument can be stated in the following slightly different form: the *average number* ("expected value" or "first moment") of winning sets completely occupied by either player is precisely

$$2|\mathcal{F}|\frac{\binom{N-n}{N/2}}{\binom{N}{N/2}}, \quad \text{which is } \textit{less than } 1.$$

Since the *minimum* is less or equal to the *average*, and the *average* is *less than 1*, there must exist a Drawing Terminal Position (i.e. no player owns a whole winning set).

This kind of "counting argument," discovered and systematically developed by Erdős, is in the same category as Euclid's proof of the existence of infinitely many primes, or Pythagoras's proof of the irrationality of $\sqrt{2}$: they are astonishingly simple and fundamentally important at the same time.

Here is a simple application: a lower bound to the van der Waerden number $W(n)$. Since the number of n-term arithmetic progressions in an interval $[N] = \{1, 2, \ldots, N\}$ is about $\frac{N^2}{2(n-1)}$, Theorem 11.3 yields the exponential lower bound

$$W(n) \geq W_{1/2}(n) \geq (1 + o(1))\sqrt{n2^n}. \tag{11.1}$$

Erdős's "counting argument" (Theorem 11.3) was later developed in two very different ways: first by Wolfgang M. Schmidt [1962], and later in a joint work by Erdős and Lovász [1975].

Notice that the family of winning sets in the n^d game has the important additional feature that the winning sets are on straight lines, and any 2 straight lines have at most 1 point in common. A hypergraph with the intersection property that any two hyperedges have at most one point in common is called **Almost Disjoint**.

Schmidt's work can be summarized in the following theorem (that we mention without proof): part (a) is a general hypergraph result; part (b) is the special case of "arithmetic progressions"; Schmidt's main goal was to improve on (11.1); Schmidt basically *squared* the lower bound in (11.1).   □

Schmidt's Theorem [1962]:

(a) Let $\mathcal{F}$ be an *n-uniform Almost Disjoint hypergraph. Assume that the Maximum Degree of $\mathcal{F}$ is less than* $2^{n-5}\sqrt{n\log n}$ *and the size* $|\mathcal{F}|$ *of the hypergraph is less than* 8^n; *then* $\mathcal{F}$ *has chromatic number two, i.e. the hypergraph has a Proper 2-Coloring.*

(b) $W(n) = W(n, 2) \geq 2^{n-5}\sqrt{n\log n}$. \tag{11.2}

Note that part (b) is not a corollary of part (a); the family of n-term arithmetic progressions in an interval $[N] = \{1, 2, \ldots, N\}$ is *not* Almost Disjoint, but it is "close enough" in the sense that the proof of (a) can be *adapted* to prove (b). Schmidt's method motivated our "game-theoretic decomposition technique" in Part D.

The following result is more general: it is about arbitrary hypergraphs (not just Almost Disjoint or "nearly" Almost Disjoint Hypergraphs), see Erdős and Lovász [1975].

Theorem 11.4 (*"Erdős–Lovász 2-Coloring Theorem"*) *If $\mathcal{F}$ is an n-uniform hyper-graph, and its Maximum Neighborhood Size is at most 2^{n-3}, then the hypergraph has a Proper 2-Coloring (i.e. the points can be colored by two colors so that no hyperedge $A \in \mathcal{F}$ is monochromatic.) In particular, if the Maximum Degree is at most $2^{n-3}/n$, then the hypergraph has a Proper 2-Coloring.*

Remark. The very surprising message of the Erdős–Lovász Theorem is that the "global size" of hypergraph $\mathcal{F}$ is irrelevant – it can even be infinite! – all what matters is the "local size."

The proof of Theorem 11.4 is strikingly short.

Proof of Theorem 11.4. The usual proof uses the Local Lemma; here we give a more direct "counting argument" the proof. It can be considered as a sophisticated generalization of the proof of Theorem 11.3.

Let $|\mathcal{F}| = M$, let $\mathcal{F} = \{A_1, A_2, \ldots, A_M\}$, let V denote the board, and let $|V| = N$. Let $\mathcal{C}$ denote the set of all 2^N possible 2-colorings of V. Let $I \subset [M]$ be an arbitrary index-set where $[M] = \{1, 2, \ldots, M\}$; then $\mathcal{C}(I : \text{proper}) \subset \mathcal{C}$ denotes the set of 2-colorings of the board V such that no $A_i, i \in I$ becomes monochromatic. For arbitrary $I \subset [M]$ and $j \in [M]$ with $j \notin I$, let $\mathcal{C}(I : \text{proper} \wedge j : \text{mono})$ denote the set of 2-colorings of the board V such that no $A_i, i \in I$ becomes monochromatic but A_j *is* monochromatic.

We actually prove a *stronger* statement; it is often easier to prove a stronger statement by induction. The proof of Theorem 11.4 is an excellent example of the principle that "to prove more may be less trouble."

Proposition: *Let $I \subset [M]$ and $j \in [M]$ with $j \notin I$. Then*

$$\frac{|\mathcal{C}(I : \text{proper} \wedge j : \text{mono})|}{|\mathcal{C}(I : \text{proper})|} \leq 2 \cdot 2^{-n+1}.$$

Remark. Note that 2^{-n+1} is the probability that in a Random 2-Coloring a given n-set becomes monochromatic.

Proof of the Proposition. We prove this *stronger* Proposition by induction on $|I|$. If $|I| = 0$ (i.e. I is the empty set), then the Proposition reduces to the following triviality:

$$\frac{|\mathcal{C}(j : \text{mono})|}{|\mathcal{C}|} = 2^{-n+1} < 2 \cdot 2^{-n+1}.$$

Next assume that index-set I is not empty. For notational convenience write $I = \{1, 2, \ldots, i\}$, and among the elements of I let $1, 2, \ldots, d$ $(d \leq i)$ denote the neighbors of A_j (i.e. $A_1, \ldots, A_d$ intersect A_j, but $A_{d+1}, \ldots, A_i$ do not intersect A_j). By hypothesis, $d \leq 2^{n-3}$. Since $A_{d+1}, \ldots, A_i$ are disjoint from A_j

$$\frac{|\mathcal{C}(\{d+1, d+2, \ldots, i\}: \text{ proper} \wedge j: \text{ mono})|}{|\mathcal{C}(\{d+1, d+2, \ldots, i\}: \text{ proper})|} = 2^{-n+1}. \tag{11.3}$$

Furthermore

$$\mathcal{C}(\{1, 2, \ldots, i\}: \text{ proper}) =$$

$$\mathcal{C}(\{d+1, d+2, \ldots, i\}: \text{ proper}) \setminus \bigcup_{k=1}^{d} \mathcal{C}(\{d+1, d+2, \ldots, i\}: \text{ proper} \wedge k: \text{ mono}), \tag{11.4}$$

and by the induction hypothesis

$$\frac{|\mathcal{C}(\{d+1, d+2, \ldots, i\}: \text{ proper} \wedge k: \text{ mono})|}{|\mathcal{C}(\{d+1, d+2, \ldots, i\}: \text{ proper})|} \leq 2 \cdot 2^{-n+1}. \tag{11.5}$$

Since $d \cdot 2 \cdot 2^{-n+1} \leq 2^{n-3} \cdot 2 \cdot 2^{-n+1} = 1/2$, by (11.4) and (11.5) we obtain

$$|\mathcal{C}(\{1, 2, \ldots, i\}: \text{ proper})| \geq \frac{1}{2}|\mathcal{C}(\{d+1, d+2, \ldots, i\}: \text{ proper})|. \tag{11.6}$$

Now the proof of the Proposition is straightforward: by (11.3) and (11.6)

$$2^{-n+1} = \frac{|\mathcal{C}(\{d+1, d+2, \ldots, i\}: \text{ proper} \wedge j: \text{ mono})|}{|\mathcal{C}(\{d+1, d+2, \ldots, i\}: \text{ proper})|} \geq$$

$$\frac{|\mathcal{C}(\{1, 2, \ldots i\}: \text{ proper} \wedge j: \text{ mono})|}{2|\mathcal{C}(\{1, 2, \ldots i\}: \text{ proper})|} = \frac{|\mathcal{C}(I: \text{ proper} \wedge j: \text{ mono})|}{2|\mathcal{C}(I: \text{ proper})|},$$

which is exactly the Proposition.

The deduction of Theorem 11.4 from the Proposition is obvious: indeed, by an *iterated application* of the Proposition

$$\frac{|\mathcal{C}(\{1, 2, \ldots M\}: \text{ proper})|}{|\mathcal{C}|} \geq \left(1 - 2^{-n+2}\right)^{|\mathcal{F}|} > 0,$$

which proves the *existence* of a Proper 2-Coloring of hypergraph $\mathcal{F}$. Since $|\mathcal{C}| = 2^N$, the total number of Proper 2-Colorings is at least

$$2^N \cdot \left(1 - 2^{-n+2}\right)^{|\mathcal{F}|} \approx 2^N \cdot e^{-|\mathcal{F}|/2^{n-2}}. \tag{11.7}$$

This completes the proof of Theorem 11.4.   □

3. Concluding remarks

(i) Finding a needle in a haystack!

The proof of the Erdös-Lovász 2-Coloring Theorem is an *existence* argument. The only way to find the existing Proper 2-Coloring is to try out all possible 2^N 2-colorings of the board, where N is the board size. This is very similar to the "combinatorial chaos" that we face dealing with the Strategy Stealing argument. Indeed, to find a winning or drawing strategy guaranteed by the Strategy Stealing argument, we have to perform a case study of size $O(N3^N)$ where N is the board size. (A systematic way to perform this "case study" is the "backward labeling of the position graph" – see Appendix C.)

In both cases, we have the same fundamental problem: can the *exponential* case study be replaced by a *polynomial* one? Observe that the proof of the Erdös–Lovász 2-Coloring Theorem does not even provide a *randomized algorithm*. Indeed, in view of (11.7), the probability that a Random 2-Coloring provides a Proper 2-Coloring is

$$\geq e^{-|\mathcal{F}|/2^{n-2}}, \tag{11.8}$$

and because in the applications of the Erdös–Lovász 2-Coloring Theorem the hypergraph-size $|\mathcal{F}|$ is *much* larger than 2^{n-2}, (11.8) is (usually) extremely small!

The Erdős–Lovász 2-Coloring Theorem implies the lower bound

$$W(n) = W(n, 2) \geq \frac{2^n}{8n}, \tag{11.9}$$

which is somewhat better than Schmidt's lower bound, but it is still in the same range $W(n) \geq (2+o(1))^n$. The elegant/short proof of the Erdős–Lovász 2-Coloring Theorem was "non-constructive": it was a *pure existence* argument, in some sense similar to Strategy Stealing; Schmidt's proof, on the other hand, is more constructive.

We improved on the van der Waerden number (lower bound) for the second time, but the best-known result belongs to the hard-core *constructivists*. We start with an example: the inequality $W(4) \geq 35$ is established by the following explicit 2-coloring of the interval $I = \{0, 1, 2, \ldots, 33\}$: the first color-class consists of 0, 11, and the quadratic non-residues (mod 11) in I

$$0, 2, 6, 7, 8, 10, 11, 13, 17, 18, 19, 21, 24, 28, 29, 30, 32,$$

and, of course, the other color-class is the complement set

$$1, 3, 4, 5, 8, 12, 14, 15, 16, 20, 22, 23, 25, 26, 27, 31, 33.$$

It is easy to see that no class contains a 4-term arithmetic progression. This algebraic construction is due to J. Folkman (the inequality is actually an equality: $W(4) = 35$).

(ii) Berlekamp's algebraic construction
A similar but more sophisticated explicit finite field construction was discovered
by Berlekamp in 1968; it gives the lower bound

$$W(n) > (n-1)2^{n-1} \text{ if } (n-1) \text{is a prime.} \qquad (11.10)$$

Berlekamp's construction, just like Folkman's example above, is a Proper *Halving*
2-Coloring (see Berlekamp [1968]). It follows that $W_{1/2}(n) > (n-1)2^{n-1}$ if $n-1$
is a *prime*, and because there is always a prime between n and $n-n^{2/3}$ if n is large
enough, Berlekamp's construction implies the lower bound $W_{1/2}(n) \geq (2+o(1))^n$
for *every* n.

(iii) What is the Local Lemma?
Let us return to Theorem 11.4: it was the motivation and the original application
of the general **Local Lemma**, see Erdős and Lovász [1975]. The *Local Lemma*
(or *Lovász Local Lemma*) is a remarkable probabilistic *sieve* argument to prove
the *existence* of certain very complicated structures that we are unable to construct
directly. To be precise, let $E_1, E_2, \ldots, E_s$ denote events in a probability space. In
the applications, the E_is are "bad" events, and we want to avoid all of them, i.e. we
wish to show that $Prob(\cup_{i=1}^s E_i) < 1$. A trivial way to guarantee this is to assume
$\sum_{i=1}^s Prob(E_i) < 1$. A completely different way to guarantee $Prob(\cup_{i=1}^s E_i) < 1$ is to
assume that $E_1, E_2, \ldots, E_s$ are *mutually independent* and all $Prob(E_i) < 1$. Indeed,
we then have $Prob(\cup_{i=1}^s E_i) = 1 - \prod_{i=1}^s (1 - Prob(E_i)) < 1$. The Local Lemma
applies in the very important case when we don't have mutual independence, but
"independence dominates" in the sense that each event is *independent* of all but a
small number of other events.

Local Lemma: *Let $E_1, E_2, \ldots, E_s$ be events in a probability space. If $Prob(E_i) \leq$
$p < 1$ holds uniformly for all i, and each event is independent of all but at most $\frac{1}{4p}$
other events, then $Prob(\cup_{i=1}^s E_i) < 1$.*
Theorem 11.4 is an easy corollary of the Local Lemma: for every $A_i \in \mathcal{F}$ let E_i
be the event "A_i is monochromatic in a Random 2-Coloring of the points of the
n-uniform hypergraph $\mathcal{F}$"; then the Local Lemma applies with $p = 2^{-n+1}$.

On the other hand, the proof of Theorem 11.4 (a special case of the Local Lemma)
can be easily adapted to prove the general result itself.

Exercise 11.1 *Prove the Local Lemma.*

(iv) More than two colors
We conclude this section with two generalizations of Theorem 11.4, where the
number of colors is more than 2.

(a) Let $k \geq 2$ be an integer, and let $\mathcal{F}$ be an n-uniform hypergraph with Maximum
    Degree of at most $k^{n-1}/4$. Then the chromatic number of $\mathcal{F}$ is $\leq k$, i.e. there

is a *Proper k-Coloring* of the vertex-set (meaning that no $A \in \mathcal{F}$ becomes monochromatic).

(b) Let $k \geq 2$ be an integer, and let $\mathcal{F}$ be an n-uniform hypergraph with Maximum Degree of at most $\frac{1}{4k} \left(\frac{k}{k-1} \right)^n$. Then there is a *Rainbow k-Coloring* of hypergraph $\mathcal{F}$. A *Rainbow k-Coloring* means that each hyperedge $A \in \mathcal{F}$ contains all k colors.

Both (a) and (b) are easy corollaries of the Local Lemma (note that we can also adapt the counting proof of Theorem 11.4).

The concepts of Proper k-Coloring and Rainbow k-Coloring are identical for $k = 2$, but they become very different for $k \geq 3$.

Finally, an almost trivial, but useful observation:

(c) Let $\mathcal{F}$ be an arbitrary finite hypergraph. If $\mathcal{F}$ has a Rainbow 3-Coloring, then it also has a Proper Halving 2-Coloring (i.e. the 2 color classes have equal size). Indeed, let C_1, C_2, C_3 be the 3 color classes of the vertex-set in a Rainbow 3-Coloring of hypergraph $\mathcal{F}$, and assume that $|C_1| \leq |C_2| \leq |C_3|$. Since C_3 is the largest color class, we can always divide it into 2 parts $C_3 = C_{3,1} \cup C_{3,2}$ such that the 2 sums $|C_1| + |C_{3,1}|$ and $|C_2| + |C_{3,2}|$ become equal (or differ by at most 1). Coloring $C_1 \cup C_{3,1}$ red and $C_2 \cup C_{3,2}$ blue gives a Proper Halving 2-Coloring of hypergraph $\mathcal{F}$:

Combining (b) and (c), we obtain:

(d) Let $\mathcal{F}$ be an n-uniform hypergraph with Maximum Degree at most $\frac{1}{12} \left(\frac{3}{2} \right)^n$; then there is a *Proper Halving* 2-Coloring of hypergraph $\mathcal{F}$.

12

Ramsey Theory and Hypercube Tic-Tac-Toe

1. The Hales–Jewett Number. In Section 11, we discussed three different lower bounds for the van der Waerden number $W(n)$ (see (11.2) and (11.9)–(11.10)); the arguments were completely different, but they gave basically the same order of magnitude (around 2^n). Unfortunately, there remains an enormous gap between the plain exponential lower bound and Gowers's 5-times iterated exponential upper bound (see (7.2)).

Next we switch from the van der Waerden number $W(n)$ to the Hales–Jewett Number $HJ(n)$, where the gap between the best-known upper and lower bounds is even much larger. The best-known upper bound is Shelah's super-tower function; what is the best-known lower bound for $HJ(n)$? We begin with the first result: in their pioneering paper Hales and Jewett [1963] proved the linear lower bound $HJ(n) \geq n$ by an explicit construction.

Theorem 12.1 ("Hales–Jewett linear lower bound") *The Hales–Jewett Number satisfies the linear lower bound $HJ(n) \geq n$.*

Proof. We can assume that $n \geq 5$. Indeed, every "reasonable" play of Tic-Tac-Toe leads to a drawing terminal position (which solves the case $n = 3$), and even if the 4^3 game is a first player win, it nevertheless *does* have a drawing terminal position, which settles the case $n = 4$.

For $n \geq 5$ we are going to define a Proper 2-Coloring of the n^{n-1} hypergraph by using an elegant explicit algebraic construction. The idea of Hales and Jewett is to add up $(n-1)$ 1-dimensional 2-colorings (i.e. 2-colorings of $[n] = \{1, 2, \ldots, n\}$), where the addition is taken (mod 2).

Let $\mathbf{v}_1, \ldots, \mathbf{v}_i = (v_{i,1}, \ldots, v_{i,n}), \ldots, \mathbf{v}_d$ be d n-dimensional 0–1 vectors, i.e. $v_{i,j} \in \{0, 1\}$ for all $1 \leq i \leq d, 1 \leq j \leq n$. Each $\mathbf{v}_i$ can be viewed as a 2-coloring of $[n] = \{1, 2, \ldots, n\}$. Now the vector sequence $\mathbf{v}_1, \ldots, \mathbf{v}_d$ defines a 2-coloring $f : [n]^d \to \{0, 1\}$ of the board of the n^d game as follows: for every $(j_1, \ldots, j_d) \in [n]^d$ let

$$f(j_1, \ldots, j_d) \equiv v_{1,j_1} + v_{2,j_2} + \cdots + v_{d,j_d} \pmod{2}. \tag{12.1}$$

Which vector sequence $\mathbf{v}_1, \ldots, \mathbf{v}_d$ defines a *Proper* 2-Coloring of the n^d hypergraph? To answer this question, consider an arbitrary winning line L. Line L can be parametrized by an "x-vector" as follows. The first coordinate of the "x-vector" is either a constant c_1, or x, or $(n+1-x)$; similarly, the second coordinate of the "x-vector" is either a constant c_2, or x, or $(n+1-x)$, and so on.

$$L : (\text{either constant} c_1 \text{ or} x \text{ or} (n+1-x), \ldots, \text{either constant} c_d \text{ or } x \text{ or } (n+1-x)), \tag{12.2}$$

and the kth point $\mathbf{P}_k$ on line L is obtained by the substitution $x = k$ in "x-vector" (12.2) ($k = 1, 2, \ldots, d$). What is the f-color – see (12.1) – of point $\mathbf{P}_k$? To answer this question, for every $i = 1, 2, \ldots, d$ write

$$\varepsilon_i = \begin{cases} 0, & \text{if the } i\text{th coordinate in (12.2) is a constant } c_i; \\ 1, & \text{otherwise.} \end{cases} \tag{12.3}$$

For an arbitrary n-dimensional vector $\mathbf{a} = (a_1, a_2, \ldots, a_n)$ define the "reverse"

$$\mathbf{a}^{(rev)} = (a_n, a_{n-1}, \ldots, a_1). \tag{12.4}$$

It follows from (12.1)–(12.4) that the f-color of the kth point $\mathbf{P}_k$ on line L is

$$f(\mathbf{P}_k) \equiv \sum_{1 \le i \le d:\ \varepsilon_i = 0} v_{i, c_i} + \left(k\text{th coordinate of } \sum_{i=1}^{d} \varepsilon_i \mathbf{w}_i \right) \pmod{2}, \tag{12.5}$$

where

$$\mathbf{w}_i = \begin{cases} \mathbf{v}_i, & \text{if the } i\text{th coordinate in (12.2) is } x; \\ \mathbf{v}_i^{(rev)}, & \text{if the } i\text{th coordinate in (12.2) is} (n+1-x). \end{cases} \tag{12.6}$$

It follows from (12.5)–(12.6) that line L is monochromatic if and only if

$$\sum_{i=1}^{d} \varepsilon_i \mathbf{w}_i \equiv \text{either } \mathbf{0} = (0, \ldots, 0) \text{ or } \mathbf{1} = (1, \ldots, 1) \pmod{2}.$$

It suffices therefore to find $(n-1)$ n-dimensional 0–1 vectors $\mathbf{v}_1, \ldots, \mathbf{v}_{n-1}$ such that for each choice of $\mathbf{w}_i \in \{\mathbf{v}_i, \mathbf{v}_i^{(rev)}\}$, $\varepsilon_i \in \{0, 1\}$, $1 \le i \le n-1$, where $(\varepsilon_1, \ldots, \varepsilon_{n-1}) \ne \mathbf{0}$, the vector

$$\varepsilon_1 \mathbf{w}_1 + \varepsilon_2 \mathbf{w}_2 + \ldots + \varepsilon_{n-1} \mathbf{w}_{n-1} \pmod{2} \text{ is neither } \mathbf{0} \text{ nor } \mathbf{1}. \tag{12.7}$$

We give the following explicit construction: For $n \geq 5$ let (watch out for the 1s)

$$
\begin{array}{l}
(1, 0, 0, 0, 0, 0, \ldots\ldots, 0, \ldots\ldots, 0, 0, 0, 0, 0, 0) \\
(0, 1, 0, 0, 0, 0, \ldots\ldots, 0, \ldots\ldots, 0, 0, 0, 0, 0, 0) \\
(0, 0, 1, 0, 0, 0, \ldots\ldots, 0, \ldots\ldots, 0, 0, 0, 0, 0, 0) \\
(0, 0, 0, 1, 0, 0, \ldots\ldots, 0, \ldots\ldots, 0, 0, 0, 0, 0, 0) \\
(0, 0, 0, 0, 1, 0, \ldots\ldots, 0, \ldots\ldots, 0, 0, 0, 0, 0, 0) \\
\rule{5cm}{0.4pt} \\
\rule{5cm}{0.4pt} \\
\rule{5cm}{0.4pt} \\
(0, 0, 0, 0, 1, 0, \ldots\ldots, 0, \ldots\ldots, 0, 1, 0, 0, 0, 0) \\
(0, 0, 0, 1, 0, 0, \ldots\ldots, 0, \ldots\ldots, 0, 0, 1, 0, 0, 0) \\
(0, 0, 1, 0, 0, 0, \ldots\ldots, 0, \ldots\ldots, 0, 0, 0, 1, 0, 0) \\
(0, 1, 0, 0, 0, 0, \ldots\ldots, 0, \ldots\ldots, 0, 0, 0, 0, 1, 0) \\
(1, 0, 0, 0, 0, 0, \ldots\ldots, 0, \ldots\ldots, 0, 0, 0, 0, 0, 1)
\end{array}
$$

That is, the first $\lfloor n/2 \rfloor$ n-dimensional vectors are $\mathbf{e}_i = (0, \ldots, 0, 1, 0, \ldots, 0)$ such that the only non-zero coordinate is 1 at the ith place, $1 \leq i \leq \lfloor n/2 \rfloor$, the rest are symmetric "self-reversed" vectors, and in each vector the $\lceil (n+1)/2 \rceil$th coordinate is 0.

It remains to show that this construction satisfies the requirement (see (12.7)). First we show that the vector

$$
\varepsilon_1 \mathbf{w}_1 + \varepsilon_2 \mathbf{w}_2 + \ldots + \varepsilon_{n-1} \mathbf{w}_{n-1} \pmod 2
$$

in (12.7) is $\neq \mathbf{1} = (1, \ldots, 1)$. Indeed, the $\lceil (n+1)/2 \rceil$th coordinate of vector (12.7) is always 0.

So assume that vector (12.7) equals $\mathbf{0} = (0, \ldots, 0)$. Then from the first and nth coordinates we see that:

either $\varepsilon_1 + \varepsilon_{n-1} \equiv \varepsilon_{n-1} \equiv 0 \pmod 2$,
or $\varepsilon_1 + \varepsilon_{n-1} \equiv \varepsilon_1 \equiv 0 \pmod 2$.

In both cases we obtain that $\varepsilon_1 = \varepsilon_{n-1} = 0$. Similarly, we have $\varepsilon_2 = \varepsilon_{n-2} = 0$, $\varepsilon_3 = \varepsilon_{n-3} = 0$, and so on. We conclude that all coefficients must be 0: $\varepsilon_1 = \varepsilon_2 = \cdots = \varepsilon_{n-1} = 0$, which is impossible. This completes the proof of Theorem 12.1. $\square$

2. Improving the linear bound. A natural idea to improve Theorem 12.1 is to apply the Erdős–Lovász 2-Coloring Theorem. The direct application doesn't work because the n^d hypergraph is very far from being Degree-Regular. The n^d hypergraph is in fact extremely *irregular*: the Average Degree of the family of winning sets in n^d is

$$
\text{AverageDegree}(n^d) = \frac{n \cdot \text{familysize}}{\text{boardsize}} = \frac{n \left((n+2)^d - n^d \right)/2}{n^d} \approx \frac{n}{2} \left(e^{2d/n} - 1 \right).
$$

This is *much* smaller than the Maximum Degree $(3^d - 1)/2$ (n odd) and $2^d - 1$ (n even), namely about (roughly speaking) the nth root of the Maximum Degree. It is natural therefore to ask the following:

Question A: Can we reduce the Maximum Degree of an arbitrary n-uniform hypergraph close to the order of the Average Degree?

The answer is an easy *yes* if we are allowed to throw out *whole* winning sets. But throwing out a whole winning set means that Breaker loses control over that set, and Maker might completely occupy it. So we cannot throw out whole sets, but we may throw out a few points from each winning set. In other words, we can *partially truncate* the winning sets, but we cannot throw them out entirely. So the right question is:

Question B: Can we reduce the Maximum Degree of an arbitrary n-uniform hypergraph, by *partially* truncating the winning sets, close to the order of the Average Degree?

The answer to Question B is *no* for general n-uniform hypergraphs (we leave it to the reader to construct an example), but it is *yes* for the special case of the n^d hypergraphs.

Theorem 12.2 (*"Degree Reduction by Partial Truncation"*)

(a) Let $\mathcal{F}_{n,d}$ denote the family of n-in-a-lines (i.e. geometric lines) in the n^d board; $\mathcal{F}_{n,d}$ is an n-uniform Almost Disjoint hypergraph. Let $0 < \alpha < 1/2$ be an arbitrary real number. Then for each geometric line $L \in \mathcal{F}_{n,d}$ there is a $2\lfloor(\frac{1}{2} - \alpha)n\rfloor$-element subset $\widetilde{L} \subset L$ such that the truncated family $\widetilde{\mathcal{F}_{n,d}} = \{\widetilde{L} : L \in \mathcal{F}_{n,d}\}$ has Maximum Degree

$$\text{MaxDegree}\left(\widetilde{\mathcal{F}_{n,d}}\right) < d + d^{\lceil d/\alpha n\rceil - 1}.$$

(b) Let $\mathcal{F}_{n,d}^c$ denote the family of combinatorial lines in the n^d board; $\mathcal{F}_{n,d}^c$ is an n-uniform Almost Disjoint hypergraph. Let $0 < \beta < 1$ be an arbitrary real number. Then for each combinatorial line $L \in \mathcal{F}_{n,d}^c$ there is a $\lfloor(1 - \beta)n\rfloor$-element subset $\widetilde{L} \subset L$ such that the truncated family $\widetilde{\mathcal{F}_{n,d}^c} = \{\widetilde{L} : L \in \mathcal{F}_{n,d}^c\}$ has Maximum Degree

$$\text{MaxDegree}\left(\widetilde{\mathcal{F}_{n,d}^c}\right) < d + d^{\lceil d/\beta n\rceil - 1}.$$

Remarks.

(1) Let $\alpha > c_0 > 0$, i.e. let α be "separated from 0." Then the upper bound $d^{O(d/n)}$ of the Maximum Degree of $\widetilde{\mathcal{F}_{n,d}}$ is not that far from the order of magnitude of the Average Degree $\frac{n}{2}\left(e^{2d/n} - 1\right)$ of $\mathcal{F}_{n,d}$. Indeed, what really matters is the exponent, and the 2 exponents $const \cdot d/n$ and $2d/n$ are the same, apart from a constant factor.

(2) In the applications of (a), we always need that $(\frac{1}{2} - \alpha)n \geq 1$, or equivalently $\alpha \leq \frac{1}{2} - \frac{1}{n}$, since otherwise the "pseudo-line" $\tilde{L}$ becomes empty.

Proof of Theorem 12.2.

Case (a): We recall that the maximum degree of the family of winning lines is $(3^d - 1)/2$ if n is *odd*, and $2^d - 1$ if n is *even*. The maximum is achieved for the center (n is odd), and for the points $(c_1, c_2, \ldots, c_d)$ such that there is a $c \in \{1, \ldots, n\}$ with $c_j \in \{c, n+1-c\}$ for every $j = 1, 2, \ldots, d$ (n is *even*). This motivates our basic idea: we define $\tilde{L} \subset L$ by throwing out the points with "large coordinate-repetition."

Let $\mathbf{P} = (a_1, a_2, a_3, \ldots, a_d)$, $a_i \in \{1, 2, \ldots, n\}$, $1 \leq i \leq d$ be an arbitrary point of the board of the n^d game. We study the *coordinate-repetitions* of $\mathbf{P}$. Let $\ell = \lfloor (n+1)/2 \rfloor$, and write $[\ell] = \{1, 2, \ldots, \ell\}$. Let $b \in [\ell]$ be arbitrary. Consider the *multiplicity* of b and $(n+1-b)$ in $\mathbf{P}$: let

$$m(\mathbf{P}, b) = |M(\mathbf{P}, b)|, \text{ where } M(\mathbf{P}, b) = \{1 \leq i \leq d : a_i = b \text{ or } (n+1-b)\}$$
$$= M(\mathbf{P}, n+1-b).$$

Observe that in the definition of multiplicity we identify b and $(n+1-b)$.

For example, let

$$\mathbf{P} = (3, 7, 3, 5, 1, 3, 5, 4, 3, 1, 5, 3, 3, 5, 5, 1, 4, 2, 6, 5, 2, 7) \in [7]^{22};$$

then $m(\mathbf{P}, 1) = 5$, $m(\mathbf{P}, 2) = 3$, $m(\mathbf{P}, 3) = 12$, $m(\mathbf{P}, 4) = 2$.

For every *n-line* L of the n^d game we choose one of the two orientations. An orientation can be described by an *x-vector* $\mathbf{v} = \mathbf{v}(L) = (v_1, v_2, v_3, \ldots, v_d)$, where the ith coordinate v_i is either a constant c_i, or variable x, or variable $(n+1-x)$, $1 \leq i \leq d$, and for at least one index i, v_i is x or $(n+1-x)$.

For example, in the ordinary 3^2 Tic-Tac-Toe

$(1, 3)$	$(2, 3)$	$(3, 3)$
$(1, 2)$	$(2, 2)$	$(3, 2)$
$(1, 1)$	$(2, 1)$	$(3, 1)$

$\{(1, 1), (2, 2), (3, 3)\}$ is a winning line defined by the *x*-vector xx, $\{(1, 2), (2, 2), (3, 2)\}$ is another winning line defined by the *x*-vector $x2$, and finally $\{(1, 3), (2, 2), (3, 1)\}$ is a winning line defined by the *x*-vector xx' where $x' = (n+1-x)$.

The kth point $\mathbf{P}_k$ ($1 \leq k \leq n$) of line L is obtained by putting $x = k$ in the *x*-vector $\mathbf{v} = \mathbf{v}(L)$ of the line. The sequence $(\mathbf{P}_1, \mathbf{P}_2, \ldots, \mathbf{P}_n)$ gives an *orientation* of line L. The second (i.e. reversed) orientation comes from *x*-vector $\mathbf{v}^*$, which is obtained from $\mathbf{v} = \mathbf{v}(L)$ by switching coordinates x and $(n+1-x)$ that are *variables*.

Let $b \in [\ell]$ ($\ell = \lfloor (n+1)/2 \rfloor$) be arbitrary, and consider the multiplicity of b and $(n+1-b)$ in x-vector $\mathbf{v} = \mathbf{v}(L)$

$$m(L, b) = |M(\mathbf{v}(L), b)|, \quad \text{where } M(\mathbf{v}(L), b) = \{1 \leq i \leq d: \ v_i = b \text{ or } (n+1-b)\}.$$

Similarly, consider the multiplicity of x and $(n+1-x)$ in x-vector $\mathbf{v} = \mathbf{v}(L)$

$$m(L, x) = |M(\mathbf{v}(L), x)|, \quad \text{where } M(\mathbf{v}(L), x) = \{1 \leq i \leq d: \ v_i = x \text{ or } (n+1-x)\}.$$

It follows that $m(\mathbf{P}_k, k) = m(L, k) + m(L, x)$, and $m(\mathbf{P}_k, b) = m(L, b)$ if $k \notin \{b, n+1-b\}$, where $\mathbf{P}_k$ is the kth point of line L in the orientation $\mathbf{v} = \mathbf{v}(L)$, and $b \in [\ell]$.

Let $m_0 = \lceil d/\alpha n \rceil$. For every line L define the "index-set"

$$B_L = \{k \in [\ell]: \ m(L, k) < m_0\}.$$

Then

$$d > \sum_{b \in [\ell]} m(L, b) \geq \sum_{b \in [\ell] \setminus B_L} m(L, b) \geq \sum_{b \in [\ell] \setminus B_L} m_0 = m_0(\ell - |B_L|),$$

and so

$$|B_L| > \ell - \frac{d}{m_0} \geq \lfloor (n+1)/2 \rfloor - \alpha n,$$

which implies that

$$|B_L| \geq \left\lfloor (\tfrac{1}{2} - \alpha)n \right\rfloor + \{1 \text{ or } 0\}$$

depending on the parity of n (1 if n is *odd*, and 0 if n is *even*). Let

$$B_L^* = \{k: \ k \in B_L \text{ or } (n+1-k) \in B_L\};$$

then clearly

$$|B_L^*| \geq 2 \left\lfloor (\tfrac{1}{2} - \alpha)n \right\rfloor + \{1 \text{ or } 0\}$$

depending on the parity of n. For every line L, the "index-set" B_L^* defines a subset $\tilde{L} \subset L$ (we call $\tilde{L}$ a *pseudo-line*) as follows: let $\tilde{L} = \{\mathbf{P}_k: \ k \in B_L^*\}$ if n is *even*, and $\tilde{L} = \{\mathbf{P}_k: \ k \in B_L^* \setminus \{\ell\}\}$ if n is *odd* (i.e. we throw out the "mid-point" when there is one). Here $\mathbf{P}_k$ is the kth point of line L in the chosen orientation.

The above-mentioned definition of the pseudo-line has one trivial formal problem: $\tilde{L}$ may have too many points, and this indeed happens for lines like the "main diagonal"; then it is sufficient to throw out arbitrary points to get to the desired size $2 \lfloor (\tfrac{1}{2} - \alpha)n \rfloor$.

Now fix an arbitrary point $\mathbf{P} = (c_1, c_2, \dots, c_d)$ of the n^d-board. We have to estimate the number of pseudo-lines through $\mathbf{P}$. To find a line L such that $\mathbf{P} = \mathbf{P}_k$ for some $k \in B_L^*$ (i.e. $\mathbf{P}$ is the kth point of line L) we must choose a subset Y of $M(\mathbf{P}, k)$ of size $y < m_0$ and for $i \in M(\mathbf{P}, k) \setminus Y$ change $c_i = k$ to x and $c_i = n+1-k$ to $n+1-x$ (here we use that $k \neq (n+1-k)$; indeed, this follows from $k \neq \ell = \lfloor (n+1)/2 \rfloor$ when n is odd).

Let $K = \{k \in [\ell] : m(\mathbf{P}, k) \neq 0\}$. Then clearly

$$|K| \le \sum_{k \in K} m(\mathbf{P}, k) \le d.$$

Thus the number of pseudo-lines through $\mathbf{P}$ is at most

$$\sum_{y=0}^{m_0-1} \sum_{k \in K} \binom{m(\mathbf{P}, k)}{y} \le |K| + \sum_{y=1}^{m_0-1} \binom{\sum_{k \in K} m(\mathbf{P}, k)}{y} < d + d^{m_0 - 1},$$

which completes the proof of Theorem 12.2 (a).

Case (b): *Combinatorial* lines. This case is even simpler than that of case (a); we leave the details to the reader. $\square$

Now we are ready to improve the Hales–Jewett linear lower bound $HJ(n) \ge n$ of the Hales–Jewett number. The improvement comes from combining Theorem 12.2 (a) with the Erdős–Lovász 2-Coloring Theorem.

Applying the Erdős–Lovász Theorem to the truncated hypergraph $\widetilde{\mathcal{F}}_{n,d}$, we get a Proper 2-Coloring of the n^d-hypergraph if $0 < c_0 < \alpha < c_1 < 1/2$ and

$$d^{d/\alpha n} \le 2^{(1 - 2\alpha)n + O(\log n)}.$$

Taking logarithms we obtain the requirement $d \log_2 d / \alpha n \le (1 - 2\alpha)n + O(\log n)$, which is equivalent to $d \log_2 d \le \alpha(1 - 2\alpha)n^2(1 + o(1))$. Since $\alpha(1 - 2\alpha)$ attains its maximum at $\alpha = 1/4$, we conclude that $d \log_2 d \le n^2(1 + o(1))/8$, which is equivalent to $d \le (\frac{\log 2}{16} + o(1))n^2 / \log n$.

How about *combinatorial* lines? Repeating the same calculations for $\widetilde{\mathcal{F}}^c_{n,d}$, we obtain the similar inequality $d \log_2 d \le \beta(1 - \beta)n^2(1 + o(1))$. Since $\beta(1 - \beta)$ attains its maximum at $\beta = 1/2$, we conclude that $d \log_2 d \le n^2(1 + o(1))/4$, which is equivalent to $d \le (\frac{\log 2}{8} + o(1))n^2 / \log n$.

This gives the following better lower bounds for the two Hales–Jewett Numbers.

Theorem 12.3 *We have the nearly quadratic lower bounds*

$$HJ(n) \ge \left(\frac{\log 2}{16} + o(1) \right) \frac{n^2}{\log n}$$

and

$$HJ^c(n) \ge \left(\frac{\log 2}{8} + o(1) \right) \frac{n^2}{\log n},$$

where the $o(1)$ in either case is in fact $O(\log \log n / \log n)$.

Unlike the proof of the linear lower bound $HJ(n) \ge n$, which was an *explicit* algebraic construction, here we cannot provide an explicit Proper 2-Coloring. Indeed, the proof of the Erdős–Lovász 2-Coloring Theorem was an existence argument: the proof didn't say a word how to find the existing Proper 2-Coloring. To try out all possible 2^N 2-colorings of the board (where N is the board-size) is intractable.

A much more efficient "polynomial time algorithmization" was developed in Beck [1991], but we pay a price for it: the local condition $(2+o(1))^n$ of the Erdős–Lovász Theorem is replaced by some c^n, where $1 < c < 2$ is a smaller constant.

Is the (nearly) quadratic lower bound in Theorem 12.3 the best that we know? The answer is no: a shockingly simple argument ("lifting colorings") leads to exponential lower bounds!

3. Separating the Weak Win Numbers from the Hales–Jewett Numbers. We show that in hypercube Tic-Tac-Toe the converse of the trivial implication

draw play is impossible $\Rightarrow$ Winning Strategy $\Rightarrow$ weak win strategy

is *totally* false!

We start the discussion with $HJ^c(n)$ ("combinatorial lines"), which is less interesting from our game-theoretic/geometric viewpoint, but more natural from a purely combinatorial viewpoint. At the end of Section 7 we applied the one-to-one mapping (see (7.1))

$$f(\mathbf{w}) = (a_1 - 1) + (a_2 - 1)n + (a_3 - 1)n^2 + \ldots + (a_d - 1)n^{d-1}, \qquad (12.8)$$

where $\mathbf{w} = (a_1, a_2, \cdots, a_d) \in [n]^d$. Notice that f maps any n-in-a-line ("geometric line") into an n-term arithmetic progression; it follows that

$$W(n, k) \leq n^{HJ(n,k)}, \qquad (12.9)$$

in particular, $W(n) \leq n^{HJ(n)}$.

Here comes the surprise: we don't really need a one-to-one mapping like (12.8), the simpler "coordinate-sum"

$$g(\mathbf{w}) = (a_1 - 1) + (a_2 - 1) + (a_3 - 1) + \ldots + (a_d - 1) \qquad (12.10)$$

leads to the new inequality

$$\frac{W(n, k) - 1}{n - 1} \leq HJ^c(n, k), \qquad (12.11)$$

which turns out to be much more efficient than (12.9).

First we prove inequality (12.11). Let $W = HJ^c(n, k) \cdot (n - 1)$, and let χ be an arbitrary k-coloring of the interval $[0, W] = \{0, 1, 2, \ldots, W\}$; we want to show that there is a monochromatic n-term arithmetic progression in $[0, W]$. Consider the d-dimensional hypercube $[n]^d$ with $d = HJ^c(n)$, where, as usual, $[n] = \{1, 2, \cdots, n\}$. Let $\mathbf{w} = (a_1, a_2, \cdots, a_d) \in [n]^d$ be an arbitrary point in the hypercube. We can define a color of point $\mathbf{w}$ as the χ-color of the coordinate-sum (see (12.10))

$$g(\mathbf{w}) = (a_1 - 1) + (a_2 - 1) + (a_3 - 1) + \ldots + (a_d - 1);$$

We refer to this particular k-coloring of hypercube $[n]^d$ as the "lift-up of χ." Since the dimension of the hypercube is $d = HJ^c(n)$, there is a monochromatic combinatorial line in $[n]^d$ (monochromatic in the "lift-up of χ"). Thus the coordinate sums of the n points on the line form a χ-monochromatic n-term arithmetic progression in $[0, W]$. This completes the proof of (12.11). $\qquad\qquad\square$

If n is a prime, then Berlekamp's bound $W(n) > (n - 1)2^{n-1}$ (see (11.10), combined with (12.11), gives

$$HJ^c(n) \geq 2^{n-1};\qquad\qquad(12.12')$$

and, in general, for arbitrary n, the Erdős–Lovász bound $W(n) \geq 2^{n-3}/n$ (see (11.9)) implies

$$HJ^c(n) \geq \frac{2^{n-3}}{n^2}.\qquad\qquad(12.12'')$$

Lower bounds ((12.12'))–((12.12'')) represent a big improvement on Theorem 12.3.

How about $HJ(n)$? Can we prove a similar exponential lower bound? The answer is yes, and we are going to employ the *quadratic* coordinate sum

$$Q(\mathbf{w}) = (a_1 - 1)^2 + (a_2 - 1)^2 + (a_3 - 1)^2 + \ldots + (a_d - 1)^2,\qquad(12.13)$$

where $\mathbf{w} = (a_1, a_2, \cdots, a_d) \in [n]^d$. Notice that the old linear function g (see (12.10)) has a handicap: it may map a whole n-in-a-line (geometric line) into a single integer (as a "degenerate n-term arithmetic progression"). The quadratic function Q in (12.13) basically solves this kind of problem, but it leads to a minor technical difficulty: the Q-image of a geometric line is a *quadratic progression* (instead of an arithmetic progression). We pay a small price for this change: the set of n-term arithmetic progressions is a 2-parameter family, but the set of n-term quadratic progressions is a 3-parameter family. Also an n-term quadratic progression is a multiset with maximum multiplicity 2 (since a quadratic equation has 2 roots), representing at least $n/2$ distinct integers (another loss of a factor of 2). After this outline, we can easily work out the details as follows.

Any geometric line can be encoded as a string of length d over the alphabet $\Lambda = \{1, 2, \cdots, n, x, x^*\}$ (where x^* represents "reverse x") with at least one x or x^*. The n points $P_1, P_2, \ldots, P_n$ constituting a geometric line can be obtained by substituting $x = 1, 2, \ldots, n$ and $x^* = n+1-x = n, n-1, \ldots, 1$. If the encoding of a geometric line L contains a occurences of symbol x and b occurences of symbol x^*, and $L = \{P_1, P_2, \ldots, P_n\}$ where P_i arises by the choice $x = i$, the sequence $Q(P_1), Q(P_2), Q(P_3), \ldots, Q(P_n)$ (see (12.13)) has the form

$$a(x-1)^2 + b(n-x)^2 + c = (a+b)x^2 - 2(a+bn)x + (c+a+bn^2) \text{ as } x = 1, 2, \ldots, n.$$
$$(12.14)$$

Let $W = HJ(n) \cdot (n-1)^2$; the quadratic sequence (12.14) falls into the interval $[0, W]$. A quadratic sequence $Ax^2 + Bx + C$ with $x = 1, 2, \ldots, n$ is called a *n-term non-degenerate quadratic progression* if A, B, C are integers and $A \neq 0$.

Motivated by van der Waerden's Theorem, we define $W_q(n)$ to be the least integer such that any 2-coloring of $[0, W_q(n) - 1] = \{0, 1, 2, \ldots, W_q(n) - 1\}$ yields a monochromatic n-term non-degenerate quadratic progression. We prove the following inequality (an analogue of (12.11))

$$\frac{W_q(n) - 1}{(n-1)^2} \leq HJ(n). \tag{12.15}$$

In order to prove (12.15), let $W = W_q(n) - 1$ and let χ be an arbitrary 2-coloring of the interval $[0, W] = \{0, 1, 2, \ldots, W\}$. We want to show that there is a monochromatic n-term non-degenerate quadratic progression in $[0, W]$. Consider the d-dimensional hypercube $[n]^d$ with $d = HJ(n)$, where $[n] = \{1, 2, \cdots, n\}$. Let $\mathbf{w} = (a_1, a_2, \cdots, a_d) \in [n]^d$ be an arbitrary point in the hypercube. We can define a color of point $\mathbf{w}$ as the χ-color of the quadratic coordinate sum (see (12.13))

$$Q(\mathbf{w}) = (a_1 - 1)^2 + (a_2 - 1)^2 + (a_3 - 1)^2 + \ldots + (a_d - 1)^2;$$

We refer to this particular 2-coloring of hypercube $[n]^d$ as the "lift-up of χ." Since the dimension of the hypercube is $d = HJ(n)$, there is a monochromatic geometric line in $[n]^d$ (monochromatic in the "lift-up of χ"). Thus the quadratic coordinate sums of the n points on the line form a χ-monochromatic n-term non-degenerate quadratic progression in $[0, W]$. This completes the proof of (12.15).

Next we need a lower bound for $W_q(n)$; the following simple bound suffices for our purposes

$$W_q(n) \geq \frac{2^{n/4}}{3n^2}. \tag{12.16}$$

Lower bound (12.16) is an easy application of Theorem 11.4 ("Erdős-Lovász 2-Coloring Theorem"). Indeed, first note that an n-term non-degenerate quadratic progression $Ax^2 + Bx + C$ represents at least $n/2$ different integers (since a quadratic polynomial has at most 2 real roots). Three different terms "almost" determine an n-term quadratic progression; they determine less than n^3 n-term quadratic progressions. Thus, any n-term non-degenerate quadratic progression contained in $[1, W]$, where $W = W_q(n)$, intersects fewer than $n^4 \cdot W^2$ other n-term non-degenerate quadratic progressions in $[1, W]$. It follows that

$$8n^4 \cdot W^2 > 2^{n/2}; \tag{12.17}$$

indeed, otherwise Theorem 11.4 applies, and yields the existence of a 2-coloring of $[1, W]$ with no monochromatic n-term non-degenerate quadratic progression, which contradicts the choice $W = W_q(n)$. (12.17) implies (12.6).

Combining (12.15) and (12.16) we obtain

$$HJ(n) \geq \frac{2^{n/4}}{3n^4}. \tag{12.18}$$

(12.18) is somewhat weaker than (12.12), but it is still exponential, representing a big improvement on Theorem 12.3.

Remark.

Inequality (12.11) (using the linear mapping (12.10)) is a short lemma in Shelah's paper (is it *folklore* in Ramsey Theory?). The similar but more complicated (12.15)–(12.17) (using the quadratic mapping (12.13)) seems to be a new result; it is an unpublished observation due to me and my Ph.D. student Sujith Vijay (Rutgers University).

Note that Berlekamp's explicit algebraic construction – $W(n) > (n-1)2^{n-1}$ if n is a prime – was a Proper *Halving* 2-Coloring. The proof of the Erdős–Lovász 2-Coloring Theorem, on the other hand, does not provide a Proper *Halving* 2-Coloring (and it is not clear at all how to modify the original proof to get a Proper Halving 2-Coloring). This raises the following natural question. Is it true that $HJ_{1/2}(n)$ (involving halving 2-colorings) is also (at least) exponentially large? The answer is, once again, "yes." One possible way to prove it is to repeat the proofs of (12.12) and (12.18) with *rainbow 3-colorings* instead of proper 2-colorings, and to apply the following (almost trivial) general fact (see Remark (c) after the Local Lemma at the end of Section 11):

Rainbow Fact: If $\mathcal{F}$ is an arbitrary finite hypergraph such that it has a rainbow 3-coloring, then it also has a Proper Halving 2-Coloring (i.e. the two color classes have equal size).

Another (more direct) way to prove an exponential lower bound for the halving Hales–Jewett number is to apply the following inequality

$$HJ_{1/2}(n) \geq HJ(n-2). \tag{12.19}$$

Inequality (12.19), due to W. Pegden, is "hypercube-specific"; it does not extend to a general hypergraph result like the Rainbow Fact above.

In fact, the following slightly stronger version of (12.19) holds

$$HJ_{1/2}^*(n) \geq HJ(n-2), \tag{12.20}$$

where $HJ_{1/2}^*(n)$ is the largest dimension d_0 such that for any $d < d_0$ the n^d hypercube has a Proper Halving 2-Coloring (*proper* means that there is no monochromatic geometric line). We recall that $HJ_{1/2}(n)$ denotes the least integer d such that in each *halving* 2-coloring of n^d there is a monochromatic geometric line (i.e. n-in-a-line). Trivially, $HJ_{1/2}(n) \geq HJ_{1/2}^*(n)$, and we cannot exclude the possibility of a strict inequality $HJ_{1/2}(n) > HJ_{1/2}^*(n)$ for some n. This means the halving Hales–Jewett number is possibly(!) a "fuzzy threshold," unlike the ordinary Hales–Jewett number $HJ(n)$ (where there is a critical dimension d_0 such that for every 2-coloring of n^d with $d \geq d_0$ there is always a monochromatic geometric line, and for every n^d with $d < d_0$ there is a 2-coloring with no monochromatic geometric line; in the halving case we cannot prove the existence of such a critical dimension).

By adding the trivial upper bound to (12.19)–(12.20) we have

$$HJ(n) \geq HJ_{1/2}(n) \geq HJ^*_{1/2}(n) \geq HJ(n-2). \tag{12.21}$$

Here is Pegden's strikingly simple proof of (12.20). The idea is to divide the $n^{HJ(n-2)-1}$ hypercube into subcubes of the form $(n-2)^j$, $j \leq HJ(n-2) - 1$, and color them independently. We make use of the Hales–Jewett linear lower bound (see Theorem 12.1)

$$HJ(n) \geq n. \tag{12.22}$$

The "large dimension" (see (12.22)) guarantees that most of the volume of the hypercube $n^{HJ(n-2)-1}$ lies on the "boundary"; this is why we can combine the proper 2-colorings of the subcubes $(n-2)^j$, $j \leq HJ(n-2) - 1$ to obtain a Proper *Halving* 2-Coloring of the whole.

The exact details go as follows. Let $H = [n]^d$ where $d = HJ(n-2) - 1$ and $[n] = \{1, 2, \ldots, n\}$; so there is a Proper 2-Coloring for the "center" $(n-2)^d \subset H$. We need to show that there is a Proper Halving 2-Coloring of H. We divide H into subcubes of the form $(n-2)^j$, $0 \leq j \leq d$: for each "formal vector"

$$\mathbf{v} = (v_1, v_2, \ldots, v_d) \in \{1, c, n\}^d$$

(here "c" stands for "center") we define the sub-hypercube $H_\mathbf{v}$ as the set of of all $(a_1, a_2, \ldots, a_d) \in H$ satisfying the following two requirements:

(1) $a_i = 1$ if and only if $v_i = 1$;
(2) $a_i = n$ if and only if $v_i = n$.

Then $H_\mathbf{v}$ is of size $(n-2)^j$, where the dimension $j = dim(H_\mathbf{v})$ is the number of coordinates of $\mathbf{v}$ equal to c, and the $H_\mathbf{v}$s form a partition of H by mimicking the binomial formula

$$n^d = ((n-2)+2)^d = (n-2)^d + \binom{d}{1} 2 \cdot (n-2)^{d-1} + \binom{d}{2} 2^2 \cdot (n-2)^{d-2} + \ldots + 2^d. \tag{12.23}$$

Notice that $H_{(c,\ldots,c)}$ is the "center" of H; by assumption $H_{(c,\ldots,c)}$ has a Proper 2-Coloring.

Call $H_\mathbf{v}$ *degenerate* if its dimension $j = 0$; these are the 2^d "corners" of hypercube H.

The following fact is readily apparent.

Simple proposition: For any geometric line L (n-in-a-line) in $H = n^d$, there is some non-degenerate subhypercube $H_v \subset H$ such that the intersection $L \cap H_v$ is a geometric line (($n-2$)-in-a-line) in H_v (considering H_v as an $(n-2)^j$ hypercube). The Simple Proposition implies that any 2-coloring of H which is *improper* (i.e. there is a monochromatic line) must be improper in its restriction to a non-degenerate sub-hypercube H_v.

As we said before, the "center" $H_{(c,...,c)} \subset H$ has a Proper 2-Coloring; we use the colors X and O (like in an ordinary Tic-Tac-Toe play). Let the proportion of Xs in the coloring be α_0; we can assume that $\alpha_0 \geq 1/2$. Considering all $(d-1)$-dimensional "slices" of the "center" $H_{(c,...,c)}$, the average proportion of Xs is α_0, so the maximum proportion of Xs, denoted by α_1, is greater or equal to α_0. It follows that the $(n-2)^{d-1}$ sub-hypercubes of H can be properly 2-colored with an α_1 fraction of Xs (or Os; we can always flip a coloring!). Thus, inductively, we find a non-decreasing sequence

$$1/2 \leq \alpha_0 \leq \alpha_1 \leq \ldots \leq \alpha_d = 1$$

of "proportions" so that for $0 \leq j \leq d$, each $(n-2)^j$ sub-hypercube can be properly 2-colored with an α_{d-j} fraction of Xs. For each $(n-2)^j$ sub-hypercube we have two options: either we keep this proper 2-coloring or we flip. By using this freedom, we can easily extend the proper 2-coloring of the "center" $H_{(c,...,c)} \subset H$ to a Proper *Halving* 2-Coloring of H as follows. Let

$$A_k = \bigcup_{dim(H_v) \geq d-k} H_v.$$

By induction on k (as $k = 0, 1, 2, \ldots, d$), we give a 2-coloring χ of H, which is proper on each of the subhypercubes $H_v \subset A_k$ and

$$disc(\chi, A_k) \leq (2\alpha_k - 1) \cdot (n-2)^{d-k}, \tag{12.24}$$

where $disc(\chi, A_k)$ denotes the *discrepancy*, i.e. the absolute value of the difference between the sizes of the color classes. Notice that $2\alpha_k - 1 = \alpha_k - (1 - \alpha_k)$ and $(n-2)^j$ is the volume of a j-dimensional H_v.

At the end, when $k = d$, coloring χ will be a Proper *Halving* 2-Coloring (indeed, the color classes on all of H will differ by at most $(2\alpha_d - 1) \cdot (n-2)^{d-d} = 1$).

(12.24) is trivial for $k = 0$: on $A_0 = H_{(c,...,c)}$ (the "center") our 2-coloring χ is the above-mentioned proper 2-coloring of the "center" with X-fraction α_0.

Next comes the general induction step: let (12.24) be satisfied for some $(k-1) \geq 0$. The number N_{d-k} of sub-hypercubes H_v of dimension $(d-k)$ is $\binom{d}{k}2^k$ (binomial theorem: see (12.23)), so by $d = HJ(n-2) - 1$ and (12.22), $N_{d-k} \geq 2d \geq n-2$. Thus since $\alpha_k \geq \alpha_{k-1}$, we have

$$N_{d-k} \cdot (2\alpha_k - 1) \cdot (n-2)^{d-k} \geq (2\alpha_{k-1} - 1) \cdot (n-2)^{d-(k-1)} \geq disc(\chi, A_{k-1}). \tag{12.25}$$

In view of inequality (12.25) we have room enough to extend χ from A_{k-1} to A_k by coloring a suitable number of the H_v of dimension $(d-k)$ with an α_k-fraction of Xs, and the rest with an α_k-fraction of Os ("we flip the coloring"). This completes the induction proof of (12.24), and (12.20) follows.

We summarize these results in a single theorem (see (12.18) and (12.20)).

Theorem 12.4 *We have*

$$HJ(n) \geq HJ_{1/2}(n) \geq HJ_{1/2}^*(n) \geq HJ(n-2) \geq \frac{2^{(n-2)/4}}{3n^4}. \tag{12.26}$$

The line in (12.26) can be extended by the "game numbers"

$$HJ(n) \geq HJ_{1/2}(n) \geq HJ_{1/2}^*(n) \geq \mathbf{w}(n\text{–line}) \geq \mathbf{ww}(n\text{–line}). \tag{12.27}$$

(12.27) is trivial, because a Strong Draw strategy of the second player – in fact, any drawing strategy! – yields the existence of a drawing terminal position, i.e. a Proper Halving 2-Coloring (indeed, the first player can "steal" the second player's strategy).

Here comes the surprise: a trivial application of Theorem 1.2 (Weak Win criterion) gives the upper bound

$$\frac{\log 2}{2} n^2 \geq \mathbf{ww}(n\text{–line}) \tag{12.28}$$

(We challenge the reader to prove this; we give the details at the beginning of Section 13), so comparing (12.26) and (12.28) we have (assuming n is large)

$$HJ_{1/2}^*(n) \geq \frac{2^{(n-2)/4}}{3n^4} > \frac{\log 2}{2} n^2 \geq \mathbf{ww}(n\text{–line}), \tag{12.29}$$

i.e. asymptotically the Ramsey threshold $HJ_{1/2}^*(n)$ is (at least) exponential and the Weak Win threshold $\mathbf{ww}(n\text{–line})$ is (at most) quadratic. Roughly speaking, Ramsey Theory has nothing to do with Weak Win!

Inequality (12.29) leads to some extremely interesting problems; the first one was already briefly mentioned at the end of Section 5.

Delicate win or delicate draw? A wonderful question! In Section 5, we introduced a classification of the family of all finite hypergraphs: we defined Classes 0–5. Perhaps the two most interesting classes are

Class 2 ("Forced win but Drawing Position exists: delicate win"): It contains those hypergraphs $\mathcal{F}$ which have a Drawing Position, but the first player can nevertheless force a win.

Class 3 ("Delicate Draw"): It contains those hypergraphs $\mathcal{F}$ for which the Positional Game is a Draw but the first player can still force a Weak Win (the Full Play Convention applies!).

The 4^3 Tic-Tac-Toe ("Cubic") is the only n^d game in Class 2 that we know, and the ordinary 3^2 Tic-Tac-Toe is the only n^d game in Class 3 that we know. Are there other examples? This is exactly Open Problem 5.4:

Is it true that each hypergraph class contains infinitely many n^d games? The unknown cases are Class 2 and Class 3.

What (12.29) implies is that the *union* of Class 2 and Class 3 is infinite. Indeed, each n^d Tic-Tac-Toe with dimension

$$HJ^*_{1/2}(n) > d \geq \mathbf{ww}(n\text{–line}) \tag{12.30}$$

belongs to either Class 2 or Class 3: if it is a first player win, the game belongs to Class 2; if it is a draw game, then it goes to Class 3. Of course, (12.29) implies that the range (12.30) is non-empty; in fact, it is a very large range (if n is large). Unfortunately, we cannot decide which class (Class 2 or Class 3) for any single game in the range (12.30); this is a wonderful open problem!

By (12.27)

$$HJ^*_{1/2}(n) \geq \mathbf{w}(n\text{–line}) \geq \mathbf{ww}(n\text{–line}),$$

and by (12.29) the Ramsey threshold $HJ^*_{1/2}(n)$ is (at least) exponential and the Weak Win threshold $\mathbf{ww}(n\text{–line})$ is (at most) quadratic. Where does the ordinary win threshold $\mathbf{w}(n\text{–line})$ fall? Is it (at least) exponential or polynomial? We don't have a clue; this is another totally hopeless question.

Open Problem 12.1

(a) *Which order is the right order of magnitude for* $\mathbf{w}(n\text{–line})$: *(at least) exponential or polynomial?*

(b) *For every* n^d *Tic-Tac-Toe, where the dimension* $d = d(n)$ *falls into range (12.30), decide whether it belongs to Class 2 or Class 3.*

4. The 3 basic categories of the book: Ramsey Theory, Game Theory (Weak Win), and Random Structures. Let's leave inequality (12.29), and compare Theorems 12.3 and 12.4: the reader is justly wondering: "why did we bother to prove Theorem 12.3 when Theorem 12.4 (and (12.12) for combinatorial lines) is so much stronger?" Well, the reason is that the proof of Theorem 12.3 can be adapted to the Weak Win threshold $\mathbf{ww}(n\text{–line})$, and it gives a nearly best possible game-theoretic result! Theorem 12.5 below is the game-theoretic analogue of Theorem 12.3 (the lower bounds).

Theorem 12.5 *We have*:

(a) $\frac{\log 2}{2} n^2 \geq \mathbf{ww}(n\text{–line}) \geq \left(\frac{\log 2}{16} + o(1)\right) \frac{n^2}{\log n}$,   (12.31)

(b) $(\log 2)n^2 \geq \mathbf{ww}(\text{comb. } n\text{–line}) \geq \left(\frac{\log 2}{8} + o(1)\right) \frac{n^2}{\log n}$.

Theorem 12.5 could not determine the true order of magnitude of the "phase transition" from Weak Win to Strong Draw, but it came pretty close to that: we proved an inequality where the upper and lower bounds differ by a mere factor of $\log n$.

Open Problem 12.2 *Which order of magnitude is closer to the truth in Theorem 12.5, $n^2/\log n$ ("the lower bound") or n^2 ("the upper bound")?*

We believe that the upper bound is closer to the truth. In fact, we guess that in the n^d hypercube Tic-Tac-Toe the phase transition from Weak Win to Strong Draw happens at $d = d(n) = ((\log 2)n^2 - n\log n)/2 + O(n)$ (see formula (9.2), and its application after that).

Now let's return to (12.26): the exponential lower bound

$$HJ^*_{1/2}(n) \geq \frac{2^{(n-2)/4}}{3n^4}$$

means that the dimension $d = d(n)$ can be exponentially large (in terms of n) and it is still possible to have a Draw Play in n^d Tic-Tac-Toe (if the two players cooperate).

On the other hand, by Theorem 12.5 Weak Win can be forced when the dimension $d = d(n)$ is roughly quadratic. If a player has a Weak Win strategy, then of course he can prevent a Draw (he can always occupy a winning set: if he does it first, he has an ordinary win; if the opponent does it first, the opponent has an ordinary win).

The third natural category is the *majority outcome:* what happens in the overwhelming majority of the n^d Tic-Tac-Toe plays? A very good approximation of the random play is the Random 2-Coloring. The expected number of monochromatic n-in-a-lines in a Random 2-Coloring of the n^d board is exactly

$$2 \cdot \frac{(n+2)^d - n^d}{2} \cdot \left(\frac{1}{2}\right)^n,$$

which undergoes a rapid change from "much less than 1" to "much larger than 1" in the sharp range $d = (1 + o(1))n/\log_2 n$. Since a "reasonable" random variable is "close" to its expected value, it is easy to prove precisely that in the range $d \geq (1 + o(1))n/\log_2 n$ the overwhelming majority of n^d Tic-Tac-Toe plays have a winner, and in the range $d \leq (1 - o(1))n/\log_2 n$ the overwhelming majority of n^d Tic-Tac-Toe plays end in a draw.

This gives us a clear-cut separation of the 3 basic categories of the book: Ramsey Theory, Game Theory (Weak Win), and Random Structures. In n^d Tic-Tac-Toe

(1) the Ramsey threshold is (at least) *exponential*;
(2) the Weak Win threshold is roughly *quadratic*; and, finally,
(3) the phase transition for the Majority Outcome happens in a *sublinear* range.

The bottom line is that Weak Win and the Majority Outcome are closely related but different, and Ramsey Theory has nothing to do with Weak Win. How about ordinary win? Is ordinary win more related to Ramsey Theory? No one knows!

Let's return to Theorem 12.5 one more time. Switching to the inverse function, inequality (12.5) can be expressed in terms of the corresponding Achievement Number as follows

$$\sqrt{2/\log 2}\sqrt{d} \le \mathbf{A}(d-\text{dim. cube; line}) \le (1+o(1))\sqrt{32/\log 2}\sqrt{d \log d}, \quad (12.32)$$

and exactly the same bounds for the Avoidance Number $\mathbf{A}(d-\text{dim. cube; line; } -)$.

5. Winning planes. Inequality (12.32) is not too elegant, certainly not as elegant as the exact results Theorem 6.4 and Theorem 8.2 about "quadratic" goal-sets. What is the quadratic version of a "winning line"? Of course, a "winning plane"! First switch from the n^d hypercube to the n^d Torus – because we want a degree-regular hypergraph! – and define the concept of a *combinatorial plane*, which is the simplest kind of a "plane." Intuitively a *combinatorial plane*, or *Comb-Plane* for short, is a "2-parameter set."

A Comb-Plane S in the n^d Torus is formally defined by a point $P \in S$ and two non-zero vectors $\mathbf{v} = (a_1, \ldots, a_d)$ and $\mathbf{w} = (b_1, \ldots, b_d)$, where each coordinate is either 0 or 1, and $a_i b_i = 0$ for $i = 1, \ldots, d$ (i.e. the non-zero coordinates in $\mathbf{v}$ and $\mathbf{w}$ form two disjoint non-empty sets); the n^2 points of Comb-Plane S are $P + k\mathbf{v} + l\mathbf{w}$, where k and l independently run through $0, 1, \ldots, n-1$.

In other words, every Comb-Plane S in the n^d Torus can be described by a single vector $\mathbf{u} = (u_1, \ldots, u_d)$, where each u_i is either a constant c_i, or $c_i + x$, or $c_i + y$, and the last two cases both occur – x and y are the two defining parameters, which independently run through $0, 1, \ldots, n-1$, and the "addition" is modulo n ("torus"). For example, the $(3+x)14(2+y)2511(2+x)$ Comb-Plane in the 5^9 Torus means the following 5×5 lattice, where in each string the single "asterisk" $*$ means 14 and the "double star" $\star\star$ means 2511:

$$3*2\star\star2 \quad 3*3\star\star2 \quad 3*4\star\star2 \quad 3*5\star\star2 \quad 3*1\star\star2$$

$$4*2\star\star3 \quad 4*3\star\star3 \quad 4*4\star\star3 \quad 4*5\star\star3 \quad 4*1\star\star3$$

$$5*2\star\star4 \quad 5*3\star\star4 \quad 5*4\star\star4 \quad 5*5\star\star4 \quad 5*1\star\star4$$

$$1*2\star\star5 \quad 1*3\star\star5 \quad 1*4\star\star5 \quad 1*5\star\star5 \quad 1*1\star\star5$$

$$2*2\star\star1 \quad 2*3\star\star1 \quad 2*4\star\star1 \quad 2*5\star\star1 \quad 2*1\star\star1$$

If $P = (c_1, \ldots, c_d)$ is an arbitrary point of the n^d Torus, then the number of Comb-Planes S containing point P equals half of the number of all $\{0, x, y\}$-strings of length d where both parameters x and y show up ("half" because the substitution $x = y$ doesn't change the plane). A simple application of the Inclusion–Exclusion principle gives the answer: $(3^d - 2^{d+1} + 1)/2$. Since the Maximum Degree is $(3^d - 2^{d+1} + 1)/2$, the total number of Comb-Planes in the n^d Torus is clearly $(3^d - 2^{d+1} + 1)n^{d-2}/2$.

Given a fixed dimension d, what is the largest value of n such that Maker can occupy a whole $n \times n$ Comb-Plane in the n^d Torus? The largest value of n is, of course, the Achievement Number $\mathbf{A}(d-\text{dim. torus; comb. plane})$, and $\mathbf{A}(d-\text{dim. torus; comb. plane; –})$ is the Avoidance Number.

Theorem 12.6 *We have*

$$\mathbf{A}(d-\text{dim. torus; comb. plane}) = \left\lfloor \sqrt{\log_2 3} \sqrt{d} + o(1) \right\rfloor,$$

and the same for the Avoidance Number.

This is another exact result.

Let me return to *ordinary win* one more time. We gave up on this concept as completely hopeless (see Sections 3–5, a collection of hopeless open problems), and switched to Weak Win. The Meta-Conjecture – the main issue of the book – is about Weak Win. Nevertheless here we discuss a new concept that is about *halfway* between ordinary win and Weak Win; we call it *second player's moral victory*. To motivate this new concept, we give first a quick application of Theorem 5.1 ("Strategy Stealing"); this simple but elegant application is the joint effort of my graduate class (the main credit goes to W. Pegden).

6. The second player can always avoid a humiliating defeat! A soccer score of 5–4 reflects a good, exciting match, but a score of 7–2 indicates a one-sided match, where one team was much better than the other ("humiliating defeat"). The score of $n\text{-}(n-2)$ (or an even bigger gap) motivates the following concept. For simplicity, assume that $\mathcal{F}$ is an n-uniform hypergraph, and the two players are playing the usual alternating (1:1) game. We say that the first player has a *Humiliating Victory Strategy* if he can occupy a whole n-element winning set in such a way that, in the moment of complete occupation, the opponent has at most $(n-2)$ marks in any other winning set. Let *Str* denote a first player's *Humiliating Victory Strategy* on a finite n-uniform hypergraph $\mathcal{F}$ with board set V; we are going to derive a contradiction as follows.

Let x_1 be the optimal opening move of the first player by using strategy *Str*. Let $\mathcal{F} \setminus \{x_1\}$ be the truncated subhypergraph with points $V \setminus \{x_1\}$ and winning sets $A \setminus \{x_1\}$, where $A \in \mathcal{F}$. Notice that if the first player has a Humiliating Victory Strategy on $\mathcal{F}$, then he has an ordinary winning strategy in the positional game on $\mathcal{F} \setminus \{x_1\}$ as a *second player*, which contradicts Theorem 5.1. This contradiction proves that strategy *Str* cannot exist.

As usual with Theorem 5.1, the argument above does not say a word about *how* to avoid a humiliating defeat.

Now we are ready to discuss the new concept which is much more than merely avoiding a humiliating defeat.

7. Second player's moral-victory. To be concrete, consider an (N, n) Van der Waerden Game. In view of Theorem 6.1 the second player has no chance to occupy an n-term A.P. first (if the first player plays rationally); similarly, the second player has no chance to occupy a *longer* A.P. than his opponent (hint: Strategy Stealing), so the natural question arises: "What is the *best* that the second player can hope for?" A suggestion for the "best" is the following concept of **moral-victory**: we say that the second player has a *moral-victory* in an (N, n) Van der Waerden Game if (1) he can occupy an n-term A.P. and at the same time (2) he can prevent the first player from occupying an $(n+1)$-term A.P. For example, in the $(5, 2)$ Van der Waerden Game the second player does *not* have a moral-victory, but he does have one in the $(4, 2)$ Van der Waerden Game. Are there infinitely many examples for moral-victory? Unfortunately we don't know the answer; this is another hopeless/depressing open problem.

Open Problem 12.3 *Are there infinitely many pairs (N, n) for which the (N, n) Van der Waerden Game is a second player's moral-victory?*

How come the Ramsey Theory doesn't help here? For simplicity assume that N is even, and also assume that the second player uses the following "reflection" pairing strategy: if first player's last move was i, then the second player replies by $(N+1-i)$ (the board is, as usual, $[N] = \{1, 2, \ldots, N\}$). This way at the end of the play the longest A.P. of the first player and the longest A.P. of the second player have the same length (reflected copies!); but what guarantees that this common length is the *given* n? Unfortunately, nothing guarantees that; the second player has no control over the play, since he is just a copycat.

The reader was promised a non-trivial result, and here it is; we can solve the 2-dimensional analogue of Open Problem 12.3. The 2-dimensional $(N \times N, n \times n)$ Van der Waerden Game is defined as follows: the board is an $N \times N$ lattice; the most natural choice is to take the subset

$$[N] \times [N] = \{(a, b) \in \mathbb{Z}^2 : 1 \le a \le N, 1 \le b \le N\}$$

of the integer lattice points $\mathbb{Z}^2$; the winning sets are the $n \times n$ Aligned Square Lattices in $N \times N$, so the game is the "Cartesian square" of the (N, n) Van der Waerden Game.

Second player's *moral-victory* in the 2-dimensional $(N \times N, n \times n)$ Van der Waerden Game means that (1) the second player can occupy an $n \times n$ Aligned Square Lattice, and at the same time (2) he can prevent the first player from occupying an $(n+1) \times (n+1)$ Aligned Square Lattice.

Theorem 12.7 *Let*

$$n_1 = \left\lfloor \sqrt{\log_2 N} + o(1) \right\rfloor \quad \text{and} \quad n_2 = \left\lceil \sqrt{\log_2 N} + o(1) \right\rceil$$

where the two "error terms" $o(1)$ in n_1 and n_2 both tend to 0 as $N \to \infty$ (the two $o(1)$s are not the same!). Playing on an $N \times N$ board, the second player can occupy an $n_1 \times n_1$ Aligned Square Lattice, and at the same time he can prevent the first player from occupying an $n_2 \times n_2$ Aligned Square Lattice.

Notice that second player's strategy in Theorem 12.7 is a *moral-victory* (i.e. n_1 and n_2 are consecutive integers), unless $\sqrt{\log_2 N}$ is "very close to an integer," so close that, due to the effect of the two different $o(1)$s, n_1 and n_2 are not consecutive. But this is very rare; for the overwhelming majority of Ns, the $(N \times N, n \times n)$ Van der Waerden Game is a second player's moral-victory!

A big bonus is that the proof of Theorem 12.7 supplies an explicit strategy. It is a potential strategy, not some kind of a "soft" (strategy stealing type) existential argument.

A byproduct of Theorem 12.7 is that the second player can always prevent the first player from occupying a *larger* lattice. This part alone is easy, and goes exactly like the 1-dimensional case: by using reflection. The handicap of the reflection strategy is that it says very little about the size, nothing beyond Ramsey Theory. Of course, Ramsey Theory never guarantees equality; it just gives an inequality. Unfortunately the quantitative bounds that we currently know about the 2-dimensional van der Waerden Numbers are ridiculously weak (much worse than Gowers's 5-times iterated exponential bound in the 1-dimensional case).

8. Summary. Finally, we summarize what we have done so far. The objective of Chapter II was to formulate the main results of the book. We concentrated on three "Ramseyish" games: the Clique Game, n^d Tic-Tac-Toe, and the Arithmetic Progression Game. They are the "natural examples" motivated by Ramsey Theory. The main results fall into 3 groups:

(1) the exact solutions: Theorem 6.4 ("Clique Games"), Theorem 8.2 ("Lattice Games"), Theorem 12.6 ("winning planes"), and Theorem 12.7 ("second player's moral victory");

(2) the asymptotic Theorem 8.1 ("arithmetic progression game"); and

(3) the "nearly asymptotic" Theorem 12.5 ("n^d Tic-Tac-Toe").

In addition, in Chapters IV and VI, we will discuss the discrepancy and biased versions of these theorems. We will derive all of these concrete results from general hypergraph theorems. These general hypergraph theorems ("ugly but useful criterions") represent my best efforts toward the Meta-Conjecture.

Part B

Basic potential technique – game-theoretic first and second moments

Part B is a practice session for the potential technique, demonstrating the enormous flexibility of this technique.

We look at about a dozen amusing "little" games (similar to the S-building game in Section 1). There is a large variety of results, starting with straightforward applications of Theorem 1.2 ("building") and Theorem 1.4 ("blocking"), and ending with sophisticated proofs like the 6-page-long proof of Theorem 20.3 ("Hamiltonian cycle game") and the 10-page-long proof of Theorem 15.1 ("Kaplansky's Game").

The core idea is the mysterious connection between games and randomness. By using the terms "game-theoretic first moment" and "game-theoretic second moment," we tried to emphasize this connection.

The point is to collect a lot of "easy" proofs. To get a "feel" for the subject the reader is advised to go through a lot of easy stuff. Reading Part B is an ideal warmup for the much harder Parts C–D.

A reader in a big rush focusing on the exact solutions may skip Part B entirely, and jump ahead to Sections 23–24 (where the "hard stuff" begins).

Chapter III
Simple applications

The results formulated in the previous chapter (Chapter II) will be proved in Chapters V–IX, that is, we will need 5 chapters, more than 250 pages! Chapter III plays an intermediate role: it is a preparation for the main task, and also it answers some of the questions raised in Section 4. For example, in Section 15 we discuss an interesting result related to Kaplansky's n-in-a-line game.

The main goal of Chapter III is to demonstrate the amazing flexibility of the potential technique on a wide range of simple applications.

13

Easy building via Theorem 1.2

Some of the statements formulated in Chapter II have easy proofs. So far we proved two potential criterions, both simple: (1) the Weak Win criterion Theorem 1.2, and (2) the Strong Draw criterion Theorem 1.4 ("Erdős–Selfridge"). In a few lucky cases a direct reference to Theorem 1.2 supplies the optimal result.

1. Weak Win in the Van der Waerden Game. A particularly simple example is the upper bound in Theorem 8.1 ("arithmetic progression game"). We recall the (N, n) Van der Waerden Game: the board is $[N] = \{1, 2, \ldots, N\}$ and the winning sets are the n-term A.P.s ("arithmetic progressions") in $[N]$. The Weak Win part of Theorem 8.1 is a straightforward application of Theorem 1.2. Indeed, the Max Pair-Degree Δ_2 of the hypergraph is clearly $\leq \binom{n}{2}$, the size of the hypergraph is $(1/2 + o(1))N^2/(n-1)$, so if

$$\frac{\left(\frac{1}{2} + o(1)\right)\frac{N^2}{n-1}}{N} > \binom{n}{2} 2^{n-3}, \tag{13.1}$$

then Theorem 1.2 applies, and yields a first player's Weak Win. Inequality (13.1) is equivalent to

$$N > (1 + o(1)) \, n^3 \cdot 2^{n-3}, \tag{13.2}$$

which proves the *upper bound* in Theorem 8.1.

2. Weak Win in hypercube Tic-Tac-Toe. Another straightforward application is the n^d hypercube Tic-Tac-Toe. By Theorem 1.2 and Theorem 3.4, the first player has a Weak Win if

$$\frac{(n+2)^d - n^d}{2} > 2^{n-3} n^d,$$

which is equivalent to

$$\left(1 + \frac{2}{n}\right)^d > 2^{n-2} + 1. \tag{13.3}$$

Inequality (13.3) holds if $d > \frac{1}{2}(\log 2) \cdot n^2$. This proves the upper bound in Theorem 12.5 (a).

Consider small values of n. Inequality (13.3) holds for the 3^3, 4^4, 5^7, 6^{10}, 7^{14}, 8^{19}, 9^{25}, 10^{31}, ... games, so in these games the first player can force a Weak Win. Note that 3^3 and 4^4 on the list can be replaced by 3^2 and 4^3. Indeed, ordinary Tic-Tac-Toe has an easy Weak Win, and in view of Patashnik's computer-assisted work, the 4^3 game has an ordinary Win (see Patashnik [1980]).

The list of small Weak Win n^d games: 3^2, 4^3, 5^7, 6^{10}, 7^{14}, 8^{19}, ... is complemented by the following list of known small Strong Draw games: 4^2, 8^3, 14^4, 20^5, 24^6, 26^7, ... There is a big gap between the two lists, proving that our knowledge of the small n^d games is very unsatisfactory.

The case of *combinatorial lines* goes similarly; the proof of the upper bound in Theorem 12.5 (b) is left to the reader.

3. Torus Tic-Tac-Toe.

A major difficulty of studying the n^d hypercube Tic-Tac-Toe comes from a technical difficulty: from the highly irregular nature of the n^d-hypergraph. The Average Degree of the n^d-hypergraph is about (very roughly) the nth root of the Maximum Degree. This huge difference between the Average Degree and the Maximum Degree explains why we needed so desperately Theorem 12.2 ("Degree Reduction by Partial Truncation"). Unfortunately the application of Theorem 12.2 has an unpleasant byproduct: it leads to an "error factor" of $\log n$ that we cannot get rid of.

If we switch to the n^d *Torus*, and consider the family of all "Torus-Lines" instead of all "geometric lines" in the n^d Hypercube, then we obtain a Degree-Regular hypergraph, and there is no need for any ("wasteful") degree reduction. The Degree-Regular family of "Torus-Lines" forms the family of winning sets of the n^d **Torus Tic-Tac-Toe.** The 8^2 Torus Tic-Tac-Toe is particularly interesting because it yields the amusing fact that the *Unrestricted 9-in-a-row on the plane* is a draw game (see the beginning of Section 4 and the end of Section 10). This was the "Fourth Proof" of Theorem 10.1; we employed the following 8×8 matrix

$$
n = 8: \quad
\begin{bmatrix}
\backslash & \backslash & - & - & | & | & / & / \\
- & - & \backslash & \backslash & / & / & | & | \\
- & - & / & / & \backslash & \backslash & | & | \\
/ & / & | & | & - & - & \backslash & \backslash \\
\backslash & \backslash & | & | & - & - & / & / \\
| & | & \backslash & \backslash & / & / & - & - \\
| & | & / & / & \backslash & \backslash & - & - \\
/ & / & - & - & | & | & \backslash & \backslash
\end{bmatrix}
$$

This 8 by 8 matrix represents is a direction-marking of the $4 \cdot 8 = 32$ "torus-lines" of the 8×8 torus. The direction-marks $-$, $|$, $\backslash$, and $/$ mean (respectively) "horizontal," "vertical," "diagonal of slope -1," and "diagonal of slope 1." Each one of the 32 torus-lines contains 2 marks of its own direction. The periodic extension of the 8 by 8 matrix over the whole plane (see the picture at the beginning of Section 4) gives a Pairing Strategy Draw for the unrestricted 9-in-a-row game. Either player replies to the opponent's last move by taking the *nearest* similarly marked square in the direction indicated by the mark in the opponent's last move square.

Let us return to the general case. The n^d *Torus* is an abelian group, implying that the n^d Torus-hypergraph is translation-invariant: all points look alike. The n^d torus-hypergraph is, therefore, degree-regular where every point has degree $(3^d - 1)/2$ – which is, by the way, the same as the degree of the center in the n^d hypergraph when n is odd. So the total number of winning sets ("Torus-Lines") is $(3^d - 1)n^{d-1}/2$. Of course, the board size remains n^d.

We owe the reader a formal definition of the concept of "Torus-Line". A *Torus-Line L* is formally defined by a point $P \in L$ and a vector $\mathbf{v} = (a_1, \ldots, a_d)$, where each coordinate a_i is either 0, or $+1$, or -1 ($1 \leq i \leq d$). The n points of line L are $P + k\mathbf{v} \pmod{n}$, where $k = 0, 1, \ldots, n - 1$.

The *combinatorial line* version goes as follows: A *Comb-Torus-Line L* is formally defined by a point $P \in L$ and a vector $\mathbf{v} = (a_1, \ldots, a_d)$, where each coordinate a_i is either 0 or $+1$ ($1 \leq i \leq d$). The n points of line L are $P + k\mathbf{v} \pmod{n}$, where $k = 0, 1, \ldots, n - 1$.

The n^d Comb-Torus-Hypergraph is degree-regular: every point has degree $2^d - 1$ (which is, by the way, the same as the maximum degree of the family of all combinatorial lines in the n^d hypercube). So the total number of winning sets ("Comb-Torus-Lines") is $(2^d - 1)n^{d-1}$.

A peculiarity of this new "line-concept" is that two different Torus-Lines may have more than one point in common! The figure below shows two different Torus-Lines with *two* common points in the 4^2 torus game.

We prove that this "pair-intersection" cannot happen when n is *odd*, and if n is *even*, then it can happen only under very special circumstances.

For Comb-Torus-Lines, however, there is no surprise.

Lemma on Torus-Lines

(a) Any two different Torus-Lines have at most *one* common point if n is *odd*, and at most *two* common points if n is *even*. In the second case the distance between the two common points along either Torus-Line containing both is always $n/2$.

(b) Any two different Comb-Torus-Lines have at most *one* common point.

The **proof** of (a) goes as follows. Let L_1 and L_2 be two different Torus-Lines with (at least) two common points **P** and **Q**. Then there exist $k, l, \mathbf{v}, \mathbf{w}$ with $1 \leq k, l \leq n-1, \mathbf{v} = (a_1, \ldots, a_d), \mathbf{w} = (b_1, \ldots, b_d)$ (where a_i and b_i are either 0, or $+1$, or -1 $(1 \leq i \leq d)$), $\mathbf{v} \neq \pm\mathbf{w}$, such that $\mathbf{Q} \equiv \mathbf{P} + k\mathbf{v} \equiv \mathbf{P} + l\mathbf{w}$ (mod n). It follows that $k\mathbf{v} \equiv l\mathbf{w}$ (mod n), or equivalently, $ka_i \equiv lb_i$ (mod n) for every $i = 1, 2, \ldots, d$. Since a_i and b_i are either 0, or $+1$, or -1 $(1 \leq i \leq d)$, the only solution is $k = l = n/2$ (no solution if n is *odd*).

We leave the proof of (b) to the reader. $\qquad\square$

What can we say about the two-dimensional n^2 torus game? Well, we know a lot; we just demonstrated that the 8^2 torus game is a Pairing Strategy Draw, and this is a sharp result (since the Point/Line ratio of the 7^2 torus is $49/28 = 7/4$; i.e. less than 2).

The Erdős–Selfridge Theorem applies if $4n + 4 < 2^n$, which gives that the n^2 torus game is a Strong Draw for every $n \geq 5$. The 4^2 torus game is also a draw (mid-size "case study"), but I don't know any elegant proof. On the other hand, the 3^2 Torus Game is an easy first player's win.

Next consider the three-dimensional n^3 Torus Game. The Erdős–Selfridge Theorem applies if $13n^2 + 13 < 2^n$, which gives that the n^3 torus game is a Strong Draw for every $n \geq 11$. We are convinced that the 10^3 Torus Game is also a draw, but we don't know how to prove it.

We have the following torus version of Theorem 12.5:

Theorem 13.1 *We have:*

(a) $\mathbf{ww}(n - \text{line in torus}) = \left(\frac{\log 2}{\log 3} + o(1)\right) n,$
(b) $\mathbf{ww}(\text{comb. } n - \text{line in torus}) = (1 + o(1))n,$

A quantitative form of lower bound (a) is $\mathbf{ww}(n-\text{line in torus}) \geq \left(\frac{\log 2}{\log 3}\right) n - O(\sqrt{n \log n})$, and the same error term for (b). This is complemented by the quantitative upper bound $\mathbf{ww}(n-\text{line in torus}) \geq \left(\frac{\log 2}{\log 3}\right) n + O(\log n)$, and the same error term for (b).

The upper bounds ("Weak Win") in (a) with *odd* n and (b) immediately follow from Theorem 1.2 (using "Almost Disjointness").

In sharp contrast, the remaining case "(a) with *even* n" is far less easy! Not only that the n^d-Torus-hypergraph is *not* Almost Disjoint, but the Max Pair-Degree of this hypergraph is *exponentially large* (namely 2^{d-1}), which makes Theorem 1.2 simply useless for this case. Indeed, let **P** be an arbitrary point of the n^d Torus, and let **Q** be another point such that the coordinates of **P** are all shifted by $n/2$ (modulo n, of course): $\mathbf{Q} = \mathbf{P} + \mathbf{n}/2$ (mod n), where $\mathbf{n}/2 = (n/2, n/2, \ldots, n/2)$; then there are exactly 2^{d-1} Torus-Lines containing both **P** and **Q**. This 2^{d-1} is in fact the Max Pair-Degree. An application of Theorem 1.2 gives that, if $|\mathcal{F}|/|V| = 3^{d-1}/n > 2^{n-3} \cdot 2^{d-1}$, then the first player can force a Weak Win; that is

$$d \geq \left(\frac{\log 2}{\log 3 - \log 2} + o(1) \right) n$$

suffices for Weak Win. This bound is clearly weaker than what we stated in (a); Theorem 1.2 is grossly insufficient.

4. Advanced Weak Win Criterion with applications. Instead of Theorem 1.2 we need to apply a much more complicated criterion (Beck [2002c]), see below.

We need some new notation. Let $\mathcal{F}$ be an arbitrary finite hypergraph; for arbitrary integer $p \geq 2$ define the "big hypergraph" $\mathcal{F}_2^p$ as follows

$$\mathcal{F}_2^p = \left\{ \bigcup_{i=1}^{p} A_i : \{A_1, \ldots, A_p\} \in \binom{\mathcal{F}}{p}, \ \left| \bigcap_{i=1}^{p} A_i \right| \geq 2 \right\}.$$

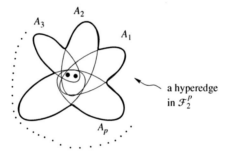

a hyperedge
in $\mathcal{F}_2^p$

In other words, $\mathcal{F}_2^p$ is the family of all union sets $\bigcup_{i=1}^{p} A_i$, where $\{A_1, \ldots, A_p\}$ runs over all unordered p-tuples of distinct elements of $\mathcal{F}$ having at least 2 points in common. Note that even if $\mathcal{F}$ is an ordinary uniform hypergraph, i.e. a set has multiplicity 0 or 1 only, and every set has the same size, $\mathcal{F}_2^p$ may become a *non-uniform multi-hypergraph* (i.e. a "big set" may have arbitrary *multiplicity*, not just 0 and 1, and the "big hypergraph" fails to remain uniform). More precisely, if $\{A_1, \ldots, A_p\}$ is an unordered p-tuple of distinct elements of $\mathcal{F}$ and $\{A_1', \ldots, A_p'\}$ is another unordered p-tuple of distinct elements of $\mathcal{F}$, $|\bigcap_{i=1}^{p} A_i| \geq 2$, $|\bigcap_{j=1}^{p} A_j'| \geq 2$, and $\bigcup_{i=1}^{p} A_i = \bigcup_{j=1}^{p} A_j'$, i.e. $\bigcup_{i=1}^{p} A_i$ and $\bigcup_{j=1}^{p} A_j'$ are equal as *sets*, then they still represent *distinct* hyperedges of the "big hypergraph" $\mathcal{F}_2^p$.

For an arbitrary hypergraph $(V, \mathcal{H})$ write

$$T(\mathcal{H}) = \sum_{A \in \mathcal{H}} 2^{-|A|}.$$

If a set has multiplicity (say) M, then, of course, it shows up M times in the summation.

Advanced Weak Win Criterion. *If there exists a positive integer $p \geq 2$ such that*

$$\frac{T(\mathcal{F})}{|V|} > p + 4p\left(T(\mathcal{F}_2^p)\right)^{1/p},$$

then in the Positional Game on hypergraph $(V, \mathcal{F})$ the first player can force a Weak Win.

At first sight this criterion is completely "out of the blue," without any motivation ("deus ex machina"), hopelessly incomprehensible. The reader is justly wondering:

(i) Where did this criterion come from?
(ii) What was the motivation to conjecture it in the first place?
(iii) How do you prove it?
(iv) Why is this complicated criterion so useful?

One thing is clear though, which gives at least a *partial* answer to question (ii): the lower index "2" in $\mathcal{F}_2^p$ is responsible for "controlling the Max Pair-Degree," and the hypothesis of the criterion means that a kind of "generalized Max Pair-Degree" is subantially less than the Average Degree. This explains why the Advanced Weak Win Criterion is a kind of "more sophisticated" version of Theorem 1.2.

To get a satisfying answer to questions (i)–(iii) above the reader is referred to Chapter V; here we answer one question only, namely question (iv), by showing an application. We apply the Advanced Weak Win Criterion to the the n^d-torus-hypergraph; we denote this hypergraph by $\mathcal{F}(t; n, d)$ (n is even). Trivially

$$T(\mathcal{F}(t; n, d)) = \frac{3^d - 1}{2} \cdot n^{d-1} \cdot 2^{-n} \text{ and } |V| = n^d.$$

To estimate the more complicated sum $T\left((\mathcal{F}(t; n, d))_2^p\right)$, where $p \geq 2$, we begin with a simple observation (we recommend the reader to study the proof of the "Lemma on Torus-Lines" again).

Observation: Let $\mathbf{P}$ and $\mathbf{Q}$ be two arbitrary points of the n^d Torus, and let k denote the number of coordinates of $\mathbf{P} - \mathbf{Q}$ (mod n), which are different from zero. If these k coordinates are all equal to $n/2$, then the number of Torus-Lines containing both $\mathbf{P}$ and $\mathbf{Q}$ is exactly 2^{k-1}; otherwise there is at most one Torus-Lines containing both $\mathbf{P}$ and $\mathbf{Q}$.

It follows from this *Observation* that

$$T\left((\mathcal{F}(t; n, d))_2^p\right) \leq n^d \left(\sum_{k=1}^{d} \binom{d}{k}\binom{2^{k-1}}{p}2^{-pn+2\binom{p}{2}}\right) <$$

$$< \frac{n^d}{2^{pn-p^2}}\left(\sum_{k=0}^{d}\binom{d}{k}(2^p)^k\right) = \frac{n^d}{2^{pn-p^2}}(1+2^p)^d < \frac{3^{pd}n^d}{2^{p(n-p)}}. \quad (13.4)$$

Let d_0 be the least integer such that $3^{d_0} \geq n^2 2^{n+1}$; this means $d_0 = (\log 2/\log 3)n + O(\log n)$. By choosing $p = p_0 = (2/\log 3)\log n + O(1)$ and $d = d_0$, in view of (13.4) we obtain that both inequalities below

$$\frac{T(\mathcal{F}(t; n, d))}{|V|} \geq n \qquad (13.5)$$

and

$$\left(T\left((\mathcal{F}(t; n, d))_2^p\right)\right)^{1/p} \leq \frac{3n^{d/p}}{2^{n-p}} < 1 \qquad (13.6)$$

hold at the same time. (13.5) and (13.6) together imply that the Advanced Weak Win Criterion applies, and guarantees a first player's Weak Win in the n^{d_0} Torus game. Finally note that, if $d > d_0$, then the application of the Advanced Weak Win Criterion is even simpler; the trivial calculations we left to the reader. This completes the proof of the upper bound in Theorem 13.1 "(a) with *even n*."

Square Lattice Games. Next consider the Weak Win part of Theorem 8.2 in the special cases (a) and (c): "aligned" and "tilted" Square Lattices. These two cases are exceptionally easy: we don't need the Advanced Weak Win Criterion, the much simpler Theorem 1.2 suffices. Indeed, the Max Pair-Degree Δ_2 of the family of $q \times q$ Aligned Square Lattices in an $N \times N$ board has the easy upper bound $\Delta_2 \leq \binom{q^2}{2} < \frac{q^4}{2}$, which is *independent* of N. So Theorem 1.2 and (16.3) imply a Weak Win if

$$\frac{N^3}{3(q-1)} > N^2 \cdot 2^{q^2-3} \cdot \Delta_2,$$

which follows from

$$N > 2^{q^2-4} \cdot 3q^5. \qquad (13.7)$$

Similarly, the Max Pair-Degree Δ_2 of the family of $q \times q$ tilted Square Lattices on an $N \times N$ board is less than q^4 ("independent of N"). To apply Theorem 1.2 we need a lower bound on the number of $q \times q$ tilted Square Lattices on $N \times N$

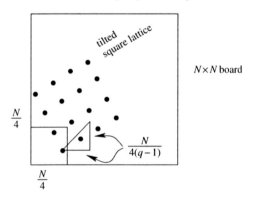

tilted square lattice

$N \times N$ board

$\frac{N}{4}$

$\frac{N}{4(q-1)}$

$\frac{N}{4}$

the lower bound

$$\left(\frac{N}{4}\right)^2 \cdot \frac{1}{2}\left(\frac{N/4}{q-1}\right)^2 = \frac{N^4}{2^9(q-1)^2}$$

is trivial from the picture. So Theorem 1.2 applies if

$$\frac{N^4}{2^9(q-1)^2} > N^2 \cdot 2^{q^2-3} \cdot q^4 > N^2 \cdot 2^{q^2-3} \cdot \Delta_2,$$

which is guaranteed by the inequality

$$N^2 > q^6 \cdot 2^{q^2+6}. \tag{13.8}$$

14

Games beyond Ramsey Theory

We show more applications of Theorem 1.2. Previously we have discussed the connections and similarities between Ramsey Theory and games such as Tic-Tac-Toe. Here the differences are studied. One possible interpretation of "games beyond Ramsey Theory" is to show games for which the Weak Win Number is much *smaller* than the corresponding Ramsey Number; we have already done this in Theorem 6.3. Another possibility, and this is what we are going to do here, is to show games for which Ramsey Theory fails to give anything, i.e. there is *no* Ramsey phenomenon whatsoever, but Theorem 1.2 (or its adaptation) still gives very interesting results. In other words, we demonstrate that the *Weak Win world* goes far beyond the *Ramsey world*.

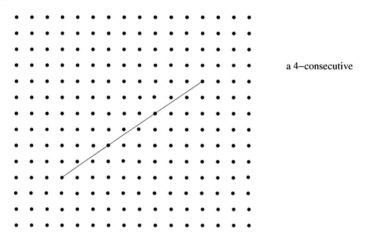

a 4–consecutive

1. *n*-in-a-row with arbitrary slopes. The game of *Unrestricted n-in-a-row on the plane* was introduced in Section 4, and was discussed in detail at the end of Section 10; here we consider the following modification: the players occupy lattice points instead of "little squares," and *n*-in-a-row means *n consecutive* lattice points on an *arbitrary* straight line; the novelty is that *every* rational slope is allowed, not just the four Tic-Tac-Toe directions. To prevent confusion "*n* consecutive lattice points on an arbitrary straight line" will be called *n-consecutive* (instead of *n*-in-a-row).

First, the reader is challenged to solve the following:

Exercise 14.1 *Show that there is no Ramsey phenomenon for n consecutives: it is possible to 2-color the set of integer lattice points in the plane in such a way that every 100 consecutive contains both colors.*

Of course, "100" is just an accidental large constant. The correct value is actually known: it is 4; this elegant result is due to Dumitrescu and Radoicic [2004], but it is easier to solve the exercise with a large constant like 100.

Even if "Ramsey phenomenon" fails here, we can still prove a "game theorem": playing on an $N \times N$ board Maker can have a whole *n*-consecutive with arbitrarily large *n* up to $n = (2 - o(1))\log_2 N$. The proof is a straightforward application of Theorem 1.2. Indeed, for simplicity assume that the lower-left corner of the $N \times N$ board is the origin. The set $\{(k + ja, l + jb) \in \mathbb{Z}^2 : j = 0, 1, 2, \ldots, n - 1\}$ gives *n consecutive* lattice points on a line *inside* the $N \times N$ square if *a* and *b* are coprime, $0 \leq k \leq N/2$, $0 \leq l \leq N/2$, $1 \leq a \leq \frac{N}{2n}$, and $1 \leq b \leq \frac{N}{2n}$; we call this *n*-set a $(k, l; a, b)$-set. Let $\mathcal{F}$ denote the family of all $(k, l; a, b)$-sets defined above. $\mathcal{F}$ is an *n*-uniform hypergraph; the board size $|V| = N^2$; and because the hyperedges are "intervals," the Max Pair-Degree $\Delta_2 = \Delta_2(\mathcal{F})$ is exactly $2(n - 2)$. (Indeed, two "intervals" on the same line may intersect at more than one point.) To estimate the size $|\mathcal{F}|$ of the hypergraph, we recall a well-known number-theoretic result: Theorem 331 from Hardy–Wright [1979], which states that the number of fractions r/s with $1 \leq r \leq s \leq m$, where *r* and *s* are coprime, is approximately $3\pi^{-2}m^2$ ("asymptotic number of Farey fractions"). It follows that

$$|\mathcal{F}| = \left(\frac{N}{2}\right)^2 \cdot 2 \cdot (1 + o(1))\frac{3}{\pi^2}\left(\frac{N}{2n}\right)^2 = (1 + o(1))\frac{3N^4}{8\pi^2 n^2},$$

so Theorem 1.2 applies if

$$(1 + o(1))\frac{3N^4}{8\pi^2 n^2} > \Delta_2 \cdot N^2 \cdot 2^{n-3} = 2(n - 2) \cdot N^2 \cdot 2^{n-3},$$

which is equivalent to

$$N^2 > \frac{4\pi^2}{3} \cdot n^2(n - 2) \cdot 2^n. \tag{14.2}$$

Inequality (14.2) is guaranteed by the choice

$$n = 2\log_2 N - O(\log \log N), \tag{14.3}$$

which completes the Weak Win part of the following:

Theorem 14.1 *Consider the Maker–Breaker game on the $N \times N$ board, where Maker's goal is to occupy n consecutive lattice points on a line ("n-consecutive"). The phase transition from Weak Win to Strong Draw happens at $n = (2 + o(1))\log_2 N$.*

The missing Strong Draw part of Theorem 14.1 can be proved by a minor modification of the proof technique of Theorem 34.1 (see Section 36); we leave it to the reader as an *exercise*. (The reason why we need to slightly modify the proof-technique of Theorem 34.1 is that the corresponding hypergraph is not Almost Disjoint: two *n*-consecutive's on the same line may intersect in more than one point.)

The two Ramsey Criterions (Theorems 6.1 and 6.2) prove that the "Weak Win world" is at least as large as the "Ramsey world." The Unrestricted *n*-in-a-row with Arbitrary Slopes was the first example to demonstrate that the "Weak Win world" is strictly bigger than the "Ramsey world." Many more examples come from:

2. Games with two-colored goals. Let us begin with two-colored arithmetic progressions (A.P.s). First a simple observation: let $N \geq W(n)$, where $W(n)$ is the van der Waerden number for *n*-term A.P.s, and consider a play in the twice as long interval $[2N] = \{1, 2, \ldots, 2N\}$. If the second player follows the "copycat strategy": for first player's x he replies by $2N + 1 - x$, then of course at the end of the play the second player will occupy an *n*-term A.P., and also the first player will occupy an *n*-term A.P. In other words, if the second player is Mr. Red, and the first player is Mr. Blue, then Mr. Red can force the appearance of *both* a monochromatic *red* and a monochromatic *blue* *n*-term A.P. This was trivial; next comes a non-trivial question.

How about if Mr. Red's goal is an *arbitrary* 2-colored *n*-term A.P.? Of course, Ramsey Theory cannot help here, but the Potential Technique works very well.

The precise definition of the "2-Colored Goal Game" goes as follows. First fix an arbitrary Red–Blue sequence of length 100: (say) $R, B, R, R, R, B, B, R, B, \ldots$; we call it the "goal sequence." Mr. Red and Mr. Blue alternately take new integers from the interval $[N] = \{1, 2, \ldots, N\}$, and color them with their own colors. Mr. Red wins if at the end of the play there is a 100-term arithmetic progression which is colored exactly like the given Red–Blue "goal sequence"; otherwise Mr. Blue wins. We call this the *Goal Sequence Game*. Is it true that Mr. Red has a winning strategy in the Goal Sequence Game if N is sufficiently large?

The answer is "yes"; of course, 100 can be replaced by any natural number.

Theorem 14.2 *Fix an arbitrary Red–Blue "goal sequence" S of length n, and consider the Goal Sequence Game on the interval $[N] = \{1, 2, \ldots, N\}$, where the "goal sequence" is S; the two players are Mr. Red and Mr. Blue. If $N \geq n^3 \cdot 2^{n-2}$, then at the end of the play Mr. Red can force the appearance of an n-term arithmetic progression which is colored exactly the same way as the given Red–Blue "goal sequence" S goes.*

Proof. For simplicity assume that Mr. Red is the first player. Assume that we are in the middle of a play: so far Mr. Red colored integers $x_1, x_2, \ldots, x_i$ ("red points") and Mr. Blue colored integers $y_1, y_2, \ldots, y_i$ ("blue points"). This defines a Partial 2-Coloring of $[N]$.

Let $\mathcal{F} = \mathcal{F}(N, n)$ be the family of all n-term A.P.s in $[N]$; clearly $|\mathcal{F}| > N^2/4(n-1)$. The Partial 2-Coloring of $[N]$ defines a Partial 2-Coloring of every n-term A.P. $A \in \mathcal{F}$. We introduce the following natural adaptation of the Power-of-Two Scoring System: if the Partial 2-Coloring of an n-term A.P. $A \in \mathcal{F}$ *contradicts* the given "goal sequence" S, then A has *zero value*, and we call it a "dead set". The rest of the elements of $\mathcal{F}$ are called "survivors"; the Partial 2-Coloring of a "survivor" $A \in \mathcal{F}$ has to be consistent with the given "goal sequence" S. The *value* of a "survivor" $A \in \mathcal{F}$ is 2^j if there are j points of A which got color in the Partial 2-Coloring.

We define the "winning chance function" in the standard way

$$C_i = \sum_{A \in \mathcal{F}} \mathrm{value}_i(A).$$

As usual, we study how the consecutive moves x_{i+1} and y_{i+1} affect the "chance function."

For an arbitrary n-term A.P. $A \in \mathcal{F}$, 2-color the elements of A copying the given "goal sequence" S. Then let $red_S(A)$ denote the set of red elements of A, and let $blue_S(A)$ denote the set of blue elements of A.

We have

$$C_{i+1} = C_i + \sum_{\substack{A \in \mathcal{F}: \\ x_{i+1} \in red_S(A)}} \mathrm{value}_i(A) - \sum_{\substack{A \in \mathcal{F}: \\ x_{i+1} \in blue_S(A)}} \mathrm{value}_i(A)$$

$$+ \sum_{\substack{A \in \mathcal{F}: \\ y_{i+1} \in blue_S(A)}} \mathrm{value}_i(A) - \sum_{\substack{A \in \mathcal{F}: \\ y_{i+1} \in red_S(A)}} \mathrm{value}_i(A)$$

$$- \sum_{\substack{A \in \mathcal{F}: \\ \{x_{i+1}, y_{i+1}\} \in red_S(A)}} \mathrm{value}_i(A) - \sum_{\substack{A \in \mathcal{F}: \\ \{x_{i+1}, y_{i+1}\} \in blue_S(A)}} \mathrm{value}_i(A)$$

$$+ \sum_{\substack{A \in \mathcal{F}: \\ x_{i+1} \in red_S(A) \\ y_{i+1} \in blue_S(A)}} \mathrm{value}_i(A) + \sum_{\substack{A \in \mathcal{F}: \\ x_{i+1} \in blue_S(A) \\ y_{i+1} \in red_S(A)}} \mathrm{value}_i(A).$$

How should Mr. Red choose his next move x_{i+1}? Well, Mr. Red's optimal move is to compute the numerical value of the function

$$f(u) = \sum_{\substack{A \in \mathcal{F}: \\ u \in red_S(A)}} \mathrm{value}_i(A) - \sum_{\substack{A \in \mathcal{F}: \\ u \in blue_S(A)}} \mathrm{value}_i(A)$$

for every unoccupied integer $u \in [N]$, and to choose that $u = x_{i+1}$ for which the *maximum* is attained. This choice implies

$$C_{i+1} \geq C_i - \binom{n}{2} 2^{n-2}.$$

Indeed, there are at most $\binom{n}{2}$ n-term arithmetic progressions containing both x_{i+1} and y_{i+1}. Therefore

$$C_{end} \geq C_{start} - \frac{N}{2}\binom{n}{2} 2^{n-2}.$$

Since $C_{start} = C_0 = |\mathcal{F}| \geq N^2/4(n-1)$, it suffices to guarantee the inequality

$$\frac{N^2}{4(n-1)} > \frac{N}{2}\binom{n}{2} 2^{n-2},$$

which is equivalent to $N > n(n-1)^2 \cdot 2^{n-2}$. Under this condition $C_{end} > 0$; that is, at the end of the play there must exist an n-term A.P., which is 2-colored by Mr. Red and Mr. Blue exactly the same way as the "goal sequence" S goes. Theorem 14.2 follows. $\qquad\square$

We leave the "clique version" of Theorem 14.2 to the reader as an exercise.

Theorem 14.3 *Fix an arbitrary 2-colored clique $K_{100}(red; blue)$ of 100 vertices ("2-colored goal-graph"). Consider the following game: Mr. Red and Mr. Blue alternately take edges of a complete graph K_n on n vertices, and color them with their own colors. Mr. Red wins if at the end of the play there is an isomorphic copy of $K_{100}(red; blue)$; otherwise Mr. Blue wins. If n is sufficiently large, then Mr. Red has a winning strategy.*

3. A highly degree-irregular hypergraph: the All-Subset Game. Here the two players alternately take *arbitrary subsets* of a ground set. More precisely, there are two players, Maker and Breaker, who alternately take previously unselected subsets of the n-element set $[n] = \{1, 2, \ldots, n\}$. Maker (the first player) wins if he has all 2^{100} subsets of some 100-element subset of $[n]$; otherwise Breaker (the second player) wins.

Can Maker win if n is sufficiently large? One thing is clear: Maker's opening move has to be the "empty set" (otherwise he loses immediately).

Observe that there is no Ramsey phenomenon here. Indeed, color the subsets depending on the parity of the size: "even" means red and "odd" means blue. This particular 2-coloring kills any chance for an "all-subset Ramsey theorem." But there is a "game theorem": We show that Maker (first player) has a winning strategy if n is sufficiently large.

How to prove the game theorem? It seems a natural idea to use Theorem 1.2. Unfortunately a direct application does *not* work. Indeed, $|\mathcal{F}| = \binom{n}{100}$ is actually *less* than $|V| = \binom{n}{0} + \binom{n}{1} + \binom{n}{2} + \cdots + \binom{n}{100}$, so Theorem 1.2 cannot give anything.

To get around this difficulty we use a technical trick. The following is proved:

Stronger Statement: *There is a 101-element subset S of* $[n]$ *such that Maker (first player) can have all* (≤ 100)*-element subsets of S, assuming n is sufficiently large.*

In fact, we can prove it for arbitrary k (not just for 100), and then the Stronger Statement goes as follows: *There is a* $(k+1)$*-element subset S of* $[n]$ *such that Maker (first player) can have all* $(\leq k)$*-element subsets of S, assuming n is sufficiently large depending on k.*

The reason why we switch to the Stronger Statement is that now $|\mathcal{F}| = \binom{n}{101}$ is roughly n-times greater than $|V| = \binom{n}{0} + \binom{n}{1} + \binom{n}{2} + \cdots + \binom{n}{100}$. This makes it possible to repeat the whole proof of Theorem 1.2. Note that a direct application of the theorem cannot work for the simple reason that the corresponding hypergraph is "very degree-irregular," and it is "bad" to work with the same Max Pair-Degree during the whole course of the play.

The proof of the Stronger Statement goes as follows. As usual, in each move Maker makes the "best choice": he chooses a subset of "maximum value," using the usual power-of-two scoring system, and if there are two *comparable* subsets, say, U and W with $U \subset W$, having the same "absolute maximum," then Maker always chooses the *smaller set*, i.e. U instead of W. (Note that if $U \subset W$, then the "value" of U is *always* at least as large as the "value" of W.) Maker's opening move is, of course, the "empty set." Let $U^{(i)}$ be an arbitrary move of Maker: $U^{(i)}$ is an i-element subset $(0 \leq i \leq k)$. Let W be Breaker's next move right after $U^{(i)}$. We know that W cannot be a subset of $U^{(i)}$–indeed, otherwise Maker would prefer W instead of $U^{(i)}$–so $|U^{(i)} \cup W| \geq i+1$. There are at most $\binom{n}{k+1-(i+1)}$ $(k+1)$-element subsets S containing *both* $U^{(i)}$ and W: this is the "actual Max Pair-Degree." Note that for $i = 0, 1, 2, \ldots, k$ Maker chooses an i-element set at most $\binom{n}{i}$ times, so by repeating the proof of Theorem 1.2 we obtain

$$C_{last} \geq C_{first} - \sum_{i=0}^{k} \binom{n}{i} \left(\frac{n}{k+1-(i+1)} \right) 2^{2^{k+1}-3},$$

where C_j is the "Chance Function" at the jth turn of the play. Clearly $C_{first} = C_0 = \binom{n}{k+1}$, so

$$C_{last} \geq \binom{n}{k+1} - \sum_{i=0}^{k} \binom{n}{i} \left(\frac{n}{k+1-(i+1)} \right) 2^{2^{k+1}-3}.$$

A trivial calculation shows that

$$\binom{n}{k+1} > \sum_{i=0}^{k} \binom{n}{i} \left(\frac{n}{k+1-(i+1)} \right) 2^{2^{k+1}-3} \text{ if } n > (k+1)! 2^{2^{k+1}}.$$

It follows that

$$C_{last} > 0 \text{ if } n > (k+1)! 2^{2^{k+1}}.$$

If the "Chance Function" is not zero at the end of a play, then Breaker could *not* block every "winning set," so there must exist a "winning set" completely occupied by Maker. In other words, there must exist a $(k+1)$-element subset S of $[n]$ such that Maker occupied all at-most-k-element subsets of S. This proves the Stronger Statement.

The last step is trivial: throwing out an arbitrary integer from S we obtain a k-element subset S' of $[n]$ such that Maker occupied all 2^k subsets of S'. This solves the "All-Subset Game."

Notice that inequality

$$n > (k+1)! 2^{2^{k+1}}$$

holds if $k = (1 + o(1))\log_2\log_2 n$.

In the other direction, we show that with a different $o(1)$ the choice $k = (1 + o(1))\log_2\log_2 n$ is impossible. Indeed, the Erdős–Selfridge criterion applies if

$$\binom{n}{k} < 2^{2^k - 1}.$$

This also holds for $k = (1 + o(1))\log_2\log_2 n$, and we obtain (see Beck [2005]):

Theorem 14.4 *The breaking point for Weak Win in the All-Subset Game on $\{1, 2, \ldots, n\}$ is the iterated binary logarithm of n:* $(1 + o(1))\log_2\log_2 n$.

Notice that the Power-Set, the goal of the All-Subset Game, is a "very rapidly changing configuration": $2^{n+1} = 2 \times 2^n$. Changing the value of n by we *double* the size of the Power-Set. This is why such a crude approach–the simplest form of the Erdős–Selfridge technique–can still give an asymptotically satisfactory answer.

4. Applying the Biased Weak Win Criterion to a Fair Game. Recall the biased Weak Win criterion Theorem 2.2: *If*

$$\sum_{A \in \mathcal{F}} \left(\frac{p+q}{p}\right)^{-|A|} > p^2 \cdot q^2 \cdot (p+q)^{-3} \cdot \Delta_2(\mathcal{F}) \cdot |V(\mathcal{F})|,$$

where $\Delta_2(\mathcal{F})$ is the Max Pair-Degree of hypergraph $\mathcal{F}$, and $V(\mathcal{F})$ is the board, then the first player can occupy a whole winning set $A \in \mathcal{F}$ in the biased $(p : q)$ play on $\mathcal{F}$ (the first player takes p new points and the second player takes q new points per move).

We show a rather surprising application of the Biased Weak Win criterion: it is applied to a fair game; in fact, to a fair tournament game. Recall that a tournament means a "directed complete graph" such that every edge of a complete graph is directed by one of the two possible orientations; it represents a tennis tournament where any two players played with each other, and an arrow points from the winner to the loser.

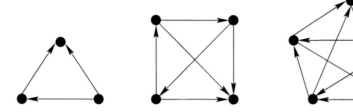

Fix an arbitrary goal tournament T_k on k vertices. The two players are Red and Blue, who alternately take new edges of a complete graph K_n, and for each new edge choose one of the two possible orientations ("arrow"). Either player colors his arrow with his own color. At the end of a play, the players create a 2-colored tournament on n vertices. Red wins if there is a red copy of T_k; otherwise Blue wins. Is it true that, if n is sufficiently large compared to k, then Red always has a winning strategy?

The answer is "yes."

Theorem 14.5 *Given an arbitrary goal tournament T_k, Maker can build a copy of T_k from his own arrows if the board clique K_n is large enough.*

Proof. It is a simple application of the Biased Weak Win Criterion. The idea is to associate with the Fair Tournament Game a biased (1:3) hypergraph game! To understand the application, the reader is recommended to study the following picture

The board $V = V(\mathcal{F})$ of the biased $(1:3)$ hypergraph game is the set of $2\binom{n}{2}$ arrows of $K_n(\uparrow\downarrow)$, where $K_n(\uparrow\downarrow)$ means that every edge of the complete graph K_n shows up twice with the two orientations. The winning sets $A \in \mathcal{F}$ are the arrow-sets of all possible copies of T_k in $K_n(\uparrow\downarrow)$. So $\mathcal{F}$ is a $\binom{k}{2}$-uniform hypergraph, and trivially $|\mathcal{F}| \geq \binom{n}{k}$. If the mth move of Red and Blue are, respectively, $i_1 \to j_1$ and $i_2 \to j_2$, then these two moves (arrows) automatically exclude the extra arrows $j_1 \to i_1$ and $j_2 \to i_2$ from $K_n(\uparrow\downarrow)$ for the rest of the play. (There may be some coincidence among i_1, j_1, i_2, j_2, but it does not make any difference.) This means 1 arrow for Maker, and 3 arrows for Breaker in the hypergraph game on $\mathcal{F}$, which explains how the biased (1:3) play enters the story.

All what is left is to apply the biased criterion to the (1:3) hypergraph game on $\mathcal{F}$: If

$$\sum_{A \in \mathcal{F}} 4^{-\binom{k}{2}} > \frac{9}{64} \cdot \Delta_2(\mathcal{F}) \cdot |V(\mathcal{F})|,$$

where $\Delta_2(\mathcal{F})$ is the Max Pair-Degree and $V(\mathcal{F})$ is the board, then Red can force a win in the Tournament Game. We have the trivial equality $|V(\mathcal{F})| = 2\binom{n}{2}$, and the less trivial inequality $|\Delta_2(\mathcal{F})| \le \binom{n-3}{k-3} \cdot k!$. Indeed, a tournament T_k *cannot* contain parallel arrows (i.e., both orientations of an edge), so a pair of red and blue arrows contained by a copy of T_k must span at least 3 vertices, and there are at most $\binom{n-3}{k-3} \cdot k!$ ways to extend an unparallel arrow-pair to a copy of T_k.

Combining these facts, it suffices to check that

$$\binom{n}{k} > \frac{9}{64} \cdot 4^{\binom{k}{2}} \cdot \binom{n-3}{k-3} \cdot k! \cdot 2\binom{n}{2}.$$

This inequality is trivial if $n \ge c_0 \cdot k^{k+3} \cdot 4^{\binom{k}{2}}$. The threshold $n_0(k) = c_0 \cdot k^{k+3} \cdot 4^{\binom{k}{2}}$ works uniformly for all goal-tournaments T_k on k vertices. $\square$

Next we switch from biased building to biased blocking.

5. Biased version of the Unrestricted n-in-a-row on the plane. Let us start with the (2:1) version: First Player claims 2 little squares and Second Player claims 1 little square per move. Unlike the Fair (1:1) case where Second Player can block every 8-in-a-row (see the "Sixth Proof" of Theorem 10.2), in the Biased case Second Player *cannot block*. Indeed, First Player can easily occupy n-in-a-row for arbitrarily large n as follows. In his first 2^{n-1} moves First Player marks 2^n distinct rows. Second Player can block at most half of the rows, so at least 2^{n-1} rows remain Second Player-free. In his next 2^{n-2} moves first player can achieve 2^{n-1} distinct rows with 2 consecutive marks in each. Second Player can block at most half of them, so there are at least 2^{n-2} rows which are all:

(1) Second Player-free, and
(2) contain 2 consecutive marks of First Player.

Repeating this argument, at the end there is at least one row which is:

(1) Second Player-free, and
(2) contains n consecutive marks of First Player.

In this strategy First Player may be forced to place his 2 points per move "very far" from each other (as far as 2^n). To make up for First Player's (2:1) advantage, we can make First Player's job harder by implementing the following *Distance Condition*: during the whole play First Player's 2 marks per move must have a distance at most (say) 1000. What happens under this Distance Condition? We show that in this case First Player *cannot* have n-in-a-row for arbitrarily large n. In other words, the Distance Condition can successfully neutralize First Player's (2:1) advantage.

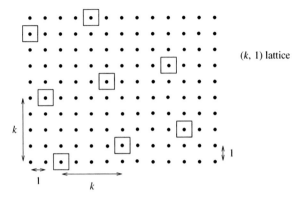

$(k, 1)$ lattice

In general, consider the Biased $(p : 1)$ version of the Unrestricted n-in-a-row on the plane with the following Distance Condition: First Player's $p(\geq 2)$ marks per move are always in a circle of radius $r(\geq 2)$. It doesn't make any difference if the players mark integer lattice points instead of little squares, so we work with the lattice point model.

We give two proofs. The first one, due to P. Csorba, uses the concept of the tilted infinite Square Lattice, see the picture above. We call the lattice on the picture above a $(k, 1)$-lattice. We can direction-mark the vertices of the $(k, 1)$-lattice by the 4×4 matrix used in the Third Proof of Theorem 10.2. We need the following simple Lemma:

Lattice Lemma: *If k is even, then the $(k, 1)$-lattice intersects every horizontal, vertical, and diagonal (of slopes ± 1) lattice-line. The intersection is of course an infinite arithmetic progression, and the gap between consecutive points is always $\leq k^2$.*

Exercise 14.2 *Prove the Lattice Lemma.*

By using the Lattice Lemma, we can repeat the "Third proof" of Theorem 10.2, and obtain the following statement: *In the Biased $(p : 1)$ version of the Unrestricted n-in-a-row with Distance Condition r, Second Player can block every n-in-a-row if $n \geq c_0 r^2$ where c_0 is an absolute constant.*

This proof has an interesting special feature: the value of the bias parameter p is irrelevant. Since a circle of radius r can contain no more than $\pi r^2 + O(r)$ lattice points, we clearly have $p \leq \pi r^2 + O(r)$.

If the value of p is much smaller than r^2, then the following alternative proof gives a much better value for n (this was my original proof).

Alternative proof using potentials. We shall employ the following biased $(p:1)$ version of the Erdős–Selfridge Theorem (see Beck [1982]): *If*

$$\sum_{A \in \mathcal{F}} 2^{-|A|/p} < 1, \quad \text{or} \ < \frac{1}{2},$$

then First Player (or Second Player) can block every winning set $A \in \mathcal{F}$ in the biased $(1:p)$ (or $(p:1)$) play on $\mathcal{F}$. (The $(f:s)$ play means that First Player takes f points and Second Player takes s points per move.)

This is a special case of a more general result – Theorem 20.1 – which will be proved at the end of Section 20.

For the application we divide the plane into $m \times m$ squares, where $m \geq 2r$. First we discuss the following:

Special Case: Assume that during the whole course of the play, in every single move, the p marks of First Player are always completely inside of some $m \times m$ square.

Then we can repeat the "First Proof" of Theorem 10.2 as follows. Since the p points per move are always completely inside of some $m \times m$ square, Second Player replies in the same $m \times m$ square by using the Biased $(p:1)$ version of the Erdős–Selfridge Theorem. If $m \geq n$, then every n-in-a-row intersects some $m \times m$ square in an "interval-piece" of length $\geq n/3$. We use the Biased Erdős–Selfridge Theorem for the $(p:1)$ game: it applies if

$$4m^2 < \frac{2^{\lceil n/3 \rceil/p}}{2}, \quad \text{where } m \geq \max\{2r, n\}. \tag{14.4}$$

If $m = \max\{2r, n\}$ and $n = c_1 \cdot p \cdot \log r$, then (14.4) holds for a sufficiently large absolute constant c_1. It follows that Second Player can block every n-in-a-row with $n = c_1 \cdot p \cdot \log r$. This solves the *Special Case*.

In the **General Case**, the p points (of First Player in any single move) are in a circle of radius r, and, if $m \geq 2r$, a circle of radius r can intersect as many as *four* $m \times m$ squares. The question is: which one of the (possibly) four $m \times m$ squares should Second Player pick to reply?

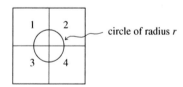

Well, a natural answer is to use a Periodic Rule: if the same situation repeats itself several times, then Second Player should respond *periodically*, e.g. $1, 2, 3, 4, 1, 2, 3, 4, 1, 2, 3, 4, \ldots$.

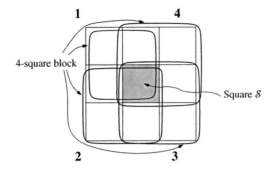

Let us see what happens to an *arbitrary but fixed* $m \times m$ square S? S is surrounded by 8 neighbors, and there are 4 "four-square-blocks" which contain S (S is the center square in the 3×3 arrangement). For each one of these "four-square-blocks" we use a Periodic Rule $\ldots, 1, 2, 3, 4, 1, 2, 3, 4, 1, 2, 3, 4, \ldots$ which tells Second Player where to reply. For a fixed "four-square-block" there are at most 3 consecutive steps in a Periodic Rule $\ldots, 1, 2, 3, 4, 1, 2, 3, 4, 1, 2, 3, 4, \ldots$ indicating *not* to reply in S. Since there are 4 of them, by the Pigeonhole Principle, in every $13 = 4 \cdot 3 + 1$ consecutive steps, when First Player puts (at least one) mark in S, at least once, the Periodic Rule tells Second Player to reply in S.

This means in the *General Case* we can repeat the proof of the *Special Case* with a minor modification: we have to use the $(13p : 1)$ version of the Biased Erdős–Selfridge Theorem instead of the $(p : 1)$ version. The same calculation works, and gives that Second Player can block every n-in-a-row with $n = c_2 \cdot p \cdot \log r$, where c_2 is an absolute constant (of course, c_2 is larger than c_1 in the Special Case).

Note that, for small p the logarithmic factor $\log r$ is necessary here. Indeed, as we have explained above, in the Biased $(2 : 1)$ version First Player can occupy n-in-a-row staying inside of a *fixed* circle of radius $r = 2^n$.

What we have just proved is summarized below:

Theorem 14.6 *In the Biased $(p : 1)$ version of the Unrestricted n-in-a-row with Distance Condition r, Second Player can block every n-in-a-row if $n \geq \min\{c_0 r^2, c_2 \cdot p \cdot \log r\}$, where c_0 and c_2 are absolute constants.*

15

A generalization of Kaplansky's game

1. The Maker-Breaker version. In Section 4 Kaplansky's k-in-a-line game was introduced. It is a Shutout Game where the two players alternately occupy new lattice points in the plane, and that player wins who owns first a k-in-a-line such that the whole line is opponent-free. Note that even the case of $k = 4$ is unsolved; Kleitman and Rothschild conjectured it to be a draw game.

The original version remains wide open, but the Maker–Breaker version was solved by Kleitman and Rotschild [1972]. They proved that, playing the usual (1:1) game, for every k Maker can build (in a finite number of moves) a Breaker-free k-in-a-line. (Needless to say, their proof cannot guarantee that Maker does it first.) Kleitman and Rotschild [1972] didn't use any potential function – instead they applied the Hales–Jewett Theorem in a very clever way. This Hales–Jewett type approach has two shortcomings:

(1) the upper bound on the Move Number is ridiculous;
(2) they used a "copycat" argument, and copycat doesn't extend to the biased case where Maker is the underdog (like the (1:2) game).

In this section, we give a completely different *potential* proof, which completely avoids the Hales–Jewett Theorem. This way we get rid of both shortcomings at once: (1) the new proof gives a nice, plain exponential upper bound on the Move Number, and (2) it automatically extends to the general biased play, including the case when Maker is the underdog.

Consider the (1:b) underdog play where the first player takes 1 new lattice point and the second player takes b (≥ 1) new lattice point(s) per move. The Kaplansky-($b; k, l$) Game is the generalization where the two players are playing the (1:b) play on the set of lattice points, first player's goal is an opponent-free k-in-a-line and second player's goal is an opponent-free l-in-a-line. Whoever does it first in a finite number of moves is declared the winner; an ω-long play without a winner means a draw.

Theorem 15.1 *If*

$$\frac{l}{k} \geq c_0, \quad \text{where the constant} \quad c_0 \text{ is defined as } c_0 = c_0(b) = \frac{64b \log(1+2b)}{\log 2},$$

then underdog first player has a winning strategy in the Kaplansky-$(b; k, l)$ Game. In fact, the first player can always win in at most $(1+2b)^{16k}$ moves.

Remark. Theorem 15.1 is about the "restricted" Kaplansky game, but we can also consider the "unrestricted" game, which has the same rules except that the players may make their choices from all the points of the plane instead of just from the lattice points. The only relevant geometric feature is the *collinearity* of subsets of points. Thus if one player has a winning strategy in the unrestricted game, then he can convert his strategy to the restricted game. Therefore, we need only consider the unrestricted game; this gives us a little extra flexibility in the notation.

Proof. We solve the unrestricted version, i.e. the players may select points anywhere in the plane. The first player uses his 1st, 3rd, 5th, 7th, ... moves for "building" and uses his 2nd, 4th, 6th, 8th, ... moves for "blocking." In other words, the first player divides the $(1:b)$ game into a "twice as biased" $(1:2b)$ Building Part and a $(1:2b)$ Blocking Part; the two parts are handled like two disjoint/non-interacting games (of course, this is in first player's mind only). First player's plan is to find an integer M such that:

(1) in his first $M/2$ "odd" moves, the first player can build an opponent-free k-in-a-line; and

(2) in his first $M/2$ "even" moves, he can prevent the second player from building an opponent-free l-in-a-line if $l \geq c_0 \cdot k$. $\qquad \square$

2. Building Part. Let V_0 denote the $(2n+1) \times (2n+1)$ Square Lattice of integer points centered at the origin

$$V_0 = \left\{ (u, v) \in \mathbb{Z}^2 : -n \leq u \leq n, \ -n \leq v \leq n \right\},$$

and also consider the following set of rational points

$$V_1 = \left\{ (u, v) \in \mathbb{R}^2 : u = \frac{p_1}{q_1}, v = \frac{p_2}{q_2}, \max\{|u|, |v|\} \leq n^{1+\varepsilon}, \max\{|q_1|, |q_2|\} \leq n^{\varepsilon} \right\},$$

$$(15.1)$$

where $\varepsilon > 0$ will be specified later as a small positive absolute constant. In (15.1) the equality "$u = \frac{p_1}{q_1}, v = \frac{p_2}{q_2}$" means that both u and v are rational numbers, p_1, q_1, p_2, q_2 are integers, p_1, q_1 are relatively prime, and p_2, q_2 are also relatively prime. Clearly $V_0 \subset V_1$.

Let $\mathcal{L}$ denote the family of all straight lines in the plane which intersect lattice V_0 in $\geq k$ lattice points.

The first player restricts himself to the hypergraph $(V_1, \mathcal{L} \cap V_1)$, i.e. V_1 is the board and $\mathcal{L} \cap V_1 = \{L \cap V_1 : L \in \mathcal{L}\}$ are the "winning sets." The second player may or may not stay in board V_1 (note that every line $L \in \mathcal{L}$ can be blocked outside of V_1). Since the first player uses his 1st, 3rd, 5th, 7th, ... moves for "building" (and the rest for "blocking"), for the first player the Building Part means a "kind of" $(1:2b)$ Maker–Breaker Shutout Game (of size k) on hypergraph $(V_1, \mathcal{L} \cap V_1)$, where the first player is the underdog Maker. Here the term "kind of" refers to the novelty that Breaker may move (and block!) outside of V_1.

In the Building Part, for simplicity, we call the first player Maker and the second player Breaker.

Assume that we are in the middle of a $(1:2b)$ play. Let $X(i) = \{x_1, \ldots, x_i\}$ $(\subset V_1)$ denote Maker's points and

$$Y(i) = \{y_1^{(1)}, y_1^{(2)}, \ldots, y_1^{(2b)}, \ldots, y_i^{(1)}, y_i^{(2)}, \ldots, y_i^{(2b)}\}$$

denote Breaker's points selected so far; of course, some of the points of Breaker may be outside of V_1. The key question is how to choose $x_{i+1} \in V_1$. Consider the potential functions

$$T_i = \sum_{L \in \mathcal{L}:\ L \cap Y(i) = \emptyset} (1 + 2b)^{|L \cap X(i)|} \quad \text{and} \quad T_i(z) = \sum_{z \in L \in \mathcal{L}:\ L \cap Y(i) = \emptyset} (1 + 2b)^{|L \cap X(i)|}. \quad (15.2)$$

It is easy to describe the effect of the $(i+1)$st moves x_{i+1} and $\{y_{i+1}^{(1)}, y_{i+1}^{(2)}, \ldots, y_{i+1}^{(2b)}\}$:

$$T_{i+1} = T_i + 2b \cdot T_i(x_{i+1}) - \alpha_i - 2b \cdot \beta_i, \quad (15.3)$$

where

$$\alpha_i = \sum_{L \in \mathcal{L}:\ L \cap Y(i) = \emptyset,\, L \cap Y(i+1) = \neq \emptyset} (1 + 2b)^{|L \cap X(i)|} \quad (15.4)$$

and

$$\beta_i = \sum_{x_{i+1} \in L \in \mathcal{L}:\ L \cap Y(i) = \emptyset,\, L \cap Y(i+1) = \neq \emptyset} (1 + 2b)^{|L \cap X(i)|}. \quad (15.5)$$

Trivially

$$\alpha_i \leq \sum_{j=1}^{2b} T_i(y_{i+1}^{(j)}),$$

so by (15.3)

$$T_{i+1} \geq T_i + 2b \left(T_i(x_{i+1}) - \frac{1}{2b} \sum_{j=1}^{2b} T_i(y_{i+1}^{(j)}) - \beta_i \right). \quad (15.6)$$

Maker selects his $(i+1)$st point x_{i+1} as the "best" new point in V_1: x_{i+1} is the new point $z \in V_1$ for which the function $f(z) = T_i(z)$ attains its *maximum*.

The following inequality may or may not hold

$$T_i(x_{i+1}) \geq \max_{1 \leq j \leq 2b} T_i(y_{i+1}^{(j)}), \quad (15.7)$$

and accordingly we distinguish several cases.

Case 1: Inequality (15.7) holds during the whole course of the play, i.e. until the two players exhaust board V_1

Then by (15.6), $T_{i+1} \geq T_i - 2b \cdot \beta_i$ holds for every i. We will now see that, at some stage of the play on V_1, Maker owns an opponent-free k-in-a-line. Assume to the contrary that $|\mathcal{L} \cap X(i)| < k$ holds whenever $\mathcal{L} \cap Y(i) = \emptyset$ during the whole play; then by (15.5)

$$T_{i+1} \geq T_i - 2b \cdot \beta_i \geq T_i - 2b \cdot 2b(1 + 2b)^{k-1}, \tag{15.8}$$

where the second factor of "$2b$" comes from the fact that each one of the $2b$ point pairs $\{x_{i+1}, y_{i+1}^{(j)}\}$, $j = 1, 2, \ldots, 2b$ uniquely determines a straight line. Let d denote the duration of the play; then by (15.8)

$$T_{end} = T_d \geq T_{start} - d \cdot (1 + 2b)^{k+1}. \tag{15.9}$$

Trivially $d \leq |V_1|$, and from the definition of V_1 (see (15.1))

$$d \leq |V_1| \leq 4(n^{1+3\varepsilon})^2 = 4n^{2+6\varepsilon}.$$

Somewhat less trivial is the lower bound

$$T_{start} = T_0 = |\mathcal{L} \cap V_1| \geq n^2 \cdot \frac{3}{\pi^2} \left(\frac{n}{k}\right)^2.$$

The last step follows from the well-known number-theoretic fact that the number of lattice points in $[1, M] \times [1, M]$ which are visible from the origin (i.e. the coordinates are relatively prime) is about

$$\frac{3}{\pi^2} M^2 = \frac{M^2}{2\zeta(2)} \quad \text{where} \quad \zeta(2) = \sum_{s=1}^{\infty} \frac{1}{s^2} = \frac{\pi^2}{6}.$$

We already used this fact in the proof of Theorem 19.1. Returning to (15.9) we have

$$T_{end} = T_d \geq T_{start} - d \cdot (1 + 2b)^{k+1} \geq \frac{3}{\pi^2} \cdot \frac{n^4}{k^2} - 4n^{2+6\varepsilon} \cdot (1 + 2b)^{k+1}. \tag{15.10}$$

Therefore, if the inequality

$$(1 + 2b)^{k+1} < \frac{3}{4\pi^2} \cdot \frac{n^{2-6\varepsilon}}{k^2} \tag{15.11}$$

holds, then by (15.10) $T_{end} > 0$, proving that Maker really owns an opponent-free k-in-a-line.

In the rest of the argument, we can assume that inequality (15.7) fails at some point of the play. Let $i = i_0$ be the first time the failure occurs

$$\max_{1 \leq j \leq 2b} T_{i_0}(y_{i_0+1}^{(j)}) \geq T_i(x_{i_0+1}). \tag{15.12}$$

Case 2: After the $(i_0 + 1)$st round the subset V_0 $(\subset V_1)$ is already completely occupied by the two players.

We recall that V_0 is a $(2n+1) \times (2n+1)$ sub-lattice of $\mathbb{Z}^2$ centered at the origin (the formal definition is right before (15.1)).

Since $i = i_0$ is the first violation of (15.7), we can still save the following analogue of inequality (15.10) from Case 1

$$T_{i_0} \geq T_{start} - i_0 \cdot (1+2b)^{k+1} \geq \frac{3}{\pi^2} \cdot \frac{n^4}{k^2} - 4n^{2+6\varepsilon} \cdot (1+2b)^{k+1}. \tag{15.13}$$

Therefore, if the inequality

$$(1+2b)^{k+1} < \frac{3}{8\pi^2} \cdot \frac{n^{2-6\varepsilon}}{k^2} \tag{15.14}$$

holds, then by (15.10)

$$T_{i_0} \geq \frac{1}{2} \cdot \frac{3}{\pi^2} \cdot \frac{n^4}{k^2}. \tag{15.15}$$

(Notice that requirement (15.14) is stronger than that of (15.11).)

Now let L be an arbitrary Breaker-free line after the i_0th round; (15.15) implies that such an L exists. By hypothesis L is completely occupied in the next round. Thus there are two possibilities:

(1) either $L \cap V_0$ is already occupied by Maker;
(2) or $L \cap V_0$ contains an unoccupied point, namely one of the $\leq 2b+1$ unoccupied points of V_0.

If (1) holds for some L, we are done. Therefore, we can assume that every Breaker-free line $L \in \mathcal{L}$ contains one of the $\leq 2b+1$ unoccupied points of V_0. Combining this with (15.15), we see that there is an unoccupied $z_0 \in V_0$ with

$$T_{i_0}(z_0) \geq \frac{1}{2b+1} \cdot \frac{1}{2} \cdot \frac{3}{\pi^2} \cdot \frac{n^4}{k^2}. \tag{15.16}$$

The number of lines $L \in \mathcal{L}$ with $z_0 \in L$ is obviously $\leq |V_0| = (2n+1)^2$, so by (15.16) there is a Breaker-free line $L_0 \in \mathcal{L}$ such that $z_0 \in L_0$ and

$$(1+2b)^{|L_0 \cap X(i_0)|} \geq \frac{1}{(2n+1)^2} \cdot \frac{3n^4}{2\pi^2(2b+1)k^2}. \tag{15.17}$$

Comparing (15.14) and (15.17), we obtain $|L_0 \cap X(i_0)| \geq k$, and we are done again.

Case 3: After the (i_0+1)st round the subset $V_0 (\subset V_1)$ still has some unoccupied point.

Since $x_{i_0+1} = z$ is the point where the function $f(z) = T_{i_0}(z)$ attains its maximum in V_1, and also because (15.7) is violated

$$T_{i_0}(x_{i_0+1}) < \max_{1 \leq j \leq 2b} T_{i_0}(y_{i_0+1}^{(j)}), \tag{15.18}$$

the maximum in the right-hand side of (15.18) is attained *outside* of V_1. Let (say)

$$T_{i_0}(y_{i_0+1}^{(1)}) > T_{i_0}(x_{i_0+1}) \quad \text{and} \quad y_{i_0+1}^{(1)} \notin V_1,$$

then

$$T_{i_0}(y_{i_0+1}^{(1)}) > T_{i_0}(x_{i_0+1}) \geq \max_{z \in V_1:\ \text{unselected}} T_{i_0}(z) \geq \max_{w \in V_0:\ \text{unselected}} T_{i_0}(w), \qquad (15.19)$$

where by hypothesis the set $\{w \in V_0 : \text{unselected}\}$ is non-empty.

Repeating the argument of Case 2, we obtain (see (15.14)–(15.16)): if the inequality

$$(1+2b)^{k+1} < \frac{3}{8\pi^2} \cdot \frac{n^{2-6\varepsilon}}{k^2} \qquad (15.20)$$

holds, then

$$\max_{w \in V_0:\ \text{unselected}} T_{i_0}(w) \geq \frac{1}{|V_0|} \cdot \frac{1}{2} \cdot \frac{3}{\pi^2} \cdot \frac{n^4}{k^2} \geq \frac{n^2}{4\pi^2 k^2}. \qquad (15.21)$$

Combining (15.19) and (15.21), under condition (15.20) we have

$$T_{i_0}(y_{i_0+1}^{(1)}) > \frac{n^2}{4\pi^2 k^2}. \qquad (15.22)$$

We need:

Lemma 1: *If a point u in the plane is outside of V_1, then the number of lines $L \in \mathcal{L}$ passing through u is $\leq 8|V_0|n^{-\varepsilon} = 8(2n+1)^2 \cdot n^{-\varepsilon}$, meaning: "relatively few."*

Proof. Let $u = (x, y)$ be an arbitrary point in the plane. Two distinct lines $L_1, L_2 \in \mathcal{L}$ ("rational lines") always intersect at a rational point, so we can assume that both coordinates of u are rational; write $x = p_1/q_1$ and $y = p_2/q_2$, where the fractions are in the standard form (the numerator and the denominator are relatively prime). There are 4 reasons why $u \notin V_1$ (see (15.1)).

Case A: $q_2 > n^\varepsilon$

Write $y = \lfloor y \rfloor + \frac{r}{q_2}$ as the sum of the integral part and the fractional part; here $1 \leq r < q_2$ and q_2 are relatively prime.

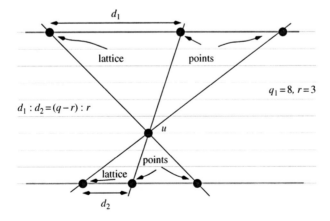

It follows from the picture that the gaps $(q_2 - r)$ ("up") and r ("down") are unavoidable, implying that there are

$$\leq \frac{1}{\max\{r, q_2 - r\}} |V_0| \leq \frac{2}{q_2}(2n+1)^2 \leq \frac{2}{n^\varepsilon}(2n+1)^2$$

lines $L \in \mathcal{L}$ passing through the given point u (which has a "large" denominator in the second coordinate).

Case B: $q_1 > n^\varepsilon$
The same bound as in Case A.

Case C: $|x| > n^{1+\varepsilon}$
Let $u \in L \in \mathcal{L}$, and let (a_1, b_1), (a_2, b_2) be 2 arbitrary points in $L \cap V_0$. Then the slope $\frac{b_2 - b_1}{a_2 - a_1}$ of line L has absolute value $< n^{-\varepsilon}$. Thus there are less than $2 \cdot (2n+1) \cdot (2n+1)n^{-\varepsilon}$ different values for the slopes of lines L with $u \in L \in \mathcal{L}$. Since different lines $u \in L \in \mathcal{L}$ have different slopes, we obtain the upper bound $2n^{-\varepsilon}(2n+1)^2$ for the number of lines $u \in L \in \mathcal{L}$.

Case D: $|y| > n^{1+\varepsilon}$
The same bound as in Case C. The proof of Lemma 1 is complete.   □

Let us return to (15.20)–(15.22). If the inequality

$$(1+2b)^{k+1} < \frac{3}{8\pi^2} \cdot \frac{n^{2-6\varepsilon}}{k^2}$$

holds, then

$$T_{i_0}(y_{i_0+1}^{(1)}) > \frac{n^2}{4\pi^2 k^2}.$$

Since $y_{i_0+1}^{(1)} \notin V_1$, by Lemma 1 the sum

$$T_{i_0}(y_{i_0+1}^{(1)}) = \sum_{y_{i_0+1}^{(1)} \in L \in \mathcal{L}:\ L \cap Y(i_0) = \emptyset} (1+2b)^{|L \cap X(i_0)|}$$

contains $\leq 8(2n+1)^2 \cdot n^{-\varepsilon} \leq 40n^{2-\varepsilon}$ terms, so by (15.22) there is a line $L_0 \in \mathcal{L}$ with $y_{i_0+1}^{(1)} \in L_0$ such that

$$(1+2b)^{|L_0 \cap X(i_0)|} \geq \frac{\frac{n^2}{4\pi^2 k^2}}{40n^{2-\varepsilon}} = \frac{n^\varepsilon}{160\pi^2 k^2}. \tag{15.23}$$

On the other hand, by (15.20)

$$(1+2b)^{k+1} < \frac{3}{8\pi^2} \cdot \frac{n^{2-6\varepsilon}}{k^2},$$

which contradicts (15.23) if $\varepsilon = 2 - 6\varepsilon$, or equivalently, if $\varepsilon = 2/7$ (note that k is small constant times $\log n$).

Conditions (15.11), (15.14), (15.20) are all satisfied if

$$(1+2b)^{k+1} < \frac{3}{8\pi^2} \cdot \frac{n^{2-6\varepsilon}}{k^2} = \frac{3}{8\pi^2} \cdot \frac{n^{2/7}}{k^2}. \tag{15.24}$$

Notice that the choice

$$k = \frac{\log n}{4\log(1+2b)} \tag{15.25}$$

satisfies (15.24). It follows that, in $\leq |V_1| \leq 4n^{2+6\varepsilon} \leq n^{4-1/7}$ moves, Maker ("the first player") can always build an opponent-free k-in-a-line, where k is defined in (15.25). By (15.25)

$$n^{4-1/7} = (1+2b)^{4(4-1/7)k} < (1+2b)^{16k}. \tag{15.26}$$

This completes the Building Part.

It remains to discuss the:

3. Blocking Part. The family of all lines in the plane is an (infinite) Almost Disjoint hypergraph. We have already proved that, playing the usual (1:1) game on an Almost Disjoint hypergraph, Shutout takes an (at least) exponentially long time (see Theorem 1.3). What we need here is just the underdog version of this simple result.

Lemma 2: *Let* $(V, \mathcal{F})$ *be an arbitrary (finite or infinite) Almost Disjoint hypergraph:* V *is the board and* $\mathcal{F}$ *is the family of hyperedges. Maker and Breaker are playing the* $(s{:}1)$ *game, where Maker takes s points per move and Breaker takes 1 point per move. Maker wants to achieve a Shutout of size l, and Breaker simply wants to stop Maker. Assume that Breaker is the first player; then he can prevent Maker from achieving a Shutout of size in* $\frac{1}{2s}2^{\frac{l}{2s}}$ *moves.*

The **proof** of Lemma 2 is a simple (underdog) adaptation of the proof of Theorem 1.3 ("(1:1) game").

Assume we are in the middle of a play, $X(i) = \{x_1, \ldots, x_i\}$ is the set of Breaker's points and

$$Y(i) = \{y_1^{(1)}, y_1^{(2)}, \ldots, y_1^{(s)}, \ldots, y_i^{(1)}, y_i^{(2)}, \ldots, y_i^{(s)}\}$$

is the set of Maker's points (Maker is the second player).

Let

$$\mathcal{F}(i) = \{A \in \mathcal{F} : A \cap X(i) \text{ and } |A \cap Y(i)| \geq 2\}, \tag{15.27}$$

where the requirement "≥ 2" in (15.27) is motivated by the fact that in an Almost Disjoint hypergraph 2 points uniquely determine a hyperedge.

Write

$$T_i = \sum_{A \in \mathcal{F}(i)} 2^{|A \cap Y(i)|/s} \text{ and } T_i(z) = \sum_{z \in A \in \mathcal{F}(i)} 2^{|A \cap Y(i)|/s}. \tag{15.28}$$

Breaker (the first player) selects the unselected $x_{i+1} = z \in V \setminus (X(i) \cup Y(i))$ for which the function $f(z) = T_i(z)$ attains its *maximum*.

It is claimed that, by using this "best-point" strategy, Breaker can force the inequality

$$T_{i+1} \le T_i + 4(s+1) \cdot s^2 \tag{15.29}$$

for the whole play.

Notice that (15.29) implies Lemma 2. Indeed, by a repeated application of (15.29) we obtain

$$T_i \le T_0 + 4s^2 (\sum_{j=1}^{i} j) = 0 + 4s^2 \binom{i+1}{2} = 2s^2 i(i+1). \tag{15.30}$$

If $i \le \frac{1}{2s} 2^{\frac{l}{2s}}$, then (15.30) gives

$$T_i \le 2i(i+1)s^2 < 2^{\frac{l}{s}}. \tag{15.31}$$

Now assume that, in the ith turn with $i \le \frac{1}{2s} 2^{\frac{l}{2s}}$ Maker can achieve a Shutout of size l in some $A_0 \in \mathcal{F}$, then

$$T_i \ge 2^{|A_0 \cap Y(i)|/s} = 2^{l/p},$$

which contradicts (15.31). This contradiction proves Lemma 2, *assuming* inequality (15.29) holds.

To prove (15.29) we describe the effect of the $(i+1)$st moves x_{i+1} and $\{y_{i+1}^{(1)}, \ldots, y_{i+1}^{(s)}\}$, we have

$$T_{i+1} = T_i - T_i(x_{i+1}) + \gamma_i + \sum_{A \in \mathcal{F}(i+1) \setminus \mathcal{F}(i)} 2^{|A \cap Y(i+1)|/s}, \tag{15.32}$$

where

$$\gamma_i = \sum_{A \in \mathcal{F}(i): \, *} \left(2^{|A \cap Y(i+1)|/s} - 2^{|A \cap Y(i)|/s} \right) \tag{15.33}$$

and property $*$ means that $x_{i+1} \notin A$ and $A \cap (Y(i+1) \setminus Y(i)) \ne \emptyset$.

First, we have an upper bound on γ_i: it is claimed that

$$\gamma_i \le T_i(x_{i+1}). \tag{15.34}$$

To prove (15.34), some notation is introduced: let

$$Y(i, j) = Y(i) \cup \{y_{i+1}^{(1)}, \ldots, y_{i+1}^{(j)}\}, \quad j = 0, 1, \ldots, s,$$

and

$$\gamma_{i,j} = \sum_{y_{i+1}^{(j)} \in A \in \mathcal{F}(i)} \left(2^{|A \cap Y(i,j)|/s} - 2^{|A \cap Y(i,j-1)|/s} \right).$$

It is easy to see that $\gamma_i \le \sum_{j=1}^{s} \gamma_{i,j}$.

Another easy inequality is the following

$$\gamma_{i,j} = \sum_{y_{i+1}^{(j)} \in A \in \mathcal{F}(i)} \left(2^{|A \cap Y(i,j)|/s} - 2^{(|A \cap Y(i,j)|-1)/s}\right) \le$$

$$\le \sum_{y_{i+1}^{(j)} \in A \in \mathcal{F}(i)} \left(2^{(j+|A \cap Y(i)|)/s} - 2^{(j-1+|A \cap Y(i)|)/s}\right) =$$

$$= \sum_{y_{i+1}^{(j)} \in A \in \mathcal{F}(i)} \left(2^{j/s} - 2^{(j-1)/s}\right) 2^{|A \cap Y(i)|/s}.$$

Thus we have

$$\gamma_i \le \sum_{j=1}^{s} \gamma_{i,j} \le \sum_{j=1}^{s} \sum_{y_{i+1}^{(j)} \in A \in \mathcal{F}(i)} \left(2^{j/s} - 2^{(j-1)/s}\right) 2^{|A \cap Y(i)|/s} =$$

$$= \sum_{j=1}^{s} \sum_{y_{i+1}^{(j)} \in A \in \mathcal{F}(i)} \left(2^{j/s} - 2^{(j-1)/s}\right) T_i(y_{i+1}^{(j)}),$$

and because $\max_j T_i(y_{i+1}^{(j)}) \le T_i(x_{i+1})$, we have

$$\gamma_i \le \sum_{j=1}^{s} \left(2^{j/s} - 2^{(j-1)/s}\right) T_i(x_{i+1})$$

$$= T_i(x_{i+1}) \sum_{j=1}^{s} \left(2^{j/s} - 2^{(j-1)/s}\right) = T_i(x_{i+1}),$$

proving (15.34).

By (15.32) and (15.34)

$$T_{i+1} = T_i + \sum_{A \in \mathcal{F}(i+1) \setminus \mathcal{F}(i)} 2^{|A \cap Y(i+1)|/s}. \tag{15.35}$$

We claim the following upper bound for the last term in (15.35)

$$\sum_{A \in \mathcal{F}(i+1) \setminus \mathcal{F}(i)} 2^{|A \cap Y(i+1)|/s} < 4(i+1)s^2. \tag{15.36}$$

If $A \in \mathcal{F}(i+1) \setminus \mathcal{F}(i)$, then:

(1) either $|A \cap Y(i)| = 1$ and $A \cap (Y(i+1) \setminus Y(i)) \ne \emptyset$,
(2) or $|A \cap (Y(i+1) \setminus Y(i))| \ge 2$.

Since 2 points uniquely determine a hyperedge ($\mathcal{F}$ is Almost Disjoint!), we have

$$|\mathcal{F}(i+1) \setminus \mathcal{F}(i)| \le is \cdot s + \binom{s}{2}.$$

If $A \in \mathcal{F}(i+1) \setminus \mathcal{F}(i)$, then $|A \cap Y(i)| \le 1$, yielding $|A \cap Y(i+1)| \le 1+s$. Combining these facts we have

$$\sum_{A \in \mathcal{F}(i+1) \setminus \mathcal{F}(i)} 2^{|A \cap Y(i+1)|/s} \le \left(is^2 + \binom{s}{2} \right) \cdot 2^{(1+s)/s} < 4(i+1)s^2,$$

proving (15.36).

Finally, (15.35) and (15.36) yield (15.29), which completes the proof of Lemma 2. □

Since the first player uses his 2nd, 4th, 6th, 8th, ... moves for blocking, we apply Lemma 2 with $s = 2b$, $V = $ the whole plane, $\mathcal{F} = $ the family of all straight lines in the plane. Combining the Building Part above (see (15.26)) with Lemma 2 we obtain: if

$$k = \frac{\log n}{4 \log(1+2b)} \quad \text{and} \quad n^4 = (1+2b)^{16k} \le 2^{\frac{l}{2s}} = 2^{\frac{l}{4b}},$$

then the first player has a winning strategy in the Kaplansky-$(b; k, l)$ Game. The inequality

$$(1+2b)^{16k} \le 2^{\frac{l}{4b}} \quad \text{is equivalent to} \quad \frac{l}{k} \ge \frac{64b \log(1+2b)}{\log 2}.$$

This completes the proof of Theorem 15.1. □

4. Concluding remark: a finite analogue of Theorem 15.1. The hypergraph of all straight lines in the plane is the most natural example of *infinite* Almost Disjoint hypergraphs. The hypergraph of all Lines in a Finite Projective Plane is the most natural example of *finite* Almost Disjoint hypergraphs.

In the infinite case we have unbounded Shutout (Theorem 15.1); what is the largest Shutout in a Finite Projective Plane? Section 15 concludes with an answer to this question.

Let q be an arbitrary prime power. By using the finite field of order q we can construct a Finite Projective Plane, which is a finite hypergraph with the following intersection properties:

(1) there are $q^2 + q + 1$ points;
(2) there are $q^2 + q + 1$ Lines where each Line consists of $q + 1$ points;
(3) any two Lines intersect in a unique point;
(4) any two points determine a unique Line.

For simplicity consider the ordinary (1:1) play on this hypergraph: what is the largest Shutout that (say) the first player can achieve?

It is proved that the first player can always achieve a Shutout of size $c \cdot \log q$, where $c > 0$ is an absolute constant ($c = 1/3$ is a good choice). This is best possible apart from the value of the constant factor c.

What makes the proof interesting is that, besides the usual potential technique, it also uses some linear algebra, in particular Pythagoras Theorem, in a novel way.

We call the first player Maker ("Shutout–Maker") and the second player Breaker. Assume that we are in the middle of a play, $X(i) = \{x_1, x_2, \ldots, x_i\}$ denotes the set of Maker's points and $Y(i) = \{y_1, y_2, \ldots, y_i\}$ denotes the set of Breaker's points selected so far. Maker uses the standard Power-of-Two Scoring System

$$T_i = \sum_{L \in \mathcal{L}: \, L \cap Y(i) = \emptyset} 2^{|L \cap X(i)|} \text{ and } T_i(z_1, \ldots, z_m) = \sum_{\{z_1, \ldots, z_m\} \subseteq L \in \mathcal{L}: \, L \cap Y(i) = \emptyset} 2^{|L \cap X(i)|},$$

where $\mathcal{L}$ is the family of $q^2 + q + 1$ Lines. Clearly

$$T_{i+1} = T_i + T_i(x_{i+1}) - T_i(y_{i+1}) - T_i(x_{i+1}, y_{i+1}).$$

If Maker chooses the "best" point in the usual way, then $T_i(x_{i+1}) \geq T_i(y_{i+1})$ and

$$T_{i+1} = T_i + T_i(x_{i+1}) - T_i(y_{i+1}) - T_i(x_{i+1}, y_{i+1}) \geq T_i - T_i(x_{i+1}, y_{i+1}). \quad (15.37)$$

Assume the contrary that Breaker can prevent Maker from achieving a Shutout of size $\delta \cdot \log_2 q$ (constant $0 < \delta < 1$ will be specified later); We want to derive a contradiction.

Since 2 points uniquely determine a Line, by (15.37) we have

$$T_{i+1} \geq T_i - T_i(x_{i+1}, y_{i+1}) \geq T_i - 2^{\delta \cdot \log_2 q} = T_i - q^{\delta}. \quad (15.38)$$

By iterated application of (15.38)

$$T_i \geq T_0 - i \cdot q^{\delta} = |\mathcal{L}| - i \cdot q^{\delta} = (q^2 + q + 1) - i \cdot q^{\delta} \geq \frac{q^2 + q + 1}{2} \quad (15.39)$$

as long as $i \leq \frac{1}{2} q^{2-\delta}$.

Next comes the Linear Algebra part. Let V denote the set of $q^2 + q + 1$ points of the Finite Projective Plane. For every Line $L \in \mathcal{L}$, let χ_L denote the characterisctic function of Line L: it is a 0-1-valued function defined on V, 1 on the Line and 0 outside of the Line. We look at these $|V| = q^2 + q + 1$ functions χ_L, $L \in \mathcal{L}$ as $|V|$-dimensional vectors, and construct an *orthonormal basis* in the $|V|$-dimensional Euclidean Space as follows: for every $L \in \mathcal{L}$ let

$$v_L = \frac{1}{\sqrt{q}} (\chi_L - \varepsilon \cdot u) \text{ where } u = (1, 1, \ldots, 1) \text{ and } \varepsilon = \frac{1}{q + \sqrt{q} + 1}.$$

This is an orthonormal basis because (1) any two different vectors v_{L_1} and v_{L_2} are orthogonal (we can compute the inner product $\langle \cdots \rangle$ by using the intersection properties)

$$\langle v_{L_1}, v_{L_2} \rangle = \frac{1}{q} \left(\langle \chi_{L_1}, \chi_{L_2} \rangle - 2\varepsilon \langle u, \chi_L \rangle + \varepsilon^2 ||u||^2 \right) =$$

$$= \frac{1}{q} \left(1 - 2\varepsilon(q+1) + \varepsilon^2(q^2+q+1) \right) = 0 \quad \text{if} \quad \varepsilon = \frac{1}{q+\sqrt{q}+1};$$

furthermore, (2) each vector has norm 1

$$||v_L||^2 = \frac{1}{q} \left(||\chi_L||^2 - 2\varepsilon \langle u, \chi_L \rangle + \varepsilon^2 ||u||^2 \right) =$$

$$= \frac{1}{q} \left(q + 1 - 2\varepsilon(q+1) + \varepsilon^2(q^2+q+1) \right) = \frac{1}{q}(q+0) = 1,$$

if $\varepsilon = 1/(q+\sqrt{q}+1)$.

We have an orthonormal basis; the idea is to define a new vector w_1 by the play, and to apply Pythagoras theorem to w_1. Vector w_1 is motivated by (15.39): let

$$V_1 = V \setminus \left(X \left(\frac{1}{2} q^{2-\delta} \right) \cup Y \left(\frac{1}{2} q^{2-\delta} \right) \right). \tag{15.40}$$

That is, V_1 denotes the unoccupied part of the board (Finite Projective Plane) after the $\frac{1}{2} q^{2-\delta}$-th turn of the play.

Let

$$\mathcal{L}_1 = \left\{ L \in \mathcal{L} : L \cap Y \left(\frac{1}{2} q^{2-\delta} \right) = \emptyset \right\}$$

denote the Lines where Maker still has a chance for a Shutout after the $\frac{1}{2} q^{2-\delta}$-th turn of the play. Then by (15.39)

$$T_{\frac{1}{2} q^{2-\delta}} = \sum_{L \in \mathcal{L}_1} 2^{|L \cap X(\frac{1}{2} q^{2-\delta})|} \geq \frac{q^2+q+1}{2}. \tag{15.41}$$

By hypothesis Breaker can prevent Maker from achieving a Shutout of size $\delta \log_2 q$, so

$$\left| L \cap X \left(\frac{1}{2} q^{2-\delta} \right) \right| \leq \delta \log_2 q \tag{15.42}$$

holds for every $L \in \mathcal{L}_1$. Thus by (15.41)–(15.42)

$$|\mathcal{L}_1| \geq \frac{1}{2} q^{2-\delta}. \tag{15.43}$$

Now let w_1 denote the characteristic function of subset V_1 ($\subset V$), see (15.40), i.e. w_1 is 1 on V_1 and 0 on $V \setminus V_1$; by Pythagoras Theorem

$$||w_1||^2 = \sum_{L \in \mathcal{L}} (\langle w_1, v_L \rangle)^2. \tag{15.44}$$

First, we compute the left-hand side of (15.44)

$$||w_1||^2 = |V_1| = (q^2+q+1) - 2 \cdot \frac{1}{2} q^{2-\delta}. \tag{15.45}$$

Next we estimate the right-hand side of (15.44) from below; we use (15.43)

$$\sum_{L \in \mathcal{L}} (\langle w_1, v_L \rangle)^2 \geq \sum_{L \in \mathcal{L}_1} (\langle w_1, v_L \rangle)^2 = \sum_{L \in \mathcal{L}_1} \left(\langle w_1, \frac{1}{\sqrt{q}} (\chi_L - \varepsilon \cdot u) \rangle \right)^2 \geq$$

$$\geq \frac{1}{2} q^{2-\delta} \left(\frac{1}{\sqrt{q}} ((q+1-\delta \log_2 q) - \varepsilon |V_1|) \right)^2 =$$

$$= \frac{1}{2} q^{2-\delta} \frac{1}{q} \left((q+1-\delta \log_2 q) - \frac{1}{q+\sqrt{q}+1} ((q^2+q+1) - 2 \cdot \frac{1}{2} q^{2-\delta}) \right)^2 =$$

$$= \frac{1}{2} q^{1-\delta} \left((q+1-\delta \log_2 q) + \frac{q^{2-\delta}}{q+\sqrt{q}+1} - (q+1-\sqrt{q}) \right)^2 =$$

$$= \frac{1}{2} q^{1-\delta} \left(\sqrt{q} - \delta \log_2 q + \frac{q^{1-\delta}}{1+q^{-1/2}+q^{-1}} \right)^2 \geq$$

$$\geq \frac{1}{2} q^{1-\delta} \cdot \frac{q^{2-2\delta}}{2}. \tag{15.46}$$

Combining (15.44)–(15.46)

$$q^2 \geq ||w_1||^2 = \sum_{L \in \mathcal{L}} (\langle w_1, v_L \rangle)^2 \geq \frac{q^{3-3\delta}}{4},$$

which is a contradiction if $\delta < 1/3$. This proves:

Theorem 15.2 *Playing the (1:1) Shutout Game on a Finite Projective Plane of q^2+q+1 points and the same number of Lines (the Lines are the winning sets), the first player can always achieve a Shutout of size*

$$\left(\frac{1}{3} - o(1) \right) \log_2 q \tag{15.47}$$

in some Line.

The Erdős–Selfridge Theorem gives an upper bound on the maximum shutout: if m denotes the maximum Line-Shutout, then

$$q^2 + q + 1 < 2^m, \text{ or equivalently, } m \geq 2 \log_2 q + O(1). \tag{15.48}$$

It seems a difficult problem to find the correct constant factor of $\log_2 q$. In view of (15.47)–(15.48) the truth is between 1/3 and 2.

Exercise 15.1 *Generalize Theorem 15.2 to the underdog play (1:s) where the first player takes one new point and the second player takes s new points per move, and the first player wants a large Line-Shutout.*

Chapter IV

Games and randomness

This chapter gives new insights to the Meta-Conjecture (see the discussion about the "Phantom Decomposition Hypothesis" at the end of Section 19). Some of the illustrations are off the main trend, but they are still very instructive and amusing, and – what is most important – the proofs are relatively short.

The results of the discrepancy sections (16–17) will be applied later in Chapter VI. Finally, Section 20 answers a question which was raised at the end of Section 4 ("biased connectivity games").

16

Discrepancy Games and the variance

1. Balancing. As far is known, the field of Discrepancy Games was initiated by the following Degree Game problem of Paul Erdős, raised in the early 1970s. Two players, called Maker and Breaker, alternately take previously unselected edges of a given complete graph K_n on n vertices. Each player takes 1 edge per move, Maker colors his edges red and Breaker colors his edges blue. At the end of the play, Maker will own half of the edges, i.e. $\binom{n}{2}/2$ red edges, so the average degree in the red graph is $\binom{n}{2}/n = (n-1)/2$, implying that the maximum degree in Maker's graph is $\geq (n-1)/2$.

Maker's goal in the Degree Game on K_n is to maximize his maximum degree, i.e. to force a red degree $\frac{n}{2} + \Delta$ with discrepancy $\Delta = \Delta(n)$ as large as possible. What is the largest discrepancy $\Delta = \Delta(n)$ that Maker can force? Notice that we want one-sided, strictly positive discrepancy.

In 1981, L. Székely proved that $\Delta = \Delta(n) \to \infty$ as $n \to \infty$, and gave the following explicit upper and lower bounds

$$\text{Breaker can always force } \Delta = \Delta(n) < c_1\sqrt{n\log n}, \qquad (16.1)$$

$$\text{Maker can always force } \Delta = \Delta(n) > c_2\log n. \qquad (16.2)$$

The upper bound (16.1) immediately follows from the following general discrepancy theorem (see Lemma 3 in Beck [1981b]).

Theorem 16.1 *Let $\mathcal{F}$ be an arbitrary finite hypergraph, and let ε with $0 < \varepsilon \leq 1$ be an arbitrary real number. There are two players, Balancer and Unbalancer, who play the (1:1) game on $\mathcal{F}$: they alternate, and each player takes one new point per move. Unbalancer's goal is to achieve a majority: he wins if he owns $\geq \frac{1+\varepsilon}{2}$ part of some $A \in \mathcal{F}$; otherwise Balancer wins. Here is a Balancer's win criterion: if*

$$\sum_{A \in \mathcal{F}} \left((1+\varepsilon)^{1+\varepsilon}(1-\varepsilon)^{1-\varepsilon}\right)^{-|A|/2} < 1, \qquad (16.3)$$

then Balancer, as the first player, has a winning strategy.

It is critical to see that the base $(1+\varepsilon)^{1+\varepsilon}(1-\varepsilon)^{1-\varepsilon}$ is greater than 1 for every $0 < \varepsilon \le 1$ (why?).

Note that in the special case $\varepsilon = 1$, where Unbalancer's goal is to occupy a whole set $A \in \mathcal{F}$, Theorem 16.1 gives back the Erdős–Selfridge Theorem.

The **proof** of Theorem 16.1 is very similar to that of the Erdős–Selfridge Theorem. Assume that we are in the middle of a play: Unbalancer already occupied $u_1, u_2, \ldots, u_t$ and Balancer occupied $b_1, b_2, \ldots, b_t$; t is the time parameter. Write $U(t) = \{u_1, u_2, \ldots, u_t\}$, $B(t) = \{b_1, b_2, \ldots, b_t\}$, and consider the "one-sided" potential function

$$P_t = \sum_{A \in \mathcal{F}} (1+\varepsilon)^{|A \cap U(t)| - \frac{1+\varepsilon}{2}|A|}(1-\varepsilon)^{|A \cap B(t)| - \frac{1-\varepsilon}{2}|A|}, \tag{16.4}$$

which is very sensitive ("exponentially sensitive") to Unbalancer's lead. The core idea, which is taken from the Erdős–Selfridge proof, is that Balancer can force the monotone *decreasing* property

$$P_0 \ge P_1 \ge P_2 \ge \cdots \ge P_{end},$$

so $P_{start} = P_0 \ge P_{end}$.

If Unbalancer wins, then by (16.4) we have $P_{end} \ge 1$; on the other hand, by hypothesis (16.3), $P_{start} = P_0 < 1$, which together imply Balancer's win. We leave the details to the reader. $\qquad\square$

Let us apply Theorem 16.1 to the star hypergraph of K_n: the hyperedges are the n stars (each star has $n-1$ edges), so $\mathcal{F}$ is an $(n-1)$-uniform hypergraph with $|\mathcal{F}| = n$. By choosing

$$\varepsilon = \sqrt{\frac{c \log n}{n}}$$

with some unspecified (yet) constant $c > 0$, criterion (16.3), applied to the star hypergraph of K_n, gives

$$\sum_{A \in \mathcal{F}} \left((1+\varepsilon)^{1+\varepsilon}(1-\varepsilon)^{1-\varepsilon}\right)^{-|A|/2} = ne^{-((1+\varepsilon)\log(1+\varepsilon)+(1-\varepsilon)\log(1-\varepsilon))\frac{n-1}{2}}$$

$$= ne^{-((1+\varepsilon)(\varepsilon - \frac{\varepsilon^2}{2} \pm \cdots) + (1-\varepsilon)(-\varepsilon - \frac{\varepsilon^2}{2} - \cdots))\frac{n-1}{2}}$$

$$= ne^{-(\varepsilon^2 + O(\varepsilon^3))\frac{n-1}{2}} = ne^{-\frac{c \log n}{2}(1+O(\varepsilon))} = n^{1-\frac{c}{2}(1+O(\varepsilon))},$$

which is less than 1 if $c > 2$ and n is sufficiently large. So Theorem 16.1 applies with

$$\varepsilon = \sqrt{\frac{(2+o(1))\log n}{n}},$$

and yields (16.1) with the explicit constant $c_1 = \sqrt{1/2 + o(1)}$.

In general we have:

Corollary of Theorem 16.1 *Let $\mathcal{F}$ be an n-uniform hypergraph, and consider the Balancer–Unbalancer game played on hypergraph $\mathcal{F}$ where Unbalancer's goal is to own at least $\frac{n+\Delta}{2}$ points from some $A \in \mathcal{F}$. If*

$$\Delta = \left(1 + O\left(\sqrt{\frac{\log|\mathcal{F}|}{n}}\right)\right) \sqrt{2n \log|\mathcal{F}|},$$

then Balancer has a winning strategy.

Proof of the Corollary. Let $\varepsilon = \Delta/n$; in view of Theorem 16.1 we have to check the inequality

$$(1+\varepsilon)^{(n+\Delta)/2} \cdot (1-\varepsilon)^{(n-\Delta)/2} \geq |\mathcal{F}|.$$

Note that

$$(1+\varepsilon)^{(n+\Delta)/2} \cdot (1-\varepsilon)^{(n-\Delta)/2} = (1-\varepsilon^2)^{n/2} \cdot \left(\frac{1+\varepsilon}{1-\varepsilon}\right)^{\Delta/2} \approx e^{-\varepsilon^2 n/2 + \varepsilon\Delta} = e^{\Delta^2/2n}.$$

More precisely, we have

$$e^{(1+O(\Delta/n))\frac{\Delta^2}{2n}} = (1+\varepsilon)^{(n+\Delta)/2} \cdot (1-\varepsilon)^{(n-\Delta)/2} \geq |\mathcal{F}| = e^{\log|\mathcal{F}|},$$

which implies

$$(1+O(\Delta/n))\frac{\Delta^2}{2n} \geq \log|\mathcal{F}|,$$

or equivalently,

$$\Delta \geq \left(1 + O\left(\sqrt{\frac{\log|\mathcal{F}|}{n}}\right)\right) \sqrt{2n \log|\mathcal{F}|},$$

which proves the Corollary. $\qquad\qquad\square$

2. Forcing a Standard Deviation size discrepancy. There is a very large gap between upper bound (16.1) and lower bound (16.2). In his paper Székely conjectured that the $\log n$ in (16.2) was the true order of magnitude; he called it "Problem 2," see Székely [1981]. In Beck [1993c] this conjecture was disproved by improving the lower bound $\log n$ to $\sqrt{n}$: it was proved that

$$\text{Maker can always force } \Delta = \Delta(n) > \frac{\sqrt{n}}{32}. \tag{16.5}$$

This result shows that upper bound (16.1) is actually pretty close to the truth, there is only a factor of $\log n$ between the best known upper and lower bounds.

In Beck [1993c] the complete bipartite graph $K_{n,n}$ was actually worked with instead of K_n (of course, the proof is the same in both cases), because the Degree Game on $K_{n,n}$ is equivalent to an elegant Row–Column Game on an $n \times n$ board.

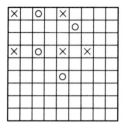

Row–Column Game
on a 9×9 board

The two players play on an $n \times n$ chessboard, and alternately mark previously unmarked little squares. Maker uses (say) X and Breaker uses (say) O, like in Tic-Tac-Toe; Maker's goal is to achieve a large *lead* in some line, where a "line" means either a row or a column. Let $\frac{n}{2} + \Delta$ denote the maximum number of Xs ("Maker's mark") in some line at the end of a play; then the difference $(\frac{n}{2} + \Delta) - (\frac{n}{2} - \Delta) = 2\Delta$ is Maker's *lead*; Maker wants to maximize $\Delta = \Delta(n)$.

Of course, there is hardly any difference between the Degree Game on $K_{n,n}$ or on K_n, but we personally find the Row–Column Game the most attractive variant.

Besides his "Problem 2," that was disproved in (16.5), Székely formulated two more problems in his paper (Székely [1981]). The first one goes as follows:

Székely's "Problem 1": *Let G be an arbitrary finite n-regular graph. Is it true that, playing the Degree Game on G, Maker can always achieve a degree $\geq \frac{n}{2} + \Delta$ with $\Delta = \Delta(n) \to \infty$ as $n \to \infty$?*

Recently this conjecture was proved. In fact, a general hypergraph theorem was proved, and the solution of "Problem 1" above follows from applying the hypergraph theorem (see Theorem 16.2 below) in the special case of the "star hypergraph of G" (the hyperedges are the "stars"). Here is the hypergraph result:

Theorem 16.2 *Let $\mathcal{F}$ be a hypergraph which is (1) n-uniform, (2) Almost Disjoint: $|A_1 \cap A_2| \leq 1$ for any two different elements of hypergraph $\mathcal{F}$, and (3) the common degree of $\mathcal{F}$ is 2: every point of the hypergraph is contained in exactly two hyperedges. Maker and Breaker play the usual (1:1) game on $\mathcal{F}$. Then, at the end of the play, Maker can occupy at least $\frac{n}{2} + c\sqrt{n}$ points from some $A \in \mathcal{F}$.*

We are going to give two different proofs for Theorem 16.2. It is very instructive to compare them. Note that Theorem 16.2 holds with $c = 1/15$.

Applying Theorem 16.2 to the star hypergraph of an n-regular graph G, we obtain the following solution to Székely's "Problem 1".

Theorem 16.3 *Consider the Maker–Breaker Degree Game on an arbitrary finite n-regular graph G. Maker can force that, at the end of the play, his maximum degree is $\geq \frac{n}{2} + c\sqrt{n}$ with $c = 1/15$.*

Notice that the constant factor $c = 1/15$ in Theorem 16.3 is better than the earlier constant factor $c = 1/32$ was in (16.5) in the special cases $G = K_n$ and $G = K_{n,n}$ (the old proof in Beck [1993c] was different).

First proof of Theorem 16.2. Assume we are in the middle of a play, Maker already occupied $x_1, x_2, \ldots, x_t$ (t is the time) and Breaker occupied $y_1, y_2, \ldots, y_t$. The claimed $\sqrt{n}$ discrepancy is familiar from Probability Theory: $\sqrt{n}$ is the "standard deviation" of the n-step random walk. The "standard deviation" is the square-root of the "variance" – this motivates the introduction of the following "game-theoretic variance." Let $X(t) = \{x_1, x_2, \ldots, x_t\}$, $Y(t) = \{y_1, y_2, \ldots, y_t\}$, and write ("V" for "variance")

$$V_t = \sum_{A \in \mathcal{F}} (|A \cap X(t)| - |A \cap Y(t)|)^2. \tag{16.6}$$

What is the effect of the $(t+1)$st moves x_{t+1} (by Maker) and y_{t+1} (by Breaker)? Well, the answer is easy; by using the trivial fact $(u \pm 1)^2 = u^2 \pm 2u + 1$, we have

$$V_{t+1} = V_t + \sum_{A \in \mathcal{F}:\ x_{t+1} \in A \not\ni y_{t+1}} \left(2(|A \cap X(t)| - |A \cap Y(t)|) + 1 \right)$$

$$+ \sum_{A \in \mathcal{F}:\ y_{t+1} \in A \not\ni x_{t+1}} \left(-2(|A \cap X(t)| - |A \cap Y(t)|) + 1 \right)$$

$$= V_t + 2 \sum_{A \in \mathcal{F}:\ x_{t+1} \in A} (|A \cap X(t)| - |A \cap Y(t)|) - 2$$

$$\sum_{A \in \mathcal{F}:\ y_{t+1} \in A} (|A \cap X(t)| - |A \cap Y(t)|) +$$

$$+ \sum_{A \in \mathcal{F}:\ x_{t+1} \in A \not\ni y_{t+1}} 1 + \sum_{A \in \mathcal{F}:\ y_{t+1} \in A \not\ni x_{t+1}} 1. \tag{16.7}$$

It follows from the hypothesis of Theorem 16.2 that

$$\sum_{A \in \mathcal{F}:\ x_{t+1} \in A \not\ni y_{t+1}} 1 + \sum_{A \in \mathcal{F}:\ y_{t+1} \in A \not\ni x_{t+1}} 1 \geq 2. \tag{16.8}$$

Since Maker chooses his x_{t+1} before Breaker chooses y_{t+1}, Maker can have the "best" point: x_{t+1} is that unoccupied point z for which the sum

$$\sum_{A \in \mathcal{F}:\ z \in A} (|A \cap X(t)| - |A \cap Y(t)|)$$

attains its *maximum*. Then

$$\sum_{A \in \mathcal{F}:\ x_{t+1} \in A} (|A \cap X(t)| - |A \cap Y(t)|) \geq \sum_{A \in \mathcal{F}:\ y_{t+1} \in A} (|A \cap X(t)| - |A \cap Y(t)|),$$

so by (16.7)–(16.8)

$$V_{t+1} \geq V_t + 2, \quad \text{implying} \quad V_T \geq V_0 + 2T, \tag{16.9}$$

where T is the "total time," i.e. $T = n|\mathcal{F}|/4$ is half of the board size. Since trivially $V_0 = 0$, we conclude

$$V_T \geq V_0 + 2T = 2T = \frac{n|\mathcal{F}|}{2}. \tag{16.10}$$

Comparing (16.6) and (16.10), we see that there must exist an $A \in \mathcal{F}$ such that

$$(|A \cap X(T)| - |A \cap Y(T)|)^2 \geq \frac{n}{2}. \tag{16.11}$$

By taking square-root of (16.11), we obtain

$$||A \cap X(T)| - |A \cap Y(T)|| \geq \sqrt{n/2}, \tag{16.12}$$

but (16.12) is not what we want! We want a one-sided estimate with $|A \cap X(T)| > |A \cap Y(T)|$, but in (16.11) "squaring kills the sign," and there is no way to guarantee $|A \cap X(T)| > |A \cap Y(T)|$.

A technical trick. To overcome the difficulty of "squaring kills the sign," we modify the "quadratic" variance in (16.6) with an extra "exponential" term as follows: let

$$S_t = \sum_{A \in \mathcal{F}} \left((|A \cap X(t)| - |A \cap Y(t)|)^2 - \alpha n (1-\lambda)^{|A \cap X(t)|} (1+\lambda)^{|A \cap Y(t)|} \right), \tag{16.13}$$

where parameters α with $0 < \alpha < 1$ and λ with $0 < \lambda < 1$ will be specified later. What motivates the introduction of the auxiliary "exponential" term

$$\alpha n (1-\lambda)^{|A \cap X(t)|} (1+\lambda)^{|A \cap Y(t)|} \tag{16.14}$$

in (16.13)? We had 2 reasons in mind.

(1) The obvious problem with variance (16.6) is that "squaring kills the sign," so it may happen that, there is a "very large" $|A \cap Y(t)|$, which is the only reason why the variance is "large" (this $|A \cap Y(t)|$ is useless for us!). But "exponential" beats "quadratic," this is how, by using (16.13) instead of (16.6), Maker can nevertheless guarantee a large one-sided discrepancy.

(2) A second reason is that in the proof of Theorem 16.1 we already used the exponential potential

$$\sum_{A \in \mathcal{F}} (1+\varepsilon)^{|A \cap U(t)| - \frac{1+\varepsilon}{2}|A|} (1-\varepsilon)^{|A \cap B(t)| - \frac{1-\varepsilon}{2}|A|}, \tag{16.15}$$

with some $0 < \varepsilon < 1$; here $B(t)$ is Balancer's set and $U(t)$ is Unbalancer's set. (16.14) and (16.15) are basically the same; the success of (16.15) is a good sign, and motivates the modification of (16.6) by (16.14).

What is the effect of the $(t+1)$st moves x_{t+1} (by Maker) and y_{t+1} (by Breaker) in the new sum (16.13)? By definition

$$S_{t+1} = S_t + 2 \sum_{A \in \mathcal{F}: \, x_{t+1} \in A} (|A \cap X(t)| - |A \cap Y(t)|) - 2 \sum_{A \in \mathcal{F}: \, y_{t+1} \in A} (|A \cap X(t)| - |A \cap Y(t)|)$$

$$+ \sum_{A \in \mathcal{F}: \, x_{t+1} \in A \ne y_{t+1}} 1 + \sum_{A \in \mathcal{F}: \, y_{t+1} \in A \ne x_{t+1}} 1 + \alpha n \lambda \sum_{A \in \mathcal{F}: \, x_{t+1} \in A} (1 - \lambda)^{|A \cap X(t)|} (1 + \lambda)^{|A \cap Y(t)|}$$

$$- \alpha n \lambda \sum_{A \in \mathcal{F}: \, y_{t+1} \in A} (1 - \lambda)^{|A \cap X(t)|} (1 + \lambda)^{|A \cap Y(t)|}$$

$$+ \alpha n \lambda^2 \cdot \delta(x_{t+1}, y_{t+1}) \cdot (1 - \lambda)^{|A_0 \cap X(t)|} (1 + \lambda)^{|A_0 \cap Y(t)|}, \tag{16.16}$$

where $\delta(x_{t+1}, y_{t+1}) = 1$ if there is an $A \in \mathcal{F}$ containing both x_{t+1} and y_{t+1}; Almost Disjointness yields that, if there is 1, then there is exactly 1: let A_0 be this uniquely determined $A \in \mathcal{F}$; finally let $\delta(x_{t+1}, y_{t+1}) = 0$ if there is *no* $A \in \mathcal{F}$ containing both x_{t+1} and y_{t+1}.

From the hypothesis of Theorem 16.2

$$\sum_{A \in \mathcal{F}: \, x_{t+1} \in A \ne y_{t+1}} 1 + \sum_{A \in \mathcal{F}: \, y_{t+1} \in A \ne x_{t+1}} 1 \ge 2. \tag{16.17}$$

If Maker chooses that previously unselected $x_{t+1} = z$ for which the function

$$g_t(z) = \sum_{A \in \mathcal{F}: \, z \in A} \left(2(|A \cap X(t)| - |A \cap Y(t)|) - \alpha n \lambda (1 - \lambda)^{|A \cap X(t)|} (1 + \lambda)^{|A \cap Y(t)|} \right)$$

attains its *maximum*, then $g_t(x_{t+1}) \ge g_t(y_{t+1})$. So by (16.15) and (16.17)

$$S_{t+1} \ge S_t + 2 + g_t(x_{t+1}) - g_t(y_{t+1}) \ge S_t + 2. \tag{16.18}$$

An iterated application of (16.17) gives that $S_{end} = S_T \ge S_0 + 2T$, where $T = n|\mathcal{F}|/4$ is the duration of the play ("half of the board size"). By definition $S_0 = S_{blank} = -\alpha n|\mathcal{F}|$, so we have

$$S_{end} = S_T \ge \frac{n|\mathcal{F}|}{2} - \alpha n|\mathcal{F}| = n|\mathcal{F}| \left(\frac{1}{2} - \alpha \right),$$

or equivalently

$$\sum_{A \in \mathcal{F}} \left((|A \cap X(T)| - |A \cap Y(T)|)^2 - \alpha n (1 - \lambda^2)^{n/2} \left(\frac{1 - \lambda}{1 + \lambda} \right)^{|A \cap X(T)| - n/2} \right)$$

$$\ge n|\mathcal{F}| \left(\frac{1}{2} - \alpha \right). \tag{16.19}$$

Let $|\mathcal{F}| = N$, and write $\mathcal{F} = \{A_1, A_2, \ldots, A_N\}$ and $|A_i \cap X(T)| = |A_i \cap X_{end}| = \frac{n}{2} + \Delta_i$, $1 \le i \le N$.

Let $\alpha = 1/8$ and $\lambda = \sqrt{2/n}$. Then $(1 - \lambda^2)^{n/2} \approx e^{-1}$, and by using the easy inequality

$$4u^2 < \alpha \cdot e^{-1 + 2\sqrt{2}u} \quad \text{for all } u \ge \sqrt{5},$$

from (16.19) we obtain the following inequality

$$\sum_{i:\ \Delta_i \ge -\sqrt{5n}} 4\Delta_i^2 \ge n|\mathcal{F}|\left(\frac{1}{2}-\alpha\right) = \frac{3nN}{8}. \tag{16.20}$$

Now we are almost done. First note that $\sum_{i=1}^{N}\Delta_i = 0$, which implies

$$\sum_{i:\ \Delta_i>0} \Delta_i = \sum_{i:\ \Delta_i\le 0} |\Delta_i| \ge \sum_{i:\ -\sqrt{5n}\le\Delta_i\le 0} |\Delta_i|, \tag{16.21}$$

and by (16.21)

$$\sqrt{5n}\left(\sum_{i:\ \Delta_i>0}\Delta_i\right) \ge \sum_{i:\ -\sqrt{5n}\le\Delta_i\le 0} \Delta_i^2 \tag{16.22}$$

By (16.20) and (16.22)

$$\sum_{i:\ \Delta_i>0}\Delta_i^2 + \sqrt{5n}\left(\sum_{i:\ \Delta_i>0}\Delta_i\right) \ge \frac{3nN}{32}, \tag{16.23}$$

i.e. in (16.23) we have positive discrepancy only!

To complete the proof, in sum (16.23) we distinguish 2 cases. Either

$$\sqrt{5n}\left(\sum_{i:\ \Delta_i>0}\Delta_i\right) \ge \frac{nN}{11},$$

which immediately gives the lower bound

$$\max \Delta_i \ge \frac{\sqrt{n}}{11\sqrt{5}} \ge \frac{\sqrt{n}}{25},$$

or, by (16.23)

$$\sum_{i:\ \Delta_i>0}\Delta_i^2 \ge \left(\frac{3}{32}-\frac{1}{11}\right)nN,$$

which immediately gives the better lower bound

$$\max \Delta_i \ge \sqrt{\left(\frac{3}{32}-\frac{1}{11}\right)n} \ge \frac{\sqrt{n}}{19}.$$

This proves Theorem 16.2 with $c = 1/25$.

Second proof of Theorem 16.2. In the first proof above we used the "exponential" expression

$$\sum_{A\in\mathcal{F}} (1-\lambda)^{|A\cap X(t)|}(1+\lambda)^{|A\cap Y(t)|} \tag{16.24}$$

to fix the shortcoming of the variance ("squaring kills the sign"), and the "exponential" expression (16.24) played a secondary, counter-balancing role only.

The new idea is to work with the "twin brother" of (16.24): consider the single-term potential function

$$S_t = \sum_{A \in \mathcal{F}} (1+\lambda)^{|A \cap X(t)|} (1-\lambda)^{|A \cap Y(t)|}. \tag{16.25}$$

The reader may find it surprising that such a simple potential will do the job.

We have the following analogue of (16.16), in fact the new identity is simpler

$$\begin{aligned}
S_{t+1} = S_t + \lambda \sum_{A \in \mathcal{F}: \, x_{t+1} \in A} (1+\lambda)^{|A \cap X(t)|} (1-\lambda)^{|A \cap Y(t)|} \\
- \lambda \sum_{A \in \mathcal{F}: \, y_{t+1} \in A} (1+\lambda)^{|A \cap X(t)|} (1-\lambda)^{|A \cap Y(t)|} \\
- \lambda^2 \cdot \delta(x_{t+1}, y_{t+1}) \cdot (1+\lambda)^{|A_0 \cap X(t)|} (1-\lambda)^{|A_0 \cap Y(t)|},
\end{aligned} \tag{16.26}$$

where $\delta(x_{t+1}, y_{t+1}) = 1$ if there is an $A \in \mathcal{F}$ containing both x_{t+1} and y_{t+1}. Almost Disjointness yields that, if there is 1, then there is exactly 1: let A_0 be this uniquely determined $A \in \mathcal{F}$. Finally, let $\delta(x_{t+1}, y_{t+1}) = 0$ if there is *no* $A \in \mathcal{F}$ containing both x_{t+1} and y_{t+1}.

Since Maker's x_{t+1} is selected before Breaker's y_{t+1}, Maker can select the "best" point: Maker chooses that $x_{t+1} = z$ for which the sum

$$\sum_{A \in \mathcal{F}: \, z \in A} (1+\lambda)^{|A \cap X(t)|} (1-\lambda)^{|A \cap Y(t)|}$$

attains its *maximum*. Then by (16.26)

$$S_{t+1} \geq S_t - \lambda^2 \cdot \delta(x_{t+1}, y_{t+1}) \cdot (1+\lambda)^{|A_0 \cap X(t)|} (1-\lambda)^{|A_0 \cap Y(t)|}. \tag{16.27}$$

Let $\Delta = \Delta(n)$ denote the largest positive discrepancy that Maker can achieve; it means $\frac{n}{2} + \Delta$ points from some $A \in \mathcal{F}$. If Δ is the maximum discrepancy, then the inequality $|A \cap X(t)| - |A \cap Y(t)| \leq 2\Delta$ must hold during the whole play (meaning every t) and for every $A \in \mathcal{F}$. Indeed, if $|A \cap X(t)| - |A \cap Y(t)| > 2\Delta$, then Maker can keep this lead for the rest of the play till the end, contradicting the maximum property of Δ. Combining this observation with (16.27) we have

$$S_{t+1} \geq S_t - \lambda^2 \cdot \delta(x_{t+1}, y_{t+1}) \cdot (1+\lambda)^{z_t + \Delta} (1-\lambda)^{z_t - \Delta}, \tag{16.28}$$

where

$$z_t = \frac{|A_0 \cap X(t)| + |A_0 \cap Y(t)|}{2}.$$

Since $S_0 = S_{start} = |\mathcal{F}|$ and "total time"$= T = nN/4$, from (16.28) we obtain

$$S_{end} = S_T \geq S_0 - \lambda^2 \sum_{t=1}^{nN/4} (1+\lambda)^{z_t+\Delta}(1-\lambda)^{z_t-\Delta}$$

$$= N - \lambda^2 \left(\frac{1+\lambda}{1-\lambda}\right)^{\Delta} \left(\sum_{t=1}^{nN/4}(1-\lambda^2)^{z_t}\right)$$

$$\geq N - \lambda^2 \left(\frac{1+\lambda}{1-\lambda}\right)^{\Delta} \frac{nN}{4}. \tag{16.29}$$

On the other hand, by definition

$$S_{end} = S_T \leq N(1+\lambda)^{n/2+\Delta}(1-\lambda)^{n/2-\Delta} =$$

$$= N \left(\frac{1+\lambda}{1-\lambda}\right)^{\Delta} (1-\lambda^2)^{n/2}. \tag{16.30}$$

Combining (16.29) and (16.30)

$$N - \lambda^2 \left(\frac{1+\lambda}{1-\lambda}\right)^{\Delta} \frac{nN}{4} \leq N \left(\frac{1+\lambda}{1-\lambda}\right)^{\Delta} (1-\lambda^2)^{n/2},$$

or equivalently

$$\left(\frac{1+\lambda}{1-\lambda}\right)^{\Delta} \geq \frac{1}{\frac{\lambda^2 n}{4} + (1-\lambda^2)^{n/2}}. \tag{16.31}$$

We want to minimize the denominator in the right-hand side of (16.31): we are looking for an optimal λ in the form $\lambda = \sqrt{2\beta/n}$, where β is an unspecified constant (yet); then

$$\frac{\lambda^2 n}{4} + (1-\lambda^2)^{n/2} \approx \frac{\beta}{2} + e^{-\beta} = \frac{1+\log 2}{2}$$

if $\beta = \log 2$. With this choice of β (16.31) becomes

$$e^{2\lambda\Delta} \approx \left(\frac{1+\lambda}{1-\lambda}\right)^{\Delta} \geq \frac{1}{\frac{\lambda^2 n}{4} + (1-\lambda^2)^{n/2}} \geq \frac{1}{\frac{1+\log 2}{2}},$$

implying

$$2\sqrt{2\log 2} \cdot \frac{\Delta}{\sqrt{n}} \geq \log\left(\frac{2}{1+\log 2}\right),$$

that is

$$\Delta \geq \frac{\log\left(\frac{2}{1+\log 2}\right)}{2\sqrt{2\log 2}} \sqrt{n} \geq \frac{\sqrt{n}}{15}.$$

This proves Theorem 16.2 with $c = 1/15$. $\square$

3. Extensions. Theorem 16.2 is about very special hypergraphs: we assumed that hypergraph $\mathcal{F}$ is (1) n-uniform, (2) Almost Disjoint, and (3) 2-regular (i.e. every point has degree 2. What happens for a general hypergraph? For example, assume that hypergraph $\mathcal{F}$ is (1) n-uniform, (3′)D-regular, i.e. every point has degree $D \geq 2$ – we skip Almost Disjointness – can we still guarantee any discrepancy? The answer is an easy "no." Indeed, let me recall the class of Strictly Even Hypergraphs from Section 9. A Strictly Even Hypergraph has an underlying "pairing," and if a hyperedge intersects a "pair," then the hyperedge must contain the whole "pair."

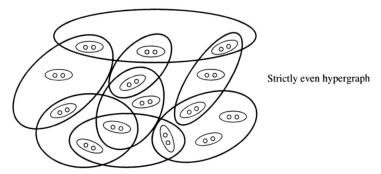

Strictly even hypergraph

In a **Strictly Even Hypergraph**, every winning set has even size, and, in general, the intersection of an arbitrary family of winning sets has even size too (explaining the name "strictly even").

The Positional Game played on an arbitrary Strictly Even Hypergraph is trivial: the first player cannot occupy a whole winning set; in fact, the first player cannot even achieve a "lead" by 1. Indeed, by using a Pairing Strategy, the second player can take the exact half from each winning set, preventing any "lead" of the opponent.

One of the many peculiarities of a Strictly Even Hypergraph is that its Max Pair-Degree D_2 equals the Max Degree D: $D_2 = D$.

A hypergraph is very different from a Strictly Even Hypergraph; Max Pair-Degree D_2 is subtantially less than the Max Degree D. Under this condition Maker can always force a large lead.

Theorem 16.4 *Let $\mathcal{F}$ be an arbitrary n-uniform D-regular hypergraph, and let $D_2 = D_2(\mathcal{F})$ denote the Max Pair-Degree of $\mathcal{F}$. Playing the usual (1:1) game on $\mathcal{F}$, at the end of the play Maker can always occupy at least*

$$\frac{n}{2} + \frac{\log\left(\frac{D}{D_2(1+\log(D/D_2))}\right)}{2\sqrt{2\log(D/D_2)}}\sqrt{n}$$

points from some $A \in \mathcal{F}$.

Remark. If the ratio D/D_2 is strictly larger than 1, then the coefficient

$$\frac{\log\left(\frac{D}{D_2(1+\log(D/D_2))}\right)}{2\sqrt{2\log(D/D_2)}}$$

of $\sqrt{n}$ is positive. Indeed, if $\log(D/D_2) = \varepsilon > 0$, then

$$\frac{D}{D_2(1+\log(D/D_2))} = \frac{e^\varepsilon}{1+\varepsilon} = \frac{1+\varepsilon+\varepsilon^2/2+\varepsilon^3/6+\cdots}{1+\varepsilon} > 1,$$

so its logarithm is positive.

If D_2 is close to D, i.e. if $\frac{D}{D_2} - 1 \approx \log(D/D_2) = \varepsilon$ is small, then

$$\frac{\log\left(\frac{D}{D_2(1+\log(D/D_2))}\right)}{2\sqrt{2\log(D/D_2)}} \approx \frac{\varepsilon^2/2}{2\sqrt{2\varepsilon}} = \frac{\varepsilon^{3/2}}{2^{5/2}}. \tag{16.32}$$

Theorem 16.4 is somewhat "ugly," but it immediately solves Székely's "Problem 3." We discuss this application right after the proof.

Proof. Let $|\mathcal{F}| = N$. We repeat the argument of the Second Proof of Theorem 16.2; we have the following perfect analogue of (16.31)

$$N - D_2 \cdot \lambda^2 \left(\frac{1+\lambda}{1-\lambda}\right)^\Delta \frac{nN}{2D} \le N \left(\frac{1+\lambda}{1-\lambda}\right)^\Delta (1-\lambda^2)^{n/2},$$

or equivalently

$$\left(\frac{1+\lambda}{1-\lambda}\right)^\Delta \ge \frac{1}{\frac{D_2\lambda^2 n}{2D} + (1-\lambda^2)^{n/2}}. \tag{16.33}$$

We want to minimize the denominator in the right-hand side of (16.33): we are looking for an optimal λ in the form $\lambda = \sqrt{2\beta/n}$, where β is an unspecified constant (yet); then

$$\frac{D_2\lambda^2 n}{2D} + (1-\lambda^2)^{n/2} \approx \frac{D_2\beta}{D} + e^{-\beta} = \frac{D_2}{D}(1+\log(D/D_2))$$

if $\beta = \log(D/D_2)$. With this choice of β (16.32) becomes

$$e^{2\lambda\Delta} \approx \left(\frac{1+\lambda}{1-\lambda}\right)^\Delta \ge \frac{1}{\frac{\lambda^2 n}{4} + (1-\lambda^2)^{n/2}} \ge \frac{D}{D_2(1+\log(D/D_2))},$$

implying

$$\Delta \ge \frac{\log\left(\frac{D}{D_2(1+\log(D/D_2))}\right)}{2\sqrt{2\log(D/D_2)}} \sqrt{n},$$

which completes the proof of Theorem 16.4. □

Theorem 16.3 solved Székely's "Problem 1." We already mentioned Székely's "Problem 2," which was about the Degree Game on the complete graph K_n. Székely's "Problem 3" is the Degree Game on complete *hyper*graphs.

Székely's "Problem 3": Let $r \ge 2$ be a fixed integer, and let K_n^r denote the r-uniform complete hypergraph on n points (K_n^r contains $\binom{n}{r}$ r-sets). Consider the

usual Maker-Breaker game on K_n^r (the players alternate, each player takes one
r-set per move). What is the largest degree that Maker can achieve? Is it true that
Maker can always achieve a degree $\geq \frac{1}{2}\binom{n-1}{r-1} + \Delta$ with $\Delta = \Delta(n) \to \infty$ as $n \to \infty$?

Theorem 16.4 immediately solves this problem. Indeed, the corresponding hyper-
graph $\mathcal{F}$ is $\binom{n-1}{r-1}$-uniform, r-regular, and the Max Pair-Degree is $r - 1$. So Theorem 16.4
gives the lower bound

$$\Delta \geq \frac{\log\left(\frac{r}{(r-1)(1+\log(D/D_2))}\right)}{2\sqrt{2\log(r/(r-1))}}\sqrt{\binom{n-1}{r-1}}. \tag{16.34}$$

In view of (16.32) the right-hand side of (16.34) is

$$\Delta \geq \frac{1}{\sqrt{32r^3}}\sqrt{\binom{n-1}{r-1}}.$$

This gives the following:

Corollary 1 of Theorem 16.4 *Székely's "Problem 3" holds with*

$$\Delta \geq \frac{1}{\sqrt{32r^3}}\sqrt{\binom{n-1}{r-1}}. \qquad \square$$

Next we apply Theorem 16.4 to a famous combinatorial structure: the Finite Projective Plane of order q (it exists if q is a prime-power). The Finite Projective Plane
of order q has $q^2 + q + 1$ points, $q^2 + q + 1$ Lines, each Line has $q + 1$ points, and
each point is contained in exactly $q + 1$ Lines. We apply Theorem 16.4: the Lines
are the winning sets, so $n = q + 1$; also $D = q + 1$ and $D_2 = 1$ (since any two Lines
intersect in one point only, so two points uniquely determine a Line); we obtain the
following:

Corollary 2 of Theorem 16.4 *Playing on a Finite Projective Plane of order q (the
winning sets are the Lines), Maker can always occupy at least*

$$\frac{q}{2} + \frac{\log(q/\log q)}{2\sqrt{2\log q}}\sqrt{q}$$

points of some Line. $\qquad \square$

How good are these lower bounds? To answer the question we apply the Corollary
of Theorem 16.1. This corollary gives the following upper bounds to the game-
theoretic discrepancy:

(i) $O(\sqrt{\log n})\sqrt{\binom{n-1}{r-1}}$ for Corollary 1 of Theorem 16.4, and
(ii) $O(\sqrt{q\log q})$ for Corollary 2 of Theorem 16.4.

This means that Corollary 2 is sharp (apart from a constant factor), and Corollary 1 is nearly sharp (apart from a factor of $\log n$).

4. A humiliating problem. How about Theorem 16.3? What upper bound can we prove? Unfortunately, the Corollary is useless: in the application the factor $\log |\mathcal{F}|$ becomes "unbounded," because the vertex-set of an n-regular graph G can be arbitrarily large. The sad truth is that we do not know a satisfying upper bound.

Very recently, Tibor Szabó told me the following upper bound result: if the underlying graph is n-regular, then Breaker can prevent Maker from achieving a degree greater than $3n/4$ (for simplicity assume that n is divisible by 4). Breaker applies the following simple pairing strategy. The "star hypergraph" has common degree 2 (since every edge belongs to exactly 2 stars determined by the 2 endpoints), so by the Bigamy version of the Marriage Theorem every star has an $n/2$-element Private Part such that the Private Parts are pairwise disjoint. Whenever Maker takes an edge from a Private Part, Breaker replies in the same Private Part. This way Breaker occupies exactly $n/4$ elements from every Private Part, preventing Maker from achieving a degree larger than $3n/4$.

Note that we don't even need the Marriage Theorem: if n is even, the n-regular graph G contains an Euler trail ("the first theorem in Graph Theory"), implying an orientation of the edges such that every vertex has $n/2$ in-edges and $n/2$ out-edges. The set of (say) out-edges in each vertex define the required Private Parts.

Unfortunately, this is the best that we currently know.

Open Problem 16.1 *Can the reader replace the upper bound $3n/4$ in the Degree Game above with some $c \cdot n$, where $c < 3/4$? Is it possible to get $c = \frac{1}{2} + o(1)$?*

In view of the Local Lemma (see Exercise 16.1 below) Open Problem 16.1 is even more frustrating.

Exercise 16.1 *Prove, by using the Local Lemma (see the end of Section 11), that it is possible to 2-color the edges of an n-regular graph such that in every vertex the difference between the number of red edges and the number of blue edges starting from that vertex is uniformly less than $2\sqrt{n \log n}$.*

17

Biased Discrepancy Games: when the extension from fair to biased works!

1. Why is the biased case more difficult? In the previous section we discussed the solution of 3 problems raised by L. Székely in his old paper (Székely [1981]). Here we discuss a related "biased" problem, which was raised very recently in 2005. It took 25 years to realize that the transition from the usual (1:1) play to the general $(p:q)$ biased play is not obvious!

In the Workshop of "Erdős Magic: Random Structures and Games," Bertinoro, Italy (April 23–29, 2005) M. Krivelevich ended his lecture by asking the following innocent-looking question. The problem is basically an "underdog" version of Erdős's Degree Game from the previous section.

Problem 17.1 *Two players, called Maker and Breaker, are playing the following biased game on K_n (complete graph on n vertices). It is a (1:2) play: in each turn Maker takes 1 new edge of K_n and colors it red, but Breaker takes 2 new edges per turn and colors them blue. Can Maker guarantee that, at the end of the play, each degree of the red graph is $\approx n/3$? More precisely, can Maker force that every red degree falls into the short interval $\frac{n}{3}(1-\varepsilon) <degree< \frac{n}{3}(1+\varepsilon)$ with $\varepsilon = \varepsilon(n) \to 0$ as $n \to \infty$?*

Note that Krivelevich and his co-authors (see Alon, Krivelevich, Spencer, and Szabó [2005]) could solve the analogous problem for the fair (1:1) play, where both Maker and Breaker take 1 edge per move. In the (1:1) play Maker can force that, at the end, every degree in his graph is $\approx \frac{n}{2}$, in fact equals $\frac{n}{2} + O(\sqrt{n \log n})$. Switching from the fair (1:1) play to the biased (1:2) play, where Maker is the "underdog," led to some unexpected technical difficulties. This is how Problem 17.1 was raised.

Our main objective in this section is to solve Problem 17.1, and to extend the solution from the (1:2) play to an arbitrary biased play. First we have to understand the "unexpected technical difficulties" caused by the unfair play. The fair (1:1) play is easy, and it was covered by Theorem 16.1; for convenience it is recalled here.

Theorem 16.1 *Let $\mathcal{F}$ be an arbitrary finite hypergraph, and let ε with $0 < \varepsilon < 1$ be an arbitrary real number. There are 2 players, Balancer and Unbalancer, who play the (1:1) game on $\mathcal{F}$: they alternate, and each player takes 1 new point per move. Unbalancer's goal is to achieve a majority: he wins if he owns $\geq \frac{1+\varepsilon}{2}$ part of some $A \in \mathcal{F}$; otherwise Balancer wins. Here is a Balancer's win criterion: if*

$$\sum_{A \in \mathcal{F}} \left((1+\varepsilon)^{1+\varepsilon} (1-\varepsilon)^{1-\varepsilon} \right)^{-|A|/2} < 1, \tag{17.1}$$

then Balancer, as the first player, has a winning strategy.

For later purposes it is necessary to briefly recall the proof. Assume that we are in the middle of a play: unbalancer already occupied $u_1, u_2, \ldots, u_t$ and Balancer occupied $b_1, b_2, \ldots, b_t$; t is the time parameter. Write $U(i) = \{u_1, u_2, \ldots, u_t\}$, $B(i) = \{b_1, b_2, \ldots, b_t\}$, and consider the "one-sided" potential function

$$P_t = \sum_{A \in \mathcal{F}} (1+\varepsilon)^{|A \cap U(t)| - \frac{1+\varepsilon}{2}|A|} (1-\varepsilon)^{|A \cap B(t)| - \frac{1-\varepsilon}{2}|A|}, \tag{17.2}$$

which is very sensitive ("exponentially sensitive") to Unbalancer's lead. The core idea, which is taken from the Erdős–Selfridge proof, is that Balancer can force the monotone *decreasing* property

$$P_0 \geq P_1 \geq P_2 \geq \cdots \geq P_{end},$$

so $P_{start} = P_0 \geq P_{end}$.

If Unbalancer wins, then by (17.2) we have $P_{end} \geq 1$; on the other hand, by hypothesis (17.1), $P_{start} = P_0 < 1$, which together imply Balancer's win.

If Balancer works with the "symmetric" potential function

$$S_t = \sum_{A \in \mathcal{F}} \Bigg((1+\varepsilon)^{|A \cap U(t)| - \frac{1+\varepsilon}{2}|A|} (1-\varepsilon)^{|A \cap B(t)| - \frac{1-\varepsilon}{2}|A|} +$$

$$+ (1+\varepsilon)^{|A \cap B(t)| - \frac{1+\varepsilon}{2}|A|} (1-\varepsilon)^{|A \cap U(t)| - \frac{1-\varepsilon}{2}|A|} \Bigg) \tag{17.3}$$

instead of the "one-sided" P_t in (17.2), then the same proof gives the following refinement of Theorem 16.1.

Theorem 17.1 *Let $\mathcal{F}$ be an arbitrary finite hypergraph, and let ε with $0 < \varepsilon < 1$ be an arbitrary real number. Balancer and Unbalancer alternate, each player takes 1 new point per move. If*

$$\sum_{A \in \mathcal{F}} \left((1+\varepsilon)^{1+\varepsilon} (1-\varepsilon)^{1-\varepsilon} \right)^{-|A|/2} < \frac{1}{2}, \tag{17.4}$$

then Balancer, as the first player, can force that, at the end of the play, for every $A \in \mathcal{F}$, Unbalancer's part in A is between $\frac{1-\varepsilon}{2}|A|$ and $\frac{1+\varepsilon}{2}|A|$.

Let us apply Theorem 17.1 to the star hypergraph of K_n: the hyperedges are the n stars (each star has $n-1$ edges), so $\mathcal{F}$ is an $(n-1)$-uniform hypergraph with $|\mathcal{F}| = n$. By choosing

$$\varepsilon = \sqrt{\frac{c\log n}{n}}$$

with some unspecified (yet) constant $c > 0$, criterion (17.4), applied to the star hypergraph of K_n, gives

$$\sum_{A \in \mathcal{F}} \left((1+\varepsilon)^{1+\varepsilon}(1-\varepsilon)^{1-\varepsilon} \right)^{-|A|/2} = ne^{-((1+\varepsilon)\log(1+\varepsilon)+(1-\varepsilon)\log(1-\varepsilon))\frac{n-1}{2}}$$

$$= ne^{-((1+\varepsilon)(\varepsilon-\frac{\varepsilon^2}{2}\pm\cdots)+(1-\varepsilon)(-\varepsilon-\frac{\varepsilon^2}{2}-\cdots))\frac{n-1}{2}}$$

$$= ne^{-(\varepsilon^2+O(\varepsilon^3))\frac{n-1}{2}} = ne^{-\frac{c\log n}{2}(1+O(\varepsilon))} = n^{1-\frac{c}{2}(1+O(\varepsilon))},$$

which is less than $1/2$ if $c > 2$ and n is sufficiently large. Thus Theorem 17.1 applies with

$$\varepsilon = \sqrt{\frac{(2+o(1))\log n}{n}},$$

and yields the following solution of the fair (1:1) version of Problem 17.1: Maker (as Balancer) can force that, at the end of the play, every vertex in his red graph has

$$\left| \text{degree} - \frac{n}{2} \right| \le \frac{\varepsilon}{2}n = \sqrt{\frac{n\log n}{2+o(1)}}. \tag{17.5}$$

Note that the proof of Alon, Krivelevich, Spencer, and Szabó [2005] was somewhat different, but they also used an Erdős–Selfridge type argument.

Now we explain how to modify my proof of (17.5) ("(1:1) game") to solve Problem 17.1 ("(1:2) game").

Assume that we are in the middle of a play in the (1:2) game played on an arbitrary finite hypergraph $\mathcal{F}$: Balancer already occupied $B(t) = \{b_1, b_2, \ldots, b_t\}$ and Unbalancer occupied the twice as large set $U(t) = \{u_1^{(1)}, u_1^{(2)}, u_2^{(1)}, u_2^{(2)}, \ldots, u_t^{(1)}, u_t^{(2)}\}$ (t is the time). To compensate for Balancer's 1:2 handicap, it is natural to introduce the following "asymmetric" version of potential function (17.3)

$$V_t = \sum_{A \in \mathcal{F}} \left((1+\frac{\varepsilon}{3})^{|A \cap U(t)| - \frac{2+\varepsilon}{3}|A|} (1-\varepsilon)^{|A \cap B(t)| - \frac{1-\varepsilon}{2}|A|} + \right.$$

$$\left. + (1+\varepsilon)^{|A \cap B(t)| - \frac{1+\varepsilon}{2}|A|} (1-\frac{\varepsilon}{2})^{|A \cap U(t)| - \frac{2-\varepsilon}{3}|A|} \right). \tag{17.6}$$

Why did we call potential (17.6) "natural"? Well, because both (17.3) and (17.6) are motivated by the so-called "Probabilistic Method." Indeed, assume that the points of hypergraph $\mathcal{F}$ are colored blue and yellow (blue for Balancer and yellow for Unbalancer) randomly: Pr[a given point is blue]$= p$ and Pr[a given point is yellow$]= q = 1 - p$; the points are colored independently of each other. Potentials (17.2) and (17.6) correspond to $p = 1/2$ and $p = 1/3$, respectively. Let $A \in \mathcal{F}$ be an arbitrary hyperedge; with $|A| = n$ we have

$$\text{Pr}\left[\text{at least}(1+\varepsilon)p \text{ part of set A is blue}\right] = \sum_{b:\ b \geq (1+\varepsilon)pn} \binom{n}{b} p^b (1-p)^{n-b}$$

$$\approx \binom{n}{(1+\varepsilon)pn} p^{(1+\varepsilon)pn}(1-p)^{n-(1+\varepsilon)pn}$$

$$\approx \frac{\left(\frac{n}{e}\right)^n \cdot p^{(1+\varepsilon)pn}(1-p)^{n-(1+\varepsilon)pn}}{\left(\frac{(1+\varepsilon)pn}{e}\right)^{(1+\varepsilon)pn} \left(\frac{n-(1+\varepsilon)pn}{e}\right)^{n-(1+\varepsilon)pn}}$$

$$= (1+\varepsilon)^{-(1+\varepsilon)|A|p}\left(1-\varepsilon\frac{p}{q}\right)^{-(1-\varepsilon p/q)|A|q},$$

$$(17.7)$$

and similarly

$$\text{Pr}\left[\text{at most}(1-\varepsilon)p \text{ part of set A is blue}\right] = \sum_{b:\ b \leq (1-\varepsilon)pn} \binom{n}{b} p^b (1-p)^{n-b}$$

$$\approx (1-\varepsilon)^{-(1-\varepsilon)|A|p}\left(1+\varepsilon\frac{p}{q}\right)^{-(1+\varepsilon p/q)|A|q}.$$

$$(17.8)$$

Note that in both (17.7) and (17.8) the approximation $\approx$ is defined in the weak sense as in the weak form of the Stirling's formula $M! \approx (M/e)^M$, where for simplicity we ignore the logarithmic factors.

Notice that (17.7)–(17.8) with $p = 1/2$ motivates potential (17.3), and (17.7)–(17.8) with $p = 1/3$ motivates potential (17.6).

It has just been explained why potential function (17.6) is a perfect candidate to work with, but here comes the "unexpected technical difficulty" mentioned at the beginning: unfortunately, potential (17.6) *cannot* work! Indeed, a hypothetical success of potential (17.6) would lead to the following analogue of Theorem 16.1 ("H(?)" stands for hypothetical).

Theorem H(?): *Let $\mathcal{F}$ be an arbitrary finite hypergraph, and let ε, with $0 < \varepsilon < 1$, be an arbitrary real number. Balancer and Unbalancer alternate: Balancer takes 1 new point per move and Unbalancer takes 2 new points per move. If*

$$\sum_{A \in \mathcal{F}} \left(\left(1+\frac{\varepsilon}{2}\right)^{\frac{2+\varepsilon}{3}}(1-\varepsilon)^{\frac{1-\varepsilon}{3}}\right)^{-|A|} = O(1), \qquad (17.9)$$

then Balancer, as the first player, can force that, at the end of the play, for every $A \in \mathcal{F}$, his part in A is strictly more than $\frac{1-\varepsilon}{3}|A|$.

Now here comes the bad news: Theorem H(?) is false! Indeed, if $\mathcal{F}$ is n-uniform, and $\varepsilon = 1$, then criterion (17.9) simplifies to

$$|\mathcal{F}| = O\left((3/2)^n\right). \tag{17.10}$$

Here is a counter-example: consider two disjoint copies of hypergraph $\mathcal{F}^*$.

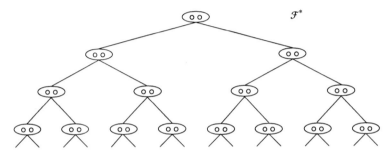

$\mathcal{F}^*$

Notice that hypergraph $\mathcal{F}^*$ arises from a binary tree of $n/2$ levels, where each vertex is replaced by 2 points. The winning sets of $\mathcal{F}^*$ are the $2^{n/2}$ full-length branches. Playing on 2 disjoint copies of hypergraph $\mathcal{F}^*$, Unbalancer, as the second player, can always occupy a full-length branch in 1 of the 2 copies of $\mathcal{F}^*$. This violates the statement of Theorem H(?) with $\varepsilon = 1$. On the other hand, in view of the trivial inequality

$$2|\mathcal{F}^*| = 2 \cdot 2^{n/2} < (3/2)^n \tag{17.11}$$

criterion (17.8) applies, which is a contradiction! (Note that inequality (17.11) is obvious from the numerical fact $\sqrt{2} = 1.414 < 3/2 = 1.5$.) This contradiction "kills" Theorem H(?).

2. Forcing the Decreasing Property. The failure of Theorem H(?) is the first indication of the kind of unpleasant technical difficulties that we face in biased games (for more examples, see Chapter VI). The collapse of Theorem H(?) forces us to abandon potential (17.6). We are looking for a "good potential" in the following more general form

$$W_t = \sum_{A \in \mathcal{F}} \left(\left(1 + \frac{\varepsilon}{2}\right)^{|A \cap U(t)| - \frac{2+\varepsilon}{3}|A|} (1 - \gamma)^{|A \cap B(t)| - \frac{1-\varepsilon}{3}|A|} + \right.$$
$$\left. + \left(1 - \frac{\varepsilon}{2}\right)^{|A \cap U(t)| - \frac{2-\varepsilon}{3}|A|} (1 + \delta)^{|A \cap B(t)| - \frac{1+\varepsilon}{3}|A|} \right), \tag{17.12}$$

where both parameters $\gamma = \gamma(\varepsilon)$ and $\delta = \delta(\varepsilon)$ will be specified later in terms of ε; note that both are between 0 and 1. "Good potential" precisely means the Decreasing Property; how can we achieve the Decreasing Property? Balancer's $(t+1)$st move is b_{t+1} and Unbalancer's $(t+1)$st move is the point pair $u_{t+1}^{(1)}, u_{t+1}^{(2)}$; how do they affect (17.12)? We clearly have

$$
\begin{aligned}
W_{t+1} = W_t &- \gamma \cdot W_t^+(b_{t+1}) + \frac{\varepsilon}{2} \cdot W_t^+(u_{t+1}^{(1)}) + \frac{\varepsilon}{2} \cdot W_t^+(u_{t+1}^{(2)}) \\
&- \frac{\varepsilon\gamma}{2} \cdot W_t^+(b_{t+1}, u_{t+1}^{(1)}) - \frac{\varepsilon\gamma}{2} \cdot W_t^+(b_{t+1}, u_{t+1}^{(2)}) \\
&+ \delta \cdot W_t^-(b_{t+1}) - \frac{\varepsilon}{2} \cdot W_t^-(u_{t+1}^{(1)}) - \frac{\varepsilon}{2} \cdot W_t^-(u_{t+1}^{(2)}) \\
&- \frac{\varepsilon\delta}{2} \cdot W_t^-(b_{t+1}, u_{t+1}^{(1)}) - \frac{\varepsilon\delta}{2} \cdot W_t^-(b_{t+1}, u_{t+1}^{(2)}) \\
&+ \frac{\varepsilon^2}{4} \cdot W_t^-(b_{t+1}, u_{t+1}^{(1)}) + \frac{\varepsilon^2\delta}{4} \cdot W_t^-(b_{t+1}, u_{t+1}^{(1)}, u_{t+1}^{(2)}),
\end{aligned} \tag{17.13}
$$

where for an arbitrary point set Z

$$
W_t^+(Z) = \sum_{A \in \mathcal{F}: \, Z \subset A} \left(1 + \frac{\varepsilon}{2}\right)^{|A \cap U(t)| - \frac{2+\varepsilon}{3}|A|} (1 - \gamma)^{|A \cap B(t)| - \frac{1-\varepsilon}{3}|A|} \tag{17.14}
$$

and

$$
W_t^-(Z) = \sum_{A \in \mathcal{F}: \, Z \subset A} (1 + \delta)^{|A \cap B(t)| - \frac{1+\varepsilon}{3}|A|} \left(1 - \frac{\varepsilon}{2}\right)^{|A \cap U(t)| - \frac{2-\varepsilon}{3}|A|}. \tag{17.15}
$$

The following inequality is trivial from definitions (17.14)–(17.15)

$$
W_t^\pm(b_{t+1}, u_{t+1}^{(1)}, u_{t+1}^{(2)}) \leq W_t^\pm(u_{t+1}^{(1)}, u_{t+1}^{(2)}) \leq \frac{W_t^\pm(u_{t+1}^{(1)}) + W_t^\pm(u_{t+1}^{(2)})}{2}.
$$

Applying this trivial inequality in (17.13), we have

$$
\begin{aligned}
W_{t+1} \leq W_t &- \gamma \cdot W_t^+(b_{t+1}) + \left(\frac{\varepsilon}{2} + \frac{\varepsilon^2}{8}\right) \cdot W_t^+(u_{t+1}^{(1)}) + \left(\frac{\varepsilon}{2} + \frac{\varepsilon^2}{8}\right) \cdot W_t^+(u_{t+1}^{(2)}) + \\
&+ \delta \cdot W_t^-(b_{t+1}) - \left(\frac{\varepsilon}{2} - \frac{\varepsilon^2}{8} - \frac{\varepsilon^2\delta}{8}\right) \cdot W_t^-(u_{t+1}^{(1)}) - \left(\frac{\varepsilon}{2} - \frac{\varepsilon^2}{8} - \frac{\varepsilon^2\delta}{8}\right) \cdot W_t^-(u_{t+1}^{(2)}).
\end{aligned} \tag{17.16}
$$

Inequality (17.16) motivates the following choice of parameters γ and δ: let

$$
\gamma = 2\left(\frac{\varepsilon}{2} + \frac{\varepsilon^2}{8}\right) = \varepsilon + \frac{\varepsilon^2}{4} \tag{17.17}
$$

and

$$
\delta = 2\left(\frac{\varepsilon}{2} - \frac{\varepsilon^2}{8} - \frac{\varepsilon^2\delta}{8}\right) = \varepsilon - \frac{\varepsilon^2}{4} - \frac{\varepsilon^2\delta}{4},
$$

or equivalently

$$
\delta = \frac{\varepsilon - \frac{\varepsilon^2}{4}}{1 + \frac{\varepsilon^2}{4}}. \tag{17.18}
$$

In view of (17.17)–(17.18), we can rewrite (17.14) in the following more transparent form

$$W_{t+1} \le W_t - g_t(b_{t+1}) + \frac{1}{2}g_t(u_{t+1}^{(1)}) + \frac{1}{2}g_t(u_{t+1}^{(2)}), \tag{17.19}$$

where for every unoccupied point z we define the function (see (17.14)–(17.15) and (17.17)–(17.18))

$$g_t(z) = \gamma \cdot W_t^+(z) - \delta \cdot W_t^-(z). \tag{17.20}$$

Since Balancer chooses his $(t+1)$st point b_{t+1} before Unbalancer's $(t+1)$st point pair $u_{t+1}^{(1)}$, $u_{t+1}^{(2)}$, Balancer can choose the "best" point as follows: Balancer chooses that unoccupied $b_{t+1} = z$ for which the function $g_t(z)$, defined in (17.20), attains its *maximum*. Then

$$g_t(b_{t+1}) - \frac{1}{2}g_t(u_{t+1}^{(1)}) - \frac{1}{2}g_t(u_{t+1}^{(2)}) \ge 0,$$

which, by (17.19), implies the desperately needed *decreasing* property: $W_{t+1} \le W_t$, assuming, of course, that Balancer chooses the "best" point.

Now we are ready to replace the false Theorem H(?) with a correct statement! The decreasing property gives $W_{end} \le W_{start}$, and trivially

$$W_{start} = W_{blank} = \sum_{A \in \mathcal{F}} \left(\left((1 + \frac{\varepsilon}{2})^{2+\varepsilon}(1-\gamma)^{1-\varepsilon} \right)^{-|A|/3} + \left((1+\delta)^{1+\varepsilon}(1-\frac{\varepsilon}{2})^{2-\varepsilon} \right)^{-|A|/3} \right). \tag{17.21}$$

On the other hand, if at the end of the play (let B_{end} denote Balancer's part) either $|A \cap B_{end}| \ge \frac{1+\varepsilon}{3}|A|$ or $|A \cap B_{end}| \le \frac{1-\varepsilon}{3}|A|$ holds for some set $A \in \mathcal{F}$, then $W_{end} \ge 1$, because in sum (17.12) there is a single term ≥ 1.

Therefore, if we assume that the sum in (17.21) is less than 1, then $W_{end} < 1$, proving the following result.

Theorem 17.2 *Let $\mathcal{F}$ be an arbitrary finite hypergraph, and let ε be an arbitrary real number with $0 < \varepsilon < 1$. Balancer and Unbalancer play the $(1:2)$ game: they alternate, Balancer takes 1 new point per move and Unbalancer takes 2 new points per move. If*

$$\sum_{A \in \mathcal{F}} \left(\left(\left(1 + \frac{\varepsilon}{2}\right)^{\frac{2+\varepsilon}{3}} \left(1 - \varepsilon - \frac{\varepsilon^2}{4}\right)^{\frac{1-\varepsilon}{3}} \right)^{-|A|} + \left(\left(1 - \frac{\varepsilon}{2}\right)^{\frac{2-\varepsilon}{3}} \left(1 + \frac{\varepsilon - \frac{\varepsilon^2}{4}}{1 + \frac{\varepsilon^2}{4}}\right)^{\frac{1+\varepsilon}{3}} \right)^{-|A|} \right) < 1, \tag{17.22}$$

then Balancer, as the first player, can force that, at the end of the play, for every $A \in \mathcal{F}$, his part in A is strictly between $\frac{1-\varepsilon}{3}|A|$ and $\frac{1+\varepsilon}{3}|A|$.

The proof argument requires $1 - \varepsilon - \frac{\varepsilon^2}{4} > 0$, so the range for ε is $0 < \varepsilon < 2(\sqrt{2} - 1) = .828427$.

Theorem 17.2 is an "ugly" result, but it is very useful: it immediately solves Problem 17.1 (mentioned at the beginning of the section). Indeed, we apply Theorem 17.2 to the "star hypergraph" of K_n with

$$\varepsilon = \sqrt{\frac{c \log n}{n}},$$

where $c > 0$ is an absolute constant (to be specified later). The "ugly" criterion (17.22) gives

$$\sum_{A \in \mathcal{F}} \left(\left((1 + \tfrac{\varepsilon}{2})^{\frac{2+\varepsilon}{3}} \left(1 - \varepsilon - \tfrac{\varepsilon^2}{4} \right)^{\frac{1-\varepsilon}{3}} \right)^{-|A|} + \left((1 - \tfrac{\varepsilon}{2})^{\frac{2-\varepsilon}{3}} \left(1 + \varepsilon - \tfrac{\varepsilon^2}{4} + O(\varepsilon^3) \right)^{\frac{1+\varepsilon}{3}} \right)^{-|A|} \right)$$

$$= n e^{-((2+\varepsilon) \log(1+\varepsilon/2) + (1-\varepsilon) \log(1-\varepsilon-\varepsilon^2/4)) \frac{n-1}{3}}$$

$$\qquad + n e^{-((2-\varepsilon) \log(1-\varepsilon/2) + (1+\varepsilon) \log(1+\varepsilon-\varepsilon^2/4+O(\varepsilon^3))) \frac{n-1}{3}}$$

$$= n e^{-((2+\varepsilon)(\varepsilon/2-\varepsilon^2/8+O(\varepsilon^3)) + (1-\varepsilon)(-\varepsilon-3\varepsilon^2/4+O(\varepsilon^3))) \frac{n-1}{3}}$$

$$\qquad + n e^{-((2-\varepsilon)(-\varepsilon/2-\varepsilon^2/8+O(\varepsilon^3)) + (1+\varepsilon)(\varepsilon-3\varepsilon^2/4+O(\varepsilon^3))) \frac{n-1}{3}}$$

$$= n e^{-(\varepsilon^2/2+O(\varepsilon^3)) \frac{n-1}{3}} + n e^{-(\varepsilon^2/2+O(\varepsilon^3)) \frac{n-1}{3}}$$

$$= n^{1 - \frac{c}{6}(1 + O(\varepsilon))},$$

which is less than 1 if $c > 6$ and n is sufficiently large. So Theorem 17.2 applies, with

$$\varepsilon = \sqrt{\frac{(6 + o(1)) \log n}{n}}, \qquad (17.23)$$

and gives the following: in Problem 17.1 underdog Maker (as Balancer) can force that, at the end of the (1:2) play, every vertex in his red graph has

$$\left| \text{degree} - \frac{n}{3} \right| \le \frac{\varepsilon}{3} n = \sqrt{\left(\frac{2}{3} + o(1) \right) n \log n}; \qquad (17.24)$$

that is, we get the same strikingly small "error term" $O(\sqrt{n \log n})$ as in the fair (1:1) case (17.5).

Theorem 17.3 *Problem 17.1 (formulated at the beginning of this section) holds with*

$$\varepsilon = \varepsilon(n) = \sqrt{\frac{(6 + o(1)) \log n}{n}} \to 0 \text{ as } n \to \infty.$$

We will see another application of Theorem 17.2 in Section 33.

In the proof of Theorem 17.2, it was explained why the simpler potential (17.6) cannot work. Nevertheless, if ε is "small," as in our application (17.23) in the proof of Theorem 17.3, then by (17.17)–(17.18)

$$\gamma = \varepsilon + \frac{\varepsilon^2}{4} \approx \varepsilon \text{ and } \delta = \frac{\varepsilon - \frac{\varepsilon^2}{4}}{1 + \frac{\varepsilon^2}{4}} \approx \varepsilon,$$

so the successful potential (17.12) is almost the same as the failure (17.6). This means, the "probabilistic intuition" turned out to be more or less correct after all.

3. Random play. The probabilistic analogue of Problem 17.1 is a routine question in Probability Theory. An easy calculation gives that, in a Random Graph $\mathbf{R}(K_n; p)$ with edge probability p (let $q = 1 - p$)

$$\Pr\left[\text{every degree is between } (n-1)p \pm (\lambda + o(1))\sqrt{npq}\right] = 1 - o(1),$$

where $\lambda = \lambda(n)$ comes from the equation $e^{-\lambda^2/2} = 1/n$, i.e. $\lambda = \sqrt{2\log n}$. Of course, the function $e^{-\lambda^2/2}$ comes from the Central Limit Theorem. With $p = 1/2$, we get the bounds

$$\frac{n-1}{2} \pm \sqrt{(2 + o(1))\log n}\sqrt{n \cdot \frac{1}{2} \cdot \frac{1}{2}} = \frac{n}{2} \pm \sqrt{\frac{n\log n}{2 + o(1)}}, \tag{17.25}$$

and

$$\frac{n-1}{3} \pm \sqrt{(2 + o(1))\log n}\sqrt{n \cdot \frac{1}{3} \cdot \frac{2}{3}} = \frac{n}{2} \pm \left(\frac{2}{3} + o(1)\right)\sqrt{n\log n}. \tag{17.26}$$

Notice that (17.25) is exactly the game-theoretic upper bound (17.5), but (17.26) is weaker than (17.24) – indeed, $2/3 < \sqrt{2/3}$ – due to the fact that potential (17.6) didn't work, and we had to switch to the slightly different (17.12) with a "second-order" difference.

In spite of this minor "second-order" difference, we still refer to our potential technique as a "fake probabilistic method."

4. Non-uniform case. Let us return to Problem 17.1: it was solved by applying Theorem 17.2 to the star hypergraph of K_n. The star hypergraph is a particular $(n-1)$-uniform hypergraph $\mathcal{F} = \mathcal{F}(K_n)$ with $|\mathcal{F}| = n$; what happens in general, for an arbitrary hypergraph $\mathcal{F}$ with $|\mathcal{F}| = n$ and $\max_{A \in \mathcal{F}}|A| \leq n$? Well, in the non-uniform case we have to be careful: we have to modify the proofs of Theorem 17.1 and Theorem 17.2 in such a way that parameter ε is not fixed anymore, the value ε_i does depend on the size $|A_i|$ as $A_i \in \mathcal{F}$.

First we discuss the non-uniform analogue of Theorem 17.1, i.e. we study the fair (1:1) case. Let $\mathcal{F} = \{A_1, A_2, \ldots, A_N\}$, and *asumme* that $n \geq |A_i| \geq (\log n)^2$ holds for every $A_i \in \mathcal{F}$. The only novelty is to work with

$$\varepsilon_i = \sqrt{(2 + o(1))\frac{\log N}{|A_i|}}, \quad 1 \leq i \leq N$$

instead of a fixed ε; by repeating the proof of Theorem 17.1 with these ε_is, we obtain:

Theorem 17.4 *Let $\mathcal{F}$ be a hypergraph with $|\mathcal{F}| = N$ and $n \geq |A| \geq (\log n)^2$ for every $A_i \in \mathcal{F}$. Balancer and Unbalancer play the (1:1) game: they alternate, and each player takes 1 new point per move. Then Balancer, as the first player, can force that, at the end of the play, his part in every $A_i \in \mathcal{F}$ is between*

$$\frac{1 \pm \varepsilon_i}{2} |A_i| = \frac{|A_i|}{2} \pm \sqrt{\left(\frac{1}{2} + o(1)\right) |A_i| \log N}.$$

In the special case $|\mathcal{F}| = n$ and $\max_{A \in \mathcal{F}} |A| \leq n$, Theorem 17.4 gives the "error term" $\sqrt{n \log n / 2}$. Note that Székely [1981] proved a much weaker "error term" $O(n^{2/3} (\log n)^{1/3})$ – he was not careful.

The "error term" $O(\sqrt{n \log n})$ is *best possible*. To prove this we will give a hypergraph $\mathcal{F}$ with $|\mathcal{F}| = n$ and $\max_{A \in \mathcal{F}} |A| \leq n$ such that Unbalancer can force a lead of size $\sqrt{n \log n}$. We need a slight generalization of the "ugly" Theorem 16.4 from the previous section. In Theorem 16.4 we assumed that the hypergraph is D-regular (i.e. every point has degree D), because this condition automatically holds in the application ("Székely's Problem 3"). The same lower bound works under the more general condition that every degree is $\geq D$ (instead of equality).

Theorem 16.4' *Let $\mathcal{F}$ be an arbitrary n-uniform hypergraph, let $D_2 = D_2(\mathcal{F})$ denote the Max Pair-Degree. and let D denote the **minimum** degree of $\mathcal{F}$. Playing the usual (1:1) game on $\mathcal{F}$, at the end of the play Maker can always occupy at least*

$$\frac{n}{2} + \frac{\log\left(\frac{D}{D_2(1 + \log(D/D_2))}\right)}{2\sqrt{2 \log(D/D_2)}} \sqrt{n}$$

points from some $A \in \mathcal{F}$.

Assume we have an n-uniform hypergraph $\mathcal{F}_0$ with $|\mathcal{F}_0| = n$, minimum degree $D = n^{1/2 + o(1)}$, and Max Pair-Degree $D_2 = n^{o(1)}$. An application of Theorem 16.4' to this hypergraph gives an Unbalancer's lead $\geq \frac{1}{4}\sqrt{n \log n}$, proving the sharpness of the "error term" $O(\sqrt{n \log n})$.

The existence of the desired hypergraph $\mathcal{F}_0$ can be proved by a routine application of the "Probabilistic Method." Indeed, let X be a set of size $n^{3/2}$, and let A_1 be a "random subset" of X where the "inclusion" probability is $p = n^{-1/2}$ (i.e. every $x \in X$ is included in A_1 with probability p, and these decisions are independent). Repeating this random construction independently n times, we get a hypergraph $\mathcal{F}_0 = \{A_1, A_2, \ldots, A_n\}$, where every hyperedge $A_i \approx |X| p = n$, every point $x \in X$ has $\mathcal{F}_0$-degree $\approx np = n^{1/2}$, and every point pair x_1, x_2 $(x_1 \neq x_2)$ has pair-degree $\approx np^2 = 1$. It is very easy to make this heuristic argument precise; the details are left to the reader.

5. General Biased Case. Theorems 17.2–17.3 were about a particular underdog play: the (1:2) play. The proof can be extended to the most general biased case as follows. Consider an arbitrary $(p:q)$ Balancer–Unbalancer play on an arbitrary finite n-uniform hypergraph $\mathcal{F}$. Assume that we are in the middle of a play, "t" is the "time," Balancer owns the point set

$$B(t) = \{b_1^{(1)}, \ldots, b_1^{(p)}, b_2^{(1)}, \ldots, b_2^{(p)}, \ldots, b_t^{(1)}, \ldots, b_t^{(p)}\}$$

and Unbalancer owns the point set

$$U(t) = \{u_1^{(1)}, \ldots, u_1^{(q)}, u_2^{(1)}, \ldots, u_2^{(q)}, \ldots, u_t^{(1)}, \ldots, u_t^{(q)}\}.$$

We are going to work with the following potential function

$$W_t = \sum_{A \in \mathcal{F}} \left((1+\alpha)^{|A \cap U(t)| - \frac{q+\varepsilon}{p+q}|A|} (1-\gamma)^{|A \cap B(t)| - \frac{p-\varepsilon}{p+q}|A|} \right.$$
$$\left. + (1-\beta)^{|A \cap U(t)| - \frac{q-\varepsilon}{p+q}|A|} (1+\delta)^{|A \cap B(t)| - \frac{p+\varepsilon}{p+q}|A|} \right),$$

with 4 unspecified (yet) parameters $\alpha = \alpha(\varepsilon)$, $\beta = \beta(\varepsilon)$, $\gamma = \gamma(\varepsilon)$, $\delta = \delta(\varepsilon)$ (they are all between 0 and 1).

The effect of the $(t+1)$st moves $\{b_{t+1}^{(1)}, \ldots, b_{t+1}^{(p)}\}$ and $\{u_{t+1}^{(1)}, \ldots, u_{t+1}^{(q)}\}$ in W_t is the following

$$W_{t+1} = W_t + \alpha \sum_{j=1}^{q} W_t^+(u_{t+1}^{(j)}) + \alpha^2 \sum_{1 \le j_1 < j_2 \le q} W_t^+(u_{t+1}^{(j_1)}, u_{t+1}^{(j_2)})$$

$$- \gamma \sum_{i=1}^{p} W_t^+(b_{t+1}^{(i)}) + \gamma^2 \sum_{1 \le i_1 < i_2 \le p} W_t^+(u_{t+1}^{(i_1)}, u_{t+1}^{(i_2)})$$

$$- \alpha\gamma \sum_{j=1}^{q} \sum_{i=1}^{p} W_t^+(u_{t+1}^{(j)}, b_{t+1}^{(i)}) \pm \cdots$$

$$- \beta \sum_{j=1}^{q} W_t^-(u_{t+1}^{(j)}) + \beta^2 \sum_{1 \le j_1 < j_2 \le q} W_t^-(u_{t+1}^{(j_1)}, u_{t+1}^{(j_2)})$$

$$+ \delta \sum_{i=1}^{p} W_t^-(b_{t+1}^{(i)}) + \delta^2 \sum_{1 \le i_1 < i_2 \le p} W_t^-(u_{t+1}^{(i_1)}, u_{t+1}^{(i_2)})$$

$$- \beta\delta \sum_{j=1}^{q} \sum_{i=1}^{p} W_t^-(u_{t+1}^{(j)}, b_{t+1}^{(i)}) \pm \cdots, \qquad (17.27)$$

where for an arbitrary point set Z

$$W_t^+(Z) = \sum_{A \in \mathcal{F}:\ Z \subset A} (1+\alpha)^{|A \cap U(t)| - \frac{q+\varepsilon}{p+q}|A|} (1-\gamma)^{|A \cap B(t)| - \frac{p-\varepsilon}{p+q}|A|}$$

and

$$W_t^-(Z) = \sum_{A \in \mathcal{F}:\; Z \subset A} (1-\beta)^{|A \cap U(t)| - \frac{q-\varepsilon}{p+q}|A|}(1+\delta)^{|A \cap B(t)| - \frac{p+\varepsilon}{p+q}|A|}.$$

We delete all the negative terms on the right-hand side of (17.27) except the 2 linear terms with "coefficients" "$-\gamma$" and "$-\beta$"

$$
\begin{aligned}
W_{t+1} \leq W_t &+ \alpha \sum_{j=1}^{q} W_t^+(u_{t+1}^{(j)}) + \alpha^2 \sum_{1 \leq j_1 < j_2 \leq q} W_t^+(u_{t+1}^{(j_1)}, u_{t+1}^{(j_2)}) \\
&- \gamma \sum_{i=1}^{p} W_t^+(b_{t+1}^{(i)}) + \gamma^2 \sum_{1 \leq i_1 < i_2 \leq p} W_t^+(u_{t+1}^{(i_1)}, u_{t+1}^{(i_2)}) + \cdots \\
&- \beta \sum_{j=1}^{q} W_t^-(u_{t+1}^{(j)}) + \beta^2 \sum_{1 \leq j_1 < j_2 \leq q} W_t^-(u_{t+1}^{(j_1)}, u_{t+1}^{(j_2)}) \\
&+ \delta \sum_{i=1}^{p} W_t^-(b_{t+1}^{(i)}) + \delta^2 \sum_{1 \leq i_1 < i_2 \leq p} W_t^-(u_{t+1}^{(i_1)}, u_{t+1}^{(i_2)}) + \cdots .
\end{aligned}
\tag{17.28}
$$

The following inequalities are trivial from the definition

$$W_t^\pm(z_1, z_2) \leq \frac{W_t^\pm(z_1) + W_t^\pm(z_2)}{2},$$

$$W_t^\pm(z_1, z_2, z_3) \leq \frac{W_t^\pm(z_1) + W_t^\pm(z_2) + W_t^\pm(z_3)}{3},$$

and so on. Applying these trivial inequalities in (17.28), we obtain

$$
\begin{aligned}
W_{t+1} \leq W_t &+ \sum_{j=1}^{q} W_t^+(u_{t+1}^{(j)})\left(\alpha + \frac{1}{2}(q-1)\alpha^2 + O(q^2)\alpha^3 + O(p^2)\alpha\gamma^2\right) \\
&- \sum_{i=1}^{p} W_t^+(b_{t+1}^{(i)})\left(\gamma - \frac{1}{2}(p-1)\gamma^2 + O(p^3)\gamma^4 + O(pq)\alpha\gamma^2\right) \\
&- \sum_{j=1}^{q} W_t^+(u_{t+1}^{(j)})\left(\beta - \frac{1}{2}(q-1)\beta^2 + O(q^3)\beta^4 + O(pq)\beta^2\delta\right) \\
&+ \sum_{i=1}^{p} W_t^+(b_{t+1}^{(i)})\left(\delta + \frac{1}{2}(p-1)\delta^2 + O(p^2)\delta^3 + O(q^2)\beta^2\delta\right).
\end{aligned}
\tag{17.29}
$$

We require the 2 equalities

$$q\left(\alpha + \frac{1}{2}(q-1)\alpha^2 + O(q^2)\alpha^3 + O(p^2)\alpha\gamma^2\right) = \sigma$$

$$= p\left(\gamma - \frac{1}{2}(p-1)\gamma^2 + O(p^3)\gamma^4 + O(pq)\alpha\gamma^2\right)
\tag{17.30}$$

and

$$q\left(\beta - \frac{1}{2}(q-1)\beta^2 + O(q^3)\beta^4 + O(pq)\beta^2\delta\right) = \tau$$

$$= p\left(\delta + \frac{1}{2}(p-1)\delta^2 + O(p^2)\delta^3 + O(q^2)\beta^2\delta\right); \tag{17.31}$$

notice that the common value in (17.30) is denoted by σ and the common value in (17.31) is denoted by τ.

Under the conditions (17.30)–(17.31), we can rewrite (17.29) in a more transparent form

$$W_{t+1} \le W_t + \frac{1}{q}\sum_{j=1}^{q} f(u_{t+1}^{(j)}) - \frac{1}{p}\sum_{i=1}^{p} f(b_{t+1}^{(i)}), \tag{17.32}$$

where

$$f(z) = \sigma \cdot W^+(z) - \tau \cdot W^-(z).$$

If Balancer is the first player, then his $(t+1)$st move $\{b_{t+1}^{(1)}, \ldots, b_{t+1}^{(p)}\}$ comes before Unbalancer's $(t+1)$st move $\{u_{t+1}^{(1)}, \ldots, u_{t+1}^{(q)}\}$. If Balancer first chooses the smallest value of $f(z)$, followed by the second smallest, and then the third smallest, and so on, then

$$\max_i f(b_{t+1}^{(i)}) \le \min_j f(u_{t+1}^{(j)}),$$

implying

$$\frac{1}{p}\sum_{i=1}^{p} f(b_{t+1}^{(i)}) \le \frac{1}{q}\sum_{j=1}^{q} f(u_{t+1}^{(j)}),$$

so by (17.32) we obtain the critical decreasing property $W_{t+1} \le W_t$.

The decreasing property gives $W_{end} \le W_{start} = W_{blank}$, and trivially

$$W_{blank} = |\mathcal{F}|\left(\left((1+\alpha)^{q+\varepsilon}(1-\gamma)^{p-\varepsilon}\right)^{-n/(p+q)} + \left((1-\beta)^{q-\varepsilon}(1+\delta)^{p+\varepsilon}\right)^{-n/(p+q)}\right). \tag{17.33}$$

On the other hand, if at the end of the play (let B_{end} denote Balancer's part) either $|A \cap B_{end}| \ge \frac{p+\varepsilon}{p+q}n$ or $|A \cap B_{end}| \le \frac{p-\varepsilon}{p+q}n$ holds for some set $A \in \mathcal{F}$, then $W_{end} \ge 1$ (because there is a term ≥ 1).

Therefore, if we assume that the sum in (17.33) is less than 1, then $W_{end} < 1$, proving that Balancer can guarantee

$$\frac{p+\varepsilon}{p+q}n \ge |A \cap B_{end}| \ge \frac{p-\varepsilon}{p+q}n \quad \text{for all} \quad A \in \mathcal{F}.$$

It remains to specify parameters $\alpha = \alpha(\varepsilon)$, $\beta = \beta(\varepsilon)$, $\gamma = \gamma(\varepsilon)$, $\delta = \delta(\varepsilon)$. The goal is to optimize either base in (17.33)

$$(1+\alpha)^{q+\varepsilon}(1-\gamma)^{p-\varepsilon} \quad \text{and} \quad (1-\beta)^{q-\varepsilon}(1+\delta)^{p+\varepsilon}. \tag{17.34}$$

First we go back to requirements (17.30)–(17.31); an easy calculation shows that (17.30) is equivalent to

$$\alpha = \frac{1}{q}\left(\sigma - \frac{q-1}{2q}\sigma^2 + O(\sigma^3)\right),\tag{17.35}$$

$$\gamma = \frac{1}{p}\left(\sigma + \frac{p-1}{2p}\sigma^2 + O(\sigma^3)\right),\tag{17.36}$$

and (17.31) is equivalent to

$$\beta = \frac{1}{q}\left(\tau + \frac{q-1}{2q}\tau^2 + O(\tau^3)\right),\tag{17.37}$$

$$\delta = \frac{1}{p}\left(\tau - \frac{p-1}{2p}\tau^2 + O(\tau^3)\right).\tag{17.38}$$

Taking the logarithm of the first terms in (17.34), and applying (17.35)–(17.36) we have

$$(q+\varepsilon)\log(1+\alpha) + (p-\varepsilon)\log(1-\gamma)$$

$$= (q+\varepsilon)\log(\alpha - \frac{1}{2}\alpha^2 + O(\alpha^3)) - (p-\varepsilon)\log(\gamma + \frac{1}{2}\gamma^2 + O(\gamma^3))$$

$$= (q+\varepsilon)\left(\frac{1}{q}\left(\sigma - \frac{q-1}{2q}\sigma^2 + O(\sigma^3)\right) - \frac{1}{2}\left(\frac{\sigma}{q}\right)^2 + O(\alpha^3)\right)$$

$$- (p-\varepsilon)\left(\frac{1}{p}\left(\sigma + \frac{p-1}{2p}\sigma^2 + O(\sigma^3)\right) + \frac{1}{2}\left(\frac{\sigma}{p}\right)^2 + O(\gamma^3)\right)$$

$$= -\sigma^2 + \sigma\varepsilon\left(\frac{p+q}{pq}\right) + O(\sigma^3).\tag{17.39}$$

The maximum of the quadratic polynomial

$$g(\sigma) = -\sigma^2 + \sigma\varepsilon\left(\frac{p+q}{pq}\right) \quad \text{is attained at} \quad \sigma = \frac{\varepsilon(p+q)}{2pq},$$

and the maximum value itself is

$$\sigma^2 = \frac{\varepsilon^2(p+q)^2}{4p^2q^2}.$$

Similar calculations work for the second term in (17.34).

Returning to (17.33) we obtain the following criterion. If

$$\exp\left(\frac{\varepsilon^2(p+q)}{4p^2q^2}n + O(\frac{\varepsilon^3(p+q)^2}{p^3q^3})n\right) > 2|\mathcal{F}|,\tag{17.40}$$

then Balancer can guarantee that, at the end of the play his part in every $A \in \mathcal{F}$ is between

$$\frac{p+\varepsilon}{p+q}n \quad \text{and} \quad \frac{p-\varepsilon}{p+q}n. \tag{17.41}$$

This implies:

Theorem 17.5 *Let $\mathcal{F}$ be an arbitrary finite n-uniform hypergraph. Balancer and Unbalancer play the (p:q) game: they alternate, Balancer takes p new points and Unbalancer takes q new points per move. Then Balancer, as the first player, can force that, at the end of the play, for every $A \in \mathcal{F}$, his part in A is strictly between $\frac{p+\varepsilon}{p+q}n$ and $\frac{p-\varepsilon}{p+q}n$, where*

$$\varepsilon = \left(1 + O\left(pq\sqrt{\frac{\log|\mathcal{F}|}{(p+q)n}}\right)\right) 2pq\sqrt{\frac{\log|\mathcal{F}|}{(p+q)\cdot n}}.$$

Applying Theorem 17.5 to the Degree Game on K_N (where the "stars" define an n-uniform hypergraph with $n = N - 1$ and $|\mathcal{F}| = N$), we can immediately extend Theorem 17.3 for the general case of $(p : q)$ biased game (with arbitrary $p \geq 1$ and $q \geq 1$).

6. A case of unexpected asymmetry: 1 direction is much easier than the other!

In Theorem 17.5 Balancer can force that, for every winning set $A \in \mathcal{F}$, at the end of the play his part in A is strictly between $\frac{p+\varepsilon}{p+q}|A|$ and $\frac{p-\varepsilon}{p+q}|A|$. If Balancer just wants the *lower bound* $\geq \frac{p-\varepsilon}{p+q}|A|$, then there is a much simpler proof. We discuss this simpler argument at the end of Section 20 – see (20.33) in the Remark after the proof of Theorem 20.1.

The simpler argument does *not* work for the upper bound.

18

A simple illustration of "randomness" (I)

The subject of the book is to discuss the surprising "randomness" associated with some classes of games of complete information. The relation is indirect/motivational; this is why we refer to our proof technique as a "fake probabilistic method." The proofs of the main results of the book – the exact solutions – are long and difficult; it is easy to get lost in the technical details. It is very beneficial, therefore, to see some other illustrations of "randomness in games," which have simple (or at least much simpler) proofs.

1. Picker–Chooser Row Imbalance Games. The board is an $n \times k$ chess-board (k rows, each of length n), a standard board, but the way the two players divide the board into two parts is not standard. It is not the Maker–Breaker play; the two players divide the board into two halves in the "Picker–Chooser way," meaning that in each turn Picker picks two previously unselected little squares on the board, Chooser chooses one of them, and the other square goes back to Picker. Chooser marks his squares with letter "C" and Picker marks his squares with letter "P."

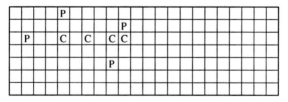

$k = 7$ rows
$n = 22$ columns

This is a Row Imbalance Game: one player wants a large "discrepancy" in a row. In other words, one player, called **Unbalancer**, wants a row where the number of Ps differs from the number of Cs by a large amount (we take absolute value!). For example, on the picture above, in the 3rd row we have a discrepancy of 3. The opponent of Unbalancer is called **Balancer**: his goal is to minimize the maximum row discrepancy.

260

There are two versions: (1) Picker is Balancer, (2) Picker is Unbalancer. Observe that the first version (1) is trivial. Indeed, if Picker proceeds "row-wise" (i.e. first taking pairs from the first row, and when the first row is exhausted, taking pairs from the second row, and when the second row is exhausted, taking pairs from the third row, and so on), then, at the end of the play, the maximum row discrepancy is 0 or 1 depending on the parity of n (Chooser's strategy is irrelevant).

The second version (2) is not trivial; we call it the Picker–Chooser Row Imbalance Game on an $n \times k$ board (k rows), where Picker is Unbalancer. Let $D(n, k)$ denote the maximum row discrepancy that Picker can achieve against a perfect opponent. How large is $D(n, k)$?

First we discuss the case when k is much smaller than n, like $k = 10$ and $n = 1000$. We show a heuristic argument, which seems to prove that Picker can achieve a row-discrepancy $\geq \lfloor k/2 \rfloor$ if k is much smaller than n. Here is the argument: if during the whole play every row has a row-discrepancy $< \lfloor k/2 \rfloor$, i.e. if every *row-sum* (meaning: the number of Ps minus the number of Cs) is between $-\lfloor k/2 \rfloor + 1$ and $\lfloor k/2 \rfloor - 1$, then by the Pigeonhole Principle there is always a pair of rows with the same row-sum: $r_{i_1}(j) = r_{i_2}(j)$. In other words, after the jth turn, the i_1st and the i_2nd rows ($i_1 \neq i_2$) have the same row-sum. Then, *assuming* the i_1st and the i_2nd rows both still have unmarked squares and Picker picks 1 unmarked square from each, the "game-theoretic variance"

$$V_j = \sum_{i=1}^{k} (r_i(j))^2 \tag{18.1}$$

increases by two

$$V_{j+1} - V_j = (r_{i_1}(j) \pm 1)^2 + (r_{i_2}(j) \mp 1)^2 - (r_{i_1}(j))^2 - (r_{i_2}(j))^2 = 2.$$

This way $V_{end} = V_{nk/2} = nk$, implying that at the end there exists a row with row-sum r, where $r^2 \geq n$, i.e. $|r| \geq \sqrt{n} \geq k/2$ if $n \geq k^2/4$.

This argument repeatedly used the assumption that "there are two rows with equal row-sums such that both rows have unmarked squares." How can we guarantee this assumption? A simple way to save the argument is to replace $k/2$ by $k/4$: we show that Picker can always achieve a row-discrepancy $\geq \lfloor k/4 \rfloor$ if k is much smaller than n. Indeed, if during the whole play every *row-sum* (the number of Ps minus the number of Cs) is between $-\lfloor k/4 \rfloor + 1$ and $\lfloor k/4 \rfloor - 1$, then by the Pigeonhole Principle there are always less than $k/2$ rows with multiplicity-one row-sum value. Assume there are l distinct row-sum values, and let $m_1 \leq m_2 \leq \cdots \leq m_l$ denote the multiplicities in increasing order. Let $m_j = 1 < 2 \leq m_{j+1}$; then $j < k/2$ and $m_{j+1} + \cdots + m_l > k/2$. The only reason why the "game-theoretic variance" does *not* increase by 2 is that there are at least

$$(m_{j+1} - 1) + \cdots + (m_l - 1) \geq \frac{m_{j+1} + \cdots + m_l}{2} > \frac{k/2}{2} = \frac{k}{4}$$

"full rows," meaning at least $nk/4$ marked cells. It follows $V_{nk/8} = nk/4$, implying that at this stage of the play there exists a row with row-sum r, where $r^2 \geq n/4$, i.e. $|r| \geq \sqrt{n}/2 \geq k/4$ if $n \geq k^2/4$. We have just proved:

Theorem 18.1 (a) *The maximum row-discrepancy $D(n, k) \geq \lfloor k/4 \rfloor$ if $n \geq k^2/4$.*

Next consider the opposite direction: How well can Chooser balance the rows? Here is a very simple balancing strategy. If the pair of squares that Picker picks to choose from are in the same row, then Chooser's move is irrelevant. The relevant moves are those where the 2 squares are from 2 different rows, say, from the ith row and the jth row where $1 \leq i < j \leq k$. For every pair $\{i, j\}$ with $1 \leq i < j \leq k$ Chooser follows a fixed alternating $+, -, +, -, \cdots$ pattern in the sense that, whenever Picker's last move gives the $\{i, j\}$ pair, Chooser acts according to the next sign in the alternating $\pm$ sequence: $+$ means Chooser chooses the ith row and $-$ means Chooser chooses the jth row. This way the discrepancy in a fixed row, say, the i_0th row, is estimated from above by the number of pairs $\{i_0, j\}$, which is clearly $k - 1$. Thus we get:

Theorem 18.1 (b) *The maximum row-discrepancy $D(n, k) < k$.*

Theorem 18.1 (a)–(b) shows that the discrepancy function $D(n, k)$ is basically linear if k is very small compared to n. If n is fixed/large and k increases beyond $\sqrt{n}$, then the function $D(n, k)$ "slows down" and picks up a completely different "square-root like behavior" – this is the message of the following result.

Theorem 18.2 *The maximum row-discrepancy $D(n, k) < \sqrt{(2 + o(1))n \log k}$. (Notice that we can assume $k > \sqrt{n}$, since otherwise Theorem 18.1 (b) beats Theorem 18.2.)*

Comparing Theorem 18.2 with Theorem 18.1 (b) we see a "phase transition" in the evolution of the Row Imbalane Game as k goes beyond $\sqrt{n}$: a simple "linear" game becomes more complex.

Theorem 18.2 is a very simple special case of the much more general:

Theorem 18.3 *Let $\mathcal{F}$ be an arbitrary finite hypergraph, and let ε with $0 < \varepsilon < 1$ be an arbitrary real number. Consider the Picker–Chooser play on $\mathcal{F}$ (i.e. in each turn Picker picks two previously unselected points, Chooser chooses one of them, and the other one goes back to Picker). Assume that Chooser is Balancer. If*

$$\sum_{A \in \mathcal{F}} \left((1 + \varepsilon)^{1+\varepsilon} (1 - \varepsilon)^{1-\varepsilon} \right)^{-|A|/2} < \frac{1}{2}, \qquad (18.2)$$

then Chooser (as Balancer) can force that, at the end of the play, for every $A \in \mathcal{F}$, his part in A is between $\frac{1-\varepsilon}{2}|A|$ and $\frac{1+\varepsilon}{2}|A|$.

Notice that Theorem 18.3 is the perfect analogue of the Maker–Breaker Theorem 17.1. The proof is basically the same: Chooser always chooses the "better point" (from the pair offered to him by Picker to choose from) according to the potential function defined in the proof of Theorem 17.1. The reader is challenged to check the details.

Let us apply Theorem 18.3 to the row hypergraph of the $n \times k$ board: the hyperedges are the k rows, $\mathcal{F}$ is an n-uniform hypergraph with $|\mathcal{F}| = k$. By choosing

$$\varepsilon = \sqrt{\frac{c \log k}{n}}$$

with some unspecified (yet) constant $c > 0$, criterion (18.2), applied to the row hypergraph of the $n \times k$ board, gives

$$\sum_{A \in \mathcal{F}} \left((1+\varepsilon)^{1+\varepsilon}(1-\varepsilon)^{1-\varepsilon} \right)^{-|A|/2} = k e^{-((1+\varepsilon)\log(1+\varepsilon)+(1-\varepsilon)\log(1-\varepsilon))\frac{n}{2}} =$$

$$= k e^{-((1+\varepsilon)(\varepsilon - \frac{\varepsilon^2}{2} \pm \cdots) + (1-\varepsilon)(-\varepsilon - \frac{\varepsilon^2}{2} - \cdots))\frac{n}{2}} =$$

$$= k e^{-(\varepsilon^2 + O(\varepsilon^3))\frac{n}{2}} = k e^{-\frac{c \log k}{2}(1+O(\varepsilon))} = k^{1-\frac{c}{2}(1+O(\varepsilon))},$$

which is less than $1/2$ if $c > 2$ and k is sufficiently large (we can assume that $k > \sqrt{n}$, since otherwise Theorem 18.1 (b) beats Theorem 18.2). So Theorem 18.3 applies with

$$\varepsilon = \sqrt{\frac{(2+o(1)) \log k}{n}},$$

and yields the following row-balancing: Chooser (as Balancer) can force that, at the end of the play, every row-sum has absolute value $\leq \sqrt{(2+o(1))n \log k}$. This proves Theorem 18.2.

In the symmetric case $k = n$ Theorem 18.2 gives the upper bound

$$D(n) = D(n, n) \leq \sqrt{(2+o(1))n \log n}.$$

How good is this upper bound? Well, we are going to see that $\sqrt{n \log n}$ is the correct order of magnitude (but not the correct constant factor). In fact, the following three results will now be proved:

Theorem 18.4 *In the $n \times n$ Picker–Chooser Row Imbalance Game:*

(a) the maximum lead that Picker (as Unbalancer) can force is around $\sqrt{n \log n}$;
(b) the L_2-norm of the leads over the n rows is around $\sqrt{n}$;
(c) the maximum shutout that Picker can force in a row is exactly $\lfloor \log_2 n \rfloor$.

Of course, we have to define the concept of **shutout** (see (c)). The maximum shutout is the largest lead that Picker (as Unbalancer) can achieve in a row in such a way that Chooser has not yet put any mark into that row. Moreover, the vague term of "around" in (a)–(b) means "apart from a constant factor."

Let us return to the shutout. On the diagram below, Picker has a shutout of 3 in the 5th row: Picker has 3 marks and Chooser has none.

$\bigcirc$ = Chooser

Notice that the "3" comes from the exponent $8 = 2^3$ of the board size. In general, if $n \geq 2^s$, then Picker can force a shutout of s as follows: first Picker picks 2^{s-1} pairs from the first column, and throws out the rows with "C"; he then picks 2^{s-2} pairs from the second column, and throws out the rows with "C"; he then picks 2^{s-3} pairs from the third column, and throws out the rows with "C"; and so on. This proves one direction of (c); the other direction is left to the reader as an easy exercise.

Theorem 18.4 (a) shows that the max-lead is much larger than the max-shutout.

Before proving Theorem 18.4 (a) and (b), we will formulate another result (see Theorem 18.5 below) and compare it with Theorem 18.4. This new result is about the first variant where Picker is the Balancer. The first variant is trivial for the Row Game (Picker can force row-discrepancy ≤ 1), but the Row–Column version is far from trivial: the maximum discrepancy jumps from 1 to $\sqrt{n}$. We are going to prove the following precise result.

Theorem 18.5 *In the $n \times n$ Picker–Chooser Row–Column Imbalance Game where Chooser is the Unbalancer*

(a) *the maximum lead that Chooser (as Unbalancer) can force is around $\sqrt{n}$;*
(b) *the L_2-norm of the leads over the n rows and n columns is around $\sqrt{n}$;*
(c) *the maximum shutout that Chooser can force over the $2n$ lines (n rows and n columns) is ≤ 3.*

2. Phantom Decomposition. What kind of intuitive explanation can we give for Theorems 18.4 and 18.5? Well, the case of Theorem 18.4 is self-explanatory: a standard "random model" predicts the 3 statements (a)–(b)–(c) surprisingly well. What we mean by a "random model" is the Random 2-Coloring of the n^2 little squares of the $n \times n$ board. Of course, the little squares are 2-colored independently of each other, and each color has probability 1/2. In view of the Central Limit Theorem, the maximum row-discrepancy is around $\sqrt{n \log n}$ (see (a)), and the L_2-norm of the n row-discrepancies is around the standard deviation $\sqrt{n}$ (see (b)). Notice that in (a) the extra factor $\sqrt{\log n}$ comes from the inverse $\sqrt{\log x}$ of the familiar function e^{-x^2} in the integral form of the normal distribution.

These "explain" (a) and (b), but (c) is even simpler. Indeed, if we color the little squares sequentially, then the "pure heads" equation $n \cdot 2^{-x} = 1$ has the solution $x = \log_2 n$, which "explains" (c). (The precise proof of (c) is trivial anyway.)

Thus we can say: the standard "random model" successfully predicts Theorem 18.4, but what kind of "randomness" predicts Theorem 18.5? This is a non-standard randomness that we call "local randomness." What does "local randomness" mean? It requires a longer explanation.

The hypergraph of n rows in the $n \times n$ board consists of *disjoint* hyperedges, but the hypergraph of $2n$ "lines" (n rows and n columns) has degree 2: every cell is covered by 2 "lines." The Row–Column hypergraph does *not* fall apart into disjoint hyperedges; nevertheless, my interpretation of "local randomness" assumes an imaginary "phantom decomposition," where the hypergraph falls apart into n *disjoint* row–column pairs (see the diagram below), and Theorem 18.5 is "explained" by the randomization of a single row–column pair.

Phantom Decomposition of 5×5

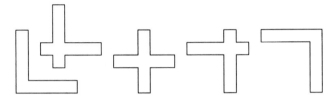

Indeed, if we randomly 2-color a row–column pair consisting of $(2n - 1)$ little squares, then the row–column discrepancy is typically around the standard deviation $\sqrt{n}$ (see Theorem 18.5 (a)–(b)); also if we color the little squares sequentially, then the longest "pure heads" sequence starting at the corner has typical length $O(1)$ (see (c)).

In other words, my interpretation of Theorem 18.5 involves a non-existing "phantom decomposition," meaning that the hypergraph of winning sets "pretends to fall apart into disjoint neighborhoods," and we predict the game-theoretic outcome by "randomizing the largest neighborhood" (meaning: two *random* players are playing in the largest neighborhood, ignoring the rest of the hypergraph).

Of course, this "Phantom Decomposition Hypothesis" is just an alternative interpretation of the Meta-Conjecture, but we find this new form very instructive.

We have the symbolic equality:

Local Randomness in Games $=$ Phantom Decomposition Hypothesis $=$ MetaConjecture

3. The proofs. This concludes my interpretation of "local randomness"; we return to this issue later at the end of Section 19. It remains to prove Theorems 18.4–18.5.

Proof of Theorem 18.4 (a). We now prove one-sided discrepancy, namely that Picker can force a lead of $c\sqrt{n \log n}$ in some row ($c > 0$ is an absolute constant).

Notice that once a lead is established in the middle of a play, Picker can sustain it till the end of the play – by picking pairs from that row so long as he can – except of a possible loss of 1 due to "parity reasons."

The obvious problem with game-theoretic variance (let $P_i(j)$ denote the number of Picker's marks and $C_i(j)$ denote the number of Chooser's marks in the ith row after the j turn)

$$V_j = \sum_{i=1}^{n} (r_i(j))^2, \text{ where } r_i(j) = P_i(j) - C_i(j)$$

is that "squaring kills the sign," and a large row-discrepancy is not necessarily Picker's lead. We can overcome this technical difficulty by applying a trick: we replace variance V_j above with the modified sum

$$V_j^* = \sum_{i=1}^{n} \left(\left(r_i(j) + \frac{\sqrt{n}}{8} \right)^+ \right)^2, \text{ where } (x)^+ = \max\{x, 0\}.$$

Write $\varrho_i(j) = \left(r_i(j) + \frac{\sqrt{n}}{8} \right)^+$; we refer to $\varrho_i(j)$ as the *modified ith row-sum* (after the jth turn).

One new idea is to work with the modified variance V_j^* instead of V_j; the second new idea is *iteration*: this is how we supply the extra factor $\sqrt{\log n}$ next to the standard deviation $\sqrt{n}$.

The iteration requires a more general setting: instead of the square-shaped $n \times n$ board we switch to a general rectangle shape $m \times k$, meaning k rows of length m each. The corresponding "modified variance" is

$$V_j^* = \sum_{i=1}^{k} \left(\left(r_i(j) + \frac{\sqrt{m}}{8} \right)^+ \right)^2;$$

we write $\varrho_i(j) = \left(r_i(j) + \frac{\sqrt{m}}{8} \right)^+$ for the *modified ith row-sum* (after the jth turn).

The idea is the following: we divide the $n \times n$ board into $\log n$ sub-boards of size $(n/\log n) \times n$ each, and accordingly the play is divided into $\log n$ stages.

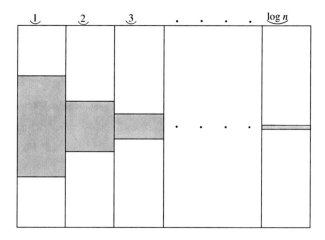

At the end of the first stage we find $\geq n/2$ rows in the first (left-most) sub-board, where Picker=Unbalancer has a lead $\approx \sqrt{n/\log n}$ in each. In the second stage Picker moves to the second (left-most) sub-board; among the $\geq n/2$ "good" rows he finds $\geq n/4$ for which he can force an additional $\approx \sqrt{n/\log n}$ lead in each (altogether $\approx 2\sqrt{n/\log n}$). In the third stage, Picker moves to the third (left-most) sub-board; among the $\geq n/4$ "good" rows he finds $\geq n/8$ for which he can force an additional $\approx \sqrt{n/\log n}$ lead in each (altogether $\approx 3\sqrt{n/\log n}$), and so on. We keep repeating this $c\log n$ times, which gives a total lead of

$$c\log n \cdot \sqrt{n/\log n} = c\sqrt{n\log n}.$$

How to handle a typical stage; that is, what is Picker's strategy in a sub-board? Consider a sub-board of size $m \times k$, where $m = n/\log n$.

Case 1: There are $\geq 2\sqrt{m}\log m$ different values among the modified row-sums $\varrho_i(j) = \left(r_i(j) + \frac{\sqrt{m}}{8}\right)^+$, $i = 1, 2, \ldots, k$.
Then $\max_i r_i(j) \geq 2\sqrt{m}\log m - \sqrt{m}/8 \geq \sqrt{m}\log m$, and we are done.

Case 2: There are $\geq k/10$ rows with $\varrho_i(j) = 0$.
Then

$$\left| \sum_{i:r_i(j)<0} r_i(j) \right| \geq \frac{k}{10} \cdot \frac{\sqrt{m}}{8},$$

and combining this with the trivial fact $\sum_{i=1}^{k} r_i(j) = 0$, we conclude

$$\left| \sum_{i:r_i(j)>0} r_i(j) \right| \geq \frac{k}{10} \cdot \frac{\sqrt{m}}{8}.$$

Then there is an integer $r \geq 2$ such that the lower bound $r_i(j) \geq r\sqrt{m}$ holds for at least $\dfrac{k}{r^2}$ values of i.

Case 3: There are $\geq k/10$ rows which are fully occupied.
Then, of course, $j \geq km/20$.

Case 4: Neither one of Cases 1 and 2 and $j < km/20$.
Then there are $k - k/10 = 9k/10$ rows with $\varrho_i(j) \geq 1$. Assume that there are exactly l different row-sum values, and the l values show up with multiplicities $\mu_1, \mu_2, \ldots, \mu_l$. We have $\mu_1 + \mu_2 + \cdots + \mu_l \geq 9k/10$ and $l \leq 2\sqrt{m}\log m$; for notational convenience let $\mu_1 \leq \mu_2 \leq \ldots \leq \mu_l$. If $k \geq 10\sqrt{m}\log m$ and $j < km/20$, then we can always find two rows with the same non-zero modified row-sum such that neither row is fully occupied yet. If Picker picks 1 new square from each one of these two rows, the modified variance V_j^* increases by 2 in the usual way

$$V_{j+1}^* = V_j^* + (\varrho \pm 1)^2 + (\varrho \mp 1)^2 - 2\varrho^2 = V_j^* + 2,$$

so

$$\frac{km}{10} = V^*_{km/20} = \sum_{i=1}^{k} \left(\left(r_i(j) + \frac{\sqrt{m}}{8} \right)^+ \right)^2,$$

which implies that there is an integer $r \geq 2$ such that the lower bound $r_i(j) \geq r\sqrt{m}$ holds for at least $\frac{k}{r^2}$ values of i.

We summarize Cases 1–4 in a single statement:

Lemma 1: *Playing on an $m \times k$ board (k rows of length m each), where $k \geq 10\sqrt{m}\log m$, Picker can force that, at some stage of the play:*

(1) either there is a row where Picker has a lead $\geq \sqrt{m}\log m$,

(2) or there is an integer r with $\log m \geq r \geq 2$ such that Picker has a lead $\geq r\sqrt{m}$ in at least $\frac{k}{r^2}$ rows.

Theorem 18.4 (a) follows from an iterated application of Lemma 1. Indeed, let $m = n/\log n$; in the first stage, played on the left-most sub-board, we have $k = n$. By Lemma 1 there are two cases: (1) or (2). If (1) holds, we are done. If (2) holds, there is an integer r_1 with $\log m \geq r_1 \geq 2$ such that Picker has a lead $\geq r_1\sqrt{m}$ in at least $\frac{k}{r^2}$ rows, say, in the i_1st row, in the i_2nd row, in the i_3rd row, …, in the i_{k_1}th row, where $k_1 \geq \frac{k}{r^2}$.

In the second stage we move to the second sub-board; we keep the "good" rows $i_1, i_2, i_3, \ldots, i_{k_1}$, and throw out the rest of the rows. This means we apply Lemma 1 to an $m \times k_1$ board where $k_1 \geq \frac{k}{r^2}$ and again $m = n/\log n$. By Lemma 1 there are two cases: (1) or (2). If (1) holds, we are done. If (2) holds, there is an integer r_2 with $\log m \geq r_2 \geq 2$ such that Picker has a lead $\geq r_2\sqrt{m}$ in at least $\frac{k_1}{r_2^2}$ rows, say, in the j_1st row, in the j_2nd row, in the j_3rd row, …, in the j_{k_2}th row, where $k_2 \geq \frac{k_1}{r_2^2}$.

In the third stage we move to the third sub-board; we keep the "good" rows $j_1, j_2, j_3, \ldots, j_{k_2}$, and throw out the rest of the rows. This means we apply Lemma 1 to an $m \times k_2$ board where $k_2 \geq \frac{k_1}{r_2^2}$ and again $m = n/\log n$. By Lemma 1 there are two cases: (1) or (2), and so on. Repeating this argument, at the end of the sth stage we have

$$\geq \frac{n}{r_1^2 r_2^2 \cdots r_s^2} \text{ rows with Picker's lead} \geq (r_1 + r_2 + \cdots + r_s)\sqrt{m}$$

in each. Lemma 1 applies so long as

$$\frac{n}{r_1^2 r_2^2 \cdots r_s^2} < 10\sqrt{m}\log m \text{ where } m = n/\log n.$$

An easy calculation shows that the minimum occurs when $2 = r_1 = r_2 = r_3 = \cdots$, implying a Picker's lead

$$\geq c_1\sqrt{m}\log n = c_1\sqrt{n/\log n}\log n = c_1\sqrt{n\log n}.$$

This completes the proof of Theorem 18.4 (a).   □

The **proof** of Theorem 18.4 (b) is easy now: Chooser can trivially guarantee the upper bound

$$V_j = \sum_{i=1}^{n}(r_i(j))^2 \leq 2j, \quad j = 1, 2, \ldots, n/2.$$

Indeed, if Picker's two little squares are in the same row, then always $V_{j+1} = V_j$. If Picker's two little squares are in two different rows i_1 and i_2, and the row-sums are equal: $r_{i_1}(j) = r_{i_2}(j)$, then there is always an increase by 2: $V_{j+1} = 2 + V_j$. Finally, if Picker's two little squares are in two different rows i_1 and i_2 and the row-sums are different: $r_{i_1}(j) \neq r_{i_2}(j)$, then Chooser can force $V_{j+1} \leq V_j$. So the L_2-norm is bounded from above by

$$\left(\frac{1}{n}\sum_{i=1}^{n}(r_i(j))^2\right)^{1/2} \leq \sqrt{n^2/n} = \sqrt{n}$$

if Chooser plays rationally.

The other direction is trivial from the proof argument of Lemma 1. Indeed, repeating the proof of Lemma 1 we can prove the following:

Proposition: Playing on an $n \times n$ board, Picker can force that, at some stage of the play:

(1) *either there are at least $n/10$ different positive row-sum values,*
(2) *or there is an integer $r \geq 2$ such that Picker has a lead $\geq r\sqrt{n}$ in at least $\frac{n}{r^2}$ rows.*

In both cases we have $\sqrt{\frac{1}{n}V_j} \geq c_2\sqrt{n}$, and Theorem 18.4 (b) follows.   □
Since we already proved Theorem 18.4 (c) (which was trivial anyway), the proof of Theorem 18.4 is complete.

19

A simple illustration of "randomness" (II)

1. Proof of Theorem 18.5 (a). The easy part of (a) is to show that Chooser=
Unbalancer can force a lead of $c\sqrt{n}$.

Lemma 1: *Let $\mathcal{F}$ be a hypergraph which is (1) n-uniform, (2) Almost Disjoint: $|A_1 \cap A_2| \leq 1$ for any two different elements of hypergraph $\mathcal{F}$, and (3) the common degree of $\mathcal{F}$ is 2: every point of the hypergraph is contained in exactly two hyperedges. Playing the Picker–Chooser game on $\mathcal{F}$ where Chooser=Unbalancer, at the end of the play Chooser can occupy at least $\frac{n}{2} + c\sqrt{n}$ points from some $A \in \mathcal{F}$ ($c = 1/15$ is a good choice).*

Lemma 1 is the perfect analogue of Theorem 18.2 in the "Discrepancy Section."
Two proofs were given for Theorem 18.2, and both can be trivially adapted to
get Lemma 1 (Chooser chooses the "better" point by applying the same potential
function).

The hard part is the converse: Picker=Balancer can force the upper bound $\leq c \cdot \sqrt{n}$
for the line-discrepancy ("line" means *row* or *column*). Picker's strategy consists
of several phases.

1st Phase: Picker picks his cell-pairs by employing a potential function F_1; function
F_1 will be specified later. A *line* (row or column) becomes "dangerous" in a play
when the absolute value of the line-discrepancy equals $10\sqrt{n}$. The unoccupied part
of a "dangerous" line, called an *emergency set*, becomes part of the *Emergency
Room*, or simply the E.R. The E.R., a "growing" set, is exactly the union of all
emergency sets arising in the course of the 1st Phase. In the 1st Phase Picker
completely avoids the E.R.; the 1st Phase is over when the complement of the E.R.
is completely exhausted by Picker.

The board of the 2nd Phase is the E.R. The key step in the proof is to confirm:

Lemma 2: *In the 1st Phase, Picker can force that there are less than n/2 emergency
sets in the 2nd Phase.*

270

Proof of Lemma 2. Let $\{L_1, L_2, \ldots, L_{n/4}\}$ be an arbitrary set of $n/4$ lines (rows and columns). Since the hypergraph of all $2n$ lines has degree 2, we have the formal equality

$$L_1 + L_2 + \ldots + L_{n/4} = H_0 + H_2,$$

where $H_0 = H_1 \cup H_2$, H_1 is the set of degree-one elements and H_2 is the set of degree-two elements of the union set $L_1 \cup L_2 \cup \ldots \cup L_{n/4}$.

We define an auxiliary hypergraph $\mathcal{H}$: the hyperedges are the sets $H_0 = H_0(L_1, L_2, \ldots, L_{n/4})$ and $H_2 = H_2(L_1, L_2, \ldots, L_{n/4})$ for all possible sets $\{L_1, L_2, \ldots, L_{n/4}\}$ of $n/4$ lines. Clearly $|\mathcal{H}| = 2\binom{2n}{n/4}$.

Picker's goal in the 1st Phase: Picker's goal is to guarantee that, during the whole phase, every $H \in \mathcal{H}$ has discrepancy of absolute value $< 1/2 \cdot n/4 \cdot 10\sqrt{n}$.

Notice that, if Picker can achieve this goal in the 1st Phase, then Lemma 2 holds. Indeed, assume that in the 1st Phase there are $\geq n/2$ emergency sets; then there are $n/4$ emergency sets where the discrepancy has the same sign (all positive or all negative); let $\{L_1, L_2, \ldots, L_{n/4}\}$ denote the super-sets of these $n/4$ emergency sets, and write

$$L_1 + L_2 + \ldots + L_{n/4} = H_0 + H_2,$$

where $H_0 = H_1 \cup H_2$, H_1 is the set of degree-one elements and H_2 is the set of degree-two elements of the union set $L_1 \cup L_2 \cup \ldots \cup L_{n/4}$. Then either $H_0 \in \mathcal{H}$ or $H_2 \in \mathcal{H}$ has a discrepancy $\geq 1/2 \cdot n/4 \cdot 10\sqrt{n} = 5n^{3/2}/4$, which contradicts Picker's goal.

As said before, Picker achieves his goal by applying a potential function F_1 (to be defined below in (19.1)). Assume that we are in the middle of the 1st Phase, $X(i) = \{x_1, \ldots, x_i\}$ denotes Picker's squares and $Y(i) = \{y_1, \ldots, y_i\}$ denotes Chooser's squares selected so far. The question is what new pair $\{x_{i+1}, y_{i+1}\}$ should Picker pick in his $(i+1)$st move (of course, Picker does not know in advance which element of the pair will be chosen by Chooser as y_{i+1}). For every previously unselected square z write

$$F_1(i; z) = \sum_{z \in L} \left((1+\varepsilon)^{|X(i) \cap L|}(1-\varepsilon)^{|Y(i) \cap L|} - (1+\varepsilon)^{|Y(i) \cap L|}(1-\varepsilon)^{|X(i) \cap L|} \right), \quad (19.1)$$

where parameter $\varepsilon(1 > \varepsilon > 0)$ will be specified later. Since z is contained by 1 row and 1 column, $F_1(z)$ is the sum of 4 terms only. Picker picks that unselected pair $\{z_1, z_2\} = \{x_{i+1}, y_{i+1}\}$ outside of the E.R. for which the difference $|F_1(i; z_1) - F_1(i; z_2)|$ attains its *minimum*.

Let

$$F_1(i) = \sum_{L: \, 2n \text{ lines}} \left((1+\varepsilon)^{|X(i) \cap L|}(1-\varepsilon)^{|Y(i) \cap L|} + (1+\varepsilon)^{|Y(i) \cap L|}(1-\varepsilon)^{|X(i) \cap L|} \right). \quad (19.2)$$

Then, after Chooser made his $(i+1)$st move, we have

$$F_1(i+1) = F_1(i) + \varepsilon F_1(i; x_{i+1}) - \varepsilon F_1(i; y_{i+1})$$

$$- \varepsilon^2 \sum_{\{x_{i+1}, y_{i+1}\} \subset L} \left((1+\varepsilon)^{|X(i) \cap L|} (1-\varepsilon)^{|Y(i) \cap L|} + (1+\varepsilon)^{|Y(i) \cap L|} (1-\varepsilon)^{|X(i) \cap L|} \right)$$

$$\leq F_1(i) + \varepsilon(F_1(i; x_{i+1}) - F_1(i; y_{i+1}))$$

$$\leq F_1(i) + \varepsilon \min_{z_1 \neq z_2} |F_1(i; z_1) - F_1(i; z_2)|. \tag{19.3}$$

Let $U(i)$ denote the set of unselected squares outside of the E.R. From the trivial inequality

$$|F_1(i; z_1)| \leq F_1(i) \text{ for all } z \in U(i),$$

by the Pigeonhole Principle we have

$$\min_{z_1 \neq z_2} |F_1(i; z_1) - F_1(i; z_2)| \leq \frac{F_1(i)}{|U| - 1}.$$

Returning to (19.3)

$$F_1(i+1) \leq F_1(i) + \frac{\varepsilon}{|U(i)| - 1} F_1(i) = \left(1 + \frac{\varepsilon}{|U(i)| - 1} \right) F_1(i) \leq e^{\frac{\varepsilon}{|U(i)| - 1}} F_1(i),$$

and by iterated application, we have

$$F_1(i) \leq F_1(start) \cdot e^{\varepsilon \sum_i \frac{1}{|U(i)| - 1}} \leq F_1(start) \cdot e^{\varepsilon \sum_{i=1}^{n^2} \frac{1}{i}}$$

$$\leq F_1(start) \cdot e^{\varepsilon \cdot 2 \log n} \text{ holds for all } i.$$

Since

$$F_1(start) = |\mathcal{H}| = 2 \binom{2n}{n/4},$$

we have

$$F_1(i) \leq 2 \binom{2n}{n/4} e^{\varepsilon \cdot 2 \log n}$$

holds for the whole course of the 1st Phase. Assume that, at some point in the 1st Phase, one player leads by $5n^{3/2}/4$ in some $H \in \mathcal{H}$. Then for some i and $t(5n^{3/2}/4 \leq t \leq n^2/8)$

$$(1+\varepsilon)^{t + 5n^{3/2}/4} (1-\varepsilon)^{t - 5n^{3/2}/4} \leq F_1(i) \leq 2 \binom{2n}{n/4} e^{\varepsilon \cdot 2 \log n}. \tag{19.4}$$

We have

$$(1+\varepsilon)^{t+5n^{3/2}/4}(1-\varepsilon)^{t-5n^{3/2}/4} = \left(\frac{1+\varepsilon}{1-\varepsilon}\right)^{5n^{3/2}/4} \cdot (1-\varepsilon^2)^t$$

$$\geq \left(\frac{1+\varepsilon}{1-\varepsilon}\right)^{5n^{3/2}/4} \cdot (1-\varepsilon^2)^{n^2/8}$$

$$= e^{(2\varepsilon+O(\varepsilon^2))5n^{3/2}/4 - \varepsilon^2 n^2/8 + O(\varepsilon^4)n^2}, \tag{19.5}$$

where the inequality $t \leq n^2/8$ follows from the fact that $n/4$ lines cover at most $n^2/4$ cells.

By choosing $\varepsilon = 10/\sqrt{n}$ in (19.5), we obtain

$$(1+\varepsilon)^{t+5n^{3/2}/4}(1-\varepsilon)^{t-5n^{3/2}/4} \geq e^{(25/2-o(1))n},$$

and returning to (19.4)

$$e^{(25/2-o(1))n} \leq 2\binom{2n}{n/4}e^{10/\sqrt{n}\cdot 2\log n}. \tag{19.6}$$

Finally, (19.6) is an obvious contradiction, since

$$\binom{2n}{n/4} \leq \binom{2n}{n} \leq 2^{2n} < e^{25n/2}.$$

This completes the proof of Lemma 2. □

Now we are ready to discuss the:

2nd Phase. The board is the E.R. at the end of the 1st Phase; from Lemma 2 we know that there are $n_1 \leq n/2$ emergency sets. Of course, there may exist a large number of lines which are not dangerous yet and intersect the E.R.: we call these intersections *secondary sets*. Since the E.R. is the union of the emergency sets and any two lines intersect in ≤ 1 cell, every secondary set has size $\leq n_1 \leq n/2$.

At the end of the 1st Phase, a similar "halving" cannot be guaranteed for the emergency sets; the size of an emergency set may remain very close to n. The objective of the 2nd Phase is exactly to achieve a size-reduction in every single emergency set (close to "halving").

Let $E_1, E_2, \ldots, E_{n_1}$ denote the $n_1(\leq n/2)$ emergency sets, and consider the sub-hypergraph $\{E_1, E_2, \ldots, E_{n_1}\}$. Every E_j has a "private part," meaning: the degree-one cells of E_j in the sub-hypergrap, we denote the "private part" by $PR(E_j)$, $j = 1, 2, \ldots, n_1$. If Picker always picks a pair from a some "private part" $PR(E_j)$, then there is no change in the discrepancy of any emergency set, but, of course, the discrepancy of a secondary set may change (in fact, may increase).

This is exactly what Picker does: in the 2nd Phase he exhausts every "private part" $PR(E_j)$, $j = 1, 2, \ldots, n_1$ by picking pairs from them so long as he can "legally" do it. In the 2nd Phase the discrepancy of an emergency set doesn't change; a secondary set becomes "dangerous" when its discrepancy (restricted to the board of the 2nd Phase) has absolute value $10 \cdot \log(6n/n_1) \cdot \sqrt{n_1}$. The unoccupied part of a "dangerous" secondary set, called a *new emergency set*, becomes part of the 2nd E.R. The 2nd E.R., a growing set, is exactly the union of the new emergency sets arising during the 2nd Phase:

Lemma 3: *In the 2nd Phase Picker can force that, the number of new emergency sets is less than $n_1/10$.*

Proof of Lemma 3. We proceed somewhat similarly to Lemma 2. Let $\{S_1, S_2, \ldots, S_{n_1/20}\}$ be an arbitrary set of $n_1/20$ secondary sets. Again we have the formal equality

$$S_1 + S_2 + \ldots + S_{n_1/20} = H_0 + H_2,$$

where $H_0 = H_1 \cup H_2$, H_1 is the set of degree-one elements and H_2 is the set of degree-two elements of the union set $S_1 \cup S_2 \cup \ldots \cup S_{n_1/20}$.

We define an auxiliary hypergraph $\mathcal{H}$: the hyperedges are the sets $H_0 = H_0(S_1, S_2, \ldots, S_{n_1/20})$ and $H_2 = H_2(S_1, S_2, \ldots, S_{n_1/20})$ for all possible sets $\{S_1, S_2, \ldots, S_{n_1/20}\}$ of $n_1/20$ secondary sets. Clearly $|\mathcal{H}| = 2\binom{2n}{n_1/20}$.

Picker's goal in the 2nd Phase: Picker's goal is to guarantee that, during the whole phase, every $H \in \mathcal{H}$ has discrepancy of absolute value $< 1/2 \cdot n_1/20 \cdot 10 \log(6n/n_1)\sqrt{n_1}$.

Notice that, if Picker can achieve this goal in the 2nd Phase, then Lemma 3 holds. Indeed, assume that in the 2nd Phase there are $\geq n_1/10$ new emergency sets; then there are $n_1/20$ new emergency sets, where the discrepancy has the same sign (all positive or all negative); let $\{S_1, S_2, \ldots, S_{n_1/20}\}$ denote the super-sets of these $n_1/20$ emergency sets, and write

$$S_1 + S_2 + \ldots + S_{n_1/20} = H_0 + H_2,$$

where $H_0 = H_1 \cup H_2$, H_1 is the set of degree-one elements and H_2 is the set of degree-two elements of the union set $S_1 \cup S_2 \cup \ldots \cup S_{n_1/20}$. Then either $H_0 \in \mathcal{H}$ or $H_2 \in \mathcal{H}$ has a discrepancy $\geq 1/2 \cdot n_1/20 \cdot 10 \log(6n/n_1)\sqrt{n_1}$, which contradicts Picker's goal.

Picker achieves his goal by applying a potential function F_2. Function F_2 is the perfect analogue of F_1 in Lemma 2: for every previously unselected square z write

$$F_2(i; z) = \sum_{z \in S} \left((1+\varepsilon)^{|X(i) \cap S|}(1-\varepsilon)^{|Y(i) \cap S|} - (1+\varepsilon)^{|Y(i) \cap S|}(1-\varepsilon)^{|X(i) \cap S|} \right),$$

where S runs over the secondary sets, and parameter ε ($1 > \varepsilon > 0$) will be specified later (it will have a different value than in Lemma 2). Since z is contained by 1 row and 1 column, $F_2(z)$ is the sum of 4 terms only.

The only novelty is that here Picker picks his pairs from the "private parts" $PR(E_j)$, $j = 1, 2, \ldots, n_1$ of the old emergency sets. Picker exhausts $PR(E_j)$ as long as he can; the only restriction is that he must avoid the 2nd E.R. If the "private parts" $PR(E_j)$ contains at least 2 unselected cells outside of the 2nd E.R., then Picker picks that unselected pair $\{z_1, z_2\} = \{x_{i+1}, y_{i+1}\}$ from $PR(E_j)$ outside of the new E.R. for which the difference $|F_2(i; z_1) - F_2(i; z_2)|$ attains its *minimum*.

Picker keeps doing this for every "private part" $PR(E_j)$ until he has no legal move left. The analogue of (19.4) goes as follows

$$(1+\varepsilon)^{t+n_1^{3/2}\log(6n/n_1)/4}(1-\varepsilon)^{t-n_1^{3/2}\log(6n/n_1)/4} \le F_2(i) \le 2\binom{2n}{n_1/20}\left(e^{\varepsilon \cdot \log n}\right)^{n_1}. \quad (19.7)$$

We have

$$(1+\varepsilon)^{t+n_1^{3/2}\log(6n/n_1)/4}(1-\varepsilon)^{t-n_1^{3/2}\log(6n/n_1)/4} = \left(\frac{1+\varepsilon}{1-\varepsilon}\right)^{n_1^{3/2}\log(6n/n_1)/4} \cdot (1-\varepsilon^2)^t$$

$$\ge \left(\frac{1+\varepsilon}{1-\varepsilon}\right)^{n_1^{3/2}\log(6n/n_1)/4} \cdot (1-\varepsilon^2)^{n_1^2/40}$$

$$= e^{(2\varepsilon+O(\varepsilon^2))n_1^{3/2}\log(6n/n_1)/4-(\varepsilon^2+O(\varepsilon^4))n_1^2},$$

$$(19.8)$$

where the inequality $t \le n_1^2/40$ follows from the fact that $n_1/20$ lines cover at most $n_1^2/20$ cells.

By choosing $\varepsilon = 10\log(6n/n_1)/\sqrt{n_1}$ in (19.8), we obtain

$$(1+\varepsilon)^{t+n_1^{3/2}\log(6n/n_1)/4}(1-\varepsilon)^{t-n_1^{3/2}\log(6n/n_1)/4} \ge e^{(5/2-o(1))n_1\log(6n/n_1)},$$

and returning to (19.7)

$$e^{(5/2-o(1))n_1\log(6n/n_1)} \le 2\binom{2n}{n_1}e^{10\log(6n/n_1)/\sqrt{n_1}\cdot\log n \cdot n_1}. \quad (19.9)$$

Finally, (19.9) is a contradiction; indeed

$$e^{(5/2-o(1))n_1\log(6n/n_1)} = \left(\frac{6n}{n_1}\right)^{(5/2-o(1))n_1}$$

and

$$\binom{2n}{n_1}e^{10\log(6n/n_1)/\sqrt{n_1}\cdot\log n \cdot n_1}$$

$$\le \left(\frac{2en}{n_1}\right)^{n_1}\left(\frac{n}{n_1}\right)^{10\sqrt{n_1}\cdot\log n}.$$

This contradiction proves Lemma 3. □

The 2nd Phase terminates when every "private part" $PR(E_j)$, $j = 1, 2, \ldots, n_1$ has ≤ 1 unselected square outside of the 2nd E.R.. By Lemma 3, the 2nd E.R. consists of less than $n_1/10$ new emergency sets. It follows that, at the end of the 2nd Phase, every new emergency set has unoccupied part

$$\le n_1 + \frac{n_1}{10} = \frac{11n_1}{10} \le \frac{11n}{20}.$$

We already know that every new emergency set has size $\le n_1 \le n/2$ (a byproduct of Lemma 2). Summarizing, at the end of the 2nd Phase, every emergency set, old and new, has an unoccupied part of size $\le 11n/20$, and, so far in the play, every line has discrepancy

$$\le 10\sqrt{n} + 10\log(6n/n_1)\sqrt{n_1}, \tag{19.10}$$

coming from the 1st Phase and the 2nd Phase.

The board of the 3rd Phase is the union of the emergency sets (old and new). Of course, there may exist a large number of lines which are not "dangerous" yet and still have an unoccupied part. The (non-empty) unoccupied parts are called *secondary sets*; every secondary set has size $\le 11n/20$.

Notice that we already made a big progress: the set-size is reduced from the original n to $\le 11n/20$ (a 45% reduction), and the maximum total discrepancy so far is $\le 30\sqrt{n}$ (see (19.10)). The rest of the argument is plain *iteration*.

Let n_2 denote the number of emergency sets (old and new) at the end of the 2nd Phase. The argument above gives that:

(α) each emergency set has size $\le n_2 \le 11n/20$;

(β) every secondary set has size $\le n_2$;

(γ) the only bad news is that the total number of sets may remain very close to the initial $2n$.

The 3rd Phase is the analogue of the 1st Phase and the 4th Phase is the analogue of the 2nd Phase (that is, we have a periodicity in the argument where the length of the period is 2). Let us focus on the 3rd Phase: a set (emergency or secondary) becomes "dangerous" when its discrepancy has absolute value $10\log(6n/n_2)\sqrt{n_2}$.

Warning: when we speak about discrepancy in a particular phase, the set (row or column) is restricted to the particular sub-board, in this case to the board of the 3rd Phase, and the discrepancy outside doesn't count. When "everything counts," we will call it the *total discrepancy*.

We need:

Lemma 4: *In the 3rd Phase Picker can force that the number of "dangerous sets" remains less than $n_2/2$.*

The proof is very similar to that of Lemma 2. The critical part is the following inequality (analogue of (19.6) and (19.9))

$$e^{2\varepsilon \cdot 1/2 \cdot n_2/4 \cdot 10\log(6n/n_2)\sqrt{n_2} - \varepsilon^2 \cdot n_2^2/8} \le \binom{2n}{n_2}, \tag{19.11}$$

which provides the necessary contradiction in the *reductio ad absurdum* proof. Indeed, choosing $\varepsilon = 10\log(6n/n_2)/\sqrt{n_2}$, the left-hand side of (19.11) becomes much larger than the right-hand side

$$\left(\frac{6n}{n_2}\right)^{5n_2/2} > \left(\frac{2n}{n_2}\right) \approx \left(\frac{2en}{n_2}\right)^{n_2}.$$

The 4th Phase is the perfect analogue of the 2nd Phase. Indeed, let $E_1, E_2, \ldots, E_{n_3}$ denote the emergency sets at the end of the 3rd Phase; Lemma 4 yields $n_3 \le n_2/2$. Again Picker picks his pairs from the "private parts" $PR(E_j)$, $j = 1, 2, \ldots, n_3$. This way the discrepancy of the emergency sets doesn't change. A secondary set becomes "dangerous" in the 4th Phase when its discrepancy has absolute value $10\log(6n/n_3)\sqrt{n_3}$.

We need:

Lemma 5: *In the 4th Phase Picker can force that the number of "dangerous sets" remains less than $n_3/10$.*

The proof is exactly the same as that of Lemma 3.

Let me summarize the progress we have made at the end of the 4th Phase:

(α) every emergency set has size $\le n_3 + n_3/10 \le 11n_2/20 = (11/20)^2 n$;
(β) every secondary set has size $\le (11/20)^2 n$;
(γ) the maximum total discrepancy so far is

$$\le 30\sqrt{n} + 30\log(6n/n_2)\sqrt{n_2};$$

(δ) the only bad news is that the total number of sets may remain very close to the initial $2n$.

A k-times iteration of this argument gives that the set-size is reduced to $\le (11/20)^k n$, and the maximum total discrepancy so is estimated from above by the sum

$$\le 30\sqrt{n} + 30\log(6n/n_2)\sqrt{n_2} + 30\log(6n/n_4)\sqrt{n_4} + 30\log(6n/n_6)\sqrt{n_6} + \ldots,$$
(19.12)

where $n > n_2 > n_4 > n_6 > \cdots$ and $n_{2j} \le (11/20)^j n$, $j = 1, 2, 3, \ldots$. Sum (19.13) represents a rapidly convergent series (like a geometric series) where the sum is less than constant times the first term $O(\sqrt{n})$. This completes the proof of Theorem 18.5 (a). □

Lemma 1 (at the beginning of this section) was a Picker–Chooser version, where Chooser=Unbalancer, of the Maker–Breaker result Theorem 16.2. Two proofs were given for Theorem 16.2, and a straightforward adaptation of the first proof solves one-half of Theorem 16.5 (b). Indeed, let Δ_i, $i = 1, 2, \ldots, 2n$ denote Chooser's lead in the $2n$ lines at the end (n rows and n columns; $\Delta_i > 0$ means that Chooser has more

marks in that line). The first proof of Theorem 16.2 gives the alternative result that
(1) either the L_1-norm is "large"; (2) or the L_2-norm is "large." Formally, (1) either

$$\frac{1}{2n} \sum_{i:\ \Delta_i>0} \Delta_i > c\sqrt{n},$$

(2) or

$$\left(\frac{1}{2n} \sum_{i:\ \Delta_i>0} \Delta_i^2\right)^{1/2} > c\sqrt{n}.$$

Since L_2-norm $\geq L_1$-norm, in both cases the L_2-norm of Chooser's lead is $> c\sqrt{n}$
(if Chooser plays rationally).

The converse is trivial from (a); indeed, L_∞-norm $\geq$ L_2-norm. This
completes (b). $\qquad\square$

Finally, Theorem 18.5 (c) is very easy. Every domino in the picture represents a
Picker's move. Each row and each column contains a whole domino.

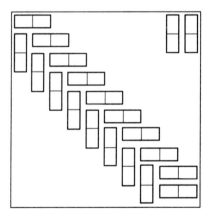

Picker's strategy to prevent
a shutout ≥ 4

It follows that the largest shutout that Chooser can force is ≤ 3. The proof of
Theorem 18.5 is complete. $\qquad\square$

Note that Theorem 18.5 (c) has the following far-reaching generalization. Let G be
an arbitrary d-regular graph; for simplicity assume that the common degree d is even.
Then by a classical theorem of Petersen, G is the union of $d/2$ perfect 2-factors. A
perfect 2-factor means a sub-graph which contains every vertex of G and every degree
is exactly 2; it is the union of vertex-disjoint cycles. Picker takes two perfect 2-factors
from G, and gives an orientation to each cycle. Then every vertex of G has two out-
edges. In each turn Picker offers to Chooser 2 out-edges with a common starting point.
Then, of course, Chooser cannot make a Shutout larger than 2.

Theorem 18.5 (c) is the special case when G is the $n \times n$ complete bipartite
graph.

2. Explaining games with probabilistic models. We recall that $\sqrt{n}$ is the *standard deviation* of the n-step random walk (with steps ± 1), and the quadratic sum V_j defined above can be interpreted as a "game-theoretic variance" (and V_j^* is a "*modified* variance"). How far can we go with this kind of **probabilistic analogy**? Consider a *random process* where for each one of the n^2 little squares of the board we toss a fair coin: Heads means we write a "P" and Tails means we write a "C" in the little square. This random process gives a "Random 2-Coloring" of the $n \times n$ board; what is the largest lead that Picker can achieve in the family of all rows? Well, with probability tending to 1 as $n \to \infty$, the maximum lead is $\sqrt{(2+o(1))\log n}\sqrt{n}$. Indeed, let $\lambda = \sqrt{(2+\varepsilon)\log n}$ with $\varepsilon > 0$; then by the Central Limit Theorem

$$Pr[\text{max lead} \geq \lambda\sqrt{n}] \leq n\frac{1}{\sqrt{2\pi}}\int_\lambda^\infty e^{-u^2/2}du \leq \frac{n}{n^{1+\varepsilon}} = n^{-\varepsilon} \to 0.$$

Next let $\lambda = \sqrt{(2-\varepsilon)\log n}$ with $\varepsilon > 0$; then

$$Pr[\text{max lead} \leq \lambda\sqrt{n}] \leq \left(1 - \frac{1}{\sqrt{2\pi}}\int_\lambda^\infty e^{-u^2/2}du\right)^n \leq (1-n^{-1+\varepsilon})^n \leq e^{-n^\varepsilon} \to 0,$$

proving what we wanted.

How far can we go with the **probabilistic analogy**? Is it true that that the largest lead that Picker can force in the $n \times n$ Picker–Chooser Row Discrepancy Game is the same $\sqrt{(2+o(1))\log n}\sqrt{n}$ (as $n \to \infty$)? In one direction we could really prove this: Chooser can always prevent Picker from achieving a lead of $\sqrt{(2+o(1))\log n}\sqrt{n}$; in the other direction we missed by a constant factor: Picker can always achieve a lead of $c\sqrt{\log n}\sqrt{n}$, where $c > 0$ is an absolute constant. Even if the upper and lower bounds differ by a constant factor, the appearance of the same (strange!) quantity $\sqrt{\log n}\sqrt{n}$ is a remarkable coincidence.

The extra factor $\sqrt{\log n}$ – the solution of the equation $e^{-x^2} = 1/n$ – next to $\sqrt{n}$ ("standard deviation") is an unmistakable sign of Randomness (Central Limit Theorem).

Notice that, inspecting the n rows in the "Random 2-Coloring," the longest "pure P sequence" at the beginning of the n rows (like PP...P) has length $\log_2 n + O(1)$ with probability tending to 1 as $n \to \infty$ (why?). A "pure P sequence" at the beginning of a row corresponds to a "shutout"; the appearance of the same $\log_2 n$ is another interesting sign of a deeper analogy between the Picker–Chooser game and the "Random 2-Coloring."

Actually, the analogy went deeper: it was converted into the proofs! For example, to prove that Chooser can always prevent Picker from achieving a lead of $\sqrt{(2+o(1))\log n}\sqrt{n}$, we employed a Potential Function, which was clearly motivated by the probabilistic analogy as follows. Restricting the "Random 2-Coloring" to a fixed row (say, the first row), we have

$$Pr\,[\text{Picker has a lead } \geq l \text{ in the first row}] = \sum_{j=(n+l)/2}^{n} \binom{n}{j} 2^{-n}.$$

The dominant term on the right-hand side is the first term; by using the weak form $k! \approx (k/e)^k$ of Stirling's formula we get the approximation

$$\binom{n}{(n+l)/2} 2^{-n} = \frac{n!}{\frac{n+l}{2}!\frac{n-l}{2}!2^n} \approx \frac{(n/e)^n}{((n+l)/2e)^{(n+l)/2}((n-l)/2e)^{(n-l)/2}2^n}$$

$$= \left(1+\frac{l}{n}\right)^{-(n+l)/2} \cdot \left(1-\frac{l}{n}\right)^{-(n-l)/2}.$$

The approximation

$$Pr\,[\text{Picker has a lead } \geq l \text{ in the first row}] \approx \left(1+\frac{l}{n}\right)^{-(n+l)/2} \cdot \left(1-\frac{l}{n}\right)^{-(n-l)/2}$$

of the "large deviation probability" motivates the introduction of the following Potential Function: let $l = \sqrt{(2+o(1))\log n}\sqrt{n}$, and consider the sum

$$T_i = \sum_{j=1}^{n} \left(1+\frac{l}{n}\right)^{\#P_j-(n+l)/2} \cdot \left(1-\frac{l}{n}\right)^{\#C_j-(n-l)/2}.$$

We now see what this means: we are right after the ith turn where Picker has i marks "P" and Chooser has i marks "C"; $\#P_j$ is the number of marks of Picker in the jth row; $\#C_j$ is the number of marks of Chooser in the jth row; the present danger of the event "at the end of the play Picker achieves a lead $\geq l$ in the jth row" is measured by the product

$$\left(1+\frac{l}{n}\right)^{\#P_j-(n+l)/2} \cdot \left(1-\frac{l}{n}\right)^{\#C_j-(n-l)/2};$$

the sum T_i above, extended over all rows, measures the "total danger."

3. Picker–Chooser is a "global" game and Chooser–Picker is a "local" game.
We tried to "explain" the Chooser–Picker game with the "randomized neighborhood" model, but where does the concept of "neighborhood" come from? Why is it so important? Well, the Chooser–Picker game (similarly to the Maker–Breaker game) is a *local game* in the following sense. Assume that a particular Chooser–Picker game falls apart into *components* $C_1, C_2, C_3, \ldots$ with disjoint sub-boards; then Chooser cannot take advantage of the (possibly) large number of components: the best that he can achieve in the *whole game* is the same that he can achieve in the "largest" component. Indeed, Picker can proceed component-wise by exhausting first (say) the first component, next the second component, next then the third component, and so on. This means Chooser's best play is to *focus on one component*, preferably on the "largest" component, and to ignore the rest. My way of understanding a Chooser–Picker game is to visualize a *Phantom Decomposition* where

the game falls apart into disjoint components; a component is a "neighborhood" in the narrow sense: it is the family of all winning sets containing a *fixed element* of the board. Once we have the components, we pick the "largest one" – because this is all what Chooser can do: namely to pick 1 component and focus on it – and *randomize* the "largest component" exactly as in the Picker–Chooser game. This particular probabilistic model that we call the "randomized neighborhood" is our guide to understanding the Chooser–Picker game.

It is important to emphasize that the *Phantom Decomposition* mentioned above is an "imaginary" decomposition: it is not real; in fact the game remains a coherent entity during the whole course of the play.

The *Phantom Decomposition* is not real, but in the last part (Part D) of the book we are going to discuss several (real!) *decomposition techniques*; these decomposition techniques support my *Phantom Decomposition* "philosophy."

Note that the "emergency room" idea in the proof of Theorem 18.5 (a) was already a decomposition technique.

Theorems 18.4 and 18.5 illustrate that the Picker–Chooser game and the Chooser–Picker game can both be explained by some *probabilistic models*: the Picker–Chooser game is explained by the "randomized game" model and the Chooser–Picker game is explained by the "randomized neighborhood" model. The bounds in Theorems 18.4–18.5 were not exact (with the lucky exception of Theorem 18.4 (c)), so most likely the reader has his/her strong reservations about my "probabilistic model explanation," finding the results presented so far less than convincing. The most natural objection against a probabilistic model is that it has a built-in "fluctuation," raising the question of how on earth can it predict an exact/deterministic outcome, namely the outcome of an optimal play between perfect players? For example, the *standard deviation* of an n-step symmetric random walk is exactly $\sqrt{n}$, but what is a "typical" deviation? A deviation of $\sqrt{n}/2$ is just as "typical" as that of $\sqrt{n}$; in fact, a deviation of $\sqrt{n}/2$ is more likely than a deviation of $\sqrt{n}$, and the most likely deviation is in fact the zero(!) deviation.

Then we would probably say: "Let us take an *average* of the deviations!"; but which average is the "natural" average? The *linear* average and the *quadratic* average are equally "natural"; from the Central Limit Theorem and the 2 integrals below

$$\frac{1}{\sqrt{2\pi}} \int_{-\infty}^{\infty} |u| e^{-u^2/2} du = \sqrt{2/\pi} \quad \text{and} \quad \frac{1}{\sqrt{2\pi}} \int_{-\infty}^{\infty} u^2 e^{-u^2/2} du = 1$$

we see that the *linear* average of the deviations is $\sqrt{2/\pi}\sqrt{n}$ and the *quadratic* average of the deviations is $\sqrt{n}$. There is a slightly more than 20 percent difference between the 2 averages; which one (if any) predicts the truth in the Chooser–Picker game? Unfortunately we cannot decide on the question, because we don't know the right constant factor in the solution of the game.

It seems that a probabilistic model cannot predict an exact value at all, but here comes the surprise: there are some "natural" probabilistic models in which the key parameter – a random variable! – happens to be almost single-valued (i.e. almost deterministic, so the problem of "which average is the right one" disappears). For example, in a Random Graph on K_N with edge probability $1/2$ (which means that for each one of the $\binom{N}{2}$ edges of the complete graph K_N on N vertices we toss a fair coin; we keep the edge if the outcome is Heads and discard the edge if the outcome is Tails) the largest clique K_q has size

$$q = \lfloor 2\log_2 N - 2\log_2\log_2 N + 2\log_2 e - 1\rfloor$$

with probability tending to 1 as $N \to \infty$. Here the probabilistic model gives a (basically) deterministic prediction, which leads to the following precise question: is it true that, in the Picker–Chooser game played on the edges of K_N, the largest clique K_q that Picker can occupy has $q = \lfloor 2\log_2 N - 2\log_2\log_2 N + 2\log_2 e - 1\rfloor$ vertices? The answer is "yes" (at least for the overwhelming majority of the values of N) – this is proved in Section 22.

Now how about the Chooser–Picker version? What the "randomized neighborhood" model suggests is the following: fix an (arbitrary) edge $\{u, v\}$ of K_N, and find the largest clique in the Random Graph with edge probability $1/2$ which *contains* edge $\{u, v\}$; let this largest clique be a copy of K_s. The value of s is again a random variable, and, similarly to q, s is also concentrated on a single value: this time the value is

$$\lfloor 2\log_2 N - 2\log_2\log_2 N + 2\log_2 e - 3\rfloor$$

with probability tending to 1 as $N \to \infty$ (for clarity this was over-simplified a little bit). Because the probabilistic model gave a (basically) deterministic value, it is natural to ask the following question: is it true that, in the Chooser–Picker game played on the edges of K_N, the largest clique K_q that Chooser can occupy has $q = \lfloor 2\log_2 N - 2\log_2\log_2 N + 2\log_2 e - 3\rfloor$ vertices? Again the answer is "yes"! (At least for the overwhelming majority of the values of N.)

Next switch to the lattices games introduced in Section 8: consider a Random 2-Coloring of the $N \times N$ board with marks "P" and "C" (involving N^2 coin tosses); what is the size of the largest Aligned Square Lattice which is exclusively marked "P" (the gap of the lattice is arbitrary)? Let the largest Square Lattice be of size $q \times q$; parameter q is a random variable (depending on the N^2 coin tosses). A simple calculation shows that it is concentrated on the single value

$$\lfloor\sqrt{3\log_2 N}\rfloor$$

with probability tending to 1 as $N \to \infty$. We ask the usual question: is it true that, in the Picker–Chooser game played on the $N \times N$ board, the largest Aligned Square Lattice that Picker can occupy has size $q \times q$ with

$$q = \lfloor \sqrt{3 \log_2 N} \rfloor$$

Again the answer is "yes" (at least for the overwhelming majority of the values of N).

How about the Chooser–Picker version? What the "randomized neighborhood" model suggests here is the following: fix a little square of the $N \times N$ chess-board (say, the lower left corner), and find the largest Aligned Square Lattice in the random P-C coloring which *contains* the fixed little square and is exclusively marked "C"; let this lattice be of size $s \times s$. The value of s is again a random variable, and, similarly to q, s is also concentrated on a single value: this time the value is

$$\lfloor \sqrt{\log_2 N} \rfloor$$

with probability tending to 1 as $N \to \infty$ (again for clarity this was over-simplified a little bit). Because the probabilistic model gave a (basically) deterministic value, it is natural to ask: Is it true that, in the Chooser–Picker game played on the $N \times N$ chess-board, the largest Aligned Square Lattice that Chooser can occupy has size $s \times s$ with $s = \lfloor \sqrt{\log_2 N} \rfloor$? Again the answer is "yes"! (At least for the overwhelming majority of the values of N.)

Comparing the "Clique Game" with the "Square Lattice game" we can see a major difference: in the "Clique Game" r differs from s by an additive constant only; in the "aligned square game" r is $\sqrt{3} = 1.732$ times larger than s.

There is also a big difference in the difficulty of the proofs: the "extreme concentration" of the random variables can be proved in 1 page; the Picker–Chooser results can be proved on 5–6 pages; but the proof of the Chooser–Picker results needs about 80 pages! This justifies our intuition that the probabilistic model is "easy," the Picker–Chooser game is "more difficult," and the Chooser–Picker game is "very difficult" (because Chooser has much less control than Picker does).

These sporadic results can be extended to a whole scale of results; 1 extension is the discrepancy version. Let α be an arbitrary fixed real in the interval $1/2 < \alpha \le 1$; what happens if the goal is relaxed in such a way that it suffices to occupy a given majority, say the α part of the original goal (instead of complete occupation)? For simplicity we just discuss the "Square Lattice game"; in the α version of the "Square Lattice game" the goal is to occupy the α part of a large Aligned Square Lattice in the $N \times N$ board; given α and N, what is the largest size $q \times q$ that is still α-achievable?

First, as usual, consider a Random 2-Coloring of the $N \times N$ board with marks "P" and "C" (involving N^2 coin tosses); what is the size of the largest Aligned Square Lattice which has at least α-part "P"? Let the largest such Square Lattice be of size $q \times q$. Parameter q is a random variable (depending on the N^2 coin tosses); it is concentrated on the single value

$$\left\lfloor \sqrt{\frac{3}{1-H(\alpha)}} \log_2 N \right\rfloor$$

with probability tending to 1 as $N \to \infty$. Here the novelty is the appearance of the Shannon entropy $H(\alpha) = -\alpha \log \alpha - (1-\alpha) \log(1-\alpha)$, which is a positive constant less than 1. We ask the usual question: Is it true that, in the Picker–Chooser game played on the $N \times N$ board, the largest Aligned Square Lattice that Picker can "α-occupy" has size $q \times q$ with

$$q = \left\lfloor \sqrt{\frac{3}{1-H(\alpha)}} \log_2 N \right\rfloor$$

Again the answer is "yes" (at least for the overwhelming majority of the values of N).

How about the Chooser–Picker version? What the "randomized neighborhood" model suggests here is the following: fix a little square of the $N \times N$ chess-board (say, the lower left corner), and find the largest Aligned Square Lattice in the random P-C coloring which *contains* the fixed little square and has at least α-part "C"; let this lattice be of size $s \times s$. The value of s is again a random variable, and, similarly to q, s is also concentrated on a single value: the value is

$$\left\lfloor \sqrt{\frac{1}{1-H(\alpha)}} \log_2 N \right\rfloor$$

with probability tending to 1 as $N \to \infty$ (again for clarity this was over-simplified a little bit). Again the usual question: is it true that, in the Chooser–Picker game played on the $N \times N$ chess-board, the largest Aligned Square Lattice that Chooser can α-occupy has size $s \times s$ with

$$s = \left\lfloor \sqrt{\frac{1}{1-H(\alpha)}} \log_2 N \right\rfloor$$

The answer is, one more time, "yes"! (At least for the overwhelming majority of the values of N.)

These *exact* results – holding for all real numbers α in $1/2 < \alpha \le 1$ and for all large enough N – represent strong evidence in favor of the "probabilistic model explanation."

Changing q to $q+1$, the clique K_q increases by q edges and the $q \times q$ lattice increases by $2q+1$ elements. This square-root size increase provides a comfortable "safety cushion" for our (rather complicated) potential calculations.

Note that the Maker–Breaker game has a striking similarity to the Chooser–Picker game – we can prove (almost) identical(!) results – but we cannot give any *a priori* explanation for this coincidence (except the complicated proof itself).

The basic philosophy is repeated. The probabilistic model is "easy" to work with; it is easy to compute (or estimate) the relevant probabilities. This is why we are using the probabilistic model as a guide. The Picker–Chooser game is "more difficult." This is why we are using it as a "warmup" and a motivation. The Chooser–Picker game is "very difficult"; and the same goes for the Maker–Breaker game; and, finally, the "who-does-it-first" games are hopeless.

20

Another illustration of "randomness" in games

1. When Achievement and Avoidance are very different. We recall Lehman's remarkable theorem from Section 4. Consider the (1:1) Maker–Breaker Connectivity Game on a finite multi-graph G where Maker's goal is to own a spanning tree in G. When can Maker win? The shockingly simple answer goes as follows: Maker, as the second player, has a winning strategy if and only if G contains 2 edge-disjoint spanning trees.

What happens in the (1:b) underdog play with $b \geq 2$ if Maker is the underdog? Of course, the (1:b) play with $b \geq 2$ means that Maker takes 1 new edge and Breaker takes b new edges per move. It is very tempting to believe in the following generalization of Lehman's theorem:

Question 20.1 *Is it true that, if multi-graph G contains $b+1$ edge-disjoint spanning trees, then at the end of the (1:b) play underdog Maker can own a whole spanning tree?*

For example, the complete graph K_n contains $\lfloor n/2 \rfloor$ edge-disjoint spanning trees (hint: if n is even, K_n is the union of $n/2$ edge-disjoint Hamiltonian paths – see the very end of the section). Is it true that, playing the (1:b) game on K_n with $b = \lfloor n/2 \rfloor - 1$, underdog Maker can own a whole spanning tree? Well, it is very tempting to say "yes," especially since the answer is really "yes" for the Avoidance version, see Theorem 4.2.

In spite of this, the correct answer for the Achievement game is "no." What is more, even if b is $o(n)$, such as $b = (1 + o(1))n/\log n$, Breaker can still prevent Maker from occupying a whole spanning tree. This result, due to Erdős and Chvátal [1978], completely "kills" Question 20.1.

Breaker's strategy goes as follows: if $b = (1 + \varepsilon)n/\log n$ with some $\varepsilon > 0$, then Breaker can occupy a star of $(n-1)$ edges; a Breaker's star means an isolated point in Maker's graph, and an isolated point trivially prevents a spanning tree.

A good motivation for the "isolated point" is:

Erdős's Random Graph intuition. The game-theoretic result of Erdős and Chvátal shows a striking similarity with the following well-known *Random Graph* result. The duration of a $(1:b)$ play allows for approximately $n^2/2(b+1)$ Maker's edges. In particular, if $b = cn/\log n$, then Maker will have the time to create a graph with $n\log n/2c$ edges. It is well known that a *Random Graph* with n vertices and $n\log n/2c$ edges is almost certainly disconnected for $c > 1$; in fact, it has many isolated vertices(!). This is an old result of Erdős and Rényi.

By using this "Random Graph intuition" Erdős suspected that the "breaking point" for the Connectivity Game (and also for the Hamilton Cycle Game, where Maker wants to own a whole Hamiltonian cycle, see below) should come around $b = n/\log n$.

Erdős's "Random Graph intuition," together with the Erdős–Selfridge Theorem, are the two pioneering results of the "fake probabilistic method."

After this short *detour*, we return to Breaker's "isolated-point-forcing" strategy mentioned above: if $b = (1+\varepsilon)n/\log n$ with some $\varepsilon > 0$, then Breaker can occupy a star of $(n-1)$ edges, forcing an isolated point in Maker's graph. How can Breaker do it? Well, Breaker proceeds in two stages. In the First Stage, Breaker claims a K_m with $m \le b/2$ in the strong sense that no Maker's edge has a common endpoint with this K_m. In the Second Stage, Breaker turns one of the m vertices of K_m into an isolated vertex in Maker's graph.

First Stage. It goes by a simple induction in (at most) m moves. Assume that, at some point in the first $(i-1)$ moves, Breaker has created a K_{i-1} such that none of Maker's edges has an endpoint in $V(K_{i-1})$ ("vertex set of K_{i-1}"). At that point of the play, Maker owns $\le (i-1)$ edges, so if $i < n/2$, there are at least two vertices u, v in the complement of $V(K_{i-1})$ that are incident with none of Maker's edges. If $i \le b/2$, then in his next move Breaker can take the $\{u, v\}$ edge plus the $2(i-1)$ edges joining $\{u, v\}$ to $V(K_{i-1})$, thereby enlarging K_{i-1} by 2 vertices. In his next move, Maker can kill 1 vertex from this $V(K_{i+1})$ (by claiming an edge incident with that vertex), but a clique K_i of i vertices will certainly survive. This completes the induction step.

Second Stage. At the end of the First Stage the vertices of K_m define m edge-disjoint stars; each star consists of $n - m$ edges, which are as yet completely unoccupied. Breaker's goal in the Second Stage is to own one of these m stars. In terms of hypergraphs, this is the "disjoint game," where the hyperedges are pairwise disjoint. The "disjoint game" represents the simplest possible case; the analysis of the game is almost trivial.

If Maker takes an edge from a star, the star becomes "dead" (Breaker cannot own it), and it is removed from the game. In each move Breaker divides his b edges among the "survivor" stars as evenly as possible. After i moves Breaker's part in a "survivor" star is

$$\geq \left\lfloor \frac{b}{m} \right\rfloor + \left\lfloor \frac{b}{m-1} \right\rfloor + \left\lfloor \frac{b}{m-2} \right\rfloor + \cdots + \left\lfloor \frac{b}{m-i+1} \right\rfloor \geq b \sum_{j=m-i+1}^{m} \frac{1}{j} + i$$

$$= b \log \left(\frac{m}{m-i+1} \right) - O(b+i). \tag{20.1}$$

By choosing

$$m = \frac{b}{\log n} = (1+o(1))\frac{n}{(\log n)^2} \quad \text{and} \quad i = m-1,$$

the right-hand side of (20.1) becomes $b \log m - O(b) \geq n \geq n - m$, proving that Breaker can completely occupy a star. This proves the Erdős–Chvátal Theorem (see [1978]) that, if $b = (1+\varepsilon)n/\log n$ with some $\varepsilon > 0$, then, playing the $(1:b)$ game on K_n, where n is large enough, Breaker can force an isolated point in Maker's graph (Maker is the underdog).

In the other direction, Erdős and Chvátal proved the following: if $b = \frac{1}{4}n/\log n$, then, playing the $(1:b)$ game on K_n, underdog Maker can build a spanning tree in K_n.

The "weakness" of this result is the constant factor $\frac{1}{4}$; indeed, the Random Graph intuition mentioned above suggests $1 - o(1)$ instead of $\frac{1}{4}$. Can we replace $\frac{1}{4}$ with $1 - o(1)$? Unfortunately, this innocent-looking problem has remained unsolved for nearly 30 years.

Open Problem 20.1 *Consider the $(1:b)$ Connectivity Game on the complete graph K_n. Is it true that, if $b = (1 - o(1))n/\log n$ and n is large enough, then underdog Maker can build a spanning tree?*

The constant factor $\log 2 = 0.693$ (see see Beck [1982]) is somewhat better than the Erdős–Chvátal constant $1/4$ mentioned above. The main reason to include this proof (instead of the Erdős–Chvátal proof) is to illustrate a new idea: the so-called "Transversal Hypergraph Method." In the Erdős–Chvátal proof, Maker is directly building a spanning tree; this proof is indirect: Maker prevents Breaker from occupying a whole "cut." This proof applies the following *biased* version of the Erdős–Selfridge Theorem (see Beck [1982]).

Theorem 20.1 *Playing the $(p:q)$ game on a finite hypergraph $\mathcal{F}$, where Maker takes p new points and Breaker takes q new points per move, if*

$$\sum_{A \in \mathcal{F}} (1+q)^{-|A|/p} < \frac{1}{1+q}, \tag{20.2}$$

then Breaker, as the second player, can put his mark in every $A \in \mathcal{F}$.

The proof of Theorem 20.1 is postponed to the end of this section.

2. Building via blocking: the Transversal Hypergraph Method. (20.2) is a "blocking" criterion, and we use it to build a spanning tree. How is it possible?

Building and Blocking are complementary properties; to use one to achieve the other sounds like a paradox. But, of course, there is no paradox here; the explanation is that *connectivity* is a simple graph property: it has a "good characterization," which means that both *connectivity* and its complement, *disconnectivity*, can be desribed in simple terms.

Indeed, *connectivity* means to have a spanning tree, and *disconnectivity* means to have a *cut* (i.e. a partition of the vertex set into 2 parts such that there is no edge between the parts).

The early developments of Graph Theory focused on the graph properties which have a good characterization (e.g. planarity, 1-factor). Unfortunately, the list is very short; the overwhelming majority of the interesting graph properties do not have a "good characterization" (e.g., to decide whether a graph contains a Hamiltonian cycle).

Let us return to connectivity; as said before, *disconnectivity* of a graph means the existence of a *cut*, where a *cut* means the absence of a complete bipartite graph $K_{a,b}$ with $1 \leq a, b \leq n-1$, $a+b = n$.

By using hypergraph terminology, we can say the following: the complete bipartite graphs $K_{a,b}$ with $1 \leq a, b \leq n-1$, $a+b = n$ are exactly the minimal elements of the *transversal* of the hypergraph of all spanning trees in K_n.

This statement doesn't make any sense unless the reader knows the definition of the *transversal hypergraph*. For an arbitrary finite hypergraph $\mathcal{F}$ write

$$\text{Transv}(\mathcal{F}) = \{S \subset V(\mathcal{F}) : S \cap A \neq \emptyset \text{ for all } A \in \mathcal{F}\};$$

$\text{Transv}(\mathcal{F})$ is called the *transversal hypergraph* of $\mathcal{F}$.

Consider a Maker–Breaker play on hypergraph $\mathcal{F}$. At the end of a play the two players split the board

$$V = V(\mathcal{F}) = M \cup B,$$

where M is the set of Maker's points and B is the set of Breaker's points. If Maker doesn't win, then $B \in \text{Transv}(\mathcal{F})$. It follows that, if Maker can block $\text{Transv}(\mathcal{F})$, then at the end Maker owns a whole $A \in \mathcal{F}$. This is how Maker can build via blocking. This is referred to as the Transversal Hypergraph Method.

This is a very general method; the Connectivity Game is a good example – for a few more instances (see Beck [1993c]).

In the Connectivity Game, Maker wants to prevent Breaker from occupying a cut. Here are the details. Let $\mathcal{H}_n$ denote the family of all spanning complete bipartite sub-graphs $K_{t,n-t}$, $1 \leq t \leq (n-1)/2$ of K_n. As said before, $\mathcal{H}_n$ is the set of minimal elements of the transversal of the hypergraph of all spanning trees in K_n. Maker applies the Transversal Hypergraph Method: it suffices to block hypergraph $\mathcal{H}_n$. By the underdog blocking criterion (20.2) with $p = b$ and $q = 1$, where Maker is the second player, we just have to check the inequality

$$\sum_{K_{t,n-t}\in\mathcal{H}_n} 2^{-|K_{t,n-t}|/b} = \sum_{t=1}^{(n-1)/2} 2^{-t(n-t)/b} < \frac{1}{2} \tag{20.3}$$

with $b = (\log 2 - \varepsilon)n/\log n$ and $n > n_0(\varepsilon)$ ("n is large enough").

Inequality (20.3) is just a routine calculation. Indeed, by using the easy fact $\binom{n}{t} \leq (en/t)^t$ ("Stirling formula"), we have

$$\sum_{t=1}^{(n-1)/2} 2^{-t(n-t)/b} \leq \sum_{t=1}^{(n-1)/2} \left(\frac{en}{t} 2^{-(n-t)/b}\right)^t. \tag{20.4}$$

To evaluate the right-hand side of (20.4) we distinguish 2 cases. If t is in the range $1 \leq t \leq \sqrt{n}$ and $n > n_1(\varepsilon)$, then

$$\frac{n-t}{b} \geq \frac{n-\sqrt{n}}{(\log 2 - \varepsilon)n/\log n} \geq \frac{(1+\varepsilon)\log n}{\log 2},$$

and so

$$\frac{en}{t} 2^{-(n-t)/b} \leq en \cdot 2^{-(1+\varepsilon)\log n/\log 2} = \frac{en}{n^{1+\varepsilon}} \leq \frac{1}{5}$$

holds for $n > n_2(\varepsilon)$. Therefore, if $n > \max\{n_1(\varepsilon), n_2(\varepsilon)\}$, then by (20.4)

$$\sum_{t=1}^{\sqrt{n}} 2^{-t(n-t)/b} \leq \sum_{t=1}^{\sqrt{n}} \left(\frac{en}{t} 2^{-(n-t)/b}\right)^t$$

$$\leq \sum_{t=1}^{\sqrt{n}} 5^{-t} < \sum_{t=1}^{\infty} 5^{-t} = \frac{1}{4}. \tag{20.5}$$

Next consider the range $\sqrt{n} \leq t \leq (n-1)/2$; then

$$\frac{n-t}{b} \geq \frac{n}{2b} = \frac{n}{2(\log 2 - \varepsilon)n/\log n} \geq \frac{(1+\varepsilon)\log n}{2\log 2},$$

and so

$$\frac{en}{t} 2^{-(n-t)/b} \leq \frac{en}{\sqrt{n}} \cdot 2^{-(1+\varepsilon)\log n/2\log 2} = \frac{en}{\sqrt{n} \cdot n^{(1+\varepsilon)/2}} = \frac{e}{n^{\varepsilon/2}} \leq \frac{1}{5}$$

holds for $n > n_3(\varepsilon)$. Thus for $n > n_3(\varepsilon)$,

$$\sum_{t=\sqrt{n}}^{(n-1)/2} 2^{-t(n-t)/b} \leq \sum_{t=\sqrt{n}}^{(n-1)/2} \left(\frac{en}{t} 2^{-(n-t)/b}\right)^t$$

$$\leq \sum_{t=\sqrt{n}}^{(n-1)/2} 5^{-t} < \sum_{t=1}^{\infty} 5^{-t} = \frac{1}{4}. \tag{20.6}$$

(20.5) and (20.6) imply (20.3) for $n > \max\{n_1(\varepsilon), n_2(\varepsilon), n_3(\varepsilon)\}$. This completes the proof of part (2) in the following (see Erdős and Chvátal [1978] and Beck [1982]):

Theorem 20.2 *Consider the (1:b) Connectivity Game on K_n, where Maker is the underdog:*

(1) If $b = (1 + o(1))n/\log n$, then Breaker has a winning strategy.
(2) If $b = (\log 2 - o(1))n/\log n$, then Maker has a winning strategy.

It is rather disappointing that the constant factors in (1) and (2) do not coincide, especially that *connectivity* is an "easy" property.

How about a typical "hard" property like to have a Hamiltonian cycle? In other words, what happens if Maker wants a Hamiltonian cycle ("round-trip") instead of a spanning tree (of course, a Hamiltonian cycle contains a spanning tree)? Where is the game-theoretic breaking point? Erdős's *Random Graph intuition* suggests the game-theoretic breaking point to be around the very same $b = n/\log n$. Indeed, a Random Graph with n vertices and $(\frac{1}{2} + o(1))n \log n$ edges almost certainly contains a Hamiltonian cycle. This is a famous theorem in the theory of Random Graphs, due to the joint effort of several mathematicians like Pósa, Korshunov, Komlós–Szemerédi, and Bollobás. The main difficulty is that we cannot describe the non-existence of a Hamiltonian cycle in simple terms.

The best that we can prove is a constant factor weaker than the conjectured truth. Unfortunately, the constant factor $c = \frac{\log 2}{27} = .02567$ is rather weak (see Beck [1985]).

Theorem 20.3 *If $b = (\frac{\log 2}{27} - o(1))n/\log n = (.02567 - o(1))n/\log n$, then playing the (1:b) Hamiltonian Cycle Game on K_n, underdog Maker can build a Hamiltonian cycle.*

Proof. We need 4 ideas: (1) the so-called Pósa Lemma, (2) the Longer-Path Argument, (3) the Transversal Hypergraph Method, and (4) the Trick of Fake Moves.

The first two ideas, the Pósa Lemma and the Longer-Path Argument, are simply borrowed from Graph Theory; the other two – the Transversal Hypergraph Method and the Trick of Fake Moves – are the new, game-specific ideas. The Transversal Hypergraph Method guarantees that Maker's graph possesses some fundamental properties of Random Graphs, like "expander" type properties (provided Maker plays rationally). The Trick of Fake Moves is the real novelty here; the price that we pay for it is the poor constant factor $c = \frac{\log 2}{27} = .02567$.

We begin with the so-called Pósa Lemma: it says (roughly speaking) that every *expander* graph has a long path. We need a notation: given a simple and undirected graph G, and an arbitrary subset S of the vertex-set $V(G)$ of G, denote by $\Gamma_G(S)$ the set of vertices in G adjacent to at least 1 vertex of S. Let $|S|$ denote the number of elements of a set S.

The following lemma is in fact a technical refinement of the Pósa Lemma, see Pósa [1976]. The figure below illustrates the "Pósa deformation." $\square$

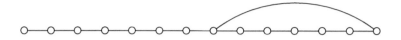

Pósa-deformation

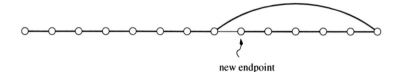

new endpoint

Lemma 1: *Let G be a non-empty graph, $v_0 \in V(G)$ and consider a path $P = (v_0, v_1, \ldots, v_m)$ of maximum length, which starts from vertex v_0. If $\{v_i, v_m\} \in G$ ($1 \le i \le m - 1$), then we say that the path $(v_0, \ldots, v_i, v_m, v_{m-1}, \ldots, v_{i+1})$ arises by a Pósa-deformation from P. Let $End(G, P, v_0)$ denote the set of all endpoints of paths arising by repeated Pósa-deformations from P, keeping the starting point v_0 fixed. Assume that for every vertex-set $S \subset V(G)$ with $|S| \le k$, $|\Gamma_G(S) \setminus S| \ge 2|S|$. Then $|End(G, P, v_0)| \ge k + 1$.*

Proof of Lemma 1 (due to Pósa). Let $X = End(G, P, v_0) = \{v_{i_1}, v_{i_2}, \ldots, v_{i_r}, v_m\}$ where $0 \le i_1 < i_2 < \ldots < i_r < m$. The main difficulty is to show that

$$\Gamma_G(X) \subset \{v_{m-1}, v_{i_l \pm 1} : l = 1, 2, \ldots, r\} \cup X. \tag{20.7}$$

To prove (20.7), let $y \in \Gamma_G(X)$ but $y \notin X$. Then y is adjacent to a point $v \in X$. By definition v is an endpoint of a maximum path whose points are the points of P (in different order), so y must belong to P; let $y = v_\nu$.

By the definition of vertex v, there is a sequence of paths $P_0 = P, P_1, P_2, \ldots, P_s$ such that P_{i+1} arises from P_i by a Pósa-deformation ($i = 0, 1, \ldots, s - 1$) and v is an endpoint of P_s.

Case 1: If both edges $\{v_\nu, v_{\nu-1}\}$ and $\{v_\nu, v_{\nu+1}\}$ belong to path P_s, then let (say) $\{v_\nu, v_{\nu+1}\}$ be the first edge on the $(y = v_\nu, v)$-arc of P_s. Then $v_{\nu+1}$ is an endpoint of a path arising from P_s by a Pósa-deformation, implying $v_{\nu+1} \in X$. This proves (20.7) in Case 1.

Case 2: Suppose that (say) edge $\{v_\nu, v_{\nu+1}\}$ does not belong to path P_s. Then there is a largest index j in $0 \le j \le s - 1$ such that edge $\{v_\nu, v_{\nu+1}\}$ belongs to path P_j. Since P_{j+1} arises from P_j by a Pósa-deformation, this can only happen if one of the two vertices $y = v_\nu, v_{\nu+1}\}$ is the endpoint of P_{j+1}. Since by hypothesis $y \notin X$, we must have $v_{\nu+1} \in X$, which proves (20.7) in Case 2 as well.

Now by (20.7)

$$|\Gamma_G(X) \setminus X| < 2|X|,$$

which implies that $|X| = |End(G, P, v_0)| \ge k + 1$, and Lemma 1 follows. $\square$

It is worth while mentioning the following nice corollary of (the rather technical) Lemma 1: If Maker's graph is expanding by a factor of 2 up to size k, then the longest path has length $\geq 3k$.

Now we are ready to discuss the:

Basic idea of the proof: the Longer-Path Argument. Assume that we are in the middle of a play, let G denote Maker's graph (G is "growing"), and let $P = \{v_0, v_1, \ldots, v_m\}$ be a maximum length path in G. Also assume that Maker's graph G satisfies the following two properties:

(α) for every vertex-set $S \subset V(G)$ with $|S| \leq k$, $|\Gamma_G(S) \setminus S| \geq 2|S|$;
(β) G is connected on n vertices.

The common idea in most Hamiltonian cycle proofs is to produce a new path *longer* than P. Repeating this argument several times Maker will end up with a Hamiltonian cycle.

How can Maker produce a *longer* path? Well, let $End(G, P, v_0) = \{x_1, x_2, \ldots, x_q\}$, where $q \geq k+1$ (see Lemma 1), and denote by $P(x_i)$ a path such that (1) $P(x_i)$ arises from P by a sequence of Pósa-deformations, and (2) x_i is an endpoint of $P(x_i)$. Again by Lemma 1 $|End(G, P(x_i), x_i)| \geq k+1$ for every $x_i \in End(G, P, v_0)$.

Consider the following critical sub-graph:

$$Close(G, P) = \Big\{\{x_i, y\} : x_i \in End(G, P, v_0), y \in End(G, P(x_i), x_i), i = 1, 2, \ldots, q\Big\};$$

trivially $|Close(G, P)| \geq (k+1)^2/2$.

Notice that sub-graph $Close(G, P)$ consists of the "closing edges": the edges which turn a maximum path into a cycle on the same set of vertices.

Case 1: Maker already owns an edge from $Close(G, P)$.

Then Maker's graph contains a cycle C of $m+1$ edges, where $m = |P|$ is the number of edges in the longest path P in G. If $m+1 = n$, then C is a Hamiltonian cycle, and Maker wins the play. If $m+1 < n$, then there is a vertex w outside of C, and by the connectivity of G (see property (β)), there is a path P^* outside of C joining vertex w to C. Let v be the endpoint of path P^* on C; deleting an edge of C with endpoint v and adding path P^* to C, we obtain a new path in G, which is longer than P, a contradiction.

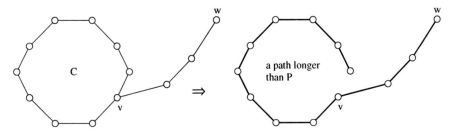

a path longer than P

This contradiction proves that in Case 1 Maker's graph already contains a Hamiltonian cycle.

Case 2: There is an edge in $Close(G, P)$, which is unoccupied in the play yet.
Let $e \in Close(G, P)$ be an unoccupied edge; Maker's next move is to take edge e. Then, repeating the argument of Case 1, we obtain a path *longer* than P.

There are exactly 2 reasons why this Making-a-Longer-Path procedure may stop: either (1) Maker's graph already contains a Hamiltonian cycle (and Maker wins!), or (2) Case 3 (see below) holds.

Case 3: Every edge in $Close(G, P)$ is already occupied by Breaker.
Then, of course, Breaker owns at least $|Close(G, P)| \geq (k+1)^2/2$ edges, requesting at least $\frac{(k+1)^2}{2b}$ moves.

This way we have proved:

Lemma 2: *If Maker can guarantee that after the tth move with $t = \left(\frac{(k+1)^2}{2b} - n \right)$ his graph satisfies properties (α) and (β), then with at most n extra moves he will always own a Hamiltonian cycle.* $\qquad\square$

Notice that Lemmas 1–2 are pure graph-theoretic results; the rest of the proof is the game-specific argument.

Maker is going to force properties (α) and (β) by using the Transversal Hypergraph Method, i.e. Maker builds via blocking.

Let $k \leq n/3$ be a parameter to be specified later, and let $\mathcal{G}_{n,k}$ be the set of all complete bipartite graphs of type

$$K_{j,n-3j+1} \subset K_n \text{ where } j = 1, 2, \ldots, k.$$

Hypergraph $\mathcal{G}_{n,k}$ is clearly motivated by property (α): If Maker can occupy at least 1 edge from each $K_{j,n-3j+1} \in \mathcal{G}_{n,k}$ during the first $t = \left((k+1)^2/2b - n \right)$ moves, then property (α) is clearly satisfied.

Notice that property (α) implies property (β). Indeed, every spanning complete bipartite sub-graph $K_{l,n-l}$ with $1 \leq l \leq (n-1)/2$ clearly contains a whole $K_{j,n-3j+1} \in \mathcal{G}_{n,k}$, and disconnectivity means the absence of a $K_{l,n-l}$.

Maker's goal is, therefore, to block hypergraph $\mathcal{G}_{n,k}$ during the first $t = \left((k+1)^2/2b - n \right)$ moves (with an appropriate choice of parameter $k \leq n/3$). Of course, Maker can try applying the underdog blocking criterion (20.2), but there is a novel **technical difficulty** with the application: the potential technique guarantees blocking at the end only, but what Maker needs is blocking at a relatively early stage of the play, namely after the tth move with $t = \left((k+1)^2/2b - n \right)$.

We overcome this technical difficulty by using a trick: We employ a large number of "fake moves." The auxiliary "fake moves" speed up the time, and turn the early stage into the end stage. This way we can successfully apply the underdog blocking criterion (20.2).

The Trick of Fake Moves. We discuss the trick in a general setup: let $\mathcal{G}$ be an arbitrary finite hypergraph with board $V = V(\mathcal{G})$. Consider the $(1:b)$ play on $\mathcal{G}$, where the first player takes 1 new point and the second player takes b new points per move. Let $q > 1$; the first player wants to block every $A \in \mathcal{G}$ at an early stage of the play, say when the (topdog) second player owns $\leq |V|/q$ points of board V. When can the first player achieve this early stage blocking? Here is a sufficient criterion: if

$$\sum_{A \in \mathcal{G}} 2^{-\frac{|A|}{bq}} < 1, \tag{20.8}$$

then the first player can block every $A \in \mathcal{G}$ at an early stage of the play when the (topdog) second player owns $\leq |V|/q$ points of board V.

The *proof* of criterion (20.8) goes as follows. The first player defines a fake $(1:bq)$ play, and uses the ordinary potential function blocking strategy Str (see (20.2)) for this $(1:bq)$ play on hypergraph $\mathcal{G}$.

Let $x1$ and $Y(1) = \left\{ y_1^{(1)}, y_1^{(2)}, \ldots, y_1^{(b)} \right\}$ denote the first moves of the two players; these are "real" moves. To define the fake $(1:bq)$ play, the first player selects an arbitrary $(q-1)b$-element subset $Z(1)$ from $V \setminus (\{x_1\} \cup Y(1))$, and adds $Z(1)$ to $Y(1)$

$$W(1) = Y(1) \cup Z(1).$$

Of course, the second player does not know about set $Z(1)$, it is in first player's mind only. $W(1)$ is a qb-element set, so the first player can apply potential strategy Str. Strategy Str gives first player's second move $x_2 = Str(\{x_1\}; W(1))$; in other words, the "real" game is a $(1:b)$ game, but the first player pretends to play a $(1:bq)$ game, and chooses his next moves accordingly.

Let $Y(2) = \left\{ y_2^{(1)}, y_2^{(2)}, \ldots, y_2^{(b)} \right\}$ denote second player's second move; the set $Y(2)$ may or may not intersect $W(1)$; in any case the first player selects an arbitrary $(qb - |W(1) \cap Y(2)|)$-element subset $Z(2)$ from $V \setminus (\{x_1, x_2\} \cup Y(1) \cup Y(2))$, and adds $Z(2)$ to $Y(2)$

$$W(2) = Y(2) \cup Z(2).$$

$W(2)$ is a qb-element set, disjoint from $\{x_1, x_2\} \cup W(1)$, so the first player can apply potential strategy Str. Strategy Str gives first player's third move $x_3 = Str(\{x_1, x_2\}; W(1), W(2))$; in other words, the first player pretends to play a $(1:bq)$ game, and chooses his next moves accordingly.

Let $Y(3) = \left\{ y_3^{(1)}, y_3^{(2)}, \ldots, y_3^{(b)} \right\}$ denote second player's third move; the set $Y(3)$ may or may not intersect $W(1) \cup W(2)$. In any case the first player selects an arbitrary $(qb - |(W(1) \cup W(2)) \cap Y(3)|)$-element subset $Z(3)$ from $V \setminus (\{x_1, x_2, x_3\} \cup Y(1) \cup Y(2) \cup Y(3))$, and adds $Z(3)$ to $Y(3)$

$$W(3) = Y(3) \cup Z(3).$$

$W(3)$ is a qb-element set, disjoint from $\{x_1, x_2, x_3\} \cup W(1) \cup W(2)$, so the first player can apply potential strategy *Str*, and so on.

The whole point is that, because of the artificially constructed fake sets $Z(1), Z(2), \ldots$, the fake $(1{:}bq)$ play will end at the early stage of the real $(1{:}b)$ play, where the second player owns $\leq |V|/q$ points. An application of criterion (20.2) proves criterion (20.8).

Conclusion of the proof. Before applying criterion (20.8) it is worth while noticing that parameter q is not necessarily an integer; all what we need in the proof is that qb is an integer.

Now we are ready for the application: let $\varepsilon > 0$ be arbitrarily small but fixed; let

$$k = \left\lfloor \left(\frac{1}{3} - \varepsilon\right) n \right\rfloor \quad \text{and} \quad b = \left\lfloor \left(\frac{\log 2}{27} - \varepsilon\right) \frac{n}{\log n} \right\rfloor$$

(lower integral part), and define integer qb as the upper integral part

$$qb = \left\lceil \binom{n}{2} / t \right\rceil \quad \text{where} \quad t = \frac{(k+1)^2}{2b} - n.$$

Then $t \geq 1/qb\binom{n}{2}$, and in view of Lemma 2 and criterion (20.8), it suffices to check the inequality

$$\sum_{K_{j,n-3j+1} \in \mathcal{G}_{n,k}:\ 1 \leq j \leq k} 2^{-|K_{j,n-3j+1}|/qb} < 1. \tag{20.9}$$

The proof of (20.9) is just a routine calculation. Indeed, the left-hand side of (20.9) equals

$$\sum_{j=1}^{k} \binom{n}{j} \binom{n-j}{2j-1} 2^{-j(n-3j+1)/r}, \tag{20.10}$$

where

$$r = qb = \left\lceil \binom{n}{2} / t \right\rceil \leq \frac{(1-\varepsilon)\log 2}{3} \cdot \frac{n}{\log n}.$$

By using the elementary fact $\binom{n}{j} \leq (en/j)^j$, we can estimate (20.10) from above as follows

$$\sum_{j=1}^{k} \binom{n}{j} \binom{n}{2j} 2^{-j(n-3j+1)/r} \leq \sum_{1 \leq j \leq (1/3-\varepsilon)n} \left(\left(\frac{en}{j}\right) \cdot \left(\frac{en}{2j}\right)^2 \cdot 2^{-(n-3j+1)/r} \right)^j$$

$$\leq \sum_{1 \leq j \leq (1/3-\varepsilon)n} \left(\frac{e^3 n^3}{2j^3} \cdot n^{-(3-9j/n)/(1-\varepsilon)} \right)^j. \tag{20.11}$$

A trivial calculation shows that

$$\max_{1 \leq j \leq (1/3-\varepsilon)n} \frac{n^3}{j^3} \cdot n^{-(3-9j/n)/(1-\varepsilon)} \leq \frac{1}{100} \quad \text{if } n > n_0(\varepsilon). \tag{20.12}$$

Thus by (20.11)–(20.12)

$$\sum_{j=1}^{k} \binom{n}{j}\binom{n}{2j} 2^{-j(n-3j+1)/r} \leq \sum_{j=1}^{k} 5^{-j} \leq \sum_{j=1}^{\infty} 5^{-j} = \frac{1}{4}, \qquad (20.13)$$

if $n > n_0(\varepsilon)$. Finally, (20.13) and (20.10) trivially imply (20.9), which completes the proof of Theorem 20.3. □

Concluding remark. The basic idea of the proof of Theorem 20.3 was the Longer-Path Argument: taking an unoccupied edge from the sub-graph $Close(G, P)$ produced a longer path. Of course, Maker is eager to take an edge from $Close(G, P)$, but how about the Reverse Hamiltonian Game? Reluctant Avoider may want to stay away from $Close(G, P)$ as long as he can, and then the argument above seems to collapse. Can the reader still save the proof for the Reverse Hamiltonian Game?

Open Problem 20.2 *Consider the Reverse Hamiltonian Game, played on K_n, where Avoider takes 1 and Forcer takes f new edges per move; Forcer wins if, at the end, Avoider's graph contains a Hamiltonian cycle. Is it true that, if $f = c_0 n / \log n$ for some positive absolute constant and n is large enough, then Forcer can force Avoider to own a Hamiltonian cycle?*

By the way, apart from the "Random Graph intuition" we don't know any *a priore* reason why the Achievement and the Avoidance (i.e. Reverse) Hamiltonian Games should have the same breaking point. Is it true that they have exactly the same breaking point?

Recently Hefetz, Krivelevich, and Szabó [2007] came very close to solving Open Problem 20.2: they could prove the weaker version with

$$f = c_0 \frac{n}{(\log n)(\log \log n)}, \qquad (20.14)$$

i.e. it falls short of a factor of $\log \log n$. Very recently Hefetz, Krivelevich, and Szabó announced a positive solution of Open Problem 20.2.

3. Proofs of the biased criterions. As promised at the beginning of the section, there follows a proof of:

Theorem 20.1 *Playing the $(p{:}q)$ Achievement game on a finite hypergraph $\mathcal{F}$, where Maker takes p new points and Breaker takes q new points per move, if*

$$\sum_{A \in \mathcal{F}} (1+q)^{-|A|/p} < \frac{1}{1+q},$$

then Breaker, as the second player, can put his mark in every $A \in \mathcal{F}$.

Note in advance that the proof of Theorem 20.1 can be easily adapted in the $(a{:}1)$ Avoidance version (the general Avoidance case, however, remains a mystery!).

Theorem 20.4 *Let $a \geq 2$ be an arbitrary integer, and let $\mathcal{F}$ be an n-uniform hypergraph with size*

$$|\mathcal{F}| < \left(\frac{a+1}{a}\right)^n.$$

Then, playing the $(a:1)$ Avoidance Game on $\mathcal{F}$, where Forcer is the underdog, Avoider (as the first player) has a winning strategy.

Remark. We learned very recently that Theorem 20.4 was independently discovered and proved in Hefetz, Krivelevich, and Szabó [2007]. They used it to prove bound (20.14) for the Reverse Hamiltonian cycle game. We are going to use it to prove the exact solution for the $(a:1)$ Avoidance versions of the Lattice Games (see Theorem 9.2).

Exercise 20.1 *By using the proof technique of Theorem 20.1 prove Theorem 20.4.*

We will return to Theorem 20.4 in Section 30, where 2 proofs are included and some applications are shown.

Now I give a:

Proof of Theorem 20.1. We basically repeat the proof of the Erdős–Selfridge Theorem, but we work with the powers of a suitable $(1 + \mu)$ where parameter $\mu > 0$ will be specified later. It is also very important to have a good notation.

Given a hypergraph $\mathcal{G}$ and two disjoint subsets X and Y of the board $V = V(\mathcal{G})$, write

$$\phi(X, Y, \mathcal{G}) = \sum_{A \in \mathcal{G}:\ A \cap Y = \emptyset} (1 + \mu)^{-|A \setminus X|}. \tag{20.15}$$

For an arbitrary $z \in V(\mathcal{G})$, write

$$\phi(X, Y, \mathcal{G}, z) = \sum_{z \in A \in \mathcal{G}:\ A \cap Y = \emptyset} (1 + \mu)^{-|A \setminus X|}. \tag{20.16}$$

We are going to repeatedly use the following two completely trivial inequalities

$$\phi(X, Y \cup \{y_1\}, \mathcal{G}, y_2) \leq \phi(X, Y, \mathcal{G}, y_2), \tag{20.17}$$

$$\phi(X \cup \{x_1\}, Y, \mathcal{G}, x_2) \leq (1 + \mu)\phi(X, Y, \mathcal{G}, x_2). \tag{20.18}$$

Assume we are in the middle of a $(p:q)$ play, Maker (the first player) already occupied

$$X_i = \left\{ x_1^{(1)}, \ldots, x_1^{(p)}, x_2^{(1)}, \ldots, x_2^{(p)}, \ldots, x_i^{(1)}, \ldots, x_i^{(p)} \right\} \tag{20.19}$$

and Breaker (the second player) occupied

$$Y_i = \left\{ y_1^{(1)}, \ldots, y_1^{(q)}, y_2^{(1)}, \ldots, y_2^{(q)}, \ldots, y_i^{(1)}, \ldots, y_i^{(q)} \right\}. \tag{20.20}$$

Let

$$X_{i,j} = X_i \cup \{x_{i+1}^{(1)}, \ldots, x_{i+1}^{(j)}\} \quad \text{and} \quad Y_{i,j} = Y_i \cup \{y_{i+1}^{(1)}, \ldots, y_{i+1}^{(j)}\}.$$

After the ith move of the first player the actual play defines a truncation of our hypergraph $\mathcal{F}$ as follows

$$\mathcal{F}_i = \{A \setminus X_i : A \in \mathcal{F} \text{ and } A \cap Y_{i-1} = \emptyset\}.$$

That is, we throw away the winning sets $A \in \mathcal{F}$, which are blocked by the second player, and from the unblocked winning sets we remove the first player's points. Write

$$\psi(\mathcal{F}_i) = \phi(X_i, Y_{i-1}, \mathcal{F}) = \sum_{B \in \mathcal{F}_i} (1 + \mu)^{-|B|}.$$

First player's Weak Win is equivalent to the fact that for some i the truncated hypergraph $\mathcal{F}_i$ contains the empty-set; in this case the contribution of the empty-set alone is $(1 + \mu)^0 = 1$, so if the second player can enforce the inequality $\psi(\mathcal{F}_i) < 1$ for the whole course of the play, then at the end of the play the second player blocks ever $A \in \mathcal{F}$.

Here is second player's blocking strategy: at his ith move for every $k = 1, \ldots, q$ he computes the value of $\phi(X_i, Y_{i-1}, \mathcal{F}, y)$ for each unoccupied $y \in V = V(\mathcal{F})$, and picks that $y = y_i^{(k)}$ for which the maximum is attained.

Let μ be defined by the equality $(1 + \mu) = (1 + q)^{1/p}$. We claim that by making this choice of parameter μ the inequality below holds

$$\psi(\mathcal{F}_{i+1}) \leq \psi(\mathcal{F}_i), \tag{20.21}$$

independently of first player's $(i+1)$th move. As usual, decreasing property (20.21) is the key step, the rest is standard.

To prove (20.21) note that

$$\psi(\mathcal{F}_{i+1}) = \psi(\mathcal{F}_i) - \sum_{k=1}^{q} \phi(X_i, Y_{i-1,k-1}, \mathcal{F}, y_i^{(k)}) + \mu \sum_{j=1}^{p} \phi(X_{i,j-1}, Y_i, \mathcal{F}, x_{i+1}^{(j)}). \tag{20.22}$$

By (20.17) for $k = 1, \ldots, q - 1$

$$\phi(X_i, Y_{i-1,k}, \mathcal{F}, y_i^{(k+1)}) \leq \phi(X_i, Y_{i-1,k-1}, \mathcal{F}, y_i^{(k+1)}),$$

and by the maximum property of $y_i^{(k)}$

$$\phi(X_i, Y_{i-1,k-1}, \mathcal{F}, y_i^{(k+1)}) \leq \phi(X_i, Y_{i-1,k-1}, \mathcal{F}, y_i^{(k)}),$$

so combining the last 2 inequalities we obtain

$$\phi(X_i, Y_{i-1,k}, \mathcal{F}, y_i^{(k+1)}) \leq \phi(X_i, Y_{i-1,k-1}, \mathcal{F}, y_i^{(k)}) \tag{20.23}$$

for $k = 1, \ldots, q - 1$. Similarly, for $j = 0, \ldots, p - 1$

$$\phi(X_i, Y_i, \mathcal{F}, x_{i+1}^{(j+1)}) \leq \phi(X_i, Y_{i-1,q-1}, \mathcal{F}, y_i^{(q)}), \tag{20.24}$$

and by (20.18), for $j = 1, \ldots, p$ and for any z

$$\phi(X_{i,j}, Y_i, \mathcal{F}, z) \leq (1+\mu)\phi(X_{i,j-1}, Y_i, \mathcal{F}, z). \tag{20.25}$$

By repeated application of (20.23), for $j = 1, \ldots, q$

$$\phi(X_i, Y_{i-1,q-1}, \mathcal{F}, y_i^{(q)}) \leq \phi(X_i, Y_{i-1,j-1}, \mathcal{F}, y_i^{(j)}). \tag{20.26}$$

By (20.24) and (20.25), for every $j = 0, \ldots, p-1$

$$\phi(X_{i,j}, Y_i, \mathcal{F}, x_{i+1}^{(j+1)}) \leq (1+\mu)^j \phi(X_i, Y_{i-1,q-1}, \mathcal{F}, y_i^{(q)}). \tag{20.27}$$

Returning to (20.22), by (20.26) and (20.27) we conclude

$$\psi(\mathcal{F}_{i+1}) \leq \psi(\mathcal{F}_i) - \left(q - \sum_{j=0}^{p-1} \mu(1+\mu)^j \right) \phi(X_i, Y_{i-1,q-1}, \mathcal{F}, y_i^{(q)}). \tag{20.28}$$

The choice $(1+\mu) = (1+q)^{1/p}$ gives

$$\sum_{j=0}^{p-1} \mu(1+\mu)^j = \mu \frac{(1+\mu)^p - 1}{(1+\mu)-1} = q,$$

i.e. (20.28) yields (20.21). The proof of Theorem 20.1 is complete. $\qquad\square$

Remark. An adaptation of the last proof works for the one-sided discrepancy case where Breaker wants more than just 1 point from every winning set: Breaker = Balancer in fact wants "his proportional share" from every winning set (it is not a problem if he has more than his share). This nice adaptation is due to András Pluhár; the details go as follows.

Let $\mathcal{F}$ be a finite n-uniform hypergraph. There are 2 players, Balancer and Unbalancer, who play a general $(b : u)$ game on $\mathcal{F}$: Balancer takes b points per move and Unbalancer takes u points per move. Balancer's goal is to force that, at the end of the play, he owns at least $b - \varepsilon/b + u$ part from every winning set $A \in \mathcal{F}$. Here $\varepsilon > 0$ is, of course, "small" – the smaller, the better.

Assume we are in the middle of a $(b : u)$ play: Balancer – for simplicity assume that he is the first player – already occupied

$$Y_i = \{y_1^{(1)}, \ldots, y_1^{(b)}, y_2^{(1)}, \ldots, y_2^{(b)}, \ldots, y_i^{(1)}, \ldots, y_i^{(b)}\},$$

and Unbalancer (the second player) occupied

$$X_i = \{x_1^{(1)}, \ldots, x_1^{(u)}, x_2^{(1)}, \ldots, x_2^{(u)}, \ldots, x_i^{(1)}, \ldots, x_i^{(u)}\}.$$

At this stage of the play, for every winning set $A \in \mathcal{F}$ define the "weight"

$$w_i(A) = (1-\lambda)^{|Y_i \cap A| - \frac{bn}{b+u}} \cdot (1+\tau)^{|X_i \cap A| - \frac{un}{b+u}},$$

and define the "total weight"

$$T_i(\mathcal{F}) = \sum_{A \in \mathcal{F}} w_i(A).$$

Note that λ and τ are unspecified (yet) parameters; we will optimize them later. We just note that $0 < \lambda < 1$ and $\tau > 0$.

By repeating the proof of Theorem 20.1, we obtain the following perfect analogue of inequality (20.28): Balancer can force that

$$T_{i+1}(\mathcal{F}) \le T_i(\mathcal{F}) - \left(b \cdot \lambda - \left((1+\tau)^u - 1\right)\right) \cdot \sum_{A:\ y_{i+1}^{(b)} \in A \in \mathcal{F}} (1-\lambda)^{|Y_{i,b-1} \cap A| - \frac{bn}{b+u}} \cdot (1+\tau)^{|X_i \cap A| - \frac{un}{b+u}},$$

where $Y_{i,b-1} = Y_i \cup \{y_{i+1}^{(1)}, \ldots, y_{i+1}^{(b-1)}\}$. It follows that, by choosing

$$b \cdot \lambda = (1+\tau)^u - 1, \tag{20.29}$$

we can enforce the critical "decreasing property" $T_{i+1}(\mathcal{F}) \le T_i(\mathcal{F})$.

Therefore

$$T_{end}(\mathcal{F}) \le T_{start}(\mathcal{F}) = |\mathcal{F}| \cdot (1-\lambda)^{-\frac{bn}{b+u}} \cdot (1+\tau)^{-\frac{un}{b+u}}. \tag{20.30}$$

If at the end of the play Unbalancer can own $\ge un/b+u+\Delta$ points from some $A_0 \in \mathcal{F}$, then, of course

$$(1-\lambda)^{-\Delta} \cdot (1+\tau)^{\Delta} \le T_{end}(\mathcal{F}). \tag{20.31}$$

Combining (20.30)–(20.31) we obtain

$$\Delta \le \frac{\log |\mathcal{F}| - \frac{n}{b+u}(b \log(1-\lambda) + u \log(1+\tau))}{\log(1+\tau) - \log(1-\lambda)}. \tag{20.32}$$

The last (routine!) step is to minimize the right-hand side of (20.32) under the side condition (20.29); this gives an upper bound on the one-sided discrepancy Δ.

To minimize the right-hand side of (20.32), we can use the approximations $\log(1 \pm \varepsilon) = \pm\varepsilon - (\varepsilon^2)/2 + O(\varepsilon^3)$ and $(1+\varepsilon)^u - 1 = u\varepsilon + \binom{u}{2}\varepsilon^2 + O(\varepsilon^3)$, where ε is "small," and after some long but routine calculations (that is left to the reader) we obtain the following upper bound: at the end of the play Balancer can own $\ge bn/b+u-\Delta$ points from all $A \in \mathcal{F}$ with

$$\Delta = 3\sqrt{n \log |\mathcal{F}|} \frac{b+u}{(b+u)^{3/2}}. \tag{20.33}$$

This result can be interpreted as the "easy direction" of Theorem 17.5.

Finally, we include a detailed proof of the "biased building criterion" Theorem 2.2 (which was introduced at the end of Section 2). The notation applied in the previous proof of Theorem 20.1 comes very handy here.

Theorem 2.2 ("biased building") *If*

$$\sum_{A \in \mathcal{F}} \left(\frac{p+q}{p}\right)^{-|A|} > p^2 \cdot q^2 \cdot (p+q)^{-3} \cdot \Delta_2(\mathcal{F}) \cdot |V(\mathcal{F})|,$$

where $\Delta_2(\mathcal{F})$ is the Max Pair-Degree of hypergraph $\mathcal{F}$, and $V(\mathcal{F})$ is the board, then the first player can occupy a whole winning set $A \in \mathcal{F}$ in the biased $(p:q)$ play on $\mathcal{F}$ (the first player takes p new points and the second player takes q new points per move).

Proof of Theorem 2.2. We use the notation $x_i^{(j)}$, $y_i^{(k)}$, X_i, Y_i, ϕ introduced at the beginning of the previous proof (see (20.15)–(20.16) and (20.19)–(20.20)). Let $X_0 = Y_0 = \emptyset$. Here is first player's building strategy: at his $(i+1)$st move $(i \geq 0)$ the first player computes the value of $\phi(X_i, Y_i, \mathcal{F}, x)$ for each unoccupied point $x \in V \setminus (X_i \cup Y_i)$, and sequentially picks $x_{i+1}^{(1)}, x_{i+1}^{(2)}, \ldots, x_{i+1}^{(p)}$ which have the maximum value.

Now let $\psi_i = \phi(X_i, Y_i, \mathcal{F})$. We want to show that $\psi_{end} > 0$, which clearly implies a Weak Win.

To prove $\psi_{end} > 0$ we estimate $\psi_{i+1} - \psi_i$ from below, i.e. we want to control the decrease. We have

$$\psi_{i+1} - \psi_i = \sum_{A \in \mathcal{F}:\ A \cap Y_{i+1} = \emptyset} \left\{ (1+\mu)^{-|A \setminus X_{i+1}|} - (1+\mu)^{-|A \setminus X_i|} \right\} - \sum_A^* (1+\mu)^{-|A \setminus X_i|},$$

$$(20.34)$$

where the summation $\sum^*$ is taken over all $A \in \mathcal{F}$ for which $A \cap Y_{i+1} \neq \emptyset$ but $A \cap Y_i = \emptyset$.

The first sum on the right-hand side of (20.34) equals

$$F = \sum_{j=1}^{p} \left\{ (1+\mu)^j - 1 \right\} \sum_A^{(j)} (1+\mu)^{-|A \setminus X_i|},$$

where the summation $\sum^{(j)}$ is taken over all $A \in \mathcal{F}$ for which $A \cap Y_{i+1} = \emptyset$ but $|A \cap (X_{i+1} \setminus X_i)| = j$. Using the trivial fact $(1+\mu)^j \geq 1 + j\mu$, we have

$$F \geq \sum_{j=1}^{p} j\mu \sum_A^{(j)} (1+\mu)^{-|A \setminus X_i|} = \mu \sum_{l=1}^{p} \phi(X_i, Y_{i+1}, \mathcal{F}, x_{i+1}^{(l)}). \qquad (20.35)$$

Let us introduce the sub-hypergraph $\mathcal{F}_i^j \subset \mathcal{F}$

$$\mathcal{F}_i^j = \left\{ A \in \mathcal{F} : x_{i+1}^{(j)} \in A, A \cap Y_i = \emptyset \text{ but } A \cap (Y_{i+1} \setminus Y_i) \neq \emptyset \right\}.$$

Obviously $|\mathcal{F}_i^j| \leq |Y_{i+1} \setminus Y_i| \Delta_2 = q \cdot \Delta_2$, and by definition

$$\phi(X_i, Y_{i+1}, \mathcal{F}, x_{i+1}^{(j)}) = \phi(X_i, Y_i, \mathcal{F}, x_{i+1}^{(j)}) - \sum_{A \in \mathcal{F}_i^j} (1+\mu)^{-|A \setminus X_i|}$$

$$\geq \phi(X_i, Y_i, \mathcal{F}, x_{i+1}^{(j)}) - |\mathcal{F}_i^j|(1+\mu)^{-2}$$

$$\geq \phi(X_i, Y_i, \mathcal{F}, x_{i+1}^{(j)}) - q \cdot \Delta_2 \cdot (1+\mu)^{-2}.$$

Therefore, by (20.35)

$$F \geq \mu \sum_{j=1}^{p} \phi(X_i, Y_i, \mathcal{F}, x_{i+1}^{(j)}) - pq \cdot \Delta_2 \cdot \mu(1+\mu)^{-2}. \tag{20.36}$$

On the other hand, the second sum on the right-hand side of (20.34) is less or equal to

$$\sum_{k=1}^{q} \phi(X_i, Y_i, \mathcal{F}, y_{i+1}^{(k)}).$$

Thus by (20.36) we have

$$\psi_{i+1} - \psi_i \geq \mu \sum_{j=1}^{p} \phi(X_i, Y_i, \mathcal{F}, x_{i+1}^{(j)})$$

$$- \sum_{k=1}^{q} \phi(X_i, Y_i, \mathcal{F}, y_{i+1}^{(k)}) - pq \cdot \Delta_2 \cdot \mu(1+\mu)^{-2}. \tag{20.37}$$

Since $x_{i+1}^{(j)}$ was picked before $y_{i+1}^{(k)}$, we have

$$\phi(X_i, Y_i, \mathcal{F}, x_{i+1}^{(j)}) \geq \phi(X_i, Y_i, \mathcal{F}, y_{i+1}^{(k)}) \text{ for every } 1 \leq j \leq p, \ 1 \leq k \leq q.$$

It follows that

$$\frac{q}{p} \sum_{j=1}^{p} \phi(X_i, Y_i, \mathcal{F}, x_{i+1}^{(j)}) \geq \sum_{k=1}^{q} \phi(X_i, Y_i, \mathcal{F}, y_{i+1}^{(k)}),$$

so by choosing $\mu = p/q$ in (20.37), we obtain

$$\psi_{i+1} - \psi_i \geq -\left(\frac{pq}{p+q}\right)^2 \Delta_2. \tag{20.38}$$

By repeated application of (20.38) we get the following lower bound for ψ_i

$$\psi_i \geq \psi_0 - i \cdot \left(\frac{pq}{p+q}\right)^2 \Delta_2. \tag{20.39}$$

Since

$$\psi_0 = \psi_{blank} = \sum_{A \in \mathcal{F}} (1+\mu)^{-|A|} = \sum_{A \in \mathcal{F}} \left(1+\frac{p}{q}\right)^{-|A|},$$

and the length of the play is $\leq \frac{|V|}{p+q}$, by (20.39)

$$\psi_{end} \geq \sum_{A \in \mathcal{F}} \left(\frac{p+q}{p}\right)^{-|A|} - \frac{|V|}{p+q} \cdot \left(\frac{pq}{p+q}\right)^2 \Delta_2 > 0, \tag{20.40}$$

where the positivity of the right-hand side of (20.40) is exactly the hypothesis of Theorem 2.2. This completes the proof. □

Part C

Advanced Weak Win – game-theoretic higher moment

Here is a nutshell summary of what we did in Part A: the goal of the first chapter was to introduce the basic concepts such as Positional Game, Weak Win, Strong Draw, and to demonstrate the power of the potential technique on several amusing examples. The goal of the second chapter was to formulate the main results such as Theorem 6.4 and Theorem 8.2 ("exact solutions"), and also the Meta-Conjecture, the main issue of the book.

Part B was a practice session for the potential technique.

In the forthcoming Parts C–D, we discuss the most difficult proofs, in particular the exact solutions of our Ramseyish games with 2-dimensional goals. Part C is the building part and Part D is (mainly) the blocking part.

In Part A, we introduced two simple "linear" criterions (Theorem 1.2 and Theorem 1.4), and gave a large number of applications. Here, in Part C, we develop some more sophisticated "higher moment" criterions. The motivation for "higher moment" comes from Probability Theory. The "higher moment" criterions are applied in a way very similar to how some of the main results of classical Probability Theory – such as the central limit theorem and the law of the iterated logarithm – are all based on higher moment techniques.

Note in advance that the last part of the book (Part D) also has a strong probabilistic flavor: Part D is about how to "sequentialize" the global concept of *statistical independence*.

A common technical feature of Parts C–D is to involve big auxiliary hypergraphs. The big auxiliary hypergraphs explain the "higher moment" in Part C, and explain why it is possible to find the exact solutions for large board size.

Parts C–D (together with Chapter IV) justify the name *fake probabilistic method*. The word "fake" refers to the fact that, when we actually define an optimal strategy, the "probabilistic" part completely disappears. The probabilistic argument is converted into a perfectly deterministic potential strategy such as – metaphorically speaking – how a caterpillar turns into a butterfly. This "metamorphosis" is a regular feature of the proofs.

The deep connection with Probability Theory is definitely surprising, but the most surprising fact is that the upper and lower bound techniques often *coincide*, supplying *exact* solutions for infinitely many natural games. This is why we dared to use the term "theory" in the title of the book. The author discovered this *coincidence* quite recently, only a couple of years ago. (The original, pre-exact-solution version of the book had the less of imposing title "Positional Games"; without the exact solutions it didn't deserve to be called a "theory.")

Chapter V
Self-improving potentials

In Chapter IV, we start to explore the connection between randomness and games. A more systematic study is made of the probabilistic approach, that is actually refered to as a "fake probabilistic method."

The main ingredients of the "fake probabilistic method" are:

(1) the two linear criterions ("Part A") – for some applications see Part B;

(2) the advanced Weak Win criterion together with the *ad hoc* method of Section 23 ("Part C");

(3) the BigGame–SmallGame Decomposition and its variants ("Part D").

The main result in Chapter V is (2): the Advanced Weak Win Criterion, a complicated "higher moment" criterion. It is complicated in many different ways:

(i) the form of the criterion is already rather complicated;

(ii) the proof of the criterion is long and complicated;

(iii) the application to the Clique Game requires complicated calculations.

This criterion basically solves the building part of the Meta-Conjecture (see Section 9).

21
Motivating the probabilistic approach

Let us return to Section 6: consider the Maker–Breaker version of the (K_N, K_q) Clique Game (we don't use the notation $[K_N, K_q]$ any more). How do we prove lower bound (6.1)? How can Maker build such a large clique?

1. Halving Argument. The Ramsey criterion Theorem 6.2, combined with the Erdős–Szekeres bound, gives the size $q = \frac{1}{2}\log_2 N$, which is roughly $\frac{1}{4}$ of the truth. We can easily obtain a factor of 2 improvement by using the Ramsey proof technique ("ramification") instead of the theorem itself. "Ramification" is an iterated "halving process," which has the natural limitation of the binary logarithm of N. In other words, a ramification strategy cannot build a clique K_q with q larger than $\log_2 N + O(1)$. In the following theorem we achieve the bound $\log_2 N + O(1)$, the natural limitation of the Ramsey technique. (The second "factor of 2 improvement" will be accomplished later in Sections 24–25.)

Theorem 21.1 *Consider the (K_N, K_q) Clique Game. If $N \geq 2^{q+2}$, then Maker can force a Weak Win; in fact, Maker can build a K_q of his own in less than 2^{q+2} moves.*

Remark. This result was discovered by Beck, and independently, by Pekec and Tuza. The following proof is due to Beck [2002a]. As a byproduct, Theorem 21.1 happens to give the best-known upper bound on the following "Move Number" question: "How long does it take to build a clique?" We will return to this question at the end of Section 25.

Proof. The trick is to combine the standard "Ramification" argument with the following "Sparse Subgraph Lemma." □

Lemma 1: *Let $G = (V, E)$ be a simple graph (i.e. no loops and there is at most 1 edge between 2 vertices). Two players, We call them First Player and Second Player, alternately occupy the vertices of G: at the end of the play they split the vertex-set V into 2 parts V' (First Player's points) and V'' (Second Player's points). Let $G(V'')$ denote the restriction of G to the vertex-set V'' (i.e. the induced* sub-graph). *First Player can always force that the number of edges of G(V′) is at most* 1/4 *of the number of edges of G.*

Proof of Lemma 1. Actually the following much more general statement is true.

Lemma 2: *Let V be a finite set, and let $\mathcal{F}$ be an n-uniform family of subsets of V. Two players, First Player and Second Player, alternately occupy the points of V. First Player can force that, at the end of the play, the number of sets $A \in \mathcal{F}$ completely occupied by him is at most $|\mathcal{F}|2^{-n}$.*

The **proof of Lemma 2** is almost identical to the proof of the Erdős–Selfridge Theorem. The only difference is that instead of picking a point of *maximum* weight, First Player picks his next point as a point of *minimum* weight. □

Lemma 1 is the special case of Lemma 2 with $n = 2$. □

Consider the complete graph K_{2^q}, and let V_0 be its vertex set: $|V_0| = 2^q$. Let $u_1 \in V_0$ be an arbitrary vertex. Then playing on K_{2^q} Maker (as the first player) can pick 2^{q-1} edges incident with u_1. Let $V_1 (\subset V_0)$ denote the set of other endpoints of these 2^{q-1} edges of Maker: $|V_1| = 2^{q-1}$. Consider the complete graph K_{V_1} on vertex-set V_1. The graph K_{V_1} doesn't have any edge of Maker, but it may contain some edges of Breaker: let E_1 denote the set of edges of Breaker in K_{V_1}. Clearly $|E_1| \leq |V_1| = 2^{q-1}$. Let G_1 be the graph (V_1, E_1). The average degree $\overline{d_1}$ of G_1 is

$$\overline{d_1} = \frac{2|E_1|}{|V_1|} \leq 2 \cdot \frac{2^{q-1}}{2^{q-1}} = 2.$$

Let $u_2 \in V_1$ be a point with minimum degree in G_1. So the degree of u_2 in G_1 is ≤ 2.

By playing on K_{V_1} and choosing edges from point u_2 Maker (as the first player) can *trivially* pick

$$\left\lceil \frac{|V_1| - \overline{d_1}}{2} \right\rceil \quad \text{(upper integral part)}$$

edges. These edges are all incident with $u_2 \in V_1$, and let $V_2 (\subset V_1)$ denote the set of other endpoints. Clearly

$$|V_2| = \left\lceil \frac{|V_1| - \overline{d_1}}{2} \right\rceil.$$

So

$$2^{q-2} \geq |V_2| \geq \left\lceil \frac{2^{q-1} - 2}{2} \right\rceil = 2^{q-2} - 1.$$

The complete graph K_{V_2} with vertex-set V_2 doesn't have any edge of Maker, but it may contain some edges of Breaker: let E_2 denote the set of edges of Breaker in K_{V_2}. Clearly

$$|E_2| \le |E_1| + |V_2|.$$

But this trivial upper bound can be substantially improved if Maker picks his

$$\left\lceil \frac{|V_1| - \overline{d_1}}{2} \right\rceil$$

edges incident with $u_2 \in V_1$ (i.e. the set V_2) in a *clever* way, namely by using First Player's strategy in Lemma 1. Then Maker can guarantee the stronger inequality

$$|E_2| \le \frac{|E_1|}{4} + |V_2|$$

instead of the trivial one

$$|E_2| \le |E_1| + |V_2|.$$

So Maker can force the upper bound

$$|E_2| \le \frac{|E_1|}{4} + |V_2| \le \frac{|V_1|}{4} + |V_2|.$$

Let G_2 be the graph (V_2, E_2). The average degree $\overline{d_2}$ of G_2 is

$$\overline{d_2} = \frac{2|E_2|}{|V_2|} \le 2\left(\frac{|V_1|}{4|V_2|} + \frac{|V_2|}{|V_2|} \right) \le 2 + \frac{2^{q-1}}{2(2^{q-2}-1)} \le 4$$

if $q \ge 3$. Let $u_3 \in V_2$ be a point with minimum degree in G_2. So the degree of u_3 in G_2 is ≤ 4.

By playing on K_{V_2} and choosing edges from point u_3 Maker (as the first player) can *trivially* pick

$$\left\lceil \frac{|V_2| - \overline{d_2}}{2} \right\rceil$$

edges. Let V_3 ($\subset V_2$) denote the set of other endpoints of these edges of Maker

$$|V_3| = \left\lceil \frac{|V_2| - \overline{d_2}}{2} \right\rceil.$$

So

$$2^{q-3} \ge |V_3| \ge \left\lceil \frac{2^{q-2} - 4}{2} \right\rceil = 2^{q-3} - 2.$$

The complete graph K_{V_3} doesn't have any edge of Maker, but it may contain some edges of Breaker: let E_3 denote the set of edges of Breaker in K_{V_3}. Clearly

$$|E_3| \le |E_2| + |V_3|.$$

But again this trivial upper bound can be substantially improved if Maker picks his

$$\left\lceil \frac{|V_2| - \overline{d_2}}{2} \right\rceil$$

edges incident with $u_3 \in V_2$ (i.e. the set V_3) by using First Player's strategy in Lemma 1. Then Maker can guarantee the stronger inequality

$$|E_3| \leq \frac{|E_2|}{4} + |V_3|$$

instead of the trivial one

$$|E_3| \leq |E_2| + |V_3|.$$

So Maker can force the upper bound

$$|E_3| \leq \frac{|E_2|}{4} + |V_3| \leq \frac{|V_1|}{4^2} + \frac{|V_2|}{4} + |V_3|.$$

Let G_3 be the graph (V_3, E_3). The average degree $\overline{d_3}$ of G_3 is

$$\overline{d_3} = \frac{2|E_3|}{|V_3|} \leq 2\left(\frac{|V_1|}{4^2|V_3|} + \frac{|V_2|}{4|V_3|} + \frac{|V_3|}{|V_3|}\right),$$

and so on. By iterating this argument, we have the following inequalities in general

$$|V_i| \geq \frac{|V_{i-1}| - \overline{d_{i-1}}}{2},$$

and

$$\overline{d_i} = \frac{2|E_i|}{|V_i|} \leq 2\left(\frac{|V_1|}{4^{i-1}|V_i|} + \frac{|V_2|}{4^{i-2}|V_i|} + \frac{|V_3|}{4^{i-3}|V_i|} + \cdots + \frac{|V_i|}{|V_i|}\right).$$

We are going to prove *by induction* that if $1 \leq i \leq q-4$, then $2^{q-i} \geq |V_i| \geq 2^{q-i} - 6$ and $\overline{d_i} \leq 6$.

We have already proved the cases $i = 1$ and $i = 2$.

Now assume that the inequalities hold for all $1 \leq j \leq i-1$, and want to show that they hold for $j = i$ as well. But this is just trivial calculations. Indeed, by hypothesis

$$|V_i| \geq \frac{|V_{i-1}| - \overline{d_{i-1}}}{2} \geq \frac{(2^{q-i+1} - 6) - 6}{2} = 2^{q-i} - 6.$$

Note that the upper bound $2^{q-i} \geq |V_i|$ is trivial.

On the other hand, we have

$$\overline{d_i} = \frac{2|E_i|}{|V_i|} \leq 2\left(\frac{|V_1|}{4^{i-1}|V_i|} + \frac{|V_2|}{4^{i-2}|V_i|} + \frac{|V_3|}{4^{i-3}|V_i|} + \cdots + \frac{|V_i|}{|V_i|}\right)$$

$$\leq 2 + \frac{2^{q-i+1}}{2(2^{q-i} - 6)}\left(1 + \frac{1}{2} + \frac{1}{2^2} + \frac{1}{2^3} + \cdots\right) \leq 4 + \frac{12}{2^{q-i} - 6} \leq 6$$

if $q - i \geq 4$. This completes the proof of the inequalities $2^{q-i} \geq |V_i| \geq 2^{q-i} - 6$ and $\overline{d_i} \leq 6$ if $1 \leq i \leq q-4$.

Let $u_{q-3} \in V_{q-4}$ be a point with minimum degree in graph $G_{q-4} = (V_{q-4}, E_{q-4})$. So the degree of u_{q-3} in G_{q-4} is $\leq \overline{d_{q-4}} \leq 6$. Now playing on $K_{V_{q-4}}$ Maker (as the first player) can *trivially* pick

$$\left\lceil \frac{|V_{q-4}| - \overline{d_{q-4}}}{2} \right\rceil \geq \frac{(2^4 - 6) - 6}{2} = 2$$

edges incident with $u_{q-3} \in V_{q-4}$. Let V_{q-3} ($\subset V_{q-4}$) denote the set of other endpoints of these edges of Maker, and let $u_{q-2} \in V_{q-3}$ be an arbitrary point. It follows from the construction (which is a slight modification of the standard proof of the graph Ramsey Theorem) that any two of the $(q-2)$ vertices $u_1, u_2, u_3, \cdots, u_{q-2}$ are joined by an edge of Maker. This means that playing on K_{2^q} Maker can build a K_{q-2} of his own in less than 2^q moves. This proves Theorem 21.1. $\square$

2. How to beat the Halving Argument? Approaching the Optimal Play via the Majority Play. Theorem 21.1 is certainly an improvement, but it is still just "half" of the truth. With any kind of halving strategy we are hopelessly stuck at $\log_2 N$. How can we go beyond $\log_2 N$, and "double" it? It is very instructive to solve first a much easier question: "What is the *majority outcome* of the game?"

In Section 6 "random" players was mentioned, and whole of Chapter IV focused on the topic of "randomness and games." Here a more detailed, systematic treatment of this angle is given.

What the *majority outcome* means is the following most naive approach to solve a game. Assume we possess the supernatural power to go through and analyze all possible plays at once (needless to say most of the plays are hopelessly dull, or even dumb), and if the overwhelming majority of the plays ends with a Weak Win (or Strong Draw), then we suspect that the *outcome of the game* is the same: a Weak Win Strategy (or a Strong Draw Strategy). The naive intuition is that the outcome of the game (which is the outcome of the optimal play between two perfect players) is the same as the majority outcome.

A terminal position of a play in the Clique Game gives a Halving 2-Coloring of the edges of K_N, so we have to study the Random Halving 2-Colorings of K_N. A technical simplification (in fact a slight "cheating") is to consider the simpler *Random Graph* model: we include or exclude each edge of K_N with probability $1/2$, and we do this inclusion-or-exclusion for all $\binom{N}{2}$ edges of K_N independently of each other. This way we get a *Random 2-Coloring* of the edges of K_N, which is not exactly Halving, but Nearly Halving (with a "square-root error" by the Central Limit Theorem).

We want to understand the mysterious relationship between deterministic Graph Games (of complete information) and Random Graphs. The first step in exploring this idea is the study of the

Clique Number of the Random Graph. Since Maker occupies half of the edges of K_N, we study the Clique Number of the Random Graph $G = \mathbf{R}(K_N, 1/2)$ where the edge probability is $1/2$. The Clique Number means the number of vertices in the largest clique. It turns out that the Clique Number of the Random Graph has an extremely sharp concentration: "almost all" graphs on N vertices have the *same* Clique Number. The common value of the Clique Number is *not* $\log_2 N$ (the natural limitation of the "ramification"), but the (almost) twice as large $2\log_2 N - 2\log_2\log_2 N + O(1)$. The point is that the Clique Number of the Random Graph does go beyond $\log_2 N + O(1)$, and this gives us a hope for a similar factor 2 improvement in the Clique Game.

The "sharp concentration of the Clique Number," already mentioned in Section 6, is one of the most surprising results in the theory of Random Graphs. It is based on the combined efforts of several authors: Erdős, Rényi, Bollobás, and Matula. The statement is very surprising, but the proof itself is shockingly simple: it is an almost routine application of the probabilistic *First* and *Second Moment Method*. We use the standard notation $\omega(G)$ for the Clique Number of a graph G, and $\mathbf{R}(K_N, 1/2)$ denotes the Random Graph with N vertices and edge probability $1/2$.

Sharp Concentration of the Clique Number: $\omega(\mathbf{R}(K_N, 1/2))$ *is basically concentrated on a single value (depending, of course, on N) with probability tending to one as N tends to infinity.*

Remark. This is a pure probabilistic statement which seemingly has absolutely nothing to do with the optimal play. The reason why it was decided to still include an outline of the proof (see below) was to demonstrate how the critical probabilistic calculations here will *unexpectedly* show up again later in the game-theoretic considerations. This is an excellent illustration of how the "Probabilistic Method" is converted into a "Fake Probabilistic Method" (the method of the book).

The concentration of random variables in general: We start our discussion with the vague intuition that "a random variable is close to its expected value." For example, if one tosses a fair coin N times, the number of Heads is typically in the range $N/2 \pm c\sqrt{N}$. The typical fluctuation is described by the Central Limit Theorem: the norming factor is the square root of N (which is "small" compared to $N/2$ if N is large).

Square-root-of-N size fluctuation is very typical, but many "natural" random variables are far more concentrated – exhibit constant-size fluctuation. For example, instead of counting the number of Heads, we may be interested in the length of the *longest run* of consecutive Heads in N trials; let $L = L(N)$ denote the *longest run* random variable. As far as is known, it was Erdős and Rényi who made the first systematic study of the *longest run*. For simplicity assume that $N = 2^n$ where n is an integer. Erdős and Rényi proved the following elegant result: if c is a fixed integer and $N = 2^n$ tends to infinity, then

$$\Pr[L(N) = \log_2 N + c] \to e^{-2^{-c-2}} - e^{-2^{-c-1}}. \tag{21.1}$$

In (21.1) the choice $c = -1$ gives the maximum probability $e^{-1/2} - e^{-1} = .2387$; the other choices $c = 0, 1, 2$ give the probabilities, respectively, $e^{-1/4} - e^{-1/2} = .1723$, $e^{-1/8} - e^{-1/4} = .1037$, $e^{-1/16} - e^{-1/8} = .057$.

(21.1) shows that the *longest run* random variable is concentrated on a few values $\log_2 N + O(1)$ (centered at $\log_2 N$) with probability nearly one.

The proof of (21.1) is based on the Inclusion–Exclusion Principle combined with the trick of *non-overlapping extensions*. The term *non-overlapping extensions* means that if $H \cdots H$ is a block of consecutive Heads, then we extend it in both directions into a block $TH \cdots HT$ (or possibly $TH \cdots H(last)$, $(first)H \cdots HT$). The critical property of the extended patterns $TH \cdots HT$, $TH \cdots H(last)$, $(first)H \cdots HT$ is that they cannot overlap! Working with the extended patterns, the application of the Inclusion–Exclusion formula becomes particularly simple, and (21.1) follows easily.

Another example of a "type (21.1) concentration" is the length of the longest monochromatic arithmetic progression in a Random 2-Coloring of $[1, N] = \{1, 2, \ldots, N\}$. Then the *longest monochromatic length*, as a random variable, is concentrated on a few values $2 \log_2 N - \log_2 \log_2 N + O(1)$ with probability nearly 1. The reader is challenged to prove both (21.1) and the last statement.

The third type of concentration is so extreme that there is no fluctuation; the random variable is deterministic (or almost deterministic). The best illustration is exactly what we are interested in, namely the clique number of the Random Graph.

Outline of the proof of the sharp concentration of the clique number. Let

$$f(q) = f_N(q) = \binom{N}{q} 2^{-\binom{q}{2}}$$

denote the *expected number* of q-cliques in the Random Graph $\mathbf{R}(K_N, 1/2)$. The function $f(q)$ is monotone decreasing and drops under one around $q \approx 2 \log_2 N$. Let q_0 be the last integer q such that $f(q) > 1$, i.e. $f(q_0) > 1 \geq f(q_0 + 1)$. The "real solution" of the equation $f(q) = 1$ is $q = 2 \log_2 N - 2 \log_2 \log_2 N + 2 \log_2 e - 1 + o(1)$, and q_0 is the lower integral part of this real number. The *crucial* fact is that $f(q)$ is a very *rapidly* changing function

$$\frac{f(q)}{f(q+1)} = \frac{q+1}{N-q} 2^q = N^{1+o(1)}$$

if $q \approx 2 \log_2 N$. The reason why $f(q)$ is rapidly changing is pretty simple: the complete graph K_q has $\binom{q}{2}$ edges, and changing q by one makes a large – in fact, square-root size – increase in the number of edges: $\binom{q+1}{2} = \binom{q}{2} + q$.

In view of the rapid change, it is very unlikely that either $N^\varepsilon \geq f(q_0) > 1$ or $1 \geq f(q_0 + 1) \geq N^{-\varepsilon}$ occurs, where $\varepsilon > 0$ is an arbitrarily small but fixed constant.

Case 1: $f(q_0) > N^\varepsilon > N^{-\varepsilon} > f(q_0 + 1)$ holds for some $\varepsilon > 0$

In this "typical" case the Clique Number is concentrated on a *single* value, namely

$$\Pr\left\{ \omega(\mathbf{R}(K_N, 1/2)) = q_0 \right\} \to 1$$

as $N \to \infty$. Indeed, for each q-element vertex-set S of K_N let χ_S denote the indicator random variable of the event that "S is the vertex-set of a clique in $\mathbf{R}(K_N, 1/2)$", i.e. χ_S is 1 if "S spans a clique" and 0 if "not." Let

$$\Omega = \Omega_q = \sum_{|S|=q} \chi_S.$$

The expected value $\mathbf{E}(\Omega) = f(q)$. The *variance* of Ω is as follows

$$Var(\Omega) = \sum_{|S_1|=q} \sum_{|S_2|=q} \left(\mathbf{E}(\chi_{S_1} \chi_{S_2}) - 2^{-2\binom{q}{2}} \right),$$

where

$$\mathbf{E}(\chi_{S_1} \chi_{S_2}) = \Pr\{\chi_{S_1} = 1 = \chi_{S_2}\} = 2^{-2\binom{q}{2}}$$

if $|S_1 \cap S_2| \leq 1$, and

$$\mathbf{E}(\chi_{S_1} \chi_{S_2}) = \Pr\{\chi_{S_1} = 1 = \chi_{S_2}\} = 2^{-2\binom{q}{2}+\binom{i}{2}}$$

if $2 \leq i = |S_1 \cap S_2| \leq q$. It follows that

$$\frac{Var(\Omega)}{\mathbf{E}^2(\Omega)} = \sum_{i=2}^{q} \frac{\binom{q}{i}\binom{N-q}{q-i}}{\binom{N}{q}} \left(2^{\binom{i}{2}} - 1 \right) = \sum_{i=2}^{q} g(i), \qquad (21.2)$$

where

$$g(i) = g_{N,q}(i) = \frac{\binom{q}{i}\binom{N-q}{q-i}}{\binom{N}{q}} \left(2^{\binom{i}{2}} - 1 \right). \qquad (21.3)$$

If $q \approx 2 \log_2 N$, then $\sum_{i=2}^{q} g(i) \approx g(2) + g(q)$. Indeed, $g(2) \approx \frac{q^4}{2N^2} = c\frac{(\log N)^4}{N^2}$, $g(3) \approx \frac{q^6}{6N^3} = c\frac{(\log N)^6}{N^3}$, and, on the other end, $g(q) \approx \frac{1}{f(q)}$, $g(q-1) \approx \frac{q}{Nf(q)}$, and so on. For $\Omega = \Omega_q$ with $q = q_0$ by Chebyshev's inequality

$$\Pr\left\{\Omega = 0\right\} \leq \Pr\left\{|\Omega - \mathbf{E}(\Omega)| \geq \mathbf{E}(\Omega)\right\} \leq \frac{Var(\Omega)}{\mathbf{E}^2(\Omega)} \approx c\frac{(\log N)^4}{N^2} + \frac{1}{f(q_0)} < N^{-\varepsilon} \to 0$$

as $N \to \infty$. So $K_{q_0} \subset \mathbf{R}(K_N, 1/2)$ with probability tending to 1.

On the other hand, for $\Omega = \Omega_q$ with $q = q_0 + 1$ Markov's inequality yields

$$\Pr\left\{\Omega \geq 1\right\} \leq \mathbf{E}(\Omega) = f(q_0 + 1) < N^{-\varepsilon} \to 0$$

as $N \to \infty$. Therefore, $K_{q_0+1} \subset \mathbf{R}(K_N, 1/2)$ with probability tending to 0.

Case 2: Either $N^{\varepsilon} \geq f(q_0) > 1$ or $1 \geq f(q_0 + 1) \geq N^{-\varepsilon}$

Repeating the argument of Case 1, we conclude that either

$$\Pr\left\{\omega(\mathbf{R}(K_N, 1/2)) = q_0 + 1 \text{ or } q_0\right\} \to 1$$

or

$$\Pr\left\{\omega(\mathbf{R}(K_N, 1/2)) = q_0 \text{ or } q_0 - 1\right\} \to 1$$

as $N \to \infty$. This completes the outline of the proof of the sharp concentration of the clique number of the Random Graph $G = \mathbf{R}(K_N, 1/2)$.   □

3. How to conjecture the Advanced Weak Win Criterion? As said before, the main idea is to convert the probabilistic First and Second Moment argument above into a deterministic game-theoretic proof via potentials. Guessing the right criterion ("Advanced Weak Win Criterion") is already a serious challenge; the guess work is nearly as difficult as the proof itself. We already formulated the Advanced Weak Win Criterion in Section 13, but here we pretend not to know about it. The point is to see how the probabilistic considerations give us a strong hint/motivation to discover the advanced criterion.

Note that the advanced criterion is a "Game-Theoretic Higher Moment Method"; in fact, a pth Moment Method, where in the (majority of) applications the parameter $p \to \infty$.

Let's start with the first moment; what is the Game-Theoretic First Moment? This job is already done: "First Moment" means the two linear criterions together, the Erdős–Selfridge theorem for Strong Draw (Theorem 1.4) and Theorem 1.2 for Weak Win (explaining the title of Part A). To justify this claim, we recall the non-uniform version of the Erdős–Selfridge Theorem.

Non-uniform Erdős–Selfridge: *If*

$$\sum_{A \in \mathcal{F}} 2^{-|A|} < 1 \quad (< 1/2),$$

then Breaker, as the first (second) player, has an explicit Strong Draw strategy in the positional game on $(V, \mathcal{F})$.

Consider the Random 2-Coloring of the points of hypergraph $(V, \mathcal{F})$, where the points are colored red and blue independently of each other with probability $1/2$. Then the *expected number* of monochromatic winning sets is precisely $\sum_{A \in \mathcal{F}} 2^{-|A|+1}$. If this expected number is less than 1, then there exists at least one Drawing Terminal Position. By the non-uniform Erdős–Selfridge Theorem this Drawing Position can be "upgraded" to a Strong Draw Strategy.

A probabilistic interpretation of the Erdős–Selfridge criterion has just been given: it is about the "expected number of monochromatic winning sets in a Random 2-Coloring." (Note that the "expected value" is often called the "first moment" in probability theory.) But the probabilistic interpretation goes much deeper! It goes far beyond the criterion alone: the Strong Draw Strategy itself also has a natural probabilistic interpretation. Indeed, the *strategy* is to minimize the *defeat probability in the randomized game*. To explain what it means, assume that we are in the middle of a play where Maker (the first player) already occupied $x_1, x_2, \ldots, x_i$, Breaker (the second player) occupied $y_1, y_2, \ldots, y_{i-1}$, and the question is how to choose Breaker's next point y_i. Again we use the concepts of "dead set" (i.e. blocked by Breaker) and "survivor." A "survivor" $A \in \mathcal{F}$ has a chance to eventually be completely occupied by Maker, so each one represents some "danger." What is the total "danger" of the whole position? Consider the *randomized game* starting from this given position, i.e. we color the unoccupied points of the board V red and blue independently with probability $1/2$, red for Maker and blue for Breaker. Let $A_s \in \mathcal{F}$ be a "survivor" winning set, and let E_s denote the event that A_s becomes monochromaticly red in the randomized game. The probability of E_s is clearly 2^{-u_s}, where u_s is the number of unoccupied points of A_s. Let $\{A_s : s \in S_i\}$ be the family of all survivor sets at this stage of the play, i.e. S_i is the index-set for the "survivor" winning sets. We are interested in the "defeat-probability" $\Pr\{\bigcup_{s \in S_i} E_s\}$. By the inclusion–exclusion formula

$$\Pr\left\{\bigcup_{s \in S_i} E_s\right\} = \sum_{s \in S_i} \Pr\{E_s\} - \sum_{s < t} \Pr\{E_s \cap E_t\} + \sum_{s < t < v} \Pr\{E_s \cap E_t \cap E_v\} \mp \cdots .$$

This expression is obviously far too complicated, so we need to approximate. A natural choice is the "linear approximation," i.e. to keep the first term $\sum_{s \in S_i} \Pr\{E_s\}$ on the right-hand side and ignore the rest. The first term is nothing else than the *expected number!* This convinces us to evaluate the given position by the following "total danger" $T_i = \sum_{s \in S_i} 2^{-u_s}$, where u_s is the number of unoccupied points of the "survivor" A_s ($s \in S_i$) and the index i indicates that we are at the stage of choosing the ith point y_i of Breaker. What is the effect of the next two moves y_i and x_{i+1}? The familiar equality

$$T_{i+1} = T_i - \sum_{s \in S_i: \, y_i \in A_s} 2^{-u_s} + \sum_{s \in S_i: \, x_{i+1} \in A_s} 2^{-u_s} - \sum_{s \in S_i: \, \{y_i, x_{i+1}\} \subseteq A_s} 2^{-u_s}$$

explains why the natural choice for y_i is the "maximum decrease," which minimizes the "total danger" T_{i+1}.

For later application note that hypergraph $\mathcal{F}$ can be a *multi-hypergraph*; a winning set may have arbitrarily large multiplicity (not just 0 and 1); the proof remains the same: in every sum a winning set is counted with its own multiplicity.

Random 2-Coloring of a hypergraph. Let $\psi : V \to \{red, blue\}$ be a Random 2-Coloring of the board V of hypergraph $\mathcal{F}$ such that the points are colored red and blue independently of each other with probability $1/2$. Let $\Omega = \Omega(\psi)$ denote the number of red (i.e. monochromatic red) hyperedges $A \in \mathcal{F}$ in 2-coloring ψ. Write

$$T(\mathcal{F}) = \sum_{A \in \mathcal{F}} 2^{-|A|}. \tag{21.4}$$

(Of course, if a hyperedge has multiplicity (say) M, then it shows up M times in the summation above.) Notice that $T(\mathcal{F})$ is, in fact, the *expected value* $\mathbf{E}\Omega$ of the random variable Ω.

Next we compute the *variance* $Var(\Omega)$ of Ω. Let χ_A denote the indicator variable of the event that "A is red," i.e. χ_A is 1 if "A is red" and 0 otherwise. Clearly $\Omega = \sum_{A \in \mathcal{F}} \chi_A$, and we have

$$Var(\Omega) = E(\Omega^2) - E^2(\Omega) = \sum_{A \in \mathcal{F}} \sum_{B \in \mathcal{F}} \Big(E(\chi_A \chi_B) - E(\chi_A) E(\chi_B) \Big).$$

If A and B are disjoint sets, then χ_A and χ_B are independent random variables, so $E(\chi_A \chi_B) = E(\chi_A) E(\chi_B)$. It follows that

$$Var(\Omega) = \sum_{A \in \mathcal{F}} \Big(2^{-|A|} - 2^{-2|A|} \Big) + 2 \sum_{\{A,B\} \in \binom{\mathcal{F}}{2} : |A \cap B| \geq 1} \Big(2^{-|A \cup B|} - 2^{-|A|-|B|} \Big).$$

Here we employed the (standard) *notation* that, if X is a set, then $\binom{X}{k}$ denotes the family of k-element subsets of X.

Since $|A \cup B| = |A| + |B| - |A \cap B| \leq |A| + |B| - 1$, we have

$$\frac{1}{2} S \leq Var(\Omega) \leq S \quad \text{where} \quad S = T(\mathcal{F}) + 2T(\mathcal{F}_1^2) \tag{21.5}$$

and

$$\mathcal{F}_1^2 = \left\{ A \cup B : \{A, B\} \in \binom{\mathcal{F}}{2}, \ |A \cap B| \geq 1 \right\}. \tag{21.6}$$

That is, $\mathcal{F}_1^2$ is the family of all union-sets (counted with multiplicity) of the unordered pairs of distinct non-disjoint elements of $\mathcal{F}$.

By Chebyshev's inequality, the "red set counting" random variable Ω falls in the interval

$$E(\Omega) - O(\sqrt{Var(\Omega)}) < \Omega < E(\Omega) + O(\sqrt{Var(\Omega)}) \tag{21.7}$$

with probability close to 1. We want to guarantee that $\Omega \geq 1$ holds with probability close to 1, i.e. the Random 2-Coloring provides a (monochromatic) red winning set ("Maker wins") with probability close to 1. To apply the Chebyshev's inequality we need the Double Requirement that "$E(\Omega)$ is large" and "$\sqrt{Var(\Omega)}/E(\Omega)$ is small" hold at the same time. This kind of Double Requirement (we express (21.5)–(21.7) in terms of $T(\mathcal{F})$ and $T(\mathcal{F}_1^2)$):

(1) $T(\mathcal{F})$ is "large" and
(2) $\sqrt{T(\mathcal{F}_1^2)}/T(\mathcal{F})$ is "small",

is that we may expect to be the new Advanced Weak Win Criterion. Well, the Advanced Weak Win Criterion ("Theorem 24.2") looks more complicated than (1)–(2), but there is a variant of the Clique Game in which the "builder's win" criterion looks exactly like (1)–(2)! This variant is the *Picker–Chooser Game*, and this is what we are going to discuss in the next section. The *Picker–Chooser Game* plays an auxiliary role in our study; it is an ideal "warm up" for the real challenge (Sections 23–24).

We already saw the sharp ("single-valued") concentration of the Clique Number of the Random Graph $R(K_N, 1/2)$ – this justifies the notion of Majority-Play Clique Number. Similarly, we can talk about the Majority-Play Lattice Number for each type listed in Section 8. We know the following Majority-Play Numbers (we use the notation $\mathbf{M}(\cdots)$ similarly to the notation $\mathbf{A}(\cdots)$ for the Achievement Number)

$$\mathbf{M}(K_N; \text{clique}) = \lfloor 2\log_2 N - 2\log_2\log_2 N + 2\log_2 e - 1 + o(1)\rfloor,$$

$$\mathbf{M}(N \times N; \text{ square lattice}) = \left\lfloor \sqrt{3\log_2 N} + o(1)\right\rfloor,$$

$$\mathbf{M}(N \times N; \text{ rectangle lattice}) = \left\lfloor 2\sqrt{\log_2 N} + o(1)\right\rfloor,$$

$$\mathbf{M}(N \times N; \text{ tilted square lattice}) = \left\lfloor 2\sqrt{\log_2 N} + o(1)\right\rfloor,$$

$$\mathbf{M}(N \times N; \text{ tilted rectangle lattice}) = \left\lfloor 2\sqrt{\log_2 N} + o(1)\right\rfloor,$$

$$\mathbf{M}(N \times N; \text{ rhombus lattice}) = \left\lfloor 2\sqrt{\log_2 N} + o(1)\right\rfloor,$$

$$\mathbf{M}(N \times N; \text{ parallelogram lattice}) = \left\lfloor \sqrt{6\log_2 N} + o(1)\right\rfloor,$$

$$\mathbf{M}(N \times N; \text{ area one lattice}) = \left\lfloor 2\sqrt{\log_2 N} + o(1)\right\rfloor.$$

Exercise 21.1 *By using (21.5)–(21.7) prove the Majority-Play Number formulas above.*

22

Game-theoretic second moment: application to the Picker–Chooser game

This section is a *detour* (but an instructive one!); a reader in a rush may skip it. Here we study the

1. Picker–Chooser and Chooser–Picker games. These games are motivated by the well-known "I-Cut-You'll-Choose" way of dividing (say) a piece of cake between two people (it is, in fact, a sequential variant). We have already studied the discrepancy version of these games in Sections 18–19.

First, the special case of *Clique Game* is discussed; the generalization to arbitrary hypergraphs will be straightforward.

In the *Clique Game* the board is, as always, the complete graph K_N. In each round of the play Picker picks two previously unselected edges of K_N, Chooser chooses one of them, and the other one goes back to Picker. There are 2 variants. In the first variant Picker colors his edges blue, Chooser colors his edges red, and Chooser wins if he can occupy all the $\binom{q}{2}$ edges of some complete sub-graph K_q of K_N; otherwise Picker wins. In other words, Chooser wins if and only if there is a red copy of K_q. This is the (K_N, K_q) **Chooser Picker Clique Game**. This is obviously a *Ramsey type* game. Indeed, it is a sequential version of the following *1-round game* ("Ramsey Theorem in disguise"): Picker divides the edge-set of K_n into 2 parts, Chooser chooses 1 part, and the other part goes back to Picker; Chooser wants a K_q in his own part. Chooser has a winning strategy in this *1-round game* if and only if $N \geq R(q)$, where is the standard (diagonal) Ramsey Number.

In the second variant Picker wants a K_q; we call it the Picker–Chooser game. In the (K_N, K_q) **Picker–Chooser Clique Game** the board is the complete graph K_N, and just like before, in each round of the play Picker picks 2 previously unselected edges of K_N, Chooser chooses one of them, and the other one goes back to Picker. In this variant Chooser colors his edges blue, Picker colors his edges red, and Picker wins if he occupies all the $\binom{q}{2}$ edges of some complete sub-graph K_q of K_N; otherwise Chooser wins. In other words, Picker wins if and only if there is a red copy of K_q in K_N.

There is a surprising similarity between the game-theoretic breaking points of these Clique Games and the Clique Number of the *Random Graph* on N vertices with edge probability $1/2$. We show that the game-theoretic threshold for the Picker–Chooser game is $2\log_2 N - 2\log_2\log_2 N + 2\log_2 e - 1 + o(1)$, which is exactly the Clique Number of the Random Graph on N vertices with edge probability $1/2$. (For the Chooser–Picker game the threshold is *two less*.)

Of course, one can play the Picker–Chooser and Chooser–Picker games on an arbitrary finite hypergraphs. Let $(V, \mathcal{F})$ be a finite hypergraph; the rules are the same: in each round of the play Picker picks two previously unselected points of the board V, Chooser chooses one of them, and the other one goes back to Picker. In the Chooser–Picker version Chooser colors his points red and Picker colors his points blue. Chooser wins if he can occupy all the points of some winning set $A \in \mathcal{F}$; otherwise Picker wins.

In the Picker–Chooser version Chooser colors his points blue and Picker colors his points red. Picker wins if he occupies all the points of some winning set $A \in \mathcal{F}$; otherwise Chooser wins.

Observe that in both games there is a *builder*, who wants to occupy a whole winning set. It is the *builder* whose name comes fist: Picker in the Picker–Chooser game and Chooser in the Chooser–Picker game.

The opponent of *builder* is a *blocker* – his name comes second – who simply wants to prevent *builder* from occupying a whole winning set.

Builder colors his points red, and *blocker* colors his points blue. Each play has two possible outcomes only: either *builder* wins, or *blocker* wins.

Let us return to the Clique Game. It is clear that Picker has much more control in the Picker–Chooser version than Maker does in the Maker–Breaker version (or Chooser does in the Chooser–Picker version). In view of the next exercise, Picker can occupy a whole winning set even if the hypergraph is *not* dense.

Exercise 22.1 *Show that, if $\mathcal{F}$ contains (at least) 2^k pairwise disjoint k-sets, then Picker can occupy a whole k-set.*

We challenge the reader to solve this (really simple) exercise.

The relative simplicity of the Picker–Chooser game explains why it is a good idea to understand the "building part" in the Picker–Chooser game first, and discuss the more challenging/interesting Maker–Breaker and Chooser–Picker games later.

Here is a general Picker's Building Criterion.

Theorem 22.1 *Consider the Picker–Chooser game on hypergraph $(V, \mathcal{F})$. Assume that*

$$T(\mathcal{F}) \geq 10^{14}\|\mathcal{F}\|^{14}\left(\sqrt{T(\mathcal{F}_1^2)} + 1\right),$$

where $T(\mathcal{F})$ is defined in (21.4) and $\|\mathcal{F}\| = \max\{|A| : A \in \mathcal{F}\}$ is the rank of hypergraph $\mathcal{F}$. Then Picker (i.e. builder) has an explicit winning strategy.

Remark. The hypothesis is a *Double Requirement* (as we predicted in Section 21):

(1) $T(\mathcal{F})$ has to be "large"; and
(2) $\sqrt{T(\mathcal{F}_1^2)}/T(\mathcal{F})$ has to be "small."

The large factor $10^{14}\|\mathcal{F}\|^{14}$ is accidental. It is (almost) irrelevant in the applications; this is why we do not make an effort to find better constants.

If $\mathcal{F}$ consists of pairwise disjoint sets, then $\mathcal{F}_1^2$ is empty. In that trivial case the large factor can be dropped, and the condition can be reduced to the much simpler $T(\mathcal{F}) \geq \|\mathcal{F}\|$ (we leave the proof to the reader).

First we show how Theorem 22.1 applies to the Picker–Chooser Clique Game, and postpone the proof of Theorem 22.1 to the end of the section.

2. Application. Let $\mathbf{P_A}(K_N; \text{clique})$ be the largest value of $q = q(N)$ such that Picker has a winning strategy in the (K_N, K_q) Picker–Chooser Clique Game. I call it the Clique Picker-Achievement Number.

Exercise 22.2 *If $\mathcal{F}$ is n-uniform and $|\mathcal{F}| < 2^n$, then Chooser can always prevent Picker from occupying a whole $A \in \mathcal{F}$ in the Picker-Chooser Game.*

In particular, if the inequality

$$\binom{N}{q} 2^{-\binom{q}{2}} < 1 \tag{22.1}$$

holds, then Chooser (i.e. "blocker") has a winning strategy in the (K_N, K_q) Picker–Chooser Clique Game.

Next we apply Picker's win criterion: Theorem 22.1. Let $[N] = \{1, 2, \ldots, N\}$ be the vertex-set of K_N, and define

$$\mathcal{F} = \{K_S : S \subset [N], |S| = q\}.$$

We have

$$\mathcal{F}_1^2 = \{K_{S_1} \cup K_{S_2} : S_i \subset [N], |S_i| = q, |K_{S_1} \cap K_{S_2}| \geq 1\},$$

where $\{S_1, S_2\}$ runs over the unordered pairs of distinct q-element subsets of $[N]$. We have to check that

$$T(\mathcal{F}) \geq 10^{14} \|\mathcal{F}\|^{14} \left((T(\mathcal{F}_1^2))^{1/2} + 1 \right).$$

Clearly $\|\mathcal{F}\| = \binom{q}{2}$, $T(\mathcal{F}) = \binom{N}{q} 2^{-\binom{q}{2}}$, and

$$T(\mathcal{F}_1^2) = \sum_{j=2}^{q-1} \frac{1}{2} \binom{N}{q} \binom{q}{j} \binom{N-q}{q-j} 2^{-2\binom{q}{2} + \binom{j}{2}},$$

where $j = |S_1 \cap S_2|$. Therefore

$$\frac{T(\mathcal{F}_1^2)}{(T(\mathcal{F}))^2} = \frac{1}{2} \sum_{j=2}^{q-1} \frac{\binom{q}{j}\binom{N-q}{q-j}}{\binom{N}{q}} 2^{\binom{j}{2}} = \frac{1}{2} \sum_{j=2}^{q-1} g(j),$$

where

$$g(j) = \frac{\binom{q}{j}\binom{N-q}{q-j}}{\binom{N}{q}} 2^{\binom{j}{2}}.$$

Note that this is the same as (21.2)–(21.3) in the calculation of the variance of the Clique Number of the Random Graph $\mathbf{R}(K_N, 1/2)$ (see Section 21). Thus we have

$$\frac{T(\mathcal{F}_1^2)}{(T(\mathcal{F}))^2} = \frac{1}{2} \sum_{j=2}^{q-1} g(j) \approx \frac{g(2) + g(q-1)}{2} \approx \frac{q^4}{2N^2} + \frac{q}{2NT(\mathcal{F})}.$$

Therefore, it remains to check that

$$T(\mathcal{F}) = \binom{N}{q} 2^{-\binom{q}{2}} > cq^{28}, \tag{22.2}$$

where c is a sufficiently large absolute constant. By Stirling's formula the two different inequalities (see (22.1) and (22.2))

$$\binom{N}{q} 2^{-\binom{q}{2}} < 1 \quad \text{and} \quad \binom{N}{q} 2^{-\binom{q}{2}} > cq^{28}$$

determine the *same* threshold

$$2 \log_2 N - 2 \log_2 \log_2 N + 2 \log_2 e - 1 + o(1)$$

(with different $o(1)$'s, but $o(1)$ tends to zero anyway). Thus we obtain:

Theorem 22.2 *Consider the Picker–Chooser Clique Game; the Clique Picker-Achievement Number equals*

$$\mathbf{P_A}(K_N; \text{clique}) = \lfloor 2 \log_2 N - 2 \log_2 \log_2 N + 2\log_2 e - 1 + o(1) \rfloor. \qquad \square$$

It means that for the Clique Game on K_N the Picker-Achievement Number equals the Majority-Play Number (well, at least for the overwhelming majority of Ns). The same holds for the Lattice Games.

Theorem 22.3 *Consider the Picker–Chooser Lattice Game on an $N \times N$ board. The Lattice Picker-Achievement Number equals*

(a) $\mathbf{P_A}(N \times N; \text{square lattice}) = \lfloor \sqrt{3 \log_2 N} + o(1) \rfloor$,

(b) $\mathbf{P_A}(N \times N; \text{rectangle lattice}) = \lfloor 2\sqrt{\log_2 N} + o(1) \rfloor$,

(c) $\mathbf{P_A}(N \times N; \text{tilted square lattice}) = \lfloor 2\sqrt{\log_2 N} + o(1) \rfloor$,

(d) $\mathbf{P_A}(N \times N; \text{tilted rectangle lattice}) = \lfloor 2\sqrt{\log_2 N} + o(1) \rfloor$,

(e) $\mathbf{P_A}(N \times N; \text{ rhombus lattice}) = \left\lfloor 2\sqrt{\log_2 N} + o(1) \right\rfloor$,

(f) $\mathbf{P_A}(N \times N; \text{ parallelogram lattice}) = \left\lfloor \sqrt{6\log_2 N} + o(1) \right\rfloor$,

(g) $\mathbf{P_A}(N \times N; \text{ area one lattice}) = \left\lfloor 2\sqrt{\log_2 N} + o(1) \right\rfloor$.

Exercise 22.3 *Prove Theorem 22.3 above.*

3. The proof. An important ingredient of the proof of Theorem 22.1 is the *Variance Lemma* below. It will be used as a sort of "game-theoretic Chebyshev's inequality."

We begin with the notation: we use (21.4), that is, for an arbitrary hypergraph $(V, \mathcal{H})$ write

$$T(\mathcal{H}) = \sum_{A \in \mathcal{H}} 2^{-|A|},$$

and for any point $x \in V$ of the board write

$$T(\mathcal{H}; x) = \sum_{A \in \mathcal{H}:\ x \in A} 2^{-|A|}.$$

(Of course, if a set has multiplicity ≥ 1, then it shows up that many times in the summations above.)

Variance Lemma. *For any point $x \in V$ of the board*

$$T(\mathcal{H}; x) < 2\left(\left(T(\mathcal{H}_1^2) \right)^{1/2} + 1 \right).$$

We postpone the proof of the Variance Lemma to Section 24 (because the Variance Lemma is a special case of the more general Theorem 24.1).

Proof of Theorem 22.1. Assume we are at a stage of the play where Picker has the points $x_1, x_2, \ldots, x_i$, and Chooser has $y_1, y_2, \ldots, y_i$. The question is how to find Picker's next 2-element set $\{v, w\}$, from which Chooser will choose y_{i+1}, and the other one (i.e. x_{i+1}) will go back to Picker.

Let $X_i = \{x_1, x_2, \ldots, x_i\}$ and $Y_i = \{y_1, y_2, \ldots, y_i\}$. Let $V_i = V \setminus (X_i \cup Y_i)$. Clearly $|V_i| = |V| - 2i$.

Let $\mathcal{F}(i)$ be that truncated sub-family of $\mathcal{F}$, which consists of the unoccupied parts of the "survivors"

$$\mathcal{F}(i) = \{A \setminus X_i :\ A \in \mathcal{F},\ A \cap Y_i = \emptyset\}.$$

If Picker can guarantee that $T(\mathcal{F}(i)) > 0$ during the whole course of the play, in particular if $T(\mathcal{F}(end)) > 0$, then Picker wins. Let x_{i+1} and y_{i+1} denote, respectively, the $(i+1)$st points of Picker and Chooser. We have

$$T(\mathcal{F}(i+1)) = T(\mathcal{F}(i)) + T(\mathcal{F}(i); x_{i+1}) - T(\mathcal{F}(i); y_{i+1}) - T(\mathcal{F}(i); x_{i+1}, y_{i+1})$$

$$\geq T(\mathcal{F}(i)) - |T(\mathcal{F}(i); x_{i+1}) - T(\mathcal{F}(i); y_{i+1})| - T(\mathcal{F}(i); x_{i+1}, y_{i+1}),$$

(22.3)

and similarly, but in the other direction

$$T(\mathcal{F}_1^2(i+1)) \leq T(\mathcal{F}_1^2(i)) + |T(\mathcal{F}_1^2(i); x_{i+1}) - T(\mathcal{F}_1^2(i); y_{i+1})|. \qquad (22.4)$$

We want to *maximize* $T(\mathcal{F}(i+1))$ and *minimize* $T(\mathcal{F}_1^2(i+1))$ at the same time. We are going to apply the following lemma.

Lemma 1: *Let* $r_{j,\ell}$, $1 \leq j < \ell \leq m$ *be* $\binom{m}{2}$ *non-negative reals, let* r_j, $1 \leq j \leq m$ *be* m *non-negative reals, and, finally, let* t_j, $1 \leq j \leq m$ *be* m *non-negative reals, where*

$$\sum_{1 \leq j < \ell \leq m} r_{j,\ell} \leq z, \quad \sum_{j=1}^{m} r_j \leq u, \quad \sum_{j=1}^{m} t_j \leq s.$$

Assume that $m \geq 10^7$. *Then there exists a pair* $\{j_1, j_2\}$ *with* $1 \leq j_1 \neq j_2 \leq m$ *such that*

$$r_{j_1, j_2} \leq \frac{6z}{m^{8/7}}, \quad |r_{j_1} - r_{j_2}| \leq \frac{2u}{m^{8/7}}, \quad |t_{j_1} - t_{j_2}| \leq \frac{3s}{m^{8/7}}.$$

To prove the rather complicated Lemma 1, we need the following much simpler:

Lemma 2: *If* $t_1, t_2, \ldots, t_m$ *are non-negative real numbers and* $t_1 + t_2 + \cdots + t_m \leq s$, *then*

$$\min_{1 \leq j < \ell \leq m} |t_j - t_\ell| \leq \frac{s}{\binom{m}{2}}.$$

Proof. of Lemma 2. We can assume that $0 \leq t_1 < t_2 < \cdots < t_m$. Write $g = \min_{1 \leq j < \ell \leq m} |t_j - t_\ell|$. Then $t_{j+1} - t_j \geq g$ for every j, and

$$\binom{m}{2} g = g + 2g + \ldots + (m-1)g \leq t_1 + t_2 + \ldots + t_m \leq s.$$

This completes the proof of Lemma 2. □

Now we are ready to prove Lemma 1.

Proof of Lemma 1. Consider the graph

$$G_m = \left\{ \{j, \ell\} \in \binom{[m]}{2} = K_m : r_{j,\ell} \geq \frac{6z}{m^{8/7}} \right\}$$

on m points. Since $\sum_{1 \leq j < \ell \leq m} r_{j,\ell} \leq z$, it follows that the number of edges $|G_m|$ of graph G_m satisfies $|G_m| \leq m^{8/7}/6$. The average degree $d = d(G_m)$ of graph G_m is $2|G_m|/m = m^{1/7}/3$, so by Turán's well-known theorem there is an *independent point-set* $J \subseteq \{1, 2, \ldots, m\}$ in graph G_m with $|J| \geq \frac{m}{d+1} \geq m(m^{1/7}/3 + 1)^{-1} \geq 2m^{6/7}$. Let K_J denote the complete graph on point-set J; then

$$r_{j,\ell} < \frac{6z}{m^{8/7}} \text{ for all pairs } \{j, \ell\} \in K_J = \binom{J}{2}. \tag{22.5}$$

Next consider the inequality $\sum_{j\in J} r_j \leq \sum_{j=1}^{m} r_j \leq u$. It follows that there is a subset J_1 of J such that $|J_1| \geq |J|/2 \geq m^{6/7}$ and $r_j \leq um^{-6/7}$ for all $j \in J_1$. Let $c = um^{-6/7}$ and $m_1 = \lfloor m^{2/7} \rfloor$ (lower integral part), and divide the interval $[0, c]$ into m_1 equal sub-intervals $[0, c] = \bigcup_{k=1}^{m_1} I_k$, where $I_k = [(k-1)c/m_1, kc/m_1]$. By the Pigeonhole Principle there is a sub-interval I_k containing at least $|J_1|/m_1 \geq |J_1|m^{-2/7} \geq m^{4/7}$ r_j's. In other words, there is a subset J_2 of J_1 such that $|J_2| \geq m^{4/7}$ and

$$|r_j - r_\ell| \leq \frac{2u}{m^{8/7}} \text{ for all pairs } \{j, \ell\} \subseteq J_2. \tag{22.6}$$

Finally, consider the inequality $\sum_{j\in J_2} t_j \leq \sum_{j=1}^{m} t_j \leq s$. By Lemma 2 there is a pair $\{j_1, j_2\} \subseteq J_2$ with $j_1 \neq j_2$ such that

$$|t_{j_1} - t_{j_2}| \leq \frac{s}{\binom{|J_2|}{2}} \leq \frac{3s}{|J_2|^2} \leq \frac{3s}{m^{8/7}}. \tag{22.7}$$

Since $\{j_1, j_2\} \subseteq J_2 \subseteq J$, Lemma 1 follows from (22.5), (22.6), and (22.7). $\square$

Clearly

$$\sum_{\{v,w\}\subset V_i : v\neq w} T(\mathcal{F}(i); v, w) \leq \binom{\|\mathcal{F}\|}{2} T(\mathcal{F}(i)),$$

$$\sum_{v\in V_i} T(\mathcal{F}(i); v) \leq \|\mathcal{F}\| T(\mathcal{F}(i)),$$

$$\sum_{v\in V_i} T(\mathcal{F}_1^2(i); v) \leq 2\|\mathcal{F}\| T(\mathcal{F}_1^2(i)).$$

If $|V_i| \geq 10^7$, then by Lemma 2 there is a pair $\{v_0, w_0\} \subset V_i$ ($v_0 \neq w_0$) of previously unselected points such that

$$T(\mathcal{F}(i); v_0, w_0) \leq \frac{6}{|V_i|^{8/7}} \binom{\|\mathcal{F}\|}{2} T(\mathcal{F}(i)),$$

$$|T(\mathcal{F}(i); v_0) - T(\mathcal{F}(i); w_0)| \leq \frac{2}{|V_i|^{8/7}} \|\mathcal{F}\| T(\mathcal{F}(i)),$$

$$|T(\mathcal{F}_1^2(i); v_0) - T(\mathcal{F}_1^2(i); w_0)| \leq \frac{6}{|V_i|^{8/7}} \|\mathcal{F}\| T(\mathcal{F}_1^2(i)).$$

It follows that if Picker offers the pair $\{v_0, w_0\} \subset V_i$ for Chooser to choose from, then by (22.3) and (22.4)

$$T(\mathcal{F}(i+1)) \geq T(\mathcal{F}(i)) \left\{ 1 - \frac{3\|\mathcal{F}\|^2 + 2\|\mathcal{F}\|}{|V_i|^{8/7}} \right\}; \tag{22.8}$$

and

$$T(\mathcal{F}_1^2(i+1)) \leq T(\mathcal{F}_1^2(i)) \left\{ 1 + \frac{6\|\mathcal{F}\|}{|V_i|^{8/7}} \right\}. \tag{22.9}$$

Since $|V_i| = |V| - 2i$, for any $M > 0$ we have

$$\sum_{i:|V_i|>M} \frac{1}{|V_i|^{8/7}} \leq \int_M^\infty x^{-8/7} dx = \frac{7}{M^{1/7}}. \tag{22.10}$$

We use the trivial inequalities $1 + x \leq e^x$ and $1 - x \geq e^{-2x}$ for $0 \leq x \leq 1/2$; then by iterated applications of (22.8) and (22.9), and also by (22.10) it follows that

$$T(\mathcal{F}(i+1)) \geq \frac{1}{2} T(\mathcal{F}) \text{ and } T(\mathcal{F}_1^2(i+1)) \leq 2T(\mathcal{F}_1^2)$$

for all i with $|V_i| > M = (70\|\mathcal{F}\|^2)^7$. Let i_0 be the *last* index with $|V_i| > M = (70\|\mathcal{F}\|^2)^7$. Then

$$|V_{i_0+1}| \leq M, \ T(\mathcal{F}(i_0+1)) \geq \frac{1}{2} T(\mathcal{F}) \text{ and } T(\mathcal{F}_1^2(i_0+1)) \leq 2T(\mathcal{F}_1^2).$$

Note that $B \in \mathcal{F}(i_0+1)$ implies $B \subseteq V_{i_0+1}$. We distinguish two cases.

Case 1: The empty set is an element of $\mathcal{F}(i_0+1)$.
Then Picker already completed a winning set, and wins the play.

Case 2: $B \in \mathcal{F}(i_0+1)$ implies $|B| \geq 1$
We show that Case 2 is *impossible*. Indeed, in Case 2

$$\sum_{v \in V_{i_0+1}} T(\mathcal{F}(i_0+1); v) \geq T(\mathcal{F}(i_0+1)) \geq \frac{1}{2} T(\mathcal{F}),$$

so by the Pigeonhole Principle there is a point $x \in V_{i_0+1}$ such that

$$T(\mathcal{F}(i_0+1); x) \geq \frac{T(\mathcal{F})}{2|V_{i_0+1}|} \geq \frac{T(\mathcal{F})}{2M}. \tag{22.11}$$

On the other hand, by the Variance Lemma (mentioned above) we obtain the following inequality

$$T(\mathcal{F}(i_0+1); x) \leq 2 \left(T \left((\mathcal{F}(i_0+1))_1^2 \right) \right)^{1/2} + 2.$$

By using the trivial fact $(\mathcal{F}(i))_1^2 \subseteq \mathcal{F}_1^2(i)$, we conclude that

$$T(\mathcal{F}(i_0+1); x) \leq 2 \left(T(\mathcal{F}_1^2(i_0+1)) \right)^{1/2} + 2. \tag{22.12}$$

Comparing (22.11) and (22.12)

$$T(\mathcal{F}) \leq 2(70\|\mathcal{F}\|^2)^7 \left(2 \left(2T(\mathcal{F}_1^2) \right)^{1/2} + 2 \right),$$

which contradicts the hypothesis of Theorem 22.1. Theorem 22.1 follows. $\qquad\square$

A brief summary of Sections 21–22 goes as follows

Majority Play Number = Picker Achievement Number,

which holds for our "Ramseyish" games with quadratic goal sets (for the overwhelming majority of the board size parameter N).

23

Weak Win in the Lattice Games

In this section the "board" is very restricted: it always means the $N \times N$ lattice in the plane. Weak Win in the Square Lattice Games, both aligned and tilted, are easy: Theorem 1.2 applies and gives the correct lower bounds

$$\mathbf{A}(N \times N; \text{square lattice}) \geq \left\lfloor \sqrt{\log_2 N} + o(1) \right\rfloor$$

$$\mathbf{A}(N \times N; \text{tilted square lattice}) \geq \left\lfloor \sqrt{2 \log_2 N} + o(1) \right\rfloor$$

for the Achievement Numbers, see Section 8. The reason why the simple Theorem 1.2 can give the truth is that any 2 given points "nearly" determine a $q \times q$ Square Lattice; in fact, 2 points determine at most $\binom{q^2}{2}$ $q \times q$ Square Lattices. This $\leq \binom{q^2}{2}$ is polylogarithmic in terms of N; of course, "polylogarithmic" is asymptotically negligible because the goal set size is a rapidly changing quadratic function. We refer to this property as *2-determinedness*.

1. An *ad hoc* Higher Moment technique. Next we discuss the much more challenging case of $q \times q$ parallelogram lattices. The family $\mathcal{F}$ of all $q \times q$ parallelogram lattices in an $N \times N$ board is *3-determined*: 3 non-collinear points "nearly" determine a $q \times q$ parallelogram lattice: the multiplicity is $\leq \binom{q^2}{3}$, which is again polylogarithmic in terms of N. Here we develop an *ad hoc* Higher Moment technique, which works for 3-determined hypergraphs (unlike Theorem 1.2, which heavily requires 2-determinedness).

Assume we are in the middle of a Parallelogram Lattice Game on an $N \times N$ board, Maker owns the points $X(i) = \{x_1, \ldots, x_i\}$ and Breaker owns the points $Y(i) = \{y_1, \ldots, y_i\}$. The main question is how to choose Maker's next move x_{i+1}. Let $\mathcal{F}(i)$ be a truncated sub-family of $\mathcal{F}$, namely the family of the unoccupied parts of the "survivors"

$$\mathcal{F}(i) = \{A \setminus X(i) : A \in \mathcal{F}, \ A \cap Y(i) = \emptyset\}.$$

329

$\mathcal{F}(i)$ is a multi-hypergraph: if A and A' are distinct elements of $\mathcal{F}$ and $A \setminus X(i) = A' \setminus X(i)$, then $A \setminus X(i)$ and $A' \setminus X(i)$ are considered different elements of $\mathcal{F}(i)$. Note that the empty set can be an element of $\mathcal{F}(i)$; it simply means that Maker already occupied a winning set, and wins the play in the ith round or before.

As usual, for an arbitrary hypergraph $(V, \mathcal{H})$ write

$$T(\mathcal{H}) = \sum_{A \in \mathcal{H}} 2^{-|A|},$$

and for any m-element subset of points $\{x_1, \ldots, x_m\} \subseteq V$ ($m \geq 1$) write

$$T(\mathcal{H}; x_1, \ldots, x_m) = \sum_{A \in \mathcal{H}: \, \{x_1, \ldots, x_m\} \subseteq A} 2^{-|A|}.$$

Of course, if a set has multiplicity (say) M, then it shows up M times in every summation.

We call $T(\mathcal{F}(i))$ the *Winning Chance Function*; note that the Chance Function can be much larger than 1 (it is *not* a probability!), but it is always non-negative. If Maker can guarantee that the Chance Function remains non-zero during the whole play, i.e. $T(\mathcal{F}(i)) > 0$ for all i, then Maker obviously wins. Let x_{i+1} and y_{i+1} denote the $(i+1)$st moves of the 2 players; they affect the Chance Function as follows

$$T(\mathcal{F}(i+1)) = T(\mathcal{F}(i)) + T(\mathcal{F}(i); x_{i+1}) - T(\mathcal{F}(i); y_{i+1}) - T(\mathcal{F}(i); x_{i+1}, y_{i+1}).$$

Indeed, x_{i+1} doubles the value of each $B \in \mathcal{F}(i)$ with $x_{i+1} \in B$; on the other hand, y_{i+1} "kills" all $B \in \mathcal{F}(i)$ with $y_{i+1} \in B$. Of course, we have to make a correction due to those $B \in \mathcal{F}(i)$ which contain *both* x_{i+1} and y_{i+1}: these $B \in \mathcal{F}(i)$ were "doubled" first and "killed" second, so we have to subtract their values one more time.

"Optimizing the Chance Function" was the basic idea of the proof of the Linear Weak Win Criterion Theorem 1.2. Unfortunately, Theorem 1.2 gave very poor result for the (K_N, K_q) Clique Game: it applies and yields a Weak Win if

$$\binom{N}{q} > 2^{\binom{q}{2}-3} \binom{N}{2} \Delta_2.$$

For this particular family of winning sets the Max Pair-Degree $\Delta_2 \leq \binom{N}{q-3}$ (indeed, two distinct edges determine at least 3 different vertices); therefore, we have

$$\binom{N}{q} > 2^{\binom{q}{2}-3} \binom{N}{2} \binom{N}{q-3}.$$

This yields $N \geq 2^{q^2/2}$, or taking the inverse, $q < \sqrt{2 \log_2 N}$, which is roughly square-root of the truth. A very disappointing bound!

From this example we learn that optimizing the Chance Function alone is not enough. The new idea is to *maximize* the Chance Function $T(\mathcal{F}(i))$, and, at the same time, *minimize* the "one-sided discrepancy" $T(\mathcal{F}(i); x_{i+1}, y_{i+1})$. How do we handle these two jobs at the same time? Well, we are going to introduce a single *Potential Function*, which is a linear combination of the Chance Function $T(\mathcal{F}(i))$ (with positive weight 1), and an "auxiliary part" related to the "discrepancy" $T(\mathcal{F}(i); x_{i+1}, y_{i+1})$. The "auxiliary part" requires some preparations.

The plan is to involve a key inequality, which estimates the "one-sided discrepancy" $T(\mathcal{F}(i); x_{i+1}, y_{i+1})$ from above (see inequality (23.7) below). Select an arbitrary $A_1 \in \mathcal{F}$ with $\{x_{i+1}, y_{i+1}\} \subset A_1$ and $A_1 \cap Y(i) = \emptyset$ (which means a "survivor" $q \times q$ parallelogram lattice in the $N \times N$ board such that the lattice contains the given point pair $\{x_{i+1}, y_{i+1}\}$; such an $A_1 \in \mathcal{F}$ has a non-zero contribution in the sum $T(\mathcal{F}(i); x_{i+1}, y_{i+1}))$, and select an arbitrary point $z \in A_1$, which is not on the $x_{i+1}y_{i+1}$-line (the $x_{i+1}y_{i+1}$-line means the straight line joining the two lattice points x_{i+1}, y_{i+1}). The property of 3-determinedness gives that there are $\leq \binom{q^2}{3} \leq q^6/6$ winning sets $A_2 \in \mathcal{F}$, which contain the non-collinear triplet $\{x_{i+1}, y_{i+1}, z\}$; as z runs, we obtain that there are $\leq q^8/6$ winning sets $A_2 \in \mathcal{F}$, which contain the pair $\{x_{i+1}, y_{i+1}\}$ such that the intersection $A_1 \cap A_2$ is not contained by the $x_{i+1}y_{i+1}$-line.

For $k = 0, 1, 2, \ldots$ write

$$\mathcal{F}(x_{i+1}, y_{i+1}; k) = \{A \in \mathcal{F}: \{x_{i+1}, y_{i+1}\} \subset A, A \cap Y(i) = \emptyset, |A \cap X(i)| = k\}$$

and $|\mathcal{F}(x_{i+1}, y_{i+1}; k)| = M_k$. Clearly

$$T(\mathcal{F}(i); x_{i+1}, y_{i+1}) = \sum_{k=0}^{q^2} T(\mathcal{F}(x_{i+1}, y_{i+1}; k)) \tag{23.1}$$

and

$$T(\mathcal{F}(x_{i+1}, y_{i+1}; k)) = M_k \cdot 2^{k-q^2}. \tag{23.2}$$

Assume that $M_k \geq 3q^8$; then there are at least

$$\frac{M_k \cdot (M_k - q^8/6) \cdot (M_k - 2q^8/6) \cdot (M_k - 3q^8/6) \cdot (M_k - 4q^8/6)}{5!} \geq \frac{M_k^5}{3^5} = \frac{M_k^5}{243} \tag{23.3}$$

5-tuples

$$\{A_1, A_2, A_3, A_4, A_5\} \in \binom{\mathcal{F}(x_{i+1}, y_{i+1}; k)}{5}$$

such that, deleting the points of the $x_{i+1}y_{i+1}$-line, the remaining parts of the 5 sets $A_1, \ldots, A_5$ become *disjoint*. This motivates the following auxiliary "big hypergraph" (we apply it with $p = 5$) on $N \times N$: given a finite hypergraph $\mathcal{H}$ on $N \times N$, for arbitrary integer $p \geq 2$ write

$$\mathcal{H}_{2*}^{p} = \left\{ \bigcup_{i=1}^{p} A_i : \{A_1, \ldots, A_p\} \in \binom{\mathcal{H}}{p}, \ \bigcap_{i=1}^{p} A_i \text{ is a collinear set of cardinality } \geq 2, \right.$$

$$\left. \text{the truncated sets } A_j \setminus (\bigcap_{i=1}^{p} A_i - line), j = 1, \ldots, p \text{ are disjoint} \right\},$$

$$(23.4)$$

where "$\bigcap_{i=1}^{p} A_i - line$" means the uniquely determined straight line containing the (≥ 2)-element collinear set $\bigcap_{i=1}^{p} A_i$.

$\mathcal{H}_{2*}^{p}$ is the family of all union sets $\bigcup_{i=1}^{p} A_i$, where $\{A_1, \ldots, A_p\}$ runs over all unordered p-tuples of distinct elements of $\mathcal{H}$ having a collinear intersection of at least 2 points such that deleting this common part from the sets they become disjoint.

Trivially

$$\left| \left(\bigcup_{i=1}^{p} A_i \right)_{u.o.p.} \right| \leq \sum_{i=1}^{p} |(A_i)_{u.o.p}|,$$

where "u.o.p." stands for *unoccupied part*. By using this trivial inequality and (23.3)–(23.4), we have

$$T(\mathcal{F}_{2*}^{5}(i)) \geq \frac{M_k^5}{3^5} \cdot 2^{5(k-q^2)} \text{ if } M_k \geq 3q^8. \tag{23.5}$$

(23.5) implies

$$3 \left(T(\mathcal{F}_{2*}^{5}(i)) \right)^{1/5} + 3q^8 \cdot 2^{k-q^2} \geq M_k \cdot 2^{k-q^2}, \tag{23.6}$$

so by (23.1)–(23.2) and (23.6) we obtain the:

Key inequality

$$6q^8 + 3q^2 \left(T(\mathcal{F}_{2*}^{5}(i)) \right)^{1/5} \geq \sum_{k=0}^{q^2} M_k \cdot 2^{k-q^2} = T(\mathcal{F}(i); x_{i+1}, y_{i+1}). \tag{23.7}$$

We are sure the reader is wondering "Why did we pick $p = 5$ in (23.4)?" This natural question will be answered later.

After this preparation we are ready to define our *Potential Function*

$$L_i = T(\mathcal{F}(i)) - \lambda \cdot T(\mathcal{F}_{2*}^{5}(i)), \tag{23.8}$$

where the positive constant λ is specified by the side condition

$$L_0 = \frac{1}{2} T(\mathcal{F}), \text{ that is, } \lambda = \frac{T(\mathcal{F})}{2T(\mathcal{F}_{2*}^{5})}. \tag{23.9}$$

By definition

$$L_{i+1} = T(\mathcal{F}(i+1)) - \lambda \cdot T(\mathcal{F}_{2*}^5(i+1)) =$$

$$= T(\mathcal{F}(i)) + T(\mathcal{F}(i); x_{i+1}) - T(\mathcal{F}(i); y_{i+1}) - T(\mathcal{F}(i); x_{i+1}, y_{i+1})$$

$$- \lambda \Big(T(\mathcal{F}_{2*}^5(i)) + T(\mathcal{F}_{2*}^5(i); x_{i+1}) - T(\mathcal{F}_{2*}^5(i); y_{i+1}) - T(\mathcal{F}_{2*}^5(i); x_{i+1}, y_{i+1}) \Big)$$

$$= L_i + L_i(x_{i+1}) - L_i(y_{i+1}) - T(\mathcal{F}(i); x_{i+1}, y_{i+1}) + \lambda \cdot T(\mathcal{F}_{2*}^5(i); x_{i+1}, y_{i+1}),$$
$$(23.10)$$

where

$$L_i(z) = T(\mathcal{F}(i); z) - \lambda \cdot T(\mathcal{F}_{2*}^5(i); z)$$

for any $z \in V \setminus (X(i) \cup Y(i))$.

A natural choice for x_{i+1} is to maximize the Potential L_{i+1}. The best choice for x_{i+1} is that unoccupied $z \in V \setminus (X(i) \cup Y(i))$ for which $L_i(z)$ attains its *maximum*. Then clearly $L_i(x_{i+1}) \geq L_i(y_{i+1})$, so by (23.10)

$$L_{i+1} \geq L_i - T(\mathcal{F}(i); x_{i+1}, y_{i+1}).$$

By using the key inequality (23.7) we have

$$L_{i+1} \geq L_i - \left(6q^8 + 3q^2 \left(T(\mathcal{F}_{2*}^5(i)) \right)^{1/5} \right), \qquad (23.11)$$

or equivalently

$$T(\mathcal{F}(i+1)) - \lambda \cdot T(\mathcal{F}_{2*}^5(i+1)) = L_{i+1} \geq L_i - \left(6q^8 + 3q^2 \left(T(\mathcal{F}_{2*}^5(i)) \right)^{1/5} \right), \quad (23.11')$$

that we will call the Critical Inequality.

Before applying the Critical Inequality, we need an upper bound for $T(\mathcal{F}_{2*}^5)$; this is estimated it in a trivial way

$$T(\mathcal{F}_{2*}^5) \leq (N^2)^2 \cdot (N^2)^5 \cdot 2^{-5(q^2-q)} = N^{14} \cdot 2^{-5q^2+5q}. \qquad (23.12)$$

We also need a lower bound for $T(\mathcal{F})$; the picture below

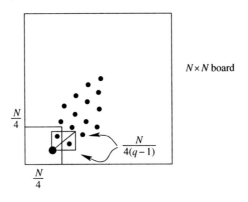

explains the easy lower bound

$$|\mathcal{F}| \geq \left(\frac{N}{4}\right)^2 \cdot \left(\frac{1}{2}\left(\frac{N/4}{q-1}\right)^2\right)^2 = \frac{N^6}{2^{14}(q-1)^4},$$

which implies

$$T(\mathcal{F}) \geq \frac{N^6}{2^{14}(q-1)^4} 2^{-q^2}. \tag{23.13}$$

2. Conclusion of the proof. We distinguish two cases: either there is or there is not an index i such that

$$T(\mathcal{F}_{2*}^5(i)) > N^{5.5} \cdot T(\mathcal{F}_{2*}^5). \tag{23.14}$$

Case 1: There is no index i such that (23.14) holds.
Then we finish the proof exactly like in Theorem 1.2. Indeed, by (23.9), the Critical Inequality (23.11), and (23.14)

$$L_{end} = L_{N^2/2} \geq L_0 - \sum_{i=0}^{\frac{N^2}{2}-1} \left(6q^8 + 3q^2 \left(T(\mathcal{F}_{2*}^5(i))\right)^{1/5}\right)$$

$$\geq L_0 - \frac{N^2}{2} \cdot \left(6q^8 + 3q^2 \cdot N^{1.1} \cdot \left(T(\mathcal{F}_{2*}^5(i))\right)^{1/5}\right)$$

$$= \frac{1}{2}T(\mathcal{F}) - \frac{N^2}{2} \cdot \left(6q^8 + 3q^2 \cdot N^{1.1} \cdot \left(T(\mathcal{F}_{2*}^5)\right)^{1/5}\right). \tag{23.15}$$

By (23.12), (23.13), and (23.15)

$$L_{end} \geq \frac{1}{2} \cdot \frac{N^6}{2^{14}(q-1)^4} 2^{-q^2} - \frac{N^2}{2} \cdot \left(6q^8 + 3q^2 \cdot N^{1.1} \cdot N^{14/5} \cdot 2^{-q^2+q}\right). \tag{23.16}$$

By choosing $q = (2+o(1))\sqrt{\log_2 N}$ we have $2^{q^2} = N^{4+o(1)}$, so by (23.16)

$$L_{end} \geq \frac{N^6}{2^{15}(q-1)^4} 2^{-q^2} - 3N^2 q^8 - N^{1.9+o(1)}. \tag{23.17}$$

The right-hand side of (23.17) is strictly positive if $q = \lfloor 2\sqrt{\log_2 N} - o(1) \rfloor$ (with an appropriate $o(1)$) and N is large enough.

If $L_{end} > 0$, then $T(\mathcal{F}(end)) \geq L_{end} > 0$ implies the existence of a "survivor" (i.e. Breaker-free) $A_0 \in \mathcal{F}$ at the end of the play. This $A_0 \in \mathcal{F}$ is completely occupied by Maker, proving the Weak Win.

Case 2: There is an index i such that (23.14) holds.
We will prove that Case 2 is impossible! Indeed, let $i = j_1$ denote the *first* index such that (23.14) holds. Then we can save the argument of (23.15)–(23.17) in Case 1, and obtain the inequality $L_{j_1} > 0$ if $q = \lfloor 2\sqrt{\log_2 N} - o(1) \rfloor$ (with an appropriate

$o(1)$) and N is large enough. By definition (see (23.8))

$$L_{j_1} = T(\mathcal{F}(j_1)) - \lambda \cdot T(\mathcal{F}^5_{2*}(j_1)) > 0,$$

implying

$$T(\mathcal{F}(j_1)) > \lambda \cdot T(\mathcal{F}^5_{2*}(j_1)) > \lambda \cdot N^{5.5} \cdot T(\mathcal{F}^5_{2*})$$

$$= \frac{T(\mathcal{F})}{2T(\mathcal{F}^5_{2*})} \cdot N^{5.5} \cdot T(\mathcal{F}^5_{2*}) = \frac{1}{2} N^{5.5} \cdot T(\mathcal{F}). \qquad (23.18)$$

By (23.13) and (23.18)

$$T(\mathcal{F}(j_1)) > N^{5.5} \cdot \frac{N^6}{2^{15}(q-1)^4} 2^{-q^2} = N^{7.5 - o(1)}, \qquad (23.19)$$

which is clearly false! Indeed, a corner point and its two neighbors uniquely determine a $q \times q$ parallelogram lattice, implying the trivial upper bound

$$T(\mathcal{F}(j_1)) \le |\mathcal{F}(j_1)| \le |\mathcal{F}| \le (N^2)^2 = N^6,$$

which contradicts (23.19). This contradiction proves that Case 2 is impossible. This completes the proof of the following lower bound on the Achievement Number ("Weak Win" part).

Theorem 23.1 *We have*

$$\mathbf{A}(N \times N; \text{ parallelogram lattice}) \ge \left\lfloor 2\sqrt{\log_2 N} - o(1) \right\rfloor$$

with an appropriate $o(1)$ tending to zero as $N \to \infty$. □

Notice that the proof technique works if $(2 - \frac{4}{p})p + 2 > 6$ holds, i.e. if $p > 4$, explaining the (at first sight) accidental choice $p = 5$ in (23.4).

3. Another illustration. The argument applies, with minor modifications, for all lattices in Section 8 (see Theorem 8.2). Just one more example will be discussed, the *aligned rectangle* lattice, and the rest left to the reader.

An obvious novelty is that in the aligned rectangle lattice the *horizontal* and *vertical* directions play a special role, which leads to some natural changes in the argument.

Assume we are in the middle of an Aligned Rectangle Lattice Game on an $N \times N$ board, Maker owns the points $X(i) = \{x_1, \ldots, x_i\}$ and Breaker owns the points $Y(i) = \{y_1, \ldots, y_i\}$. The main question is how to choose Maker's next move x_{i+1}. Again we have the equality

$$T(\mathcal{F}(i+1)) = T(\mathcal{F}(i)) + T(\mathcal{F}(i); x_{i+1}) - T(\mathcal{F}(i); y_{i+1}) - T(\mathcal{F}(i); x_{i+1}, y_{i+1}),$$

and again the plan is to involve a key inequality, which estimates the "one-sided discrepancy" $T(\mathcal{F}(i); x_{i+1}, y_{i+1})$ from above.

Case a: the $x_{i+1}y_{i+1}$-line is neither horizontal, nor vertical.
Then trivially

$$T(\mathcal{F}(i); x_{i+1}, y_{i+1}) \le |\{A \in \mathcal{F} : \{x_{i+1}, y_{i+1}\} \subset A\}| \le \binom{q^2}{2} \le \frac{q^4}{2}. \qquad (23.20)$$

Case b: the $x_{i+1}y_{i+1}$-line is (say) horizontal (the vertical case goes similarly).
Select an arbitrary $A_1 \in \mathcal{F}$ with $\{x_{i+1}, y_{i+1}\} \subset A_1$ and $A_1 \cap Y(i) = \emptyset$ (which means
a "survivor" $q \times q$ aligned rectangle lattice in the $N \times N$ board such that the lattice
contains the given point pair $\{x_{i+1}, y_{i+1}\}$; such an $A_1 \in \mathcal{F}$ has a non-zero contribution
in the sum $T(\mathcal{F}(i); x_{i+1}, y_{i+1}))$. Again there are $\le q^8/6$ winning sets $A_2 \in \mathcal{F}$ which
contain the pair $\{x_{i+1}, y_{i+1}\}$ such that the intersection $A_1 \cap A_2$ is not contained by
the $x_{i+1}y_{i+1}$-line.
For $k = 0, 1, 2, \ldots$ write

$$\mathcal{F}(x_{i+1}, y_{i+1}; k) = \{A \in \mathcal{F} : \{x_{i+1}, y_{i+1}\} \subset A, A \cap Y(i) = \emptyset, |A \cap X(i)| = k\}$$

and $|\mathcal{F}(x_{i+1}, y_{i+1}; k)| = M_k$. Clearly

$$T(\mathcal{F}(i); x_{i+1}, y_{i+1}) = \sum_{k=0}^{q^2} T(\mathcal{F}(x_{i+1}, y_{i+1}; k)) \qquad (23.21)$$

and

$$T(\mathcal{F}(x_{i+1}, y_{i+1}; k)) = M_k \cdot 2^{k-q^2}. \qquad (23.22)$$

The first change is that we need 6-tuples (instead of 5-tuples): assume that $M_k \ge 4q^8$;
then there are at least

$$\frac{M_k \cdot (M_k - q^8/6) \cdot (M_k - 2q^8/6) \cdots (M_k - 5q^8/6)}{6!} \ge \frac{M_k^6}{4^6} \qquad (23.23)$$

6-tuples

$$\{A_1, \ldots, A_6\} \in \binom{\mathcal{F}(x_{i+1}, y_{i+1}; k)}{6}$$

such that, deleting the points of the $x_{i+1}y_{i+1}$-line, the remaining parts of the 6 sets
$A_1, \ldots, A_6$ become *disjoint*. This motivates the following auxiliary "big hypergraph"
(we apply it with $p = 6$); notice the change that here "line" means horizontal or
vertical only: given an finite hypergraph $\mathcal{H}$ on $N \times N$, for arbitrary integer $p \ge 2$
write

$$\mathcal{H}_{2**}^p = \left\{ \bigcup_{i=1}^{p} A_i : \{A_1, \ldots, A_p\} \in \binom{\mathcal{H}}{p}, \bigcap_{i=1}^{p} A_i \text{ is a collinear set of cardinality} \ge 2, \right.$$

where the supporting line is either horizontal or vertical

$$\left. \text{and the truncated sets } A_j \setminus (\bigcap_{i=1}^{p} A_i - line), j = 1, \ldots, p \text{ are disjoint} \right\}, \qquad (23.24)$$

where "$\bigcap_{i=1}^{p} A_i - line$" means the uniquely determined horizontal or vertical line containing the (≥ 2)-element collinear set $\bigcap_{i=1}^{p} A_i$.

Trivially

$$\left| \left(\bigcup_{i=1}^{p} A_i \right)_{u.o.p.} \right| \leq \sum_{i=1}^{p} |(A_i)_{u.o.p}|,$$

where "u.o.p." stands for *unoccupied part*. By using this trivial inequality and (23.23)–(23.24), we have

$$T(\mathcal{F}_{2**}^6(i)) \geq \frac{M_k^6}{4^6} \cdot 2^{6(k-q^2)} \text{ if } M_k \geq 4q^8. \tag{23.25}$$

(23.25) implies

$$4 \left(T(\mathcal{F}_{2**}^6(i)) \right)^{1/6} + 4q^8 \cdot 2^{k-q^2} \geq M_k \cdot 2^{k-q^2}, \tag{23.26}$$

so by (23.21)–(23.22) and (23.26) we obtain the new:

Key inequality

$$8q^8 + 4q^2 \left(T(\mathcal{F}_{2*}^6(i)) \right)^{1/6} \geq \sum_{k=0}^{q^2} M_k \cdot 2^{k-q^2} = T(\mathcal{F}(i); x_{i+1}, y_{i+1}). \tag{23.27}$$

In view of (23.20), inequality (23.27) holds in both cases (a) and (b).

After this preparation we are ready to define our *Potential Function*

$$L_i = T(\mathcal{F}(i)) - \lambda \cdot T(\mathcal{F}_{2**}^6(i)), \tag{23.28}$$

where the positive constant λ is specified by the side condition

$$L_0 = \frac{1}{2} T(\mathcal{F}), \text{ i.e. } \lambda = \frac{T(\mathcal{F})}{2T(\mathcal{F}_{2**}^6)}. \tag{23.29}$$

Let

$$L_i(z) = T(\mathcal{F}(i); z) - \lambda \cdot T(\mathcal{F}_{2**}^6(i); z)$$

for any $z \in V \setminus (X(i) \cup Y(i))$. The best choice for x_{i+1} is that unoccupied $z \in V \setminus (X(i) \cup Y(i))$ for which $L_i(z)$ attains its *maximum*; then we get

$$L_{i+1} \geq L_i - T(\mathcal{F}(i); x_{i+1}, y_{i+1}). \tag{23.30}$$

By using the key inequality (23.27) in (23.30), we obtain

$$L_{i+1} \geq L_i - \left(8q^8 + 4q^2 \left(T(\mathcal{F}_{2**}^6(i)) \right)^{1/6} \right), \tag{23.31}$$

or equivalently

$$T(\mathcal{F}(i+1)) - \lambda \cdot T(\mathcal{F}_{2**}^6(i+1)) = L_{i+1} \geq L_i - \left(8q^8 + 4q^2 \left(T(\mathcal{F}_{2**}^6(i)) \right)^{1/6} \right),$$
$$\tag{23.31'}$$

that we call the (new) Critical Inequality.

Before applying the Critical Inequality, we need an upper bound for $T(\mathcal{F}_{2**}^6)$; this is estimated it in a most trivial way

$$T(\mathcal{F}_{2**}^6) \leq (N^2) \cdot N \cdot N^6 \cdot 2^{-6(q^2-q)} = N^9 \cdot 2^{-6q^2+6q}, \qquad (23.32)$$

We also need a lower bound for $T(\mathcal{F})$; clearly

$$T(\mathcal{F}) \approx \frac{N^4}{4(q-1)^2} 2^{-q^2}. \qquad (23.33)$$

Conclusion. We distinguish two cases: either there is or there is not an index i such that

$$T(\mathcal{F}_{2**}^6(i)) > N^{2.5} \cdot T(\mathcal{F}_{2**}^6). \qquad (23.34)$$

Case 1: There is no index i such that (23.34) holds.
Then we have the following analogue of (23.16)

$$L_{end} \geq \frac{1}{2} \cdot \frac{N^4}{4(q-1)^2} 2^{-q^2} - \frac{N^2}{2} \cdot \left(8q^8 + 4q^2 \cdot N^{5/12} \cdot N^{9/6} \cdot 2^{-q^2+q}\right). \qquad (23.35)$$

By choosing $q = (1+o(1))\sqrt{2\log_2 N}$, we have $2^{q^2} = N^{2+o(1)}$, so by (23.35)

$$L_{end} \geq \frac{N^4}{8(q-1)^2} 2^{-q^2} - 4N^2 q^8 - N^{23/12+o(1)}. \qquad (23.36)$$

The right-hand side of (23.36) is strictly positive if $q = \lfloor \sqrt{2\log_2 N} - o(1) \rfloor$ (with an appropriate $o(1)$) and N is large enough.

If $L_{end} > 0$, then $T(\mathcal{F}(end)) \geq L_{end} > 0$ implies Maker's Weak Win.

Case 2: There is an index i such that (23.34) holds.
Again we will prove that Case 2 is impossible! Indeed, let $i = j_1$ denote the *first* index such that (23.14) holds. Then we can save the argument of Case 1 and obtain the inequality $L_{j_1} > 0$ if $q = \lfloor \sqrt{2\log_2 N} - o(1) \rfloor$ (with an appropriate $o(1)$) and N is large enough. By definition

$$L_{j_1} = T(\mathcal{F}(j_1)) - \lambda \cdot T(\mathcal{F}_{2**}^6(j_1)) > 0,$$

implying

$$T(\mathcal{F}(j_1)) > \lambda \cdot T(\mathcal{F}_{2**}^6(j_1)) > \lambda \cdot N^{2.5} \cdot T(\mathcal{F}_{2**}^6)$$

$$= \frac{T(\mathcal{F})}{2T(\mathcal{F}_{2**}^6)} \cdot N^{2.5} \cdot T(\mathcal{F}_{2**}^6) = \frac{1}{2} N^{2.5} \cdot T(\mathcal{F}). \qquad (23.37)$$

By (23.33) and (23.37)

$$T(\mathcal{F}(j_1)) > N^{2.5} \cdot \frac{N^4}{8(q-1)^2} 2^{-q^2} = N^{4.5-o(1)}, \qquad (23.38)$$

which is clearly false! Indeed, a corner point and its two non-collinear neighbors uniquely determine a $q \times q$ aligned rectangular lattice, implying the trivial upper bound

$$T(\mathcal{F}(j_1)) \leq |\mathcal{F}(j_1)| \leq |\mathcal{F}| \leq (N^2) \cdot N \cdot N = N^4,$$

which contradicts (23.38). This contradiction proves that Case 2 is impossible. This completes the proof of the following lower bound on the Achievement Number ("Weak Win" part).

Theorem 23.2 *We have*

$$\mathbf{A}(N \times N; \text{ aligned rectangular lattice}) \geq \left\lfloor \sqrt{2 \log_2 N} - o(1) \right\rfloor,$$

with an appropriate o(1) tending to zero as $N \to \infty$. □

Again the reader is wondering "Why did we pick $p = 6$ in (23.24)?" The reader is challenged to explain why the (at first sight) accidental choice $p = 6$ is the right choice.

24
Game-theoretic higher moments

The proof technique of the last section does *not* apply for the Clique Game. Indeed, 3 non-collinear points nearly determine a $q \times q$ parallelogram lattice, but 3 edges determine at most 6 vertices, and the remaining $(q - 6)$ vertices of a clique K_q (i.e. the overwhelming majority!) remain completely "free."

A straightforward application of Section 23 doesn't work, but we can *develop* the proof technique one step further. This is exactly what we are going to do here.

1. Big auxiliary hypergraphs. The basic idea of Section 23 was to work with a Potential Function

$$L_i = T(\mathcal{F}(i)) - \lambda \cdot T(\mathcal{F}_{2*}^p(i))$$

with $p = 5$ for the parallelogramm lattice, and

$$L_i = T(\mathcal{F}(i)) - \lambda \cdot T(\mathcal{F}_{2**}^p(i))$$

with $p = 6$ for the aligned rectangle lattice, involving big "auxiliary" hypergraphs $\mathcal{H}_{2*}^p$ (see (23.4)) and $\mathcal{H}_{2**}^p$ (see (23.24)). These auxiliary hypergraphs are defined for the $N \times N$ board only: the definitions included geometric concepts like "collinear," "horizontal," "vertical"; these concepts do not generalize for arbitrary hypergraphs. What we use in this section is a simpler concept: a straightforward generalization of the "second moment" hypergraph $\mathcal{H}_1^2$ (see Sections 21–22) to a "higher moment" hypergraph $\mathcal{H}_2^p$ (to be defined below). Note in advance that, for an optimal result in the Clique Game, parameter p has to tend to infinity! (In Section 23 it was enough to work with finite constants like $p = 5$ and $p = 6$.)

In general, for an arbitrary finite board (not just $N \times N$), for an arbitrary finite hypergraph $\mathcal{H}$, and for arbitrary integers $p \geq 2$ and $m \geq 1$ define the "big" hypergraph $\mathcal{H}_m^p$ as follows

$$\mathcal{H}_m^p = \left\{ \bigcup_{i=1}^p A_i : \{A_1, \ldots, A_p\} \in \binom{\mathcal{H}}{p}, \left| \bigcap_{i=1}^p A_i \right| \geq m \right\}.$$

In other words, $\mathcal{H}_m^p$ is the family of all union sets $\bigcup_{i=1}^p A_i$, where $\{A_1, \ldots, A_p\}$ runs over all unordered p-tuples of distinct elements of $\mathcal{H}$ having at least m points in common. Note that even if $\mathcal{H}$ is an ordinary hypergraph, i.e. a set has multiplicity 0 or 1 only, $\mathcal{H}_m^p$ can become a *multi-hypergraph* (i.e. a set may have arbitrary *multiplicity*, not just 0 and 1). More precisely, if $\{A_1, \ldots, A_p\}$ is an unordered p-tuple of distinct elements of $\mathcal{H}$ and $\{A_1', \ldots, A_p'\}$ is another unordered p-tuple of distinct elements of $\mathcal{H}$, $|\bigcap_{i=1}^p A_i| \geq m$, $|\bigcap_{j=1}^p A_j'| \geq m$, and $\bigcup_{i=1}^p A_i = \bigcup_{j=1}^p A_j'$, i.e. $\bigcup_{i=1}^p A_i$ and $\bigcup_{j=1}^p A_j'$ are equal as *sets*, then they still represent *distinct* hyperedges of hypergraph $\mathcal{H}_m^p$.

As usual, for an arbitrary hypergraph $(V, \mathcal{H})$ (where V is the union set), write

$$T(\mathcal{H}) = \sum_{A \in \mathcal{H}} 2^{-|A|},$$

and for any m-element subset of points $\{x_1, \ldots, x_m\} \subseteq V$ $(m \geq 1)$, write

$$T(\mathcal{H}; x_1, \ldots, x_m) = \sum_{A \in \mathcal{H}: \{x_1, \ldots, x_m\} \subseteq A} 2^{-|A|}.$$

Of course, if a set has multiplicity (say) M, then it shows up M times in every summation.

We shall employ the following generalization of the Variance Lemma (see Section 22). We shall apply Theorem 24.1 below in the special cases of $m = 1$ and $m = 2$ only, but we formulate it for arbitrary m anyway.

Theorem 24.1 *("Generalized Variance Lemma") For arbitrary integers $p \geq 2$ and $m \geq 1$, and for arbitrary points $x_1, \ldots, x_m$ of the board*

$$T(\mathcal{H}; x_1, \ldots, x_m) < p\left(\left(T(\mathcal{H}_m^p)\right)^{1/p} + 1\right).$$

Proof. Let $\psi: V \to \{red, blue\}$ be a Random 2-Coloring of the board V. Let $\Omega = \Omega(\psi)$ denote the number of *red* hyperedges $A \in \mathcal{H}$ with $\{x_1, \ldots, x_m\} \subseteq A$. The *expected value* of Ω is equal to $\mathbf{E}(\Omega) = T(\mathcal{H}; x_1, \ldots, x_m)$.

On the other hand, the expected number of (monochromatic) red unordered p-tuples $\{A_1, A_2, \ldots, A_p\}$ such that A_i, $i = 1, 2, \ldots, p$ are distinct elements of $\mathcal{H}$ and $\bigcap_{i=1}^p A_i \supseteq \{x_1, \ldots, x_m\}$, is equal to

$$\mathbf{E}\left(\binom{\Omega}{p}\right) = \sum_{\{A_1, \ldots, A_p\}}^* 2^{-|\bigcup_{i=1}^p A_i|},$$

where $*$ indicates that $A_i \in \mathcal{H}$, $i = 1, \ldots, p$ are distinct elements and $\bigcap_{i=1}^p A_i \supseteq \{x_1, \ldots, x_m\}$. It follows that

$$T(\mathcal{H}_m^p) \geq \mathbf{E}\left(\binom{\Omega}{p}\right).$$

We distinguish 2 cases.

Case 1: $p = 2$ and $m = 1$, i.e. the Variance Lemma in Section 22.
Note that $g(x) = \binom{x}{2}$ is a *convex* function for every x. By using the general inequality

$$\mathbf{E}\Big(g(\Omega)\Big) \geq g\Big(\mathbf{E}(\Omega)\Big)$$

which holds for any convex $g(x)$, we conclude

$$T(\mathcal{H}_1^2) \geq \binom{T(\mathcal{H}; x)}{2},$$

and Case 1 (i.e. the Variance Lemma) easily follows:

Case 2: The general case.
The general case is technically a little bit more complicated because for $p \geq 3$ the function $g(x) = \binom{x}{p} = x(x-1)\cdots(x-p+1)/p!$ is *not* convex for every x. But luckily $g(x)$ *is* convex if $x > p - 1$, which implies the following inequality for *conditional expectations*

$$\mathbf{E}\left[\binom{\Omega}{p} \Big| \Omega \geq p\right] \geq \binom{\mathbf{E}[\Omega | \Omega \geq p]}{p}.$$

Since

$$\binom{x}{p} \geq \left(\frac{x}{p}\right)^p$$

if $x \geq p$, we obtain

$$p\left(\mathbf{E}\left[\binom{\Omega}{p} \Big| \Omega \geq p\right]\right)^{1/p} \geq \mathbf{E}[\Omega | \Omega \geq p].$$

By definition

$$T(\mathcal{H}_m^p) \geq \mathbf{E}\left(\binom{\Omega}{p}\right) = \mathbf{E}\left[\binom{\Omega}{p} \Big| \Omega \geq p\right]\Pr\{\Omega \geq p\} + \mathbf{E}\left[\binom{\Omega}{p} \Big| \Omega < p\right]\Pr\{\Omega < p\}$$

$$= \mathbf{E}\left[\binom{\Omega}{p} \Big| \Omega \geq p\right]\Pr\{\Omega \geq p\},$$

since random variable Ω has integer values ≥ 0 and $\binom{\Omega}{p} = 0$ if $\Omega < p$. Therefore

$$p\Big(T(\mathcal{H}_m^p)\Big)^{1/p}\Big(\Pr\{\Omega \geq p\}\Big)^{-\frac{1}{p}} \geq \mathbf{E}[\Omega | \Omega \geq p].$$

Summarizing, we have

$$T(\mathcal{H}; x_1, \ldots, x_m) = \mathbf{E}(\Omega) = \mathbf{E}[\Omega | \Omega \geq p]\Pr\{\Omega \geq p\} + \mathbf{E}[\Omega | \Omega < p]\Pr\{\Omega < p\}$$

$$\leq p\Big(T(\mathcal{H}_m^p)\Big)^{1/p}\Big(\Pr\{\Omega \geq p\}\Big)^{1-\frac{1}{p}} + p \cdot \Pr\{\Omega < p\} < p\Big(T(\mathcal{H}_m^p)\Big)^{1/p} + p,$$

completing the proof of Theorem 24.1. ∎

2. Advanced Weak Win Criterion. The main objective of this section is to prove:

Theorem 24.2 ["Advanced Weak Win Criterion"] *If there exists a positive integer* $p \geq 2$ *such that*

$$\frac{T(\mathcal{F})}{|V|} > p + 4p\left(T(\mathcal{F}_2^p)\right)^{1/p},$$

then the first player has an explicit Weak Win strategy in the positional game on hypergraph $(V, \mathcal{F})$.

Remarks

(a) Note that under the same condition Chooser (i.e. "builder") has an explicit winning strategy in the Chooser–Picker Game on hypergraph $(V, \mathcal{F})$.

(b) Observe that Theorem 24.2 means a Double Requirement that (1) $T(\mathcal{F})/|V|$ is "large"; and (2) $\left(T(\mathcal{F}_2^p)\right)^{1/p}/T(\mathcal{F})$ is "small" (in fact, less than $1/|V|$, where $|V|$ is the *duration* of the play).

(c) If $\mathcal{F}$ is n-uniform, then $T(\mathcal{F})/|V|$ equals $\text{AverDeg}(\mathcal{F}) \cdot 2^{-n}/n$.

(d) If $\mathcal{F}$ is *Almost Disjoint*, then $\mathcal{F}_2^p$ is empty. In this special case Theorem 24.2 is essentially equivalent to the old Linear Weak Win Criterion Theorem 1.2.

In both Weak Win Criterions – Theorems 1.2 and 24.2 – the *Average Degree* is the key parameter; on the other hand, in the "Neighborhood Conjecture" about Strong Draw (see Open Problem 9.1) the *Maximum Degree* is the key parameter. If the hypergraph is *degree-regular*, then the Maximum Degree equals the Average Degree. This is the reason why for *degree-regular* hypergraphs, like the clique-hypergraph and the lattice-hypergraphs (the latter are nearly degree-regular), we can prove exact results.

Unfortunately, the n^d-hypergraph ("multidimensional n^d Tic-Tac-Toe") is very far from being degree-regular.

3. Technique of self-improving potentials. An alternative (very instructive!) name for this proof technique is "sliding potentials," see Figure on p. 346.

Proof of Theorem 24.2. Assume we are at a stage of the play where Maker, as the first player, already occupied $x_1, x_2, \ldots, x_i$, and Breaker, the second player, occupied $y_1, y_2, \ldots, y_i$. The question is how to choose Maker's next point x_{i+1}.

Let $X_i = \{x_1, x_2, \ldots, x_i\}$ and $Y_i = \{y_1, y_2, \ldots, y_i\}$. Let $\mathcal{F}(i)$ be a truncated sub-family of $\mathcal{F}$, namely the family of the unoccupied parts of the "survivors"

$$\mathcal{F}(i) = \{A \setminus X_i : A \in \mathcal{F}, \ A \cap Y_i = \emptyset\}.$$

$\mathcal{F}(i)$ is a multi-hypergraph: if A and A' are distinct elements of $\mathcal{F}$ and $A \setminus X_i = A' \setminus X_i$, then $A \setminus X_i$ and $A' \setminus X_i$ are considered different elements of $\mathcal{F}(i)$. Note that the empty set can be an element of $\mathcal{F}(i)$: it simply means that Maker already occupied a winning set, and wins the play in the ith round or before.

We call $T(\mathcal{F}(i))$ the *Winning Chance Function*; note that the Chance Function can be much larger than 1 (it is *not* a probability!), but it is always non-negative. If Maker can guarantee that the Chance Function remains non-zero during the whole play, i.e. $T(\mathcal{F}(i)) > 0$ for all i, then Maker obviously wins. Let x_{i+1} and y_{i+1} denote the $(i+1)$st moves of the two players; it is easy to describe their effect

$$T(\mathcal{F}(i+1)) = T(\mathcal{F}(i)) + T(\mathcal{F}(i); x_{i+1}) - T(\mathcal{F}(i); y_{i+1}) - T(\mathcal{F}(i); x_{i+1}, y_{i+1}).$$
(24.1)

Indeed, x_{i+1} doubles the value of each $B \in \mathcal{F}(i)$ with $x_{i+1} \in B$; on the other hand, y_{i+1} "kills" all $B \in \mathcal{F}(i)$ with $y_{i+1} \in B$. Of course, we have to make a correction due to those $B \in \mathcal{F}(i)$ that contain *both* x_{i+1} and y_{i+1}: these $B \in \mathcal{F}(i)$ were "doubled" first and "killed" second, so we have to subtract their values one more time.

Following Section 23, the new idea is to *maximize* the Chance Function $T(\mathcal{F}(i))$, and, at the same time, *minimize* the "one-sided discrepancy" $T(\mathcal{F}(i); x_{i+1}, y_{i+1})$. To handle these two jobs at the same time we introduce a single *Potential Function*, which is a linear combination of the Chance Function $T(\mathcal{F}(i))$ (with positive weight 1) and an "auxiliary part" $T(\mathcal{F}_2^p(i))$ related to the "discrepancy" $T(\mathcal{F}(i); x_{i+1}, y_{i+1})$ via Theorem 24.1 with negative weight. Naturally

$$\mathcal{F}_2^p(i) = \{A \setminus X_i : A \in \mathcal{F}_2^p, \ A \cap Y_i = \emptyset\}.$$
(24.2)

Note that the relation $(\mathcal{F}(i))_2^p \subseteq \mathcal{F}_2^p(i)$ is trivial; it will be used repeatedly in the proof.

The Potential Function is defined as follows:

$$L_i = T(\mathcal{F}(i)) - \lambda \cdot T(\mathcal{F}_2^p(i)),$$
(24.3)

where the positive constant λ is specified by the side condition $L_0 = \frac{1}{2}T(\mathcal{F})$. In other words

$$\frac{1}{2}T(\mathcal{F}) = \lambda \cdot T(\mathcal{F}_2^p).$$
(24.4)

Note that the Potential Function L_i is *less* than the Chance Function $T(\mathcal{F}(i))$. If the Chance Function is 0, then every winning set is blocked by Breaker, and Breaker wins. If the Chance Function is positive, then Maker hasn't lost yet (he still has a "chance" to win). Therefore, if the Potential Function is positive, then Maker still has a"chance" to win; if the Potential Function is negative, then we call it "inconclusive."

By (24.2)–(24.3)

$$L_{i+1} = T(\mathcal{F}(i+1)) - \lambda \cdot T(\mathcal{F}_2^p(i+1))$$

$$= T(\mathcal{F}(i)) + T(\mathcal{F}(i); x_{i+1}) - T(\mathcal{F}(i); y_{i+1}) - T(\mathcal{F}(i); x_{i+1}, y_{i+1})$$

$$- \lambda\Big(T(\mathcal{F}_2^p(i)) + T(\mathcal{F}_2^p(i); x_{i+1}) - T(\mathcal{F}_2^p(i); y_{i+1}) - T(\mathcal{F}_2^p(i); x_{i+1}, y_{i+1})\Big)$$

$$= L_i + L_i(x_{i+1}) - L_i(y_{i+1}) - T(\mathcal{F}(i); x_{i+1}, y_{i+1}) + \lambda \cdot T(\mathcal{F}_2^p(i); x_{i+1}, y_{i+1}),$$
$$\tag{24.5a}$$

where

$$L_i(z) = T(\mathcal{F}(i); z) - \lambda \cdot T(\mathcal{F}_2^p(i); z) \tag{24.5b}$$

for any $z \in V \setminus (X_i \cup Y_i)$.

A natural choice for x_{i+1} is to maximize the Potential L_{i+1}. The best choice for x_{i+1} is that unoccupied $z \in V \setminus (X_i \cup Y_i)$ for which $L_i(z)$ attains its *maximum*. Then clearly $L_i(x_{i+1}) \ge L_i(y_{i+1})$, so by (24.5a)

$$L_{i+1} \ge L_i - T(\mathcal{F}(i); x_{i+1}, y_{i+1}).$$

Since $(\mathcal{F}(i))_2^p \subseteq \mathcal{F}_2^p(i)$, by Theorem 24.1 we obtain the following **critical inequality**

$$L_{i+1} \ge L_i - p\Big(T(\mathcal{F}_2^p(i))\Big)^{1/p} - p.$$

The last step was a *kind of* "Chebyshev's inequality": we estimated the "one-sided discrepancy" $T(\mathcal{F}(i); x_{i+1}, y_{i+1})$ with a "standard deviation" $\Big(T(\mathcal{F}_2^p(i))\Big)^{1/p}$ ("Theorem 24.1").

In view of (24.3), we can rewrite the *critical inequality* as follows

$$T(\mathcal{F}(i+1)) - \lambda \cdot T(\mathcal{F}_2^p(i+1)) = L_{i+1} \ge L_i - p\Big(T(\mathcal{F}_2^p(i))\Big)^{1/p} - p. \tag{24.6}$$

The reason why the proof (to be described below) works is that the "consecutive" terms $T(\mathcal{F}_2^p(i+1))$ and $T(\mathcal{F}_2^p(i))$ show up in *different powers* at the two ends of inequality (24.6).

We divide the whole course of the play into several *phases*. In each *phase* we use a *new* Potential Function, which will lead to a surprising "self-improvement." Right before the old Potential Function turns into negative ("inconclusive"), we switch to a new Potential Function which will be "large positive" again. At the end of each *phase* we employ formula (24.6), and as a result we gain a multiplicative factor: a "p-power." In other words, the more phases we have, the better for Maker; this is how we increase Maker's "chances" to win.

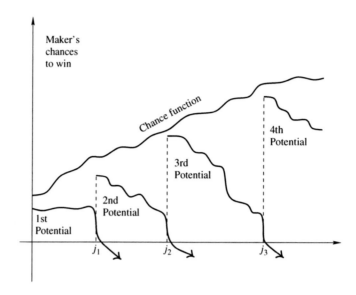

The picture above should help the reader to visualize the machinery of the proof.

The *first phase* of the play is $0 \leq i < j_1$, i.e. the "time index" i runs in an interval $0 \leq i < j_1$, where j_1 is defined by

$$T(\mathcal{F}_2^p(i)) < 4^p T(\mathcal{F}_2^p) \quad for \ 0 \leq i < j_1 \tag{24.7}$$

and

$$T(\mathcal{F}_2^p(j_1)) \geq 4^p T(\mathcal{F}_2^p). \tag{24.8}$$

That is, inequality $T(\mathcal{F}_2^p(i)) \geq 4^p T(\mathcal{F}_2^p)$ happens for first time at $i = j_1$.

We claim that, for every i with $0 \leq i < j_1$, $L_{i+1} > 0$. Indeed, by (24.4) and (24.6)–(24.7)

$$L_{i+1} \geq L_0 - \sum_{\ell=0}^{j_1-1} p\left(\left(T(\mathcal{F}_2^p(\ell))\right)^{1/p} + 1 \right)$$

$$\geq \frac{1}{2}T(\mathcal{F}) - \frac{|V|}{2}p\left(4\left(T(\mathcal{F}_2^p)\right)^{1/p} + 1 \right)$$

$$= \frac{1}{2}\left(T(\mathcal{F}) - 4p|V|\left(\left(T(\mathcal{F}_2^p)\right)^{1/p} + \frac{1}{4} \right) \right) > 0$$

by the hypothesis of Theorem 24.2.

In particular, we get the inequality $L_{j_1} > 0$. If the first phase is the *whole* play, then we are clearly done. Indeed, $L_i > 0$ implies $T(\mathcal{F}(i)) \geq L_i > 0$, i.e. the Chance Function is always positive, and Maker wins. The idea is to show that, if there is a next phase, that just *helps* Maker, i.e. Maker has a *better* "chance" to win!

By (24.8) there is a real number $t_1 \geq 1$ such that we have the equality

$$T(\mathcal{F}_2^p(j_1)) = (4t_1)^p T(\mathcal{F}_2^p). \tag{24.9}$$

Since $L_{j_1} = T(\mathcal{F}(j_1)) - \lambda \cdot T(\mathcal{F}_2^p(j_1)) > 0$, it follows by (24.4) and (24.9) that

$$T(\mathcal{F}(j_1)) > \lambda T(\mathcal{F}_2^p(j_1)) = (4t_1)^p \lambda T(\mathcal{F}_2^p) = \frac{4^p}{2} t_1^p T(\mathcal{F}). \tag{24.10}$$

"Time index" j_1 is the beginning of the **second phase** $j_1 \leq i < j_2$, where the "endpoint" j_2 will be defined by (24.13)–(24.14) below. In the *second phase* $j_1 \leq i < j_2$ we modify the Potential Function L_i by *halving* the value of λ: let

$$L_i^{(2)} = T(\mathcal{F}(i)) - \frac{\lambda}{2} T(\mathcal{F}_2^p(i)). \tag{24.11}$$

Comparing (24.10) and (24.11), we see that in (24.11) we "lost" a factor of 2, but in (24.10) we "gained" much more: we gained a factor of $4^p/2$, meaning that Maker's "chance" to win is definitely *improved*.

Since $L_{j_1} > 0$, we have

$$L_{j_1}^{(2)} = \frac{1}{2} T(\mathcal{F}(j_1)) + \frac{1}{2} L_{j_1} > \frac{1}{2} T(\mathcal{F}(j_1)). \tag{24.12}$$

The *second phase* of the play is the "time interval" $j_1 \leq i < j_2$, where j_2 is defined by

$$T(\mathcal{F}_2^p(i)) < \left(\frac{4^p}{2}\right)^p T(\mathcal{F}_2^p(j_1)) \quad \text{for} \quad j_1 \leq i < j_2 \tag{24.13}$$

and

$$T(\mathcal{F}_2^p(j_2)) \geq \left(\frac{4^p}{2}\right)^p T(\mathcal{F}_2^p(j_1)). \tag{24.14}$$

By (24.9) and (24.13), for every i with $j_1 \leq i < j_2$

$$\left(T(\mathcal{F}_2^p(i))\right)^{1/p} < 4 \left(\frac{4^p}{2}\right) t_1 \left(T(\mathcal{F}_2^p)\right)^{1/p}. \tag{24.15}$$

In the second phase we use the *new* Potential Function $L_i^{(2)}$, so the analogue of (24.5a) goes like this

$$L_{i+1}^{(2)} = L_i^{(2)} + L_i^{(2)}(x_{i+1}) - L_i^{(2)}(y_{i+1})$$

$$- T(\mathcal{F}(i); x_{i+1}, y_{i+1}) + \frac{\lambda}{2} T(\mathcal{F}_2^p(i); x_{i+1}, y_{i+1}), \tag{24.5a'}$$

where

$$L_i^{(2)}(z) = T(\mathcal{F}(i); z) - \frac{\lambda}{2} T(\mathcal{F}_2^p(i); z) \tag{24.5b'}$$

for any $z \in V \setminus (X_i \cup Y_i)$.

In the second phase Maker's $(i+1)$st move x_{i+1} is that unoccupied $z \in V \setminus (X_i \cup Y_i)$ for which $L_i^{(2)}(z)$ attains its maximum. Then clearly $L_i^{(2)}(x_{i+1}) \geq L_i^{(2)}(y_{i+1})$, so by (24.5a′)

$$L_{i+1}^{(2)} \geq L_i^{(2)} - T(\mathcal{F}(i); x_{i+1}, y_{i+1}).$$

By Theorem 24.1 we obtain the analogue of 24.6

$$L_{i+1}^{(2)} \geq L_i^{(2)} - p\left(T(\mathcal{F}_2^p(i))\right)^{1/p} - p. \tag{24.6′}$$

We claim that for every i with $j_1 \leq i < j_2$, $L_{i+1}^{(2)} > 0$. Indeed, by (24.4), (24.6′), (24.10), (24.12) and (24.15)

$$L_{i+1}^{(2)} \geq L_{j_1}^{(2)} - \sum_{\ell=j_1}^{j_2-1} p\left(\left(T(\mathcal{F}_2^p(\ell))\right)^{1/p} + 1\right)$$

$$\geq \frac{1}{2} T(\mathcal{F}(j_1)) - \frac{|V|}{2} p\left(4\left(\frac{4^p}{2}\right) t_1 \left(T(\mathcal{F}_2^p)\right)^{1/p} + 1\right)$$

$$\geq 4^{p-1} t_1 \left(t_1^{p-1} T(\mathcal{F}) - 4p|V|\left(\left(T(\mathcal{F}_2^p)\right)^{1/p} + \frac{1}{4\left(\frac{4^p}{2}\right) t_1}\right)\right)$$

$$\geq 4^{p-1} t_1 \left(T(\mathcal{F}) - 4p|V|\left(\left(T(\mathcal{F}_2^p)\right)^{1/p} + \frac{1}{4}\right)\right) > 0$$

by the hypothesis of Theorem 24.2. In particular, $L_{j_2}^{(2)} > 0$. If the second phase is the last one then we are done again. Indeed, $L_i^{(2)} > 0$ implies $T(\mathcal{F}(i)) \geq L_i^{(2)} > 0$, and Maker wins. We show that if there is a next phase, then again Maker's chances to win are improving.

By (24.14) there is a real number $t_2 \geq 1$ such that

$$T(\mathcal{F}_2^p(j_2)) = \left(\frac{4^p}{2} t_2\right)^p T(\mathcal{F}_2^p(j_1)). \tag{24.16}$$

Since $L_{j_2}^{(2)} = T(\mathcal{F}(j_2)) - \frac{\lambda}{2} T(\mathcal{F}_2^p(j_2)) > 0$, it follows by (24.4), (24.9), and (24.16) that

$$T(\mathcal{F}(j_2)) > \frac{\lambda}{2} T(\mathcal{F}_2^p(j_2)) = \frac{4^p}{2} \frac{\left(\frac{4^p}{2}\right)^p}{2} (t_1 t_2)^p T(\mathcal{F}). \tag{24.17}$$

The general step of this "self-refining procedure" goes exactly the same way. Let

$$\beta_0 = 4, \beta_1 = \frac{4^p}{2} = \frac{\beta_0^p}{2}, \beta_2 = \frac{\beta_1^p}{2}, \dots, \beta_{k+1} = \frac{\beta_k^p}{2}, \dots.$$

We have

$$\beta_k = 2^{2p^k - (p^k - 1)/(p-1)} > 2^{p^k}. \tag{24.18}$$

i.e., β_k, $k = 1, 2, 3, \ldots$ is a very rapidly increasing sequence. This is how rapidly Maker's chances are improving – see (24.20) below.

We are going to prove by induction on the "phase-index" k ($k \geq 1$) that there are positive integers j_k ($j_0 = 0$) and real numbers $t_k \geq 1$ such that for every i with $j_{k-1} \leq i < j_k$ ("kth phase")

$$L_{i+1}^{(k)} = T(\mathcal{F}(i+1)) - \frac{\lambda}{2^{k-1}} T(\mathcal{F}_2^p(i+1)) > 0, \tag{24.19}$$

and *if* the play is not over in the kth phase yet, then

$$T(\mathcal{F}(j_k)) > \beta_1 \beta_2 \cdots \beta_k (t_1 t_2 \cdots t_k)^p T(\mathcal{F}), \tag{24.20}$$

and

$$T(\mathcal{F}_2^p(j_k)) = \left(\beta_0 \beta_1 \cdots \beta_{k-1} t_1 t_2 \cdots t_k \right)^p T(\mathcal{F}_2^p). \tag{24.21}$$

Statements (24.19)–(24.21) are true for $k = 1$ and 2. Assume that they hold for k, and we prove them for $k+1$.

If the play is over in the kth phase, i.e. $j_k = |V|/2$, then let $j_{k+1} = j_k$ and $t_{k+1} = t_k$. Otherwise "time index" j_k is the beginning of the $(k+1)$th **phase** $j_k \leq i < j_{k+1}$, where the "endpoint" j_{k+1} will be defined by (24.24)–(24.25) below. In the $(k+1)$st phase $j_k \leq i < j_{k+1}$ we modify the kth potential function $L_i^{(k)}$ by halving λ

$$L_i^{(k+1)} = T(\mathcal{F}(i)) - \frac{\lambda}{2^k} T(\mathcal{F}_2^p(i)). \tag{24.22}$$

By induction $L_{j_k}^{(k)} > 0$, so

$$L_{j_k}^{(k+1)} = \frac{1}{2} T(\mathcal{F}(j_k)) + \frac{1}{2} L_{j_k}^{(k)} > \frac{1}{2} T(\mathcal{F}(j_k)). \tag{24.23}$$

The $(k+1)$st *phase* of the play is the "time interval" $j_k \leq i < j_{k+1}$, where j_{k+1} is defined by

$$T(\mathcal{F}_2^p(i)) < \beta_k^p T(\mathcal{F}_2^p(j_k)) \quad \text{for} \quad j_k \leq i < j_{k+1} \tag{24.24}$$

and

$$T(\mathcal{F}_2^p(j_{k+1})) \geq \beta_k^p T(\mathcal{F}_2^p(j_k)). \tag{24.25}$$

If there is no j_{k+1} satisfying (24.25), then the play is over in the $(k+1)$st phase and we write $j_{k+1} = |V|/2$. By (24.21) and (24.24), for every i with $j_k \leq i < j_{k+1}$

$$\left(T(\mathcal{F}_2^p(i)) \right)^{1/p} < \beta_0 \beta_1 \cdots \beta_k t_1 t_2 \cdots t_k \left(T(\mathcal{F}_2^p) \right)^{1/p}. \tag{24.26}$$

In the $(k+1)$st phase we use the new Potential Function $L_i^{(k+1)}$, so the analogue of (24.5a) goes like this

$$L_{i+1}^{(k+1)} = L_i^{(k+1)} + L_i^{(k+1)}(x_{i+1}) - L_i^{(k+1)}(y_{i+1})$$

$$- T(\mathcal{F}(i); x_{i+1}, y_{i+1}) + \frac{\lambda}{2^k} T(\mathcal{F}_2^p(i); x_{i+1}, y_{i+1}), \tag{24.5a''}$$

where

$$L_i^{(k+1)}(z) = T(\mathcal{F}(i); z) - \frac{\lambda}{2^k} T(\mathcal{F}_2^p(i); z) \qquad (24.5b'')$$

for any $z \in V \setminus (X_i \cup Y_i)$.

In the $(k+1)$st phase, Maker's $(i+1)$st move x_{i+1} is that unoccupied $z \in V \setminus (X_i \cup Y_i)$ for which $L_i^{(k+1)}(z)$ attains its maximum. Then clearly $L_i^{(k+1)}(x_{i+1}) \geq L_i^{(k+1)}(y_{i+1})$, so by (24.5a'')

$$L_{i+1}^{(k+1)} \geq L_i^{(k+1)} - T(\mathcal{F}(i); x_{i+1}, y_{i+1}).$$

By Theorem 24.1 we obtain the analogue of (24.6)

$$L_{i+1}^{(k+1)} \geq L_i^{(k+1)} - p\left(T(\mathcal{F}_2^p(i))\right)^{1/p} - p. \qquad (24.6'')$$

We claim that, for every i with $j_k \leq i < j_{k+1}$, $L_{i+1}^{(k+1)} > 0$. Indeed, by (24.4), (24.6''), (24.20), (24.23), and (24.26)

$$L_{i+1}^{(k+1)} \geq L_{j_k}^{(k+1)} - \sum_{\ell=j_k}^{j_{k+1}-1} p\left(\left(T(\mathcal{F}_2^p(\ell))\right)^{1/p} + 1\right)$$

$$\geq \frac{1}{2} T(\mathcal{F}(j_k)) - \frac{|V|}{2} p\left(\beta_0\beta_1 \cdots \beta_k t_1 t_2 \cdots t_k \left(T(\mathcal{F}_2^p)\right)^{1/p} + 1\right)$$

$$\geq \frac{1}{2}\beta_1 \cdots \beta_k t_1 t_2 \cdots t_k \left((t_1 \cdots t_k)^{p-1} T(\mathcal{F}) - 4p|V|\left(\left(T(\mathcal{F}_2^p)\right)^{1/p} + \frac{1}{4}\right)\right)$$

$$\geq \frac{1}{2}\beta_1 \cdots \beta_k t_1 t_2 \cdots t_k \left(T(\mathcal{F}) - 4p|V|\left(\left(T(\mathcal{F}_2^p)\right)^{1/p} + \frac{1}{4}\right)\right) > 0$$

by the hypothesis of Theorem 24.2. This proves (24.19) for $k+1$. In particular, we have $L_{j_{k+1}}^{(k+1)} > 0$.

Now assume that the play is not over in the $(k+1)$st phase yet, i.e. $j_{k+1} < |V|/2$. By (24.25) there is a real number $t_{k+1} \geq 1$ such that

$$T(\mathcal{F}_2^p(j_{k+1})) = (\beta_k t_{k+1})^p T(\mathcal{F}_2^p(j_k)). \qquad (24.27)$$

Combining (24.21) and (24.27), we obtain (24.21) for $k+1$.

Since $L_{j_{k+1}}^{(k+1)} > 0$, it follows by (24.4), and by (24.21) for $k+1$, that

$$T(\mathcal{F}(j_{k+1})) > \frac{\lambda}{2^k} T(\mathcal{F}_2^p(j_{k+1}))$$

$$= 2^{-k-1}(\beta_0\beta_1 \cdots \beta_k t_1 t_2 \cdots t_{k+1})^p T(\mathcal{F})$$

$$= \beta_1\beta_2 \cdots \beta_{k+1}(t_1 t_2 \cdots t_{k+1})^p T(\mathcal{F}),$$

since $\beta_{i+1} = \frac{\beta_i^p}{2}$. This proves (24.20) for $k+1$, and the induction step is complete.

In view of (24.19), $T(\mathcal{F}(i)) > 0$ during the whole course of the play, and Maker wins. This completes the proof of Theorem 24.2. $\square$

The proof of the *Chooser–Picker* version of Theorem 24.2 is exactly the same except one natural change: instead of choosing a point of *maximum value* (as Maker does), Chooser chooses the one from the two points offered to him (by Picker) that has the larger value.

25

Exact solution of the Clique Game (I)

We recall that $A(K_N; \text{clique})$ denotes the largest value of $q = q(N)$ such that Maker has a winning strategy in the (K_N, K_q) Clique Game ("Weak Win"). The Erdős–Selfridge Strong Draw Criterion gives that if

$$\binom{N}{q} 2^{-\binom{q}{2}} < \frac{1}{2}, \tag{25.1}$$

then Breaker can force a Strong Draw in the (K_N, K_q) Clique Game. By Stirling's formula this means $N < e^{-1} q 2^{(q-1)/2}$. Let q_0 be the smallest integer q for which this inequality holds; q_0 is the upper integral part

$$q_0 = \lceil 2 \log_2 N - 2 \log_2 \log_2 N + 2 \log_2 e - 1 + o(1) \rceil, \tag{25.2}$$

and so the Clique Achievement Number has the upper bound

$$A(K_N; \text{clique}) \leq q_0 - 1 = \lfloor 2 \log_2 N - 2 \log_2 \log_2 N + 2 \log_2 e - 1 + o(1) \rfloor. \tag{25.3}$$

Upper bound (25.3) is a good starting point: it is not optimal, but it is very close to that. Next comes:

1. The lower bound. It is a direct application of Theorem 24.2. Unfortunately, the calculations are rather tiresome. Let $[N] = \{1, 2, \ldots, N\}$ be the vertex-set of K_N. Define the hypergraph

$$\mathcal{F} = \Big\{ K_S : S \subset [N], \ |S| = q \Big\},$$

where K_S denotes the complete graph on vertex-set S. We have to check that

$$T(\mathcal{F}) > 4p \binom{N}{2} \left(\left(T(\mathcal{F}_2^p) \right)^{1/p} + \frac{1}{4} \right)$$

holds for some integer $p \geq 2$. For simplicity consider the first "useful choice" of parameter p: let $p = 4$ (later it will be explained why $p \leq 3$ doesn't give anything).

We clearly have

$$T(\mathcal{F}) = \binom{N}{q} 2^{-\binom{q}{2}}.$$

Technically it is much more complicated to find a similar exact formula for $T(\mathcal{F}_2^4)$, but luckily we don't need that: we are perfectly satisfied with a "reasonably good" upper bound. By definition

$$\mathcal{F}_2^4 = \left\{ \bigcup_{i=1}^{4} K_{S_i} : \{K_{S_1}, \ldots, K_{S_4}\} \in \binom{\mathcal{F}}{4}, \left| \bigcap_{i=1}^{4} K_{S_i} \right| \geq 2 \right\}.$$

What it means is that $\mathcal{F}_2^4$ is the family of all clique-unions $\bigcup_{i=1}^{4} K_{S_i}$, where $\{S_1, \ldots, S_4\}$ runs over the unordered 4-tuples of distinct q-element subsets of $[N]$ for which $\left| \bigcap_{i=1}^{4} K_{S_i} \right| \geq 2$. Two edges of a graph are either vertex-disjoint or have a common vertex, i.e. two edges span either 4 or 3 vertices. It is easy to guess that the main contribution of $T(\mathcal{F}_2^4)$ comes from the two *extreme cases:* (1) from the "3-core sunflowers," which represent the most sparse case, and (2) from the *most crowded case.*

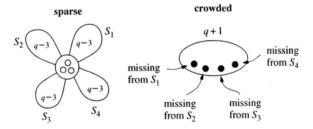

(1) The "3-core sunflowers" are those elements $\bigcup_{i=1}^{4} K_{S_i}$ of family $\mathcal{F}_2^4$ for which $\bigcap_{i=1}^{4} S_i$ is a 3-element set and the sets $S_\ell \setminus (\bigcap_{i=1}^{4} S_i)$, $\ell = 1, \ldots, 4$ are *pairwise disjoint.*

The corresponding contribution in $T(\mathcal{F}_2^4)$ is that we call the *first main term*; it is precisely

$$f(N, q) = \frac{1}{4!} \binom{N}{3} \binom{N-3}{q-3} \binom{N-q}{q-3} \binom{N-2q+3}{q-3} \binom{N-3q+6}{q-3} 2^{-4\binom{q}{2}+9}.$$

It is easily seen

$$f(N, q) \approx \frac{2^9}{4!3!} \frac{q^{12}}{N^9} \left(\binom{N}{q} 2^{-\binom{q}{2}} \right)^4 = \frac{2^9}{4!3!} \frac{q^{12}}{N^9} (T(\mathcal{F}))^4, \qquad (25.4)$$

so

$$(f(N, q))^{1/4} < T(\mathcal{F}) N^{-9/4 + o(1)}.$$

The *critical fact* is that $9/4 > 2$: this is why we need $p \geq 4$ (see later).

(2) The *most crowded* elements of $\mathcal{F}_2^4$ are those elements $\bigcup_{i=1}^4 K_{S_i}$ for which $\bigcup_{i=1}^4 S_i$ is a $(q+1)$-element set. The corresponding contribution in $T(\mathcal{F}_2^4)$ is what we call the *second main term*; it is precisely

$$h(N, q) = \binom{N}{q+1}\binom{q+1}{4}2^{-\binom{q}{2}-q}.$$

It is easily seen

$$h(N, q) \approx \frac{q^3}{4!}T(\mathcal{F})\frac{N}{2^q} = \frac{T(\mathcal{F})}{N^{1+o(1)}}$$

if $q = (2+o(1))\log_2 N$. Clearly

$$T(\mathcal{F}_2^4) \geq f(N, q) + h(N, q),$$

and *assume* that

$$T(\mathcal{F}_2^4) \approx f(N, q) + h(N, q). \tag{25.5}$$

Then

$$\left(T(\mathcal{F}_2^4)\right)^{1/4} \leq \left(f(N, q)\right)^{1/4} + \left(h(N, q)\right)^{1/4}$$

$$\leq T(\mathcal{F})N^{-9/4+o(1)} + \left(T(\mathcal{F})N^{-1+o(1)}\right)^{1/4}. \tag{25.6}$$

To apply Theorem 24.2 with $p = 4$ we have to check

$$T(\mathcal{F}) > 16\binom{N}{2}\left(\left(T(\mathcal{F}_2^4)\right)^{1/4} + \frac{1}{4}\right). \tag{25.7}$$

Inequality (25.7) follows from

$$T(\mathcal{F}) > O\left(N^2 q^3 N^{-9/4}T(\mathcal{F})\right) + O\left(N^2 N^{-\frac{1}{4}+\epsilon}(T(\mathcal{F}))^{1/4}\right) + 2N^2 =$$

$$= O\left(N^{-1/4+o(1)}T(\mathcal{F})\right) + O\left(N^{7/4+o(1)}(T(\mathcal{F}))^{1/4}\right) + 2N^2, \tag{25.8}$$

(see (25.6)) where $N \geq e^{-1}q2^{(q-1)/2}$ and $q = (2+o(1))\log_2 N$.

The first part

$$T(\mathcal{F}) > O\left(N^{-1/4+o(1)}T(\mathcal{F})\right)$$

of (25.8) is trivial. The second part of (25.8)

$$T(\mathcal{F}) > O\left(N^{7/4+o(1)}(T(\mathcal{F}))^{1/4}\right)$$

holds if $T(\mathcal{F}) > N^{7/3+o(1)}$, which, of course, implies the entire (25.8). The condition

$$T(\mathcal{F}) = \binom{N}{q}2^{-\binom{q}{2}} > N^{7/3+o(1)}$$

is equivalent to (Stirling's formula)

$$q \leq 2\log_2 N - 2\log_2 \log_2 N + 2\log_2 e - \frac{10}{3} + o(1). \tag{25.9}$$

If q is the lower integral part of the right-hand side of (25.9), then (25.7) holds, and Theorem 24.2 yields a Maker's win ("Weak Win") in the Clique Game (K_N, K_q). Looking back to the calculations of the proof, now we see why it works: two edges determine at least 3 vertices, and N^3 is much larger than $\binom{N}{2}$, the *duration* of a play.

Of course, we are not done yet: we have to justify the vague assumption (25.5). Also we hope that working with $\mathcal{F}_2^p$, where $p \to \infty$, instead of $\mathcal{F}_2^4$, will lead to an improvement; note that the "3-core sunflowers" (the most sparse case) have the following contribution in $\mathcal{F}_2^p$

$$f_p(N, q) = \frac{1}{p!}\binom{N}{3}\binom{N-3}{q-3}\binom{N-q}{q-3}\cdots\binom{N-(p-1)q+3(p-2)}{q-3}2^{-p\binom{q}{2}+3(p-1)},$$

and

$$h_p(N, q) = \binom{N}{q+1}\binom{q+1}{p}2^{-\binom{q}{2}-q}$$

represents the most crowded case (we assume $p \le q-2$). The analogue of (25.5) is

$$T(\mathcal{F}_2^p) \approx f_p(N, q) + h_p(N, q). \tag{25.10}$$

Under what condition can we prove (25.10)? To answer this question we divide $\mathcal{F}_2^p$ into 3 parts

$$\mathcal{F}_2^p = \mathcal{G}_1 \cup \mathcal{G}_2 \cup \mathcal{G}_3,$$

where (parameter $A = A(q)$ below will be specified later):

(1) $\mathcal{G}_1$ is the family of those $\bigcup_{i=1}^p K_{S_i} \in \mathcal{F}_2^p$ for which there is a pair i_1, i_2 of indices with $1 \le i_1 < i_2 \le p$ such that $A < |S_{i_1} \cap S_{i_2}| < q - A$ ("irrelevant case");
(2) $\mathcal{G}_2$ is the family of those $\bigcup_{i=1}^p K_{S_i} \in \mathcal{F}_2^p$ for which $|S_i \cap S_j| \le A$ for all $1 \le i < j \le p$ ("sparse case," the main contribution is expected to be $f_p(N, q)$); and, finally,
(3) $\mathcal{G}_3$ is the family of those $\bigcup_{i=1}^p K_{S_i} \in \mathcal{F}_2^p$ for which $|S_i \cap S_j| \ge q - A$ for all $1 \le i < j \le A$ ("crowded case," the main contribution is expected to be $h_p(N, q)$).

To estimate $T(\mathcal{G}_1)$ from above, let $i_1, i_2, \ldots, i_p$ be a permutation of $1, 2, \ldots, p$ such that $|S_{i_1} \cap S_{i_2}| = m_1$ with $A < m_1 < q - A$

$$\left|\left(\bigcup_{j=1}^2 S_{i_j}\right) \cap S_{i_3}\right| = m_2, \ldots, \left|\left(\bigcup_{j=1}^{p-1} S_{i_j}\right) \cap S_{i_p}\right| = m_{p-1},$$

and let $\mathcal{G}_1(m_1, m_2, \ldots, m_p)$ denote the corresponding part of $\mathcal{G}_1$.

We have

$$\mathcal{G}_1 = \bigcup_{m_1=A}^{q-A} \bigcup_{m_2=3}^{q} \cdots \bigcup_{m_{p-1}=3}^{q} \mathcal{G}_1(m_1, m_2, \ldots, m_{p-1}),$$

so

$$T(\mathcal{G}_1) \le \sum_{m_1=A}^{q-A} \sum_{m_2=3}^{q} \cdots \sum_{m_{p-1}=3}^{q} T(\mathcal{G}_1(m_1, m_2, \ldots, m_{p-1})).$$

It is easy to see that

$$T(\mathcal{G}_1(m_1, m_2, \ldots, m_{p-1})) \le \binom{N}{3}\binom{N-3}{q-3}\binom{q-3}{m_1-3}\binom{N-q}{q-m_1}\binom{2q-m_1-3}{m_2-3}$$

$$\cdot \binom{N-2q+m_1}{q-m_2} \cdots \binom{(p-1)q-m_1-\cdots-m_{p-2}-3}{m_{p-1}-3}$$

$$\cdot \binom{N-(p-1)q+m_1+\cdots+m_{p-2}}{q-m_{p-1}} \cdot 2^{-p\binom{q}{2}+\binom{m_1}{2}+\binom{m_2}{2}+\cdots+\binom{m_{p-1}}{2}}.$$

Comparing this upper bound of $T(\mathcal{G}_1(m_1, m_2, \ldots, m_{p-1}))$ to the *main term* $f_p(N, q)$, after some easy computations we have

$$\frac{T(\mathcal{G}_1(m_1, m_2, \ldots, m_{p-1}))}{f_p(N, q)} < \left(p!\binom{q}{m_1-3}\left(\frac{q}{N-q}\right)^{m_1-3}2^{\binom{m_1}{2}+3}\right)$$

$$\cdot \prod_{j=2}^{p-1}\left(\binom{jq}{m_j-3}\left(\frac{q}{N-jq}\right)^{m_j-3}2^{\binom{m_j}{2}+3}\right).$$

Therefore

$$\frac{T(\mathcal{G}_1)}{f_p(N, q)} \le p!S_1S_2\cdots S_{p-1},$$

where

$$S_1 = \sum_{m=A}^{q-A}\binom{q}{m-3}\left(\frac{q}{N-q}\right)^{m-3}2^{\binom{m}{2}+3}$$

and

$$S_j = \sum_{m=3}^{q}\binom{jq}{m-3}\left(\frac{q}{N-jq}\right)^{m-3}2^{\binom{m}{2}+3}$$

with $2 \le j \le p-1$. Let $j \in \{2, 3, \ldots, p-1\}$; then by $N \ge e^{-1}q2^{(q-1)/2}$ we have

$$S_j \le 2\binom{jq}{q-3}\left(\frac{q}{N-jq}\right)^{q-3}2^{\binom{q}{2}+3}$$

(indeed, the sum behaves like a rapidly convergent geometric series, and it roughly equals the *last* term), so

$$S_j \le 2\binom{jq}{q-3}\left(\frac{q}{N-jq}\right)^{q-3}2^{\binom{q}{2}+3}$$

$$\le 2^7\left(\frac{ejq^22^{q/2+1}}{(q-3)N}\right)^{q-3} \le 2^7\left(2\sqrt{2}e^2j\right)^{q-3} \le 2^{5q+7}j^q,$$

and so

$$S_2 S_3 \cdots S_{p-1} \le \prod_{j=2}^{p-1} 2^{5q+7} j^q$$

$$\le 2^{7p+5pq} (p!)^q \le \left(2^{7/q} 32 p\right)^{pq}.$$

On the other hand

$$p! S_1 \le 2p! \left(\binom{q}{A-3} \left(\frac{q}{N-q}\right)^{A-3} 2^{\binom{A}{2}+3} + \binom{q}{q-A-3} \left(\frac{q}{N-q}\right)^{q-A-3} 2^{\binom{q-A}{2}+3} \right)$$

$$\le p! 2^{-qA/4}.$$

It follows that $T(\mathcal{G}_1) \le f_p(N, q)$ if

$$p! \left(2^{7/q} 32 p\right)^{pq} \le 2^{qA/4},$$

which follows from

$$\left(2^{7/q} p^{1/q} 32 p\right)^p \le 2^{A/4}. \tag{25.11}$$

Next we estimate $T(\mathcal{G}_2)$; in this case

$$\left| \left(\bigcup_{i=1}^{j} S_i\right) \cap S_{j+1} \right| \le jA,$$

for $j = 1, 2, \ldots, p-1$, so, exactly the same way as we estimated $T(\mathcal{G}_1)$, we obtain (we *separate* the suspected main term $f_p(N, q)$ from the sum)

$$\frac{T(\mathcal{G}_2) - f_p(N, q)}{f_p(N, q)} \le p! \prod_{1 \le j \le p-1}^{*} \left(\sum_{m_j=3}^{jA} \binom{jq}{m_j - 3} \left(\frac{q}{N-jq}\right)^{m_j-3} 2^{\binom{m_j}{2}+3} \right),$$

where the asterisk $*$ indicates that in the expansion of the product at least one $m_j \ge 4$ (due to the fact that $f_p(N, q)$ is *separated* from $T(\mathcal{G}_2)$). It follows that

$$\frac{T(\mathcal{G}_2) - f_p(N, q)}{f_p(N, q)} \le p! 2^{6p} \frac{1}{\sqrt{N}};$$

noting that in the argument above we used the inequality $pA \le q/4$, i.e. $A \le q/4p$.

Finally, consider $T(\mathcal{G}_3)$. If $\bigcup_{i=1}^{p} K_{S_i} \in \mathcal{F}_2^p$, then $|S_i \cap S_j| \ge q - A$ for all $1 \le i < j \le p$. Write

$$|S_1 \cap S_2| = m_1, \quad \left| \left(\bigcup_{i=1}^{2} S_i\right) \cap S_3 \right| = m_2, \cdots, \left| \left(\bigcup_{i=1}^{p-1} S_i\right) \cap S_p \right| = m_{p-1}.$$

Clearly

$$q - A \le m_1 \le q - 1, \quad q - A \le m_i \le q \quad \text{for } i = 2, \ldots, p-1, \quad \text{and}$$

$$q + 1 \le \left| \bigcup_{i=1}^{p} S_i \right| \le q + (p-1)A.$$

Let $\mathcal{G}_3(m_1, m_2, \ldots, m_{p-1})$ denote the corresponding part of $\mathcal{G}_3$. Since m_i is almost equal to q, it is a good idea to introduce a new variable: let $d_i = q - m_i$, $1 \le i \le p-1$. Clearly, $1 \le d_1 \le A$ and $0 \le d_i \le A$ for $i = 2, \ldots, p-1$. If $h_p(N, q)$ is subtracted from $T(\mathcal{G}_3)$, then $d_2 + \ldots + d_{p-1} \ge 1$ (we already know $d_1 \ge 1$). We have

$$T(\mathcal{G}_3) - h_p(N, q) \le \sum_{d_1=1}^{A} \sum_{d_2=0}^{A} \cdots \sum_{0 \le m_{p-1} \le A}^{*} T(\mathcal{G}_3(q - d_1, q - d_2, \ldots, q - d_{p-1})),$$

where the $*$ indicates the above-mentioned restriction $d_2 + \ldots + d_{p-1} \ge 1$.

It is easy to see that

$$T(\mathcal{G}_3(q - d_1, q - d_2, \ldots, q - d_{p-1})) \le \binom{N}{3}\binom{N-3}{q-3}\binom{q-3}{d_1}\binom{N-q}{d_1}\binom{q+d_1-3}{q-d_2-3}$$

$$\cdot \binom{N-q-d_1}{d_2}\binom{q+d_1+d_2-3}{q-d_3-3}\binom{N-q-d_1-d_2}{d_3} \cdots \binom{q+d_1+\cdots+d_{p-2}-3}{q-d_{p-1}-3}$$

$$\cdot \binom{N-q-d_1-\cdots-d_{p-2}}{d_{p-1}} 2^{-\binom{q}{2}-d_1(q-d_1)-d_2(q-d_2)-\cdots-d_{p-1}(q-d_{p-1})}.$$

Since

$$\binom{q+d_1+\cdots+d_{k-1}-3}{q-d_k-3} = \binom{q+d_1+\cdots+d_{k-1}-3}{d_1+\cdots+d_k},$$

we obtain

$$T(\mathcal{G}_3) - h_p(N, q) \le \sum_{d_1=1}^{A} \sum_{d_2=0}^{A} \cdots \sum_{0 \le m_{p-1} \le A}^{*} \binom{N}{3}\binom{N}{q-3} 2^{-\binom{q}{2}} \cdot$$

$$\binom{N}{d_1} q^{d_1} 2^{-d_1(q-d_1)} \cdot \binom{N}{d_2} q^{d_1+d_2} 2^{-d_1(q-d_1)} \cdots$$

$$\binom{N}{d_{p-1}} q^{d_1+d_2+\cdots+d_{p-1}} 2^{-d_{p-1}(q-d_{p-1})} \le T(\mathcal{F}) N^{-2+o(1)}$$

(noting that $*$ indicates the above-mentioned restriction $d_2 + \ldots + d_{p-1} \ge 1$).

Summarizing, we have just proved the following:

Key Technical Lemma. *If $p = \frac{1}{4}\sqrt{\frac{q}{\log q}} \ge 4$, then*

$$T(\mathcal{F}_2^p) \le 2f_p(N, q) + 2h_p(N, q).$$

Remark. The condition $p = \frac{1}{4}\sqrt{\frac{q}{\log q}} \geq 4$ comes from (25.11). The *Key Technical Lemma* is the precise form of the vague (25.10).

Repeating steps (25.7)–(25.9) we conclude: *if q is the lower integral part of*

$$2\log_2 N - 2\log_2 \log_2 N + 2\log_2 e - 3 + o(1), \qquad (25.12)$$

then Maker can force a Weak Win in the (K_N, K_q) *Clique Game*. This proves the Weak Win part of Theorem 6.4 (a).

We already have a remarkable result: the gap between the Strong Draw bound (25.2)–(25.3) and the Weak Win bound (25.12) is an additive constant, namely 2, which is independent of N. The last step is to eliminate constant 2; this will be done in Section 38 (first we need to develop the technique of BigGame–SmallGame Decomposition).

Note that the implicit error term "$o(1)$" in (25.12) is in fact less than the explicit bound $10/(\log N)^{1/6}$ if $N \geq 2^{1000}$ – we challenge the reader to prove this technical fact by inspecting the proofs here and in Section 38.

2. Game-theoretic law of large numbers – is it an accidental phenomenon? The condition $p = \frac{1}{4}\sqrt{\frac{q}{\log q}} \geq 4$ in the Key Technical Lemma holds for $q/\log q \geq 256$, i.e. when q is in the range $q \geq 2000$. Since $q \approx 2\log_2 N$, Theorem 24.2 works and gives a very good Weak Win result in the range $N \geq 2^{1000}$. For small Ns such as in the range $N \leq 2^{100}$ the proof technique of Theorem 24.2 does *not* seem to work. Then the only result we can use is Theorem 21.1, which leads to a large amount of uncertainty.

For example, if $N = 20$, then the Erdős–Selfridge Theorem implies that $\mathbf{A}(K_{20}; clique) < 7$, so $\mathbf{A}(K_{20}; clique) = 4$ or 5 or 6, and we do not know the truth.

If $N = 100$, then the Erdős–Selfridge Theorem implies that $\mathbf{A}(K_{100}; clique) < 10$, so $\mathbf{A}(K_{20}; clique) = 5$ or 6 or 7 or 8 or 9, and we do not know the truth.

If $N = 2^{100}$, then Theorem 21.1 yields a Weak Win if $q \leq 98$; on the other hand, the Erdős–Selfridge Theorem (see (25.1)–(25.2)) implies that the (K_N, K_{189}) Clique Game is a Strong Draw (this is the best that we know in this range). This means, in the (K_N, K_q) Clique Game with $N = 2^{100}$ and $q = 99, 100, \ldots, 187, 188$ that we don't know whether the game is a Weak Win or a Strong Draw; i.e. with this particular "mid-size" N there are 90 values of q for which we don't know whether Maker or Breaker can force a win.

In the range $N \geq 2^{1000}$, Theorem 24.2 begins to work well: for a fixed N there are at most 3 values of $q = q(N)$ for which we don't know the outcome of the Clique Game ("Weak Win or Strong Draw").

Finally, in the range $N \geq 2^{1000000}$, we (typically) know the exact threshold between Weak Win and Strong Draw, or – very rarely – there is (at most) one value of $q = q(N)$ which we don't know. In other words, the larger the N, the smaller the uncertainty. This sounds like a "paradox," since larger N means "larger complexity." We refer to this "paradox" as a "game-theoretic law of large numbers."

Of course, there is no real paradox here; the "game-theoretic law of large numbers" is simply the shortcoming of this proof technique. This is why it is called an "accidental phenomenon." We are convinced that (25.12) holds with an error term $o(1)$, which is actually uniformly ≤ 2 for all N, including small Ns. For example, if $N = 2^{100}$, then

$$2\log_2 N - 2\log_2\log_2 N + 2\log_2 e - 3 = 200 - 13.29 + 2.88 - 3 = 186.59,$$

and we are convinced that the Achievement (Avoidance) Number is equal to one of the following four consecutive integers: 185, 186, 187, 188. Of course, these are the closest neighbors of 186.59 in (25.13). Can anyone prove this?

3. How long does it take to build a clique? Bound (25.12) is sharp (see Theorem 6.4 (a)), but it does not mean that every question about cliques is solved. Far from it! For example, the Move Number remains a big mystery.

What can we say about the Move Number of K_q? Well, an equivalent form of (25.12) is the following (see also (6.11)): playing on K_N with $N = \frac{\sqrt{2}}{e}q2^{q/2}(1+o(1))$, Maker can build a clique K_q. This way it takes at most

$$\frac{1}{2}\binom{N}{2} = \frac{1+o(1)}{2e^2}q^2 2^q$$

moves for Maker to get a K_q.

Now here comes the surprise: (the otherwise weak) Theorem 21.1 gives the slightly better bound $\leq 2^{q+2}$ for the Clique Move Number. This is in fact the best known to the author. Can the reader improve on the bound $O(2^q)$?

Open Problem 25.1 *For simplicity assume that the board is the infinite complete graph K_∞ (or at least a "very large" finite K_N); playing the usual (1:1) game, how long does it take to build a K_q?*

In the other direction the best known to the author is formulated in:

Exercise 25.1 *Show that, playing the usual (1:1) game on K_∞, Breaker can prevent Maker from building a K_q in $2^{q/2}$ moves (let $q \geq 20$).*

Which one is closer to the truth, upper bound 2^q or lower bound $2^{q/2}$?

Next we replace the usual board K_N with a "typical" graph on N vertices. What is the largest achievable clique? We can adapt the proof above, and obtain:

Exercise 25.2 *Show that, playing the usual (1:1) game on the symmetric Random Graph $\mathbf{R}(K_N, 1/2)$, with probability tending to 1 as N tends to infinity, Maker can occupy a clique K_q with*

$$q = \lfloor \log_2 N - \log_2 \log_2 N + \log_2 e - 1 + o(1) \rfloor . \qquad (25.13)$$

Note that (25.13) is the best possible (this follows from the technique of Section 38). Another adaptation is:

Exercise 25.3 *Prove the Maker's part in formulas (8.7) and (8.8).*

Exercise 25.4 *What is the Avoidance version of Theorem 21.1? That is, what is the largest clique that Forcer can force Avoider to build in K_N, assuming N is not too large (i.e. N is in the range before Theorem 24.2 begins to work)?*

26

More applications

1. The Weak Win part of Theorem 6.4 (b)–(c)–(d). The good thing about Theorem 24.2 is that it extends from (ordinary) graphs to k-graphs ($k \geq 3$) without any problem. We illustrate this briefly in the case $k = 3$.

First an easy Strong Draw bound: the Erdős–Selfridge Theorem applies and yields a Strong Draw if

$$\binom{N}{q} 2^{-\binom{q}{3}} < \frac{1}{2};$$

the upper integral part of $\sqrt{6 \log_2 N} + o(1)$ is a good choice for q.

Next the other direction: to get a Weak Win we apply Theorem 24.2 to the hypergraph

$$\mathcal{F} = \left\{ \binom{S}{3} : S \subset [N], \ |S| = q \right\},$$

noting that $\binom{S}{3}$ denotes a *set* – in fact a "complete 3-uniform hypergraph" – and not a binomial coefficient. We have to check that

$$T(\mathcal{F}) > 4p \binom{N}{3} \left(\left(T(\mathcal{F}_2^p) \right)^{\frac{1}{p}} + \frac{1}{4} \right)$$

holds for some integer $p \geq 2$. We clearly have

$$T(\mathcal{F}) = \binom{N}{q} 2^{-\binom{q}{3}}.$$

Let $p = 5$; we need a good upper bound for $T(\mathcal{F}_2^5)$. By definition

$$\mathcal{F}_2^5 = \left\{ \bigcup_{i=1}^{5} \binom{S_i}{3} : S_i \subset [N], \ |S_i| = q, \ 1 \leq i \leq 5, \ \left| \bigcap_{i=1}^{5} \binom{S_i}{3} \right| \geq 2 \right\},$$

where $\{S_1, \ldots, S_5\}$ runs over the unordered 5-tuples of distinct q-element subsets of $[n]$.

Two 3-edges of $\binom{N}{3}$ span either 4, or 5, or 6 points. It is easy to guess that the main contribution of $T(\mathcal{F}_2^5)$ The most sparse case is the "4-core sunflowers": the

362

elements $\bigcup_{i=1}^{5} \binom{S_i}{3}$ of family $\mathcal{F}_2^5$ for which $\bigcap_{i=1}^{5} S_i$ is a 4-element set and the sets $S_\ell \setminus (\bigcap_{i=1}^{5} S_i)$, $\ell = 1, \dots, 5$ are *pairwise disjoint*. The corresponding contribution of $T(\mathcal{F}_2^5)$ is what we call the *first main term*; it is precisely

$$f^{(3)}(N, q) = \frac{1}{5!}\binom{N}{4}\binom{N-4}{q-4}\binom{N-q}{q-4}\binom{N-2q+4}{q-4}\binom{N-3q+8}{q-4}$$
$$\cdot \binom{N-4q+12}{q-4}2^{-4\binom{q}{3}+16}.$$

We have

$$f^{(3)}(N, q) \approx \frac{2^{16}}{5!4!}\left(\frac{q}{N}\right)^{16}\left(\binom{N}{q}2^{-\binom{q}{3}}\right)^5 = \frac{2^{16}}{5!4!}\left(\frac{q}{N}\right)^{16}(T(\mathcal{F}))^5.$$

So

$$\left(f^{(3)}(N, q)\right)^{\frac{1}{5}} < \frac{T(\mathcal{F})}{N^{16/5+o(1)}}.$$

The *critical fact* is that $N^{16/5}$ is much larger than $\binom{N}{3}$, the duration of a play.

The *second main term* comes from the "most crowded" configurations

$$h^{(3)}(N, q) = \binom{N}{q+1}\binom{q+1}{5}2^{-\binom{q+1}{3}}.$$

Clearly

$$h^{(3)}(N, q) \approx \frac{T(\mathcal{F})}{N^{1+o(1)}} \text{ if } q \approx \sqrt{6\log_2 N}.$$

Again we have to show that

$$T(\mathcal{F}_2^5) \approx f^{(3)}(N, q) + h^{(3)}(N, q).$$

The choice $p = 5$ was just an illustration; it is good enough to approach the truth by an additive error $O(1)$. Again the optimal result comes from applying Theorem 24.2 with $p \to \infty$. The technical details are very much the same as in the case of ordinary graphs. We stop here and leave the rest of the proof to the reader.

2. Weak Win part of Theorem 12.6. Let $\mathcal{F}(n, d)$ denote the family of all Comb-Planes ("2-parameter sets") of the n^d Torus. $\mathcal{F}(n, d)$ is an n^2-uniform hypergraph of size $\frac{3^d - 2^{d+1}+1}{2}n^{d-2}$, and its board size is, of course, n^d. Assume that

$$\frac{T(\mathcal{F}(n, d))}{|V|} = \frac{\frac{3^d - 2^{d+1}+1}{2}n^{d-2} \cdot 2^{-n^2}}{n^d} = \frac{3^d - 2^{d+1} + 1}{2n^2}2^{-n^2} > 2p. \tag{26.1}$$

To apply Theorem 24.2 we have to estimate $T((\mathcal{F}(n, d))_2^p)$ from above ($p \geq 2$). Fix 2 points $\mathbf{P}$ and $\mathbf{Q}$, and consider the difference vector $\mathbf{P} - \mathbf{Q}$ (mod n). If the coordinates of the vector $\mathbf{P} - \mathbf{Q}$ (mod n) have at least two different non-zero values, then there is exactly one Comb-Plane containing both $\mathbf{P}$ and $\mathbf{Q}$. This case, therefore, has no contribution to the sum $T((\mathcal{F}(n, d))_2^p)$.

It remains to study the case when the coordinates of the vector $\mathbf{P}-\mathbf{Q}$ (mod n) have precisely one non-zero value; let j denote the multiplicity of the non-zero value, and let J denote the set of places where the non-zero coordinate shows up in vector $\mathbf{P}-\mathbf{Q}$ (mod n). Clearly $|J| = j \geq 1$. If S is a Comb-Plane containing both $\mathbf{P}$ and $\mathbf{Q}$, then S has the form $\mathbf{P}+k\mathbf{u}+l\mathbf{w}$ and the equation $\mathbf{Q} = \mathbf{P}+k\mathbf{u}+l\mathbf{w}$ has an integral solution (k, l). Let U denote the set of places where vector $\mathbf{u}$ has coordinate 1, and let W denote the set of places where vector $\mathbf{w}$ has coordinate 1 (U and W are disjoint non-empty sets). It is easily seen that there are three alternatives only:

(a) either $J = U$,
(b) or $J = W$,
(c) or $J = U \cup W$.

This explains the following upper bound

$$T((\mathcal{F}(n, d))_2^p) \leq \sum_{j=1}^{d} n^d \cdot \binom{d}{j} n \cdot \binom{2^j + 2^{d-j}}{p} \cdot 2^{-pn^2 + (p-1)n};$$

indeed, n^d is the number of ways to choose $\mathbf{P}$, $\binom{d}{j}n$ is the number of ways to choose $\mathbf{Q}$; $\binom{2^j + 2^{d-j}}{p}$ is an upper bound on the number of ways to choose p Comb-Planes each, containing $\mathbf{P}$ and $\mathbf{Q}$; and, finally, the exponent $-(pn^2 - (p-1)n)$ of 2 is motivated by the fact that the p Comb-Planes, all containing the $\mathbf{PQ}$-line, are in fact disjoint apart from the $\mathbf{PQ}$-line ("n points").

Taking pth root

$$\left(T((\mathcal{F}(n, d))_2^p)\right)^{1/p} \leq n^{2d/p} \cdot 2^d \cdot 2^{-n^2 + (1 - 1/p)n}.$$

By choosing p around constant times $\log n$, the critical term

$$\left(T((\mathcal{F}(n, d))_2^p)\right)^{1/p}$$

becomes much smaller than $T(\mathcal{F}(n, d))/|V|$, i.e. Theorem 24.2 applies and yields a Weak Win if (26.1) holds.   □

We already settled the issue of "Weak Win in the Lattice Games" in Section 23 by using an *ad hoc* Higher Moment technique. It is worth knowing that our Advanced Weak Win Criterion (Theorem 24.2) applies, too. In the next two applications we give alternative proofs for the *aligned rectangle* lattice and the *parallelogram* lattice games (they were covered in Theorems 23.1–23.2).

3. Weak Win part of Theorem 8.2: case (b). The case of *aligned rectangle lattices* follows from Theorem 24.2; luckily the calculations are much simpler than in the Clique Game (Theorem 6.4 (a), see Section 25). Let P_1 and P_2 be two points of the $N \times N$ board. The *typical* case is when P_1 and P_2 are neither on the same horizontal, nor on the same vertical line; then the number of $q \times q$ aligned rectangle lattices containing both P_1 and P_2 is estimated from above by the same bound $\binom{q^2}{2} < \frac{q^4}{2}$

("log N-power") as in Case (a) ("Square Lattices"). However, if P_1 and P_2 are on the same horizontal or vertical line – non-typical case! – then we may have *much more*, namely (around) N $q \times q$ aligned rectangle lattices containing both P_1 and P_2. This is why Theorem 24.2 works better than Theorem 1.2: in Theorem 24.2 the non-typical case has a very small "weight." The actual application of Theorem 24.2 goes as follows: by (8.4)

$$\frac{T(\mathcal{F})}{|V|} \approx \frac{\frac{N^4}{4(q-1)^2} 2^{-q^2}}{N^2} = \frac{N^2}{(q-1)^2 2^{q^2+2}},$$

where, of course, $\mathcal{F}$ denotes the family of all $q \times q$ aligned rectangle lattices in the $N \times N$ board. Next we estimate the more complicated $T(\mathcal{F}_2^p)$ from above. We have the *typical* and the *non-typical* parts

$$T(\mathcal{F}_2^p) = T_{typ}(\mathcal{F}_2^p) + T_{no-typ}(\mathcal{F}_2^p).$$

Here is an easy upper bound for the *typical* part

$$T_{typ}(\mathcal{F}_2^p) \le N^2 \cdot (N^2 - 2N + 1) \cdot \binom{q^4/2}{p} \cdot 2^{-q^2},$$

where N^2 is the number of ways to choose P_1 and $N^2 - 2N + 1$ is the number of ways to choose P_2 (they are in "general position").

To estimate the *non-typical* part $T_{no-typ}(\mathcal{F}_2^p)$, assume that P_1 and P_2 are on the same (say) horizontal line, and consider p distinct $q \times q$ aligned rectangle lattices $L_1, L_2, \ldots, L_p$, each containing both P_1 and P_2. The projection of an aligned $q \times q$ rectangle lattice on the vertical axis is a q-term A.P. ("arithmetic progression"); I call it the *vertical A.P. of the lattice*. Consider now the *vertical A.P.s* of the p lattices $L_1, L_2, \ldots, L_p$; they all have a common element: the projection of the $P_1 P_2$-line. Throwing out this common element, we obtain p $(q-1)$-element sets ("almost A.P.s"). Here comes a key concept: how many pairwise disjoint sets can be selected from this collection of p $(q-1)$-element sets ("almost A.P.s")? The answer is denoted by k; it is a key parameter – note that k can be anything between $1 \le k \le p$.

Parameter k is a key concept, but we also need the following simple fact. Let P_3 be a point of the $N \times N$ board which is not on the (horizontal) $P_1 P_2$-line; then the number of $q \times q$ aligned rectangle lattices containing the non-collinear triplet $\{P_1, P_2, P_3\}$ is at most $\binom{q}{2}^2 \le q^4/4$.

Now we are ready to estimate. For a fixed point pair P_1, P_2 on the same (say) horizontal line, and for a fixed value of parameter k introduced above, the corresponding contribution $T_{no-typ,P_1,P_2,k}(\mathcal{F}_2^p)$ in $T_{no-typ}(\mathcal{F}_2^p)$ is at most

$$T_{no-typ,P_1,P_2,k}(\mathcal{F}_2^p) \le \left(N \cdot \binom{q}{2}\right)^k \cdot \binom{kq^2}{p-k} \cdot \left(\frac{q^4}{4}\right)^{p-k} \cdot 2^{-kq^2+(k-1)q}. \qquad (26.2)$$

Since there are N^2 ways to choose P_1 and $(2N-1)$ ways to choose P_2, by (26.2), we have

$$T_{no-typ}(\mathcal{F}_2^p) \le N^2(2N-1) \sum_{k=1}^{p} \left(N \cdot \binom{q}{2}\right)^k \cdot \binom{kq^2}{p-k} \cdot \left(\frac{q^4}{4}\right)^{p-k} \cdot 2^{-kq^2+(k-1)q}.$$

Using the trivial fact

$$(a_1 + a_2 + \ldots + a_p)^{1/p} \le a_1^{1/p} + a_2^{1/p} + \ldots + a_p^{1/p},$$

we have

$$\left(T_{no-typ}(\mathcal{F}_2^p)\right)^{1/p} \le 2 \cdot N^{3/p} \sum_{k=1}^{p} \left(\frac{N \cdot \binom{q}{2}}{2^{q^2-q}}\right)^{k/p} \cdot kq^2 \cdot \frac{q^4}{4}.$$

Summarizing, if

$$\frac{T(\mathcal{F})}{|V|} \approx \frac{N^2}{(q-1)^2 2^{q^2+2}} \ge 100q^{20} \quad \text{and} \quad p = q^2, \tag{26.3}$$

then Theorem 24.2 applies, and yields a Weak Win. Indeed, by (26.3)

$$\left(T_{no-typ}(\mathcal{F}_2^p)\right)^{1/p} \le N^{3/p} \cdot q^8 \sum_{k=1}^{p} N^{-\frac{k}{2p}}$$

and

$$\left(T_{typ}(\mathcal{F}_2^p)\right)^{1/p} \le N^{2/p} \cdot \frac{q^4}{2} \cdot \left(\frac{N^2}{2^{q^2}}\right)^{1/p},$$

implying the desired inequality

$$\left(T(\mathcal{F}_2^p)\right)^{1/p} \le \left(T_{typ}(\mathcal{F}_2^p)\right)^{1/p} + \left(T_{no-typ}(\mathcal{F}_2^p)\right)^{1/p} < \frac{1}{8p} \frac{T(\mathcal{F})}{|V|}.$$

4. Weak Win part of Theorem 8.2: case (f). The case of *parallelogram lattices* is similar to case (b) discussed above. Let $\mathcal{F}$ denote the family of all $q \times q$ parallelogram lattices in the $N \times N$ board. The figure below

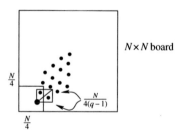

explains the easy lower bound

$$|\mathcal{F}| \ge \left(\frac{N}{4}\right)^2 \cdot \left(\frac{1}{2}\left(\frac{N/4}{q-1}\right)^2\right)^2 = \frac{N^6}{2^{14}(q-1)^4}.$$

Therefore

$$\frac{T(\mathcal{F})}{|V|} \geq \frac{N^6}{N^2 \cdot 2^{14}(q-1)^4 \cdot 2^{q^2}} = \frac{N^4}{2^{14}(q-1)^4 2^{q^2}}.$$

Next we estimate the more complicated $T(\mathcal{F}_2^p)$ from above. We begin with an easy observation: if P_1 and P_2 are two different points in the $N \times N$ board, and L_1, L_2 are two different $q \times q$ paralelogram lattices, each containing both P_1 and P_2, then the intersection $L_1 \cap L_2$ is either the $P_1 P_2$-line alone ("collinear intersection"), or it is a larger non-collinear set.

Let $L_1, L_2, \ldots, L_p$ be a collection of p $q \times q$ parallelogram lattices, each containing both P_1 and P_2. Here comes a key concept: select the maximum number of members among $L_1, L_2, \ldots, L_p$ such that any two intersect in the $P_1 P_2$-line only ("collinear intersection"). Let k denote the maximum; it is a key parameter – note that k can be anything between $1 \leq k \leq p$.

We use the following simple fact: given 3 non-collinear points P_1, P_2, P_3 of the $N \times N$ board, the total number of $q \times q$ parallelogram lattices each containing triplet P_1, P_2, P_3 is less than $(q^2)^3$.

Since there are N^2 ways to choose P_1 and $N^2 - 1$ ways to choose P_2, we have

$$T(\mathcal{F}_2^p) \leq \sum_{k=1}^{p} N^2 (N^2 - 1) \left(N^2 \cdot (q^2)^3\right)^k \cdot \binom{kq^2}{p-k} \cdot ((q^2)^3)^{p-k} \cdot 2^{-kq^2 + (k-1)q}$$

$$\leq N^4 \sum_{k=1}^{p} \left(\frac{q^6 N^2}{2^{q^2 - q}}\right)^k \cdot (kq^8)^{p-k}.$$

It follows that

$$\left(T(\mathcal{F}_2^p)\right)^{1/p} \leq N^{4/p} \sum_{k=1}^{p} \left(\frac{q^6 N^2}{2^{q^2 - q}}\right)^{k/p} \cdot (kq^8)^{1 - \frac{k}{p}}.$$

If

$$\frac{T(\mathcal{F})}{|V|} \geq \frac{N^4}{(q-1)^4 2^{q^2 + 14}} \geq 100 q^{20} \quad \text{and} \quad p = q^2, \tag{26.4}$$

then Theorem 24.2 applies, and yields a Weak Win. Indeed, by (26.4)

$$\left(T(\mathcal{F}_2^p)\right)^{1/p} \leq N^{4/p} \cdot q^{10} \sum_{k=1}^{p} N^{-k/p} < \frac{1}{8p} \frac{T(\mathcal{F})}{|V|}.$$

5. Weak Win part of Theorem 8.2: case (g). The case of *area one parallelogram lattices* is similar to cases (b) and (f) discussed above. Let $\mathcal{F}$ denote the family of all $q \times q$ area one parallelogram lattices in the $N \times N$ board. A $q \times q$ area one (parallelogram) lattice can be parametrized as follows

$$\{\mathbf{u} + k\mathbf{v} + l\mathbf{w} \in [N] \times [N] : 0 \leq k \leq q - 1, 0 \leq l \leq q - 1\},$$

where $[N] = \{1, 2, \ldots, N\}$, $\mathbf{u} \in \mathbb{Z}^2$, $\mathbf{v} = (a, b) \in \mathbb{Z}^2$, $\mathbf{w} = (c, d) \in \mathbb{Z}^2$ with $ad - bc = \pm 1$. To estimate $|\mathcal{F}|$ from below, let $\mathbf{u} \in [N/2] \times [N/2]$, and let $1 \leq a \leq b \leq \frac{N}{2(q-1)}$ be coprime integers: there are

$$\frac{3}{\pi^2}\left(\frac{N}{2(q-1)}\right)^2$$

integral vectors $\mathbf{v} = (a, b)$ with this property. Since a and b are coprime, the equation $ax - by = 1$ has an integral solution $(x, y) = (c, d)$ with the same bounds $1 \leq x \leq y \leq \frac{N}{2(q-1)}$ as for a and b above. It follows that

$$|\mathcal{F}| \geq \left(\frac{N}{2}\right)^2 \cdot \frac{3}{\pi^2}\left(\frac{N}{2(q-1)}\right)^2 = \frac{3N^4}{16\pi^2(q-1)^2},$$

and so

$$\frac{T(\mathcal{F})}{|V|} \geq \frac{N^2}{16\pi^2(q-1)^2 2^{q^2}}.$$

Next we estimate the more complicated $T(\mathcal{F}_2^p)$ from above. Fix two distinct points P_1 and P_2 in the $N \times N$ board, and let $L \in \mathcal{F}$ be a $q \times q$ area one paralelogram lattice containing both P_1 and P_2. Let P_3 be a third point of lattice L which is *not* on the $P_1 P_2$-line; then the area of the $P_1 P_2 P_3$ triangle is between $1/2$ and $q^2/2$. It follows that P_3 has to be in a *strip* around the $P_1 P_2$-line of height $h = q^2/|P_1 P_2|$, where $|P_1 P_2|$ is the distance of the two points.

Since every lattice triangle has area $\geq 1/2$, this *strip* contains at most $O(Nq^2/|P_1 P_2|)$

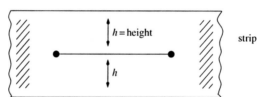

lattice points of the $N \times N$ board; i.e. there are at most $O(Nq^2/|P_1 P_2|)$ options for a third point $P_3 \in L$.

The rest of the argument is similar to case (f). Let $L_1, L_2, \ldots, L_p$ be a collection of p $q \times q$ area one lattices, each containing both P_1 and P_2. Here comes a key concept: select the maximum number of members among $L_1, L_2, \ldots, L_p$ such that any two intersect in the $P_1 P_2$-line only ("collinear intersection"). Let k denote the maximum; it is a key parameter – note that k can be anything between $1 \leq k \leq p$.

We use the following simple fact: given 3 non-collinear points P_1, P_2, P_3 of the $N \times N$ board, the total number of $q \times q$ parallelogram lattices each containing triplet P_1, P_2, P_3 is less than $(q^2)^3 = q^6$. Now we are ready to estimate $T(\mathcal{F}_2^p)$ from above. Assume that $|P_1 P_2|$ is around 2^j; then there are N^2 ways to choose P_1 and $O(4^j)$

ways to choose P_2, so we have

$$T(\mathcal{F}_2^p) \le \sum_{k=1}^{p} N^2 \sum_{j=1}^{\log_2 N} O(4^j) \left(\frac{q^2 N}{2^j} \cdot q^6 \right)^k \cdot \binom{kq^2}{p-k} \cdot (q^6)^{p-k} \cdot 2^{-kq^2+(k-1)q}$$

$$\le N^4 \sum_{k=1}^{p} \left(\frac{q^8 N}{2^{q^2-q}} \right)^k \cdot (kq^8)^{p-k}.$$

It follows that

$$\left(T(\mathcal{F}_2^p) \right)^{1/p} \le N^{4/p} \sum_{k=1}^{p} \left(\frac{q^8 N}{2^{q^2-q}} \right)^{k/p} \cdot (kq^8)^{1-\frac{k}{p}}.$$

If

$$\frac{T(\mathcal{F})}{|V|} \ge \frac{3N^2}{16\pi^2(q-1)^2 2^{q^2}} \ge 100q^{20} \quad \text{and} \quad p = q^2, \tag{26.5}$$

then Theorem 24.2 applies, and yields a Weak Win. Indeed, by (26.4)

$$\left(T(\mathcal{F}_2^p) \right)^{1/p} \le N^{4/p} \cdot q^{10} \sum_{k=1}^{p} N^{\frac{-k}{2p}} < \frac{1}{8p} \frac{T(\mathcal{F})}{|V|},$$

and the hypothesis of Theorem 24.2 is satisfied.

6. Concluding remarks. The reader must be wondering, "Why did we skip cases (d) and (e)?". Recall that case (d): *tilted rectangle lattices* and case (e): *rhombus lattices* give two different kinds of extensions of case (c): *tilted Square Lattices*, but these extensions are *negligible*. What does *negligible* mean? These extensions are very minor in the quantitative sense that the Max Degree of the hypergraph hardly increases. To justify the point, we are going to discuss the following three questions:

Question I: What is the Max Degree of the family of all $q \times q$ tilted Square Lattices in the $N \times N$ board?

Question II: What is the Max Degree of the family of all $q \times q$ tilted rectangle lattices in the $N \times N$ board?

Question III: What is the Max Degree of the family of all $q \times q$ rhombus lattices in the $N \times N$ board?

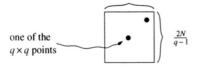

one of the
$q \times q$ points

$\frac{2N}{q-1}$

In view of the picture above the answer to Question I is

$$q^2 O((N/(q-1)^2) = O(N^2). \tag{26.6}$$

The answer to Question II is

$$O(q^2) \sum_{0 \le a \le N/(q-1), 0 \le b \le N/(q-1)}^{*} O(N/(q-1)) \frac{g.c.d.(a,b)}{a+b}, \tag{26.7}$$

where the asterisk $*$ in the summation indicates that $(0,0)$ is excluded, and g.c.d.(a,b) denotes, as usual, the *greatest common divisor* of integers a and b. If g.c.d.$(a,b) = d$, then $a = kd$ and $b = ld$ with some integers k, l, and (26.7) can be estimated from above by the sum

$$O(q \cdot N) \sum_{d=1}^{N/(q-1)} \sum_{0 \le k \le N/d(q-1), 0 \le l \le N/d(q-1)}^{*} \frac{1}{k+l}. \tag{26.8}$$

Since

$$\frac{1}{k+l} \le \frac{2}{\sqrt{1+k}} \frac{2}{\sqrt{1+l}},$$

(26.8) can be estimated from above by

$$O(q \cdot N) \sum_{d=1}^{n/(q-1)} \left(\sum_{1 \le j \le 2N/dq} \frac{2}{\sqrt{j}} \right)^2,$$

which gives the following answer to Question II

$$O(q \cdot N) \sum_{d=1}^{N} \frac{O(N)}{qd} = O(N^2) \sum_{d=1}^{N} = O(N^2 \log N). \tag{26.9}$$

Comparing (26.6) and (26.9) we see that case (d) ("tilted rectangle lattices") represents a larger hypergraph than case (c) ("tilted Square Lattices"), but the increase in the Max Degree is a small factor of $\log N$, which has a "negligible" effect in the "advanced Strong Draw criterion" later. This explains why the Achievement Number is the same for cases (c) and (d) (well, *almost* the same: the same apart from $o(1)$).

The answer to Question III is estimated from above by

$$O(q^2) \sum_{d=1}^{N/(q-1)} \sum_{0 \le a \le N/(q-1), 0 \le b \le N/(q-1)}^{*} \tau(a^2+b^2), \tag{26.10}$$

where again the asterisk $*$ in the summation indicates that $(0,0)$ is excluded, and $\tau(a,b)$ denotes, as usual, the *divisor function*; $\tau(m)$ denotes the number of divisors of integer m. In the upper bound (26.10) we used the well-known number-theoretic fact that the number of solutions of the equation $m = a^2 + b^2$ is at most $4\tau(m)$.

Sum (26.10) is estimated from above by

$$O(q^2) \sum_{1 \le m \le 2(N/(q-1))^2} \tau^2(m). \tag{26.11}$$

A theorem of Ramanujan states that

$$\sum_{1 \leq m \leq M} \tau^2(m) = O(M(\log M)^3); \tag{26.12}$$

see e.g. in Section 18.2 of Hardy-Wright: *An introduction to the theory of numbers.*
Formulas (26.10)–(26.12) lead to the following upper bound in Question III

$$O(q^2) \cdot O(N^2/(q-1)^2)) \cdot (\log N)^3 = O(N^2 \cdot (\log N)^3). \tag{26.13}$$

This time the increase in the Max Degree is a factor of $(\log N)^3$, which is still "negligible."

27

Who-scores-more games

There are two large classes of sports and games: the class of who-does-it-first and the class of who-scores-more. Every race sport (running, swimming, cycling, etc.) belongs to the class of who-does-it-first, and most of the team sports (basketball, soccer, ice-hockey, etc.) belong to the class of who-scores-more. The same applies for games: Chess is a who-does-it-first game (who gives the first checkmate) and Go is clearly a who-scores-more type of game (who captures more of the opponent's stones). Tic-Tac-Toe, and every other Positional Game, belongs to the class of who-does-it-first (the winner is the one who occupies a whole winning set first). Unfortunately, we know almost nothing about ordinary win; this is why we had to shift our attention from ordinary win to Weak Win. So far everything (well, *nearly* everything) was about Weak Win (and its complementary concept: Strong Draw) – it is time now to expand our horizons and do something different. A natural extension of the class of Positional Games is the class of "who-scores-more games." As usual, the players alternately occupy new points, but we have to clarify the notion of "scoring." In a Positional Game the most natural way of "scoring" is to occupy a whole winning set, so "who-scores-more" actually means "which player owns more winning sets at the end of the play." The symmetric version, when the two players share the same hypergraph, seems to be hopelessly complicated (just like "ordinary win," i.e. the "who-does-it-first" version); what we discuss in this section is:

1. The asymmetric version. Here the two players have two different hypergraphs – say, First Player has hypergraph $\mathcal{F}$ and Second Player has hypergraph $\mathcal{G}$ (the board is the same!) – and each player wants to occupy as many hyperedges of his own hypergraph as possible. The player who owns the most hyperedges is declared the winner; equality means a draw.

Here is an illustration. Two players are playing on the $N \times N$ board, First Player's goal is to occupy a maximum number of $q_1 \times q_1$ Aligned *Square* Lattices and Second Player's goal is to occupy a maximum number of $q_2 \times q_2$ aligned *rectangle* lattices.

In view of Theorem 8.2, the case

$$q_1 > \left\lfloor \sqrt{\log_2 N} + o(1) \right\rfloor \quad \text{and} \quad q_2 > \left\lfloor \sqrt{2\log_2 N} + o(1) \right\rfloor \quad (27.1)$$

is a boring scoreless draw (if the players play rationally); to have some action we assume that

$$q_1 \leq \left\lfloor \sqrt{\log_2 N} + o(1) \right\rfloor \quad \text{and} \quad q_2 \leq \left\lfloor \sqrt{2\log_2 N} + o(1) \right\rfloor. \quad (27.2)$$

How about the first interesting case itself

$$q_1 = \left\lfloor \sqrt{\log_2 N} + o(1) \right\rfloor \quad \text{and} \quad q_2 = \left\lfloor \sqrt{2\log_2 N} + o(1) \right\rfloor. \quad (27.3)$$

Which player ends up with more copies of his own? More precisely, which player has a winning strategy in the "square-lattice vs. rectangle-lattice who-scores-more game with (27.3)"?

Before answering this question, let me briefly recall where the "critical values" in (27.3) come from. Let $\mathcal{F} = \mathcal{F}_{N,s}$ denote the family of all $s \times s$ Aligned *Square* Lattices in the $N \times N$ board, and let $\mathcal{G} = \mathcal{G}_{N,r}$ denote the family of all $r \times r$ aligned *rectangle* lattices in the same $N \times N$ board. In $\mathcal{F}_{N,s}$ we keep the value of N fixed and s is a *variable*; then $s = \left\lfloor \sqrt{\log_2 N} + o(1) \right\rfloor$ is the largest value of integral variable s such that

$$\frac{1}{|V|}\left(\sum_{A \in \mathcal{F}_{N,s}} 2^{-|A|}\right) = \frac{1}{N^2}|\mathcal{F}_{N,s}|2^{-s^2} > 1. \quad (27.4)$$

Similarly, if N is fixed, then $r = \left\lfloor \sqrt{2\log_2 N} + o(1) \right\rfloor$ is the largest value of integral variable r such that

$$\frac{1}{|V|}\left(\sum_{B \in \mathcal{G}_{N,r}} 2^{-|B|}\right) = \frac{1}{N^2}|\mathcal{G}_{N,r}|2^{-r^2} > 1. \quad (27.5)$$

Looking at (27.4) and (27.5), we see that the factor $1/N^2$ is the same, so it is natural to come up with the *conjecture* that what really matters in the $(\mathcal{F}_{N,s}, \mathcal{G}_{N,r})$ who-scores-more game is the relation of the two sums

$$T(\mathcal{F}_{N,s}) = \left(\sum_{A \in \mathcal{F}_{N,s}} 2^{-|A|}\right) \quad \text{and} \quad T(\mathcal{G}_{N,r}) = \left(\sum_{B \in \mathcal{G}_{N,r}} 2^{-|B|}\right).$$

We would guess that if $T(\mathcal{F}_{N,s}) > T(\mathcal{G}_{N,r})$, then First Player ("Square Lattice guy") should have more copies of his own goal, and if $T(\mathcal{F}_{N,s}) < T(\mathcal{G}_{N,r})$, then Second Player ("rectangle lattice guy") should have more copies of his own goal. This vague conjecture can be justified by the following variant of Theorem 24.2.

Theorem 27.1 (("Who-Scores-More Criterion")) *Let $\mathcal{F}$ and $\mathcal{G}$ be two finite hypergraphs sharing the same board set V.*

(a) If the inequality

$$\frac{T(\mathcal{F}) - 100T(\mathcal{G})}{|V|} > p + 4p\left(T(\mathcal{F}_2^p)\right)^{1/p}$$

holds for some integer $p \geq 2$, then First Player can force the following: at the end of the play the number of As with $A \in \mathcal{F}$ completely occupied by him (i.e. First Player) is more than 100 times the number of Bs, with $B \in \mathcal{G}$ completely occupied by Second Player.

(b) Assume that $\mathcal{G}$ is m-uniform; if

$$\frac{T(\mathcal{G}) - \mathrm{MaxDeg}(\mathcal{G})2^{-m} - 100T(\mathcal{F})}{|V|} > p + 4p\left(T(\mathcal{G}_2^p)\right)^{1/p}$$

holds for some integer $p \geq 2$, then Second Player can force the following: at the end of the play the number of Bs with $B \in \mathcal{G}$ completely occupied by him (i.e. Second Player) is more than 100 times the number of As, with $A \in \mathcal{F}$ completely occupied by First Player.

Remarks. The factor of "100" in (a) and (b) was accidental: we just wanted to make sure that 1 player overwhelmingly dominates. Of course, "100" and "100 times" can be replaced by any constant (like "2" and "twice").

The proof of the Who-Scores-More Criterion above is exactly the same as that of Theorem 24.2; we leave it to the reader as an exercise.

Let's return to the the "square-lattice vs. rectangle-lattice who-scores-more game with (27.1)." How large is

$$\frac{T(\mathcal{F}_{N,q_1})}{|V|} = \frac{1}{N^2}|\mathcal{F}_{N,q_1}| \cdot 2^{-q_1^2},$$

where $q_1 = \left\lfloor \sqrt{\log_2 N} \right\rfloor$? Writing ($\{x\}$ is the fractional part of x)

$$q_1 = \left\lfloor \sqrt{\log_2 N} \right\rfloor = \sqrt{\log_2 N} - \left\{ \sqrt{\log_2 N} \right\}$$

we have

$$q_1^2 = \log_2 N - 2\left\{ \sqrt{\log_2 N} \right\}\sqrt{\log_2 N} + O(1).$$

Since

$$|\mathcal{F}_{N,q_1}| \approx \frac{N^3}{3(q_1 - 1)},$$

we obtain

$$\frac{1}{N^2}|\mathcal{F}_{N,q_1}|2^{-q_1^2} \approx \frac{N}{3(q_1 - 1)}2^{-\log_2 N + 2\left\{ \sqrt{\log_2 N} \right\}\sqrt{\log_2 N} + O(1)}$$

$$= \frac{O(1)}{q_1}2^{2\left\{ \sqrt{\log_2 N} \right\}\sqrt{\log_2 N}}. \tag{27.6}$$

Similarly

$$\frac{1}{N^2}|\mathcal{G}_{N,q_1}|2^{-q_1^2} = \frac{O(1)}{q_2}2^{2\left\{\sqrt{2\log_2 N}\right\}}\sqrt{2\log_2 N}$$

$$= \frac{O(1)}{q_2}2^{2\sqrt{2}\left\{\sqrt{2}\sqrt{\log_2 N}\right\}}\sqrt{\log_2 N}. \tag{27.7}$$

Combining the Who-Scores-More Criterion with (27.6)–(27.7), the "Square Lattice vs. rectangle lattice who-scores-more game with (27.3)" converts into an easy *fractional part* problem.

2. Fractional Part Problem.

(a) Which fractional part is larger

$$\left\{\sqrt{\log_2 N}\right\} \quad \text{or} \quad \sqrt{2}\left\{\sqrt{2}\sqrt{\log_2 N}\right\}?$$

Another natural question is:
(b) If we "randomly select" a positive integer N, then what is the "chance" that

$$\left\{\sqrt{\log_2 N}\right\} > \sqrt{2}\left\{\sqrt{2}\sqrt{\log_2 N}\right\}?$$

For typical N the fractional parts $\left\{\sqrt{\log_2 N}\right\}$ and $\sqrt{2}\left\{\sqrt{2}\sqrt{\log_2 N}\right\}$ are not too close, implying that the ratio

$$\frac{T(\mathcal{F}_{N,q_1})}{T(\mathcal{G}_{N,q_2})} = \frac{|\mathcal{F}_{N,q_1}|2^{-q_1^2}}{|\mathcal{G}_{N,q_2}|2^{-q_2^2}}$$

$$= O(1)2^{\left(\left\{\sqrt{\log_2 N}\right\}-\sqrt{2}\left\{\sqrt{2}\sqrt{\log_2 N}\right\}\right)\sqrt{\log_2 N}}$$

is typically either very large (much larger than 100) or very small (much smaller than 1/100). Therefore, by the Who-Scores-More Criterion, for a typical N one player overwhelmingly dominates the "who-scores-more" game.

Randomly choosing an N, what is the chance that the "Square Lattice over-scores the rectangle lattice"? By the Who-Scores-More Criterion this question is equivalent to part (b) of the Fractional Part Problem. Notice that the problem is not well-stated (what does "randomly chosen" mean) unless we introduce a *density concept* for integers. To guarantee that $\{\sqrt{\log_2 N}\}$ is uniformly distributed in the unit interval, the simple fact

$$\sqrt{\log_2(N+1)} - \sqrt{\log_2 N} \approx \frac{1}{2N\sqrt{\log_2 N}}$$

makes it plausible to work with the following *sub-logarithmic density*: given an infinite subset $A \subset \mathbf{N}$ of the natural numbers, we say that A has *density* α if the limit below

$$\lim_{x \to \infty} \frac{\sum_{a \in A:\ a \leq x} \frac{1}{a\sqrt{\log a}}}{\sum_{n \leq x} \frac{1}{n\sqrt{\log n}}} \tag{27.8}$$

exists and equals to α. Working with density (27.8) the following facts are clear:

(1) $\left\{\sqrt{\log_2 N}\right\}$ is uniformly distributed in the unit interval;

(2) $\left\{\sqrt{2}\sqrt{\log_2 N}\right\}$ is also uniformly distributed in the unit interval;

(3) $\left\{\sqrt{\log_2 N}\right\}$ and $\left\{\sqrt{2}\sqrt{\log_2 N}\right\}$ form independent random variables.

Statement (3) easily follows from the well-known fact that the "$n\alpha$ sequence" $\alpha, 2\alpha, 3\alpha, \ldots$ (mod 1) is uniformly distributed in the unit interval if α is irrational (we apply it with $\alpha = \sqrt{2}$).

(1)–(3) imply that part (b) of the Fractional Part Problem above is equivalent to the following elementary problem in Probability Theory: if X and Y are independent, uniformly distributed random variables in the unit interval, then what is the probability of $X > \sqrt{2}Y$?

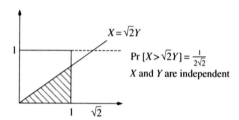

Since X and Y are independent, we can visualize the problem on the product space $[0, 1] \times [0, 1]$ (see figure), which is, of course, the unit square, and then the answer is simply the area of the shaded triangle: $1/2\sqrt{2} = .35355$.

Theorem 27.2 (a) *Consider the "aligned square-lattice vs. aligned rectangle lattice who-scores-more game" on the $N \times N$ board with the largest achievable sizes (27.3). If we randomly select the board-size parameter N – we work with the sublogarithmic density (27.8) – then the "Square Lattice" over-scores the "rectangle lattice" with probability $1/2\sqrt{2}$.*

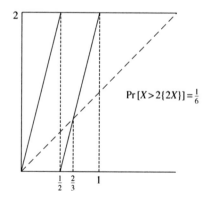

3. More Fractional Part Problems. What happens if "Aligned Square Lattice vs. aligned rectangle lattice" is replaced by the "Aligned Square Lattice vs. parallelogram lattice"? This case is even simpler. A straightforward application of the Who-Scores-More Criterion reduces the problem to the following very simple fractional part question: If X is a uniformly distributed random variable in the unit interval, then what is the probability of the event $X > 2\{2X\}$? Comparing the graphs of the two functions $y = x$ and $y = 2\{2x\}$ in $0 \le x \le 1$, the answer is clear: the probability is $1/6$.

Theorem 27.2 (b) *Consider the "Aligned Square Lattice vs. parallelogram lattice who-scores-more game" on the $N \times N$ board with the largest achievable sizes: $q_1 \times q_1$ Aligned Square Lattice with $q_1 = \lfloor \sqrt{\log_2 N} \rfloor$ and $q_2 \times q_2$ parallelogram lattice with $q_2 = \lfloor 2\sqrt{\log_2 N} \rfloor$. If we randomly select the board-size parameter N – we work with the sub-logarithmic density (27.6) – then the "Square Lattice" over-scores the "parallelogram lattice" with probability $1/6$.*

What happens if the "square shape" is replaced by arbitrary lattice polygons (see Section 8 around formula (8.5))? For example, how about the "aligned pentagon vs. aligned triangle who-scores-more game" on the $N \times N$ board if from each shape we take the largest achievable size? In view of (8.5) the largest achievable size is

$$\left\lfloor \frac{1}{\sqrt{A}} \sqrt{\log_2 N} - \frac{B}{4A} + o(1) \right\rfloor, \qquad (27.9)$$

where A is the area of the initial shape $S = S(1)$ and B is the number of lattice points on the boundary of S. The Who-Scores-More Criterion reduces this problem to the following fractional part question: Given two lattice polygons S_1 and S_2, which one is larger

$$\left\{ \frac{1}{\sqrt{A_1}} \sqrt{\log_2 N} - \frac{B_1}{4A_1} \right\} \left(\frac{1}{\sqrt{A_1}} \sqrt{\log_2 N} - \frac{B_1}{4A_1} \right)$$

or

$$\left\{ \frac{1}{\sqrt{A_2}} \sqrt{\log_2 N} - \frac{B_2}{4A_2} \right\} \left(\frac{1}{\sqrt{A_2}} \sqrt{\log_2 N} - \frac{B_2}{4A_2} \right)?$$

Here A_1 is the area of the initial shape $S_1 = S_1(1)$, B_1 is the number of lattice points on the boundary of S_1, and, similarly, A_2 is the area of the initial shape $S_2 = S_2(1)$, and B_2 is the number of lattice points on the boundary of S_2. If A_1/A_2 is not a rational square number – this happens, for example, if S_1 is the "pentagon" of area 3 and S_2 is the "triangle" of area $1/2$ – then we can repeat the argument of Theorem 27.2, and obtain the following simple answer to the "probability question" (notice that the probability is irrational).

Theorem 27.2 (c) *Consider the "aligned lattice polygon S_1 vs. aligned lattice polygon S_2 who-scores-more game" on the $N \times N$ board with the largest achievable* sizes (27.9). *Assume that the ratio* A_1/A_2 *of the areas is not a rational square number. Let* $A_1 > A_2$; *if we randomly select the board-size parameter* N – *we work with the sub-logarithmic density* (27.8) – *then shape* S_1 *over-scores shape* S_2 *with probability* $\sqrt{A_2}/2\sqrt{A_1}$.

Theorem 27.2 does not cover the case when A_1/A_2 is a rational square number. The simplest way to make A_1/A_2 a rational square number is to assume $A_1 = A_2$; then we can repeat the argument of Theorem 27.2, and get the following simple result (notice that the probability is rational).

Theorem 27.2 (d) *Consider the "aligned lattice polygon* S_1 *vs. aligned lattice polygon* S_2 *who-scores-more game" on the* $N \times N$ *board with the largest achievable sizes* (27.9). *Assume that the areas are equal:* $A_1 = A_2 = A$, *and* $B_1 > B_2$ (*i.e. the boundary of* S_1 *contains more lattice points*). *If we randomly select the board-size parameter* N – *we work with the sub-logarithmic density* (27.8) – *then shape* S_1 *over-scores shape* S_2 *with probability* $(B_1 - B_2)/4A$.

Of course, cases (c) and (d) do not cover everything; for example, in the "fish vs. octagon who-scores-more game" the ratio $28/7 = 4 = 2^2$ of the areas is a square number, so case (c) doesn't apply. We discussed cases (c) and (d) because the answer to the probability question was particularly simple. By the way, the "fish" over-scores the "octagon" with probability 13/84 (why?).

The last example is a *graph game* played on K_N. First Player's goal is to occupy a maximum number of copies of $K_{b,b}$ ("complete bipartite graph") and Second Player's goal is to occupy a maximum number of copies of $K_{t,t,t}$ ("complete tripartite graph").

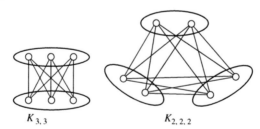

$$K_{3,3} \qquad\qquad K_{2,2,2}$$

Let

$$b = b(N) = \lfloor 2\log_2 N - 2\log_2 \log_2 N + 2\log_2 e - 3 \rfloor \qquad (27.10)$$

and

$$t = t(N) = \left\lfloor \log_2 N - \log_2 \log_2 N + \log_2 e - \frac{2}{3} \right\rfloor; \qquad (27.11)$$

note that taking the *upper* integral part in (27.10)–(27.11) would lead to a boring scoreless game: (27.10)–(27.11) represent the largest achievable values (see (8.7)). Which player wins the who-scores-more game? A straightforward application of the Who-Scores-More Criterion reduces the problem to the following elementary question: consider the expression ($\{y\}$ denotes, as usual, the fractional part of real number y)

$$2\{2\log_2 N - 2\log_2 \log_2 N + 2\log_2 e - 3\} - 3\{\log_2 N - \log_2 \log_2 N + \log_2 e - 2/3\}$$

$$= 3\left(\frac{2}{3}\{2\lambda(N) + 1/3\} - \{\lambda(N)\}\right)$$

where

$$\lambda(N) = \log_2 N - \log_2 \log_2 N + \log_2 e - 2/3; \qquad (27.12)$$

is $\frac{2}{3}\{2\lambda + 1/3\} - \{\lambda\}$ positive or negative? If $\frac{2}{3}\{2\lambda + 1/3\} - \{\lambda\}$ is positive, where $\lambda = \lambda(N)$, then First Player has more bipartite graphs $K_{b,b}$; if $\frac{2}{3}\{2\lambda + 1/3\} - \{\lambda\}$ is negative, then Second Player has more tripartite graphs $K_{t,t,t}$.

Note that working with the usual *logarithmic density*

$$\lim_{x \to \infty} \frac{\sum_{a \in A: \, a \le x} \frac{1}{a}}{\sum_{n \le x} \frac{1}{n}} \qquad (27.13)$$

the following fact is clear: $\{\lambda(N)\}$ is uniformly distributed in the unit interval as $N \to \infty$.

It follows that the fractional part problem above is equivalent to the following elementary problem in Probability Theory: if X is a uniformly distributed random variable in the unit interval, what is the probability of

$$X > \frac{2}{3}\left\{2X + \frac{1}{3}\right\}?$$

Comparing the graphs of the two functions $y = x$ and $y = \frac{2}{3}\{2x + 1/3\}$ in $0 \le x \le 1$, the answer is clear: the probability is 2/3.

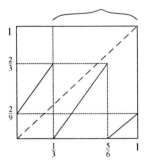

Theorem 27.2 (e) *Consider the "bipartite vs. tripartite who-scores-more game" on the complete graph K_N with the largest achievable sizes (27.10) and (27.11). If we randomly select the vertex-size parameter N – we work with the logarithmic density (27.13) – then $K_{b,b}$ over-scores $K_{t,t,t}$ with probability 2/3.*

Chapter VI

What is the Biased Meta-Conjecture, and why is it so difficult?

There are two natural ways to generalize the concept of Positional Game: one way is the (1) discrepancy version, where Maker wants (say) 90% of some hyperedge instead of 100%. Another way is the (2) biased version like the $(1:2)$ play, where underdog Maker claims 1 point per move and Breaker claims 2 points per move.

Chapter VI is devoted to the discussion of these generalizations.

Neither generalization is a perfect success, but there is a big difference. The *discrepancy version* generalizes rather smoothly; the *biased version*, on the other hand, leads to some unexpected tormenting(!) technical difficulties.

The main issue here is to formulate and prove the Biased Meta-Conjecture. The biased case is *work in progress*; what we currently know is a bunch of (very interesting!) sporadic results, but the general case remains wide open.

We don't see any *a priori* reason why the biased case should be more difficult than the fair (1:1) case. No one understands why the general biased case is still unsolved.

The Biased Meta-Conjecture is the most exciting research project that the book can offer. We challenge the reader to participate in the final solution.

The biased Maker–Breaker and Avoider–Forcer games remain mostly unsolved, but we are surprisingly successful with the biased $(1:s)$ Chooser–Picker game where Chooser is the underdog (in each turn Picker picks $(s+1)$ new points, Chooser chooses one of them, and the rest goes back to Picker). In this case we come very close to the perfect analogue of Theorem 6.4 ("Clique Games") and Theorem 8.2 ("Lattice Games") with the natural change that the base 2 logarithm $\log_2$ is replaced by the base $(s+1)$ logarithm $\log_{s+1}$ (see Theorem 33.4).

380

28

Discrepancy games (I)

1. What is the right conjecture? Recall that the Clique Achievement Number $A(K_N; clique)$ is the largest integer $q = q(N)$ such that Maker can always occupy a whole sub-clique K_q in the usual (1:1) play on K_N. The exact value of $A(K_N; clique)$ is the lower integral part of the function

$$f(N) = 2\log_2 N - 2\log_2 \log_2 N + 2\log_2 e - 3; \tag{28.1}$$

more precisely, this is true for the overwhelming majority of Ns, and to cover every N we usually add an extra $o(1)$ (which tends to 0 as $N \to \infty$) to (28.1) (see Theorem 6.4 (a)).

Function (28.1) is explained by the Phantom Decomposition Hypothesis ("Neighborhood Conjecture" and "local randomness" are alternative names for the same thing, see Sections 8, 18, and 19). The Phantom Decomposition Hypothesis says that there is a virtual decomposition into "neighborhoods": for a fixed edge $e_0 \in K_N$, there are $\binom{N-2}{q-2}$ copies of K_q in K_N containing edge e_0, and (28.1) is the ("real") solution of the equation

$$\binom{N-2}{q-2} 2^{-\binom{q}{2}} = 1 \tag{28.2}$$

in variable $q = q(N)$. Here the term $\binom{N-2}{q-2} 2^{-\binom{q}{2}}$ has a probabilistic interpretation: it is the expected number of monochromatic red copies of K_q with $e_0 \in K_q$ in a Random 2-Coloring of the edges of K_N (using colors red and blue).

Next we modify Maker's goal: instead of owning a whole K_q, Maker just wants a given majority, (say) $\geq 90\%$ of the edges from some K_q, or, in general, he wants $\geq \alpha\binom{q}{2}$ edges from some K_q, where $\frac{1}{2} < \alpha \leq 1$ is a given constant. Fix an $\frac{1}{2} < \alpha \leq 1$; the largest $q = q(N)$ such that, playing the usual (1:1) game on K_N, Maker can always own at least $\alpha\binom{q}{2}$ edges from some K_q in K_N, is called the α-Clique Achievement Number, denoted by $A(K_N; clique; \alpha)$. How large is $A(K_N; clique; \alpha)$? The Phantom Decomposition Hypothesis suggests to solve the equation

$$\binom{N-2}{q-2} \left(\sum_{m \geq \alpha\binom{q}{2}} \binom{\binom{q}{2}}{m} 2^{-\binom{q}{2}} \right) = 1 \tag{28.3}$$

in variable $q = q(N, \alpha)$; observe that (28.3) is the perfect discrepancy version of (28.3): the left-hand side of (28.3) is the expected number of $(\geq \alpha)$-red copies of K_q with $e_0 \in K_q$ in a Random 2-Coloring of the edges of K_N (using red and blue). Of course, "$(\geq \alpha)$-red" means that at least α part is red and at most $(1 - \alpha)$ part is blue.

To find a "real" solution of (28.3) we use the weak form $k! \approx (k/e)^k$ of the Stirling formula (the ignored factor $\sqrt{2\pi k}$ gives a negligible contribution in $q = q(N, \alpha)$ anyway)

$$1 = \binom{N-2}{q-2} \left(\sum_{m \geq \alpha\binom{q}{2}} \binom{\binom{q}{2}}{2} 2^{-\binom{q}{2}} \right) \approx \binom{N-2}{q-2} \binom{\binom{q}{2}}{\alpha\binom{q}{2}} 2^{-\binom{q}{2}}$$

$$\approx \binom{N-2}{q-2} \left(2\alpha^\alpha (1-\alpha)^{1-\alpha} \right)^{-\binom{q}{2}} \approx \binom{N-2}{q-2} 2^{-(1-H(\alpha))\binom{q}{2}}, \tag{28.4}$$

where $H(\alpha) = -\alpha \log_2 \alpha - (1-\alpha) \log_2(1-\alpha)$ is the well-known Shannon's entropy. Notice that, if α goes from $1/2$ to 1, the term $(1 - H(\alpha))$ goes from 0 to 1.

Returning to (28.3)–(28.4), we have

$$1 = \binom{N-2}{q-2} 2^{-(1-H(\alpha))\binom{q}{2}},$$

which is equivalent to

$$(1 - H(\alpha))\binom{q}{2} = \log_2 \binom{N-2}{q-2} = \log_2 \left(\frac{eN}{q} \right)^{q-2} = (q-2)\log_2(eN/q),$$

which is equivalent to

$$(1 - H(\alpha))\frac{q}{2} = \left(1 - \frac{1}{q-1} \right)(\log_2 N - \log_2 q + \log_2 e) =$$

$$= \log_2 N - \log_2 q - \frac{\log_2 N}{q-1} + \log_2 e + o(1),$$

which is equivalent to

$$q = \frac{2}{1 - H(\alpha)} \left(\log_2 N - \log_2 q - \frac{\log_2 N}{q-1} + \log_2 e + o(1) \right) =$$

$$= \frac{2}{1 - H(\alpha)} \left(\log_2 N - \log_2 \left(\frac{2}{1 - H(\alpha)} \log_2 N \right) - \frac{\log_2 N}{\frac{2}{1-H(\alpha)}\log_2 N} + \log_2 e + o(1) \right) =$$

$$= \frac{2}{1 - H(\alpha)} \left(\log_2 N - \log_2 \log_2 N + \log_2(1 - H(\alpha)) + \log_2 e - 1 \right) - 1 + o(1).$$

That is, the real solution of the Phantom Decomposition Hypothesis equation (28.3) is (α and N are fixed)

$$q = q(N) = \frac{2}{1-H(\alpha)}\left(\log_2 N - \log_2 \log_2 N + \log_2(1-H(\alpha)) + \log_2 e - 1\right) - 1 + o(1).$$
(28.5)

Notice that choosing $\alpha = 1$ in (28.5) we get back (28.1).

Is it true that $\mathbf{A}(K_N; clique; \alpha)$ is the lower integral part of (28.5)? Let's see how far we can go by adapting the proof of the case $\alpha = 1$. In the proof of the case $\alpha = 1$ we used the self-improving Potential Function

$$L_i = T(\mathcal{F}(i)) - \lambda \cdot T(\mathcal{F}_2^p(i))$$

with an appropriate positive constant λ, where T indicates the usual Power-of-Two Scoring System

$$T(\mathcal{H}) = \sum_{A \in \mathcal{H}} 2^{-|A|} \quad \text{and} \quad T(\mathcal{H}; u_1, \ldots, u_m) = \sum_{A \in \mathcal{H}: \ \{u_1, \ldots, u_m\} \subseteq A} 2^{-|A|}. \tag{28.6}$$

Switching from $\alpha = 1$ to a general $\frac{1}{2} < \alpha \leq 1$ we have to change scoring system (28.6). Assume we are in the middle of a play where Maker already occupies $X(i) = \{x_1, \ldots, x_i\}$ and Breaker already occupies $Y(i) = \{y_1, \ldots, y_i\}$, and with $\alpha = (1+\varepsilon)/2$ define

$$T_{i,\varepsilon}(\mathcal{H}) = \sum_{A \in \mathcal{H}}(1+\varepsilon)^{|A \cap X(i)| - (1+\varepsilon)|A|/2}(1-\varepsilon)^{|A \cap Y(i)| - (1-\varepsilon)|A|/2}; \tag{28.7}$$

we use the special cases $\mathcal{H} = \mathcal{F}$ and $\mathcal{H} = \mathcal{F}_2^p$. We employ the Potential Function

$$L_i = T_{i,\varepsilon}(\mathcal{F}(i)) - \lambda \cdot T_{i,\varepsilon}(\mathcal{F}_2^p(i))$$

with the usual side condition $L_0 = \frac{1}{2}T_{0,\varepsilon}(\mathcal{F})$, or equivalently

$$\lambda = \frac{T_{0,\varepsilon}(\mathcal{F})}{2T_{0,\varepsilon}(\mathcal{F}_2^p)}.$$

If Maker chooses that unoccupied $z = x_{i+1}$ for which the function

$$L_i(z) = T_{i,\varepsilon}(\mathcal{F}(i); z) - \lambda \cdot T_{i,\varepsilon}(\mathcal{F}_2^p(i); z)$$

attains its maximum, then we obtain the usual inequality

$$L_{i+1} \geq L_i + \varepsilon \cdot L_i(x_{i+1}) - \varepsilon \cdot L_i(y_{i+1}) - \varepsilon^2 \cdot T_{i,\varepsilon}(\mathcal{F}(i); x_{i+1}, y_{i+1}) \geq$$
$$\geq L_i - \varepsilon^2 \cdot T_{i,\varepsilon}(\mathcal{F}(i); x_{i+1}, y_{i+1}). \tag{28.8}$$

So far so good! Next we need a discrepancy analogue of the Generalized Variance Lemma (Theorem 24.1). A straightforward adaptation of the proof of Theorem 24.1 would require an inequality such as

$$(1+\varepsilon)^{\sum_{j=1}^{p}(|A_j \cap X(i)|-(1+\varepsilon)|A_j|/2)}(1-\varepsilon)^{\sum_{j=1}^{p}(|A_j \cap Y(i)|-(1-\varepsilon)|A_j|/2)} \leq$$

$$\leq (1+\varepsilon)^{|\bigcup_{j=1}^{p} A_j \cap X(i)|-(1+\varepsilon)|\bigcup_{j=1}^{p} A_j|/2}(1-\varepsilon)^{|\bigcup_{j=1}^{p} A_j \cap Y(i)|-(1-\varepsilon)|\bigcup_{j=1}^{p} A_j|/2}. \qquad (28.9)$$

Unfortunately, we get stuck here: inequality (28.9) is *false* in general! Here is a counter-example: assume that:

(1) $|A_j| = n, \ j = 1, \ldots, p$;
(2) the p intersections $A_j \cap X(i), \ j = 1, \ldots, p$ are equal to the same set B with $|B| = (1+\varepsilon)n/2$;
(3) the p intersections $A_j \cap Y(i), \ j = 1, \ldots, p$ are pairwise disjoint;
(4) $|A_j \cap Y(i)| = (1-\varepsilon)n/2, \ j = 1, \ldots, p$.

Then (28.9) simplifies to

$$(1+\varepsilon)^{(1+\varepsilon)(|\bigcup_{j=1}^{p} A_j|-|A_1|)}\left(\frac{1}{1-\varepsilon}\right)^{(1-\varepsilon)\left(\sum_{j=1}^{p}|A_j|-|\bigcup_{j=1}^{p} A_j|\right)} \leq 1, \qquad (28.10)$$

which is clearly non-sense. Indeed, inequality (28.10) cannot hold in general, because both bases, $(1+\varepsilon)$ and $1/(1-\varepsilon)$, are > 1, and their exponents can be arbitrarily large, so the left-hand side of (28.10) can be arbitrarily large.

The failure of (28.9) explains why we cannot solve the following:

Open Problem 28.1 *Is it true that the α-Clique Achievement Number $A(K_N; \text{clique}; \alpha)$ is the lower integral part of*

$$q = q(N, \alpha) = \frac{2}{1 - H(\alpha)}\left(\log_2 N - \log_2 \log_2 N + \log_2(1 - H(\alpha)) + \log_2 e - 1\right)$$

$$- 1 + o(1),$$

or at least the distance between the two quantities is $O(1)$? Here the function $H(\alpha) = -\alpha \log_2 \alpha - (1-\alpha)\log_2(1-\alpha)$ is the well-known Shannon's entropy.

2. Nearly perfect solutions. Inequality (28.9) is false in general, but it holds (with equality!) for any family of *disjoint* A_js. This is very good news, because the auxiliary big hypergraphs $\mathcal{F}_{2*}^p$ and $\mathcal{F}_{2**}^p$ in Section 23 ("*ad hoc* argument") had some "near-disjointness." This gives us a hope to solve the analogue of Open Problem 28.1 for at least the Lattice Games; and, indeed, we are going to prove the following precise statement (see Theorem 9.1 below, which was already formulated in Section 9). For every α with $1/2 < \alpha \leq 1$, let $A(N \times N; \text{square lattice}; \alpha)$ denote the largest value of

q such that Maker can always occupy $\geq \alpha$ part of some $q \times q$ Aligned Square Lattice in the $N \times N$ board. We will call it the α-*Discrepancy Achievement Number*. We can similarly define the α-Discrepancy Achievement Number for the rest of the Lattice Games in Section 8, and also for the Avoidance version.

Theorem 9.1 *Consider the $N \times N$ board; this is what we know about the α-Discrepancy Achievement Numbers of the Lattice Games:*

(a) $\left\lceil \sqrt{\frac{\log_2 N}{1-H(\alpha)}} + o(1) \right\rceil \geq A(N \times N; \text{ square lattice}; \alpha) \geq \left\lfloor \sqrt{\frac{\log_2 N}{1-H(\alpha)}} - c_0(\alpha) - o(1) \right\rfloor$,

(b) $\left\lceil \sqrt{\frac{2\log_2 N}{1-H(\alpha)}} + o(1) \right\rceil \geq A(N \times N; \text{ rectangle lattice}; \alpha) \geq \left\lfloor \sqrt{\frac{2\log_2 N}{1-H(\alpha)}} - c_0(\alpha) - o(1) \right\rfloor$,

(c) $\left\lceil \sqrt{\frac{2\log_2 N}{1-H(\alpha)}} + o(1) \right\rceil \geq A(N \times N; \text{ tilted square latt.}; \alpha) \geq \left\lfloor \sqrt{\frac{2\log_2 N}{1-H(\alpha)}} - c_0(\alpha) - o(1) \right\rfloor$,

(d) $\left\lceil \sqrt{\frac{2\log_2 N}{1-H(\alpha)}} + o(1) \right\rceil \geq A(N \times N; \text{ tilt. rect. lattice}; \alpha) \geq \left\lfloor \sqrt{\frac{2\log_2 N}{1-H(\alpha)}} - c_0(\alpha) - o(1) \right\rfloor$,

(e) $\left\lceil \sqrt{\frac{2\log_2 N}{1-H(\alpha)}} + o(1) \right\rceil \geq A(N \times N; \text{ rhombus lattice}; \alpha) \geq \left\lfloor \sqrt{\frac{2\log_2 N}{1-H(\alpha)}} - c_0(\alpha) - o(1) \right\rfloor$,

(f) $\left\lceil 2\sqrt{\frac{\log_2 N}{1-H(\alpha)}} + o(1) \right\rceil \geq A(N \times N; \text{ parall. lattice}; \alpha) \geq \left\lfloor 2\sqrt{\frac{\log_2 N}{1-H(\alpha)}} - c_0(\alpha) - o(1) \right\rfloor$,

(g) $\left\lceil \sqrt{\frac{2\log_2 N}{1-H(\alpha)}} + o(1) \right\rceil \geq A(N \times N; \text{ area one lattice}; \alpha) \geq \left\lfloor \sqrt{\frac{2\log_2 N}{1-H(\alpha)}} - c_0(\alpha) - o(1) \right\rfloor$,

and the same for the corresponding Avoidance Number. Here the function $H(\alpha) = -\alpha \log_2 \alpha - (1-\alpha) \log_2(1-\alpha)$ is the Shannon's entropy, and

$$c_0(\alpha) = \sqrt{\frac{\log 2 \cdot \alpha(1-\alpha)}{2}} \log_2\left(\frac{\alpha}{1-\alpha}\right)$$

is a constant (depending only on α) which tends to 0 as $\alpha \to 1$.

If α is close to 1, then the additive constant $c_0(\alpha)$ is small, so the upper and lower bounds coincide. Thus for the majority of the board size N we know the exact value of the α-Discrepancy Achievement (and Avoidance) Numbers for the Lattice Games.

Here and in the next section we prove the lower bounds; the upper bounds require the techniques of Part D.

We begin the lower bound **proof** with cases (a) and (c), i.e. the two:

3. Simplest lattice games: the Aligned and the Tilted Square Lattice Games. For these two games the Weak Win part of the special case $\alpha = 1$ was solved by a trivial application of the "linear" criterion Theorem 1.2 (see Section 13); we didn't need any sophisticated "higher moment hypergraph." The most natural idea is to try to develop an α-Discrepancy version of Theorem 1.2. Is the extension trivial? The answer is "no"; at some point we need a technical trick. To understand the necessary modifications, first we briefly recall the proof of Theorem 1.2. The simple basic idea was to study the Opportunity Function

$$T_i(\mathcal{F}) = \sum_{A \in \mathcal{F}:\ A \cap Y(i) = \emptyset} 2^{|A \cap X(i)| - |A|}, \quad i = 0, 1, 2, \ldots \tag{28.11}$$

and to guarantee that $T_{end}(\mathcal{F}) > 0$. If $T_{end}(\mathcal{F}) > 0$, then at the end of play there still exists an $A_0 \in \mathcal{F}$ which is not blocked by Breaker – this A_0 is clearly occupied by Maker, and the Weak Win is confirmed.

Let $\alpha = (1 + \varepsilon)/2$; the natural analogue of (28.11) is the sum (see (28.7))

$$T_{i,\varepsilon}(\mathcal{F}) = \sum_{A \in \mathcal{F}} (1 + \varepsilon)^{|A \cap X(i)| - (1+\varepsilon)|A|/2} (1 - \varepsilon)^{|A \cap Y(i)| - (1-\varepsilon)|A|/2}. \tag{28.12}$$

Notice that if $\alpha = 1$, (28.12) simplifies to (28.11).

Now assume $T_{end,\varepsilon}(\mathcal{F}) > 0$; does this imply that at the end of play there exists an $A_0 \in \mathcal{F}$ from which Maker owns $\geq \alpha$ part? Well, not necessarily! Even if Maker can force a "large" lower bound such as

$$T_{end,\varepsilon}(\mathcal{F}) \geq \frac{1}{2} T_{start,\varepsilon}(\mathcal{F}) = \text{large},$$

that still implies nothing. Unfortunately, the sum $T_{end,\varepsilon}(\mathcal{F})$ can be large for the "wrong reason": that there are an unusually large number of sets $A \in \mathcal{F}$ from which Maker owns β part with some $\beta < \alpha$, and α part may not occur at all! We refer to this "wrong reason" as the "concentration on small discrepancy."

4. A technical trick. We can prevent the "concentration on small discrepancy" by a technical trick that is very similar to what we did in the first proof of Theorem 16.2 (see formula (16.13)). We modify (28.12) as follows, let

$$R_i = T_{i,\gamma}(\mathcal{F}) - \sum_{j \geq 0} T_{i,\beta_j}(\mathcal{F}) =$$

$$= \sum_{A \in \mathcal{F}} \left((1 + \gamma)^{|A \cap X(i)| - \eta|A|} (1 - \gamma)^{|A \cap Y(i)| - \eta|A|} - \right.$$

$$\left. \sum_{j \geq 0} \delta_j \cdot (1 + \beta_j)^{|A \cap X(i)| - \alpha_j|A|} (1 - \beta_j)^{|A \cap Y(i)| - \alpha_j|A|} \right), \tag{28.13}$$

where $\eta = (1 + \gamma)/2$

$$\alpha = \alpha_0 = \frac{1 + \beta_0}{2}, \quad \alpha_j = \frac{1 + \beta_j}{2} \text{ and } \beta_j = \beta_0 - \frac{2jc_\alpha}{q} \text{ for } j = 1, 2, 3, \ldots, \quad \gamma = \beta_0 + \frac{2c_\alpha}{q}. \tag{28.14}$$

Here c_α is a positive constant, depending only on α, and $\delta_j \to 0$ very rapidly; both will be specified later.

It will be enough to show that $R_{end} > 0$ (the explanation comes later). The proof argument of Theorem 1.2 gives the usual inequality (the analogue of (1.18)–(1.19))

$$R_{end} \geq R_{start} - \gamma^2 \cdot \frac{|V|}{2} \cdot \Delta_2, \tag{28.15}$$

where $|V| = N^2$ is the board size, and $\Delta_2 = \Delta_2(\mathcal{F})$ is the Max Pair-Degree of the q^2-uniform hypergraph $\mathcal{F}$; here $\mathcal{F}$ is either the family of all $q \times q$ Aligned Square Lattices, or the family of all $q \times q$ tilted Square Lattices in $N \times N$. Clearly

$$R_{start} = R_0 = |\mathcal{F}| \left(2^{-(1-H(\eta))q^2} - \sum_{j \geq 0} \delta_j \cdot 2^{-(1-H(\alpha_j))q^2} \right). \tag{28.16}$$

In view of (28.15) we want R_{start} to be "large."

We claim that the choice of parameters

$$\delta_0 = \left(\frac{1-\gamma}{1+\gamma} \right)^{(\eta-\alpha)q^2} = \left(\frac{1-\gamma}{1+\gamma} \right)^{c_\alpha q}, \tag{28.17}$$

and, in general

$$\delta_j = \left(\frac{1-\gamma}{1+\gamma} \right)^{(\eta-\alpha_j)q^2} = \left(\frac{1-\gamma}{1+\gamma} \right)^{c_\alpha(j+1)q}, \tag{28.18}$$

where δ_j, $j \geq 0$ will prevent the "concentration on small discrepancy" (the proof comes later).

By (28.16)–(28.18)

$$R_{start} = |\mathcal{F}| \left(2^{-(1-H(\eta))q^2} - \sum_{j \geq 0} 2^{-((1-H(\alpha_j)+(\eta-\alpha_j)\log_2((1+\gamma)/(1-\gamma)))q^2} \right). \tag{28.19}$$

In order to guarantee that R_{start} is "large," by (28.19) we certainly need the inequality

$$1 - H(\eta) < 1 - H(\alpha_j) + (\eta - \alpha_j) \log_2 \left(\frac{1+\gamma}{1-\gamma} \right),$$

which is equivalent to

$$\frac{(1-H(\eta)) - (1-H(\alpha_j))}{\eta - \alpha_j} < \log_2 \left(\frac{1+\gamma}{1-\gamma} \right) = \log_2 \left(\frac{\eta}{1-\eta} \right)$$

$$= g'(\eta), \quad \text{where } g(x) = 1 - H(x). \tag{28.20}$$

In (28.20) $g'(x)$ denotes, of course, the derivative of function $g(x) = 1 - H(x)$.

Inequality (28.20) follows from *convexity*: in the figure below the chord is below the tangent line

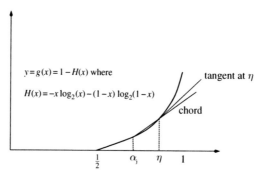

$$y = g(x) = 1 - H(x) \text{ where}$$

$$H(x) = -x \log_2(x) - (1 - x) \log_2(1 - x)$$

tangent at η

chord

$\frac{1}{2}$ α_j η 1

By Taylor's formula

$$g(x - \varepsilon) \approx g(x) - \varepsilon \cdot g'(x) + \frac{\varepsilon^2}{2} \cdot g''(x),$$

so by computing the two derivatives

$$g'(x) = (1 - H(x))' = \log_2\left(\frac{x}{1 - x}\right) \text{ and } g''(x) = \left(\frac{\log x - \log(1 - x)}{\log 2}\right)' = \frac{1}{\log 2 \cdot x(1 - x)}$$

we get

$$(1 - H(\alpha_j)) \approx (1 - H(\eta)) - (\eta - \alpha_j) \log_2\left(\frac{\eta}{1 - \eta}\right) + \frac{(\eta - \alpha_j)^2}{2} \cdot \frac{1}{\log 2 \cdot \eta(1 - \eta)},$$

or equivalently, by (28.14)

$$(1 - H(\alpha_j)) - (\eta - \alpha_j) \log_2\left(\frac{\eta}{1 - \eta}\right) - (1 - H(\eta)) \approx \frac{(\eta - \alpha_j)^2}{2} \cdot \frac{1}{\log 2 \cdot \eta(1 - \eta)} =$$

$$= \frac{c_\alpha^2 \cdot (j + 1)^2}{2 \log 2 \cdot \alpha(1 - \alpha)q^2}.$$

$$(28.21)$$

By (28.21) and (28.19)

$$R_{start} \approx |\mathcal{F}| \cdot 2^{-(1 - H(\eta))q^2}\left(1 - \sum_{j \geq 0} 2^{-\frac{c_\alpha^2 \cdot (j+1)^2}{2 \log 2 \cdot \alpha(1 - \alpha)}}\right),$$

$$(28.22)$$

so by choosing

$$c_\alpha = \sqrt{2 \log 2 \cdot \alpha(1 - \alpha)}$$

$$(28.23)$$

in (28.22) we have

$$R_{start} \approx |\mathcal{F}| \cdot 2^{-(1 - H(\eta))q^2}\left(1 - \sum_{j \geq 0} 2^{-(j+1)^2}\right) \geq \frac{1}{3}|\mathcal{F}| \cdot 2^{-(1 - H(\eta))q^2}.$$

$$(28.24)$$

Now let's return to (28.13). *Assume* that

$$R_{start} > \gamma^2 |V| \Delta_2 = \gamma^2 N^2 \Delta_2 \tag{28.25}$$

holds; then by (28.13) and (28.24)

$$R_{end} \geq \frac{1}{2} R_{start} \approx \frac{1}{2} R_{start} \geq \frac{1}{6} |\mathcal{F}| \cdot 2^{-(1-H(\eta))q^2} > 0. \tag{28.26}$$

We claim that (28.26) implies α-Discrepancy, i.e. We are going to derive from (28.26) the existence of an $A_0 \in \mathcal{F}$ such that Maker owns at least α part from A_0. Indeed, if there is no α-Discrepancy, then from every $A \in \mathcal{F}$ Maker owns less than α part. In other words, $|A \cap X(end)| < \alpha q^2$, implying that $\alpha_{j+1} q^2 \leq |A \cap X(end)| < \alpha_j q^2$ holds for some $j = j(A) \geq 0$. Then trivially (see (28.17)–(28.18) and (28.14))

$$(1+\gamma)^{|A \cap X(end)|-\eta q^2} \cdot (1-\gamma)^{|A \cap Y(end)|-(1-\eta)q^2} =$$

$$= (1+\gamma)^{|A \cap X(end)|-\alpha_j q^2 - (\eta-\alpha_j)q^2} \cdot (1-\gamma)^{|A \cap Y(end)|-(1-\alpha_j)q^2 + (\eta-\alpha_j)q^2} =$$

$$= \delta_j \cdot (1+\gamma)^{|A \cap X(end)|-\alpha_j q^2} \cdot (1-\gamma)^{|A \cap Y(end)|-(1-\alpha_j)q^2} \leq$$

$$\leq \delta_j \cdot (1+\beta_j)^{|A \cap X(end)|-\alpha_j q^2} \cdot (1-\beta_j)^{|A \cap Y(end)|-(1-\alpha_j)q^2}. \tag{28.27}$$

(28.27) immediately gives that

$$(1+\gamma)^{|A \cap X(end)|-\eta q^2} \cdot (1-\gamma)^{|A \cap Y(end)|-(1-\eta)q^2} \leq$$

$$\leq \sum_{j\geq 0} \delta_j \cdot (1+\beta_j)^{|A \cap X(end)|-\alpha_j q^2} \cdot (1-\beta_j)^{|A \cap Y(end)|-(1-\alpha_j)q^2}, \tag{28.28}$$

holds for every $A \in \mathcal{F}$. Combining (28.28) and (28.13) we conclude that $R_{end} \leq 0$, which contradicts (28.26). This contradiction proves that, if (28.25) holds, then Maker forces an α-Discrepancy.

It remains to guarantee inequality (28.5)

$$R_{start} > \gamma^2 |V| \Delta_2 = \gamma^2 N^2 \Delta_2.$$

We recall (28.24)

$$R_{start} \geq \frac{1}{3} |\mathcal{F}| \cdot 2^{-(1-H(\eta))q^2};$$

so it suffices to check

$$\frac{1}{3} |\mathcal{F}| \cdot 2^{-(1-H(\eta))q^2} \geq N^2 \Delta_2, \tag{28.29}$$

where by (28.13) and (28.23)

$$\eta = \alpha + \frac{c_\alpha}{q} = \alpha + \frac{\sqrt{2\log 2 \cdot \alpha(1-\alpha)}}{q}. \tag{28.30}$$

Let $g(x) = 1 - H(x)$; by using (28.30) and the linear approximation

$$g(\eta) \approx g(\alpha) + (\eta - \alpha)g'(\alpha) = g(\alpha) + \frac{\sqrt{2\log 2 \cdot \alpha(1-\alpha)}}{q} \cdot \log_2\left(\frac{\alpha}{1-\alpha}\right), \quad (28.31)$$

in (28.29), we have

$$\frac{|\mathcal{F}|}{N^2} \geq 3\Delta_2 \cdot 2^{(1-H(\eta))q^2} =$$

$$= 3\Delta_2 \cdot 2^{q^2\left(\frac{\sqrt{2\log 2 \cdot \alpha(1-\alpha)}}{q} \cdot \log_2\left(\frac{\alpha}{1-\alpha}\right)\right)} =$$

$$= 3\Delta_2 \cdot 2^{q^2 + 2q \cdot c_0(\alpha)}, \quad (28.32)$$

where

$$c_0(\alpha) = \sqrt{\frac{\log 2 \cdot \alpha(1-\alpha)}{2}} \log_2\left(\frac{\alpha}{1-\alpha}\right). \quad (28.33)$$

(28.32) clearly follows from the (slightly stronger) inequality

$$\frac{|\mathcal{F}|}{N^2} \geq 3\Delta_2 \cdot 2^{(q + c_0(\alpha))^2}. \quad (28.34)$$

(28.34) is equivalent to the key inequality

$$\sqrt{\frac{1}{1 - H(\alpha)} \log_2\left(\frac{|\mathcal{F}|}{3\Delta_2 \cdot N^2}\right)} - c_0(\alpha) \geq q. \quad (28.35)$$

Now we are ready to prove the lower bounds in Theorem 9.1 for the two Square Lattices.

5. Case (a) in Theorem 9.1: Here $\mathcal{F}$ is the family of all $q \times q$ Aligned Square Lattices in $N \times N$

Then

$$|\mathcal{F}| \approx \frac{N^3}{3q} \text{ and } \Delta_2(\mathcal{F}) \leq \binom{q^2}{2} \leq \frac{q^4}{2},$$

so (28.35) is equivalent to

$$\sqrt{\frac{1}{1 - H(\alpha)} \log_2\left(\frac{2N^3}{9q^5 \cdot N^2}\right)} - c_0(\alpha) = \sqrt{\frac{1}{1 - H(\alpha)} \log_2 N} - c_0(\alpha) - o(1) \geq q. \quad (28.36)$$

Therefore, if q is the lower integral part of

$$\sqrt{\frac{1}{1 - H(\alpha)} \log_2 N} - c_0(\alpha) - o(1),$$

then Maker can always force an α-Discrepancy in the Aligned Square Lattice Game. This proves the lower bound in Theorem 9.1 (a).

6. Case (c) in Theorem 9.1: $\mathcal{F}$ is the family of all $q \times q$ tilted Square Lattices in $N \times N$

Then

$$|\mathcal{F}| \approx N^4 \text{ and } \Delta_2(\mathcal{F}) \le \binom{q^2}{2} \le \frac{q^4}{2},$$

so (28.35) is equivalent to

$$\sqrt{\frac{1}{1-H(\alpha)} \log_2\left(\frac{N^2}{N^2}\right)} - c_0(\alpha) - o(1) = \sqrt{\frac{2}{1-H(\alpha)} \log_2 N} - c_0(\alpha) - o(1) \ge q.$$

$$(28.36)$$

Therefore, if q is the lower integral part of

$$\sqrt{\frac{2}{1-H(\alpha)} \log_2 N} - c_0(\alpha) - o(1),$$

then Maker can always force an α-Discrepancy in the Tilted Square Lattice Game. This proves the lower bound in Theorem 9.1 (c).

29

Discrepancy games (II)

In the last section we proved Theorem 9.1 for the two Square Lattice games; how about the rest? We consider next:

1. Case (f) in Theorem 9.1: $\mathcal{F}$ is the family of all $q \times q$ parallelogram lattices in $N \times N$.

Then the Max Pair-Degree $\Delta_2 = \Delta_2(\mathcal{F}) = N^{2-o(1)}$ is very large, so Theorem 1.2 becomes inefficient. We can save the day by developing a discrepancy version of the *ad hoc* Higher Moment method of Section 23; in fact, we are going to combine the techniques of Sections 28 (preventing the "concentration on small discrepancy") and Sections 23 (involving the auxiliary hypergraphs $\mathcal{F}_{2*}^p$ and $\mathcal{F}_{2**}^p$).

Assume that we are in the middle of a play where Maker owns $X(i) = \{x_1, \ldots, x_i\}$ and Breaker owns $Y(i) = \{y_1, \ldots, x_i\}$. We want to prevent the "concentration on small discrepancy" by a technical trick of Section 28; by using the notation of Section 28, let

$$L_i = \left(T_{i,\gamma}(\mathcal{F}) - \sum_{j \geq 0} T_{i,\beta_j}(\mathcal{F}) \right) - \lambda \cdot T_{i,\gamma}(\mathcal{F}_{2*}^p), \qquad (29.1)$$

where parameter p will be specified later (note in advance that $p = 9$ will be a good choice) and λ is determined by the natural side condition

$$L_0 = \frac{1}{2} \left(T_{0,\gamma}(\mathcal{F}) - \sum_{j \geq 0} T_{0,\beta_j}(\mathcal{F}) \right),$$

which is equivalent to

$$\lambda = \frac{T_{0,\gamma}(\mathcal{F}) - \sum_{j \geq 0} T_{0,\beta_j}(\mathcal{F})}{2 T_{0,\gamma}(\mathcal{F}_{2*}^p)}. \qquad (29.2)$$

Similarly to Section 28, it will be enough to show that $L_{end} > 0$.

If Maker chooses that unoccupied $x_{i+1} = z$ for which the function

$$L_i(z) = \left(T_{i,\gamma}(\mathcal{F}; z) - \sum_{j\geq 0} T_{i,\beta_j}(\mathcal{F}; z) \right) - \lambda \cdot \gamma \cdot T_{i,\gamma}(\mathcal{F}^p_{2*}; z)$$

attains its maximum, then

$$L_{i+1} \geq L_i + L_i(x_{i+1}) - L_i(y_{i+1}) - \gamma^2 \cdot T_{i,\gamma}(\mathcal{F}; x_{i+1}, y_{i+1}) =$$

$$\geq L_i - \gamma^2 \cdot T_{i,\gamma}(\mathcal{F}; x_{i+1}, y_{i+1}). \tag{29.3}$$

The plan is to estimate $T_{i,\gamma}(\mathcal{F}; x_{i+1}, y_{i+1})$ from above. We follow the *ad hoc* technique of Section 23.

For $k = 0, 1, 2, \ldots, q^2$ and $l = 0, 1, 2, \ldots, q^2$ write

$$\mathcal{F}(x_{i+1}, y_{i+1}; k, l) = \{A \in \mathcal{F}: \{x_{i+1}, y_{i+1}\} \subset A, A \cap Y(i)$$

$$= \emptyset, |A \cap X(i)| = k, |A \cap Y(i)| = l\}$$

and $|\mathcal{F}(x_{i+1}, y_{i+1}; k, l)| = M_{k,l}$.

We develop a key inequality (see (29.5) below), which estimates the "one-sided discrepancy" $T(\mathcal{F}(i); x_{i+1}, y_{i+1})$ from above. Select an arbitrary $A_1 \in \mathcal{F}(x_{i+1}, y_{i+1}; k, l)$ and an arbitrary point $z \in A_1$ which is not on the $x_{i+1}y_{i+1}$-line (the $x_{i+1}y_{i+1}$-line means the straight line joining the two lattice points x_{i+1}, y_{i+1}). The property of 3-determinedness gives that there are $\leq \binom{q^2}{3} \leq q^6/6$ winning sets $A_2 \in \mathcal{F}$, which contain the non-collinear triplet $\{x_{i+1}, y_{i+1}, z\}$; as z runs, we obtain that there are $\leq q^8/6$ winning sets $A_2 \in \mathcal{F}$, which contain the pair $\{x_{i+1}, y_{i+1}\}$ such that the intersection $A_1 \cap A_2$ is not contained by the $x_{i+1}y_{i+1}$-line.

Assume that $M_{k,l} \geq p^2q^8$; then there are at least

$$\frac{M_{k,l} \cdot (M_{k,l} - q^8/6) \cdot (M_{k,l} - 2q^8/6) \cdots (M_{k,l} - (p-1)q^8/6)}{p!} \geq \left(\frac{M_{k,l}}{p} \right)^p$$

p-tuples

$$\{A_1, \ldots, A_p\} \in \binom{\mathcal{F}(x_{i+1}, y_{i+1}; k, l)}{p}$$

such that, deleting the points of the $x_{i+1}y_{i+1}$-line, the remaining parts of the p sets $A_1, \ldots, A_p$ become *disjoint*.

It follows that, if $M_{k,l} \geq p^2q^8$, then

$$T_{i,\gamma}(\mathcal{F}^p_{2*}) \geq \left(\frac{M_{k,l}}{p} \right)^p \cdot (1+\gamma)^{p(k-(1+\gamma)q^2/2)-(p-1)q} \cdot (1-\gamma)^{p(l-(1-\gamma)q^2/2)+(p-1)q}, \tag{29.4}$$

where the extra factors $(1+\gamma)^{-(p-1)q}$ and $(1-\gamma)^{(p-1)q}$ are due to a possible over-lapping on the $x_{i+1}y_{i+1}$-line (note that a straight line intersects a given $q \times q$ parallelogram lattice in at most q points). By (29.4)

$$p^2 q^8 + p \left(\frac{1+\gamma}{1-\gamma} \right)^q \cdot \left(T_{i,\gamma}(\mathcal{F}_{2*}^p) \right)^{1/p}$$

$$\geq M_{k,l} \cdot (1+\gamma)^{(k-(1+\gamma)q^2/2)} \cdot (1-\gamma)^{(l-(1-\gamma)q^2/2)},$$

which implies the **Key inequality**

$$T_{i,\gamma}(\mathcal{F}; x_{i+1}, y_{i+1}) = \sum_{k+l \leq q^2: \ k \geq 0, l \geq 0} M_{k,l} \cdot (1+\gamma)^{(k-(1+\gamma)q^2/2)} \cdot (1-\gamma)^{(l-(1-\gamma)q^2/2)}$$

$$\leq p^2 q^{12} + p q^4 \left(\frac{1+\gamma}{1-\gamma} \right)^q \cdot \left(T_{i,\gamma}(\mathcal{F}_{2*}^p) \right)^{1/p}. \tag{29.5}$$

By (29.3) and (29.5)

$$L_{i+1} \leq L_i - \gamma^2 \cdot \left(p^2 q^{12} + p q^4 \left(\frac{1+\gamma}{1-\gamma} \right)^q \cdot \left(T_{i,\gamma}(\mathcal{F}_{2*}^p) \right)^{1/p} \right). \tag{29.6}$$

2. Distinguishing two cases. Similarly to Section 23 we distinguish 2 cases: either there is or there is no index i such that

$$T_{i,\gamma}(\mathcal{F}_{2*}^p) > N^{2p-5} \cdot T_{0,\gamma}(\mathcal{F}_{2*}^p). \tag{29.7}$$

Case 1: There is no index i such that (29.7) holds.
Then by (29.6)–(29.7)

$$L_{end} = L_{N^2/2} \geq L_0 - \left(p^2 q^{12} + p q^4 \left(T_{i,\gamma}(\mathcal{F}_{2*}^p) \right)^{1/p} \right)$$

$$\geq L_0 - p^2 q^{12} \frac{N^2}{2} + p q^4 \frac{N^2}{2} \left(\frac{1+\gamma}{1-\gamma} \right)^q \cdot N^{2-\frac{5}{p}} \left(T_{i,\gamma}(\mathcal{F}_{2*}^p) \right)^{1/p}. \tag{29.8}$$

By (29.1)–(29.2)

$$L_0 = \frac{1}{2} |\mathcal{F}| \left((1+\gamma)^{-(1+\gamma)q^2/2} (1-\gamma)^{-(1-\gamma)q^2/2} - \sum_{j \geq 0} \delta_j \cdot (1+\beta_j)^{-(1+\beta_j)q^2/2} (1-\beta_j)^{-(1-\beta_j)q^2/2} \right)$$

$$= \frac{1}{2} |\mathcal{F}| \left(2^{-(1-H(\eta))q^2} - \sum_{j \geq 0} \delta_j \cdot 2^{-(1-H(\alpha_j))q^2} \right),$$

where $\eta = (1+\gamma)/2$

$$\alpha = \alpha_0, \quad \alpha_j = \frac{1+\beta_j}{2} \quad \text{and} \quad \alpha_j = \alpha - \frac{jc_\alpha}{q} \quad \text{for } j = 1, 2, 3, \ldots, \quad \eta = \alpha + \frac{c_\alpha}{q}; \tag{29.9}$$

here $c_\alpha = \sqrt{2 \log 2 \cdot \alpha(1-\alpha)}$, and $\delta_j \to 0$ very rapidly as follows

$$\delta_j = \left(\frac{1-\gamma}{1+\gamma} \right)^{(\eta - \alpha_j)q^2} = \left(\frac{1-\gamma}{1+\gamma} \right)^{c_\alpha(j+1)q}, \tag{29.10}$$

$j = 0, 1, 2, \ldots$.

By repeating the calculations in Section 28 we obtain the analogue of (28.24)

$$L_{start} = L_0 = \frac{1}{2}|\mathcal{F}|\left(2^{-(1-H(\eta))q^2} - \sum_{j\geq 0}\delta_j \cdot 2^{-(1-H(\alpha_j))q^2}\right)$$

$$\geq \frac{1}{6}|\mathcal{F}|\cdot 2^{-(1-H(\eta))q^2}. \tag{29.11}$$

Trivially

$$T_{0,\gamma}(\mathcal{F}_{2*}^p) \leq (N^2)^2 \cdot (N^2)^p \cdot (1+\gamma)^{-p\eta q^2+p\eta q} \cdot (1-\gamma)^{-p(1-\eta)q^2},$$

where the extra factor $(1+\gamma)^{p\eta q}$ comes from the possible overlappings on the intersection line $\bigcap_{i=1}^p A_i$-line (see (23.4)), which means the uniquely determined straight line containing the (≥ 2)-element collinear set $\bigcap_{i=1}^p A_i$. Thus we have

$$\left(T_{0,\gamma}(\mathcal{F}_{2*}^p)\right)^{1/p} \leq N^{2+\frac{4}{p}} \cdot (1+\gamma)^{-\eta q^2+\eta q} \cdot (1-\gamma)^{-(1-\eta)q^2}$$

$$= N^{2+\frac{4}{p}} \cdot (1+\gamma)^{\eta q} \cdot 2^{-(1-H(\eta))q^2}. \tag{29.12}$$

We have the analogue of (23.13)

$$T(\mathcal{F}) \geq \frac{N^6}{2^{14}(q-1)^4}2^{-(1-H(\eta))q^2},$$

so by (29.8), (29.11), and (29.12)

$$L_{end} \geq \frac{N^6}{2^{14}(q-1)^4}2^{-(1-H(\eta))q^2} - \frac{p^2}{2}q^{12}\cdot N^2$$

$$- pq^4\frac{N^2}{2}\left(\frac{1+\gamma}{1-\gamma}\right)^q \cdot N^{2-\frac{5}{p}}N^{2+\frac{4}{p}}\cdot (1+\gamma)^{\eta q}\cdot 2^{-(1-H(\eta))q^2}$$

$$= \left(\frac{N^6}{3\cdot 2^{15}(q-1)^4} - \frac{pq^4}{2}N^{6-\frac{1}{p}}\frac{(1+\gamma)^{2q}}{(1-\gamma)^q}\right)\cdot 2^{-(1-H(\eta))q^2} - \frac{p^2}{2}q^{12}\cdot N^2. \tag{29.13}$$

By choosing $2^{(1-H(\eta))q^2} = N^{4+o(1)}$, that is

$$q = (2+o(1))\sqrt{\frac{1}{1-H(\eta)}\log_2 N}, \tag{29.14}$$

(29.13) simplifies to

$$L_{end} \geq \frac{N^6}{2^{17}(q-1)^4}2^{-(1-H(\eta))q^2} - \frac{p^2}{2}q^{12}\cdot N^2. \tag{29.15}$$

Assume that $L_{end} > 0$. Then the choice of parameters δ_j, $j \geq 0$ prevents the "concentration on small discrepancy" exactly the same way as in Section 28, and implies the existence of an α-Discrepancy.

It remains to guarantee $L_{end} > 0$. By (29.15) and (29.10) we have to check

$$\frac{N^6}{2^{17}(q-1)^4} 2^{-(1-H(\eta))q^2} > \frac{p^2}{2} q^{12} \cdot N^2,$$

which is equivalent to

$$\frac{N^4}{3 \cdot 2^{16}(q-1)^4 q^{12}} > 2^{-(1-H(\eta))q^2}, \qquad (29.16)$$

where $\eta = \alpha + \frac{c_\alpha}{q}$ with $c_\alpha = \sqrt{2 \log 2 \cdot \alpha(1-\alpha)}$.

Repeating the calculations at the end of Section 28 we see that (29.16) follows from

$$2\sqrt{\frac{1}{1-H(\alpha)}} \log_2 N - c_0(\alpha) - o(1) \geq q, \qquad (29.17)$$

where

$$c_0(\alpha) = \sqrt{\frac{\log 2 \cdot \alpha(1-\alpha)}{2}} \log_2 \left(\frac{\alpha}{1-\alpha}\right). \qquad (29.18)$$

This completes Case 1. It remains to study Case 2.

Case 2: There is an index i such that (29.7) holds.

Again we show that Case 2 is impossible. Let $i = j_1$ be the first index such that (29.7) holds. Repeating the argument of Case 1 we can still save the following weaker version of (29.15): $L_{j_1} > 0$. By definition (see (29.1))

$$T_{j_1,\gamma}(\mathcal{F}) - \lambda \cdot T_{j_1,\gamma}(\mathcal{F}_{2*}^p) \geq L_{j_1} > 0,$$

so by (29.2) and (29.7)

$$T_{j_1,\gamma}(\mathcal{F}) > \lambda \cdot T_{j_1,\gamma}(\mathcal{F}_{2*}^p) > N^{2p-5} \cdot \lambda \cdot T_{0,\gamma}(\mathcal{F}_{2*}^p),$$

which by the side condition (29.2) equals

$$= N^{2p-5} \cdot \frac{1}{2}\left(T_{0,\gamma}(\mathcal{F}) - \sum_{j\geq 0} T_{0,\beta_j}(\mathcal{F})\right) = N^{2p-5} \cdot L_0. \qquad (29.19)$$

Combining this with (29.11) we have

$$T_{j_1,\gamma}(\mathcal{F}) > N^{2p-5} \cdot L_0 \geq N^{2p-5} \cdot \frac{1}{6}|\mathcal{F}| \cdot 2^{-(1-H(\eta))q^2}$$

$$= N^{2p-5} \cdot |\mathcal{F}| \cdot N^{-4+o(1)} = N^{2p-9+o(1)} \cdot |\mathcal{F}|. \qquad (29.20)$$

By definition

$$T_{j_1,\gamma}(\mathcal{F}) = \sum_{A\in\mathcal{F}} (1+\gamma)^{|A\cap X(j_1)|-\eta|A|} (1-\gamma)^{|A\cap Y(j_1)|-(1-\eta)|A|}. \qquad (29.21)$$

We can assume that

$$|A\cap X(j_1)| - |A\cap Y(j_1)| \leq 2\gamma|A|;$$

indeed, otherwise there is a lead $> 2\gamma|A_0| > 2\beta|A_0|$ in some $A_0 \in \mathcal{F}$ that Maker can keep for the rest of the play, and ends up with

$$> \frac{1+\gamma}{2}|A_0| > \frac{1+\beta}{2}|A_0| = \alpha|A_0|$$

points in some $A_0 \in \mathcal{F}$. That is, Maker can achieve an α-Discrepancy.

We can assume, therefore, that

$$|A \cap X(j_1)| - \left(\frac{1+\gamma}{1-\gamma}\right)|A \cap Y(j_1)| \le 2\gamma|A| \text{ holds for every } A \in \mathcal{F},$$

which implies the inequality

$$(1+\gamma)^{|A \cap X(j_1)| - (1+\gamma)|A|/2}(1-\gamma)^{|A \cap Y(j_1)| - (1-\gamma)|A|/2}$$

$$= (1+\gamma)^{|A \cap X(j_1)| - \frac{1+\gamma}{1-\gamma}|A \cap Y(j_1)|} \cdot \left((1+\gamma)^{-(1+\gamma)/2}(1-\gamma)^{-(1-\gamma)/2}\right)^{|A| - \frac{2}{1-\gamma}|A \cap Y(j_1)|}$$

$$\le (1+\gamma)^{2\gamma|A|} \cdot 2^{-(1-H(\eta))\left(|A| - \frac{2}{1-\gamma}|A \cap Y(j_1)|\right)}. \tag{29.22}$$

We distinguish two cases:

(1) If $|A \cap Y(j_1)| > (1-\gamma)|A|/2$, then $|A \cap X(j_1)| < (1+\gamma)|A|/2$, and so

$$(1+\gamma)^{|A \cap X(j_1)| - (1+\gamma)|A|/2}(1-\gamma)^{|A \cap Y(j_1)| - (1-\gamma)|A|/2} \le 1.$$

(2) If $|A \cap Y(j_1)| \le (1-\gamma)|A|/2$, then $|A| \ge \frac{2}{1-\gamma}|A \cap Y(j_1)|$, and so by (29.22) $|A \cap X(j_1)| < (1+\gamma)|A|/2$, and so

$$(1+\gamma)^{|A \cap X(j_1)| - (1+\gamma)|A|/2}(1-\gamma)^{|A \cap Y(j_1)| - (1-\gamma)|A|/2} \le (1+\gamma)^{2\gamma|A|}.$$

Summarizing, in both cases we have

$$(1+\gamma)^{|A \cap X(j_1)| - (1+\gamma)|A|/2}(1-\gamma)^{|A \cap Y(j_1)| - (1-\gamma)|A|/2} \le (1+\gamma)^{2\gamma|A|}. \tag{29.23}$$

We need the following elementary inequality.

Lemma 1: *For every* $1/2 \le x \le 1$

$$(2x)^{2x-1} \le 2^{2(1-H(x))}$$

where $H(x) = -x\log_2 x - (1-x)\log_2(1-x)$.

To **prove Lemma 1** we take binary logarithm of both sides

$$(2x-1)\log_2(2x) \le 2(1 + x\log_2 x + (1-x)\log_2(1-x)),$$

which is equivalent to

$$2x - \log_2 x - 2(1-x)\log_2(1-x) \le 3. \tag{29.24}$$

Substituting $y = 1 - x$ in (29.24), we have to check that

$$2y + 2y\log_2 y + \log_2(1-y) \ge -1 \text{ holds for all } 0 \le y \le 1/2. \tag{29.25}$$

But (29.25) is trivial from the fact that in the interval $0 \leq y \leq 1/2$ the function $f(y) = 2y + 2y \log_2 y + \log_2(1 - y)$ is monotone decreasing. Indeed, computing the first two derivatives

$$f'(y) = 2 + \frac{2 \log y}{\log 2} + \frac{2}{\log 2} - \frac{1}{\log 2(1 - y)},$$

$$f''(y) = \frac{2}{\log 2y} - \frac{1}{\log 2(1 - y)^2} = \frac{1}{\log 2} \frac{2 - 5y + 2y^2}{y(1 - y)^2} \geq 0$$

if $0 \leq y \leq 1/2$. So $f'(y)$ is monotone increasing, and

$$\max_{0 \leq y \leq 1/2} f'(y) = f'(1/2) = 0,$$

completing the proof of Lemma 1. $\square$

By choosing $2^{(1 - H(\eta))q^2} = N^{4 + o(1)}$, that is

$$q = (2 + o(1)) \sqrt{\frac{1}{1 - H(\eta)} \log_2 N},$$

and applying Lemma 1, we have

$$(1 + \gamma)^{2\gamma|A|} = (1 + \gamma)^{2\gamma q^2} \leq 2^{2(1 - H((1 + \gamma)/2))q^2}$$

$$= 2^{2(1 - H(\eta))q^2} = N^{8 + o(1)},$$

so by (29.21) and (29.23)

$$T_{j_1, \gamma}(\mathcal{F}) \leq |\mathcal{F}| \cdot N^{8 + o(1)}. \tag{29.26}$$

On the other hand, by (29.20)

$$T_{j_1, \gamma}(\mathcal{F}) > N^{2p - 9 + o(1)} \cdot |\mathcal{F}|,$$

which contradicts (29.26) if $p = 9$. This contradiction proves that Case 2 is impossible with $p = 9$.

Returning to Case 1, we have just proved the following result: if q is the lower integral part of

$$2 \sqrt{\frac{1}{1 - H(\alpha)} \log_2 N} - c_0(\alpha) - o(1),$$

then Maker can always force an α-Discrepancy in the Parallelogram Lattice Game. This proves the lower bound in Theorem 9.1 (f).

3. Case (b) in Theorem 9.1: $\mathcal{F}$ is the family of all $q \times q$ aligned rectangle lattices in $N \times N$

We proceed similarly to Case (f) ("Parallelogram Lattice Game"), the only minor difference is to work with the *other* auxiliary hypergraph $\mathcal{F}^p_{2**}$, defined in (23.24) (instead of $\mathcal{F}_{2*}^{p}$, defined in (23.4)). This way we obtain the following result: if q is the lower integral part of

$$\sqrt{\frac{2}{1-H(\alpha)}} \log_2 N - c_0(\alpha) - o(1),$$

then Maker can always force an α-Discrepancy in the Parallelogram Lattice Game. This proves the lower bound in Theorem 9.1 (b).

The rest of the lattice games go similarly.

Summarizing, the Discrepancy Problem remains unsolved for the Clique Game, but has an almost perfect solution for the Lattice Games (see Theorem 9.1), where the only minor "defect" is the appearance of an extra additive constant $c_0(\alpha)$, which – luckily(!) – tends to zero as $\alpha \to 1$.

30

Biased Games (I): Biased Meta-Conjecture

Section 17 was a triumph for the biased case: we could successfully generalize a fair (1:1) game (Theorem 17.1) to the general $(p:q)$ biased game (Theorem 17.5). This kind of a "triumph" is very rare; the following Sections 30–33 are more like "defeats"; we discuss sporadic results where the upper and lower bounds rarely coincide.

1. When do we have a chance for an exact solution? We begin our discussion of the biased games with recalling the two simplest – but still fundamentally important – results in the book: the two "linear" criterions.

Linear Weak Win criterion (Theorem 1.2): If hypergraph $\mathcal{F}$ is n-uniform, Almost Disjoint, and

$$\frac{|\mathcal{F}|}{|V|} \left(\frac{1}{2}\right)^n > \frac{1}{8}, \tag{30.1}$$

then playing the (1:1) game on $\mathcal{F}$ Maker, as the first player, can force a Weak Win.

Linear Strong Draw criterion (special case of Theorem 1.4): If hypergraph $\mathcal{F}$ is n-uniform and

$$|\mathcal{F}| \left(\frac{1}{2}\right)^n < \frac{1}{2}, \tag{30.2}$$

then playing the (1:1) game on $\mathcal{F}$ Breaker, as the second player, can force a Strong Draw. Both criterions contain the same factor "$\left(\frac{1}{2}\right)^n$"; the ratio $\frac{1}{2} = \frac{1}{1+1}$ is clearly explained by the (1:1) play.

Next consider the $(m:b)$ play, where Maker (as the first player) takes m points and Breaker (the second player) takes b points per move. We know the following biased versions of (30.1)–(30.2).

Biased Weak Win criterion (Theorem 2.2): If hypergraph $\mathcal{F}$ is n-uniform, Almost Disjoint, and

$$\frac{|\mathcal{F}|}{|V|} \left(\frac{m}{m+b}\right)^n > \frac{m^2 b^2}{(m+b)^3}, \tag{30.3}$$

then playing the $(m : b)$ game on $\mathcal{F}$ Maker (the first player) can force a Weak Win.

Biased Strong Draw criterion (Theorem 20.1): If hypergraph $\mathcal{F}$ is n-uniform and

$$|\mathcal{F}| \left((1+b)^{-1/m}\right)^n < \frac{1}{1+b}, \tag{30.4}$$

then playing the $(m : b)$ game on $\mathcal{F}$ Breaker (the second player) can force a Strong Draw.

Notice that (30.2) and (30.4) are sharp. Indeed, for simplicity, assume that n is divisible by m, and consider the regular "tree" of n/m levels, where every vertex represents m points, and the out-degree is $b+1$ (except at the bottom level):

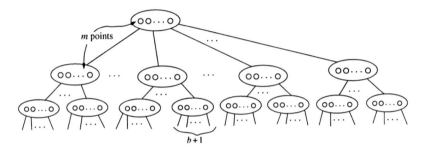

The full-length branches of the "tree" represent the winning sets of hypergraph $\mathcal{F}$; $\mathcal{F}$ is clearly $m \cdot (n/m) = n$-uniform and has $(1+b)^{(n/m)-1}$ winning sets. By "stepping down" on the tree the first player can always occupy a full-length branch, proving that this $\mathcal{F}$ is an Economical Winner.

It is good to know that this Economical system is the *only* extremal system for Theorem (30.4) with $b \geq 2$. This result was proved by Kruczek and Sundberg; the proof is long and complicated.

In sharp contrast with the biased case $b \geq 2$, in the fair (1:1) game Theorem (30.2) has many other extremal systems. Here is an extremal system which is not an Economical Winner: it is a 4-uniform hypergraph with $2^3 = 8$ winning sets, where Maker (the first player) needs at least 5 turns to win

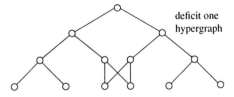

deficit one hypergraph

The players take vertices, and the winning sets are the 8 full-length branches. The reader is challenged to check that this is not an Economical Winner.

Comparing the pair (30.1)–(30.2) with (30.3)–(30.4) we see a major difference: in (30.3)–(30.4) the critical factors

$$\left(\frac{m}{m+b}\right)^n \quad \text{and} \quad \left((1+b)^{-1/m}\right)^n$$

are *different*. In fact, we have a strict inequality

$\frac{m}{m+b} < (1+b)^{-1/m}$, which holds for all pairs $(m:b)$ except the special case $(1:b)$

when we have equality. Indeed, if $m \geq 2$, then by the binomial theorem

$$\left(\frac{m+b}{m}\right)^m = \left(1+\frac{b}{m}\right)^m = 1+\binom{m}{1}\frac{b}{m}+\binom{m}{2}\left(\frac{b}{m}\right)^2+\cdots > 1+b.$$

The ultimate reason why we can determine the exact value of the Achievement and Avoidance Numbers for the usual $(1:1)$ play is the appearance of the same factor "$\left(\frac{1}{2}\right)^n$" (where the fraction $\frac{1}{2} = \frac{1}{1+1}$ comes from the $(1:1)$ play).

The exponentially large difference between

$$\left(\frac{m}{m+b}\right)^n \quad \text{and} \quad \left((1+b)^{-1/m}\right)^n \quad \text{when} \quad m \geq 2$$

prevents us from doing any kind of straightforward adaptation of the machinery of the $(1:1)$ case to the general $(m:b)$ biased case. The $(1:s)$ case is the only solution of the equation $\frac{m}{m+b} = (1+b)^{-1/m}$. The importance of the ratio $\frac{m}{m+b}$ is obvious: it is the probability that a given point of the board is taken by the first player (Maker).

With the available tools (30.3)–(30.4), the best that we can *hope* for is to find the exact solution for the $(1:b)$ play (i.e. when Maker, the first player, takes 1 point and Breaker, the second player, takes b points per move; $b \geq 2$ can be arbitrarily large).

Question 30.1 *Can we prove* **exact results** *for the underdog* $(1:b)$ *Achievement Game* $(b \geq 2)$ *where Breaker is the topdog?*

For the *Avoidance* version the "hopeful" biased case is the $(a:1)$ play, where Avoider takes $a (\geq 2)$ points and Forcer takes 1 point per move. To justify this claim, we formulate first the simplest $(1:1)$ Avoidance version of the Erdős–Selfridge Theorem.

(1:1) Avoidance Erdős–Selfridge: *If hypergraph* $\mathcal{F}$ *is n-uniform and* $|\mathcal{F}| < 2^n$, *then playing the $(1:1)$ game on* $\mathcal{F}$, *Avoider, as the first player, can always avoid occupying a whole winning set* $A \in \mathcal{F}$.

The proof of the "$(1:1)$ Avoidance Erdős–Selfridge" is exactly the same as that of the Erdős–Selfridge Theorem itself, the only minor difference is that Avoider *minimizes* the potential function (where in Section 10 Breaker was *maximizing* the potential function).

2. Avoidance Erdős–Selfridge: the (a:1) play (the rest is hopeless!). It is not too difficult to extend the "(1:1) Avoidance Erdős–Selfridge" for the $(a:1)$ play with arbitrary $a \geq 2$ (Forcer is the underdog), but to find a useful criterion for the general $(a:f)$ case with $f \geq 2$ remains unsolved! Why is the general case so hard?

The "$(a:1)$ Avoidance Erdős–Selfridge" for the $a \geq 2$ case was already formulated at the end of Section 20 without proof, see Theorem 20.4. Here we recall the statement and give two different proofs. The second proof is longer, but it supplies an extra information that we need later (to prove Theorem 30.1).

Theorem 20.4 *Let $a \geq 2$ be an arbitrary integer, and let $\mathcal{F}$ be an n-uniform hypergraph with size*

$$|\mathcal{F}| < \left(\frac{a+1}{a}\right)^n.$$

Then, playing the $(a:1)$ Avoidance Game on $\mathcal{F}$, where Forcer is the underdog, Avoider (as the first player) has a winning strategy.

First Proof. This is the "natural" proof: a straightforward adaptation of the proof technique of Theorem 20.1. Assume we are in the middle of a play, Avoider (the first player) owns the points

$$X(i) = \left\{ x_1^{(1)}, \ldots, x_1^{(a)}, x_2^{(1)}, \ldots, x_2^{(a)}, \ldots, x_i^{(1)}, \ldots, x_i^{(a)} \right\}, \text{ and } Y(i) = \{y_1, y_2, \ldots, y_i\}$$

is the set of Forcer's points. The question is how to choose Avoider's $(i+1)$st move $x_{i+1}^{(1)}, \ldots, x_{i+1}^{(a)}$. Write

$$\mathcal{F}(i) = \{A \setminus X(i) : A \in \mathcal{F}, \ A \cap Y(i) = \emptyset\},$$

i.e. $\mathcal{F}(i)$ is the family of the unoccupied parts of the "survivors." In the $(a:1)$ play, it is natural to switch from the Power-of-Two Scoring System to the Power-of-$(\frac{a+1}{a})$ Scoring System (since the $(a:1)$ play gives the ratio $\frac{a+1}{a}$)

$$T(\mathcal{H}) = \sum_{B \in \mathcal{H}} \left(\frac{a+1}{a}\right)^{-|B|} \text{ and } T(\mathcal{H}; u_1, \ldots, u_m) = \sum_{B \in \mathcal{H} : \{u_1, \ldots, u_m\} \subset B} \left(\frac{a+1}{a}\right)^{-|B|}.$$

For every $1 \leq j \leq a$, write

$$X(i, j) = X(i) \cup \{x_{i+1}^{(1)}, \ldots, x_{i+1}^{(j)}\}$$

and

$$\mathcal{F}(i, j) = \{A \setminus X(i, j) : A \in \mathcal{F}, \ A \cap Y(i) = \emptyset\}.$$

The effect of the $(i+1)$st moves $x_{i+1}^{(1)}, \ldots, x_{i+1}^{(a)}$ and y_{i+1} can be described by an inequality as follows

$$T(\mathcal{F}(i+1)) \leq T(\mathcal{F}(i)) + \frac{1}{a}T(\mathcal{F}(i); x_{i+1}^{(1)}) + \frac{1}{a}T(\mathcal{F}(i,1); x_{i+1}^{(2)}) +$$

$$+ \frac{1}{a}T(\mathcal{F}(i,2); x_{i+1}^{(3)}) + \cdots + \frac{1}{a}T(\mathcal{F}(i, a-1); x_{i+1}^{(a)}) - T(\mathcal{F}(i,a); y_{i+1}).$$
(30.5)

Notice that (30.5) is an inequality due to the effect of the sets $B \in \mathcal{F}(i)$, which (1) contain y_{i+1} and also (2) have a non-empty intersection with the set $\{x_{i+1}^{(1)}, x_{i+1}^{(2)}, \ldots, x_{i+1}^{(a)}\}$.

Let $x_{i+1}^{(1)}$ be defined as the $z \in V \setminus (X(i) \cup Y(i))$ (where V is the union set of hypergraph $\mathcal{F}$) for which the function $T(\mathcal{F}(i); z)$ attains its minimum; let $x_{i+1}^{(2)}$ be defined as the $z \in V \setminus (X(i,1) \cup Y(i))$ for which the function $T(\mathcal{F}(i,1); z)$ attains its minimum; let $x_{i+1}^{(3)}$ be defined as the $z \in V \setminus (X(i,2) \cup Y(i))$ for which the function $T(\mathcal{F}(i,2); z)$ attains its minimum; and so on. We have the inequality

$$T(\mathcal{F}(i); x_{i+1}^{(1)}) \leq T(\mathcal{F}(i); y_{i+1}) \leq T(\mathcal{F}(i,a); y_{i+1}),$$

where the first half is due to the minimum property of $x_{i+1}^{(1)}$, and the second half is trivial from the definition. Similarly we have

$$T(\mathcal{F}(i,1); x_{i+1}^{(2)}) \leq T(\mathcal{F}(i,1); y_{i+1}) \leq T(\mathcal{F}(i,a); y_{i+1}),$$

$$T(\mathcal{F}(i,2); x_{i+1}^{(3)}) \leq T(\mathcal{F}(i,2); y_{i+1}) \leq T(\mathcal{F}(i,a); y_{i+1}),$$

and so on. Substituting these inequalities back to (30.5), we obtain

$$T(\mathcal{F}(i+1)) \leq T(\mathcal{F}(i)) + \frac{1}{a}T(\mathcal{F}(i,a); y_{i+1}) + \frac{1}{a}T(\mathcal{F}(i,a); y_{i+1})$$

$$+ \frac{1}{a}T(\mathcal{F}(i,a); y_{i+1}) + \cdots + \frac{1}{a}T(\mathcal{F}(i,a); y_{i+1}) - T(\mathcal{F}(i,a); y_{i+1})$$

$$= T(\mathcal{F}(i)),$$

which is the key decreasing property. The rest of the proof is standard. □

Second Proof. The starting point of the first proof was inequality (30.5); here the starting point is an equality (see (30.6) below).

To illustrate the idea on the simplest case, we begin with the (2:1) play, i.e. $a = 2$. Assume that Avoider (the first player) owns the points

$$X(i) = \left\{ x_1^{(1)}, x_1^{(2)}, x_2^{(1)}, x_2^{(2)}, \ldots, x_i^{(1)}, x_i^{(2)} \right\}, \text{ and } Y(i) = \{y_1, y_2, \ldots, y_i\}$$

is the set of Forcer's points. The question is how to choose Avoider's $(i+1)$st move $x_{i+1}^{(1)}, x_{i+1}^{(2)}$. As usual write

$$\mathcal{F}(i) = \{A \setminus X(i) : A \in \mathcal{F}, A \cap Y(i) = \emptyset\}.$$

In the (2:1) play it is natural to switch from the Power-of-Two Scoring System to the Power-of-$(\frac{3}{2})$ Scoring System

$$T(\mathcal{H}) = \sum_{B \in \mathcal{H}} \left(\frac{3}{2}\right)^{-|B|} \quad \text{and} \quad T(\mathcal{H}; u_1, \ldots, u_m) = \sum_{B \in \mathcal{H}: \{u_1, \ldots, u_m\} \subset B} \left(\frac{3}{2}\right)^{-|B|}.$$

Let $T_i = T(\mathcal{F}(i))$; the effect of the $(i+1)$st moves $x_{i+1}^{(1)}, x_{i+1}^{(2)}$ and y_{i+1} goes as follows

$$T_{i+1} = T_i + \frac{1}{2} T(\mathcal{F}(i); x_{i+1}^{(1)}) + \frac{1}{2} T(\mathcal{F}(i); x_{i+1}^{(2)})$$

$$+ \frac{1}{4} T(\mathcal{F}(i); x_{i+1}^{(1)}, x_{i+1}^{(2)}) - T(\mathcal{F}(i); y_{i+1}) - \frac{1}{2} T(\mathcal{F}(i); x_{i+1}^{(1)}, y_{i+1})$$

$$- \frac{1}{2} T(\mathcal{F}(i); x_{i+1}^{(2)}, y_{i+1}) - \frac{1}{4} T(\mathcal{F}(i); x_{i+1}^{(1)}, x_{i+1}^{(2)}, y_{i+1}). \tag{30.6}$$

Identity (30.6) is rather long, but the underlying pattern is very simple: it is described by the expansion of the product

$$\left(1 + \frac{1}{2} x^{(1)}\right)\left(1 + \frac{1}{2} x^{(2)}\right)(1 - y) - 1 = \frac{1}{2} x^{(1)} + \frac{1}{2} x^{(2)}$$

$$+ \frac{1}{4} x^{(1)} x^{(2)} - y - \frac{1}{2} x^{(1)} y - \frac{1}{2} x^{(2)} y - \frac{1}{4} x^{(1)} x^{(2)} y. \tag{30.7}$$

For every unoccupied pair $\{u_1, u_2\} \in V \setminus (X(i) \cup Y(i))$ (where V is the union set of hypergraph $\mathcal{F}$), define the sum

$$f(u_1, u_2) = T(\mathcal{F}(i); u_1) + T(\mathcal{F}(i); u_2) + T(\mathcal{F}(i); u_1, u_2). \tag{30.8}$$

By using function (30.8) we are ready to explain how Avoider chooses his $(i+1)$st move $x_{i+1}^{(1)}, x_{i+1}^{(2)}$. Let $\{x_{i+1}^{(1)}, x_{i+1}^{(2)}\}$ be that unoccupied pair $\{u_1, u_2\}$ for which the the function $f(u_1, u_2)$ attains its *minimum*. Since y_{i+1} is selected *after* $\{x_{i+1}^{(1)}, x_{i+1}^{(2)}\}$, we have

$$f(x_{i+1}^{(1)}, x_{i+1}^{(2)}) \leq f(x_{i+1}^{(1)}, y_{i+1}) \quad \text{and} \quad f(x_{i+1}^{(1)}, x_{i+1}^{(2)}) \leq f(x_{i+1}^{(2)}, y_{i+1}). \tag{30.9}$$

By using notation (30.8) we can rewrite equality (30.6) as follows

$$T_{i+1} = T_i + f(x_{i+1}^{(1)}, x_{i+1}^{(2)}) - \frac{1}{2} f(x_{i+1}^{(1)}, y_{i+1}) - \frac{1}{2} f(x_{i+1}^{(2)}, y_{i+1})$$

$$- \frac{3}{4} T(\mathcal{F}(i); x_{i+1}^{(1)}, x_{i+1}^{(2)}) - \frac{1}{4} T(\mathcal{F}(i); x_{i+1}^{(1)}, x_{i+1}^{(2)}, y_{i+1}). \tag{30.10}$$

Combining (30.9) and (30.10) we obtain the *decreasing* property $T_{i+1} \leq T_i$, which implies $T_{end} \leq T_{start}$. By hypothesis $T_{start} = T(\mathcal{F}) < 1$, so $T_{end} < 1$. This proves that at the end of the play Avoider cannot own a whole winning set; indeed, otherwise $T_{end} \geq (3/2)^0 = 1$, a contradiction. This solves the special case $a = 2$.

A similar idea works for every $a \geq 2$. We apply the Power-of-$(\frac{a+1}{a})$ Scoring System (since the $(a{:}1)$ play gives the ratio $\frac{a+1}{a}$)

$$T(\mathcal{H}) = \sum_{B \in \mathcal{H}} \left(\frac{a+1}{a} \right)^{-|B|} \quad \text{and} \quad T(\mathcal{H}; u_1, \ldots, u_m) = \sum_{B \in \mathcal{H}: \{u_1, \ldots, u_m\} \subset B} \left(\frac{a+1}{a} \right)^{-|B|}.$$

Let $T_i = T(\mathcal{F}(i))$. The analogue of (30.7) is the expansion

$$\left(1 + \frac{1}{a} x^{(1)}\right)\left(1 + \frac{1}{a} x^{(2)}\right) \cdots \left(1 + \frac{1}{a} x^{(a)}\right)(1 - y) - 1 = \frac{1}{a} \sum_{j=1}^{a} x^{(j)}$$

$$+ \frac{1}{a^2} \sum_{1 \leq j_1 < j_2 \leq a} x^{(j_1)} x^{(j_2)} + \frac{1}{a^3} \sum_{1 \leq j_1 < j_2 < j_3 \leq a} x^{(j_1)} x^{(j_2)} x^{(j_3)} + \cdots$$

$$- y - \frac{y}{a} \sum_{j=1}^{a} x^{(j)} - \frac{y}{a^2} \sum_{1 \leq j_1 < j_2 \leq a} x^{(j_1)} x^{(j_2)}$$

$$- \frac{y}{a^3} \sum_{1 \leq j_1 < j_2 < j_3 \leq a} x^{(j_1)} x^{(j_2)} x^{(j_3)} + \cdots. \tag{30.11}$$

We are ready to explain how Avoider chooses his $(i+1)$st move $x_{i+1}^{(1)}, \ldots, x_{i+1}^{(a)}$. For every unoccupied a-tuple $\{u_1, \ldots, u_a\} \in V \setminus (X(i) \cup Y(i))$, define the sum

$$f(u_1, \ldots, u_a) = \frac{1}{a} \sum_{j=1}^{a} T(\mathcal{F}(i); u_j) + \frac{1}{a(a-1)} \sum_{1 \leq j_1 < j_2 \leq a} T(\mathcal{F}(i); u_{j_1}, u_{j_2})$$

$$+ \frac{1}{a^2(a-2)} \sum_{1 \leq j_1 < j_2 < j_3 \leq a} T(\mathcal{F}(i); u_{j_1}, u_{j_2}, u_{j_3})$$

$$+ \frac{1}{a^3(a-3)} \sum_{1 \leq j_1 < j_2 < j_3 < j_4 \leq a} T(\mathcal{F}(i); u_{j_1}, u_{j_2}, u_{j_3}, u_{j_4}) + \cdots. \tag{30.12}$$

Let $\{x_{i+1}^{(1)}, \ldots, x_{i+1}^{(a)}\}$ be that unoccupied a-tuple $\{u_1, \ldots, u_a\}$ for which the the function $f(u_1, \ldots, u_a)$ attains its *minimum*. Since y_{i+1} is selected *after* $\{x_{i+1}^{(1)}, \ldots, x_{i+1}^{(a)}\}$, we have

$$f(x_{i+1}^{(1)}, \ldots, x_{i+1}^{(a)}) \leq f(y_{i+1}, x_{i+1}^{(2)}, \ldots, x_{i+1}^{(a)}),$$

$$f(x_{i+1}^{(1)}, \ldots, x_{i+1}^{(a)}) \leq f(x_{i+1}^{(1)}, y_{i+1}, x_{i+1}^{(3)}, \ldots, x_{i+1}^{(a)}),$$

$$f(x_{i+1}^{(1)}, \ldots, x_{i+1}^{(a)}) \leq f(x_{i+1}^{(1)}, x_{i+1}^{(2)}, y_{i+1}, x_{i+1}^{(4)}, \ldots, x_{i+1}^{(a)}),$$

and so on, where the last one is

$$f(x_{i+1}^{(1)}, \ldots, x_{i+1}^{(a)}) \leq f(x_{i+1}^{(1)}, \ldots, x_{i+1}^{(a-1)}, y_{i+1}).$$

Adding up these inequalities, we get a new inequality, which has the "short form"

$$\sum_{j=1}^{a} x^{(j)} + \frac{1}{a-1} \sum_{1 \le j_1 < j_2 \le a} x^{(j_1)} x^{(j_2)} + \frac{1}{a(a-2)} \sum_{1 \le j_1 < j_2 < j_3 \le a} x^{(j_1)} x^{(j_2)} x^{(j_3)} + \cdots$$

$$\le \frac{a-1}{a} \sum_{j=1}^{a} x^{(j)} + y + \frac{a-2}{a(a-1)} \sum_{1 \le j_1 < j_2 \le a} x^{(j_1)} x^{(j_2)} + \frac{a-1}{a(a-1)} y \sum_{j=1}^{a} x^{(j)}$$

$$+ \frac{a-3}{a^2(a-2)} \sum_{1 \le j_1 < j_2 < j_3 \le a} x^{(j_1)} x^{(j_2)} x^{(j_3)} + \frac{a-2}{a^2(a-2)} y \sum_{1 \le j_1 < j_2 \le a} x^{(j_1)} x^{(j_2)} + \cdots,$$

which is equivalent to

$$f(x_{i+1}^{(1)}, \ldots, x_{i+1}^{(a)}) - \frac{1}{a} f(y_{i+1}, x_{i+1}^{(2)}, \ldots, x_{i+1}^{(a)})$$

$$- \frac{1}{a} f(x_{i+1}^{(1)}, y_{i+1}, x_{i+1}^{(3)}, \ldots, x_{i+1}^{(a)}) - \frac{1}{a} f(x_{i+1}^{(1)}, x_{i+1}^{(2)}, y_{i+1}, x_{i+1}^{(4)}, \ldots, x_{i+1}^{(a)}) - \cdots$$

$$- \frac{1}{a} f(x_{i+1}^{(1)}, \ldots, x_{i+1}^{(a-1)}, y_{i+1})$$

$$= \frac{1}{a} \sum_{j=1}^{a} x^{(j)} - y + \frac{2}{a(a-1)} \sum_{1 \le j_1 < j_2 \le a} x^{(j_1)} x^{(j_2)} - \frac{y}{a} \sum_{j=1}^{a} x^{(j)} +$$

$$+ \frac{3}{a^2(a-2)} \sum_{1 \le j_1 < j_2 < j_3 \le a} x^{(j_1)} x^{(j_2)} x^{(j_3)} - \frac{y}{a^2} \sum_{1 \le j_1 < j_2 \le a} x^{(j_1)} x^{(j_2)} \pm \cdots \le 0. \tag{30.13}$$

Combining (30.11)–(30.13) we obtain an equality that will be used later (in the proof of Theorem 30.1)

$$T(\mathcal{F}(i+1)) = T(\mathcal{F}(i)) + f(x_{i+1}^{(1)}, \ldots, x_{i+1}^{(a)}) - \frac{1}{a} f(y_{i+1}, x_{i+1}^{(2)}, \ldots, x_{i+1}^{(a)})$$

$$- \frac{1}{a} f(x_{i+1}^{(1)}, y_{i+1}, x_{i+1}^{(3)}, \ldots, x_{i+1}^{(a)}) - \frac{1}{a} f(x_{i+1}^{(1)}, x_{i+1}^{(2)}, y_{i+1}, x_{i+1}^{(4)}, \ldots, x_{i+1}^{(a)}) - \cdots$$

$$- \frac{1}{a} f(x_{i+1}^{(1)}, \ldots, x_{i+1}^{(a-1)}, y_{i+1}) - \left(\frac{2}{a(a-1)} - \frac{1}{a^2} \right) \sum_{1 \le j_1 < j_2 \le a} T(\mathcal{F}(i); x_{i+1}^{(j_1)}, x_{i+1}^{(j_2)})$$

$$- \left(\frac{3}{a^2(a-2)} - \frac{1}{a^3} \right) \sum_{1 \le j_1 < j_2 < j_3 \le a} T(\mathcal{F}(i); x_{i+1}^{(j_1)}, x_{i+1}^{(j_2)}, x_{i+1}^{(j_3)})$$

$$- \left(\frac{4}{a^3(a-3)} - \frac{1}{a^4} \right) \sum_{1 \le j_1 < j_2 < j_3 < j_4 \le a} T(\mathcal{F}(i); x_{i+1}^{(j_1)}, x_{i+1}^{(j_2)}, x_{i+1}^{(j_3)}, x_{i+1}^{(j_4)}) - \cdots$$

$$- \frac{1}{a^a} T(\mathcal{F}(i); x_{i+1}^{(1)}, \ldots, x_{i+1}^{(a)}, y_{i+1}). \tag{30.14}$$

Notice that the following coefficients in (30.14) are all positive

$$\left(\frac{2}{a(a-1)} - \frac{1}{a^2}\right) > 0, \quad \left(\frac{3}{a^2(a-2)} - \frac{1}{a^3}\right) > 0, \quad \left(\frac{4}{a^3(a-3)} - \frac{1}{a^4}\right) > 0, \dots$$

Combining (30.13)–(30.14) we obtain the crucial *decreasing property* $T_{i+1} \le T_i$, and Theorem 20.4 follows. $\square$

Theorem 20.4 is complemented by the Avoidance analogue of (30.3).

Biased Forcer's Win criterion: If hypergraph $\mathcal{F}$ is n-uniform, Almost Disjoint, and

$$\frac{|\mathcal{F}|}{|V|} \left(\frac{a}{a+f}\right)^n > \frac{a^2 f^2}{(a+f)^3}, \tag{30.15}$$

then playing the $(a{:}f)$ Avoidance game on $\mathcal{F}$, Forcer (the second player) can force Avoider to occupy a whole $A \in \mathcal{F}$.

The proof of (30.15) is very similar to that of (30.3), the only minor difference is that Forcer minimizes the potential function (where Maker was maximizing the potential function).

In the $(a{:}1)$ Avoidance Game (where Avoider takes a points and Forcer takes 1 point per move) both Theorem 20.4 and the Biased Forcer's Win criterion (30.15) have the same critical factor $\left(\frac{a}{a+1}\right)^n$, where $\frac{a}{a+1}$ is the probability that a given point is taken by Avoider. This coincidence gives us a hope to answer the following:

Question 30.2 *Can we prove* **exact results** *for the $(a{:}1)$ Avoidance Game $(a \ge 2)$, where Avoider is the topdog?*

Questions 30.1 and 30.2 lead to sporadic cases when we have exact solutions in the biased game. We will return to these questions later in Sections 32–33, but first we make a detour and study the opposite case

3. When the exact solution seems to be out of reach. How about the $(2{:}1)$ Achievement Game (where Breaker is the underdog)? The ratio $(2{:}1)$ does not belong to the class $(1{:}b)$ mentioned in Question 30.1 (see the beginning of this section), due to the fact that we cannot prove the "blocking part" of the Biased Meta-Conjecture. Can we at least do the "building part" of the Biased Meta-Conjecture? The answer is "yes," but it doesn't make us perfectly happy. Why is that? The good news is that for the $(2{:}1)$ play we can prove the perfect analogue of the Advanced Weak Win Criterion (Theorem 24.2), which implies one direction of the $(2{:}1)$ Meta-Conjecture; the bad news is that the $(2{:}1)$ Meta-Conjecture may fail to give the truth! Indeed, in the $(2{:}1)$ play (where Maker is the topdog) there is a trivial, alternative way for Maker to occupy a whole winning set, and this trivial way – described by the Cheap Halving Lemma below – can occasionally beat the sophisticated (2:1) Advanced Weak Win Criterion, disproving the (2:1) Meta-Conjecture.

Cheap Halving Lemma: *Let $\mathcal{H}$ be an n-uniform hypergraph which contains $\geq 2^{n-2}$ pairwise disjoint sets. Then playing the (2:1) game on $\mathcal{H}$, topdog Maker (as the first player) can occupy a whole $A \in \mathcal{H}$ in $\leq 2^{n-2}$ moves.*

We leave the easy proof of the Cheap Halving Lemma to the reader.

We will discuss the applications of the Cheap Halving Lemma later. First we prove the (2:1) Advanced Weak Win Criterion. The proof is a combination of the proof technique of Section 24 with identitities (3.10) (the special case $a = 2$) and (30.14) ("the general case").

Assume we are in the middle of a play, Maker (the first player) owns the points

$$X(i) = \left\{x_1^{(1)}, x_1^{(2)}, x_2^{(1)}, x_2^{(2)}, \ldots, x_i^{(1)}, x_i^{(2)}\right\}, \text{ and } Y(i) = \{y_1, y_2, \ldots, y_i\}$$

is the set of Breaker's points. The question is how to choose Maker's $(i+1)$st move $x_{i+1}^{(1)}, x_{i+1}^{(2)}$. Write

$$\mathcal{F}(i) = \{A \setminus X(i): \ A \in \mathcal{F}, \ A \cap Y(i) = \emptyset\},$$

i.e. $\mathcal{F}(i)$ is the family of the unoccupied parts of the "survivors." We use the usual Potential Function (see the proof of Theorem 24.2)

$$L_i = T(\mathcal{F}(i)) - \lambda \cdot T(\mathcal{F}_2^p(i)),$$

and also

$$L_i(u_1, \ldots, u_m) = T(\mathcal{F}(i); u_1, \ldots, u_m) - \lambda \cdot T(\mathcal{F}_2^p(i); u_1, \ldots, u_m),$$

where the Power-of-Two Scoring System is replaced by the Power-of-$(3/2)$ Scoring System

$$T(\mathcal{H}) = \sum_{B \in \mathcal{H}} \left(\frac{3}{2}\right)^{-|B|} \text{ and } T(\mathcal{H}; u_1, \ldots, u_m) = \sum_{B \in \mathcal{H}: \ \{u_1, \ldots, u_m\} \subset B} \left(\frac{3}{2}\right)^{-|B|}.$$

The effect of the $(i+1)$st moves $x_{i+1}^{(1)}, x_{i+1}^{(2)}$ and y_{i+1} is described by the following perfect analogue of (30.10) (in the application we take $\mathcal{H} = \mathcal{F}$ and $\mathcal{H} = \mathcal{F}_2^p$)

$$T(\mathcal{H}(i+1)) = T(\mathcal{H}(i)) + f_{\mathcal{H}}(x_{i+1}^{(1)}, x_{i+1}^{(2)})$$

$$- \frac{1}{2} f_{\mathcal{H}}(x_{i+1}^{(1)}, y_{i+1}) - \frac{1}{2} f_{\mathcal{H}}(x_{i+1}^{(2)}, y_{i+1})$$

$$- \frac{3}{4} T(\mathcal{H}(i); x_{i+1}^{(1)}, x_{i+1}^{(2)}) - \frac{1}{4} T(\mathcal{H}(i); x_{i+1}^{(1)}, x_{i+1}^{(2)}, y_{i+1}),$$

where (see (30.7))

$$f_{\mathcal{H}}(u_1, u_2) = T(\mathcal{H}(i); u_1) + T(\mathcal{H}(i); u_2) + T(\mathcal{H}(i); u_1, u_2).$$

Combining these facts, and using the notation

$$g(u_1, u_2) = f_{\mathcal{F}}(u_1, u_2) - \lambda \cdot f_{\mathcal{F}_2^p}(u_1, u_2),$$

we obtain

$$L_{i+1} \geq L_i + g(x_{i+1}^{(1)}, x_{i+1}^{(2)}) - \frac{1}{2}g(x_{i+1}^{(1)}, y_{i+1}) - \frac{1}{2}g(x_{i+1}^{(2)}, y_{i+1})$$

$$- \frac{3}{4}T(\mathcal{F}(i); x_{i+1}^{(1)}, x_{i+1}^{(2)}) - \frac{1}{4}T(\mathcal{F}(i); x_{i+1}^{(1)}, x_{i+1}^{(2)}, y_{i+1})$$

$$\geq L_i + g(x_{i+1}^{(1)}, x_{i+1}^{(2)}) - \frac{1}{2}g(x_{i+1}^{(1)}, y_{i+1}) - \frac{1}{2}g(x_{i+1}^{(2)}, y_{i+1})$$

$$- T(\mathcal{F}(i); x_{i+1}^{(1)}, x_{i+1}^{(2)}). \tag{30.16}$$

Following the second proof of Theorem 20.4, Maker chooses his $(i+1)$st move $x_{i+1}^{(1)}, x_{i+1}^{(2)}$ to be that unoccupied pair $\{u_1, u_2\}$ for which the the function $g(u_1, u_2)$ attains its *maximum*. Since y_{i+1} is selected *after* $\{x_{i+1}^{(1)}, x_{i+1}^{(2)}\}$, by (30.16) we have

$$L_{i+1} \geq L_i - T(\mathcal{F}(i); x_{i+1}^{(1)}, x_{i+1}^{(2)}),$$

which guarantees the success of the technique of self-improving potentials (see section 24). Just like in the second proof of Theorem 20.4, we can extend the argument from the (2:1) game to every $(m : 1)$ game where $m \geq 2$ by replacing the analogue of equality (30.10) with the analogue of equality (30.14). This way we obtain the following biased version of Theorem 24.2.

Theorem 30.1 (*"(m : 1) Advanced Weak Win Criterion"*) *Let $m \geq 2$ be an integer, and for every finite hypergraph $\mathcal{H}$ we use the notation*

$$T(\mathcal{H}) = \sum_{B \in \mathcal{H}} \left(\frac{m+1}{m}\right)^{-|B|}.$$

If there exists a positive integer $p \geq 2$ such that

$$\frac{T(\mathcal{F})}{|V|} > p + 4p\left(T(\mathcal{F}_2^p)\right)^{1/p},$$

then, playing the (m:1) Achievement Game on hypergraph $\mathcal{F}$ with board V, topdog Maker (the first player) can occupy a whole $A \in \mathcal{F}$. □

Let's apply Theorem 30.1 to the (2:1) Aligned Square Lattice Game. Playing on an $N \times N$ board topdog Maker can always build a $q \times q$ Aligned Square Lattice with $(c = 3/2)$

$$q = \left\lfloor \sqrt{\log_c N} + o(1) \right\rfloor = \left\lfloor \sqrt{\frac{\log N}{\log(3/2)}} + o(1) \right\rfloor. \tag{30.17}$$

Next apply the trivial Cheap Halving Lemma mentioned above. There are N^2/q^2 pairwise disjoint winning sets, so, by the Cheap Halving Lemma, Maker can trivially ocupy a $q \times q$ Aligned Square Lattice with

$$q^2 = \log_2 \left(\frac{N^2}{q^2} \right),$$

which is equivalent to

$$q = \sqrt{\frac{2 \log N}{\log 2}} - o(1). \tag{30.18}$$

The surprising fact is that bound (30.17) (which was proved by a sophisticated potential technique) is weaker(!) than (30.18) (which has a trivial proof). Indeed

$$2.885 = \frac{2}{\log 2} > \frac{1}{\log(3/2)} = 2.466,$$

demonstrating the failure of the Biased Meta-Conjecture for a (2:1) Achievement Lattice Game (Maker is the topdog).

Open Problem 30.1 *Playing the (2:1) game in an $N \times N$ board, what is the largest $q \times q$ Aligned Square Lattice that topdog Maker can always build?*

4. The correct form of the Biased Meta-Conjecture. Open Problem 30.1 is a special case of a much more general question: "What is the correct form of the Biased Meta-Conjecture in the achievement case $(m : b)$ with $m > b$?" The unique feature of the biased case $(m : b)$ with $m > b$ is that topdog Maker has a "cheap" way of building. The following lemma is a straightforward generalization of the Cheap Halving Lemma above.

Maker's Cheap Building Lemma. *Consider the $(m{:}b)$ achievement play with $m > b$ on an n-uniform hypergraph $\mathcal{F}$, which contains at least $(\frac{m}{m-b})^{n-m}$ pairwise disjoint winning sets. Then Maker (the first player) can always occupy a whole $A \in \mathcal{F}$.*

Proof. Let $\mathcal{G} \subset \mathcal{F}$ be a family of $(\frac{m}{m-b})^{n-m}$ pairwise disjoint winning sets. Maker's opening move is to put m marks into m different elements of $\mathcal{G}$ (1 in each). Breaker may block as many as b of them, but at least $m - b$ remain unblocked. Repeating this step several times with new sets, at the end we obtain a sub-family $\mathcal{G}_1 \subset \mathcal{G}$ such that

(1) $|\mathcal{G}_1| \ge \left(\frac{m}{m-b}\right)^{n-m-1}$,
(2) Maker has 1 mark in every element of $\mathcal{G}_1$,
(3) Breaker has no mark in any element of $\mathcal{G}_1$.

Next working with $\mathcal{G}_1$ instead of $\mathcal{G}$, we obtain a sub-family $\mathcal{G}_2 \subset \mathcal{G}_1$ such that

(1) $|\mathcal{G}_2| \ge \left(\frac{m}{m-b}\right)^{n-m-2}$,
(2) Maker has 2 marks in every element of $\mathcal{G}_2$,
(3) Breaker has no mark in any element of $\mathcal{G}_2$.

We keep doing this; at the end of the process we obtain a sub-family $\mathcal{G}_{n-m}$ such that

(1) $|\mathcal{G}_{n-m}| \geq \left(\frac{m}{m-b}\right)^{n-m-(n-m)} = 1$,

(2) Maker has $n - m$ marks in every element of $\mathcal{G}_{n-m}$,

(3) Breaker has no mark in any element of $\mathcal{G}_{n-m}$.

Finally, Maker selects an arbitrary element $A \in \mathcal{G}_{n-m}$, and marks the last m unmarked points of A. $\square$

In the fair (1:1) play the Meta-Conjecture (see Section 9) says that, if $\mathcal{F}$ is a "nice" n-uniform hypergraph and V is the board, then $n > \log_2(|\mathcal{F}|/|V|)$ yields a Strong Draw, and $n < \log_2(|\mathcal{F}|/|V|)$ yields a Weak Win.

An intuitive explanation for the threshold $n = \log_2(|\mathcal{F}|/|V|)$ is the following "Random Play plus Pairing Strategy" heuristic. Consider the inequality

$$n \cdot |\mathcal{F}| \cdot \left(\frac{1}{2}\right)^{n-2} \leq |V|, \tag{30.19}$$

which is "almost" the same as $n \geq \log_2(|\mathcal{F}|/|V|)$. In (30.19) the factor $(1/2)^{n-2}$ means the probability that in a random play $(n - 2)$ points of a fixed n-set are occupied, and they are all taken by Maker (none by Breaker; 2 points remain unmarked). Such an n-set represents a "mortal danger" for Breaker; the product $|\mathcal{F}| \cdot (1/2)^{n-2}$ is the expected number of winning sets in "mortal danger." Inequality (30.19) means that there is room for the winning sets in "mortal danger" to be pairwise disjoint. Assuming that they are *really* pairwise disjoint – we refer to it as the *Disjointness Condition* – Breaker can easily block the unmarked point-pairs of the winning sets in "mortal danger" by a Pairing Strategy. The crucial point here is that, even if there are a huge number of winning sets in mortal danger, Breaker can still block them in the last minute!

In the biased $(a{:}f)$ avoidance game (Avoider takes a and Forcer takes f points per move) the corresponding threshold is

$$n = \log_{\frac{a+f}{a}}\left(\frac{|\mathcal{F}|}{|V|}\right).$$

Similarly, in the biased $(m{:}b)$ achievement game (Maker takes m points and Breaker takes b points per move) with $m \leq b$ (i.e. Maker is *not* a topdog), the corresponding threshold is the "same"

$$n = \log_{\frac{m+b}{m}}\left(\frac{|\mathcal{F}|}{|V|}\right),$$

i.e. there is no surprise.

The surprise comes in the $(m:b)$ achievement game with $m > b$, i.e. when Maker is the topdog: in this case we conjecture that the corresponding threshold is

$$n = \log_{\frac{m+b}{m}} \left(\frac{|\mathcal{F}|}{|V|} \right) + \log_{\frac{m}{m-b}} |V|. \tag{30.20}$$

The best intuitive explanation for (30.20) is the following "Random Play plus Cheap Building" heuristic. We divide the whole play into two stages. In the First Stage we assume that Maker and Breaker play randomly. The First Stage ends when the number of Breaker-free winning sets ("survivors") becomes less than $\frac{1}{n}|V|$. The equation

$$\left(\frac{m}{m+b} \right)^y \cdot |\mathcal{F}| = \frac{1}{n}|V| \tag{30.21}$$

tells us that at the end of the First Stage there are about $\frac{1}{n}|V|$ Breaker-free winning sets such that Maker has y marks in each. The number $\frac{1}{n}|V|$ of these n-element survivors leaves enough room for disjointness: it seems perfectly reasonable to assume that the $\frac{1}{n}|V|$ n-element survivors are pairwise disjoint. Then the end-play is obvious: the Second Stage is a straightforward application of Maker's Cheap Building Lemma above applied to the (roughly) $\frac{1}{n}|V|$ pairwise disjoint n-element survivors ("Disjointness Condition"). If

$$\left(\frac{m}{m-b} \right)^{n-y-m} \geq \frac{1}{n}|V|, \tag{30.22}$$

then Maker can occupy a whole winning set (y points in the First Stage and $n - y$ points in the Second Stage). Note that (30.21) is equivalent to

$$y = \log_{\frac{m+b}{m}} \left(\frac{|\mathcal{F}|}{|V|} \right) + O(\log n), \tag{30.23}$$

and (30.22) is equivalent to

$$n - y = \log_{\frac{m}{m-b}} |V| + O(\log n). \tag{30.24}$$

Adding up (30.23) and (30.24), we obtain (30.20) with an additive error term $O(\log n)$, which is negligible compared to the "legitimate error" $o(\sqrt{n})$, which corresponds to the fact that in the exact solutions $n = \binom{q}{2}$ or q^2, i.e. the goal sets are 2-dimensional. Two-dimensional goals mean that the crucial step is to "break the square-root barrier."

In the $(m:1)$ play, when we have Theorem 30.1, it is surprisingly easy to make the "Random Play plus Cheap Building" intuition precise; well, at least half of it is doable: the Maker's part. The critical step in the proof is to enforce the Disjointness Condition, which can be done by involving an extra auxiliary hypergraph. This section is already too long, so we postpone the (somewhat ugly) details to Section 46 (another good reason why it is postponed to the end is that in Section 45 we will use very similar ideas; in fact, the arguments in Section 45 will be more complicated).

What we cannot solve is Breaker's part.

The "correct" form of the Biased Meta-Conjectured, applied in the special case of the clique-hypergraph, goes as follows:

Open Problem 30.2 ("Biased Clique Game") *Is it true that, in the* $(m : b)$ *Biased Clique Achievement Game with* $m > b$, *played on* K_N, *the corresponding Clique Achievement Number is*

$$\mathbf{A}(K_N; \text{clique}; m : b) = \lfloor 2\log_c N - 2\log_c \log_c N + 2\log_c e - 2\log_c 2 - 1$$
$$+ \frac{2\log c}{\log c_0} + o(1) \rfloor,$$

where $c = \frac{m+b}{m}$ *and* $c_0 = \frac{m}{m-b}$?

Is it true that, in the $(m : b)$ *Biased Clique Achievement Game with* $m \leq b$, *played on* K_N, *the corresponding Clique Achievement Number is*

$$\mathbf{A}(K_N; \text{clique}; m : b) = \lfloor 2\log_c N - 2\log_c \log_c N + 2\log_c e - 2\log_c 2 - 1 + o(1) \rfloor,$$

where $c = \frac{m+b}{m}$?

Is it true that the Avoidance Number

$$\mathbf{A}(K_N; \text{clique}; a : f; -) = \lfloor 2\log_c N - 2\log_c \log_c N + 2\log_c e - 2\log_c 2 - 1 + o(1) \rfloor,$$

where the base of logarithm is $c = \frac{a+f}{a}$?

Note that the "logarithmic expression" in Open Problem 30.2 comes from the equation ("Neighborhood Conjecture for the clique-hypergraph")

$$\left(\frac{m+b}{m}\right)^{\binom{q}{2}} = \frac{\binom{N}{q}}{\binom{N}{2}} \approx \binom{N}{q-2}, \tag{30.25}$$

which has the "real" solution

$$q = 2\log_c N - 2\log_c \log_c N + 2\log_c e - 2\log_c 2 - 1 + o(1), \tag{30.26}$$

where the base of the logarithm is $c = \frac{m+b}{m}$.

The deduction of (30.26) from (30.25) is easy. First take base $c = \frac{m+b}{m}$ logarithm of (30.25)

$$\binom{q}{2} = \log_c \binom{N}{q-2}.$$

Then apply Stirling's formula, and divide by $(q-1)$

$$\frac{q}{2} = \frac{q-2}{q-1}\log_c\left(\frac{eN}{q-2}\right) = \log_c N - \frac{\log_c N}{q-1} - \log_c q + \log_c e.$$

Then multiply by 2, and apply the approximation $q \approx 2 \log_c N$

$$q = 2 \log_c N - 2 \frac{\log_c N}{2 \log_c N} - 2 \log_c (2 \log_c N) + 2 \log_c e + o(1),$$

which gives (30.26).

Open Problem 30.2 remains unsolved; what we can prove is the following lower bound result.

Theorem 30.2 *In the (m:1) case the Biased Clique Achievement Number is at least as large as the logarithmic expression in Open Problem 30.2.*

The proof of Theorem 30.2 applies Theorem 30.1 and the precise version of the "Random Play plus Cheap Building" intuition; for the details see Section 46.

What we *cannot* prove is whether or not the logarithmic expression in Open Problem 30.2 is the best possible. In other words, is it true that Breaker can always prevent Maker from building a clique K_q, where q is the upper integral part of the logarithmic expression in Open Problem 30.2.

Next consider the Lattice Games of Section 8 (but not the case of "complete bipartite graph"). The analogue of Open Problem 30.2 (in fact, a generalization of Open Problem 30.1) goes as follows:

Open Problem 30.3 ("Biased Lattice Games") *Is it true that, in the (m:b) Biased Lattice Achievement Game with m > b played on an N × N board, the corresponding Lattice Achievement Number is:*

(1a) $\mathbf{A}(N \times N;$ square lattice; $m:b) = \lfloor \sqrt{\log_c N + 2 \log_{c_0} N} + o(1) \rfloor$,

(1b) $\mathbf{A}(N \times N;$ rectangle lattice; $m:b) = \lfloor \sqrt{2 \log_c N + 2 \log_{c_0} N} + o(1) \rfloor$,

(1c) $\mathbf{A}(N \times N;$ tilted square lattice; $m:b) = \lfloor \sqrt{2 \log_c N + 2 \log_{c_0} N} + o(1) \rfloor$,

(1d) $\mathbf{A}(N \times N;$ tilted rectangle lattice; $m:b) = \lfloor \sqrt{2 \log_c N + 2 \log_{c_0} N} + o(1) \rfloor$,

(1e) $\mathbf{A}(N \times N;$ rhombus lattice; $m:b) = \lfloor \sqrt{2 \log_c N + 2 \log_{c_0} N} + o(1) \rfloor$,

(1f) $\mathbf{A}(N \times N;$ parallelogram lattice; $m:b) = \lfloor 2 \sqrt{\log_c N + 2 \log_{c_0} N} + o(1) \rfloor$,

(1g) $\mathbf{A}(N \times N;$ area one lattice; $m:b) = \lfloor \sqrt{2 \log_c N + 2 \log_{c_0} N} + o(1) \rfloor$,

where $c = \frac{m+b}{m}$ *and* $c_0 = \frac{m}{m-b}$?

Is it true that, in the (m:b) Biased Lattice Achievement Game with m ≤ b played on an N × N board, the corresponding Lattice Achievement Number is:

(2a) $\mathbf{A}(N \times N;$ square lattice; $m:b) = \lfloor \sqrt{\log_c N} + o(1) \rfloor$,

(2b) $\mathbf{A}(N \times N;$ rectangle lattice; $m:b) = \lfloor \sqrt{2 \log_c N} + o(1) \rfloor$,

(2c) $\mathbf{A}(N \times N;$ tilted square lattice; $m:b) = \lfloor \sqrt{2 \log_c N} + o(1) \rfloor$,

(2d) $\mathbf{A}(N \times N;$ tilted rectangle lattice; $m:b) = \lfloor \sqrt{2 \log_c N} + o(1) \rfloor$,

(2e) $\mathbf{A}(N \times N;$ rhombus lattice; $m:b) = \lfloor \sqrt{2 \log_c N} + o(1) \rfloor$,

(2f) $\mathbf{A}(N \times N;$ parallelogram lattice; $m : b) = \left\lfloor 2\sqrt{\log_c N} + o(1) \right\rfloor$,

(2g) $\mathbf{A}(N \times N;$ area one lattice; $m : b) = \left\lfloor \sqrt{2 \log_c N} + o(1) \right\rfloor$,

where $c = \frac{m+b}{m}$?

Is it true that the Lattice Avoidance Number in the $(a:f)$ play is the same as in (2a)–(2g), except that $c = \frac{a+f}{a}$?

Applying Theorem 30.1 and the precise version of the "Random Play plus Cheap Building" intution (see Section 46) we obtain:

Theorem 30.3 *In the (m:1) Achievement Game the Biased Lattice Achievement Number is at least as large as the right-hand sides in Open Problem 30.3.*

Theorems 30.1–30.3 were about the $(m:1)$ Biased Achievement Game (where Maker was the topdog); we can prove the same results about the $(1:f)$ Biased Avoidance Game, where Forcer is the topdog. (Note there is a hidden duality here)

Theorem 30.4 ("Advanced Forcer'win Criterion in the $(1:f)$ Avoidance Game") *Let $f \geq 2$ be an integer, and for every finite hypergraph $\mathcal{H}$ write*

$$T(\mathcal{H}) = \sum_{B \in \mathcal{H}} (f+1)^{-|B|}.$$

If there exists a positive integer $p \geq 2$ such that

$$\frac{T(\mathcal{F})}{|V|} > p + 4p \left(T(\mathcal{F}_2^p) \right)^{1/p},$$

then, playing the (1:f) Avoidance Game on hypergraph $\mathcal{F}$, topdog Forcer (the first player) can force Avoider to occupy a whole $A \in \mathcal{F}$.

Proof. It is similar to the proof of Theorem 30.1. For simplicity we start with the case $f = 2$. The analogue of expansion (30.7) is the following

$$\begin{aligned}
(1+2x)(1 - y^{(1)})(1 - y^{(2)}) - 1 &= 2x - y^{(1)} - y^{(2)} \\
&\quad - 2xy^{(1)} - 2xy^{(2)} + y^{(1)}y^{(2)} + 2xy^{(1)}y^{(2)} \\
&= f(x, y^{(1)}) + f(x, y^{(2)}) - 2f(y^{(1)}, y^{(2)}) \\
&\quad - \frac{3}{2}x(y^{(1)} + y^{(2)}) + 2xy^{(1)}y^{(2)}, \qquad (30.27)
\end{aligned}$$

where

$$f(u_1, u_2) = u_1 + u_2 - \frac{1}{2}u_1 u_2. \qquad (30.28)$$

Equality (30.27) leads to the following analogue of (30.10)

$$T_{i+1} = T_i + f(x_{i+1}, y_{i+1}^{(1)}) + f(x_{i+1}, y_{i+1}^{(1)}) - 2f(y_{i+1}^{(1)}, y_{i+1}^{(2)})$$

$$- \frac{3}{2}\left(T(\mathcal{F}(i); x_{i+1}, y_{i+1}^{(1)}) + T(\mathcal{F}(i); x_{i+1}, y_{i+1}^{(2)})\right) + 2T(\mathcal{F}(i); x_{i+1}, y_{i+1}^{(1)}, y_{i+1}^{(2)}).$$

$$(30.29)$$

By using the trivial inequality

$$T(\mathcal{F}(i); x_{i+1}, y_{i+1}^{(1)}) + T(\mathcal{F}(i); x_{i+1}, y_{i+1}^{(2)}) \geq 2T(\mathcal{F}(i); x_{i+1}, y_{i+1}^{(1)}, y_{i+1}^{(2)})$$

in (30.29), we obtain the key inequality ("Decreasing Property")

$$T_{i+1} \leq T_i + f(x_{i+1}, y_{i+1}^{(1)}) + f(x_{i+1}, y_{i+1}^{(1)}) - 2f(y_{i+1}^{(1)}, y_{i+1}^{(2)}),$$

which, together with (30.29), guarantees the success of the methods of Section 24 ("self-improving potentials"). The rest of the proof of Theorem 30.4 with $f = 2$ is standard. □

The proof of the general case $f \geq 2$ requires more calculations, and goes very similarly to another proof in Section 33. This is why we postpone the proof of the general case to Section 33.

Applying Theorem 30.4 to the Clique Avoidance Game, we obtain:

Theorem 30.5 *In the (1:f) case the Biased Clique Avoidance Number is at least as large as the right-hand side in Open Problem 30.2. In other words, playing on K_N topdog Forcer can always force Avoider to build a clique K_q where q is the lower integral part of*

$$2\log_c N - 2\log_c \log_c N + 2\log_c e - 2\log_c 2 - 1 - o(1)$$

with $c = f + 1$.

What we *cannot* prove is that whether or not this lower bound is best possible.

Applying Theorem 30.4 to the Lattice Avoidance Game we obtain:

Theorem 30.6 *In the (1:f) Avoidance Game the Biased Lattice Avoidance Number is at least as large as the right-hand sides in Open Problem 30.3.*

Again what we don't know is whether or not this lower bound is the best possible.

Note that in the Avoidance Game there is no analogue of the Cheap Building Lemma. However, there is a very good chance that Open Problems 30.2 and 30.3 are all true.

31

Biased games (II): Sacrificing the probabilistic intuition to force negativity

1. The (2:2) game. The reader is probably wondering: "How come that the (2:2) game is not included in Questions 30.1-2?" The (2:2) game is a fair game just like the ordinary (1:1) game, so we would expect exactly the same results as in the (1:1) game (see e.g. Theorems 6.4 and 8.2). Let's start with Theorem 6.4: "What is the largest clique that Maker can always build in the (2:2) game on K_N?" Of course, everyone would expect a K_q with the usual

$$q = 2\log_2 N - 2\log_2 \log_2 N + O(1). \tag{31.1}$$

What could be more natural than this! The bad news is that we cannot prove (31.1).

Open Problem 31.1

(a) *Is it true that, playing the (2:2) Achievement Game on K_N, Maker can always build a clique K_q with $q = 2\log_2 N - 2\log_2 \log_2 N + O(1)$?*

(b) *Is it true that, playing the (2:2) Achievement Game on K_N, Breaker can always prevent Maker from building a clique K_q with $q = 2\log_2 N - 2\log_2 \log_2 N + O(1)$?*

What is the unexpected technical difficulty that prevents us from solving Open Problem 31.1 (a)? Well, naturally we try to adapt the proof of the Advanced Weak Win Criterion (Theorem 24.2). Since the (2:2) game is fair, we keep the Power-of-Two Scoring System: for an arbitrary hypergraph $(V, \mathcal{H})$ (V is the union set) write

$$T(\mathcal{H}) = \sum_{A \in \mathcal{H}} 2^{-|A|},$$

and for any m-element subset of points $\{u_1, \ldots, u_m\} \subseteq V$ ($m \geq 1$) write

$$T(\mathcal{H}; u_1, \ldots, u_m) = \sum_{A \in \mathcal{H}:\ \{u_1, \ldots, u_m\} \subseteq A} 2^{-|A|},$$

of course, counted with multiplicity.

Assume that we are in the middle of a play where Maker, as the first player, already occupied

$$X(i) = \left\{ x_1^{(1)}, x_1^{(2)}, x_2^{(1)}, x_2^{(2)}, \ldots, x_i^{(1)}, x_i^{(2)} \right\},$$

and Breaker, the second player, occupied

$$Y(i) = \left\{ y_1^{(1)}, y_1^{(2)}, y_2^{(1)}, y_2^{(2)}, \ldots, y_i^{(1)}, y_i^{(2)} \right\}.$$

The question is how to choose Maker's next move $x_{i+1}^{(1)}, x_{i+1}^{(2)}$.

Let

$$\mathcal{F}(i) = \{A \setminus X(i) : \ A \in \mathcal{F}, \ A \cap Y(i) = \emptyset\}.$$

$\mathcal{F}(i)$ is a multi-hypergraph; the empty set can be an element of $\mathcal{F}(i)$: it simply means that Maker already occupied a winning set, and wins the play in the ith round or before.

We can describe the effect of the $(i+1)$st moves $x_{i+1}^{(1)}, x_{i+1}^{(2)}$ and $y_{i+1}^{(1)}, y_{i+1}^{(2)}$ as follows

$$
\begin{aligned}
T(\mathcal{F}(i+1)) &= T(\mathcal{F}(i)) + T(\mathcal{F}(i); x_{i+1}^{(1)}) + T(\mathcal{F}(i); x_{i+1}^{(2)}) \\
&\quad + T(\mathcal{F}(i); x_{i+1}^{(1)}, x_{i+1}^{(2)}) - T(\mathcal{F}(i); y_{i+1}^{(1)}) - T(\mathcal{F}(i); y_{i+1}^{(2)}) \\
&\quad + T(\mathcal{F}(i); y_{i+1}^{(1)}, y_{i+1}^{(2)}) - T(\mathcal{F}(i); x_{i+1}^{(1)}, y_{i+1}^{(1)}) \\
&\quad - T(\mathcal{F}(i); x_{i+1}^{(1)}, y_{i+1}^{(2)}) - T(\mathcal{F}(i); x_{i+1}^{(2)}, y_{i+1}^{(1)}) \\
&\quad - T(\mathcal{F}(i); x_{i+1}^{(2)}, y_{i+1}^{(2)}) - T(\mathcal{F}(i); x_{i+1}^{(1)}, x_{i+1}^{(2)}, y_{i+1}^{(1)}) + T(\mathcal{F}(i); x_{i+1}^{(1)}, y_{i+1}^{(1)}, y_{i+1}^{(2)}) \\
&\quad + T(\mathcal{F}(i); x_{i+1}^{(2)}, y_{i+1}^{(1)}, y_{i+1}^{(2)}) + T(\mathcal{F}(i); x_{i+1}^{(1)}, x_{i+1}^{(2)}, y_{i+1}^{(1)}, y_{i+1}^{(2)}).
\end{aligned}
$$

$$(31.2)$$

Identity (31.2) is rather long, but the underlying pattern is very simple: it is described by the expansion of the product

$$
\begin{aligned}
(1+x^{(1)})(1+x^{(2)})(1-y^{(1)})(1-y^{(2)}) - 1 &= \sum_{j=1}^{2} x^{(j)} - \sum_{j=1}^{2} y^{(j)} + \\
&\quad + x^{(1)}x^{(2)} + y^{(1)}y^{(2)} - \sum_{j=1}^{2}\sum_{k=1}^{2} x^{(j)}y^{(k)} \\
&\quad - x^{(1)}x^{(2)}\left(\sum_{j=1}^{2} y^{(j)}\right) + y^{(1)}y^{(2)}\left(\sum_{j=1}^{2} x^{(j)}\right) \\
&\quad + x^{(1)}x^{(2)}y^{(1)}y^{(2)}.
\end{aligned}
$$

$$(31.3)$$

Maker wants to guarantee that $T(\mathcal{F}(i)) > 0$ for all i. At first sight the positive terms

$$x^{(1)}x^{(2)} + y^{(1)}y^{(2)} + y^{(1)}y^{(2)}\left(\sum_{j=1}^{2} x^{(j)}\right) + x^{(1)}x^{(2)}y^{(1)}y^{(2)} \tag{31.4}$$

in (31.3) seem to help, *but* the whole Potential Function has the positive-negative form

$$L_i = T(\mathcal{F}(i)) - \lambda \cdot T(\mathcal{F}_2^p(i)), \tag{31.5}$$

and the terms which correspond to (31.4) in the "negative Big part" $-\lambda \cdot T(\mathcal{F}_2^p(i))$ are, of course, negative, which makes it extremely difficult to estimate L_i from below. (We want to show that $L_i > 0$ for all i, which implies that $T(\mathcal{F}(i)) > 0$ for all i.)

In other words, the positive terms in (31.4) become "Big Bad Negative" terms in (31.5), and consequently the proof technique of Section 24 collapses. We have to modify the original proof technique: we have to get rid of (most of) the positive terms in (31.4). We attempt to do it in a surprising way by:

2. Sacrificing the probabilistic intuition. Let $\mathcal{H}$ be either $\mathcal{F}$ or $\mathcal{F}_2^p$; We will modify the Power-of-Two Scoring System

$$T(\mathcal{H}) = \sum_{A \in \mathcal{H}} 2^{-|A|}. \tag{31.6}$$

Note that (31.6) has a clear-cut probabilistic motivation. We replace (31.6) with the following unconventional "one-counts-only" weight

$$T_i(\mathcal{H}) = \sum_{A \in \mathcal{H}} w_i(A) \text{ and} \tag{31.7}$$

$$T_i(\mathcal{H}; u_1, \ldots, u_m) = \sum_{A \in \mathcal{H}: \{u_1, \ldots, u_m\} \subset A} w_i(A),$$

where $w_i(A) = 0$ if $A \cap Y(i) \neq \emptyset$ and

$$w_i(A) = 2^{|\{1 \leq j \leq i: \, A \cap \{x_j^{(1)}, x_j^{(2)}\} \neq \emptyset\}| - |A|} \tag{31.8}$$

if $A \cap Y(i) = \emptyset$.

Note that the weight of a "survivor" $A \in \mathcal{F}$ in $T(\mathcal{F}(i))$ (see (31.6)) is

$$2^{|A \cap X(i)| - |A|} \quad \text{if} \quad A \cap Y(i) = \emptyset, \tag{31.9}$$

and we have the trivial inequality

$$|\{1 \leq j \leq i: \, A \cap \{x_j^{(1)}, x_j^{(2)}\} \neq \emptyset\}| \leq |A \cap X(i)|.$$

A strict inequality can easily occur here; this explains why we use the phrase "one-counts-only" for the weight defined in (31.7)–(31.8).

The switch from (31.6) to (31.7)–(31.8) clearly affects identity (31.2). For notational simplicity we just write down the new form of (31.3)

$$\left(1 + x^{(1)} + x^{(2)} - x^{(1)}x^{(2)}\right)\left(1 - y^{(1)}\right)\left(1 - y^{(2)}\right) - 1$$

$$= \left(x^{(1)} + x^{(2)} - x^{(1)}x^{(2)}\right)$$

$$- \left(y^{(1)} + y^{(2)} - y^{(1)}y^{(2)}\right)$$

$$- \sum_{j=1}^{2}\sum_{k=1}^{2} x^{(j)}y^{(k)} + x^{(1)}x^{(2)}\left(\sum_{j=1}^{2} y^{(j)}\right)$$

$$+ y^{(1)}y^{(2)}\left(\sum_{j=1}^{2} x^{(j)}\right) - x^{(1)}x^{(2)}y^{(1)}y^{(2)}. \tag{31.10}$$

Notice that the first 2 rows in (31.10) have the same pattern with opposite signs; this fact motivates the introduction of the new weight (31.7)–(31.8). This pattern will define the function that Maker optimizes to get his $(i+1)$st move (see (31.17) later).

By using the trivial inequality

$$T_i(\mathcal{H}; x_{i+1}^{(1)}, x_{i+1}^{(2)}, y_{i+1}^{(j)}) \leq \sum_{k=1}^{2} T_i(\mathcal{H}; x_{i+1}^{(k)}, y_{i+1}^{(j)}) \tag{31.11}$$

(and also its symmetric version where x and y switch roles) in the "short form" (31.10), we obtain the "long form" inequality ("Decreasing Property")

$$T_{i+1}(\mathcal{F}) = T_i(\mathcal{F}) + \left(T_i(\mathcal{F}; x_{i+1}^{(1)}) + T_i(\mathcal{F}; x_{i+1}^{(2)}) - T_i(\mathcal{F}; x_{i+1}^{(1)}, x_{i+1}^{(2)})\right)$$

$$- \left(T_i(\mathcal{F}; y_{i+1}^{(1)}) + T_i(\mathcal{F}; y_{i+1}^{(2)}) - T_i(\mathcal{F}; y_{i+1}^{(1)}, y_{i+1}^{(2)})\right)$$

$$- \sum_{j=1}^{2}\sum_{k=1}^{2} T_i(\mathcal{F}; x_{i+1}^{(j)}, y_{i+1}^{(k)}) + \sum_{j=1}^{2} T_i(\mathcal{F}; x_{i+1}^{(1)}, x_{i+1}^{(2)}, y_{i+1}^{(j)})$$

$$+ \sum_{j=1}^{2} T_i(\mathcal{F}; x_{i+1}^{(j)}, y_{i+1}^{(1)}, y_{i+1}^{(2)}) - T_i(\mathcal{F}; x_{i+1}^{(1)}, x_{i+1}^{(2)}, y_{i+1}^{(1)}, y_{i+1}^{(2)})$$

$$\leq T_i(\mathcal{F}) + \left(T_i(\mathcal{F}; x_{i+1}^{(1)}) + T_i(\mathcal{F}; x_{i+1}^{(2)}) - T_i(\mathcal{F}; x_{i+1}^{(1)}, x_{i+1}^{(2)})\right)$$

$$- \left(T_i(\mathcal{F}; y_{i+1}^{(1)}) + T_i(\mathcal{F}; y_{i+1}^{(2)}) - T_i(\mathcal{F}; y_{i+1}^{(1)}, y_{i+1}^{(2)})\right). \tag{31.12}$$

Inequality (31.12) represents the desired "Negativity" mentioned in the title of the section.

We define our *Potential Function* as follows: let

$$L_i = T_i(\mathcal{F}) - \lambda \cdot T_i(\mathcal{F}_2^p), \tag{31.13}$$

where the positive constant λ is specified by the side condition

$$L_0 = \frac{1}{2}T(\mathcal{F}), \quad \text{that is, } \frac{1}{2}T(\mathcal{F}) = \lambda \cdot T(\mathcal{F}_2^p). \tag{31.14}$$

Clearly

$$
\begin{aligned}
L_{i+1} &= T_{i+1}(\mathcal{F}) - \lambda \cdot T_{i+1}(\mathcal{F}_2^p) \geq T_i(\mathcal{F}(i)) \\
&\quad + \left(T_i(\mathcal{F}; x_{i+1}^{(1)}) + T_i(\mathcal{F}; x_{i+1}^{(2)}) - T_i(\mathcal{F}; x_{i+1}^{(1)}, x_{i+1}^{(2)}) \right) \\
&\quad - \left(T_i(\mathcal{F}; y_{i+1}^{(1)}) + T_i(\mathcal{F}; y_{i+1}^{(2)}) - T_i(\mathcal{F}; y_{i+1}^{(1)}, y_{i+1}^{(2)}) \right) \\
&\quad - \sum_{j=1}^{2}\sum_{k=1}^{2} T_i(\mathcal{F}; x_{i+1}^{(j)}, y_{i+1}^{(k)}) \\
&\quad - T_i(\mathcal{F}; x_{i+1}^{(1)}, x_{i+1}^{(2)}, y_{i+1}^{(1)}, y_{i+1}^{(2)}) - \lambda \cdot T_{i+1}(\mathcal{F}_2^p),
\end{aligned}
$$

and applying (31.12) with the substitution "$\mathcal{F} = \mathcal{F}_2^p$"

$$
\begin{aligned}
L_{i+1} &\geq T_i(\mathcal{F}) + \left(T_i(\mathcal{F}; x_{i+1}^{(1)}) + T_i(\mathcal{F}; x_{i+1}^{(2)}) - T_i(\mathcal{F}; x_{i+1}^{(1)}, x_{i+1}^{(2)}) \right) \\
&\quad - \left(T_i(\mathcal{F}; y_{i+1}^{(1)}) + T_i(\mathcal{F}; y_{i+1}^{(2)}) - T_i(\mathcal{F}; y_{i+1}^{(1)}, y_{i+1}^{(2)}) \right) \\
&\quad - \sum_{j=1}^{2}\sum_{k=1}^{2} T_i(\mathcal{F}; x_{i+1}^{(j)}, y_{i+1}^{(k)}) \\
&\quad - T_i(\mathcal{F}; x_{i+1}^{(1)}, x_{i+1}^{(2)}, y_{i+1}^{(1)}, y_{i+1}^{(2)}) \\
&\quad - \lambda \cdot \left(T_i(\mathcal{F}_2^p) + \left(T_i(\mathcal{F}_2^p; x_{i+1}^{(1)}) + T_i(\mathcal{F}_2^p; x_{i+1}^{(2)}) - T_i(\mathcal{F}_2^p; x_{i+1}^{(1)}, x_{i+1}^{(2)}) \right) \right. \\
&\quad \left. - \left(T_i(\mathcal{F}_2^p; y_{i+1}^{(1)}) + T_i(\mathcal{F}_2^p; y_{i+1}^{(2)}) - T_i(\mathcal{F}_2^p; y_{i+1}^{(1)}, y_{i+1}^{(2)}) \right) \right).
\end{aligned}
$$

Thus we have

$$
\begin{aligned}
L_{i+1} &\geq L_i + \left(L_i(x_{i+1}^{(1)}) + L_i(x_{i+1}^{(2)}) - L_i(x_{i+1}^{(1)}, x_{i+1}^{(2)}) \right) \\
&\quad - \left(L_i(y_{i+1}^{(1)}) + L_i(y_{i+1}^{(2)}) - L_i(y_{i+1}^{(1)}, y_{i+1}^{(2)}) \right) \\
&\quad - \sum_{j=1}^{2}\sum_{k=1}^{2} T_i(\mathcal{F}; x_{i+1}^{(j)}, y_{i+1}^{(k)}) \\
&\quad - T_i(\mathcal{F}; x_{i+1}^{(1)}, x_{i+1}^{(2)}, y_{i+1}^{(1)}, y_{i+1}^{(2)}), \tag{31.15}
\end{aligned}
$$

where

$$L_i(u_1, \ldots, u_m) = T_i(\mathcal{F}; u_1, \ldots, u_m) - \lambda \cdot T_i(\mathcal{F}_2^p; u_1, \ldots, u_m) \tag{31.16}$$

for any set of unselected points $\{u_1, \ldots, u_m\}$.

Maker's $(i+1)$st move is that unoccupied point pair

$$\{x_{i+1}^{(1)}, x_{i+1}^{(1)}\} = \{u_1, u_2\}$$

for which the sum

$$L_i(u_1) + L_i(u_2) - L_i(u_1, u_2) \tag{31.17}$$

attains its *maximum*; then

$$L_i(x_{i+1}^{(1)}) + L_i(x_{i+1}^{(2)}) - L_i(x_{i+1}^{(1)}, x_{i+1}^{(2)})$$
$$\geq L_i(y_{i+1}^{(1)}) + L_i(y_{i+1}^{(2)}) - L_i(y_{i+1}^{(1)}, y_{i+1}^{(2)}),$$

so by (31.15)

$$L_{i+1} \geq L_i - \sum_{j=1}^{2} \sum_{k=1}^{2} T_i(\mathcal{F}; x_{i+1}^{(j)}, y_{i+1}^{(k)})$$
$$- T_i(\mathcal{F}; x_{i+1}^{(1)}, x_{i+1}^{(2)}, y_{i+1}^{(1)}, y_{i+1}^{(2)}). \tag{31.18}$$

HYPOTHESIS: Assume that the following analogue of the Generalized Variance Lemma (GVL, see Theorem 24.1) holds:

GVL: *For arbitrary integers $p \geq 2$ and $m \geq 1$, and for arbitrary points $u_1, \ldots, u_m$ of the board*

$$T_i(\mathcal{H}; u_1, \ldots, u_m) < p\left(\left(T_i(\mathcal{H}_m^p)\right)^{1/p} + 1\right). \tag{31.19}$$

An application of GVL in (31.18) would lead to the Critical Inequality

$$L_{i+1} \geq L_i - 5p\left(T_i(\mathcal{F}_2^p(i))\right)^{1/p} - 5p, \tag{31.20}$$

which seems to prove that the technique of self-improving potentials ("Section 24") works! It looks like there is no problem with proving Open Problem 31.1 (a), but here comes the bad news: unfortunately (31.19) is *false*, and the Hypothesis above collapses.

How come that the innocent-looking (31.19) is false? Well, this failure is a "side-effect" of the unconventional "one-counts-only" weight. The proof of Theorem 24.1 (i.e. (31.19) with the standard Power-of-Two Scoring System) is based on the trivial inequality

$$2^{-\sum_{i=1}^{p} |B_i|} \leq 2^{-|\bigcup_{i=1}^{p} B_i|}, \tag{31.21}$$

which follows from the even more elementary fact $\sum_{i=1}^{p} |B_i| \geq |\bigcup_{i=1}^{p} B_i|$. The switch from the standard Power-of-Two Scoring System to the "one-counts-only" weight cures one technical problem: it eliminates the "Big Bad Negative" terms in the Potential Function, but is has a serious "side-effect": "one-counts-only" means a "deficit" in (31.21), and due to the "deficit" the Generalized Variance Lemma (GVL) is not true any more.

A counter-example to (31.19). Recall the notation that

$$X(i) = \left\{ x_1^{(1)}, x_1^{(2)}, x_2^{(1)}, x_2^{(2)}, \ldots, x_i^{(1)}, x_i^{(2)} \right\}$$

denotes the set of points of Maker (right after the ith round). Let $\{u_1, u_2\}$ be a point pair disjoint from $X(i)$, and assume that $i \geq n$. Let $\mathcal{H}$ denote the n-uniform hypergraph satisfying the following properties:

(1) the given point pair $\{u_1, u_2\}$ is contained in every $A \in \mathcal{H}$;
(2) every $A \in \mathcal{H}$ has the form $\{u_1, u_2, z_1, z_2, \ldots, z_{n-2}\}$ where $z_j \in \{x_j^{(1)}, x_j^{(2)}\}$, $j = 1, 2, \ldots, n-2$.

There are 2^{n-2} ways to choose 1 element from each 1 of the $n-2$ pairs, so $|\mathcal{H}| = 2^{n-2}$. By (31.7)–(31.8)

$$T_{n-2}(\mathcal{H}; u_1, u_2) = |\mathcal{H}| \cdot 2^{-2} = 2^{n-4} \text{ and} \tag{31.22}$$

$$T_{n-2}(\mathcal{H}_2^2) = |\mathcal{H}| \cdot \sum_{k=0}^{n-2} \binom{n-2}{k} 2^{-k} \cdot 2^{-2}; \tag{31.23}$$

in (31.23) k denotes the number of pairs $\{x_j^{(1)}, x_j^{(2)}\}$, where *both* elements are contained in a union $A_1 \cup A_2$ with $A_1, A_2 \in \mathcal{H}$ (see the definition $\mathcal{H}_2^2$). This k leads to a "deficit" of k – because of "one-counts-only" – explaining the extra factor 2^{-k} in (31.23).

Returning to (31.23),

$$T_{n-2}(\mathcal{H}_2^2) = |\mathcal{H}| \cdot \sum_{k=0}^{n-2} \binom{n-2}{k} 2^{-k} \cdot 2^{-2}$$

$$= 2^{n-4} \left(\sum_{k=0}^{n-2} \binom{n-2}{k} \left(\frac{1}{2}\right)^k \right) = 2^{n-4} \left(1 + \frac{1}{2}\right)^{n-2} = \frac{3^{n-2}}{4},$$

and so

$$\left(T_{n-2}(\mathcal{H}_2^2)\right)^{1/2} = \frac{1}{2} \cdot 3^{\frac{n}{2}-1}. \tag{31.24}$$

The point is that $\sqrt{3} = 1.732 < 2$, so (31.24) is much less ("exponentially less") than (31.22), which kills our chances for any general inequality like (31.19).

The failure of (31.19) is a major roadblock. This is why we have no idea how to solve the perfectly natural Open Problem 31.1 (a).

The failure of (31.19) is similar to the failure of (28.9) in the Discrepancy Game. Again we can hope that the *ad hoc* Higher Moment Method in Section 23 will save the day, at least for the Lattice Games. We work out the details of this idea in the next section. We begin with the special case of the most general lattice: the Parallelogram Lattice.

3. The (2:2) Parallelogram Lattice Game.

We adapt the argument of Section 23. The counter-example to (31.19) (see the end of Section 31) has a very large "deficit," and by using the *ad hoc* method of Section 23 we can avoid the "deficit problem."

We keep the unconventional "one-counts-only" weight

$$T_i(\mathcal{H}) = \sum_{A \in \mathcal{H}} w_i(A) \text{ and} \tag{31.25}$$

$$T_i(\mathcal{H}; u_1, \ldots, u_m) = \sum_{A \in \mathcal{H}:\ \{u_1, \ldots, u_m\} \subset A} w_i(A),$$

where $w_i(A) = 0$ if $A \cap Y(i) \neq \emptyset$ and

$$w_i(A) = 2^{|\{1 \leq j \leq i:\ A \cap \{x_j^{(1)}, x_j^{(2)}\} \neq \emptyset\}| - |A|} \text{ if } A \cap Y(i) = \emptyset, \tag{31.26}$$

first introduced in (31.7)–(31.8). $\mathcal{F}$ denotes the family of all $q \times q$ parallelogram lattices in the $N \times N$ board, and the choice of weight (31.25)–(31.26) leads to (see inequality (31.18))

$$L_{i+1} \geq L_i - \sum_{j=1}^{2} \sum_{k=1}^{2} T_i(\mathcal{F}; x_{i+1}^{(j)}, y_{i+1}^{(k)})$$

$$- T_i(\mathcal{F}; x_{i+1}^{(1)}, x_{i+1}^{(2)}, y_{i+1}^{(1)}, y_{i+1}^{(2)}). \tag{31.27}$$

In view of (31.27) we have to estimate the terms

$$T_i(\mathcal{F}; x_{i+1}^{(\mu)}, y_{i+1}^{(\nu)}),\ \mu = 1, 2,\ \nu = 1, 2 \tag{31.28}$$

from above. Fix a pair $\{x_{i+1}^{(\mu)}, y_{i+1}^{(\nu)}\}$; write

$$\mathcal{F}(x_{i+1}^{(\mu)}, y_{i+1}^{(\nu)}) = \left\{ A_1 \in \mathcal{F}:\ \{x_{i+1}^{(\mu)}, y_{i+1}^{(\nu)}\} \subset A_1, A_1 \cap Y(i) = \emptyset \right\},$$

and for every $A_1 \in \mathcal{F}(x_{i+1}^{(\mu)}, y_{i+1}^{(\nu)})$ define

$$I_2(A_1) = \{1 \leq j \leq i:\ \{x_j^{(1)}, x_j^{(2)}\} \subset A_1\} \text{ (full pairs)}$$

and

$$I_1(A_1) = \{1 \leq j \leq i:\ |\{x_j^{(1)}, x_j^{(2)}\} \cap A_1| = 1\} \text{ (half pairs)}.$$

We have learned from the counter-example to (31.19) that set $I_2(A_1)$ is "harmless" but $I_1(A_1)$ is "dangerous": $I_1(A_1)$ may cause the violation of (31.19) in the "big hypergraph" $\mathcal{F}_2^p$; the quantitative measure of the violation is the "deficit."

Fix an arbitrary $A_1 \in \mathcal{F}(x_{i+1}^{(\mu)}, y_{i+1}^{(\nu)})$; let

$$X(i, A_1) = \left\{ \{x_j^{(1)}, x_j^{(2)}\} \setminus A_1 : i \in I_1(A_1) \right\},$$

i.e. $X(i, A_1)$ denotes the set of *other members* of the half-occupied pairs $\{x_j^{(1)}, x_j^{(2)}\}$. This is the set of candidates for the possible "deficit."

Let $z \in X(i, A_1)$ be an arbitrary point which is *not* on the $x_{i+1}^{(\mu)} y_{i+1}^{(\nu)}$-line: if a set $A^* \in \mathcal{F}(x_{i+1}^{(\mu)}, y_{i+1}^{(\nu)})$ contains z, then the union $A_1 \cup A^*$ has a "deficit" caused by z and its "twin-brother" (i.e. a pair $\{x_j^{(1)}, x_j^{(2)}\}$) – because of the "one-counts-only" weight (31.25)–(31.26) – but the good news is that the triplet $\{x_{i+1}^{(\mu)}, y_{i+1}^{(\nu)}, z\}$ is non-collinear, so there are $\leq \binom{q^2}{3}$ "deficit-dangerous" sets like A^*. Taking all points $z \in X(i, A_1)$ outside of the $x_{i+1}^{(\mu)} y_{i+1}^{(\nu)}$-line, we see that there are $\leq q^2 \binom{q^2}{3} \leq q^8/6$ winning sets $A_2 \in \mathcal{F}(x_{i+1}^{(\mu)}, y_{i+1}^{(\nu)})$ such that the union $A_1 \cup A_2$ contains a complete pair $\{x_j^{(1)}, x_j^{(2)}\}$ (representing "deficit") such that the point pair is not covered by the $x_{i+1}^{(\mu)} y_{i+1}^{(\nu)}$-line.

Exactly the same argument proves that there are $\leq q^8/6$ winning sets $A_2 \in \mathcal{F}(x_{i+1}^{(\mu)}, y_{i+1}^{(\nu)})$ such that the intersection $A_1 \cap A_2$ is not covered by the $x_{i+1}^{(\mu)} y_{i+1}^{(\nu)}$-line; the two cases together give $\leq q^8/6 + q^8/6 = q^8/3$.

Write

$$\mathcal{F}(x_{i+1}^{(\mu)}, y_{i+1}^{(\nu)}; a, b) = \left\{ A \in \mathcal{F}(x_{i+1}^{(\mu)}, y_{i+1}^{(\nu)}) : I_2(A) = a \text{ and } I_1(A) = b \right\}$$

and

$$|\mathcal{F}(x_{i+1}^{(\mu)}, y_{i+1}^{(\nu)}; a, b)| = M_{a,b}.$$

By (31.25)–(31.26) we have

$$T_i \left(\mathcal{F}(x_{i+1}^{(\mu)}, y_{i+1}^{(\nu)}) \right) = \sum_{2a+b \leq q^2: \, a \geq 0, b \geq 0} M_{a,b} \cdot 2^{a+b-q^2}. \tag{31.29}$$

For notational simplicity, write $\mathcal{F}(x_{i+1}^{(\mu)}, y_{i+1}^{(\nu)}; a, b) = \mathcal{F}(a, b)$.

Assume that $|\mathcal{F}(a, b)| = M_{a,b} \geq 10q^8$; then there are at least

$$\frac{M_{a,b} \cdot (M_{a,b} - q^8/3) \cdot (M_{a,b} - 2q^8/3) \cdot (M_{a,b} - 3q^8/3) \cdot (M_{a,b} - 4q^8/3)}{5!} \geq \frac{M_{a,b}^5}{3^5}$$

5-tuples

$$\{A_1, A_2, A_3, A_4, A_5\} \in \binom{\mathcal{F}(a, b)}{5}$$

such that:

(1) deleting the points of the $x_{i+1}^{(\mu)}y_{i+1}^{(\nu)}$-line, the remaining parts of the 5 sets $A_1, \ldots, A_5$ become *disjoint*;
(2) there is no "deficit" outside of the $x_{i+1}^{(\mu)}y_{i+1}^{(\nu)}$-line; in other words, there is no complete pair

$$\{x_j^{(1)}, x_j^{(2)}\} \subset A_1 \cup A_2 \cup A_3 \cup A_4 \cup A_5,$$

which is disjoint from the $x_{i+1}^{(\mu)}y_{i+1}^{(\nu)}$-line.

It follows from the definition of the auxiliary "big hypergraph" $\mathcal{F}_{2*}^p$ (see (23.4)) that

$$T(\mathcal{F}_{2*}^5(i)) \geq \frac{M_{a,b}^5}{3^5} \cdot 2^{5(a+b-q^2)-2q} \text{ if } M_{a,b} \geq 10q^8, \tag{31.30}$$

where the "loss factor" 2^{-2q} comes from the possibility that on the $x_{i+1}^{(\mu)}y_{i+1}^{(\nu)}$-line "deficits" may occur (a straight line intersects a $q \times q$ parallelogram lattice in at most q points).

We reformulate (31.30): if $M_{a,b} \geq 10q^8$, then

$$3 \cdot 2^{2q/5} \left(T(\mathcal{F}_{2*}^5(i))\right)^{1/5} \geq M_{a,b} \cdot 2^{a+b-q^2},$$

which implies

$$3 \cdot 2^{2q/5} \cdot \left(T(\mathcal{F}_{2*}^5(i))\right)^{1/5} + 10q^8 \cdot 2^{a+b-q^2} \geq M_{a,b} \cdot 2^{a+b-q^2}. \tag{31.31}$$

By (31.29) and (31.31) we obtain the

$$T_i\left(\mathcal{F}(x_{i+1}^{(\mu)}, y_{i+1}^{(\nu)})\right) = \sum_{2a+b\leq q^2:\ a\geq 0, b\geq 0} M_{a,b} \cdot 2^{a+b-q^2}$$

$$\leq 20q^{10} + 3q^4 \cdot 2^{2q/5} \cdot \left(T(\mathcal{F}_{2*}^5(i))\right)^{1/5}. \tag{31.32}$$

We refer to inequality (31.32) as the **Key inequality**. The point is that Key Inequality (31.32) can successfully substitute for the (false in general) inequality (31.19)!

We work with the usual Potential Function

$$L_i = T_i(\mathcal{F}) - \lambda \cdot T_i(\mathcal{F}_{2*}^5), \tag{31.33}$$

where the positive constant λ is specified by the side condition

$$L_0 = \frac{1}{2}T(\mathcal{F}), \text{ that is, } \lambda = \frac{T(\mathcal{F})}{2T(\mathcal{F}_{2*}^5)}. \tag{31.34}$$

Then by (31.27) and (31.32)

$$L_{i+1} \geq L_i - \sum_{j=1}^{2}\sum_{k=1}^{2} T_i(\mathcal{F}; x_{i+1}^{(j)}, y_{i+1}^{(k)}) - T_i(\mathcal{F}; x_{i+1}^{(1)}, x_{i+1}^{(2)}, y_{i+1}^{(1)}, y_{i+1}^{(2)})$$

$$\geq L_i - 5\left(20q^{10} + 3q^4 \cdot 2^{2q/5} \cdot \left(T(\mathcal{F}_{2*}^5(i))\right)^{1/5}\right). \tag{31.35}$$

In the rest of the argument we simply repeat Section 23. We recall (23.12) and (23.13)

$$T(\mathcal{F}_{2*}^5) \le N^{14} \cdot 2^{-5q^2+5q}, \tag{31.36}$$

$$T(\mathcal{F}) \ge \frac{N^6}{2^{14}(q-1)^4} 2^{-q^2}. \tag{31.37}$$

We distinguish two cases: either there is or there is no index i such that

$$T(\mathcal{F}_{2*}^5(i)) > N^{5.5} \cdot T(\mathcal{F}_{2*}^5). \tag{31.38}$$

Case 1: There is no index i such that (31.38) holds.
By (31.34), the Critical Inequality (31.32), and (31.38)

$$L_{end} = L_{N^2/2} \ge L_0 - \sum_{i=0}^{\frac{N^2}{2}-1} \left(100q^{10} + 15q^4 2^{2q/5} \left(T(\mathcal{F}_{2*}^5(i)) \right)^{1/5} \right)$$

$$\ge L_0 - \frac{N^2}{2} \cdot \left(100q^{10} + 15q^4 2^{2q/5} \cdot N^{1.1} \cdot \left(T(\mathcal{F}_{2*}^5(i)) \right)^{1/5} \right)$$

$$= \frac{1}{2}T(\mathcal{F}) - \frac{N^2}{2} \cdot \left(100q^{10} + 15q^4 2^{2q/5} \cdot N^{1.1} \cdot \left(T(\mathcal{F}_{2*}^5) \right)^{1/5} \right). \tag{31.39}$$

By (31.36), (31.37), and (31.39)

$$L_{end} \ge \frac{1}{2} \cdot \frac{N^6}{2^{14}(q-1)^4} 2^{-q^2} - \frac{N^2}{2} \cdot \left(100q^{10} + 15q^4 2^{2q/5} \cdot N^{1.1} \cdot N^{14/5} \cdot 2^{-q^2+q} \right). \tag{31.40}$$

By choosing $q = (2+o(1))\sqrt{\log_2 N}$ we have $2^{q^2} = N^{4+o(1)}$, so by (31.40)

$$L_{end} \ge \frac{N^6}{2^{15}(q-1)^4} 2^{-q^2} - 50N^2 q^{10} - N^{1.9+o(1)}. \tag{31.41}$$

The right-hand side of (31.41) is strictly positive if $q = \lfloor 2\sqrt{\log_2 N} - o(1) \rfloor$ (with an appropriate $o(1)$) and N is large enough.

If $L_{end} > 0$, then $T(\mathcal{F}(end)) \ge L_{end} > 0$ implying that Maker has a Weak Win in the (2:2) game.

Case 2: There is an index i such that (31.38) holds.
Again we prove that Case 2 is impossible! Indeed, let $i = j_1$ denote the *first* index such that (31.38) holds. Then we can save the argument of (31.29)–(31.41) in Case 1, and obtain the inequality $L_{j_1} > 0$ if $q = \lfloor 2\sqrt{\log_2 N} - o(1) \rfloor$ (with an appropriate $o(1)$) and N is large enough. By definition (see (31.33))

$$L_{j_1} = T(\mathcal{F}(j_1)) - \lambda \cdot T(\mathcal{F}_{2*}^5(j_1)) > 0,$$

implying

$$T(\mathcal{F}(j_1)) > \lambda \cdot T(\mathcal{F}_{2*}^5(j_1)) > \lambda \cdot N^{5.5} \cdot T(\mathcal{F}_{2*}^5)$$

$$= \frac{T(\mathcal{F})}{2T(\mathcal{F}_{2*}^5)} \cdot N^{5.5} \cdot T(\mathcal{F}_{2*}^5) = \frac{1}{2} N^{5.5} \cdot T(\mathcal{F}). \qquad (31.42)$$

By (31.37) and (31.42)

$$T(\mathcal{F}(j_1)) > N^{5.5} \cdot \frac{N^6}{2^{15}(q-1)^4} 2^{-q^2} = N^{7.5-o(1)}, \qquad (31.43)$$

which is clearly false! Indeed, a corner point and its two neighbors uniquely determine a $q \times q$ parallelogram lattice, implying the trivial upper bound

$$T(\mathcal{F}(j_1)) \le |\mathcal{F}(j_1)| \le |\mathcal{F}| \le (N^2)^2 = N^6,$$

which contradicts (31.43). This contradiction proves that Case 2 is impossible. This completes the proof of part (1) in the following lower bound result.

Theorem 31.1 *In the (2:2) Lattice Achievement Games the Achievement Numbers are at least as large as the right-hand sides in Open Problem 30.3. For example, playing the (2:2) game on an $N \times N$ board, at the end of the play Maker can always occupy a*

(1) a $q \times q$ parallelogram lattice with

$$q = \left\lfloor 2\sqrt{\log_2 N} - o(1) \right\rfloor;$$

(2) a $q \times q$ aligned rectangle lattice with

$$q = \left\lfloor \sqrt{2 \log_2 N} - o(1) \right\rfloor.$$

The same holds for the (2:2) Lattice Avoidance Games.

The same argument works, with natural modifications, for all lattices in Section 8 (see Theorem 8.2). For example, in the proof of part (2) an obvious novelty is that in the aligned rectangle lattice the *horizontal* and *vertical* directions play a special role. We can overcome this by replacing the auxiliary hypergraph $\mathcal{F}_{2*}^p$ with $\mathcal{F}_{2**}^p$, see (23.24); otherwise the proof is identical.

The big unsolved problem is always the "other direction."

Open Problem 31.2 *Is it true that Theorem 31.1 is best possible? For example, is it true that, given any constant $c > 2$, playing the (2:2) game on an $N \times N$, Breaker can prevent Maker from building a $q \times q$ parallelogram lattice with $q = c\sqrt{\log_2 N}$ if N is large enough?*

Is it true that, given any constant $c > \sqrt{2}$, playing the (2:2) game on an $N \times N$, Breaker can prevent Maker from building a $q \times q$ aligned rectangle lattice with $q = c\sqrt{\log_2 N}$ if N is large enough?

32

Biased games (III): Sporadic results

1. Exact solutions in the biased case. Unfortunately Theorem 31.1 is just a partial result: one direction is clearly missing; is there any biased play for which we can prove an *exact result*? Questions 30.1–30.2 describe the cases where we have a "chance." Which one is solvable? We succeeded with Question 30.2: the $(a : 1)$ Avoidance Game where Avoider is the topdog. We have already formulated this result in Section 9 as Theorem 9.2.

Let $a \geq 2$; recall that the $(a{:}1)$ Avoidance Game means that Avoider takes a points per move and Forcer takes 1 point per move. $A(N \times N; \ \text{square lattice}; \ a{:}1; \ -)$ denotes the largest value of q such that Forcer can always force Avoider to occupy a whole $q \times q$ Aligned Square Lattice in the $N \times N$ board. This is called the *Avoidance Number of the biased* $(a{:}1)$ *game where Avoider is the topdog*.

Theorem 9.2 *Consider the $N \times N$ board; let $a \geq 2$ and consider the $(a{:}1)$ Avoidance Game where Forcer is the underdog. We know the biased Avoidance Numbers:*

(a) $A(N \times N; \ \text{square lattice}; \ a{:}1; \ -) = \left\lfloor \sqrt{\frac{\log N}{\log(1+\frac{1}{a})}} + o(1) \right\rfloor,$

(b) $A(N \times N; \ \text{rectangle lattice}; \ a{:}1; \ -) = \left\lfloor \sqrt{\frac{2\log N}{\log(1+\frac{1}{a})}} + o(1) \right\rfloor,$

(c) $A(N \times N; \ \text{tilted square lattice}; \ a{:}1; \ -) = \left\lfloor \sqrt{\frac{2\log N}{\log(1+\frac{1}{a})}} + o(1) \right\rfloor,$

(d) $A(N \times N; \ \text{tilted rectangle lattice}; \ a{:}1; \ -) = \left\lfloor \sqrt{\frac{2\log N}{\log(1+\frac{1}{a})}} + o(1) \right\rfloor,$

(e) $A(N \times N; \ \text{rhombus lattice}; \ a{:}1; \ -) = \left\lfloor \sqrt{\frac{2\log N}{\log(1+\frac{1}{a})}} + o(1) \right\rfloor,$

(f) $A(N \times N; \ \text{parallelogram lattice}; \ a{:}1; \ -) = \left\lfloor 2\sqrt{\frac{\log N}{\log(1+\frac{1}{a})}} + o(1) \right\rfloor,$

(g) $A(N \times N; \ \text{area one lattice}; \ a{:}1; \ -) = \left\lfloor \sqrt{\frac{2\log N}{\log(1+\frac{1}{a})}} + o(1) \right\rfloor.$

Of course, this is nothing else other than Open Problem 30.3 in the special case of the $(a{:}1)$ avoidance play.

The proof of the upper bounds in Theorem 9.2 combines Theorem 20.4 with the BigGame–SmallGame Decomposition technique (see Part D). What needs to be explained here is how to prove the lower bounds in Theorem 9.2. The proof is an extension of the proof technique of Theorem 31.1. First we briefly recall the basic idea of the proof of Theorem 31.1.

Using the standard Power-of-Two Scoring System, the effect of the $(i+1)$st moves $x_{i+1}^{(1)}$, $x_{i+1}^{(2)}$ and $y_{i+1}^{(1)}$, $y_{i+1}^{(2)}$ is described by a long identity (see (31.2)), which has the "abstract form"

$$(1+x^{(1)})(1+x^{(2)})(1-y^{(1)})(1-y^{(2)}) - 1 = \sum_{j=1}^{2} x^{(j)} - \sum_{j=1}^{2} y^{(j)}$$

$$+ x^{(1)}x^{(2)} + y^{(1)}y^{(2)} - \sum_{j=1}^{2}\sum_{k=1}^{2} x^{(j)}y^{(k)}$$

$$- x^{(1)}x^{(2)} \left(\sum_{j=1}^{2} y^{(j)} \right) + y^{(1)}y^{(2)} \left(\sum_{j=1}^{2} x^{(j)} \right)$$

$$+ x^{(1)}x^{(2)}y^{(1)}y^{(2)}. \tag{32.1}$$

For simplicity consider only the "linear and quadratic" terms in (32.1)

$$\sum_{j=1}^{2} x^{(j)} + x^{(1)}x^{(2)} - \sum_{j=1}^{2} y^{(j)} + y^{(1)}y^{(2)}. \tag{32.2}$$

Expression (32.1) is *not* symmetric, but the expression below

$$\sum_{j=1}^{2} x^{(j)} - x^{(1)}x^{(2)} - \sum_{j=1}^{2} y^{(j)} + y^{(1)}y^{(2)} \tag{32.3}$$

is symmetric in the sense that we can rewrite it in the compact form

$$f(x^{(1)}, x^{(2)}) - f(y^{(1)}, y^{(2)}) \text{ where } f(u_1, u_2) = u_1 + u_2 - u_1 u_2.$$

We sacrifice the Power-of-Two Scoring System, and switch to the "one-counts-only" system

$$T_i(\mathcal{H}) = \sum_{A \in \mathcal{H}} w_i(A) \text{ and} \tag{32.4}$$

$$T_i(\mathcal{H}; u_1, \dots, u_m) = \sum_{A \in \mathcal{H}: \{u_1,\dots,u_m\} \subset A} w_i(A),$$

where $w_i(A) = 0$ if $A \cap Y(i) \neq \emptyset$ and

$$w_i(A) = 2^{|\{1 \le j \le i: A \cap \{x_j^{(1)}, x_j^{(2)}\} \neq \emptyset\}| - |A|} \tag{32.5}$$

if $A \cap Y(i) = \emptyset$.

Switching to (32.4)–(32.5) means that the product $(1+x^{(1)})(1+x^{(2)})$ in (32.1) is replaced by the quadratic polynomial $(1+x^{(1)}+x^{(2)} - x^{(1)}x^{(2)})$, which leads to the following analogue of (32.1)

$$\left(1+x^{(1)}+x^{(2)}-x^{(1)}x^{(2)}\right)\left(1-y^{(1)}\right)\left(1-y^{(2)}\right)-1$$

$$=\left(x^{(1)}+x^{(2)}-x^{(1)}x^{(2)}\right)-\left(y^{(1)}+y^{(2)}-y^{(1)}y^{(2)}\right)$$

$$-\sum_{j=1}^{2}\sum_{k=1}^{2}x^{(j)}y^{(k)}+x^{(1)}x^{(2)}\left(\sum_{j=1}^{2}y^{(j)}\right)$$

$$+y^{(1)}y^{(2)}\left(\sum_{j=1}^{2}x^{(j)}\right)-x^{(1)}x^{(2)}y^{(1)}y^{(2)}. \tag{32.6}$$

Remark. In the (2:2) play, the "trick" was to replace $1+x^{(1)}+x^{(2)}+x^{(1)}x^{(2)}$ with $1+x^{(1)}+x^{(2)}-x^{(1)}x^{(2)}$. In contrast, the (1:2) Achievement play, where Maker is the underdog, is completely unsolved; we have no "room" for any trick. In the (1:2) expansion

$$(1+2x)(1-y^{(1)})(1-y^{(2)})-1=2x-y^{(1)}-y^{(2)}$$

$$+y^{(1)}y^{(2)}-2x(y^{(1)}+y^{(2)})+2xy^{(1)}y^{(2)}$$

the product $y^{(1)}y^{(2)}$ is the "troublemaker." We do not know how to get rid of it. For example, to prove anything remotely non-trivial about the (1:2) Clique Achievement Game, we have to go back to "Ramseyish" arguments, see the Corollary of Theorem 33.7 (we can prove only half of the conjectured truth). For the (1:2) Lattice Achievement Game we can only use the linear criterion (30.3).

A closely related Coalition Game concept is defined at the end of Section 48, including a non-trivial result.

Let's now return to (32.6): we use the "formal inequalities"

$$x^{(1)}x^{(2)}\sum_{j=1}^{2}y^{(j)}\le\frac{1}{2}\sum_{j=1}^{2}\sum_{k=1}^{2}x^{(j)}y^{(k)}$$

and its "twin brother"

$$y^{(1)}y^{(2)}\sum_{j=1}^{2}x^{(j)}\le\frac{1}{2}\sum_{j=1}^{2}\sum_{k=1}^{2}x^{(j)}y^{(k)}.$$

Of course, these "formal inequalities" have obvious "real" meaning like the trivial inequality

$$T_i(\mathcal{H};x_{i+1}^{(1)},x_{i+1}^{(2)},y_{i+1}^{(j)})\le\frac{1}{2}\sum_{k=1}^{2}T_i(\mathcal{H};x_{i+1}^{(k)},y_{i+1}^{(j)}).$$

Now applying the "formal inequalities" in (32.6) we obtain the "formal" upper bound

$$\left(1+x^{(1)}+x^{(2)}-x^{(1)}x^{(2)}\right)\left(1-y^{(1)}\right)\left(1-y^{(2)}\right)-1$$

$$\le f(x^{(1)},x^{(2)})-f(y^{(1)},y^{(2)}) \quad \text{where } f(u_1,u_2)=u_1+u_2-u_1u_2. \tag{32.7}$$

Consider the *Potential Function*

$$L_i = T_i(\mathcal{F}) - \lambda \cdot T_i(\mathcal{F}_{2*}^p)$$

(or $T_i(\mathcal{F}_{2**}^p)$) with the usual side condition

$$L_0 = \frac{1}{2}T(\mathcal{F}), \quad \text{that is,} \quad \frac{1}{2}T(\mathcal{F}) = \lambda \cdot T(\mathcal{F}_{2*}^p).$$

Then by (32.6)–(32.7) we have the inequality

$$L_{i+1} \geq L_i + g(x_{i+1}^{(1)}, x_{i+1}^{(2)}) - g(y_{i+1}^{(1)}, y_{i+1}^{(2)})$$

$$- \sum_{j=1}^{2}\sum_{k=1}^{2} T_i(\mathcal{F}; x_{i+1}^{(j)}, y_{i+1}^{(k)}) - T_i(\mathcal{F}; x_{i+1}^{(1)}, x_{i+1}^{(2)}, y_{i+1}^{(1)}, y_{i+1}^{(2)})$$

(notice the Negativity of the second line!), where

$$g(u_1, u_2) = L_i(u_1) + L_i(u_2) - L_i(u_1, u_2) \tag{32.8}$$

and for $m = 1, 2$

$$L_i(u_1, \ldots, u_m) = T_i(\mathcal{F}; u_1, \ldots, u_m) - \lambda \cdot T_i(\mathcal{F}_{2*}^p; u_1, \ldots, u_m)$$

for any set $\{u_1, \ldots, u_m\}$ of unselected points.

Maker's $(i+1)$st move is that unoccupied point pair

$$\{x_{i+1}^{(1)}, x_{i+1}^{(1)}\} = \{u_1, u_2\}$$

for which the function $g(u_1, u_2)$ attains its *maximum*; then from (32.8) we obtain the critical inequality

$$L_{i+1} \geq L_i - \sum_{j=1}^{2}\sum_{k=1}^{2} T_i(\mathcal{F}; x_{i+1}^{(j)}, y_{i+1}^{(k)})$$

$$- T_i(\mathcal{F}; x_{i+1}^{(1)}, x_{i+1}^{(2)}, y_{i+1}^{(1)}, y_{i+1}^{(2)}), \tag{32.9}$$

and the technique of Section 24 works without any difficulty.

Now we leave the (2:2) Achievement Game (Theorem 31.1), and turn to the proof of Theorem 9.2. In the $(a:1)$ Avoidance Game with $a \geq 2$ the "natural" scoring is the Power-of-$(\frac{a+1}{a})$ Scoring System, which is described by the "abstract identity" below (the analogue of (32.1))

$$\left(1+\frac{1}{a}x^{(1)}\right)\left(1+\frac{1}{a}x^{(2)}\right)\cdots\left(1+\frac{1}{a}x^{(a)}\right)(1-y)-1=\frac{1}{a}\sum_{j=1}^{a}x^{(j)}$$

$$+\frac{1}{a^2}\sum_{1\le j_1<j_2\le a}x^{(j_1)}x^{(j_2)}+\frac{1}{a^3}\sum_{1\le j_1<j_2<j_3\le a}x^{(j_1)}x^{(j_2)}x^{(j_3)}+\cdots$$

$$-y-\frac{y}{a}\sum_{j=1}^{a}x^{(j)}-\frac{y}{a^2}\sum_{1\le j_1<j_2\le a}x^{(j_1)}x^{(j_2)}$$

$$-\frac{y}{a^3}\sum_{1\le j_1<j_2<j_3\le a}x^{(j_1)}x^{(j_2)}x^{(j_3)}+\cdots . \tag{32.10}$$

Identity (32.10) is very complicated, there is little hope of obtaining an upper bound such as (32.7). We make a drastic simplification in (32.10) by replacing the product

$$\left(1+\frac{1}{a}x^{(1)}\right)\left(1+\frac{1}{a}x^{(2)}\right)\cdots\left(1+\frac{1}{a}x^{(a)}\right) \tag{32.11}$$

with the linear polynomial $(1+\frac{1}{a}\sum_{j=1}^{a}x^{(j)})$

$$\left(1+\frac{1}{a}\sum_{j=1}^{a}x^{(j)}\right)(1-y)-1=\frac{1}{a}\sum_{j=1}^{a}x^{(j)}-y-\frac{y}{a}\sum_{j=1}^{a}x^{(j)}. \tag{32.12}$$

Replacing product (32.11) with a linear polynomial (see (32.12)) means that the Power-of-$(\frac{a+1}{a})$ Scoring System is replaced by a new "linear" scoring system: if $A\cap Y(i)=\emptyset$ then

$$w_i(A)=\prod_{1\le j\le i}\left(1+\frac{1}{a}|\{x_j^{(1)},x_j^{(2)},\ldots,x_j^{(a)}\}\cap A|\right), \tag{32.13}$$

and $w_i(A)=0$ if $A\cap Y(i)\ne\emptyset$.

In view of (32.12) the Potential Function

$$L_i=T_i(\mathcal{F})-\lambda\cdot T_i(\mathcal{F}_{2*}^p)$$

with the usual side condition

$$L_0=\frac{1}{2}T_0(\mathcal{F}), \quad\text{that is,}\quad \frac{1}{2}T_0(\mathcal{F})=\lambda\cdot T_0(\mathcal{F}_{2*}^p),$$

satisfies the inequality ("Decreasing Property")

$$L_{i+1}\ge L_i+\frac{1}{a}\sum_{k=1}^{a}h(x_{i+1}^{(k)})-h(y_{i+1})-\frac{1}{a}\sum_{k=1}^{a}T_i(\mathcal{F};x_{i+1}^{(k)},y_{i+1}), \tag{32.14}$$

where

$$h(u)=T_i(\mathcal{F};u)-\lambda\cdot T_i(\mathcal{F}_2^p;u). \tag{32.15}$$

Assume that Forcer is the first player, so y_{i+1} is selected before $x_{i+1}^{(k)}$, $k = 1, 2, \ldots, a$. Forcer's $(i+1)$st move is the unoccupied u for which the function $h(u)$ attains its *minimum*. Then we obtain the analogue of (32.9)

$$L_{i+1} \geq L_i - \frac{1}{a} \sum_{k=1}^{a} T_i(\mathcal{F}; x_{i+1}^{(k)}, y_{i+1}), \tag{32.16}$$

and the technique of Section 24 works without any difficulty.

Again we have a problem with finding an analogue of the Generalized Variance Lemma with the unusual "linear" weight (32.13). Unfortunately this is a major roadblock, but we can get around this technical difficulty by working with the "nearly disjoint" auxiliary hypergraph $\mathcal{F}_{2*}^p$ for the parallelogram lattice, working with $\mathcal{F}_{2**}^p$ for the aligned rectangle lattice, and so on. By repeating the proof of Theorem 31.1 we can routinly complete the proof of the Forcer's win part of Theorem 9.2.

The missing Avoider's win part of Theorem 9.2 comes from combining Theorem 20.4 with the decomposition techniques of Part D.

Theorem 9.2 gives an affirmative answer to Question 30.2 (well, at least for the Lattice Games). How about Question 30.1? What happens in the $(1:b)$ Achievement Game with $b \geq 2$ where Breaker is the topdog? Unfortunately in this case our machinery breaks down; the only case we can handle is the case of "easy lattices": the aligned and tilted Square Lattices. The corresponding hypergraphs are "nearly" Almost Disjoint: given 2 points in the $N \times N$ board, there are $\leq \binom{q^2}{2}$ $q \times q$ Square Lattices containing the two points; $\binom{q^2}{2}$ is polylogarithmic in N. We just need the following slight generalization of (30.3) involving the Max Pair-Degree.

If hypergraph $\mathcal{F}$ is n-uniform and

$$\frac{|\mathcal{F}|}{|V|} \left(\frac{f}{f+s} \right)^n > \frac{f^2 s^2}{(f+s)^3} \cdot \Delta_2, \tag{32.17}$$

where $\Delta_2 = \Delta_2(\mathcal{F})$ is the Max Pair-Degree of $\mathcal{F}$, then playing the $(m:b)$ game on $\mathcal{F}$ the first player (Maker) can occupy a whole $A \in \mathcal{F}$.

A straightforward application of (32.17) gives the Maker's win part of:

Theorem 32.1 *Consider the $N \times N$ board; let $b \geq 2$ and consider the $(1:b)$ Achievement Game where Breaker is the topdog. We know the following two biased Achievement Numbers:*

(a) $\mathbf{A}(N \times N; \text{ square lattice}; 1:b) = \left\lfloor \sqrt{\frac{\log N}{\log(1+b)}} + o(1) \right\rfloor$,

(b) $\mathbf{A}(N \times N; \text{ tilted square lattice}; 1:b) = \left\lfloor \sqrt{\frac{2 \log N}{\log(1+b)}} + o(1) \right\rfloor$.

The missing Breaker's win part of Theorem 32.1 comes from combining (30.4) with the decomposition techniques of Part D.

Theorems 9.2 and 32.1 represent the only known exact solutions in the biased case. How come that we can solve these cases but cannot solve the rest of the cases? What has been overlooked here? What an interesting research problem!

We conclude this section with an extension of Theorem 31.1 to the general fair case $(k : k)$. We begin with:

2. The (3:3) game. Theorem 31.1 was about the fair (2:2) play, and for the Lattice Games we could prove exactly the same lower bounds as in the usual (1:1) play. How about the (3:3) play? There is no surprise: we can prove the perfect analogue of Theorem 31.1, and the proof works for every other fair $(k:k)$ game. The analogue of (32.6) is the tricky "formal equality"

$$(1 + x^{(1)} + x^{(2)} + x^{(3)} - x^{(1)}x^{(2)} - x^{(1)}x^{(3)} - x^{(2)}x^{(3)} + x^{(1)}x^{(2)}x^{(3)})(1 - y^{(1)})(1 - y^{(2)})$$

$$(1 - y^{(3)}) - 1 = \left(1 + \left(1 - (1 - x^{(1)})(1 - x^{(2)})(1 - x^{(3)})\right)\right)$$

$$\left(1 - \left(1 - (1 - y^{(1)})(1 - y^{(2)})(1 - y^{(3)})\right)\right)$$

$$-1 = f(x^{(1)}, x^{(2)}, x^{(3)}) - f(y^{(1)}, y^{(2)}, y^{(3)}) \tag{32.18}$$

$$- \left(1 - (1 - x^{(1)})(1 - x^{(2)})(1 - x^{(3)})\right)\left(1 - (1 - y^{(1)})(1 - y^{(2)})(1 - y^{(3)})\right),$$

where

$$f(u_1, u_2, u_3) = 1 - (1 - u_1)(1 - u_2)(1 - u_3)$$

$$= u_1 + u_2 + u_3 - u_1 u_2 - u_1 u_3 - u_2 u_3 + u_1 u_2 u_3.$$

Note that the inequality

$$1 - (1 - x^{(1)})(1 - x^{(2)})(1 - x^{(3)}) \geq 0 \tag{32.19}$$

holds for for any choice of $x^{(j)} \in \{0, 1\}$, $j = 1, 2, 3$, and the same for

$$1 - (1 - y^{(1)})(1 - y^{(2)})(1 - y^{(3)}) \geq 0. \tag{32.20}$$

Combining (32.18)–(32.20) we obtain the "formal inequality"

$$(1 + x^{(1)} + x^{(2)} + x^{(3)} - x^{(1)}x^{(2)} - x^{(1)}x^{(3)} - x^{(2)}x^{(3)} + x^{(1)}x^{(2)}x^{(3)})(1 - y^{(1)})(1 - y^{(2)})$$

$$(1 - y^{(3)}) - 1 = \left(1 + \left(1 - (1 - x^{(1)})(1 - x^{(2)})(1 - x^{(3)})\right)\right)$$

$$\left(1 - \left(1 - (1 - y^{(1)})(1 - y^{(2)})(1 - y^{(3)})\right)\right) - 1 \leq f(x^{(1)}, x^{(2)}, x^{(3)}) - f(y^{(1)}, y^{(2)}, y^{(3)}). \tag{32.21}$$

Identity (32.18) means that we use the following "one-counts-only" scoring system

$$T_i(\mathcal{H}) = \sum_{A \in \mathcal{H}} w_i(A) \quad \text{and} \tag{32.22}$$

$$T_i(\mathcal{H}; u_1, \ldots, u_m) = \sum_{A \in \mathcal{H}: \{u_1, \ldots, u_m\} \subset A} w_i(A),$$

where $w_i(A) = 0$ if $A \cap Y(i) \neq \emptyset$ and

$$w_i(A) = 2^{|\{1 \leq j \leq i: \ A \cap \{x_j^{(1)}, x_j^{(2)}, x_j^{(3)}\} \neq \emptyset\}| - |A|} \tag{32.23}$$

if $A \cap Y(i) = \emptyset$.

The rest of the proof goes exactly like in the proof of Theorem 31.1. For example, consider the Parallelogram Lattice Game. We use the usual Potential Function

$$L_i = T_i(\mathcal{F}) - \lambda \cdot T_i(\mathcal{F}_{2*}^p)$$

with the side condition

$$L_0 = \frac{1}{2} T_0(\mathcal{F}), \text{ that is, } \lambda = \frac{T_0(\mathcal{F})}{2 T_0(\mathcal{F}_{2*}^p)}.$$

Then by (32.21) we obtain the "Decreasing Property"

$$L_{i+1} \geq L_i + h(x^{(1)}, x^{(2)}, x^{(3)}) - h(y^{(1)}, y^{(2)}, y^{(3)})$$

$$- \sum_{j=1}^{3} \sum_{k=1}^{3} T_i(\mathcal{F}; x_{i+1}^{(j)}, y_{i+1}^{(k)}) - \sum_{j=1}^{3} T_i(\mathcal{F}; x_{i+1}^{(1)}, x_{i+1}^{(2)}, x_{i+1}^{(3)}, y_{i+1}^{(j)})$$

$$- \sum_{j=1}^{3} T_i(\mathcal{F}; x_{i+1}^{(j)}, y_{i+1}^{(1)}, y_{i+1}^{(2)}, y_{i+1}^{(3)}) -$$

$$\sum_{1 \leq j_1 < j_2 \leq 3} \sum_{1 \leq k_1 < k_2 \leq 3} T_i(\mathcal{F}; x_{i+1}^{(j_1)}, x_{i+1}^{(j_2)}, y_{i+1}^{(k_1)}, y_{i+1}^{(k_2)})$$

$$- T_i(\mathcal{F}; x_{i+1}^{(1)}, x_{i+1}^{(2)}, x_{i+1}^{(3)}, y_{i+1}^{(1)}, y_{i+1}^{(2)}, y_{i+1}^{(3)})$$

(notice the Negativity of the 2nd, 3rd, and 4th lines!) where

$$h(z_1, z_2, z_3) = L_i(z_1) + L_i(z_2) + L_i(z_3) - L_i(z_1) L_i(z_2)$$
$$- L_i(z_1) L_i(z_3) - L_i(z_2) L_i(z_3) + L_i(z_1) L_i(z_2) L_i(z_3)$$

and for $m = 1, 2, 3$

$$L_i(u_1, \ldots, u_m) = T_i(\mathcal{F}; u_1, \ldots, u_m) - \lambda \cdot T_i(\mathcal{F}_{2*}^p; u_1, \ldots, u_m).$$

Maker chooses that unoccupied triplet $\{z_1, z_2, z_3\} = \{x_{i+1}^{(1)}, x_{i+1}^{(2)}, x_{i+1}^{(3)}\}$ for which the function $h(z_1, z_2, z_3)$ attains its *maximum*. We leave the rest of the proof to the reader.

Identity (32.18) clearly extends from 3-products to arbitrary k-products, and this way we get the following extension of Theorem 31.1.

Theorem 31.1 *In the fair* $(k{:}k)$ *Lattice Achievement Games the Achievement Numbers are at least as large as the right-hand sides in Open Problem 30.3. For example, playing the* $(k{:}k)$ *Achievement Lattice Game on an* $N \times N$ *board* $(k \geq 2)$, *at the end of the play:*

(1) Maker can always own a $q \times q$ parallelogram lattice with

$$q = \left\lfloor 2\sqrt{\log_2 N} - o(1) \right\rfloor;$$

(2) Maker can always own a $q \times q$ aligned rectangle lattice with

$$q = \left\lfloor \sqrt{2 \log_2 N} - o(1) \right\rfloor.$$

The same holds for every fair (k:k) Avoidance Lattice Game.

The same argument works, with minor modifications, for all lattices in Section 8 (see Theorem 8.2). For example, in the proof of part (2) an obvious novelty is that in the aligned rectangle lattice the *horizontal* and *vertical* directions play a special role. We can overcome this by replacing the auxiliary hypergraph $\mathcal{F}_{2*}^p$ with $\mathcal{F}_{2**}^p$, see (23.24); otherwise the proof is the same.

The big unsolved problem is always the "other direction."

Open Problem 32.1 *Is it true that Theorem 31.1′ is best possible? For example, is it true that, given any constant $c > 2$, playing the (k:k) game on an $N \times N$, Breaker can prevent Maker from building a $q \times q$ parallelogram lattice with $q = c\sqrt{\log_2 N}$ if N is large enough?*

33

Biased games (IV): More sporadic results

1. When Maker (or Avoider) is the topdog. Theorem 31.1′ was a building result, supporting one direction of the Meta-Conjecture for the general $(k:k)$ fair game. Its proof technique can be easily extended to all biased Avoidance games where Avoider is the topdog: the $(a:f)$ play with $a > f \geq 1$. For notational simplicity, we discuss the $(3:2)$ play only, and leave the general case to the reader. The $(3:2)$ avoidance analogue of (32.18) and (32.21) goes as follows ("Decreasing Property")

$$
\left(1 + \frac{1}{3}\left(1 - (1 - x^{(1)})(1 - x^{(2)})\right) + \frac{1}{3}\left(1 - (1 - x^{(1)})(1 - x^{(3)})\right) \right.
$$

$$
\left. + \frac{1}{3}\left(1 - (1 - x^{(2)})(1 - x^{(3)})\right) \right) \cdot (1 - y^{(1)})(1 - y^{(2)}) - 1
$$

$$
= \left(1 + \frac{1}{3}\left(1 - (1 - x^{(1)})(1 - x^{(2)})\right) + \frac{1}{3}\left(1 - (1 - x^{(1)})(1 - x^{(3)})\right) \right.
$$

$$
\left. + \frac{1}{3}\left(1 - (1 - x^{(2)})(1 - x^{(3)})\right) \right) \cdot \left(1 - \left(1 - (1 - y^{(1)})(1 - y^{(2)})\right)\right) - 1
$$

$$
= \frac{1}{3}f(x^{(1)}, x^{(2)}) + \frac{1}{3}f(x^{(1)}, x^{(3)}) + \frac{1}{3}f(x^{(2)}, x^{(3)}) - f(y^{(1)}, y^{(2)})
$$

$$
- \frac{1}{3}f(y^{(1)}, y^{(2)})\left(f(x^{(1)}, x^{(2)}) + f(x^{(1)}, x^{(3)}) + f(x^{(2)}, x^{(3)})\right) \leq
$$

$$
\leq \frac{1}{3}\left(f(x^{(1)}, x^{(2)}) + f(x^{(1)}, x^{(3)}) + f(x^{(2)}, x^{(3)})\right) - f(y^{(1)}, y^{(2)}),
$$

where

$$
f(u_1, u_2) = 1 - (1 - u_1)(1 - u_2).
$$

We get a strictly positive weight here because

$$
\left(1 - (1 - x^{(1)})(1 - x^{(2)})\right) + \left(1 - (1 - x^{(1)})(1 - x^{(3)})\right) + \left(1 - (1 - x^{(2)})(1 - x^{(3)})\right) \geq 0
$$

439

for every choice of $x^{(1)} \in \{0, 1\}$, $x^{(2)} \in \{0, 1\}$, $x^{(3)} \in \{0, 1\}$.

Forcer's obvious optimal move is to minimize function $f(u_1, u_2)$. The rest is standard.

How about the biased Achievement Game $(m:b)$ with $m > b \geq 1$? Again we just discuss the (3:2) play, and leav e the general case to the reader. The (3:2) achievement analogue of (32.18) and (32.21) goes as follows ("Decreasing Property")

$$\left(\frac{1}{3}(1 - x^{(1)}) + \frac{1}{3}(1 - x^{(2)}) + \frac{1}{3}(1 - x^{(3)}) + \left(1 - (1 - x^{(1)})(1 - x^{(2)})(1 - x^{(3)})\right) \right)$$

$$\cdot (1 - y^{(1)})(1 - y^{(2)}) - 1$$

$$= \left(\frac{1}{3}(1 - x^{(1)}) + \frac{1}{3}(1 - x^{(2)}) + \frac{1}{3}(1 - x^{(3)}) + \left(1 - (1 - x^{(1)})(1 - x^{(2)})(1 - x^{(3)})\right) \right)$$

$$\cdot \left(1 - \left(1 - (1 - y^{(1)})(1 - y^{(2)})\right)\right) - 1 =$$

$$= h(x^{(1)}, x^{(2)}, x^{(3)}) - \frac{1}{3}h(y^{(1)}, y^{(2)}, x^{(1)}) - \frac{1}{3}h(y^{(1)}, y^{(2)}, x^{(2)})$$

$$- \frac{1}{3}h(y^{(1)}, y^{(2)}, x^{(3)}) - h(x^{(1)}, x^{(2)}, x^{(3)}) \cdot g(y^{(1)}, y^{(2)})$$

$$\leq h(x^{(1)}, x^{(2)}, x^{(3)}) - \frac{1}{3}\left(h(y^{(1)}, y^{(2)}, x^{(1)}) + h(y^{(1)}, y^{(2)}, x^{(2)}) + h(y^{(1)}, y^{(2)}, x^{(3)})\right),$$

where

$$h(u_1, u_2, u_3) = 1 - (1 - u_1)(1 - u_2)(1 - u_3)$$

and

$$g(u_1, u_2) = 1 - (1 - u_1)(1 - u_2).$$

Again we get a strictly positive weight because

$$\left(\frac{1}{3}(1 - x^{(1)}) + \frac{1}{3}(1 - x^{(2)}) + \frac{1}{3}(1 - x^{(3)}) + \left(1 - (1 - x^{(1)})(1 - x^{(2)})(1 - x^{(3)})\right) \right) > 0$$

for every choice of $x^{(1)} \in \{0, 1\}$, $x^{(2)} \in \{0, 1\}$, $x^{(3)} \in \{0, 1\}$.

Maker's obvious optimal move is to maximize function $h(u_1, u_2, u_3)$. The rest is routine. This way we obtain:

Theorem 33.1 *In the* $(a{:}f)$ *Avoidance Lattice Games on an* $N \times N$ *board with* $a > f \geq 1$, *the Avoidance Numbers are at least as large as the right-hand sides in Open Problem 30.3. For example, at the end of the play:*

(1) Forcer can always force Avoider to own a $q \times q$ parallelogram lattice with

$$q = \left\lfloor 2\sqrt{\frac{\log N}{\log\left(\frac{a+f}{a}\right)}} - o(1) \right\rfloor ;$$

(2) Forcer can always force Avoider to own a $q \times q$ aligned rectangle lattice with

$$q = \left\lfloor \sqrt{\frac{2\log N}{\log\left(\frac{a+f}{a}\right)}} - o(1) \right\rfloor .$$

Similarly, in the (m:b) Achievement Lattice Games on an $N \times N$ board with $m > b \geq 1$, the Achievement Numbers are at least as large as the right-hand sides in Open Problem 30.3. For example, at the end of the play:

(1) Maker can always own a $q \times q$ parallelogram lattice with

$$q = \left\lfloor \sqrt{\frac{4\log N}{\log\left(\frac{m+b}{m}\right)} + \frac{2\log N}{\log\left(\frac{m}{m-b}\right)}} - o(1) \right\rfloor ;$$

(2) Maker can always own a $q \times q$ aligned rectangle lattice with

$$q = \left\lfloor \sqrt{\frac{2\log N}{\log\left(\frac{m+b}{m}\right)} + \frac{2\log N}{\log\left(\frac{m}{m-b}\right)}} - o(1) \right\rfloor .$$

To prove the Achievement version of Theorem 33.1, we need the precise version of the "Random Play plus Cheap Building" intuition (see Section 46).

Next consider the case when Maker is the underdog. Is there a game where we can prove a building result supporting the underdog Meta-Conjecture? The answer is "yes"; we demonstrate it with:

2. When Maker is the underdog: the (2:3) game is a partial success. We start with the following asymmetric analogue of (32.18)

$$\left(1 + \frac{3}{2}x^{(1)} + \frac{3}{2}x^{(2)} - 3x^{(1)}x^{(2)}\right)(1 - y^{(1)})(1 - y^{(2)})(1 - y^{(3)}) - 1$$

$$= \left(1 + \frac{3}{2}(x^{(1)} + x^{(2)} - 2x^{(1)}x^{(2)})\right)\left(1 - \left(1 - (1 - y^{(1)})(1 - y^{(2)})(1 - y^{(3)})\right)\right) - 1$$

$$= \frac{3}{2}f(x^{(1)}, x^{(2)}) - \frac{1}{2}\left(f(y^{(1)}, y^{(2)}) + f(y^{(1)}, y^{(3)}) + f(y^{(2)}, y^{(3)})\right)$$

$$- \frac{3}{2}(x^{(1)} + x^{(2)} - 2x^{(1)}x^{(2)})\left(1 - (1 - y^{(1)})(1 - y^{(2)})(1 - y^{(3)})\right), \qquad (33.1)$$

where

$$f(u_1, u_2) = u_1 + u_2 - 2u_1 u_2.$$

Note that the inequality

$$x^{(1)} + x^{(2)} - 2x^{(1)}x^{(2)} = (x^{(1)} - x^{(2)})^2 \geq 0 \qquad (33.2)$$

is trivial, and

$$1 - (1 - y^{(1)})(1 - y^{(2)})(1 - y^{(3)}) \geq 0 \qquad (33.3)$$

holds for for any choice of $y^{(j)} \in \{0, 1\}$, $j = 1, 2, 3$. Combining (33.1)–(33.3) we obtain the "formal inequality"

$$(1 + \frac{3}{2}x^{(1)} + \frac{3}{2}x^{(2)} - 3x^{(1)}x^{(2)})(1 - y^{(1)})(1 - y^{(2)})(1 - y^{(3)}) - 1$$

$$\leq \frac{3}{2}f(x^{(1)}, x^{(2)}) - \frac{1}{2}\left(f(y^{(1)}, y^{(2)}) + f(y^{(1)}, y^{(3)}) + f(y^{(2)}, y^{(3)})\right). \qquad (33.4)$$

(33.4) (i.e. the "Decreasing Property") is the analogue of (32.40), and repeating the proof of Theorem 31.1′ we obtain:

Theorem 33.2 *In the (2:3) Achievement Lattice Games on an $N \times N$ board the Achievement Numbers are at least as large as the right-hand sides in Open Problem 30.3. For example*

$$\mathbf{A}(N \times N; \text{ rectangle lattice}; 2:3) \geq \left\lfloor \sqrt{\frac{2\log N}{\log(5/2)}} + o(1) \right\rfloor,$$

and

$$\mathbf{A}(N \times N; \text{ parallelogram lattice}; 2:3) \geq \left\lfloor 2\sqrt{\frac{\log N}{\log(5/2)}} + o(1) \right\rfloor.$$

The role of factor $\log(5/2)$ is clear: $2/5$ is the probability that a given point will be picked by Maker.

We have succeeded with the (2:3) achievement game, at least with the "building part." What happens if we try the same approach for the (2:4) game?

3. The (2:4) achievement game is a breakdown! The difficulty with the (1:2) achievement game has already been mentioned in section 32, see the Remark after (32.6). In the (2:4) game we face the same difficulty. Indeed, consider the equality

$$1 - (1 - y^{(1)})(1 - y^{(2)})(1 - y^{(3)})(1 - y^{(4)}) = y^{(1)} + y^{(2)} + y^{(3)} + y^{(4)}$$

$$- y^{(1)}y^{(2)} - y^{(1)}y^{(3)} - y^{(1)}y^{(4)} - y^{(2)}y^{(3)} - y^{(2)}y^{(4)} - y^{(3)}y^{(4)} \pm \ldots$$

$$= \frac{1}{3}\left(f(y^{(1)}, y^{(2)}) + f(y^{(1)}, y^{(3)})\right.$$

$$\left. + f(y^{(1)}, y^{(4)}) + f(y^{(2)}, y^{(3)}) + f(y^{(2)}, y^{(4)}) + f(y^{(3)}, y^{(4)})\right) \pm \ldots,$$

where $f(u_1, u_2) = u_1 + u_2 - 3u_1u_2$. This suggests the following starting point

$$\left(1 + 2x^{(1)} + 2x^{(2)} - 6x^{(1)}x^{(2)}\right)(1 - y^{(1)})(1 - y^{(2)})(1 - y^{(3)})(1 - y^{(4)}) - 1$$

$$= \left(1 + 2f(x^{(1)}, x^{(2)})\right)\left(1 - \frac{1}{3}\sum_{1 \le k_1 < k_2 \le 4} f(y^{(k_1)}, y^{(k_2)}) - *\right) - 1$$

$$= 2f(x^{(1)}, x^{(2)}) - \frac{1}{3}\sum_{1 \le k_1 < k_2 \le 4} f(y^{(k_1)}, y^{(k_2)}) \pm \cdots,$$

where

$$* = y^{(1)}y^{(2)}y^{(3)} + y^{(1)}y^{(2)}y^{(4)} + y^{(1)}y^{(3)}y^{(4)} + y^{(2)}y^{(3)}y^{(4)} - y^{(1)}y^{(2)}y^{(3)}y^{(4)}.$$

There are two difficulties with the (2:4) game. The first one is that we need the inequality

$$\left(1 + 2x^{(1)} + 2x^{(2)} - 6x^{(1)}x^{(2)}\right)(1 - y^{(1)})(1 - y^{(2)})(1 - y^{(3)})(1 - y^{(4)}) - 1$$

$$\le 2f(x^{(1)}, x^{(2)}) - \frac{1}{3}\sum_{1 \le k_1 < k_2 \le 4} f(y^{(k_1)}, y^{(k_2)}),$$

but we don't know how to prove it. But there is a perhaps even bigger problem: the factor $1 + 2x^{(1)} + 2x^{(2)} - 6x^{(1)}x^{(2)}$ leads to negative(!) weights. Indeed, the corresponding weight is the product

$$w_i(A) = \prod_{1 \le j \le i} g(x_j^{(1)}, x_j^{(2)}), \tag{33.5}$$

and the polynomial

$$g(x^{(1)}, x^{(2)}) = 1 + 2x^{(1)} + 2x^{(2)} - 6x^{(1)}x^{(2)} = -1 \quad \text{if} \quad x^{(1)} = x^{(2)} = 1. \tag{33.6}$$

Here we used the slightly confusing notation $x_j^{(l)} = 1$ if $x_j^{(l)} \in A$ and $x_j^{(l)} = 0$ if $x_j^{(l)} \notin A$ (i.e. $x_j^{(l)}$ denotes a point and a 0-1-function at the same time).

(33.5)–(33.6) allow the appearance of negative weights! Unfortunately, negative weights "kill" the proof technique, and we don't know how to prevent it. The (2:4) achievement game is a breakdown for this approach. Of course, there are infinitely many similar breakdowns. Can the reader help me out here?

Here is one more sporadic example: another underdog achievement game.

4. The (4:5) achievement game: a partial success.

How should we start? The equality

$$1 - (1 - y^{(1)})(1 - y^{(2)})(1 - y^{(3)})(1 - y^{(4)})(1 - y^{(5)}) = \sum_{k=1}^{5} y^{(k)}$$

$$- \sum_{1 \le k_1 < k_2 \le 5} y^{(k_1)} y^{(k_2)} + \sum_{1 \le k_1 < k_2 < k_3 \le 5} y^{(k_1)} y^{(k_2)} y^{(k_3)}$$

$$- \sum_{1 \le k_1 < k_2 < k_3 < k_4 \le 5} y^{(k_1)} y^{(k_2)} y^{(k_3)} y^{(k_4)} + y^{(1)} y^{(2)} y^{(3)} y^{(4)} y^{(5)}$$

$$= \frac{1}{4} \Big(f(y^{(1)}, y^{(2)}, y^{(3)}, y^{(4)}) + f(y^{(2)}, y^{(3)}, y^{(4)}, y^{(5)}) + f(y^{(1)}, y^{(3)}, y^{(4)}, y^{(5)})$$

$$+ f(y^{(1)}, y^{(2)}, y^{(4)}, y^{(5)}) + f(y^{(1)}, y^{(2)}, y^{(3)}, y^{(5)}) \Big) + y^{(1)} y^{(2)} y^{(3)} y^{(4)} y^{(5)},$$

where

$$f(u_1, u_2, u_3, u_4) = u_1 + u_2 + u_3 + u_4$$

$$- \frac{4}{3} (u_1 u_2 + u_1 u_3 + u_1 u_4 + u_2 u_3 + u_2 u_4 + u_3 u_4) +$$

$$+ 2 (u_1 u_2 u_3 + u_2 u_3 u_4 + u_1 u_3 u_4 + u_1 u_2 u_4)$$

$$- 4 u_1 u_2 u_3 u_4$$

suggests the following starting point

$$\left(1 + \frac{5}{4} f(x^{(1)}, x^{(2)}, x^{(3)}, x^{(4)})\right) \left(1 - \frac{1}{4} \sum_{1 \le k_1 < k_2 < k_3 < k_4 \le 5} f(y^{(k_1)}, y^{(k_2)}, y^{(k_3)}, y^{(k_4)}) - *\right) - 1,$$

where

$$* = y^{(1)} y^{(2)} y^{(3)} y^{(4)} y^{(5)}.$$

Multiplying out the starting point product above we get

$$\frac{5}{4} f(x^{(1)}, x^{(2)}, x^{(3)}, x^{(4)}) - \frac{1}{4} \sum_{1 \le k_1 < k_2 < k_3 < k_4 \le 5} f(y^{(k_1)}, y^{(k_2)}, y^{(k_3)}, y^{(k_4)})$$

$$- \frac{5}{16} f(x^{(1)}, x^{(2)}, x^{(3)}, x^{(4)}) \left(\sum_{1 \le k_1 < k_2 < k_3 < k_4 \le 5} f(y^{(k_1)}, y^{(k_2)}, y^{(k_3)}, y^{(k_4)}) \right)$$

$$- \left(1 + \frac{5}{4} f(x^{(1)}, x^{(2)}, x^{(3)}, x^{(4)})\right) y^{(1)} y^{(2)} y^{(3)} y^{(4)} y^{(5)} \le$$

$$\le \frac{5}{4} f(x^{(1)}, x^{(2)}, x^{(3)}, x^{(4)}) - \frac{1}{4} \sum_{1 \le k_1 < k_2 < k_3 < k_4 \le 5} f(y^{(k_1)}, y^{(k_2)}, y^{(k_3)}, y^{(k_4)})$$

(note that we check the inequality "$f \geq 0$" later), which is the critical inequality ("Decreasing Property") that we need!

We also have to check that the factor

$$\left(1 + \frac{5}{4}f(x^{(1)}, x^{(2)}, x^{(3)}, x^{(4)})\right) = 1 + \frac{5}{4}\left(x^{(1)} + x^{(2)} + x^{(3)} + x^{(4)}\right.$$

$$- \frac{4}{3}(x^{(1)}x^{(2)} + x^{(1)}x^{(3)} + x^{(1)}x^{(4)} + x^{(2)}x^{(3)} + x^{(2)}x^{(4)} + x^{(3)}x^{(4)})$$

$$\left. + 2(x^{(1)}x^{(2)}x^{(3)} + x^{(2)}x^{(3)}x^{(4)} + x^{(1)}x^{(3)}x^{(4)} + x^{(1)}x^{(2)}x^{(4)}) - 4x^{(1)}x^{(2)}x^{(3)}x^{(4)}\right)$$

is strictly positive for every choice of $x^{(j)} \in \{0, 1\}$, $j = 1, 2, 3, 4$. Indeed, the polynomial

$$g(x^{(1)}, x^{(2)}, x^{(3)}, x^{(4)}) = \left(1 + \frac{5}{4}f(x^{(1)}, x^{(2)}, x^{(3)}, x^{(4)})\right)$$

shows up in the definition of the weight

$$w_i(A) = \prod_{1 \leq j \leq i} g(x_j^{(1)}, x_j^{(2)}, x_j^{(3)}, x_j^{(4)}).$$

Here we used the slightly confusing notation $x_j^{(l)} = 1$ if $x_j^{(l)} \in A$ and $x_j^{(l)} = 0$ if $x_j^{(l)} \notin A$, i.e. $x_j^{(l)}$ denotes a point and a 0-1-function at the same time. The success of our proof technique requires strictly positive weights, and in this case we really have strictly positive weights. Indeed, with $x^{(j)} \in \{0, 1\}$, $j = 1, 2, 3, 4$ we have:

(1) if $\sum_{k=1}^{4} x^{(k)} = 1$ then $f(x^{(1)}, x^{(2)}, x^{(3)}, x^{(4)}) = 1 > 0$;
(2) if $\sum_{k=1}^{4} x^{(k)} = 2$ then $f(x^{(1)}, x^{(2)}, x^{(3)}, x^{(4)}) = 2 - \frac{4}{3} > 0$;
(3) if $\sum_{k=1}^{4} x^{(k)} = 3$ then $f(x^{(1)}, x^{(2)}, x^{(3)}, x^{(4)}) = 3 - \frac{4}{3}3 + 2 > 0$;
(4) if $\sum_{k=1}^{4} x^{(k)} = 4$ then $f(x^{(1)}, x^{(2)}, x^{(3)}, x^{(4)}) = 4 - \frac{4}{3}6 + 2 \cdot 4 - 4 = 0$.

Thus we are ready to repeat the proof of Theorem 31.1', and obtain:

Theorem 33.3 *In the (4:5) Achievement Lattice Games on an $N \times N$ board the Achievement Numbers are at least as large as the right-hand sides in Open Problem 30.3. For example*

$$A(N \times N; \text{ rectangle lattice}; 4:5) \geq \left\lfloor \sqrt{\frac{2 \log N}{\log(9/4)}} + o(1) \right\rfloor,$$

and

$$A(N \times N; \text{ parallelogram lattice}; 4:5) \geq \left\lfloor 2\sqrt{\frac{\log N}{\log(9/4)}} + o(1) \right\rfloor.$$

The role of factor $\log(9/4)$ is clear: 4/9 is the probability that in the (4:5) game a given point will be picked by Maker.

We succeeded in the (2:3) and (4:5) games, and failed in the (2:4) game. What is going on here? Is there a better way to handle the biased achievement games where Maker is the underdog? We need more ideas!

5. The biased Chooser–Picker Game: a pleasant surprise.

In the fair (1:1) play we have the unexpected equality

$$\text{Achievement Number} = \text{Avoidance Number} = \text{Chooser Achievement Number}$$

about the Maker–Breaker, the Avoider–Forcer, and the Chooser–Picker games (well, at least for our "Ramseyish" games with quadratic goal sets). Among these 3 games the Chooser–Picker version has the most satisfying "biased theory." We start the discussion with:

The (1:2) Chooser–Picker Game. Here, of course, Chooser is the underdog; how can underdog Chooser build? The simplest Power-of-Three Scoring System works. Indeed, let

$$g(x, y^{(1)}, y^{(2)}) = (1+2x)(1-y^{(1)})(1-y^{(2)}) - 1$$
$$= 2x - (y^{(1)} + y^{(2)}) + y^{(1)}y^{(2)} - 2x(y^{(1)} + y^{(2)}) + 2xy^{(1)}y^{(2)},$$

and consider the "complete sum"

$$g(u_1; u_2, u_3) + g(u_2; u_3, u_1) + g(u_3; u_1, u_2) =$$
$$- 3(u_1 u_2 + u_1 u_3 + u_2 u_3) + 6u_1 u_2 u_3. \tag{33.7}$$

By using the trivial "formal inequality" $u_1 u_2 u_3 \le u_j u_k$ in (33.7) we obtain

$$g(u_1; u_2, u_3) + g(u_2; u_3, u_1) + g(u_3; u_1, u_2) \le 0. \tag{33.8}$$

The Power-of-Three Scoring System means to work with

$$T(\mathcal{H}) = \sum_{B \in \mathcal{H}} 3^{-|B|} \quad \text{and} \quad T(\mathcal{H}; v_1, \ldots, v_m) = \sum_{B \in \mathcal{H}: \{v_1, \ldots, v_m\} \subset b} 3^{-|B|}.$$

Consider the usual Potential Function

$$L_i = T(\mathcal{F}(i)) - \lambda \cdot T(\mathcal{F}_2^p(i)),$$

where

$$\mathcal{H}(i) = \{B \setminus X(i) : \ B \in \mathcal{H}, \ B \cap Y(i) = \emptyset\}$$

with $\mathcal{H} = \mathcal{F}$ and $\mathcal{H} = \mathcal{F}_2^p$.

In the $(i+1)$st move Picker offers 3 points u_1, u_2, u_3 to Chooser to choose from. Chooser has 3 options: $x_{i+1} = u_1$ or $x_{i+1} = u_2$ or $x_{i+1} = u_3$, and accordingly we have 3 possible values: $L_{i+1}^{(1)}$ or $L_{i+1}^{(2)}$ or $L_{i+1}^{(3)}$. By (33.7)–(33.8) we obtain the crucial inequality ("Decreasing Property")

$$\frac{1}{3}\left(L^{(1)}_{i+1} + L^{(2)}_{i+1} + L^{(3)}_{i+1}\right) \geq L_i - T(\mathcal{F}(i); u_1, u_2) - T(\mathcal{F}(i); u_1, u_3) - T(\mathcal{F}(i); u_2, u_3)$$

$$+ 2T(\mathcal{F}(i); u_1, u_2, u_3)$$

$$\geq L_i - \left(T(\mathcal{F}(i); u_1, u_2) + T(\mathcal{F}(i); u_1, u_3) + T(\mathcal{F}(i); u_2, u_3)\right).$$

$$(33.9)$$

Chooser chooses that point $x_{i+1} = u_j$ for which $L^{(j)}_{i+1}, j = 1, 2, 3$ attains its *maximum*. Then by (33.8)

$$L_{i+1} \geq L_i - \left(T(\mathcal{F}(i); u_1, u_2) + T(\mathcal{F}(i); u_1, u_3) + T(\mathcal{F}(i); u_2, u_3)\right),$$

and the rest of the proof is a routine application of Section 24. We obtain, therefore:

Theorem 33.4 *(a) In the (1:2) Chooser–Picker game Chooser can achieve at least as large goal size as the perfect analogue of Theorem 6.4 and Theorem 8.2 with the natural change that the base 2 logarithm $\log_2 N$ is replaced by the base 3 logarithm $\log_3 N$.*

We conjecture that Theorem 33.4 (a) is the best possible; what we can prove is weaker: we can show that Theorem 33.4 (a) is (at least!) asymptotically best possible. Not a word has been said about Picker's blocking yet, not even in the simplest (1:1) game. This section is already far too long, so we postpone the (1:1) case to the end of Section 38, and the general biased case to the end of Section 47.

The general (1:s) Chooser–Picker game. The (1:2) Chooser–Picker game turned out to be "easy." Let $s \geq 2$ be an arbitrary integer; how about the general (1:s) play (Picker picks $s+1$ points per move and Chooser chooses one of them for himself)? We want to come up with an analogue of inequality (33.8). Let

$$g(x, y^{(1)}, \ldots, y^{(s)}) = (1+sx)(1-y^{(1)})\cdots(1-y^{(s)}) - 1$$

$$= sx - (y^{(1)} + \ldots + y^{(s)}) + \sum_{1 \leq k_1 < k_2 \leq s} y^{(k_1)} y^{(k_2)}$$

$$- sx(y^{(1)} + \ldots + y^{(s)}) - \sum_{1 \leq k_1 < k_2 < k_3 \leq s} y^{(k_1)} y^{(k_2)} y^{(k_3)}$$

$$+ sx\left(\sum_{1 \leq k_1 < k_2 \leq s} y^{(k_1)} y^{(k_2)}\right) \pm \cdots.$$

$$(33.10)$$

Then the "complete sum" (in fact its average) equals

$$\frac{1}{s+1}\Big(g(u_1; u_2, \ldots, u_{s+1}) + g(u_2; u_1, u_3, \ldots, u_{s+1}) + \cdots$$

$$+ \cdots + g(u_{s+1}; u_1, \ldots, u_s)\Big)$$

$$= -A_2 + 2A_3 - 3A_4 + 4A_5 - 5A_6 \pm \cdots + (-1)^s \cdot sA_{s+1}, \qquad (33.11)$$

where

$$A_2 = \sum_{1 \le j_1 < j_2 \le s+1} u_{j_1} u_{j_2}, \quad A_3 = \sum_{1 \le j_1 < j_2 < j_3 \le s+1} u_{j_1} u_{j_2} u_{j_3},$$

$$A_4 = \sum_{1 \le j_1 < j_2 < j_3 < j_4 \le s+1} u_{j_1} u_{j_2} u_{j_3} u_{j_4}, \quad \cdots. \qquad (33.12)$$

To get the analogue of (33.8), we have to check the Negativity (see (33.11))

$$-A_2 + 2A_3 - 3A_4 + 4A_5 - 5A_6 \pm \cdots + (-1)^s \cdot sA_{s+1} \le 0. \qquad (33.13)$$

The Inclusion–Exclusion Principle gives in a "formal sense"

$$u_1 u_2 \ge \text{union of } u_1 u_2 u_k, \quad k = 3, 4, \ldots, s+1$$

$$= u_1 u_2 \left(\sum_{k=3}^{s+1} u_k - \sum_{3 \le k_1 < k_2 \le s} u_{k_1} u_{k_2} \right.$$

$$+ \sum_{3 \le k_1 < k_2 < k_3 \le s+1} u_{k_1} u_{k_2} u_{k_3} - \sum_{3 \le k_1 < k_2 < k_3 < k_4 \le s+1} u_{k_1} u_{k_2} u_{k_3} u_{k_4} \pm \cdots \Bigg),$$

which by (33.12) implies

$$A_2 \ge \frac{\binom{s+1}{2}(s+1-2)}{\binom{s+1}{3}} A_3 - \frac{\binom{s+1}{2}\binom{s+1-2}{2}}{\binom{s+1}{4}} A_4 +$$

$$+ \frac{\binom{s+1}{2}\binom{s+1-2}{3}}{\binom{s+1}{5}} A_5 - \frac{\binom{s+1}{2}\binom{s+1-2}{4}}{\binom{s+1}{6}} A_6 \pm \cdots$$

$$= 3A_3 - \binom{4}{2} A_4 + \binom{5}{2} A_5 - \binom{6}{2} A_6 \pm \cdots. \qquad (33.14)$$

Similarly

$$u_1 u_2 u_3 \geq \text{union of } u_1 u_2 u_3 u_k, \quad k = 4, 5, \ldots, s+1$$

$$= u_1 u_2 u_3 \left(\sum_{k=4}^{s+1} u_k - \sum_{4 \leq k_1 < k_2 \leq s} u_{k_1} u_{k_2} \right.$$

$$\left. + \sum_{4 \leq k_1 < k_2 < k_3 \leq s+1} u_{k_1} u_{k_2} u_{k_3} - \sum_{4 \leq k_1 < k_2 < k_3 < k_4 \leq s+1} u_{k_1} u_{k_2} u_{k_3} u_{k_4} \pm \cdots \right),$$

which by (33.12) implies

$$A_3 \geq \frac{\binom{s+1}{3}(s+1-3)}{\binom{s+1}{4}} A_4 - \frac{\binom{s+1}{3}\binom{s+1-3}{2}}{\binom{s+1}{5}} A_5 +$$

$$+ \frac{\binom{s+1}{3}\binom{s+1-3}{3}}{\binom{s+1}{6}} A_6 - \frac{\binom{s+1}{3}\binom{s+1-3}{4}}{\binom{s+1}{7}} A_7 \pm \cdots$$

$$= 4A_4 - \binom{5}{3} A_5 + \binom{6}{3} A_6 - \binom{7}{3} A_7 \pm \cdots . \tag{33.15}$$

We can rewrite (33.14) and (33.15) as follows

$$- A_2 + \binom{3}{2} A_3 - \binom{4}{2} A_4 + \binom{5}{2} A_5 - \binom{6}{2} A_6 \pm \cdots + (-1)^{s-2} \binom{s+1}{2} A_{s+1} \leq 0, \tag{33.16}$$

$$- A_3 + \binom{4}{3} A_4 - \binom{5}{3} A_5 + \binom{6}{3} A_6 - \binom{7}{3} A_7 \pm \cdots (-1)^{s-3} \binom{s+1}{3} A_{s+1} \leq 0. \tag{33.17}$$

Similarly, we have

$$- A_4 + \binom{5}{4} A_5 - \binom{6}{4} A_6 + \binom{7}{4} A_7 - \binom{8}{4} A_8 \pm \cdots (-1)^{s-4} \binom{s+1}{4} A_{s+1} \leq 0, \tag{33.18}$$

$$- A_5 + \binom{6}{5} A_5 - \binom{7}{5} A_7 + \binom{8}{5} A_8 - \binom{9}{5} A_9 \pm \cdots (-1)^{s-5} \binom{s+1}{5} A_{s+1} \leq 0, \tag{33.19}$$

and so on.

Adding up inequalities (33.16), (33.17), (33.18), and so on, we obtain exactly the desired (33.13). Indeed, the coefficient $(k-1)$ of A_k in (33.12) equals the sum

$$\binom{k}{2} - \binom{k}{3} + \binom{k}{4} - \binom{k}{5} \pm \ldots + (-1)^k \binom{k}{k} = k - 1, \tag{33.20}$$

due to the binomial formula $0 = (1-1)^k = \sum_{j=0}^{k} (-1)^j \binom{k}{j}$.

Inequality (33.13) is the key ingredient; the rest of the proof is a routine application of Section 24, just as in the (1:2) play (Theorem 33.4 (a)). We obtain, therefore, the following extension of Theorem 33.4 (a).

Theorem 33.4 *(b) Let $s \geq 2$ be an arbitrary integer. In the (1:s) Chooser–Picker game Chooser can achieve a goal size at least as large as the perfect analogues of Theorems 6.4 and 8.2 with the natural change that the base 2 logarithm $\log_2 N$ is replaced by the base $(s+1)$ logarithm $\log_{s+1} N$.*

At the end of Section 47 we will see that Theorem 33.4 is (at least!) asymptotically best possible.

At the end of Section 30 we formulated Theorem 30.4: it was an advanced building criterion in the $(1:s)$ Avoidance Game where Avoider is the underdog. The special case $s = 2$ was proved there, and here we complete the missing general case $s \geq 2$.

Proof of Theorem 30.4. We closely follow the proof of Theorem 33.4. The starting point is the same (see (33.10))

$$(1 + sx)(1 - y^{(1)}) \cdots (1 - y^{(s)}) - 1$$

$$= sx - (y^{(1)} + \ldots + y^{(s)}) + \sum_{1 \leq k_1 < k_2 \leq s} y^{(k_1)} y^{(k_2)}$$

$$- sx(y^{(1)} + \ldots + y^{(s)}) - \sum_{1 \leq k_1 < k_2 < k_3 \leq s} y^{(k_1)} y^{(k_2)} y^{(k_3)} +$$

$$+ sx \left(\sum_{1 \leq k_1 < k_2 \leq s} y^{(k_1)} y^{(k_2)} \right) \pm \cdots$$

$$= f(x, y^{(2)}, \ldots, y^{(s)}) + f(y^{(1)}, x, y^{(3)}, \ldots, y^{(s)}) + f(y^{(1)}, y^{(2)}, x, y^{(4)}, \ldots, y^{(s)}) +$$

$$+ \ldots + f(y^{(1)}, \ldots, y^{(s-1)}, x) - s \cdot f(y^{(1)}, y^{(2)}, \ldots, y^{(s)})$$

$$- \left(\frac{1}{2} A_1 - \frac{2}{3} A_2 + \frac{3}{4} A_3 - \frac{4}{5} A_4 \pm \cdots + (-1)^{s+1} \frac{s}{s+1} A_s \right) \cdot x \cdot (s+1), \quad (33.21)$$

where

$$f(u_1, \ldots, u_s) = \sum_{j=1}^{s} u_j - \frac{1}{2} \left(\sum_{1 \leq j_1 < j_2 \leq s} u_{j_1} u_{j_2} \right)$$

$$+ \frac{1}{3} \left(\sum_{1 \leq j_1 < j_2 < j_3 \leq s} u_{j_1} u_{j_2} u_{j_3} \right)$$

$$- \frac{1}{4} \left(\sum_{1 \leq j_1 < j_2 < j_3 < j_4 \leq s} u_{j_1} u_{j_2} u_{j_3} u_{j_4} \right) \pm \cdots,$$

and

$$A_1 = \sum_{j=1}^{s} y^{(j)}, \quad A_2 = \sum_{1 \le j_1 < j_2 \le s} y^{(j_1)} y^{(j_2)},$$

$$A_3 = \sum_{1 \le j_1 < j_2 < j_3 \le s} y^{(j_1)} y^{(j_2)} y^{(j_3)},$$

$$A_4 = \sum_{1 \le j_1 < j_2 < j_3 < j_4 \le s} y^{(j_1)} y^{(j_2)} y^{(j_3)} y^{(j_4)}, \quad \cdots . \tag{33.22}$$

The critical part of the proof is the inequality (see (33.21))

$$\frac{1}{2} A_1 - \frac{2}{3} A_2 + \frac{3}{4} A_3 - \frac{4}{5} A_4 \pm \cdots + (-1)^{s+1} \frac{s}{s+1} A_s \ge 0, \tag{33.23}$$

which plays the role of (33.13).

How to prove (33.23)? Well, we have the perfect analogue of (33.16)

$$A_2 - \binom{3}{2} A_3 + \binom{4}{2} A_4 - \binom{5}{2} A_5 + \binom{6}{2} A_6 \mp \cdots + (-1)^{s-2} \binom{s}{2} A_s \ge 0, \tag{33.24}$$

and we have the perfect analogue of (33.17)

$$A_3 - \binom{4}{3} A_4 + \binom{5}{3} A_5 - \binom{6}{3} A_6 + \binom{7}{3} A_7 \mp \cdots + (-1)^{s-3} \binom{s}{3} A_s \ge 0, \tag{33.25}$$

and we have the perfect analogue of (33.18)

$$A_4 - \binom{5}{4} A_5 + \binom{6}{4} A_6 - \binom{7}{4} A_7 + \binom{8}{4} A_8 \mp \cdots + (-1)^{s-4} \binom{s}{4} A_s \ge 0, \tag{33.26}$$

and we have the perfect analogue of (33.19)

$$A_5 - \binom{6}{5} A_5 + \binom{7}{5} A_7 - \binom{8}{5} A_8 + \binom{9}{5} A_9 \mp \cdots + (-1)^{s-5} \binom{s}{5} A_s \ge 0, \tag{33.27}$$

and so on.

The reader is probably wondering about the missing "first" inequality following the same pattern

$$A_1 - 2A_2 + 3A_3 - 4A_4 + 5A_5 \mp \cdots + (-1)^{s-1} s A_s \ge 0. \tag{33.28}$$

Of course, inequality (33.28) is true, and it can be proved in the same way by using the Inclusion–Exclusion formula.

Now we are ready to prove inequality (33.23). Indeed, adding up inequality (33.28) with weight $\frac{1}{2}$, and inequality (33.24) with weight $\frac{1}{3}$, and inequality (33.25) with weight $\frac{1}{4}$, and inequality (33.26) with weight $\frac{1}{5}$, and so on, at the end the sum is exactly the desired (33.23). Indeed, this follows from the identity

$$\frac{1}{2} k - \frac{1}{3} \binom{k}{2} + \frac{1}{4} \binom{k}{3} - \frac{1}{5} \binom{k}{4} \pm \cdots = \frac{k}{k+1}. \tag{33.29}$$

Identity (33.29) is equivalent to (33.20). Indeed, multiplying (33.29) with $(k+1)$ we obtain (33.20) with "$k = k+1$."

Inequality (33.20) is the key ingredient; the rest of the proof is a routine application of Section 24. This completes the proof of Theorem 30.4.

6. (1:1) games with extra difficulties. We conclude this section with two (1:1) games where we face the same kind of technical problem as in the biased games. The first one is a game about **2-colored goals**, and it is a common generalization of the Clique Achievement and Clique Avoidance Games. The board is the usual complete graph K_N; the two players are called Red and Blue, who alternately occupy new edges of K_N; in each turn Red colors his new edge red and Blue colors his new edge blue. In the old Achievement version, Red=Maker wants to build a monochromatic (red) sub-clique K_q as large as possible; in the Avoidance version Red=Forcer wants to force the reluctant Blue=Avoider to build a monochromatic (blue) sub-clique K_q as large as possible. In the two versions Red has two opposite goals: (1) a monochromatic red K_q that he builds, and a (2) a monochromatic blue K_q that the reluctant opponent builds. How about a mixed goal, a 2-colored goal? What we mean by this is the following: fix an arbitrary 2-colored copy of K_q that we denote by $K_q(\text{red}, \text{blue})$ (of course, the edges are colored red and blue); Red wins the play on K_N if at the end there is a color-isomorphic copy of the given 2-colored $K_q(\text{red}, \text{blue})$; otherwise Blue wins. We call this game the $K_q(\text{red}, \text{blue})$-building game on K_N. Notice that Red acts like a Builder – but he cannot succeed alone, he needs the opponent's cooperation(!) – and Blue acts like a Blocker. Is it true that, given a 2-colored goal graph $K_q(\text{red}, \text{blue})$, Red has a winning strategy in K_N if N is large enough? The answer is "yes"; we already proved it in Section 16 (see Theorems 16.2–16.3). Unfortunately that proof (a straightforward adaptation of Theorem 1.2) gave a poor quantitative bound (like $q = \sqrt{\log N}$). In the two extreme cases of monochromatic red and monochromatic blue K_q (i.e. the Achievement and Avoidance Games) we could prove the best possible quantitative bound; namely, we could find the exact value

$$q = \lfloor 2\log_2 N - 2\log_2 \log_2 N + 2\log_2 e - 3 + o(1) \rfloor.$$

This raises the following natural:

Open Problem 33.1 *Let $K_q(\text{red}, \text{blue})$ be an arbitrary fixed 2-colored goal graph, and let $q = (2 - o(1))\log_2 N$. Is it true that, playing on K_N, Red has a winning strategy in the $K_q(\text{red}, \text{blue})$-building game?*

Well, this seems plausible; we can try to prove it by using the usual potential technique

$$L_i = T(\mathcal{F}(i)) - \lambda \cdot T(\mathcal{F}_2^p(i)) \tag{33.30}$$

of Section 24. Here comes the bad news: this is a $(1:1)$ game; nevertheless we face the difficulty of "Big Bad Negative Terms" the same way as we did in the biased case! Indeed, let x_{i+1} be Red's $(i+1)$st move (a red edge), let y_{i+1} be Blue's $(i+1)$st move (a blue edge), and consider an arbitrary color-isomorphic copy K_N^* of the given 2-colored goal graph K_q(red, blue) with the additional property that the edges x_{i+1}, y_{i+1} both belong to K_N^* (but not necessarily with the right color!). Then there are 4 possibilities:

(1) both x_{i+1} and y_{i+1} are color-consistent with K_N^*;
(2) x_{i+1} is consistent but y_{i+1} is inconsistent with K_N^* (i.e. y_{i+1} is red in K_N^*);
(3) y_{i+1} is consistent but x_{i+1} is inconsistent with K_N^* (i.e. x_{i+1} is blue in K_N^*);
(4) both x_{i+1} and y_{i+1} are inconsistent with K_N^*.

Notice that Case (1) and Case (4) are "new cases"; they cannot occur in the special cases of Achievement and Avoidance Games. These two "new cases" contribute extra positive terms in $T(\mathcal{F}(i))$ and contribute extra negative terms in $-T(\mathcal{F}_2^p(i))$ (see (33.30)). The appearance of extra negative terms in (33.30) kills the proof technique exactly the same way as it did in the biased case. This is why we cannot solve the "plausible" Open Problem 33.1. What we can prove is "half of the conjectured truth": Red can force the appearance of a color-isomorphic copy of any given K_q(red, blue) with $(1 - o(1)) \log_2 N$. The proof is a Ramsey type Halving Argument, a variant of the proof of Theorem 21.1. In fact, we combine the Ramsey Halving Argument with the discrepancy result Theorem 17.1, and the proof will give the stronger statement that, at the end of the play, the 2-colored K_N contains *all* 2-colored copies of K_q with $(1 - o(1)) \log_2 N$.

Red (as Builder) wants to guarantee the following two graph-theoretic properties:

(α) at the end of the play, for any two disjoint vertex sets V_1 and V_2 in K_N with $|V_1| \geq N^\varepsilon$ and $|V_2| \geq N^\varepsilon$, the number of red edges in the complete bipartite graph $V_1 \times V_2$ is between $(\frac{1}{2} - \varepsilon)|V_1| \cdot |V_2|$ and $(\frac{1}{2} + \varepsilon)|V_1| \cdot |V_2|$; and also

(β) at the end of the play, for any vertex set V_1 in K_N with $|V_1| \geq N^\varepsilon$, the number of red edges in the graph $K_{V_1} = \binom{V_1}{2}$ is between $(\frac{1}{2} - \varepsilon)\binom{|V_1|}{2}$ and $(\frac{1}{2} + \varepsilon)\binom{|V_1|}{2}$.

These two density properties can be easily guaranteed by a routine application of the discrepancy result Theorem 17.1. Indeed, the criterion of Theorem 17.1 applies if

$$\sum_{k \geq N^\varepsilon, l \geq N^\varepsilon, k+l \leq N} \binom{N}{k}\binom{N-k}{l} \left((1+\varepsilon)^{1+\varepsilon}(1-\varepsilon)^{1-\varepsilon}\right)^{-kl/2} = o(1), \qquad (33.31)$$

and (33.31) is trivial with (say) $\varepsilon = (\log N)^{-1/2}$

$$\sum_{k \geq N^\varepsilon, l \geq N^\varepsilon, k+l \leq N} \binom{N}{k}\binom{N-k}{l}\left((1+\varepsilon)^{1+\varepsilon}(1-\varepsilon)^{1-\varepsilon}\right)^{-kl/2}$$

$$\leq \sum_{k \geq N^\varepsilon, l \geq N^\varepsilon, k+l \leq N} e^{-(\varepsilon^2 + O(\varepsilon^3))\frac{kl}{2} + (k+l)\log N}$$

$$= \sum_{k \geq N^\varepsilon, l \geq N^\varepsilon, k+l \leq N} e^{-(\varepsilon^2 + O(\varepsilon^3))\frac{kl}{2}} = o(1)$$

by trivial calculations.

Next we explain how the two graph-theoretic properties (α) and (β) guarantee the success of the Ramsey Halving Argument. Fix an arbitrary 2-colored goal graph K_q(red, blue); we show that, at the end of the play, the 2-colored K_N (colored red and blue) contains a color-isomorphic copy of K_q(red, blue)

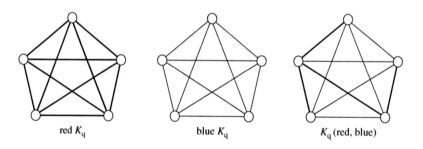

red K_q blue K_q K_q (red, blue)

Let $u_1, u_2, \ldots, u_q$ denote the vertices of the goal graph K_q(red, blue), and let $c(i, j)$ denote the color of edge $u_i u_j$ in K_q(red, blue) ($1 \leq i < j \leq q$). We will construct a color-consistent embedding f of goal graph K_q(red, blue) into the play-2-colored K_N (meaning: 2-colored by Red and Blue during the play). Of course, color-consistent means that the color of edge $f(u_i)f(u_j)$ is $c(i, j)$ for all pairs $1 \leq i < j \leq q$.

The first step is to find $f(u_1)$. Let V_1 be the set of those "bad" vertices in K_N which have red degree $\geq (\frac{1}{2} + 2\varepsilon)N$ in the play-2-colored K_N. Graph-theoretic properties (α) and (β) above guarantee that $|V_1| < N^\varepsilon$. Similarly, there are less than N^ε vertices which have red degree $\leq (\frac{1}{2} - 2\varepsilon)N$ in the play-2-colored K_N. Throwing out less than $2N^\varepsilon$ "bad" vertices, the remaining "good" vertices all have the property that the red degree is between $(\frac{1}{2} - 2\varepsilon)N$ and $(\frac{1}{2} + 2\varepsilon)N$ in the play-2-colored K_N. Let $f(u_1)$ be one of these "good" vertices.

We proceed by induction: assume that we already defined $f(u_1), \ldots, f(u_i)$, and we have to find a proper $f(u_{i+1})$. For every $k = i+1, i+2, \ldots, q$ define the set

$$W_k = W_{i,k} = \left\{ u \in V(K_N) \setminus \{f(u_1), \ldots, f(u_i)\} : \text{the color of edge } f(u_j)u \right.$$

$$\left. \text{in the play is } c(j,k), \ j = 1, \ldots, i \right\}.$$

Note that the $(q-i)$ sets W_k, $k = i+1, i+2, \ldots, q$ have the property that any two of them are either disjoint or identical. For every $k = i+1, i+2, \ldots, q$ we have

$$|W_k| \geq N \left(\frac{1}{2} - 2\varepsilon \right)^i.$$

We are looking for a proper vertex $f(u_{i+1})$ in set W_{i+1}.

Let V_1 be the set of those "bad" vertices in W_{i+1} which have red degree $\geq (\frac{1}{2} + 2\varepsilon)|W_{i+1}|$ in the complete graph $K_{W_{i+1}} = \binom{W_{i+1}}{2}$ in the play-2-colored K_N. Graph-theoretic properties (α) and (β) above guarantee that $|V_1| < N^\varepsilon$. Similarly, there are less than N^ε "bad" vertices which have red degree $\leq (\frac{1}{2} - 2\varepsilon)|W_{i+1}|$ in the complete graph $K_{W_{i+1}} = \binom{W_{i+1}}{2}$ in the play-2-colored K_N.

Similarly, there are less than N^ε "bad" vertices which have red degree $\geq (\frac{1}{2} + 2\varepsilon)|W_k|$ in the complete bipartite graph $W_{i+1} \times W_k$ in the play-2-coloring (we assume that W_{i+1} and W_k are disjoint); and the same for the red degree $\leq (\frac{1}{2} - 2\varepsilon)|W_k|$ in the complete bipartite graph $W_{i+1} \times W_k$ in the play-2-coloring.

Throwing out altogether $\leq 2(q-i)N^\varepsilon$ "bad" vertices from W_{i+1}, the remaining set of "good" vertices is still non-empty, and let $f(u_{i+1})$ be one of these "good" vertices in set W_{i+1}. The embedding works as long as

$$N \left(\frac{1}{2} - 2\varepsilon \right)^{q-1} > 2qN^\varepsilon,$$

and this inequality holds with $q = (1 - o(1)) \log_2 N$. This completes the proof of:

Theorem 33.5 *Red and Blue are playing the usual alternating (1:1) game on K_N. Then Red can force that, at the end of the play, the play-2-colored K_N contains all possible 2-colored copies of K_q with $q = (1 - o(1)) \log_2 N$.*

Theorem 33.5 was proved more than 20 years ago (it was the subject of a lecture by the author at the Balatonlelle Graph Theory Conference, Hungary, 1994, where the proof above was outlined). A special case of the result was published (see Theorem 3 in Beck [1981b]). Theorem 33.5 was later independently rediscovered by Alon, Krivelevich, Spencer, and Szabó [2005] (Krivelevich gave a talk about it in the Bertinoro Workshop, 2005; their proof was somewhat different from mine).

The sad thing about Theorem 33.5 is that (most likely) it is just a mediocre result, half of the truth. It is humiliating that we cannot solve Open Problem 33.1.

Let's go back to Section 25: recall *Exercise 25.4: What is the Avoidance version of Theorem 21.1?* The following result is a best effort at an "avoidance" analogue of Theorem 21.1; it is the special case of Theorem 33.5 where the goal graph is a monochromatic blue K_q; in fact, it is just a more quantitative form.

Theorem 33.6 *If* $N \geq q^{10} \cdot 2^q$, *then playing the Avoidance Clique Game* (K_N, K_q) *Forcer can force Avoider to build a copy of* K_q.

The good thing about Theorem 33.6 is that it holds for *every* q, not just for "sufficiently large values" like Theorem 6.4 (a). For example, if $N = 2^{250}$, then $\log_2 N - 10 \log_2 \log_2 N \approx 250 - 80 = 170$; so, by Theorem 33.6, the Clique Avoidance Number is ≥ 170. In the other direction we have the Erdős–Selfridge bound $2 \log_2 N - 2 \log_2 \log_2 N + 2 \log_2 e - 1 \approx 500 - 16 + 2 = 486$. (The Erdős–Selfridge proof has a straightforward adaptation for the avoidance game.)

Summarizing, the Clique Avoidance Number for the board size K_N with $N = 2^{250}$ is between 170 and 486. This is the best that we know, due to the fact that in the range $N = 2^{250}$ the machinery of Theorem 6.4 (a) "doesn't start to work yet."

Let's return to Theorem 33.5: what happens in the (1:2) version where Red (Builder) is the underdog? The only difference in the proof is that the Ramsey Halving Argument above is replaced by a Ramsey One-Third Argument, and Theorem 17.1 is replaced by Theorem 17.2. This way we obtain:

Theorem 33.7 *Red and Blue are playing the alternating (1:2) game on* K_N. *Then Red can force that, at the end of the play, the play-2-colored* K_N *contains all possible 2-colored copies of* K_q *with* $q = (1 - o(1)) \log_3 N$.

Corollary: *Playing the (1:2) Clique Achievement game on* K_N, *underdog Maker can always build a* K_q *with* $q = (1 - o(1)) \log_3 N$.

This Corollary is the best that we know in the (1:2) Achievement play (where Maker is the underdog).

It is very surprising that in the (1:2) Avoidance play (where Avoider is the underdog) Forcer can force Avoider to build a clique K_q with q *larger* than $(1 - o(1)) \log_3 N$. Indeed, Theorem 30.5 gives

$$q = 2 \log_3 N - 2 \log_3 \log_3 N + 2 \log_3 e - 2 \log_3 2 - 1 - o(1),$$

which is about twice as large as the Corollary above.

Of course, Theorem 33.7 can be easily extended to an arbitrary $(r : b)$ play (r edges for Red and b edges for Blue per move) The only natural change is to apply the general Theorem 17.5.

The second (1:1) game with a "biased-like attitude problem" is the **Tournament Game**, see Section 14. First we recall that a *tournament* means a "directed complete graph" such that every edge of a complete graph is directed by one of the two possible orientations; it represents a tennis tournament where any two players played with each other, and an arrow points from the winner to the loser. Fix an arbitrary goal tournament T_q on q vertices. The two players are Red and Blue, who alternately take new edges of a complete graph K_N, and for each new edge choose one of the two possible orientations ("arrow"). Either player colors his arrow with his own color. At the end of a play, the players create a 2-colored tournament on N vertices. Red wins if there is a red copy of T_q; otherwise Blue wins. Theorem 14.5 proves that, if N is sufficiently large compared to q, then Red has a winning strategy. Unfortunately Theorem 14.5 gives a poor quantitative result. An easy adaptation of the proof of Theorem 33.5 gives the following much better bound (we need the (1:3) version in Theorem 17.5, and the Halving Argument is replaced by a One-Fourth Argument).

Theorem 33.8 *Red and Blue are playing the (1:1) Tournament Game on K_N. Then Red can force that, at the end of the play, the play-2-colored tournament on N vertices contains a copy of all possible red T_qs with $q = (1 - o(1)) \log_4 N$.*

The reader is challenged to work out the details of the proof of Theorem 33.8. We are sure that Theorem 33.8 is not optimal. What is the optimal result? How can it be proved?

Part D

Advanced Strong Draw – game-theoretic independence

The objective of game-playing is winning, but very often winning is impossible for the simple reason that the game is a draw game: either player can force a draw. Blocking the opponent's winning sets is a solid way to force a draw; this is what we call a Strong Draw.

The main issue here is the Neighborhood Conjecture. The general case remains unsolved, but we can prove several useful partial results about blocking (called the Three Ugly Theorems).

Our treatment of the blocking part has a definite architecture. Metaphorically speaking, it is like a five-storied building where Theorems 34.1, 37.5, 40.1 represent the first three floors in this order, and Sections 43 and 44 represent the fourth and fifth floors; the higher floors are supported by the lower floors (there is no shortcut!).

An alternative way to look at the Neighborhood Conjecture is the Phantom Decomposition Hypothesis (see the end of Section 19), which is a kind of game-theoretic *independence*. In fact, there are two interpretations of game-theoretic *independence*: a "trivial" interpretation and a "non-trivial" one.

The "trivial" (but still very useful) interpretation is about *disjoint* games; Pairing Strategy is based on this simple observation. Disjointness guarantees that in each component either player can play independently from the rest of the components.

In the "non-trivial" interpretation the initial game does *not* fall apart into disjoint components. Instead Breaker can *force* – by playing rationally – that eventually, in a much later stage of the play, the family of unblocked (yet) hyperedges *does* fall apart into much smaller (disjoint) components. This is how Breaker can eventually finish the job of blocking the whole initial hypergraph, namely blocking componentwise in the small components.

A convincing "intuition" behind the "non-trivial" version is the Erdős–Lovász 2-Coloring Theorem. The proof of the Erdős–Lovász 2-Coloring Theorem was based on the idea of *statistical independence*; the proof is a repeated application of the simple fact that the two *events*: "the Random 2-Coloring of hypergraph $(V, \mathcal{F})$ is proper" and "the Random 2-Coloring of hypergraph $(W, \mathcal{G})$ is proper," are *independent* if the boards V and W are disjoint.

Strategy is a sequential procedure. The basic challenge of the Neighborhood Conjecture is how to *sequentialize* the global concept of statistical independence.

Chapter VII

BigGame–SmallGame Decomposition

We focus on the n^d board, and study both the hypercube and the torus versions of multi-dimensional Tic-Tac-Toe. Our main tool is a decomposition technique called BigGame–SmallGame Decomposition, combined with the Power-of-Two Scoring System. The key result of the chapter is Theorem 34.1, representing the "first floor" in the architecture of advanced blocking. By using this theorem (combined with the degree-reduction result Theorem 12.2) we can prove the so-called Hales–Jewett Conjecture (well, at least asymptotically).

Our proof technique works best for Degree-Regular hypergraphs. Unfortunately the n^d-hypergraph is very irregular, which leads to serious technical difficulties.

34

The Hales–Jewett Conjecture

1. Can we beat the Pairing Strategy? Hales and Jewett [1963] proved, by a pioneering application of Theorem 11.2, that if

$$n \geq 3^d - 1 \quad (n \text{ odd}) \quad \text{or} \quad n \geq 2^{d+1} - 2 \quad (n \text{ even}), \qquad (34.1)$$

then the n^d Tic-Tac-Toe game is a Pairing Strategy Draw. Indeed, by Theorem 3.4 (b)–(c): if n is odd, there are at most $(3^d - 1)/2$ winning lines through any point; if n is even, the maximum degree of the family of geometric lines drops to $2^d - 1$ (which is much smaller than $(3^d - 1)/2$ if d is large).

The special case $d = 2$ in (34.1) gives the bounds

$$n \geq 3^d - 1 = 8 \quad (n \text{ odd}) \quad \text{and} \quad n \geq 2^{d+1} - 2 = 6 \quad (n \text{ even}),$$

which immediately solve the n^2 game for all $n \geq 9$, n odd, and $n \geq 6$, n even: the game is a Pairing Strategy Draw (we already proved a slightly stronger result in Theorem 3.3). The missing case $n = 7$ can be easily solved by applying a "truncation trick." Since the center is the only cell with 4 winning lines passing through it, we throw out the center from these 4 lines, and also throw out an arbitrary point from each one of the rest of the lines. Then the size of the winning sets decreases to 6. Since the new maximum degree is 3, and $6 = 2 \cdot 3$, Theorem 11.2 applies to the "truncated family," and proves that the 7^2 game is also a Pairing Strategy Draw.

The following simple observation applies for any Almost Disjoint hypergraph: if a pairing strategy forces a draw, then there are at least twice as many points as winning sets. Therefore, if the Point/Line ratio in an n^d Tic-Tac-Toe game is less than 2, then it is *not* a Pairing Strategy Draw. For example, in the 4^2 game pairing strategy *cannot* exist because the number of points ("cells") is less than twice the number of winning lines: $16 = 4^2 < 2(4 + 4 + 2) = 20$. Note that 4^2 is nevertheless a draw game (we gave two proofs: one in Section 3, and an entirely different one in Section 10).

Hypercube Tic-Tac-Toe defines an Almost Disjoint hypergraph. Motivated by this fact and the "Point/Line ratio" observation above, Hales and Jewett [1963] made the following elegant conjecture:

Open Problem 34.1 *("Hales–Jewett Conjecture") (a) If there are at least twice as many points (i.e. cells) as winning lines, then the n^d Tic-Tac-Toe game is always a draw.*

(b) What is more, if there are at least twice as many points as winning lines, then the draw is actually a Pairing Strategy Draw.

Since the total number of lines is $((n+2)^d - n^d)/2$, and the number of points is n^d, the condition "there are at least twice as many points as winning lines" means

$$n^d \geq (n+2)^d - n^d,$$

which is easily seen to be equivalent to

$$n \geq \frac{2}{2^{1/d} - 1}. \tag{34.2}$$

Since $2^{1/d} - 1 = \frac{\log 2}{d} + O(d^{-2})$, (34.2) is asymptotically equivalent to

$$n \geq \frac{2d}{\log 2} + O(1) = 2.88539d + O(1). \tag{34.3}$$

Remark. Golomb and Hales [2002] made the amusing number-theoretic observation that the upper integral part

$$\left\lceil \frac{2}{2^{1/d} - 1} \right\rceil \quad \text{equals the simpler} \quad \left\lfloor \frac{2d}{\log 2} \right\rfloor \quad \text{for all integers in } 1 \leq d \leq 6.8 \cdot 10^{10}. \tag{34.4}$$

(34.4) yields that in the huge range $1 \leq d \leq 6.8 \cdot 10^{10}$ the Hales–Jewett condition (34.2) is equivalent to the simpler condition

$$n \geq \left\lfloor \frac{2d}{\log 2} \right\rfloor. \tag{34.5}$$

The "Bigamy Corollary" of the Hall's Theorem (Marriage Theorem) gives the following necessary and sufficient condition for Pairing Strategy Draw: for every sub-family of winning lines the union set has at least twice as many points as the number of lines in the sub-family. (The phrase "Bigamy" refers to the fact that "each man needs 2 wives.") Consequently, what Open Problem 34.1 (b) really says is that "the point/line ratio attains its minimum for the family of all lines, and for any proper sub-family the ratio is greater or equal". This *Ratio Conjecture* is very compelling, not only when Pairing Strategy exists, but in general for arbitrary n^d-game. The *Ratio Conjecture*, as a generalization of Open Problem 34.1 (b), was formulated in Patashnik [1980]. Unfortunately, the Ratio Conjecture is false: very recently Christofides (Cambridge) disproved it.

Let's return to Open Problem 34.1. Part (a) is nearly solved: we proved it for all but a finite number of n^d games (the drawing strategy is not a pairing strategy).

Part (b) is in a less satisfying state. R. Schroeppel proved it for all $n \geq 3d - 1$ when $d \geq 3$ is odd, and for all $n \geq 3d$ when $d \geq 2$ is even. In view of (34.3) the range $2.88539 + o(1) < n/d < 3 + o(1)$ represents infinitely many unsolved cases.

2. The first Ugly Theorem.

Notice that the Erdős–Selfridge Theorem (Theorem 1.4) is not powerful enough to settle part (a) of the Hales–Jewett Conjecture (Open Problem 34.1). Indeed, the Erdős–Selfridge criterion applies to the n^d game if

$$\frac{(n+2)^d - n^d}{2} + \{2^d - 1 \text{ or } (3^d - 1)/2\} < 2^n,$$

which implies that either player can force a Strong Draw if $n > const \cdot d \cdot \log d$. This "superlinear" bound falls short of proving the linear bound $n \geq 2d/\log 2 = 2.885d$ (which is asymptotically equivalent to part (a) of the Hales–Jewett Conjecture, see (34.3)).

Even if the Erdős–Selfridge (Theorem 1.4) is not powerful enough to "beat" the pairing strategy (in the n^d game), it is still the first step in the right direction. We are going to develop several local generalizations of Theorem 1.4, called the "Three Ugly Theorems." The first one, Theorem 34.1 below, will immediately prove part (a) of the Hales–Jewett Conjecture (at least for large dimensions). Combining Theorem 34.1 with Theorem 12.2 will lead to further improvements, far superior to what a pairing strategy can do. We have to warn the reader that Theorem 34.1 is not as elegant as Theorem 1.4. This "ugly but useful" criterion has a free parameter k, which can be freely optimized in the applications.

Theorem 34.1 *Let $\mathcal{F}$ be an m-uniform Almost Disjoint hypergraph. Assume that the Maximum Degree of $\mathcal{F}$ is at most D, i.e. every point of the board is contained in at most D hyperedges of $\mathcal{F}$ ("local size"). Moreover assume that the total number of winning sets is $|\mathcal{F}| = M$ ("global size"). If there is an integer k with $2 \leq k \leq m/2$ such that*

$$M \binom{m(D-1)}{k} < 2^{km - k(k+1) - \binom{k}{2} - 1}, \tag{34.6}$$

then the second player can force a Strong Draw in the positional game on $\mathcal{F}$.

Remarks. The part "the second player can force a Strong Draw" can be always upgraded to "either player can force a Strong Draw" (Strategy Stealing).

Theorem 34.1 is rather difficult to understand at first sight; one difficulty is the role of parameter k. What is the optimal choice for k? To answer this question, take kth roots of both sides of (34.6): we see that (34.6) holds if

$$M^{1/k} \cdot (D-1) \cdot \frac{4m}{k} < 2^{m - 3k/2 - 1}. \tag{34.7}$$

(34.7) is equivalent to

$$D - 1 < k \cdot M^{-1/k} \cdot 2^{-3k/2} \cdot \frac{2^{m-3}}{m}. \tag{34.8}$$

The product $k \cdot M^{-1/k} \cdot 2^{-3k/2}$ in (34.8) attains its maximum by choosing the integral parameter k around $(\frac{2}{3}\log_2 M)^{1/2}$, where $\log_2$ is the base two logarithm (or binary logarithm) (which assumes, in view of the requirement $k \leq m/2$, that $M < 2^{3m^2/8}$). Thus we obtain the following result:

Corollary 1 *If $\mathcal{F}$ is an m-uniform Almost Disjoint hypergraph with global size $|\mathcal{F}| < 2^{3m^2/8}$, and the maximum degree D of $\mathcal{F}$ is less than*

$$\frac{2^{m - \sqrt{6\log_2|\mathcal{F}|} - 3}}{m}, \tag{34.9}$$

then the second player can force a Strong Draw in the positional game played on $\mathcal{F}$.

Theorem 34.1 is a first step toward the Neighborhood Conjecture (Open Problem 9.1). The term "Neighborhood" emphasizes the fact that what really matters here is an exponential upper bound on the "neighborhood size," and the global size $|\mathcal{F}|$ is *almost* irrelevant (the global size can be super-exponentially large like $2^{c \cdot m^2}$, see (34.9)).

Corollary 1 is a justification of the Neighborhood Conjecture for *Almost Disjoint* hypergraphs. It raises some very natural questions such as:

(i) What happens if $|\mathcal{F}| > 2^{3m^2/8}$?

(ii) What happens if the hypergraph is *not* Almost Disjoint?

We are going to return to question (i) in Section 37, and question (ii) will be discussed much later in the last chapter.

If $|\mathcal{F}| \leq m^m$, then Corollary 1 yields:

Corollary 2 *Let $\mathcal{F}$ be an m-uniform family of Almost Disjoint sets. Assume that $|\mathcal{F}| \leq m^m$ and the Maximum Degree of $\mathcal{F}$ is at most $2^{m-3\sqrt{m\log m}}$. If $m > c_0$, i.e. if m is sufficiently large, then the second player can force a Strong Draw in the positional game on $\mathcal{F}$.*

Corollary 2 is an old result of mine; it follows (fairly) easily from Beck [1981a].

Corollary 2 applies to the n^d game if

$$\frac{(n+2)^d - n^d}{2} \leq n^n \quad \text{and} \tag{34.10}$$

$$\frac{3^d - 1}{2} \leq 2^{n-3\sqrt{n\log n}} \ (n \text{ odd}), \quad 2^d - 1 \leq 2^{n-3\sqrt{n\log n}} \ (n \text{ even}). \tag{34.11}$$

Inequalities (34.10)–(34.11) trivially hold if

$$n \geq \left(\frac{\log 3}{\log 2} + \varepsilon\right) d \quad \text{with} \quad d > c_0(\varepsilon) \quad (n \text{ odd}), \tag{34.12}$$

$$n \geq (1+\varepsilon) d \quad \text{with} \quad d > c_0(\varepsilon) \quad (n \text{ even}). \tag{34.13}$$

Observe that both (34.12) and (34.13) are asymptotically better than (34.3) (note that $\log 3 / \log 2 = 1.585$). This means that there are *infinitely many* n^d games in which Pairing Strategy Draw *cannot* exist, but the game is nevertheless a draw (in fact a Strong Draw).

According to my calculations, Theorem 34.1 proves part (a) of the Hales–Jewett Conjecture for all dimensions $d \geq 32$. For example, if $d = 32$, the Hales–Jewett Conjecture applies for n at least $2/(2^{1/32} - 1)$, which is about 91.3, so it applies for all $n \geq 92$. Theorem 34.1 settles both "border line" cases "$d = 32$ and $n = 92$" and "$d = 32$ and $n = 93$" with $k = 12$.

The case n *odd* is always harder, since the maximum degree is much larger. A low-dimensional example where Theorem 34.1 "beats" Pairing Strategy is the 44^{16} game: the Point/Line ratio is less than 2 (so Pairing Strategy cannot force a draw); on the other hand, Theorem 34.1 applies with $k = 8$, and guarantees a Strong Draw.

Now we are ready to discuss Conjectures A and B introduced at the end of Section 3; we recall them:

Conjecture A ("Gammill") *The n^d game is a draw if and only if there are more points than winning lines.*

Conjecture B ("Citrenbaum") *If $d > n$, then the first player has a winning strategy in the n^d game.*

Note that the condition of Conjecture A is $n^d > ((n+2)^d - n^d)/2$, which is asymptotically equivalent to $n \geq 2d/\log 3 = 1.82d$. Since $2/\log 3 = 1.82 > \log 3/\log 2 = 1.585 > 1$, in view of (34.12) and (34.13) there are infinitely many n^d games which *contradict* Conjecture A (infinitely many with n even, and infinitely many with n odd). This "kills" Conjecture A.

An explicit counterexample is the 144^{80}-game: since $(1+2/n)^d = (1+1/72)^{80} = 3.0146 > 3$, the 144^{80}-game has more lines than points, but Theorem 34.1 applies with $k = 20$, and yields a draw.

How about Conjecture B? Well, (34.12) and (34.13) are *not* strong enough to disprove Conjecture B. It seems very hard to find a counter-example to Conjecture B in low dimensions. In large dimensions, however, Conjecture B – similarly to Conjecture A – turns out to be completely false. Note in advance that the 214^{215}-game is the least counterexample that we know. We give a detailed discussion later.

3. Hypercube Tic-Tac-Toe. So far we have proved the following:

> *If n is odd and $n > (\log 3/\log 2 + \varepsilon)\,d$, or if n is even and $n > (1+\varepsilon)\,d$,*
> *then the n^d Tic-Tac-Toe game is a Strong Draw.*

These are two linear bounds. Can these linear bounds be improved to (nearly) quadratic? Theorem 12.5 says "yes": if $d \leq \left(\frac{\log 2}{16} + o(1)\right)\frac{n^2}{\log n}$, then (by Theorem 12.5) the n^d game is a Draw.

Are we ready to prove Theorem 12.5? A natural idea is to repeat the proof of Theorem 12.3, and to combine Theorem 12.2 with Theorem 34.1 (as a substitute of the Erdős–Lovász 2-Coloring Theorem). An application of Theorem 34.1 to the truncated hypergraph $\widetilde{\mathcal{F}}_{n,d}$ (see Theorem 12.2) leads to the requirement

$$d^{\frac{d}{\alpha n}+O(1)} \cdot 2^{3k/2} \cdot n^{d/k} \leq 2^{(1-2\alpha)n+O(1)}, \tag{34.14}$$

assuming $k = const \cdot n$. Note in advance that $d = O(n^2/\log n)$, so taking logarithms in both sides of requirement (34.14) gives

$$\frac{d\log d}{\alpha n \log 2} + \frac{3k}{2} + \frac{d}{k}\cdot\frac{\log d}{2\log 2} + O(\log n) \leq (1-2\alpha)n.$$

This inequality is satisfied if $\alpha = 2/13, k = \sqrt{d\log d/3\log 2} + O(1), n > 5.5\sqrt{d\log d}$, and $n > c_0$. Note that $n > 5.5\sqrt{d\log d}$ is asymptotically equivalent to $d < \frac{n^2}{60.5\log n}$.

The analogue of (34.14) for *combinatorial lines* goes as follows

$$d^{\frac{d}{\beta n}+O(1)} \cdot 2^{3k/2} \cdot n^{d/k} \leq 2^{(1-\beta)n+O(1)}. \tag{34.14'}$$

This inequality is satisfied if $\beta = 3/10$, $k = \sqrt{d\log d/3\log 2} + O(1)$, $n > 4.5\sqrt{d\log d}$, and $n > c_0$. Note that $n > 4.5\sqrt{d\log d}$ is asymptotically equivalent to $d < \frac{n^2}{40.5\log n}$.

This proves parts (ii) and (iv) in the following:

Theorem 34.2 *In the n^d hypercube Tic-Tac-Toe game:*

(i) *if $d \geq$ "Shelah's supertower function of n", then the first player can force an (ordinary) Win (but we don't know how he wins);*

(ii) *if $d < \frac{n^2}{60.5\log n}$ and n is sufficiently large, then the second player can force a Strong Draw;*

(iii) *if $d \leq n/4 = .25n$, then the game is a Pairing Strategy Draw, but for $d \geq (\log 2)n/2 = .34657n$ Pairing Strategy cannot exist;*

(iv) *if $d < \frac{n^2}{40.5\log n}$ and n is sufficiently large, then the second player can force a Strong Draw in the "combinatorial lines only" version of the n^d game.*

Notice that Theorem 34.2 (iii) easily follows from Theorem 12.2 and Theorem 11.2. Indeed, if $n \geq 4d$, then applying Theorem 12.2 (a) with $\alpha = 1/4$ gives

$$\text{MaxDegree}\left(\widetilde{\mathcal{F}_{n,d}}\right) \leq d \leq \lfloor n/4 \rfloor,$$

so Theorem 11.2 applies to the $2\lfloor n/4 \rfloor$-uniform $\widetilde{\mathcal{F}_{n,d}}$, and yields a *Pairing Strategy Draw*. Note that the bound $n \geq 4d$ is a big step toward part (b) of Open Problem 34.1 ("pairing strategy part of the Hales–Jewett Conjecture"). It is exponentially better than (34.1), but falls short of (34.3). In Section 48 we will improve the bound $n \geq 4d$ to $n \geq 3d$. This is the current record; it is due to Schroeppel.

It follows from parts (ii) and (iii) above that, in the big range $.34657n < d < \frac{n^2}{60.5 \log n}$ from linear to nearly quadratic, the n^d Tic-Tac-Toe is a draw game (in fact a Strong Draw) but *not* a Pairing Strategy Draw.

Part (ii) above immediately disproves Conjecture B. It gives infinitely many counterexamples. An explicit counterexample is the 214^{215}-game. Indeed, first taking $\alpha = d/7n = 215/1498 = 0.143525$ in Theorem 12.2, and then applying Theorem 34.1 to the truncated hypergraph with $k = 34$ yields a draw. Dimension $d = 215$ is very large; there should be a much smaller counterexample. Our technique doesn't work well in "low dimensions"; the case of low dimensions remains wide open.

Theorem 34.2 (ii) falls short of proving Theorem 12.5 (a) by a constant factor. The constant $1/60.5 = 0.01653$ is substantially less than $(\log 2)/16 = 0.04332$, but the nearly quadratic order of magnitude $\frac{n^2}{\log n}$ is the same.

In the next chapter we develop a more sophisticated decomposition technique (see Theorem 37.5), which will take care of the missing constant factor, and thus prove Theorem 12.5.

4. How to prove Theorem 34.1? Since it has a rather difficult proof, it helps to explain the basic idea on a simpler concrete game first. Our choice is a particular n^d torus Tic-Tac-Toe.

We have already introduced n^d torus Tic-Tac-Toe in Section 13. Recall that the 2-dimensional n^2 torus Tic-Tac-Toe is completely solved: the Erdős–Selfridge Theorem applies if $4n + 4 < 2^n$, which gives that the n^2 torus game is a Strong Draw for every $n \geq 5$; the 4^2 torus game is also a draw (mid-size case study; we don't know any elegant proof); finally, the 3^2 torus game is an easy first player's win.

Next consider the 3-dimensional n^3 torus game. The Erdős–Selfridge Theorem applies if $13n^2 + 13 < 2^n$, which gives that the n^3 torus game is a Strong Draw for every $n \geq 11$. We are convinced that the 10^3 torus game is also a draw, but don't know how to prove it.

How about the 4-dimensional n^4 torus game? The Erdős–Selfridge Theorem applies if $40n^3 + 40 < 2^n$, which gives that the n^4 torus game is a Strong Draw for every $n \geq 18$. This $n = 18$ is what we are going to improve to $n = 15$ (see Beck [2005]).

Theorem 34.3 *The 15^4 torus game is a Strong Draw.*

Theorem 34.3 is (probably) just a mediocre result, not even close to the truth, but it is a concrete low-dimensional example, the simplest illustration of the technique "BigGame-SmallGame Decomposition." We prove Theorem 34.3 first in Section 35, and prove Theorem 34.1 later in Section 36.

The 15^4 torus game is the smallest four-dimensional example that we know to be a Strong Draw. In the other direction, we know that the 5^4 torus game is a Weak Win (i.e. the first player can occupy 5 consecutive points on a torus line). This is a straightforward corollary of Theorem 1.2. There is a big gap between 5 and 15, and we don't know anything about the n^4 torus game when $6 \leq n \leq 14$.

In view of Theorem 13.1 we have a very satisfying *asymptotic* result for the n^d torus Tic-Tac-Toe: if d is a fixed "large" dimension, then the "phase transition" from Weak Win to Strong Draw happens at $n \approx (\log 3 / \log 2)d$ (see Theorem 13.1 (a)). The unsolved status of the 4-dimensional torus game in the long range $6 \leq n \leq 14$ reflects the fact that dimension 4 is just not large enough.

35

Reinforcing the Erdős–Selfridge technique (I)

1. BigGame-SmallGame Decomposition. The first player is called "Maker" and the second player is called "Breaker." The basic idea of the proofs of Theorems 34.1 and 34.3 is the same: it is a *decomposition* of the game into *two non-interacting games*. (We are motivated by a 2-coloring theorem of W. M. Schmidt mentioned in Section 11, see Schmidt [1962].) The *two non-interacting games* have disjoint boards: we call them the *Big Game* played on the big board and the *small game* played in the small board. This is in Breaker's mind only; Maker does not know anything about the decomposition whatsoever.

Non-interacting games mean that playing the *Big Game* Breaker has no knowledge of the happenings in the *small game*, and, similarly, playing the *small game* Breaker has no knowledge of the happenings in the *Big Game*. In other words, we assume that Breaker is "schizophrenic": he has two personalities, one for the *Big Game* and one for the *small game*, and the two personalities know nothing about each other. We call it the "Iron Curtain Principle." This, at first sight very weird, assumption is crucial in the proof!

Whenever Maker picks a point from the *big board* ("board of the Big Game"), Breaker responds in the big board, and, similarly, whenever Maker picks a point from the *small board* ("board of the small game"), Breaker responds in the small board. In other words, Breaker follows the "Same Board Rule."

The *small game* contains the winning sets that are "dangerous," where Maker is close to winning (i.e. the *small board* is a kind of "Emergency Room"). In the *small game*, Breaker's goal is to block the most dangerous winning sets, and he uses a straightforward *Pairing Strategy*.

Breaker's goal in the *Big Game* is to prevent too complex winning-set configurations from graduating into the *small game*. This is how the *Big Game* ensures that Breaker's Pairing Strategy in the *small game* will actually work. In the *Big Game* Breaker uses the sophisticated Erdős–Selfridge Power-of-Two Scoring System (more precisely, *not* the Erdős–Selfridge Theorem itself, but rather its Shutout version – see Lemma 1 below). The key fact is that the number of *big sets* depends primarily on the Maximum Degree D (rather than on the much larger "global size" M). The (relatively) small Maximum Degree keeps the family of *big sets* under control, and this is how Breaker can ensure, by using the Power-of-Two Scoring System, that the *small game* remains relatively simple, and a pairing strategy can indeed block every single "dangerous" winning set.

The board of the Big Game (big board) is going to shrink during a play. Consequently, the board of the small game (small board), which is the complement of the big board, keeps growing during a play. At the beginning of the play the big board is equal to the whole board V (i.e. the small game is not born yet). Let $V_{BIG}(i)$ and $V_{small}(i) = V \setminus V_{BIG}(i)$ denote the big board and the small board after Maker's *i*th move and before Breaker's *i*th move. Then we have

$$V = V_{BIG}(0) \supseteq V_{BIG}(1) \supseteq V_{BIG}(2) \supseteq V_{BIG}(3) \supseteq \cdots,$$

$$\emptyset = V_{small}(0) \subseteq V_{small}(1) \subseteq V_{small}(2) \subseteq V_{small}(3) \subseteq \cdots.$$

The Big Game is played on the family of *big sets*, and the small game is played on the family of *small sets*. What are the *big sets* and the *small sets*? Well, it is much simpler (and natural) to define the *small sets* first, and to define the *big sets* later.

We feel that the proof of Theorem 34.3 is easier to understand than that of Theorem 34.1. This why we start with the:

Proof of Theorem 34.3. Since $n = 15$ is odd, the corresponding hypergraph is Almost Disjoint (see the *Lemma on Torus-Lines* in Section 13). This fact will be used repeatedly in the proof. As said before, the basic idea is to artificially *decompose* the 15^4-torus game into *two non-interacting games* with disjoint boards: the *Big Game* and the *small game*.

The *small game* deals with the winning sets ("Lines on the Torus") that are *dangerous*, where Maker is very close to winning in the sense that all but 2 points are occupied by Maker. The *small board* is a kind of "Emergency Room": in the *small game*, Breaker's goal is to block the 2-element unoccupied parts of the most dangerous winning sets by using a trivial *Pairing Strategy*. The "Emergency Room" is Breaker's last chance before it is too late.

Breaker's goal in the *Big Game* is to prevent too complex winning-set-configurations ("Forbidden Configurations") from graduating into the *small game*. For example, Pairing Strategy works in the small game **only** if the 2-element unoccupied parts of the dangerous winning sets remain *pairwise disjoint*. The *Big Game* has to enforce, among many other things, this "disjointness property." This is how the *Big Game* guarantees that Breaker's Pairing Strategy in the *small game* will indeed work. This is why we must fully understand the *small game* first; the *Big Game* plays an auxiliary role only.

In the *Big Game* Breaker uses the Erdős–Selfridge *resource count* (*not* the Erdős–Selfridge Theorem itself, but its proof-technique) – see Lemma 1 below. The key numerical fact is that the number of *big sets* depends primarily on the Maximum Degree $D = (3^4 - 1)/2 = 40$ (rather than on the much larger Total Number $40 \cdot 15^3$ of winning sets). This is why the number of *Big Sets* is "under control," and this is how Breaker can force – by using the power-of-two scoring system – that the *small game* remains "trivial"; so simple that even a Pairing Strategy can block every "dangerous" winning set.

The board of the Big Game is going to shrink during a play (more precisely: the *unoccupied part* of the Big Game is going to shrink) due to the fact that the 2-element "emergency sets" are constantly removed from the Big Board and added to the small board.

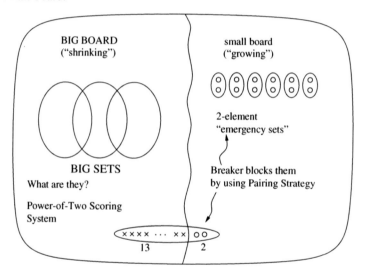

Let $x_1, x_2, \ldots, x_i, \ldots$ and $y_1, y_2, \ldots, y_i, \ldots$ denote, respectively, the points of Maker and Breaker in a particular play. At the beginning, when the board of the small game is empty (i.e. the small game is not born yet), Breaker chooses his points $y_1, y_2, y_3, \ldots$ according to an Erdős–Selfridge resource count applied to the family $\mathcal{B}$ of Big Sets ($\mathcal{B}$ will be defined later). The hypergraph of the game is the family of all Lines on the 15^4-torus; we denote it by $\mathcal{F}$. $\mathcal{F}$ is a 15-uniform hypergraph, and $|\mathcal{F}| = 40 \cdot 15^3$. In the course of a play in the Big Game an 15-element winning set $A \in \mathcal{F}$ (i.e. "Line on the Torus") becomes *dead* when it contains a mark of Breaker for the first time. Note that *dead* winning sets (i.e. elements of $\mathcal{F}$) represent no danger any more (they are marked by Breaker, so Maker cannot completely occupy them). At any time in the Big Game, the elements of $\mathcal{F}$ which are not dead yet are called *survivors*. A *survivor* $A \in \mathcal{F}$ becomes *dangerous* when Maker occupies its 13th point (all 13 points have to be in the Big Board); then the unoccupied

2-element part of this dangerous $A \in \mathcal{F}$ becomes an *emergency set*. Whenever an *emergency set* arises (in the Big Game), it is then *removed* from the board of the *Big Game* and added to the board of the *small game*. This is why the *Big Game* is shrinking. The board of the small game is precisely the union of *all 2-element emergency sets*, and, consequently, the board of the Big Game is precisely the complement of the union of all emergency sets.

If the (growing) family of 2-element emergency sets remains "disjoint" (i.e. the 2-element emergency sets never intersect during the whole course of a play), then Breaker can easily block them in the small game (on the small board) by using the following trivial Pairing Strategy: when Maker takes a member of a 2-element emergency set, Breaker then takes the other one. The Big Game is designed exactly to enforce, among other preperties, the "disjointness of the emergency sets." Therefore, Breaker must prevent the appearance of any:

Forbidden Configuration of Type 1: At some stage of the play there exist two *dangerous* sets $A_1 \in \mathcal{F}$ and $A_2 \in \mathcal{F}$ such that their 2-element emergency parts $E_1(\subset A_1)$ and $E_2(\subset A_2)$ have a common point. (Since $\mathcal{F}$ is Almost Disjoint, they cannot have more than one point in common.) See the picture on the left below (the picture on the right is a Forbidden Configuration of Type 2):

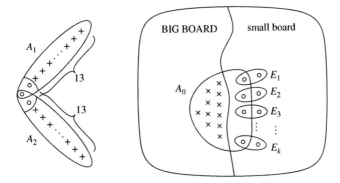

If there exists a Forbidden Configuration of Type 1, then at some stage of the play Maker occupied $13 + 13 = 26$ points of a "pair-union" $A_1 \cup A_2$ (where $A_1, A_2 \in \mathcal{F}$) in the Big Board (during the Big Game), and Breaker could not put a single mark yet in $A_1 \cup A_2$ in the Big Game (perhaps Breaker could do it in the small game, but that does not count).

Note that the total number of "intersecting pairs" $\{A_1, A_2\}$ with $|A_1 \cap A_2| = 1$ $(A_1, A_2 \in \mathcal{F})$ is exactly

$$15^4 \cdot \binom{40}{2}. \tag{35.1}$$

Indeed, the torus has 15^4 points, and each point has the same degree 40.

Another "potentially bad configuration" that Breaker is advised to prevent is any (see the picture on the right above):

Forbidden Configuration of Type 2: There exists a *survivor* $A_0 \in \mathcal{F}$:

(a) which never graduates into a *dangerous set*; and at some stage of the play,
(b) A_0's intersection with the Big Board is completely occupied by Maker, or possibly empty; and at the same time,
(c) A_0's intersection with the small board is completely covered by pairwise disjoint 2-element emergency sets.

The "danger" of (a)–(c) is obvious: since A_0 never graduates into a *dangerous set*, A_0's intersection with the small board remains "invisible" for Breaker in the whole course of the small game, so there is a real chance that A_0 will be completely occupied by Maker.

In (c) let k denote the size of the intersection of A_0 with the small board; (a) implies that the possible values of k are $3, 4, \ldots, 15$. Accordingly we can talk about Forbidden Configuration of Type $(2, k)$ where $3 \leq k \leq 15$.

Let A_0 be a Forbidden Configuration of Type $(2, k)$ $(3 \leq k \leq 15)$; then there are k disjoint 2-element emergency sets $E_1, E_2, \ldots, E_k$ which cover the k-element intersection of A_0 with the small board (see (c)). Let $A_1, A_2, \ldots, A_k$ denote the *super-sets* of $E_1, E_2, \ldots, E_k$: $E_i \subset A_i$ where every $A_i \in \mathcal{F}$ is a dangerous set (Almost Disjointness implies that for every E_i there is a unique A_i).

Then, at the particular stage of the play described by (b)–(c), Maker occupied at least $(15 - 2) + (15 - 3) + \ldots + (15 - k) + (15 - k - 1) + (15 - k) = (30 + 25k - k^2)/2$ points ("Almost Disjointness") of a union set $\bigcup_{i=0}^{k} A_i$ in the Big Board (in the Big Game), and Breaker could not put a single mark yet in $\bigcup_{i=0}^{k} A_i$ in the Big Game.

Note that the total number of configurations $A_0, \{A_1, A_2, \ldots, A_k\}$ satisfying:

(α) $A_0, A_1, \ldots, A_k \in \mathcal{F}$, and
(β) $|A_0 \cap A_i| = 1$ for $1 \leq i \leq k$, and
(γ) $A_0 \cap A_i$ with $1 \leq i \leq k$ are k distinct points

is at most

$$(40 \cdot 15^3) \cdot \binom{15}{k} \cdot (40 - 1)^k. \tag{35.2}$$

Indeed, first choose A_0; then choose the k distinct points $A_0 \cap A_i$ with $1 \leq i \leq k$, and finally choose the k sets $A_i \in \mathcal{F}$ $(i = 1, \ldots, k)$ (use "Almost Disjointness").

Note that after an arbitrary move of Maker in the Big Game, the Big Board may shrink (because of the possible appearances of new dangerous sets; the 2-element emergency parts are removed from the Big Board), but the Big Board does not change after any move of Breaker.

To prevent all Forbidden Configurations (i.e. Type 1 and Type $(2, k)$ where $k = 3, 4, \ldots, 15$), Breaker needs the following "shrinking" variant of the Erdős–Selfridge Theorem.

Lemma 1: *Let $\mathcal{B}_1, \mathcal{B}_2, \ldots, \mathcal{B}_l$ be a sequence of finite hypergraphs. Let $m_1, m_2, \ldots, m_l$ be positive integers. Consider the following "shrinking" game. Let V be the union set of the l hypergraphs; we call V the "initial board." The 2 players, called White and Black, alternate. A "move" of White is to take a previously unoccupied point of the board and at the same time White may remove an arbitrary unoccupied part from the board (if he wants any). Similarly, a "move" of Black is standard: he takes a previously unoccupied point of the board. After White's move the board may shrink, and the players are always forced to take the next point from the "available board" (which is a "decreasing" subset of the initial board). White wins, if at some stage of the play there exist $i \in \{1, \ldots, l\}$ and $B \in \mathcal{B}_i$ such that White has m_i points from B and Black has none; otherwise Black wins ("shutout game").*

Assume that

$$\sum_{i=1}^{l} (|\mathcal{B}_i| + \mathrm{MaxDeg}(\mathcal{B}_i)) \, 2^{-m_i} < 1.$$

Then Black has a winning strategy, no matter whether Black is the first or second player.

3. How to apply Lemma 1? To prevent the appearance of any Forbidden Configuration of Type 1, we have to control all possible candidates (of course, we don't know in advance which winning set will eventually become *dangerous*), so let

$$\mathcal{B}_1 = \left\{ A_1 \cup A_2 : \{A_1, A_2\} \in \binom{\mathcal{F}}{2}, \; |A_1 \cap A_2| = 1 \right\} \quad \text{and} \quad m_1 = 2(15 - 2) = 26.$$

To avoid the appearance of any Forbidden Configuration of Type $(2, 3)$, define hypergraph $\mathcal{B}_2$ as follows

$$\left\{ A_0 \cup \ldots \cup A_3 : \{A_0, \ldots, A_3\} \in \binom{\mathcal{F}}{4}, \; A_0 \cap A_i, \, 1 \leq i \leq 3 \text{ are distinct points} \right\}$$

and $m_2 = (15 - 2) + (15 - 3) + (15 - 4) + (15 - 3) = 48$. In general, to avoid the appearance of any Forbidden Configuration of Type $(2, j + 1)$, define hypergraph $\mathcal{B}_j$ as follows

$$\left\{ A_0 \cup \cdots \cup A_{j+1} : \{A_0, \ldots, A_{j+1}\} \in \binom{\mathcal{F}}{j+2}, \; A_0 \cap A_i, \, 1 \leq i \leq j+1 \right.$$

$$\left. \text{are distinct points} \right\} \qquad (35.3)$$

and

$$m_j = (15-2) + (15-3) + \cdots + (15-j-1) + (15-j-2) + (15-j-1)$$
$$= \frac{54 + 23j - j^2}{2}. \tag{35.4}$$

For technical reasons we use definition (35.3)–(35.4) for $j = 2, 3, \ldots, 6$ only. This takes care of Forbidden Configurations of Type $(2, k)$ with $3 \le k \le 7$. To prevent the appearance of any Forbidden Configuration of Type $(2, k)$ where $8 \le k \le 15$, we use a single extra hypergraph $\mathcal{B}_7$: let hypergraph $\mathcal{B}_7$ be defined as

$$\left\{ A_0 \cup A_1 \cup \cdots \cup A_8 : \{A_0, A_1, \ldots, A_8\} \in \binom{\mathcal{F}}{9}, \ A_0 \cap A_i, 1 \le i \le 8 \text{ are distinct points} \right\} \tag{35.5}$$

and

$$m_j = (15-2) + (15-3) + (15-4) + (15-5) + (15-6) + (15-7)$$
$$+ (15-8) + (15-9) = 76. \tag{35.6}$$

In the definition of m_7 we didn't include the number of marks of Maker in A_0: this is how we can deal with all Types $(2, k)$ where $8 \le k \le 15$ *at once*. Indeed, any Forbidden Configuration of Type $(2, k)$ with $8 \le k \le 15$ contains a sub-configuration described on the figure below.

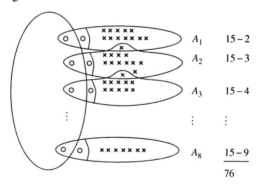

Next we estimate the sum

$$\sum_{i=1}^{7} (|\mathcal{B}_i| + \mathrm{MaxDeg}(\mathcal{B}_i)) \, 2^{-m_i}.$$

By (35.1)–(35.2) we have

$$|\mathcal{B}_1| = 15^4 \cdot \binom{40}{2}, \text{ and } |\mathcal{B}_j| = (40 \cdot 15^3) \cdot \binom{15}{j+1} \cdot (40-1)^{j+1} \text{ for } 2 \le j \le 7.$$

Since the 15^4-torus is a group, the hypergraphs $\mathcal{B}_i$ ($1 \le i \le 7$) are all degree-regular; every point has the same degree, namely the average degree. (This makes the calculations simpler.) In $\mathcal{B}_1$ every set has the same size $2 \cdot 15 - 1 = 29$, and for each $j = 2, \ldots, 7$ every $B \in \mathcal{B}_j$ has size $\geq m_j = (15 - 2) + (15 - 3) + \cdots + (15 - j - 1) + (15 - j - 2) + (15 - j - 1) = (54 + 23j - j^2)/2$. It follows that

$$(|\mathcal{B}_1| + \mathrm{MaxDeg}(\mathcal{B}_1)) = 15^4 \cdot \binom{40}{2} \left(1 + \frac{29}{15^4}\right),$$

and for $j = 2, \ldots, 7$

$$(|\mathcal{B}_j| + \mathrm{MaxDeg}(\mathcal{B}_j)) \leq (40 \cdot 15^3) \cdot \binom{15}{j+1} \cdot (40 - 1)^{j+1} \left(1 + \frac{54 + 23j - j^2}{2 \cdot 15^4}\right).$$

Therefore, easy calculations give

$$\sum_{i=1}^{7} (|\mathcal{B}_i| + \mathrm{MaxDeg}(\mathcal{B}_i)) \, 2^{-m_i} \leq 15^4 \cdot \binom{40}{2} \left(1 + \frac{29}{15^4}\right) 2^{-26}$$

$$+ \sum_{j=2}^{6} \leq (40 \cdot 15^3) \cdot \binom{15}{j+1} \cdot (40 - 1)^{j+1} \left(1 + \frac{54 + 23j - j^2}{2 \cdot 15^4}\right) 2^{-(54+23j-j^2)/2}$$

$$+ (40 \cdot 15^3) \cdot \binom{15}{8} \cdot (40 - 1)^8 \left(1 + \frac{54 + 23 \cdot 8 - 8^2}{2 \cdot 15^4}\right) 2^{-76} < \frac{9}{10} < 1.$$

Now we are ready to define the *Big Sets* (i.e. Forbidden Configurations): Let $\mathcal{B} = \mathcal{B}_1 \cup \mathcal{B}_2 \cup \cdots \cup \mathcal{B}_7$ be the family of *Big Sets*. Applying Lemma 1 (Maker is "White" and Breaker is "Black") we obtain that, in the Big Game, played on the family of Big Sets, Breaker can prevent the appearance of any Forbidden Configuration of Type 1 or Type $(2, k)$ with $3 \leq k \leq 15$. We claim that, combining this "Lemma 1 strategy" with the trivial Pairing Strategy in the family of emergency sets (small game), Breaker can block hypergraph $\mathcal{F}$, i.e. Breaker can put his mark in every Line in the 15^4 Torus Game. Indeed, an arbitrary $A \in \mathcal{F}$ ("Line on the Torus"):

(1) either eventually becomes *dangerous*, then its 2-element emergency set will be blocked by Breaker in the small board (in the small game) by the trivial Pairing Strategy ("disjointness of the emergence sets" is enforced by preventing Forbidden Configurations of Type 1);

(2) or A never becomes *dangerous*, then it will be blocked by Breaker in the Big Board: indeed, otherwise there is a Forbidden Configuration of Type $(2, k)$ with some $k \in \{3, \ldots, 15\}$.

4. The proof of Lemma 1. We basically repeat the Erdős–Selfridge proof. Black employs the following Power-of-Two Scoring System. If $B \in \mathcal{B}_i$ is marked by Black, then it scores 0. If $B \in \mathcal{B}_i$ is unmarked by Black and has w points of White, then it scores 2^{w-m_i}. The "target value" (i.e. White's win) is $\geq 2^0 = 1$; the "initial value" is less than 1 (by hypothesis). Breaker can guarantee the usual "decreasing property," which implies that no play is possible if the "target value" is larger than the "initial value."

To be more precise, suppose that we are in the middle of a play, and it is Black's turn to choose his ith point y_i. What is the "danger" of this particular position? We evaluate the position by the total sum, over all winning sets, of the *scores*; we denote it by D_i, and call it the "danger-function" (index i indicates that we are at the stage of choosing the ith point of Black). The natural choice for y_i is that unoccupied point z which makes the "biggest damage": for which the sum of the *scores* of all "survivors" $B \in \mathcal{B}$ with $z \in B$ attains the *maximum*. Loosely speaking: y_i is the "biggest damage point," so x_{i+1} is at most the "second biggest damage point." Then what we subtract from D_i is greater or equal to what we add back to it. In other words, Black can force the decreasing property $D_{start} = D_1 \geq D_2 \geq \cdots \geq D_i \geq D_{i+1} \geq \cdots \geq D_{end}$ of the "danger-function" (the "shrinking" of the unoccupied part of the board doesn't change this key property).

White wins the game if he can occupy m_i points of some set $B \in \mathcal{B}_i$ before Black could put his first mark in this B. Assume this happens right before the jth move of Breaker. Then this B alone scores $2^0 = 1$, implying $D_j \geq 1$; we call $2^0 = 1$ the "target value."

On the other hand, the "initial value" (x_1 is the first point of White)

$$D_{start} = D_1 = \sum_i \sum_{B:\ x_1 \in B \in \mathcal{B}_i} 2^1 + \sum_i \sum_{B:\ x_1 \notin B \in \mathcal{B}_i} 2^0 < 1$$

by the hypothesis of Lemma 1. By the *decreasing property* of the "danger-function," if White wins at his jth move, then

$$2^0 = 1 \leq D_j \leq D_{start} < 1,$$

and we get a contradiction. It follows that Breaker wins; Breaker's winning strategy is to keep choosing the "biggest damage point." This completes the proof of Lemma 1.   □

The proof of Theorem 34.3 is complete.   □

36

Reinforcing the Erdős–Selfridge technique (II)

1. Proof of Theorem 34.1. Again we use the *BigGame–SmallGame Decomposition* – see the beginning of Section 35. Let $x_1, x_2, \ldots, x_i, \ldots$ and $y_1, y_2, \ldots, y_i, \ldots$ denote, respectively, the points of Maker and Breaker. At the beginning of the play, when the board of the small game is empty (i.e. the small game is not born yet), Breaker chooses his points $y_1, y_2, y_3, \ldots$ according to the Erdős–Selfridge Power-of-Two Scoring System (see Lemma 5 below) applied to the family $\mathcal{B}$ of big sets (family $\mathcal{B}$ will be defined later). In the course of a play in the Big Game an n-element winning set $A \in \mathcal{F}$ is *dead* when it contains some point of Breaker (for the first time). Note that *dead* elements of $\mathcal{F}$ don't represent danger any more (they are marked by Breaker, so Maker cannot completely occupy them). At a given stage of *a* play in the Big Game those elements of $\mathcal{F}$ which are not dead yet are called *survivors*. A *survivor* $A \in \mathcal{F}$ becomes *dangerous* when Maker occupies its $(m - k - 1)$th point in the Big Game; then the unoccupied $(k + 1)$-element part of this dangerous $A \in \mathcal{F}$ becomes an *emergency set*. This is why the *Big Game* is shrinking: whenever an *emergency set* arises, its points are *removed* from the big board, and at the same time they are added to the small board. In other words, the small board is precisely the union of *all emergency sets* (and, consequently, the big board is precisely the complement of the union of all emergency sets). This means that the *small board* is a kind of "Emergency Room" where Breaker takes care of the *emergency sets* (Breaker's goal is to put his mark in every *emergency set*).

What are the *small sets*, i.e. the winning sets in the small game? Well, a set S is a *small set* if it satisfies the following two requirements:

(a) S is the intersection of a *survivor* $A \in \mathcal{F}$ and the board of the small game,

(b) $|S| \geq k + 1$.

The *emergency sets* are obviously all *small* sets. Those small sets which are *not* emergency sets are called *secondary sets*. What is the goal of the small game? Well, Breaker wins the small game if he can prevent Maker from completely occupying a

479

small set; otherwise Maker wins. (In other words, Breaker's goal is to block every small set – he will in fact employ a simple pairing strategy.)

Now we are ready to define the *big sets*. The *big sets* play an auxiliary role in the proof. We use them to guarantee that, when Breaker follows a winning strategy in the Big Game, the small game is very simple in the following sense:

(i) a secondary set doesn't exist – see Lemma 1; and
(ii) Breaker can block every emergency set by a pairing strategy employing the "private points" – see Lemma 2 below.

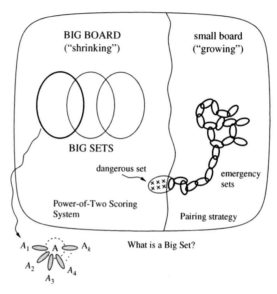

To ensure these two requirements we introduce a key definition: a *k*-element sub-family $\mathcal{G} = \{A_1, A_2, \ldots, A_k\} \subset \mathcal{F}$ is called *$\mathcal{F}$-linked* if there is a set $A \in \mathcal{F}$ with $A \notin \mathcal{G}$ such that A intersects each element of family $\mathcal{G}$, i.e. $A \cap A_i \neq \emptyset$, $1 \leq i \leq k$. (Note that parameter k is an integer between 2 and $m/2$.)

The Big Game is played on the family $\mathcal{B}$ of *big sets*. What are the *big sets*? They are the *union sets* $\bigcup_{i=1}^{k} A_i$ of all possible *$\mathcal{F}$-linked* *k*-element subfamilies $\mathcal{G} = \{A_1, A_2, \ldots, A_k\}$ of $\mathcal{F}$

$$\mathcal{B} = \Big\{ B = \bigcup_{A \in \mathcal{G}} A : \mathcal{G} \subset \mathcal{F}, \ |\mathcal{G}| = k, \ \mathcal{G} \text{ is } \mathcal{F}\text{–linked} \Big\}.$$

The total number of *big sets* is estimated from above by

$$|\mathcal{B}| \leq M \binom{m(D-1)}{k}. \tag{36.1}$$

Indeed, there are $|\mathcal{F}| = M$ ways to choose "linkage" A, there are at most $m(D-1)$ other sets intersecting A, and we have to choose k sets $A_1, A_2, \ldots, A_k$ among them.

Each big set $B = \bigcup_{i=1}^{k} A_i$ has cardinality $\geq km - \binom{k}{2}$. Indeed, since $\mathcal{F}$ is almost Disjoint

$$|B| = |\bigcup_{i=1}^{k} A_i| \geq \sum_{i=1}^{k} |A_i| - \sum_{1 \leq i < j \leq k} |A_i \cap A_j| \geq km - \binom{k}{2}. \qquad (36.2)$$

What is *Maker's goal* in the Big Game? Of course, Maker doesn't know about the "Big Game" (or the "small game"); this whole "decomposition" is in Breaker's mind only, so it is up to Breaker to define "Maker's goal" in the Big Game. The definition goes as follows: "Maker wins the Big Game" if he can occupy all but $k(k+1)$ points of some big set $B \in \mathcal{B}$ *in the big board* before Breaker could put his first mark in this B *in the big board*; otherwise Breaker wins the Big Game. The reason why we had to write "*in the big board*" twice in the previous sentence is that the big board is shrinking, and so the big board does not necessarily contain all big sets. The intersection of a big set with the small board (i.e. the part outside of the big board) is "invisible" in the Big Game: whatever happens in the small game has no effect in the Big Game. (For example, if Breaker can block a big set in the small board, that doesn't count as a "blocking in the Big Game"; this is the curse of the "Iron Curtain Principle.")

We are going to show that Breaker can win the Big Game by using the Erdős–Selfridge Power-of-Two Scoring System if the total number of big sets is not too large, namely if

$$|\mathcal{B}| < 2^{km - \binom{k}{2} - k(k+1) - 1}. \qquad (36.3)$$

Note that the board of the Big Game is *shrinking* during a play, but it is *not* going to cause any extra difficulty in the argument (see Lemma 5).

Lemma 1: *Assume that Breaker has a winning strategy in the Big Game, and he follows it in a play. Then there is no secondary set in the small game.*

Proof. Let $S^* = \{z_1, z_2, z_3, \ldots\}$ be an arbitrary secondary set. Since the small board is the union of the emergency sets, by property (a) above every point $z_i \in S^*$ is contained in some emergency set (say) S_i, $1 \leq i \leq |S^*|$. Almost disjointness implies that different points $z_i \in S^*$ are contained in different emergency sets S_i. Since S_i is an emergency set, there is a winning set $A_i \in \mathcal{F}$ such that $S_i \subset A_i$ and $A_i \setminus S_i$ was completely occupied by Maker during the play in the Big Game. Note that Breaker didn't block A_i in the Big Game, since S_i was removed from the big board (and added to the small board). Again Almost Disjointness implies that the sets A_i, $i = 1, 2, 3, \ldots$ are all different. The union set $\bigcup_{i=1}^{k} A_i$ is a *big set* since $\{A_1, A_2, \ldots, A_k\}$ is $\mathcal{F}$-*linked* by A^* where $A^* \in \mathcal{F}$ is defined by $S^* \subset A^*$ (A^* is the *ancestor* of S^*). Since for every i, $A_i \setminus S_i$ was completely occupied by Maker, Maker was able to occupy all but $k(k+1)$ points of the particular big set $B = \bigcup_{i=1}^{k} A_i$ during a play in the Big Game, and Breaker didn't block B in the Big Game, i.e. Maker wins the Big Game. This contradicts the assumption that Breaker has a winning strategy in the Big Game and follows it in the play. This contradiction shows that a secondary set cannot exist.   □

A similar argument proves:

Lemma 2: *If Breaker wins the Big Game, then every emergency set has at least two "private" points, i.e. two points which are never going to be contained in any other emergency set during the whole course of the small game. (In other words, a point of an emergency set is called "private" if it has degree one in the family of all emergency sets.)*

Let $S_1, S_2, S_3, \ldots$ be the complete list of *emergency sets* arising in this order during the course of a play (when a bunch of two or more emergency sets arise at the same time, the ordering within the bunch is arbitrary). Let $\widetilde{S}_1 = S_1$, $\widetilde{S}_2 = S_2 \setminus S_1$, $\widetilde{S}_3 = S_3 \setminus (S_1 \cup S_2)$, and in general

$$\widetilde{S}_j = S_j \setminus \left(\bigcup_{i=1}^{j-1} S_i \right).$$

We call $\widetilde{S}_j$ the "on-line disjoint part" of S_j. Of course, the "on-line disjoint part" of S_j contains its "private points," so by Lemma 2 every "on-line disjoint part" $\widetilde{S}_j$ has at least 2 elements (exactly what we need for a pairing strategy – see below).

When the first dangerous set $A \in \mathcal{F}$ arises, say, at the ith move of Maker, the whole board V splits into two non-empty parts for the *first time*. The two parts are the big board $V_{BIG}(i)$ and the small board $V_{small}(i)$, where $V = V_{BIG}(i) \cup V_{small}(i)$. Whenever Maker picks a point from the big board, Breaker then responds in the big board; whenever Maker picks a point from the small board, Breaker then responds in the small board ("Same Board Rule"). This is how the game falls apart into two non-interacting, disjoint games: the shrinking *Big Game* and the growing *small game*.

During the course of a play in the *small game* Breaker uses the following trivial pairing strategy: if Maker occupies a point of the small board which is contained in the "on-line disjoint part" $\widetilde{S}$ of an emergency set S, then Breaker picks another point of the same $\widetilde{S}$ (if he *finds* one; if he *doesn't*, then he makes an arbitrary move). In view of the remark after Lemma 2, Breaker can block every emergency set in the small game *under the condition* that he can win the Big Game. Since a secondary set cannot exist (see Lemma 1), we obtain:

Lemma 3: *If Breaker can win the Big Game, then he can win the small game, i.e. he can block every small set.*

Next we prove:

Lemma 4: *If Breaker can win the Big Game, then he can block every winning set $A \in \mathcal{F}$ either in the Big Game or in the small game.*

Proof. Indeed, assume that at the end of a play some $A_0 \in \mathcal{F}$ is completely occupied by Maker. We derive a contradiction as follows. We distinguish two cases.

Case 1: During the course of the Big Game Maker occupies $(m - k - 1)$ points of A_0, i.e. $|A_0 \cap V_{BIG}(j)| = m - k - 1$ for some j.

Let j be the first index such that $|A_0 \cap V_{BIG}(j)| = m - k - 1$, i.e. in the Big Game Maker occupied the $(m - k - 1)$th point of A_0 at his jth move. Then A_0 becomes a dangerous set, and $S_0 = A_0 \setminus V_{BIG}(j)$ goes to the small game as an emergency set. Since Breaker can block every emergency set in the small game by a "pairing strategy" (see Lemma 3), we have a contradiction.

Case 2: At the end of the Big Game Maker has less than $(m - k - 1)$ points of A_0. Then for some j, $A_0 \cap V_{small}(j)$ must become a *secondary set*, which contradicts Lemma 1. □

Therefore, the *last step* of the proof of Theorem 34.1 is to show that Breaker has a winning strategy in the Big Game. We recall that Maker wins the Big Game if he can occupy all but $k(k+1)$ points of some big set $B \in \mathcal{B}$ before Breaker could put his first mark in this B. In view of (36.3) this means (at least) $km - \binom{k}{2} - k(k+1)$ points of Maker in B (before Breaker could put his first mark in this B). What we need here is not the Erdős–Selfridge Theorem itself, but the following slightly modified version (we will apply it to the Big Game with $b = km - \binom{k}{2} - k(k+1)$, where m is from Lemma 5 below).

Lemma 5: *Let $\mathcal{B}$ be a hypergraph such that every winning set $B \in \mathcal{B}$ has at least b points. There are two players, Maker and Breaker, who alternately occupy previously unoccupied points of the board (Maker starts). Assume that after each of Maker's moves the unoccupied part of the board may shrink, but the board doesn't change after Breaker's moves. Maker wins the game if he can occupy b points of some winning set $B \in \mathcal{B}$ before Breaker could put his first mark in this B; otherwise Breaker wins ("shutout game"). Now if $|\mathcal{B}| < 2^{b-1}$, then Breaker has a winning strategy.*

Notice that Lemma 5 is a trivial special case of Lemma 1 in Section 35: it is the special case of $l = 1$ and $MaxDeg(\mathcal{B}) \le |\mathcal{B}|$.

By (36.2) each big set $B \in \mathcal{B}$ has cardinality $\ge km - \binom{k}{2}$. Therefore, we apply Lemma 5 to the Big Game with $b = km - \binom{k}{2} - k(k+1)$. We recall the upper bound on the total number of big sets (see (36.1))

$$|\mathcal{B}| \le M \binom{m(D-1)}{k}.$$

By Lemma 5 Breaker wins the Big Game if

$$|\mathcal{B}| \le M \binom{m(D-1)}{k} < 2^{km - \binom{k}{2} - k(k+1) - 1},$$

exactly as we predicted in (36.3). This completes the proof of Theorem 34.1.   □

2. A slight generalization. For a later application in the n^d Torus Tic-Tac-Toe with n is even (see the *Lemma on Torus-Lines* in Section 13) we slightly modify the intersection hypothesis of Theorem 34.1, and assume that for any two distinct hyperedges A_1 and A_2 of hypergraph $\mathcal{F}$ the intersection size $|A_1 \cap A_2|$ is ≤ 2 (instead of Almost Disjointness: $|A_1 \cap A_2| \leq 1$). Under what condition about the global size $M = |\mathcal{F}|$ and the Max Degree $D = D(\mathcal{F})$ can the second player force a Strong Draw in the positional game played on the m-uniform hypergraph $\mathcal{F}$? To answer the question we repeat the proof of Theorem 34.1 with the following minor modifications:

(1) A *survivor* $A \in \mathcal{F}$ becomes *dangerous* when Maker ("the first player") occupies its $(m-2k-1)$th point in the Big Game; then the $(2k+1)$-element part of A (unoccupied in the Big Game) becomes an *emergency set*.

(2) If an emergency set E_0 does not have a "private point pair," then at least $2k$ points of E_0 are covered by other emergency sets. Since two hyperedges intersect in at most two points, there are at least k other emergency sets $E_1, E_2, \ldots, E_k$ intersecting E_0. Let $A_0, A_1, A_2, \ldots, A_k \in \mathcal{F}$ denote the *uniquely determined* super-sets of $E_0, E_1, E_2, \ldots, E_k$ in this order (i.e. $E_i \subset A_i$ for $i = 0, 1, \ldots, k$; *uniquely determined* since $2k+1 > 2$ for $k \geq 1$).

(3) If a *survivor* $A_0 \in \mathcal{F}$ never becomes *dangerous*, then at some stage of the play the intersection of A_0 with the Big Board is fully occupied by Maker and the intersection of A_0 with the small board contains $\geq 2k+2$ points. These $\geq 2k+2$ points in the small board are covered by at least k different emergency sets.

It follows that the *old definition* of the family $\mathcal{B}$ of Big Sets works just fine and covers both cases (2) and (3) above, where the analogue of parameter "b" is $b = km - k(2k+1) - 2\binom{k}{2}$. This argument yields the following variant of Theorem 34.1, which will be applied in the next section.

Theorem 36.1 *Let $\mathcal{F}$ be an m-uniform hypergraph, and assume that $|A_1 \cap A_2| \leq 2$ holds for any two distinct hyperedges A_1 and A_2 of hypergraph $\mathcal{F}$ (instead of Almost Disjointness: $|A_1 \cap A_2| \leq 1$). Let D denote the Maximum Degree of $\mathcal{F}$, and let $M = |\mathcal{F}|$ denote the total number of winning sets. If there is an integer k with $1 \leq k \leq m/4$ such that*

$$M\binom{m(D-1)}{k} < 2^{km-k(2k+1)-2\binom{k}{2}-1},$$

then the second player can force a Strong Draw in the positional game on $\mathcal{F}$.

37

Almost Disjoint hypergraphs

1. Almost Disjoint vs. General hypergraphs. In "Tic-Tac-Toe Theory" the concept of *Almost Disjoint* hypergraph arises in a most natural way: the winning sets in hypercube Tic-Tac-Toe are collinear, and two lines intersect in (at most) 1 point.

This is a very lucky situation, because the sub-class of Almost Disjoint hypergraphs represents the simplest case for the "fake probabilistic method." Notice that the Weak Win Criterion Theorem 1.2 and the Strong Draw Criterion Theorem 34.1 are both about Almost Disjoint hypergraphs.

The general case (i.e. not necessarily Almost Disjoint) is always much more complicated; it is enough to compare Theorem 1.2 with Theorem 24.2 ("Advanced Weak Win Criterion"). The proof of Theorem 1.2 was less than 1 page; Theorem 24.2 has a 10-page proof.

The Achitecture of Advanced Blocking. The subclass of Almost Disjoint hypergraphs represents the starting point of the discussion. We develop our decomposition technique for Almost Disjoint hypergraphs first – Almost Disjoint hypergraphs are our guinea pigs! – and relax the intersection conditions later.

The whole of Part D is basically a long line of evolution of the *decomposition* idea. The highlights are

Theorem 34.1 ↪ Theorem 37.5 ↪ Theorem 40.1 ↪ Sections 43–44 : Theorem 8.2,

where the weird notation "A ↪ B" means that "criterion B is a more advanced version of criterion A." It is like a five-storied building where Theorems 34.1, 37.5, 40.1 represent the first three floors in this order, and Sections 43 and 44 are the two top floors. The three "basic floors" are all about Almost Disjoint hypergraphs, becoming increasingly difficult, and only the last two "floors" are general enough to cover the cases of the Winning Planes (Theorem 12.6) and the Lattice Games (Theorem 8.2). The plane-hypergraph and the different lattice-game hypergraphs are far from being Almost Disjoint – this is why we need the 3 preliminary steps (each one about Almost Disjoint hypergraphs) before being able to prove the Strong Draw part in Theorem 12.6 and Theorem 8.2.

After these general, introductory remarks about the "architecture of blocking," it is time now to prove something. We begin with an easy applications of Theorem 34.1 (and its variant Theorem 36.1). We discuss the:

Proof of Theorem 13.1. The statement of Theorem 13.1 is about the n^d torus Tic-Tac-Toe:

(a) $\mathbf{ww}(n\text{–line in torus}) = \left(\frac{\log 2}{\log 3} + o(1)\right) n,$ (37.1)

(b) $\mathbf{ww}(\text{comb. } n\text{–line in torus}) = (1 + o(1))n.$ (37.2)

A quantitative form of lower bound (a) is $\mathbf{ww}(n\text{–line in torus}) \geq \left(\frac{\log 2}{\log 3}\right)$ $n - O(\sqrt{n \log n})$, and (b) has the same error term. This is complemented by the quantitative upper bound $\mathbf{ww}(n\text{–line in torus}) \geq \left(\frac{\log 2}{\log 3}\right) n + O(\log n)$, and (b) has the same error term.

We already proved the upper bounds in Section 13; it remains to prove the lower bounds. In the two cases "(a) with *odd n*" and "(b)" the lower bound (i.e. Strong Draw) immediately follows from Theorem 34.1, because the corresponding hypergraphs are Almost Disjoint.

In the remaining case "(a) with *even n*" the n^d-torus-hypergraph is not Almost Disjoint, but we still have the weaker property that any two Torus-Lines have at most two common points. Thus Theorem 36.1 applies, and completes the proof.   □

Next we answer two natural questions about Almost Disjoint hypergraphs. What is the *smallest* n-uniform Almost Disjoint hypergraph in which a player can force a Weak Win? What is the *fastest way* to force a Weak Win in an Almost Disjoint hypergraph? Both questions become trivial if we drop the condition "Almost Disjoint." Indeed, the full-length branches of a binary tree of n levels (the players take vertices) form an n-uniform hypergraph of size 2^{n-1} in which the first player can win in n moves. "In n moves" is obviously the *fastest way*, and 2^{n-1} is the *smallest size* (by the Erdős–Selfridge Theorem). This hypergraph is very far from being Almost Disjoint, so for Almost Disjoint hypergraphs we would expect a rather different answer.

The first difference is that the *smallest size* jumps from 2^{n-1} up to $(4 + o(1))^n$, i.e. it is roughly "squared" (see Beck [1981a]).

Theorem 37.1 ("Almost Disjoint version of Erdős–Selfridge") *Let $\mathcal{F}$ be an n-uniform Almost Disjoint family with*

$$|\mathcal{F}| < 4^{n-4\sqrt{n \log n}}.$$

If n is sufficiently large, then, playing on $\mathcal{F}$, the second player can force a Draw, in fact a Strong Draw.

Note that Theorem 37.1 is asymptotically nearly best possible: this follows from the following result of Erdős and Lovász [1975].

Theorem 37.2 ("Erdős-Lovász Construction") *There is an n-uniform Almost Disjoint hypergraph $\mathcal{F}^*$ with at most $n^4 \cdot 4^n$ n-sets which is 3-chromatic, implying that the first player can force an ordinary win.*

What is the reason behind Theorem 37.1? Why can we replace the 2^{n-1} in the Erdős–Selfridge Theorem by its "square" $(4 + o(1))^n$? The reason is that for (relatively small) Almost Disjoint hypergraphs there is a surprising "Degree Reduction" – due to Erdős and Lovász – which makes it possible to apply Theorem 34.1. Before formulating the Erdős–Lovász Degree Reduction theorem (see [1975]), we make first a simple observation. Let $\mathcal{F}$ be an arbitrary n-uniform Almost Disjoint hypergraph, and let V denote the union set. Since $\mathcal{F}$ is Almost Disjoint, counting point pairs in two different ways gives the inequality $|\mathcal{F}|\binom{n}{2} \leq \binom{|V|}{2}$, which implies $(n-1)\sqrt{|\mathcal{F}|} < |V|$. In view of this inequality the *Average Degree* $\overline{d}$ of $\mathcal{F}$ is less than $2\sqrt{|\mathcal{F}|}$. Indeed

$$\overline{d} = \frac{1}{|V|}\sum_{x \in V} d_x = \frac{n|\mathcal{F}|}{|V|} < \frac{n|\mathcal{F}|}{(n-1)\sqrt{|\mathcal{F}|}} \leq 2\sqrt{|\mathcal{F}|}.$$

Can we prove something like this with the *Maximum Degree* instead of the Average Degree? The answer is a *yes* if we are allowed to throw out 1 point from each set $A \in \mathcal{F}$, i.e. by reducing the n-uniform hypergraph $\mathcal{F}$ to an $(n-1)$-uniform hypergraph (of course, the number of sets doesn't change).

Theorem 37.3 ("Erdős–Lovász Degree Reduction") *Let $\mathcal{F}$ be an arbitrary n-uniform Almost Disjoint hypergraph $(n \geq 2)$. Then for every n-element set $A \in \mathcal{F}$ there is an $(n-1)$-element subset $\widetilde{A} \subset A$ such that the $(n-1)$-uniform family $\widetilde{\mathcal{F}} = \{\widetilde{A} : A \in \mathcal{F}\}$ has Maximum Degree*

$$\text{MaxDegree}\left(\widetilde{\mathcal{F}}\right) \leq \sqrt{n|\mathcal{F}|}.$$

Proof. For each $A \in \mathcal{F}$ let $g(A) \in A$ be that point of A which has the largest $\mathcal{F}$-degree. From each $A \in \mathcal{F}$ throw out the corresponding point $g(A)$, i.e. let $\widetilde{A} = A \setminus \{g(A)\}$, and let $\widetilde{\mathcal{F}} = \{\widetilde{A} : A \in \mathcal{F}\}$. Let d denote the Maximum Degree of $\widetilde{\mathcal{F}}$, and let $\widetilde{A}_1, \widetilde{A}_2, \widetilde{A}_3, \ldots, \widetilde{A}_d$ be those sets of $\widetilde{\mathcal{F}}$ that contain a point (say) y of Maximum Degree in $\widetilde{\mathcal{F}}$. (The $\widetilde{\mathcal{F}}$-degree of y is d.) The d points $g(A_1)$, $g(A_2)$, $g(A_3)$, $\ldots, g(A_d)$ are all different (because y is a common point and $\mathcal{F}$ is Almost Disjoint), and all have $\mathcal{F}$-degree $\geq d$ (otherwise we would pick y instead of $g(A_i)$). It follows that

$$n|\mathcal{F}| \geq \sum_{i=1}^{d} d_i \geq d^2,$$

where d_i denotes the $\mathcal{F}$-degree of $g(A_i)$, $1 \leq i \leq d$. Theorem 37.3 follows. $\qquad \square$

Now we are ready to derive Theorem 37.1 from Theorem 34.1.

Proof of Theorem 37.1. If $|\mathcal{F}| < 4^{n-4\sqrt{n \log n}}$, then by Theorem 37.3 the maximum degree of the $(n-1)$-uniform family $\widetilde{\mathcal{F}}$ is

$$\leq \sqrt{n}2^{n-4\sqrt{n\log n}} < 2^{(n-1)-3\sqrt{(n-1)\log(n-1)}}$$

if n is sufficiently large. Since

$$|\widetilde{\mathcal{F}}| = |\mathcal{F}| < 4^{n-4\sqrt{n}} < (n-1)^{n-1},$$

Corollary 2 of Theorem 34.1 applies to $\widetilde{\mathcal{F}}$, and we are done. $\qquad \square$

Next we prove the "strong converse" of Theorem 37.1.

Proof of Theorem 37.2. We actually prove more.

Theorem 37.4 *Let* $N = 160n^4 2^n$, $M = 6400n^4 4^n$, *and* $D = 40n^2 2^n$. *Then there is an Almost Disjoint n-uniform hypergraph $\mathcal{F}$ on $2N$ points with at most M hyperedges and with Maximum Degree $\leq D$ in which each set of N points contains a hyperedge; this hypergraph is obviously at least 3-chromatic.*

Proof of Theorem 37.4. Let S be an arbitrary set of $2N$ points: this will be the union set of our hypergraph. We construct our hypergraph $\mathcal{F} = \{A_i : 1 \leq i \leq t\}$ inductively. Suppose $A_1, \ldots, A_p$ have already been chosen so that:

(1) they are Almost Disjoint;
(2) no point is contained in more than D of them.

Let $S_1, S_2, \ldots, S_{f(p)}$ be those N-element subsets of S containing no any of $A_1, \ldots, A_p$. If there is no such S_i, then we are done. Suppose $f(p) \geq 1$. Choose now the next n-set A_{p+1} in such a way that $A_1, \ldots, A_p, A_{p+1}$ satisfy (1) and (2), and A_{p+1} is contained in as many S_i, $1 \leq i \leq f(p)$ as possible. We shall show that A_{p+1} will be contained in at least $\frac{1}{20}f(p)2^{-n}$ sets from S_i, $1 \leq i \leq f(p)$ as long as $p < M$. This will imply:

(3) $f(p+1) \leq f(p)\left(1 - \frac{1}{20}2^{-n}\right)$.

Suppose we know that, if $p < M$, then (3) holds. Then

$$f(M) \leq f(0)\left(1 - \frac{1}{20}2^{-n}\right)^M < 2^{2N}e^{-M2^{-n}/20} < e^{2N - M2^{-n}/20} = 1.$$

Thus our procedure stops before the Mth step, i.e. we get a hypergraph satisfying the requirements with $< M$ hyperedges.

It remains to show (3). For every $1 \leq j \leq f(p)$ we estimate how many n-element subsets of S_j could be chosen for A_{p+1} without violating (1) and (2).

Let x be the number of points of S_j having degree D in hypergraph $\{A_1, \ldots, A_p\}$. Clearly $xD \leq np \leq nM$, so $x \leq nM/D = N/n$. Therefore, the number of points in S_j having degree $\leq (D-1)$ in hypergraph $\{A_1, \ldots, A_p\}$ is $N - x \geq (1 - \frac{1}{n})N$. Any n-element set chosen from these points will satisfy requirement (2), i.e. there are at least

$$\binom{(1 - \frac{1}{n})N}{n}$$

n-element subsets of S_j satisfying (2).

Let us see how many n-sets are excluded by requirement (1) ("Almost Disjointness"). We can describe these n-sets as those not containing any pair of points common with some A_i, $1 \leq i \leq p$. Each A_i has $\binom{n}{2}$ pairs, so there are altogether at most $p\binom{n}{2} < Mn^2/2$ excluded pairs. One excluded pair forbids at most $\binom{N-2}{n-2}$ n-sets of S_j; thus the total number of n-element subsets of S_j forbidden by requirement (1) is less than

$$\binom{N-2}{n-2}\frac{Mn^2}{2}.$$

Therefore, the number of n-element subsets of S_j that are candidates for A_{p+1}, i.e. satisfy both (1) and (2), is more than

$$\binom{(1 - \frac{1}{n})N}{n} - \binom{N-2}{n-2}\frac{Mn^2}{2}$$

$$\approx \frac{1}{e}\binom{N}{n} - \frac{Mn^4}{2N^2}\binom{N}{n} = \left(\frac{1}{e} - \frac{1}{8}\right)\binom{N}{n} > \frac{1}{5}\binom{N}{n}.$$

So the number of n-element subsets of S_j that are candidates for A_{p+1} is more than $\frac{1}{10}\binom{N}{n}$. Counting with multiplicity, there are altogether at least

$$\frac{f(p)}{10}\binom{N}{n}$$

n-sets of $S_1, S_2, \ldots, S_{f(p)}$ that can be chosen as A_{p+1}. Since the total number of n-sets is $\binom{2N}{n}$, there must be an n-set that is counted in at least

$$\frac{f(p)\binom{N}{n}}{10\binom{2N}{n}} > \frac{f(p)}{20}2^{-n}$$

different S_js. This proves (3), and the proof of Theorem 37.4 is complete. □

Notice that the board size $|V|$ of the n-uniform hypergraph $\mathcal{F}^*$ in Theorem 37.2 is "exponentially large": $|V| = 320n^4 \cdot 2^n$. This is perfectly natural, because for Almost Disjoint hypergraphs we have the general inequality $\binom{|V|}{2} \geq |\mathcal{F}| \cdot \binom{n}{2}$ by counting the point pairs in two different ways. If $\mathcal{F}$ is an n-uniform Almost Disjoint hypergraph such that the second player cannot force a strong draw, then by Theorem 37.1 $|\mathcal{F}| \geq 4^{n-4\sqrt{n \log n}}$, so the general inequality yields the lower bound $|V| \geq 2^{n-4\sqrt{n \log n}}$ for the board size. This motivates the following:

Question: Is it true that, to achieve a Weak Win in an n-uniform Almost Disjoint hypergraph, we need at least $(2 + o(1))^n$ moves?

Note that the *weaker* (but still exponential) lower bound $2^{(n-1)/2}$ was already proved in Section 1 (see Theorem 1.3). By using the BigGame-SmallGame Decomposition technique (Sections 35–36) we can easily improve the $2^{(n-1)/2}$ to $(2 + o(1))^n$. We challenge the reader to do this.

Exercise 37.1 *Let $\mathcal{F}$ be an arbitrary n-uniform Almost Disjoint hypergraph. Then Breaker can avoid losing the Maker–Breaker game on $\mathcal{F}$ in $(2 + o(1))^n$ moves.*

2. The second Ugly Theorem. So far (almost) everything was a corollary of the first Ugly Theorem (Theorem 34.1), but Theorem 34.1 failed to imply Theorem 12.5 (a). Is it true that, if $d \leq \left(\frac{\log 2}{16} + o(1) \right) \frac{n^2}{\log n}$, then the n^d game is a Draw?

By combining Theorem 12.2 with Theorem 34.1 we could prove only a somewhat weaker result which fell short by a mere constant factor.

The reason is that in the proof of Theorem 34.1 we applied the technique of "BigGame-SmallGame Decomposition," and that technique breaks down in the range $|\mathcal{F}| > 2^{3m^2/8}$ where $\mathcal{F}$ is an m-uniform Almost Disjoint hypergraph.

The following result, called the Second Ugly theorem, takes care of this problem by largely extending the range from $|\mathcal{F}| > 2^{c \cdot m^2}$ to a gigantic(!) doubly exponential bound (see Corollary 1 below).

Theorem 37.5 *If $\mathcal{F}$ is an m-uniform Almost Disjoint hypergraph such that*

$$|\mathcal{F}|^{m^2 \cdot 2^{-k/4}} \cdot \max \left\{ 2^{7k/2} \cdot D, m \cdot 2^{5k/2} \cdot D^{1 + \frac{1}{k-2}} \right\} \leq 2^m$$

holds for some integer $k \geq 8\log_2 m$ ("binary logarithm") where $D = MaxDegree(\mathcal{F})$, then the second player can force a Strong Draw in the positional game on $\mathcal{F}$.

The proof is based on a more sophisticated decomposition technique. We postpone it to Section 39.

By choosing parameter k around $\sqrt{m}$ in Theorem 37.5 (m is sufficiently large), we obtain the following special case:

Corollary 1 *If $\mathcal{F}$ is an m-uniform Almost Disjoint hypergraph*

$$MaxDeg(\mathcal{F}) < 2^{m-4\sqrt{m}} \quad \text{and} \quad |\mathcal{F}| < 2^{2^{\sqrt{m}/5}},$$

and $m > c_0$, then the second player can force a Strong Draw in the positional game on $\mathcal{F}$.

Combining Theorem 12.2 with Corollary 1 (instead of Theorem 34.1) we immediately obtain Theorem 12.5 (a). This means we can upgrade the existing Proper 2-Coloring of the n^d-hypergraph guaranteed by Theorem 12.3 to a Drawing Strategy (in fact a Strong Draw Strategy). Thus we have:

Corollary 2 *We obtain the lower bounds in Theorem 12.5*

$$(a) \quad \mathbf{ww}(n-\text{line}) \geq \left(\frac{\log 2}{16} + o(1)\right) \frac{n^2}{\log n},$$

that is, if

$$d \leq \left(\frac{\log 2}{16} + o(1)\right) \frac{n^2}{\log n},$$

then the second player can force a Strong Draw in the n^d game.

$$(b) \quad \mathbf{ww}(\text{comb. } n-\text{line}) \geq \left(\frac{\log 2}{8} + o(1)\right) \frac{n^2}{\log n},$$

that is, if

$$d \leq \left(\frac{\log 2}{8} + o(1)\right) \frac{n^2}{\log n},$$

then the second player can force a Strong Draw in the "combinatorial lines only" version of the n^d game.

The next two sections contain two long proofs. First in Section 38 we complete the proof of Theorem 6.4 (a) ("Clique Game") by supplying the Strong Draw part (the Weak Win part has already been discussed in Section 25). We develop a "clique-specific" adaptation of the methods of Sections 35–36.

In Section 39 we present the nearly 20 pages long proof of the second Ugly Theorem (Theorem 37.5).

38

Exact solution of the Clique Game (II)

1. The Strong Draw part of Theorem 6.4 (a). We prove that if

$$q \geq 2\log_2 N - 2\log_2 \log_2 N + 2\log_2 e - 3 + o(1), \tag{38.1}$$

then Breaker can prevent Maker from occupying a K_q in the (K_N, K_q) Clique Game. In view of (21.2)–(21.3) this means an additive constant 2 improvement on the Erdős–Selfridge Theorem. (38.1) basically proves the Neighborhood Conjecture (Open Problem 9.1) for the "clique hypergraph"; the improvement comes from the fact that the Max Degree $\binom{N-2}{q-2}$ of the "clique hypergraph" differs from the global size $\binom{N}{q}$ by a factor of $\approx (N/q)^2$. The "2" in the power of N/q explains the "additive constant 2" improvement!

The "clique hypergraph" is very special in the sense that the Max Degree is just a little bit smaller than the global size; in the rest of the "Ramsey type games" – multidimensional Tic-Tac-Toe, van der Waerden Game, $q \times q$ lattice games on an $N \times N$ board – the Max Degree is *much smaller* than the global size. *Much smaller* roughly means "square-root" or "a power less than 1" (or something like that). This is why the Erdős–Selfridge Theorem is so strikingly close to the truth for the Clique Game (differs from the truth only by an additive constant, namely by 2), but for the rest of the "Ramsey-type games" the Erdős–Selfridge Theorem fails by a multiplicative constant factor.

This was the good news; the bad news is that the "clique hypergraph" is very far from Almost Disjointness, in fact two K_qs may have a very large intersection: as many as $\binom{q-1}{2} = \binom{q}{2} - (q-1)$ common edges from the total $\binom{q}{2}$. This is why we need a novel variant of the technique developed in Sections 35–36 ("BigGame–SmallGame Decomposition"). Unfortunately the details are tiresome.

We assume that the reader is familiar with Sections 35–36. Following the basic idea of Sections 35–36, we are going to define several classes of *Avoidable Configurations*. By using the Power-of-Two Scoring System ("Erdős–Selfridge technique") Breaker will prevent the appearance of any *Avoidable Configuration* in the Big

Game; this is how Breaker can guarantee that the *small game* is so simple that a trivial Pairing Strategy will complete the "blocking job."

2. Avoidable Configurations of Kind One: A pair of K_qs such that:

(1*) both are Breaker-free ("survivors");
(2*) Maker owns at least $\binom{q}{2} - q^{2/3}$ edges from each K_q;
(3*) the *intersection size* of the 2 (q-element) vertex-sets is in the "middle range":
$4 \leq m \leq q - 2$.

$$K_q \qquad \qquad K_q$$

Lemma 1: *By using the Power-of-Two Scoring System in the Big Game, Breaker can prevent the appearance of any Avoidable Configuration of Kind One.*

Proof. The "target value" is at least

$$2^{2\binom{q}{2} - \binom{m}{2} - 2q^{2/3}},$$

and the number of ways to choose a pair of K_qs with m common vertices is $\binom{N}{q}\binom{q}{m}\binom{N-q}{q-m}$, so the relevant term is the product

$$f_{N,q}(m) = \binom{N}{q}\binom{q}{m}\binom{N-q}{q-m} 2^{-2\binom{q}{2} + \binom{m}{2} + 2q^{2/3}}. \tag{38.2}$$

If q satisfies (38.1), then

$$\binom{N}{q} 2^{-\binom{q}{2}} = N^{2-\varepsilon},$$

so (38.2) is approximately

$$\binom{N}{q} 2^{-\binom{q}{2}} \cdot \binom{N}{q} 2^{-\binom{q}{2}} \cdot \left(\frac{q}{N}\right)^m 2^{\binom{m}{2}} \cdot \binom{q}{m} 2^{2q^{2/3}} \approx N^{4-\varepsilon-m} \tag{38.3}$$

as long as $m = o(q)$. The ratio (see (38.2))

$$\frac{f_{N,q}(m+1)}{f_{N,q}(m)} = \frac{\binom{q}{m+1}\binom{N-q}{q-m+1} 2^{\binom{m+1}{2}}}{\binom{q}{m}\binom{N-q}{q-m} 2^{\binom{m}{2}}} = \frac{2^m (q-m)^2}{(m+1)(N-2q+m+1)} \tag{38.4}$$

decreases up to $m \approx q/2$ (since $q \approx 2\log_2 N$), and *decreases* in the rest: starting from $m \approx q/2$ and ending at $m = q - 1$. When m is close to q we write $l = q - m$; by using $\binom{q}{2} - \binom{m}{2} = l(q - \frac{l+1}{2})$ we have (see (38.2))

$$f_{N,q}(m) = f_{N,q}(q-l) = \binom{N}{q} 2^{-\binom{q}{2}} \cdot \binom{q}{l}\binom{N-q}{l} 2^{-l(q-\frac{l+1}{2})}$$

$$= N^{2-\varepsilon-l} \quad \text{if} \quad l = o(q). \tag{38.5}$$

Combining (38.3)–(38.5) we have

$$f_{N,q}(2) = N^{2-\varepsilon}, \ f_{N,q}(3) = N^{1-\varepsilon}, \ f_{N,q}(q-1) = N^{1-\varepsilon}, \quad (38.6)$$

$$f_{N,q}(4) = N^{-\varepsilon}, \ f_{N,q}(q-2) = N^{-\varepsilon}, \ \text{and} \ f_{N,q}(m) \le N^{-1-\varepsilon} \quad (38.7)$$

for all $5 \le m \le q-3$. Lemma 1 follows from (38.6)–(38.7). $\qquad\qquad\square$

We reformulate Lemma 1 as follows:

Corollary of Lemma 1: *Assume that Breaker plays rationally in the Big Game by preventing Avoidable Configurations of Kind One. During such a play, any 2 survivor K_qs with the property that Maker claims at least $\binom{q}{2} - q^{2/3}$ edges from each in the Big Game, are:*

(1) either edge-disjoint;
(2) or "nealy disjoint" having 1 or 3 common edges;
(3) or "almost identical" having $\binom{q-1}{2}$ common edges.

If a survivor K_q has at least $\binom{q}{2} - q^{2/3}$ marks ("marked edges") of Maker in the Big Game, then we call it **risky**. Sections 35–36 were about Almost Disjoint hypergraphs; the good news is that two *risky* K_qs are either Almost Disjoint, or *very close* to Almost Disjointness (see (1)–(2) in the Corollary), but unfortunately there is a new case: (3) above, which is the complete opposite of Almost Disjointness. The possibility of "almost identical" risky K_qs (see (3)) is a new difficulty which requires a new idea.

Here is the new idea: Let K_A, K_B, K_C be 3 *risky* K_qs : $|A| = |B| = |C| = q$, and assume that $K_A \equiv ai \equiv K_B$, $K_B \equiv ai \equiv K_C$, where the unusual notation $\equiv ai \equiv$ means the relation "almost identical" defined in (3). It follows that $|A \cap B| = |B \cap C| = q-1$. Now there are two possibilities:

$$\text{either} \quad |A \cup B \cup C| = q+1, \quad (38.8)$$

$$\text{or} \ |A \cup B \cup C| = q+2. \quad (38.9)$$

We claim that, by using the Power-of-Two Scoring System in the Big Game, Breaker can *exclude* the second alternative (38.9).

Avoidable Configuration of Kind Two: It means alternative (38.9). The "target value" is

$$2^{\binom{q}{2}+(q-1)+(q-1)-2q^{2/3}},$$

and the number of ways to choose triplet (A, B, C) is at most $\binom{N}{q} \cdot (N-q)q \cdot (N-q)q$, so the relevant term is the product

$$\binom{N}{q}(N-q)^2 q^2 \cdot 2^{-\binom{q}{2}-(q-1)+(q-1)+2q^{2/3}}$$

$$= \binom{N}{q} 2^{-\binom{q}{2}} \cdot \left(\frac{(N-q)q}{2^{q-1-q^{2/3}}}\right)^2 = N^{2-\varepsilon} \cdot N^{-2+o(1)} = N^{-\varepsilon+o(1)},$$

which is "small," so Breaker can indeed avoid alternative (38.9).

The other alternative (38.8) means that $K_A \equiv ai \equiv K_C$; in other words, $K_A \equiv ai \equiv K_B$ and $K_B \equiv ai \equiv K_C$ imply $K_A \equiv ai \equiv K_C$, meaning that $\equiv ai \equiv$ turns out to be an *equivalence relation* (if Breaker plays rationally); so we have just proved:

Lemma 2: *Assume that Breaker plays rationally in the Big Game by preventing Avoidable Configurations of Kind Two; then in the class of risky K_qs, the relation "almost identical" – see (3) – is an equivalence relation.*

If two risky K_qs are "almost identical," then we call them **twin brothers**.

Let k be the nearest integer to $4q^{1/3}$; a survivor K_q is called *dangerous* if Maker claims $\binom{q}{2} - k$ edges in the Big Game. Since k is less than $q^{2/3}$, every dangerous K_q is risky, but the converse is not necessarily true: a risky K_q is not necessarily dangerous (but may become dangerous *later*).

A novelty of this version of the "BigGame–SmallGame decomposition" technique is that the k-edge part of a dangerous K_q (unoccupied in the Big Game) is not necessarily added to the small board as a new *emergency set*; the answer to this question ("whether or not it is an emergency set") depends on the Max Degree and on the *twin brothers* as follows.

Let K_{A_i}, $i \in I$, where each $|A_i| = q$, be a family of *dangerous twin brothers*: by definition each K_{A_i} has a k-edge sub-graph $G_i \subset K_{A_i}$ which is unoccupied in the Big Game. We distinguish two cases:

(a) Let K_{A_1} denote the first one among them showing up in the course of a play, and assume that the Max Degree of G_1 is less than $k/3$. Then G_1 is a new *emergency set*, and (of course) it is added to the *small board*.

(b) Assume that the Max Degree of G_1 is $\geq k/3$; *then* the corresponding *star* (see the picture below) is a new *emergency set*, and (of course) it is added to the *small board*.

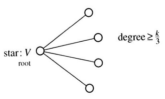

star: V root   degree $\geq \frac{k}{3}$

Among the rest of K_{A_i}s, there is at most 1 possible $j \in I \setminus \{1\}$ such that K_{A_j} does not contain the root of the star in K_{A_i} (vertex v on the picture); then G_j must have a vertex of degree $\geq k/3$ of its own, and this *star* is a second new *emergency set*, and, of course, it is also added to the *small board*.

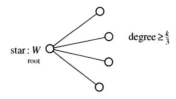

In the *small game*, played on the *small board*, Breaker applies the usual *reply on the private edges* Pairing Strategy (see Sections 35–36). Breaker can force that a survivor K_q will never be completely occupied by Maker; this is how he can do it. Consider a survivor K_A (of course, $|A| = q$); we distinguish several cases.

Case 1: Survivor K_A will eventually become dangerous, but it never has a dangerous twin brother

We claim that Breaker can block K_A in the small game. Indeed, K_A supplies a new emergency set $E = E(K_A)$, which, of course, has $\geq k/3$ edges. In Case 1 every other emergency set E^* intersects E in at most 3 edges (see (1)–(2) in the Corollary of Lemma 1). By preventing **Avoidable Configurations of Kind Three** – which will be defined below – Breaker can force that E has more than $k/6$ private edges – we just need 2! – which guarantees the blocking of E by the *reply on the private edges* Pairing Strategy (used by Breaker in the small game).

4. Avoidable Configuration of Kind Three: any "tree" formed by $s = q^{1/6}$ risky K_qs, which is twin-brother-free; see the figure below.

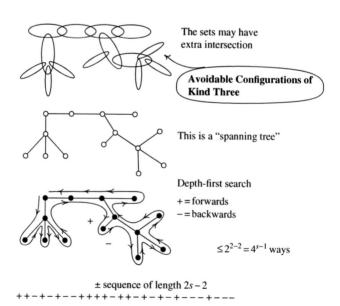

The sets may have extra intersection

Avoidable Configurations of Kind Three

This is a "spanning tree"

Depth-first search
+ = forwards
− = backwards

$\leq 2^{2-2} = 4^{s-1}$ ways

± sequence of length $2s - 2$
+ + − + − + − − + + + + − + + − + − + − + − − − + − − −

By using the Power-of-Two Scoring System, the "target value" of a "tree" is at least

$$2^s \left(\binom{q}{2} - q^{2/3} \right) - 3 \binom{s}{2},$$

noting that factor "3" comes from (2) in the Corollary of Lemma 1. There are less than 4^s *unlabeled* trees on s vertices (easy consequence of the "depth-first search" procedure; see the figure below)

So the number of ways we can build a "tree" (Avoidable Configuration of Kind Three) following the "depth-first-search" is less than

$$4^s \cdot \binom{N}{q} \cdot \left(\binom{q}{2} \binom{N-2}{q-2} \right)^{s-1}.$$

Breaker can prevent the appearance of any Avoidable Configuration of Kind Three if the product

$$4^s \binom{N}{q} \left(\binom{q}{2} \binom{N-2}{q-2} \right)^{s-1} 2^{-s \left(\binom{q}{2} - q^{2/3} \right) + 3 \binom{s}{2}} \tag{38.10}$$

is "small" like $N^{-\varepsilon}$.

If q satisfies (38.1), then $\binom{N}{q} 2^{-\binom{q}{2}} = N^{2-\varepsilon}$ and $\binom{q}{2} \binom{N-2}{q-2} 2^{-\binom{q}{2}} = N^{-\varepsilon}$, so (38.10) equals

$$\left(\frac{\binom{q}{2} \binom{N-2}{q-2} \cdot 2^{3(s-1)/2} \cdot 4}{2^{\binom{q}{2} - q^{2/3}}} \right)^{s-1} \cdot \frac{\binom{N}{q}}{2^{\binom{q}{2}}} = (N^{-\varepsilon})^s \cdot N^{2-\varepsilon} \leq N^{-\varepsilon}, \tag{38.11}$$

which is really "small" as we claimed. This proves:

Lemma 3: *By using the Power-of-Two Scoring System in the Big Game, Breaker can prevent the appearance of any Avoidable Configuration of Kind Three.*

Now we can complete Case 1: *emergency set* $E = E(K_A)$ must have at least 2 *private edges* – indeed, otherwise there is an Avoidable Configuration of Kind Three, which contradicts Lemma 3 – so Breaker can block E by Pairing Strategy.

Case 2: Survivor K_A will eventually become dangerous, it also has a dangerous twin brother, but the MaxDegree of the k-edge graph is less than $k/3$ (see case (a) above).

This is very similar to Case 1: again the *emergency set* must have at least 2 private edges (otherwise there is an Avoidable Configuration of Kind Three).

Case 3: Survivor K_A will eventually become dangerous, it also has a dangerous twin brother, but the MaxDegree of the k-edge graph is $\geq k/3$ (see case (b) above). Then the *emergency set* is a *star*; any other emergency set intersects this *star* in at most 3 edges, so the *star* must have at least 2 private edges (otherwise there is an Avoidable Configuration of Kind Three).

Case 4: Survivor K_A will never become dangerous, and at some stage of the play its intersection with the small board is $> q^{2/3}$.

The assumption "K_A's intersection with the small board is $> q^{2/3}$" implies that K_A is *not* risky, so we cannot apply the Corollary of Lemma 1: we cannot guarantee the strong intersection properties (1)–(3) ("nearly disjoint" or "almost identical"). To overcome this novel difficulty we introduce

5. Avoidable Configurations of Kind Four. We define them as follows. Consider the stage of the play when K_A's intersection with the small board is $> q^{2/3}$; consider the emergency sets intersecting K_A; for each emergency set take its super-set ("a copy of K_q"); these K_qs fall apart into *components* (see the figure below); each *component* has less than $2s = 2q^{1/6}$ sets. The last statement follows from the fact that Lemma 3 applies for the super-sets of emergency sets, noting that any family of *twin brothers*0 supplies at most 2 emergency sets (see (a) and (b)). Taking 1 super-set ("a copy of K_q") from each *component* plus K_A together form an **Avoidable Configurations of Kind Four**.

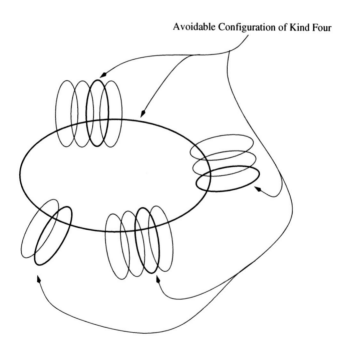

Avoidable Configuration of Kind Four

Let r denote the intersection of K_A with the small board at the end of a play; clearly $r > q^{2/3}$. Each emergency set has at most $k = 4q^{1/3}$ edges and each component size is $< 2q^{1/6}$, so if t denotes the number of components, then the inequality $t > rq^{-1/6}k^{-1}/2$ is obvious. By using the Power-of-Two Scoring System, the "target value" of an *Avoidable Configurations of Kind Four* is

$$2^{\binom{q}{2}-r} \cdot 2^{t(\binom{q}{2}-k)}.$$

The total number of ways to build an Avoidable Configuration of Kind Four with $t+1$ sets is at most

$$\binom{N}{q}\left(\binom{q}{2}\binom{N-2}{q-2}\right)^{t};$$

here we applied the trivial inequality

$$\max_{m\geq 2}\binom{q}{m}\binom{N-m}{q-m} \leq \binom{q}{2}\binom{N-2}{q-2}.$$

Breaker can prevent Case 4 if the product

$$\binom{N}{q}\left(\binom{q}{2}\binom{N-2}{q-2}\right)^{t} 2^{-\binom{q}{2}+r-t(\binom{q}{2}-k)} \tag{38.12}$$

is small like $N^{-\varepsilon}$. We rewrite and estimate (38.12) as follows (by using the obvious inequality $t > rq^{-1/6}k^{-1}/2$ mentioned above)

$$(38.12) = \binom{N}{q}2^{-\binom{q}{2}} \cdot \left(\binom{q}{2}\binom{N-2}{q-2}2^{-\binom{q}{2}+k}\right)^{t} \cdot 2^{r}$$

$$\leq \binom{N}{q}2^{-\binom{q}{2}} \cdot \left(\binom{q}{2}\binom{N-2}{q-2}2^{-\binom{q}{2}+k+2kq^{1/6}}\right)^{t} \leq N^{2-\varepsilon} \cdot (N^{-\varepsilon})^{t} < N^{-\varepsilon},$$

which is very small as we claimed. Since Breaker can prevent the appearance of any *Avoidable Configuration of Kind Four*, Case 4 does not show up at all.

The last 3 cases form the perfect analogue of Cases 1–3.

Case 5: Survivor K_A will never become dangerous, at some stage of the play its intersection with the small board is always at most $q^{2/3}$, and it never has a dangerous twin brother.

Assume that Maker can occupy all $\binom{q}{2}$ edges of K_A. When Maker owns $\binom{q}{2} - q^{2/3}$ edges of K_A in the Big Game, K_A becomes *risky* and Lemmas 1–3 all apply. We can repeat the argument of Case 1: if Maker can occupy all $\binom{q}{2}$ edges of K_A, then there is an *Avoidable Configuration of Kind Three*, which contradicts Lemma 3.

Case 6: Survivor K_A will never become dangerous, at some stage of the play its intersection with the small board is always at most $q^{2/3}$, it has a dangerous twin brother and the Max Degree of the k-edge graph is less than $k/3$ (see (a) above).

Assume that Maker can occupy all $\binom{q}{2}$ edges of K_A. When Maker owns $\binom{q}{2} - q^{2/3}$ edges of K_A in the Big Game, K_A becomes *risky*, and just as in Case 2 there is an *Avoidable Configuration of Kind Three*, which contradicts Lemma 3.

Case 7: Survivor K_A will never become dangerous, at some stage of the play its intersection with the small board is always at most $q^{2/3}$, it has a dangerous twin brother and the Max Degree of the k-edge graph is $\geq k/3$ (see (b) above).

Assume that Maker can occupy all $\binom{q}{2}$ edges of K_A. When Maker owns $\binom{q}{2} - q^{2/3}$ edges of K_A in the Big Game, K_A becomes *risky*, and just as in Case 3 there is an *Avoidable Configuration of Kind Three*, which contradicts Lemma 3.

This completes the proof of Theorem 6.4 (a).   □

The proofs of Theorem 6.4 (b)–(c)–(d) and Theorem 8.2 (h) are very similar.

Next we switch from the usual board K_N to a "typical" graph on N vertices.

Exercise 38.1 *Show that, playing the usual (1:1) game on the symmetric random graph* $R(K_N, 1/2)$, *with probability tending to 1 as N tends to infinity, Breaker can prevent Maker from occupying a clique* K_q *with*

$$q = \lceil \log_2 N - \log_2 \log_2 N + \log_2 e - 1 + o(1) \rceil .$$

Exercise 38.2 *Prove the Breaker's part in formulas (8.7) and (8.8).*

6. Chooser–Picker Game. By using Theorem 38.1 below instead of the Erdős–Selfridge Theorem, we can show, by repeating the argument of this whole section, that Chooser's Clique Achievement Number in K_N equals the ordinary (Maker's) Clique Achievement Number

$$q = \lfloor 2 \log_2 N - 2 \log_2 \log_2 N + 2 \log_2 e - 3 + o(1) \rfloor .$$

To formulate our Blocking Criterion we need the notion of *rank*: $\|\mathcal{H}\| = \max_{A \in \mathcal{H}} |A|$ is called the *rank* of hypergraph $\mathcal{H}$. The (1:1) play requests the standard Power-of-Two Scoring: $T(\mathcal{H}) = \sum_{A \in \mathcal{H}} 2^{-|A|}$.

Theorem 38.1 *If*

$$T(\mathcal{F}) \leq \frac{1}{8(\|\mathcal{F}\| + 1)}, \tag{38.13}$$

then Picker ("blocker") has an explicit winning strategy in the Chooser–Picker game on hypergraph $\mathcal{F}$.

In other words, if (38.13) holds, then Picker can prevent Chooser from completely occupying a winning set $A \in \mathcal{F}$. Theorem 38.1 is a "first moment criterion," a close relative of the Erdős–Selfridge Theorem (Theorem 1.4).

Proof of Theorem 38.1. Assume we are in the middle of a play where Chooser already selected $x_1, x_2, \ldots, x_i$, and Picker owns the points $y_1, y_2, \ldots, y_i$. The question is *how* to find Picker's next 2-element set $\{v, w\}$, from which Chooser will choose his x_{i+1} (the other one will go back to Picker).

Let $X_i = \{x_1, x_2, \ldots, x_i\}$ and $Y_i = \{y_1, y_2, \ldots, y_i\}$. Let $V_i = V \setminus (X_i \cup Y_i)$. Clearly $|V_i| = |V| - 2i$.

Let $\mathcal{F}(i)$ be that truncated subfamily of $\mathcal{F}$, which consists of the unoccupied parts of the "survivors":

$$\mathcal{F}(i) = \{A \setminus X_i : A \in \mathcal{F}, \ A \cap Y_i = \emptyset\}.$$

If Picker can guarantee that $T(\mathcal{F}(i)) < 1$ at the end of the play, i.e. $T(\mathcal{F}(end)) < 1$, then Picker wins. Let x_{i+1} and y_{i+1} denote, respectively, the $(i+1)$st points of Chooser and Picker. We have

$$T(\mathcal{F}(i+1)) = T(\mathcal{F}(i)) + T(\mathcal{F}(i); x_{i+1}) - T(\mathcal{F}(i); y_{i+1}) - T(\mathcal{F}(i); x_{i+1}, y_{i+1}).$$

It follows that

$$T(\mathcal{F}(i+1)) \leq T(\mathcal{F}(i)) + |T(\mathcal{F}(i); x_{i+1}) - T(\mathcal{F}(i); y_{i+1})|.$$

Introduce the function

$$g(v, w) = g(w, v) = |T(\mathcal{F}(i); v) - T(\mathcal{F}(i); w)|$$

which is defined for any 2-element subset $\{v, w\}$ of V_i. Picker's next move is that 2-element subset $\{v_0, w_0\}$ of V_i for which the function $g(v, w)$ achieves its *minimum*. Since $\{v_0, w_0\} = \{x_{i+1}, y_{i+1}\}$, we have

$$T(\mathcal{F}(i+1)) \leq T(\mathcal{F}(i)) + g(i), \tag{38.14}$$

where

$$g(i) = \min_{v, w : v \neq w, \{v, w\} \subseteq V_i} |T(\mathcal{F}(i); v) - T(\mathcal{F}(i); w)|. \tag{38.15}$$

We need the following simple:

Lemma 1: *If $t_1, t_2, \ldots, t_m$ are non-negative real numbers and $t_1 + t_2 + \cdots + t_m \leq s$, then*

$$\min_{1 \leq j < \ell \leq m} |t_j - t_\ell| \leq \frac{s}{\binom{m}{2}}.$$

Proof. We can assume that $0 \leq t_1 < t_2 < \cdots < t_m$. Write $g = \min_{1 \leq j < \ell \leq m} |t_j - t_\ell|$. Then $t_{j+1} - t_j \geq g$ for every j, and

$$\binom{m}{2} g = g + 2g + \cdots + (m-1)g \leq t_1 + t_2 + \ldots + t_m \leq s.$$

This completes the proof of Lemma 1. $\qquad\square$

We distinguish two *phases* of the play.

Phase 1: $|V_i| = |V| - 2i > \|\mathcal{F}\|$
Then we use the trivial fact

$$\sum_{v \in V_i} T(\mathcal{F}(i); v) \leq \|\mathcal{F}\| T(\mathcal{F}(i)).$$

By Lemma 1 and (38.15)

$$g(i) \le \frac{\|\mathcal{F}\|}{\binom{|V_i|}{2}} T(\mathcal{F}(i)),$$

so by (38.14)

$$T(\mathcal{F}(i+1)) \le T(\mathcal{F}(i)) \left\{ 1 + \frac{\|\mathcal{F}\|}{\binom{|V_i|}{2}} \right\}.$$

Since $1 + x \le e^x = \exp(x),$ we have

$$T(\mathcal{F}(i+1)) \le T(\mathcal{F}) \exp \left\{ \|\mathcal{F}\| \sum_{j=0}^{i} \frac{1}{\binom{|V_j|}{2}} \right\}.$$

It is easy to see that

$$\sum_{i:|V_i|>\|\mathcal{F}\|} \frac{1}{\binom{|V_i|}{2}} < \frac{2}{\|\mathcal{F}\|},$$

so if i_0 denotes the *last* index of the *first phase*, then

$$T(\mathcal{F}(i_0+1)) < e^2 T(\mathcal{F}) < 8T(\mathcal{F}). \tag{38.16}$$

Phase 2: $|V_i| = |V| - 2i \le \|\mathcal{F}\|$
Then we use the other trivial fact

$$\sum_{v \in V_i} T(\mathcal{F}(i); v) \le |V_i| T(\mathcal{F}(i)).$$

By Lemma 1 and (38.15)

$$g(i) \le \frac{2}{|V_i|-1} T(\mathcal{F}(i)),$$

so by (38.14)

$$T(\mathcal{F}(i+1)) \le \frac{|V_i|+1}{|V_i|-1} T(\mathcal{F}(i)). \tag{38.17}$$

By repeated application of (38.17) we have

$$T(\mathcal{F}(end)) \le T(\mathcal{F}(i_0+1)) \prod_{i:|V_i|\le\|\mathcal{F}\|} \frac{|V_i|+1}{|V_i|-1}$$

$$\le T(\mathcal{F}(i_0+1)) (\|\mathcal{F}\|+1) < 8T(\mathcal{F}) (\|\mathcal{F}\|+1). \tag{38.18}$$

In the last step we used (38.16). Combining the hypothesis of the theorem with (38.18) we conclude that $T(\mathcal{F}(end)) < 1$, so Chooser was unable to completely occupy a winning set. Theorem 38.1 follows. $\square$

A biased generalization of Theorem 38.1 will be discussed in Section 47.

This section completes the proof of the first main result: Theorem 6.4. It took us a whole section to replace an additive constant gap $O(1)$ by $o(1)$ which tends to 0, i.e. to get the *exact* solution. Similarly, if we are is satisfied with a weaker asymptotic form of Theorem 8.2 (lattice games) instead of the exact solution, then there is a major shortcut. A straightforward adaptation of the proof technique of a single section (Section 42) suffices to give at least the asymptotic form of the phase transition in Theorem 8.2.

The asymptotic form describes the truth apart from a multiplicative factor of $(1 + o(1))$. However, if we insist on having the exact value of the phase transition, then, unfortunately, we have to go through Chapters VIII and IX: an about 80 pages long argument.

Needless to say, it would be extremely desirable to reduce this 80-page-long proof to something much shorter.

Chapter VIII
Advanced decomposition

The main objective of Chapter VIII is to develop a more sophisticated version of the BigGame–SmallGame Decomposition technique (introduced in Sections 35–36).

We prove the second Ugly Theorem; We formulate and prove the third Ugly Theorem. Both are about Almost Disjoint hypergraphs. In Section 42 we extend the decomposition technique from Almost Disjoint to more general hypergraphs. We call it the RELARIN technique. These tools will be heavily used again in Chapter IX to complete the proof of Theorem 8.2.

39

Proof of the second Ugly Theorem

The Neighborhood Conjecture (Open Problem 9.1) is a central issue of the book. The first result toward Open Problem 9.1 was Theorem 34.1, or as we called it: the first Ugly Theorem (see Section 36 for the proof). The second Ugly Theorem (Theorem 37.5) is more powerful. It gives the best-known Strong Draw result for the n^d hypercube Tic-Tac-Toe (Theorem 12.5 (a)), and it is also necessary for the solution of the Lattice Games (Theorem 8.2).

1. Proof of Theorem 37.5. We assume that the reader is familiar with the proof of Theorem 34.1. In the proof of Theorem 34.1 Breaker used the Power-of-Two Scoring System in the Big Game to prevent the appearance of the "Forbidden Configurations" in the small game, and this way he could ensure the "simplicity" of the small game. The small game was so simple that Breaker could block every "emergency set" by a trivial Pairing Strategy. The new idea is very natural: we try to replace the Pairing Strategy in the small game by the Power-of-Two Scoring System. The improvement should come from the intuition that the Power-of-Two Scoring System is "exponential," far superior to the "linear" Pairing Strategy.

Note that here the "components" of the small game will be "exponentially large," so the new Big Sets will be enormous.

To have a clear understanding of the changes in the argument, we first give a brief summary of the proof of Theorem 34.1, emphasizing the key steps. Let $\mathcal{F}$ be an n-uniform Almost Disjoint hypergraph. Breaker (the second player) wants to force a Strong Draw. Breaker artificially decomposes the board (i.e. the union set) into two disjoint parts: the *Big Board* and the *Small Board*. The Decomposition Rules are the following:

(1) The decomposition is *dynamic*: the *Big Board* is shrinking, and the *small board* is growing in the course of a play.
(2) Breaker replies by the Same Board Principle: if Maker's last move was in the *Big Board* (respectively *small board*), then Breaker always replies in the *Big Board* (respectively *small board*).

(3) The *Big Game* and the *small game* are *non-interacting* in the following sense: Breaker's strategy in the *Big Game* (*small game*) does not assume any knowledge of the other game ("Breaker is schizophrenic," and the "Iron Curtain Principle" applies).

(4) At the beginning of a play the *Big Board* is the whole board (and so the complement, the *small board*, is empty). In the course of a play in the *Big Game* (played in the *Big Board*) a winning set $A \in \mathcal{F}$ is *dead* when it contains a mark of Breaker; an $A \in \mathcal{F}$, which is not *dead*, is called a *survivor*. (Of course, *survivor* is a temporary concept: a *survivor* $A \in \mathcal{F}$ may easily become *dead* later; *dead* is permanent.)

(5) A survivor $A \in \mathcal{F}$ becomes *dangerous* when Maker occupies its $(m-k-1)^{st}$ point in the *Big Game* (in the *Big Board*, of course). Then the $(k+1)$-element part of this A, unoccupied in the *Big Game*, becomes an *emergency set*. Every *emergency set* is removed from the Big Board, and added to the small board. The small board is the union of the emergency sets, and the Big Board is the complement of the small board.

(6) In the *Big Board* Breaker plays the *Big Game*. Breaker's goal in the *Big Game* is to prevent Maker from *almost* completely occupying a *Big Set* before Breaker could put his first mark in it (in the *Big Board* of course). This is how Breaker prevents "Forbidden Configurations" to graduate into the *small game*.

(7) Breaker's goal in the *small game* is to block all *emergency sets*.

(8) Breaker wins the *Big Game* by using an Erdős–Selfridge type lemma (such as "Lemma 5" in Section 36).

(9) Fact (8) implies that Breaker can win the *small game* by a trivial Pairing Strategy.

(10) Fact (8) implies that "there is no secondary danger," i.e. there is no *survivor* that never graduates into a *dangerous set*.

2. How to "beat" Theorem 34.1? As we said before, the new idea is to replace the Pairing Strategy in the small game by a more sophisticated Erdős–Selfridge type strategy. We expect the "small game" to fall apart into many disjoint "components", and the goal of the "Big Game" is to prevent the appearance of too large "components" ("too large" means exponentially large in some precise sense). If Breaker follows the Same Component Rule in the "small game" (if Maker moves to a "component" of the "small game," then Breaker replies in the same "component"), and the "components" remain at most "exponentially large" in the "small game," then Breaker should be able to block every "dangerous set" in the "small board." Of course, the "components" in the "small game" may grow, so we need a "growing version" of the Erdős–Selfridge Theorem.

Now we start to work out the details of this "heuristic outline." Again Breaker decomposes the board (i.e. the union set) into two disjoint parts: the *Big Board* and the *small board*. What happens to the Decomposition Rules (1)–(10) above? Well, (1) doesn't change:

(1) The decomposition is *dynamic*: the *Big Board* is shrinking, and the *small board* is growing in the course of a play.

(2) Will change; the Same Board Rule may be violated in the *small game* as follows:

(2′) if Maker's last move was in the *Big Board*, then Breaker always responds in the *Big Board*, *but* if Maker's last move was in the *small board*, then it may happen that Breaker replies in the *Big Board*.
   Breaker is still "schizophrenic," and the "Iron Curtain Principle" is unchanged.

(3) The *Big Game* and the *small game* are *non-interacting* in the following sense: Breaker's strategy in the *Big Game* (*small game*) does not assume any knowledge of the other game.

(4) Remains the same: at the beginning of a play the *Big Board* is the whole board (and so the complement, the *small board*, is empty). In the course of a play in the *Big Game* (played in the *Big Board*) a winning set $A \in \mathcal{F}$ is *dead* when it contains a mark of Breaker; an $A \in \mathcal{F}$, which is not *dead*, is called a *survivor*. (Of course, *survivor* is a temporary concept: a *survivor* $A \in \mathcal{F}$ may easily become *dead* later; *dead* is permanent.)

(5) Changes: the "dangerous sets" may double in size, they may have as many as $2k$ elements instead of $k+1$; more precisely,

(5′) A survivor $A \in \mathcal{F}$ becomes *dangerous* when Maker occupies its $(m-2k)$th point in the *Big Game* (in the *Big Board*, of course). Then the $2k$-element part of this A, unoccupied in the *Big Game*, is removed from the Big Board, and added to the small board. The "blank" part of this $2k$-element set, i.e. unoccupied in the *small game*, is called an *emergency set*. The small board is exactly the union of the emergency sets, and the Big Board is the complement of the small board.

(6) Remains the same: In the *Big Board* Breaker plays the *Big Game*. Breaker's goal in the *Big Game* is to prevent Maker from *almost* completely occupying a *Big Set* before Breaker could put his first mark in it (in the *Big Board* of course). This is how Breaker prevents "Forbidden Configurations" to graduate into the *small game*.

(7) Changes as follows:

(7′) Breaker's goal in the *small game* is to block all *small sets*. The *small sets* are the *emergency sets* and the *secondary sets*.

Here we have to stop and explain what the *secondary sets* are. Also we have to clarify the definition of *emergency sets*. We start with a new concept: a *survivor*

$A \in \mathcal{F}$ becomes *visible in the small game* when its intersection $A \cap V_s$ with the small board V_s has at least k elements for the first time (and remains *visible* for the rest of the play). This stage of the play is called the *birthday* of a new *small set* S ($S \subset A$), which is defined as follows. First of all, from A we remove all points occupied by Maker in the Big Game up to the *birthday* ("circumcision"); what remains from A is denoted by A'. By definition $|A'| \geq 2k$. Note that a visible survivor set A has less than k *pre-birth points* in the small board; they are all removed from A' ("circumcision" is a "security measure": it may happen that all *pre-birth points* are occupied by Maker in the small game). What remains from A' is a new *small set* S ($S \subset A$). So at its birthday a small set S is "blank," and has more than $|A'| - k \geq k$ points. If A' has exactly $2k$ points, then $S = S(A)$ is an *emergency set*. If A' has more than $2k$ points, then $S = S(A)$ is a *secondary set*. An *emergency set* is part of the small board (by definition, see (5')) but a *secondary set* may have a non-empty intersection with the Big Board. Note that a *secondary set* may later become an *emergency set,* but, if a *secondary set* is completely contained in the small board, then it will never become an *emergency set*.

(8) Remains the same: Breaker wins the *Big Game* by using "Lemma 5" in Section 36.

(9) and (10) Undergo a complete change:

(9') Fact (8) enforces that Breaker can win the *small game* by using an Erdős–Selfridge type "Growing Lemma."

(10') Breaker cannot prevent the appearance of *secondary sets*, but (8) and (9') take good care of them.

We have to explain what (9') and (10') mean.

3. The small game: applying an Erdős–Selfridge type "Growing Lemma." Let S denote the "growing" family of *small sets*, meaning the *emergency sets* and the *secondary sets* together. We have to clearly explain what "growing" means. The *basic rule of decomposition* is the following (see (2')): If Maker's last move was in the Big Board, then it is considered a "move in the Big Game," and Breaker replies in the Big Board, and his move is considered a "move in the Big Game." If Maker's last move is in the small board, then it is considered a "move in the small game," and Breaker replies in the *same component* of the hypergraph of all small sets where Maker's move was, and Breaker's move is considered a "move in the small game."

Warning: Some small sets, in fact secondary sets, are *not* necessarily contained in the small board, so it can easily happen that Breaker's reply in the small game is *outside* of the small board.

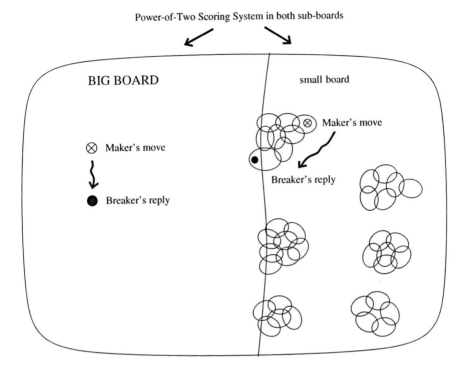

Power-of-Two Scoring System in both sub-boards

This move of Breaker, we call it an "outside move," is *not* considered a move in the Big Game (even if it is literally in the Big Board). Breaker is "schizophrenic": he has two personalities like $Break_{Big}$, who plays in the Big Game, and $Break_{small}$, who plays in the small game, and the two personalities know nothing about each other. Therefore, $Break_{Big}$ does not see an "outside move" of $Break_{small}$ as a blocking move in the Big Game. What $Break_{Big}$ can "see" here is a decrease in the set of available (unoccupied) points, i.e. a decrease in the Big Board.

This is how the original hypergraph game played on the n-uniform $\mathcal{F}$ decomposes into two disjoint ("non-interacting") games: the Big Game and the small game.

This rule of decomposition leads to the following "growing" of the small game. Assume that we are in the middle of a play. Let x^* be a move of Maker in the small board, let y^* denote Breaker's reply in the same component of the hypergraph of all simple sets $\mathcal{S}$ (y^* may be outside of the small board). Next let

$$x^{(1)}, y^{(1)}, x^{(2)}, y^{(2)}, \ldots, x^{(j)}, y^{(j)}$$

be a possible sequence of moves in the Big Game (in the Big Board, of course), and let x^{**} be Maker's first return to the small board. In the small game x^*, y^*, x^{**} are *consecutive* moves, but between y^* and x^{**} the hypergraph of small sets may *increase*. Indeed, the sequence $x^{(1)}, y^{(1)}, x^{(2)}, y^{(2)}, \ldots, x^{(j)}, y^{(j)}$ in the Big Game may lead to *new* dangerous sets $A \in \mathcal{F}$. A new dangerous set A defines a new emergency

Board

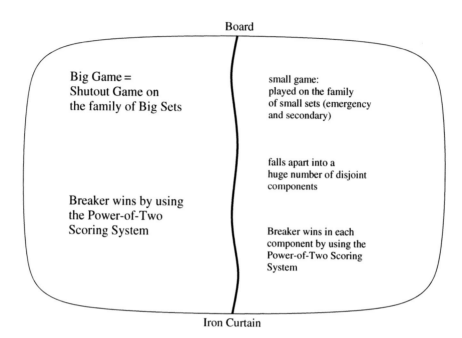

Iron Curtain

sets $E \subset A$, which is added to the small board. Larger small board means that some new survivors $B \in \mathcal{F}$ have a chance to become "visible" (i.e. have $\geq k$ points in ths small board), and give birth to some "newborn baby small sets." This is how new emergency and secondary sets may arise between the consecutive moves y^* and x^{**} in the small game.

For "security measures" we remove the "pre-birth" points from every "newborn baby small set." This way we lose less than k points (they all might be occupied by Maker in the small game). Each "newborn baby small set" is "blank:" it does not contain any point of Maker or Breaker in the small game up to that moment (i.e. before choosing x^{**}).

Warning: It is somewhat confusing to say that the hypergraph $\mathcal{S}$ of small sets is "growing." It can easily happen that between consecutive moves y^* and x^{**} in the small game some small set in fact "dies." Indeed, the moves $y^{(1)}, y^{(2)}, \ldots, y^{(j)}$ of Breaker in the Big Game may block some survivor, which is the super-set of a secondary set, and then there is no reason to keep this secondary set any longer: we could easily delete it from the hypergraph of small sets. We could, but we decided *not* to delete such a "dead secondary set" from the hypergraph of all small sets! Once a survivor becomes "visible in the small game," we keep it in hypergraph $\mathcal{S}$ forever even if it becomes blocked (i.e. "killed") by Breaker later in the Big Game.

Similarly, once a small set arises, we keep it in hypergraph $\mathcal{S}$ even if it becomes blocked later by Breaker in the small game.

Working with the "keep the small sets anyway" interpretation of S, we can really claim that the family S of all small sets is "growing" in the course of a play.

4. The Component Condition. We are going to define the family of Big Sets (the precise definition comes much later) to *ensure* the:

Component Condition: *During the whole course of the small game every component of hypergraph S ("small sets") has less than 2^k sets.*

A key ingredient of the proof is the:

Growing Erdős–Selfridge Lemma: *If the Component Condition holds, then Breaker has a strategy in the small game such that, he can prevent Maker from occupying k points from a "circumcised small set" (where the "pre-birth" points are removed) before Breaker could block it in the small game.*

Proof. Let $x_1, y_1, x_2, y_2, \ldots$ be the points of Maker and Breaker, selected in this order during a play in the small game. (Here we restrict ourselves to the small game; in other words, the Big Game does not exist, it is behind the "Iron Curtain.") We use the standard Power-of-Two Scoring System for the **small sets** (i.e. the less than k "pre-birth" points are already removed): if the small set is blocked by Breaker, then its *value* is 0; if it is Breaker-free and contains l points of Maker, then its *value* is 2^l $(l = 0, 1, 2, \ldots)$.

Let's see what happens at the very beginning. Before x_1 every small set is "blank," and has *value* $2^0 = 1$. Between x_1 and y_1 those small sets which contain x_1 have *value* $2^1 = 2$. The *value* of a component of S is, of course, the sum of the *values* of the sets in the component. Between x_1 and y_1 the *value of a component is at most twice the component-size* (i.e. the number of sets in the component).

We are going to prove, by induction on the time, that a very similar statement holds in an *arbitrary stage* of the play, see the "Overcharge Lemma" below). To formulate the "Overcharge Lemma" we have to introduce a *technical trick*: the concept of "overcharged value." To explain it, assume that we are in the middle of a particular play in the small game, and consider an arbitrary component C of the family S of all small sets. There are two possibilities:

(1) the last mark put in component C was due to Maker;
(2) the last mark put in component C was due to Breaker.

In case (1) Breaker does nothing; however, in case (2) Breaker picks the "best" unoccupied point in component C and includes it as a "fake move" of Maker. "Best" of course means that it leads to the maximum increase in the *value* of component C). By including a possible "fake move," Breaker can guarantee that, in every stage of the play, in every component of S, the *last move is always due to Maker*. Indeed, it is either a real move as in case (1), or a "fake move" as in case (2). Note that a "fake move" is always temporary: when Maker makes a real move in a component, the "fake move" is immediately erased.

The value of a *component* of S, computed including the possible "fake move," is called the **overcharged value** of the component. "Overcharged" in the sense that a possible "fake move" of Maker doubles the ordinary value of a small set.

Now we are ready to formulate the:

Overcharge Lemma: *Breaker can force that, in every stage of the small game, in every component C of all small sets S, the overcharged value of component C is $\leq 2|C|$. In other words, the overcharged value of a component is always at most twice the component size.*

Proof. We prove it by induction on the time. We study the effect of two consecutive moves y_i and x_{i+1} in the small game: y_i is the ith move of Breaker and x_{i+1} is the $(i+1)$st move of Maker. Let C_1 denote the component of S which contains y_i. Between the moves y_i and x_{i+1} family S may grow: some "newborn baby small sets" may arise. Note that the "newborn baby small sets" are always "blank": they have value $2^0 = 1$. The "newborn baby small sets" may form "bridges" to glue together some old components into a bigger one:

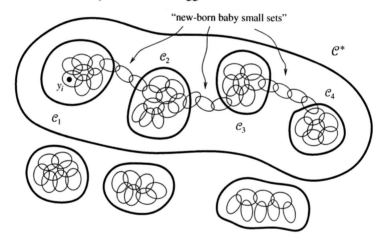

In the figure the old components $C_1, C_2, C_3, \ldots, C_r$ are glued together by some "newborn baby small sets," and form a big new component C^*.

Following this figure, we distinguish 3 cases.

Case 1: $x_{i+1} \in C_1$, that is, y_i and x_{i+1} are in the same old component of S. By the induction hypothesis, before choosing y_i, the *value* of component C_1 was at most $2|C_1|$.

Note that here the *overcharged value* is the same as the ordinary value. Indeed, x_{i-1} had to be in C_1, for Breaker's next move y_i is always in the same component, and S does *not* grow between x_{i-1} and y_i, so Maker had a real (not "fake") last move in component C_1 (namely x_{i-1}).

Since y_i comes before x_{i+1}, Breaker can pick the "best" point in the usual sense, so y_i is at least as "good" as x_{i+1}. It follows that, after x_{i+1}

$$\text{overcharged value}(C^*) \leq 2|C_1| + 2|C_2| + \cdots + 2|C_r| \tag{39.1}$$

$$+2(\text{number of newborn baby small sets in component } C^*) = 2|C^*|.$$

Note that computing the overcharged-value of C^* we drop the possible "fake moves" in $C_2, \ldots, C_r$, which explains why "inequality" may happen in (39.1) (instead of "equality").

Case 2: y_i and x_{i+1} are in different old components of C^*.
Assume that (say) $x_{i+1} \in C_2$. The last *real* move in component C_2 *before* x_{i+1} had to be a move of Breaker, so when we calculated the *overcharged value* of C_2 we had to include a "fake move" of Maker. The "fake move" is always the"best" unoccupied point, so x_{i+1} cannot be better than that. So, replacing the "fake move" by x_{i+1} in C_2 cannot increase the value, so again we get inequality (39.1).

Case 3: x_{i+1} is outside of C^*.
We can repeat the proof of Case 2. Again the idea is that a "fake move" is always the "best" available point in a component, so x_{i+1} cannot be better than the corresponding "fake move." This completes the proof of the Overcharge Lemma. $\square$

Now we are ready to complete the proof of the Growing Erdős–Selfridge Lemma. Assume that, at some stage of the small game, Maker has $k+1$ points in some small set, and Breaker has as yet no marks in it. At this stage of the play, this particular small set belongs to some component C of all small sets S, and the value of this small set alone is 2^{k+1}, so

$$\text{overcharged value}(C) = \text{value}(C) \geq 2^{k+1}. \tag{39.2}$$

On the other hand, by the Overcharge Lemma

$$\text{overcharged value}(C) \leq 2|C|,$$

and combining this inequality with (39.2) we conclude that $|C| \geq 2^k$, which contradicts the Component Condition. This contradiction proves the Growing Erdős–Selfridge Lemma. $\square$

Next we prove the:

Last-Step Lemma: *Assume that, by properly playing in the Big Game, Breaker can ensure the Component Condition in the small game. Then Breaker can block the original hypergraph $\mathcal{F}$, i.e. Breaker can put his mark in every m-element winning set $A \in \mathcal{F}$.*

Proof. If $A \in \mathcal{F}$ becomes blocked in the Big Game ("dead"), then we are done. So we can assume that $A \in \mathcal{F}$ remains a survivor in the whole course of the Big Game. We recall that in the Big Game Maker cannot occupy more than $m - 2k$ points from a survivor.

Case 1: survivor $A \in \mathcal{F}$ eventually becomes "visible" in the small game.
Then A contains a small set S. By the Growing Erdős–Selfridge Lemma, if Breaker cannot block S, then Maker can have at most k points in S, and "removing the pre-birth points" means to lose less than k points, which may all belong to Maker. It follows that Maker can have at most $(m - 2k) + k + (k - 1) = m - 1$ points of a survivor $A \in \mathcal{F}$. (Indeed, $\leq (n - 2k)$ points in the Big Game, $\leq k$ points in the small set, and $\leq (k - 1)$ "pre-birth" points.) So Maker cannot completely occupy $A \in \mathcal{F}$.

Case 2: survivor $A \in \mathcal{F}$ never becomes "visible" in the small game.
This case is even simpler: A intersects the small board in less than k points in the whole course of the play. So Maker has at most $(m - 2k) + (k - 1) < m$ points of A. The Last-Step-Lemma follows. $\qquad\square$

5. How to define the Big Sets? We have to define the family of Big Sets in such a way that, by properly playing in the Big Game, Breaker can enforce the Component Condition in the small game. To motivate the forthcoming definition of Big Sets, assume that the Component Condition is violated: at some stage of the play there is a "large" component $\mathcal{C}$ of all small sets $\mathcal{S}$: $|\mathcal{C}| \geq 2^k$. Then we can *extract* a Big Set from this "large" component $\mathcal{C}$. The precise definition of *extraction* is going to be complicated. What we can say in advance is that the "extracted Big Set" is a union of a sub-family of dangerous sets such that their emergency parts all belong to "large" component $\mathcal{C}$, and the overwhelming majority of the points in the Big Set are degree one points in the sub-family ("private points").

First we define the concept of *first, second, third, ... neighborhood* of a hyperdge (or a set of hyperedges) in an arbitrary hypergraph. Let $\mathcal{F} = \{F_1, F_2, \ldots, F_K\}$ be an arbitrary finite hypergraph. The *dependency graph* of $\mathcal{F}$ is a graph on vertices $1, 2, \ldots, K$, and two vertices i and j ($1 \leq i < j \leq K$) are joined by an edge if and only if the corresponding hypergedges F_i and F_j have non-empty intersection. Let $G = G(\mathcal{F})$ denote the *dependency graph* of hypergraph $\mathcal{F}$. Of course, $G = G(\mathcal{F})$ is not necessarily connected: it may fall apart into several *components*. For any two vertices i and j in the same component of G, the *G-distance* of i and j is the number of edges in the shortest path joining i and j. If i and j are in the different components of G, then their G-distance is ∞. For an arbitrary vertex-set $I \subset \{1, 2, \ldots, K\}$ and vertex $j \in \{1, 2, \ldots, K\} \setminus I$, the G-distance of I and j is the following: we compute the G-distance of $i \in I$ and j, and take the minimum as i runs through I.

The $\mathcal{F}$-distance of hyperedges F_i and F_j is the G-distance of i and j where $G = G(\mathcal{F})$ is the *dependency graph* of hypergraph $\mathcal{F}$. Similarly, the $\mathcal{F}$-distance of $\{F_i : i \in I\}$ and F_j is the G-distance of I and j, where $G = G(\mathcal{F})$ is the dependency graph.

If this distance happens to be d, then we say "F_j is in the dth neighborhood of F_i (or $\{F_i : i \in I\}$)."

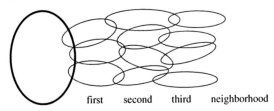

first   second   third   neighborhood

How to extract a Big Set from a large component of S? Let $\mathcal{C}$ be a component of S with $|\mathcal{C}| \geq 2^k$. Here S is the family of all small sets (up to that point of the play), and small sets are emergency sets and secondary sets. Note that every secondary set has at least k "visible" points, each one covered by emergency sets. Since the underlying hypergraph $\mathcal{F}$ is Almost Disjoint, the first $\mathcal{C}$-neighborhood of a secondary set must contain at least k distinct emergency sets. As a trivial byproduct of this argument, we obtain that $\mathcal{C}$ must contain an emergency set.

We describe a **sequential extraction process** finding $T = 2^{k/4}/m^2$ emergency sets in $\mathcal{C}$ such that the union of their *super-sets* in $\mathcal{F}$ form a Big Set. More precisely, we define a growing family $\mathcal{G}_j = \{E_1, E_2, \ldots, E_j\}$ of emergency sets, and the union $\mathcal{U}_j = \cup_{i=1}^{j} A(E_i)$ of their super-sets $(E_i \subset A(E_i) \in \mathcal{F})$ for $j = 1, 2, 3, \ldots T$, where $T = 2^{k/4}/m^2$ ("integral part").

The beginning is trivial: $\mathcal{C}$ must contain an emergency set E_1. Let $\mathcal{G}_1 = \{E_1\}$, and let $\mathcal{U}_1 = A(E_1)$ be the super-set of E_1.

To explain the general step, assume that we just completed the jth step: we have already selected a family $\mathcal{G}_{g(j)}$ of $g(j)$ distinct emergency sets, and the union of their emergency sets from $\mathcal{F}$ is denoted by $\mathcal{U}_{g(j)}$. Note that $g(j) \geq j$: indeed, in each step we shall find either 1 or $(k-2)$ new emergency sets, i.e. either $g(j+1) = g(j)+1$ or $g(j+1) = g(j)+(k-2)$. By definition $|\mathcal{U}_{g(j)}| \leq g(j) \cdot m$.

Next we describe the $(j+1)$st step. Unfortunately we have to distinguish 8 cases. The case study is going to be very "geometric" in nature, so it is absolutely necessary to fully understand the corresponding figures.

Note in advance that the family $\mathcal{B}$ of Big Sets is defined as follows

$$\mathcal{B} = \left\{ B = \mathcal{U}_T : \text{ for all possible ways one can } \mathbf{grow} \ \mathcal{U}_T \text{ of length } T = \frac{2^{k/4}}{m^2} \right.$$

$$\left. \mathbf{in\ terms\ of} \ \mathcal{F} \text{ by using Cases } 1-8 \text{ below} \right\}$$

(we shall clarify "grow" and "in terms of $\mathcal{F}$" later).

Case 1: The first $A(\mathcal{C})$-neighborhood of $\mathcal{U}_{g(j)}$ contains the super-set $A(E)$ of an emergency set E such that $|\mathcal{U}_{g(j)} \cap A(E)| = 1$. Then $\mathcal{G}_{g(j+1)} = \mathcal{G}_{g(j)} \cup \{E\}$, and so $\mathcal{U}_{g(j+1)} = \mathcal{U}_{g(j)} \cup A(E)$ satisfies $|\mathcal{U}_{g(j+1)}| = |\mathcal{U}_{g(j)}| + (m-1)$.

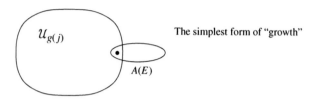

The simplest form of "growth"

We use the notation $A(\ldots)$ for the super-set in $\mathcal{F}$, and so "$A(\mathcal{C})$-neighborhood" means the family of super-sets of the members of component $\mathcal{C}$; observe that $A(\mathcal{C})$ is a connected sub-family of $\mathcal{F}$.

Case 2: The second $A(\mathcal{C})$-neighborhood of $\mathcal{U}_{g(j)}$ contains the super-set $A(E)$ of an emergency set E such that there is a neighbor A_1 of $A(E)$ from the first $A(\mathcal{C})$-neighborhood of $\mathcal{U}_{g(j)}$, which intersects $\mathcal{U}_{g(j)}$ in at least two points.

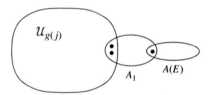

Then $\mathcal{G}_{g(j+1)} = \mathcal{G}_{g(j)} \cup \{E\}$, A_1 is an auxiliary **bond set**, and $\mathcal{U}_{g(j+1)} = \mathcal{U}_{g(j)} \cup A(E)$ satisfies $|\mathcal{U}_{g(j+1)}| = |\mathcal{U}_{g(j)}| + m$.

The name *bond set* comes from the fact that, in an Almost Disjoint hypergraph, two points determine a set. *Bond sets* have little effect in the calculation of the *total number* of Big Sets, see the **Calculations** later. To illustrate what we mean, let's calculate the number of different ways we can "grow" to $\mathcal{G}_{g(j+1)}$ from $\mathcal{G}_{g(j)}$ in Case 2. A trivial upper bound on the number of possibilities is

$$\binom{|\mathcal{U}_{g(j)}|}{2} \binom{m-2}{1} \cdot D \leq \frac{g^2(j) \cdot m^3 \cdot D}{2},$$

where $D = D(\mathcal{F})$ is the Maximum Degree of $\mathcal{F}$.

Now we interrupt our case study for a moment, and make an observation about the *secondary sets* of $\mathcal{C}$. Let $S \in \mathcal{S}$ be an arbitrary secondary set. We show that S can have 4 possible "types."

Type 1: $A(S)$ intersects $\mathcal{U}_{g(j)}$ in at least two points.

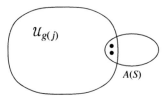

If Type 1 fails, i.e. if $|A(S) \cap \mathcal{U}_{g(j)}| \leq 1$, then there are at least $(k-1)$ "visible" points of S which are covered by distinct emergency sets of $\mathcal{C}$.

Type 2: There exists an emergency set E such that $|A(E) \cap A(S)| = 1$ and $|A(S) \cap \mathcal{U}_{g(j)}| = 1$.

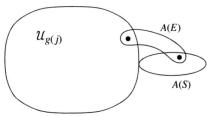

Observe that "Type 2" is covered by "Case 1," so without loss of generality we can assume that there is *no* Type 2 secondary set in $\mathcal{C}$. Also Type 1 is excluded, unless, of course, $A(S)$ is in the first $A(\mathcal{C})$-neighborhood of $\mathcal{U}_{g(j)}$

Type 3: There exist two emergency sets E_1 and E_2 such that $|A(E_i) \cap \mathcal{U}_{g(j)}| \geq 2$ and $|A(E_i) \cap A(S)| = 1$ for $i = 1, 2$.

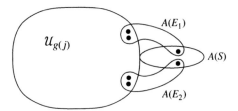

Type 4: There exist $(k-2)$ emergency sets $E_1, E_2, \ldots, E_{k-2}$ such that $|A(E_i) \cap \mathcal{U}_{g(j)}| = 0$ and $|A(E_i) \cap A(S)| = 1$ for $i = 1, 2, \ldots, k-2$; we call it the "centipede."

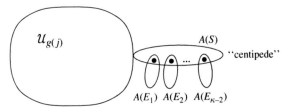

(Note that for Types 2–3–4 $|A(S) \cap \mathcal{U}_{g(j)}| = 0$ or 1.)

After the classification of "types," we can return to our case study. Assume that the second $A(\mathcal{C})$-neighborhood of $\mathcal{U}_{g(j)}$ contains the super-set $A(E)$ of an emergency set E. Let A_1 be a neighbor of $A(E)$ from the first $A(\mathcal{C})$-neighborhood of $\mathcal{U}_{g(j)}$. In view of Case 2 we can assume that A_1 intersects $\mathcal{U}_{g(j)}$ in exactly 1 point. By Case 1 we know that $A_1 = A(S)$, i.e. A_1 has to be the super-set of a secondary set S.

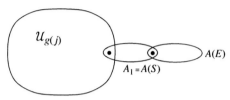

We have two possibilities: S is of Type 3 or Type 4.

Case 3: S is of Type 3. Then $\mathcal{G}_{g(j+1)} = \mathcal{G}_{g(j)} \cup \{E\}$, $A(E_1)$, $A(E_2)$, $A(S)$ are auxiliary bond sets, and clearly $|\mathcal{U}_{g(j+1)}| = |\mathcal{U}_{g(j)}| + m$.

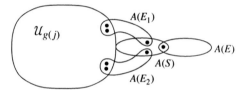

Case 4: S is of Type 4. Then $\mathcal{G}_{g(j+1)} = \mathcal{G}_{g(j)} \cup \{E_1, E_2, \ldots, E_{k-2}\}$, and $A(S)$ is an auxiliary set, but *not* a bond set. From Almost Disjointness we have

$$|\mathcal{U}_{g(j+1)}| \geq |\mathcal{U}_{g(j)}| + (k-2)m - \binom{k-2}{2}.$$

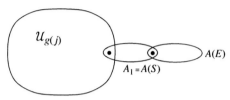

The number of ways we can "grow" to $\mathcal{G}_{g(j+1)}$ from $\mathcal{G}_{g(j)}$ in Case 4 is bounded from above by

$$|\mathcal{U}_{g(j)}| \cdot D \cdot \binom{m-1}{k-2} \cdot D^{k-2},$$

where $D = D(\mathcal{F})$ is the Maximum Degree.

In the rest of the case study we can assume that the second $A(\mathcal{C})$-neighborhood of $\mathcal{U}_{g(j)}$ consists of secondary sets only.

First assume that the second $A(\mathcal{C})$-neighborhood of $\mathcal{U}_{g(j)}$ does contain a Type 4 secondary set S. Let A_1 be a neighbor of $A(S)$, which is in the first $A(\mathcal{C})$-neighborhood of $\mathcal{U}_{g(j)}$. Then

(a) either $|A_1 \cap \mathcal{U}_{g(j)}| \ge 2$,
(b) or $|A_1 \cap \mathcal{U}_{g(j)}| = 1$.

Case 5: $|A_1 \cap \mathcal{U}_{g(j)}| \ge 2$

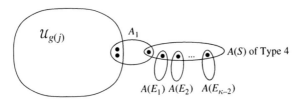

Then $\mathcal{G}_{g(j+1)} = \mathcal{G}_{g(j)} \cup \{E_1, E_2, \ldots, E_{k-2}\}$. Here A_1 and $A(S)$ are auxiliary sets: A_1 is a bond set, but $A(S)$ is *not* a bond set. From Almost Disjointness

$$|\mathcal{U}_{g(j+1)}| \ge |\mathcal{U}_{g(j)}| + (k-2)m - \binom{k-2}{2}.$$

If (a) fails, then we have (b): $|A_1 \cap \mathcal{U}_{g(j)}| = 1$. In view of Case 1 we can assume that $A_1 = A(S^*)$, where S^* is a secondary set. If S^* is of Type 4, then we have Case 4 again. So we can assume that S^* is of Type 3.

Case 6: $|A_1 \cap \mathcal{U}_{g(j)}| = 1$, where $A_1 = A(S^*)$ with a Type 3 secondary set S^*.

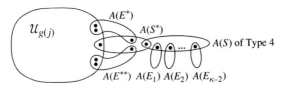

Then $\mathcal{G}_{g(j+1)} = \mathcal{G}_{g(j)} \cup \{E_1, E_2, \ldots, E_{k-2}\}$. Here $A(E^*)$, $A(E^{**})$, $A(S^*)$, $A(S)$ are auxiliary sets: all but $A(S)$ are bond sets. Clearly

$$|\mathcal{U}_{g(j+1)}| \ge |\mathcal{U}_{g(j)}| + (k-2)m - \binom{k-2}{2}.$$

In the rest of the case study we can assume that the second $A(\mathcal{C})$-neighborhood of $\mathcal{U}_{g(j)}$ contains Type 3 secondary sets only.

To complete the case study it is enough the discuss the third $A(\mathcal{C})$-neighborhood of $\mathcal{U}_{g(j)}$. This leads to the following last two cases:

Case 7: The third $A(\mathcal{C})$-neighborhood of $\mathcal{U}_{g(j)}$ contains the super-set $A(E)$ of an emergency set E. Let A_1 be a neighbor of $A(E)$, which is in the second $A(\mathcal{C})$-neighborhood of $\mathcal{U}_{g(j)}$. We know that $A_1 = A(S)$, where S is a Type 3 secondary set:

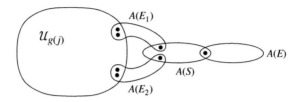

Then $\mathcal{G}_{g(j+1)} = \mathcal{G}_{g(j)} \cup \{E\}$. Here $A(E_1), A(E_2), A(S)$ are auxiliary bond sets. Clearly

$$|\mathcal{U}_{g(j+1)}| \geq |\mathcal{U}_{g(j)}| + m.$$

From now on we can assume that both the second and third $A(\mathcal{C})$-neighborhood of $\mathcal{U}_{g(j)}$ consist of secondary sets only. Let S_2 and S_3 be secondary sets such that $A(S_2)$ and $A(S_3)$ are in the second and third $A(\mathcal{C})$-neighborhood of $\mathcal{U}_{g(j)}$ and $|A(S_2) \cap A(S_3)| = 1$. We claim that S_3 cannot be of Type 3. Indeed, by the "triangle inequality" of the "distance," a neighbor of a third neighbor *cannot* be a first neighbor (i.e. intersecting $\mathcal{U}_{g(j)}$). Since Types 1 and 2 are out, the only possible type for S_3 is Type 4. Moreover, since the whole second $A(\mathcal{C})$-neighborhood of $\mathcal{U}_{g(j)}$ is of Type 3, S_2 has to be of Type 3, and we arrive at the *last case*.

Case 8:

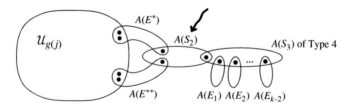

Then $\mathcal{G}_{g(j+1)} = \mathcal{G}_{g(j)} \cup \{E_1, E_2, \ldots, E_{k-2}\}$. Here $A(E^*), A(E^{**}), A(S_2), A(S_3)$ are auxiliary sets: all but $A(S_3)$ are bond sets. Clearly

$$|\mathcal{U}_{g(j+1)}| \geq |\mathcal{U}_{g(j)}| + (k-2)m - \binom{k-2}{2}.$$

6. Calculations: We go through Cases 1–8. The number of ways $\mathcal{G}_{g(j)}$ can grow to $\mathcal{G}_{g(j+1)}$ is bounded from above by

$$|\mathcal{U}_{g(j)}| \cdot D \le g(j) \cdot m \cdot D \quad \text{in Case 1 ;}$$

$$\binom{|\mathcal{U}_{g(j)}|}{2}(m-2) \cdot D \le \frac{g^2(j) \cdot m^3 \cdot D}{2} \quad \text{in Case 2;}$$

$$\binom{|\mathcal{U}_{g(j)}|}{4}\binom{4}{2}(m-2)^2(m-3) \cdot D \le \frac{g^4(j) \cdot m^7 \cdot D}{4} \quad \text{in Case 3;}$$

$$|\mathcal{U}_{g(j)}|D\binom{m-1}{k-2}D^{k-2} \le \frac{g(j) \cdot m^{k-1} \cdot D^{k-1}}{2(k-2)!} \quad \text{in Case 4;}$$

$$\binom{|\mathcal{U}_{g(j)}|}{2}(m-2)D\binom{n-1}{k-2}D^{k-2} \le \frac{g^2(j) \cdot m^{k+1} \cdot D^{k-1}}{2(k-2)!} \quad \text{in Case 5;}$$

$$\binom{|\mathcal{U}_{g(j)}|}{4}\binom{4}{2}(m-2)^2(m-3) \cdot D\binom{n-1}{k-2}D^{k-2} \le \frac{g^4(j) \cdot m^{k+5} \cdot D^{k-1}}{4(k-2)!} \quad \text{in Case 6;}$$

$$\binom{|\mathcal{U}_{g(j)}|}{4}\binom{4}{2}(m-2)^3 \cdot D \le \frac{g^4(j) \cdot m^7 \cdot D}{4} \quad \text{in Case 7;}$$

$$\binom{|\mathcal{U}_{g(j)}|}{4}\binom{4}{2}(m-2)^3 \cdot D\binom{m-1}{k-2}D^{k-2} \le \frac{g^4(j) \cdot m^{k+5} \cdot D^{k-1}}{4(k-2)!} \quad \text{in Case 8,}$$

where $D = D(\mathcal{F})$ is the Maximum Degree.

We include 1 new set in Cases 1,2,3, and 7; and include $(k-2)$ new sets in Cases 4,5,6, and 8. We have the maximum number of possibilities "per 1 inclusion" in Cases 3, 7, and in Cases 6, 8 ("centipede"). The maximum is

$$\max\left\{ \frac{g^4(j) \cdot m^7 \cdot D}{4}, \left(\frac{g^4(j) \cdot m^{k+5} \cdot D^{k-1}}{4(k-2)!} \right)^{1/(k-2)} \right\}. \tag{39.3}$$

It is clear from the case study that

$$|\mathcal{U}_{g(j)}| \ge \left(m - \frac{k}{2} \right) \cdot g(j). \tag{39.4}$$

A review of the case study will prove the following crucial:

Big Set Lemma: *If at some stage of the play there is a component C of the family S of all small sets such that $|C| \ge 2^k$, then there is a $\mathcal{G}_{g(j)} \subset C$ such that $g(j) \ge 2^{k/4}/m^2$.*

Remark. We recall that $\mathcal{G}_{g(j)}$ is a family of emergency sets from C such that in each one of the j steps the "growth" comes from one of the 8 cases described above.

Proof. Assume $g(j) < 2^{k/4}/m^2$. By going through Cases 1–8 we show how we can still "grow" family $\mathcal{G}_{g(j)}$.

If Case 1 applies, i.e. the first $A(\mathcal{C})$-neighborhood of $\mathcal{U}_{g(j)}$ ("union of the super-sets of the elements of $\mathcal{U}_{g(j)}$") contains the super-set $A(E)$ of an emergency set E such that $|A(E) \cap \mathcal{U}_{g(j)}| = 1$, then, of course, we can "grow" by adding E. So for the rest we assume that Case 1 does not apply. This implies, among many other things, that there is no Type 2 secondary set in $\mathcal{C}$.

Observe that the total number of Type 1 and Type 3 sets $A(S)$ together in $A(\mathcal{C})$ is at most

$$\binom{g(j)m}{2} + \binom{m\binom{g(j)m}{2}}{2} < \frac{g^2(j)m^2}{2} + \frac{g^4(j)m^6}{8}. \tag{39.5}$$

If the first $A(\mathcal{C})$-neighborhood of $\mathcal{U}_{g(j)}$ contains a Type 4 set $A(S)$, then there exists a set $A(E)$ ("super-set of an emergency set E") in the second $A(\mathcal{C})$-neighborhood of $\mathcal{U}_{g(j)}$, and 1 of Cases 2, 3, 4 applies. Then again we can "grow." So for the rest we assume that the first $A(\mathcal{C})$-neighborhood of $\mathcal{U}_{g(j)}$ does not contain a Type 4 secondary set $A(S)$. It follows from (39.5) that the first $A(\mathcal{C})$-neighborhood of $\mathcal{U}_{g(j)}$ has less than $\binom{g(j)m}{2} + \binom{n\binom{g(j)m}{2}}{2} < \frac{g^2(j)m^2}{2} + \frac{g^4(j)m^6}{8}$ sets. Therefore, if

$$\binom{g(j)m}{2} + \binom{m\binom{g(j)m}{2}}{2} < \frac{g^2(j)m^2}{2} + \frac{g^4(j)m^6}{8} \leq 2^k, \tag{39.6}$$

then the second $A(\mathcal{C})$-neighborhood of $\mathcal{U}_{g(j)}$ is not empty. In view of Cases 5, 6 we can assume that the second $A(\mathcal{C})$-neighborhood of $\mathcal{U}_{g(j)}$ consists of Type 3 sets $A(S)$ only. Therefore, by (39.5), if inequality (39.6) holds, the third $A(\mathcal{C})$-neighborhood of $\mathcal{U}_{g(j)}$ consists of Type 3 and Type 4 sets $A(S)$ only.

Type 3 is impossible: indeed, by the "triangle inequality," a Type 3 $A(S)$ from the third neighborhood cannot have a neighbor in the first neighborhood (in fact intersecting $\mathcal{U}_{g(j)}$ in at least 2 points). So it remains Type 4, which leads to Case 8. Summarizing, if (39.6) applies, then we can still "grow" $\mathcal{G}_{g(j)}$ inside a large component $\mathcal{C}$ of all small sets with $|\mathcal{C}| \geq 2^k$. It follows that, as long as $g(j) < 2^{k/4}/m^2$, we can still "grow" $\mathcal{G}_{g(j)}$, and the proof of the Big Set Lemma is complete. $\quad\square$

7. Finishing the proof. Now we are ready to prove:

Theorem 37.5 *If $\mathcal{F}$ is an m-uniform Almost Disjoint hypergraph such that*

$$|\mathcal{F}|^{m^2 \cdot 2^{-k/4}} \cdot \max \left\{ 2^{7k/2} \cdot D, \, m \cdot 2^{5k/2} \cdot D^{1+\frac{1}{k-2}} \right\} \leq 2^m$$

holds for for some integer $k \geq 8\log_2 m$ ("binary logarithm") where $D = \text{MaxDegree}(\mathcal{F})$, then Breaker has a winning strategy in the weak game on $\mathcal{F}$.

Proof. We have to show that Breaker can block every $A \in \mathcal{F}$. By definition, an *emergency set* E is part of a *dangerous set* $A(E) \in \mathcal{F}$. Maker had to occupy $m - 2k$ points from "survivor" $A(E)$ in the course of the Big Game; and the $2k$-element part ("emergency set E") is removed from the Big Board (and added to the small board), so Breaker cannot block $A(E)$ in the rest of the Big Game.

Now assume that, at some stage of the play, there occurs a component C of the family S of all small sets with $|C| \geq 2^k$. Then, by the Big Set Lemma, we can "extract" from component C a family $G_{g(j)} \subset C$ of emergency sets with $g(j) \geq 2^{k/4}/m^2$. The *union* $U_{g(j)}$ of the super-sets of the $g(j)$ emergency sets in $G_{g(j)}$ has the following property:

(α) Maker occupies at least $(m - \frac{5k}{2})g(j)$ points of $U_{g(j)}$ in the Big Game; and
(β) Breaker could not block $U_{g(j)}$ in the Big Game ("Shutout").

To *prevent* the appearance of a $U_{g(j)}$ with $g(j) \geq 2^{k/4}/m^2$ satisfying (α), (β), we define the family B of *Big Sets* as follows:

$$B = \left\{ B = U_T : \text{for all possible ways one can } \textbf{grow } U_T \text{ of length } T = \frac{2^{k/4}}{m^2} \right.$$

$$\left. \textbf{in terms of } \mathcal{F} \text{ by using Cases } 1-8 \text{ above} \right\}.$$

The meaning of "grow" is clear now. It remains to clarify "in terms of hypergraph $\mathcal{F}$."

What we have to clarify here is the following "confusion." In Cases 1–8 above we used the concepts of "emergency set" and "secondary set," which indirectly assume the knowledge of a particular play of the game. On the other hand, we have to define the family B of *Big Sets* **in advance, before any playing begins,** when we have no idea which winning set $A \in \mathcal{F}$ will eventually become a "dead" or "dangerous" set! In the definition of B we have to control all possible candidates, we have to prepare for the worst case scenario. In other words, we must define B in terms of the original hypergraph $\mathcal{F}$, without ever referring to the concepts of "emergency" and "secondary sets" (which are meaningless at the beginning).

Luckily the *figures* in Cases 1–8 are about *super-sets* – the elements of $\mathcal{F}$ – anyway. Therefore, a Big Set in the "broad sense" means all possible union sets U_T with $T = 2^{k/4}/m^2$ built from the elements of hypergraph $\mathcal{F}$ such that each "growth" is described by one of the **pictures** of Cases 1–8. This is how we control all possible plays at the same time.

On the other hand, the "narrow" interpretaton of Big Sets is *play-specific*: it depends on the concepts of "emergency" and "secondary sets," just as we did in our case study of Cases 1–8.

Of course, a Big Set in the "narrow sense" is a Big Set in the "broad sense." The good news is that the result of the **Calculations**, carried out for the Big Sets in the "narrow sense," works without any modification for the Big Sets in the "broad sense." It follows that the *total number* $|B|$ of Big Sets in the "broad sense" is estimated from above by (see (39.3))

$$|\mathcal{B}| \leq |\mathcal{F}| \left(\max \left\{ \frac{g^4(j) \cdot m^7 \cdot D}{4}, \left(\frac{g^4(j) \cdot m^{k+5} \cdot D^{k-1}}{4(k-2)!} \right)^{1/(k-2)} \right\} \right)^{T-1} \tag{39.7}$$

$$< |\mathcal{F}| \left(\max \left\{ 2^k \cdot D, m^2 \cdot D^{1+\frac{1}{k-2}} \right\} \right)^{T-1},$$

since $T = 2^{k/4}/m^2$ (we assume that $k \geq 8\log_2 m$).

To *prevent* the appearance of any $\mathcal{U}_T$ with $T = 2^{k/4}/m^2$ satisfying (α), (β) ("Big Set in the narrow sense"), Breaker uses "Lemma 5" in Section 36 for the family of all possible "Big Sets in the broad sense", i.e. for $\mathcal{B}$, in the Big Game. The "target value" of a "forbidden" $\mathcal{U}_T$ is $2^{(m-5k/2)T}$ (see (α)).

On the other hand, the "initial value" is $|\mathcal{B}|$, i.e. the total number of Big Sets in the "broad sense." If $|\mathcal{B}| < 2^{(m-5k/2)T-1}$ with $T = 2^{k/4}/m^2$, then, by using the strategy of "Lemma 5" in Section 36, Breaker can indeed prevent the appearance of any $\mathcal{U}_T$ with $T = 2^{k/4}/m^2$ satisfying (α), (β) ("Big Set in the narrow sense") in the Big Game. In view of the Big Set Lemma, this forces the Component Condition in the small game. Then the Last Step Lemma applies, and implies Theorem 37.5.

It remains to check the inequality $|\mathcal{B}| < 2^{(m-5k/2)T-1}$ with $T = 2^{k/4}/m^2$. In view of (39.7) it suffices to check

$$|\mathcal{F}| \left(\max \left\{ 2^k \cdot D, m^2 \cdot D^{1+\frac{1}{k-2}} \right\} \right)^{T-1} \leq 2^{(m-\frac{5k}{2})T-1}. \tag{39.8}$$

Taking the $(T-1)$st roots of both sides of (39.8), we obtain the hypothesis of Theorem 37.5, and the proof is complete. $\qquad\square$

40

Breaking the "square-root barrier" (I)

In Parts A and B we proved several Weak Win results for "Ramsey type games with quadratic goal sets" (applying either Theorem 1.2 or Theorem 24.2). The quadratic goal set size was either $n = n(q) = \binom{q}{2}$, or $n = n(q) = q^2$, or some other quadratic polynomial in q (see (8.5) and (8.6)); in each case switching the value of q to $q+1$ leads to a (roughly) $n^{1/2}$ increase in the size $n = n(q)$. This explains why the following "breaking the square-root barrier" version of the Neighborhood Conjecture (Open Problem 9.1) would suffice to prove the missing Strong Draw parts of the Lattice Games (Theorems 8.2). Breaking the "square-root barrier" refers to the small error term $n^{1/2-\varepsilon}$ in the exponent of 2, see (40.1) below.

Open Problem 9.1′ *Assume that $\mathcal{F}$ is an n-uniform hypergraph, and for some fixed positive constant $\varepsilon > 0$ the Max Degree of $\mathcal{F}$ is less than*

$$2^{n-c_1 n^{1/2-\varepsilon}} \tag{40.1}$$

where $c_1 = c_1(\varepsilon)$ is an absolute constant. Is it true that playing on $\mathcal{F}$ the second player can force a Strong Draw?

Notice that Open Problem 9.1′ is between Open Problem 9.1 (c) and (d): it follows from (c), and it implies (d).

Unfortunately we cannot solve Open Problem 9.1′, not even for Almost Disjoint hypergraphs, but we can prove a weaker version involving an additional condition.

1. The third Ugly Theorem. The additional condition that we need is a weak upper bound on the global size, which is trivially satisfied in the applications. The constants 1.1 and $2/5 = .4$ below are accidental; the main point is that 1.1 is larger than 1 and 2/5 is less than 1/2 (to break the "square-root barrier").

Theorem 40.1 *If $\mathcal{F}$ is an n-uniform Almost Disjoint hypergraph such that*

$$|\mathcal{F}| \leq 2^{n^{1.1}} \text{ and } \mathrm{MaxDeg}(\mathcal{F}) \leq 2^{n-4n^{2/5}},$$

then for $n \geq c_0$ (i.e. n is sufficiently large) the second player can force a Strong Draw in the Positional Game on $\mathcal{F}$.

Theorem 40.1 will be proved by a tricky adaptation of the proof technique of Theorem 37.5 (see the last section). The proof of Theorem 40.1 is long. At the end of this section we include a detailed outline of the proof. The rest of the proof will be discussed in the next section.

What we want to explain first is why the third Ugly Theorem (Theorem 40.1) is relevant in the Lattice Games (Theorem 8.2), i.e. how to prove the missing Strong Draw parts of Theorem 8.2. Unfortunately Theorem 40.1 itself does not apply directly: the lattice-game hypergraphs are not Almost Disjoint, but there is a way out: there is a *resemblance* to Almost Disjointness, meaning that the proof technique of Theorem 40.1 can be *adapted* to complete the proof of Theorem 8.2.

We explain what "resemblance to Almost Disjointness" in Theorem 8.2, and also in Theorem 12.6, means. The key idea is to involve a "dimension argument," namely that "intersections have smaller dimension." The best way to illustrate this idea is to introduce a new simpler game where the concept of dimension comes up naturally. We study the Finite Affine and Projective Geometry as a generalization of Theorem 12.6.

2. Illustration: planes of the Finite Affine and Projective Geometry. Note that Theorem 12.6 is about the "planes" of the n^d Torus. The n^d Torus is nothing else other than the dth power $\mathbb{Z}_n \times \cdots \times \mathbb{Z}_n = (\mathbb{Z}_n)^d$ of the additive group (mod n) $\mathbb{Z}_n$. A *combinatorial plane* (*Comb-Plane* in short) is a 2-parameter set ("plane") in $(\mathbb{Z}_n)^d$. $(\mathbb{Z}_n)^d$ is a nice algebraic structure: it is an abelian group, but the richest algebraic structure is, of course, a *field*, so it is a natural idea to consider a finite field ("Galois field") $\mathbf{F}_q$, where q is a prime power. The two most natural "plane" concepts over $\mathbf{F}_q$ are the planes of the d-dimensional Affine Geometry $AG(d, \mathbf{F}_q)$ and the planes of the d-dimensional Projective Geometry $PG(d, \mathbf{F}_q)$ (see e.g. Cameron [1994]).

The d-dimensional Affine Geometry $AG(d, \mathbf{F}_q)$ has q^d points; has $q^{d-1}(q^d - 1)/(q-1)$ lines, where each line contains q points; and has $q^{d-2}(q^d - 1)(q^{d-1} - 1)/(q^2 - 1)(q - 1)$ planes, where each plane contains q^2 points.

The d-dimensional Projective Geometry $PG(d, \mathbf{F}_q)$, on the other hand, has $(q^{d+1} - 1)/(q-1)$ points; has $(q^{d+1} - 1)(q^d - 1)/(q^2 - 1)(q-1)$ lines, where each line contains $q+1$ points; has $(q^{d+1} - 1)(q^d - 1)(q^{d-1} - 1)/(q^3 - 1)(q^2 - 1)(q-1)$ planes, where each plane contains $q^2 + q + 1$ points.

If the winning sets are the "affine planes" or the "projective planes," then we get two different positional games, and the corresponding Achievement and Avoidance Numbers are denoted by $\mathbf{A}(d\text{–dim. affine; plane})$, $\mathbf{A}(d\text{–dim. affine; plane}; -)$, and in the projective case $\mathbf{A}(d\text{–dim. projective; plane})$, $\mathbf{A}(d\text{–dim. projective; }plane; -)$; we are looking for the largest prime power q such that playing in the d-dimensional Affine (resp. Projective) Geometry Maker can occupy a whole affine (resp. projective) plane – in the Reverse version Avoider is forced to occupy a whole plane. Because it is not true that every positive integer is a prime power, the following result is not as elegant as Theorem 12.6.

Theorem 40.2 *The affine plane Achievement Number* **A**(d–dim. affine; plane) *is the largest prime power* $q = q(d)$ *such that*

$$\frac{q+o(1)}{\sqrt{2\log_2 q}} \le \sqrt{d},$$

and the same for the Avoidance Number **A**(d–dim. affine; plane; –).

The projective plane Achievement Number differs by an additive constant 1/2 in the numerator: **A**(d–dim. projective; plane) *is the largest prime power* $q = q(d)$ *such that*

$$\frac{q+1/2+o(1)}{\sqrt{2\log_2 q}} \le \sqrt{d},$$

and the same for the Avoidance Number **A**(d–dim. projective; plane;–).

The affine plane has q^2 points and the projective plane has $q^2+q+1 = (q+1/2)^2 + O(1)$ points; this extra 1/2 explains the extra 1/2 in the second part of Theorem 40.2.

There are infinitely many values of dimension d when the equations

$$\max_q : \frac{q+o(1)}{\sqrt{2\log_2 q}} \le \sqrt{d} \quad \text{and} \quad \max_q : \frac{q+1/2+o(1)}{\sqrt{2\log_2 q}} \le \sqrt{d}$$

have *different* prime power solutions q, distinguishing the *affine* case from the *projective* case (see Theorem 40.2).

Notice that the "planes" over a finite field do *not* have the "2-dimensional arithmetic progression" geometric structure any more. We included Theorem 40.2 mainly for the instructional benefits that its proof is simpler than that of Theorem 8.2 or 12.6. We begin with the easy:

3. Weak Win part of Theorem 40.2.
Case 1: d-dimensional Affine Geometry $AG(d, \mathbf{F}_q)$
The family of all affine planes in $AG(d, \mathbf{F}_q)$ is denoted by $\mathcal{F}(q, d)$; it is a q^2-uniform hypergraph, and its board size is $|V| = q^d$. Assume that

$$\frac{T(\mathcal{F}(q, d))}{|V|} \approx \frac{q^{3d-6}2^{-q^2}}{q^d} = q^{2d-6}2^{-q^2} > 2p. \tag{40.2}$$

To apply the advanced criterion Theorem 24.2 we have to estimate $T((\mathcal{F}(q, d))^p_2)$ from above. Fix two points P and Q; then a third point R and the fixed P, Q together determine an affine plane if R lies outside of the PQ-line; this gives $(q^d - q)/(q^2 - q) = (q^{d-1} - 1)/(q - 1)$ for the number of affine planes in $AG(d, \mathbf{F}_q)$ containing both P and Q. This gives the exact value

$$T((\mathcal{F}(q,d))_2^p) = \binom{q^d}{2}\binom{\frac{q^{d-1}-1}{q-1}}{p}2^{-pq^2+(p-1)q},$$

where the exponent $-(pq^2-(p-1)q)$ of 2 is explained by the fact that the p affine planes, all containing the PQ-line, are in fact disjoint apart from the PQ-line ("q points"). Therefore

$$\left(T((\mathcal{F}(q,d))_2^p)\right)^{1/p} \le q^{2d/p}\cdot 2q^{d-2}\cdot 2^{-q^2+(1-1/p)q}.$$

It follows that Theorem 24.2 applies with $p=3$ if (40.2) holds, completing the Weak Win part of the affine case.

Case 2: d-dimensional Projective Geometry $PG(d,\mathbf{F}_q)$.
The only minor difference is that the plane-size is q^2+q+1 instead of q^2, the rest of the calculations are almost identical. Again Theorem 24.2 applies with $p=3$.

$\square$

Now let's go back to Strong Draw. In Theorem 40.2 the winning sets are projective (or affine) planes, and the intersection of two planes is either a line or a point (always a lower dimensional sub-space). Two lines intersect in a point (or don't intersect at all). That is, by "repeated intersection" we go down from planes to points in two steps. This is just a little bit more complicated than the "one step way" of Almost Disjoint hypergraphs (where two hyperedges either intersect in a point or don't intersect at all). This is what we mean by a "resemblance" to Almost Disjointness.

The same argument applies for the *combinatorial planes* ("2-parameter sets") in Theorem 12.6. Theorem 8.2 ("Lattice Games") is a different case: for the $q\times q$ lattices in Theorem 8.2 – regarded as "planes" – the principle of "intersections have smaller dimension" is clearly violated. Indeed, two different $q\times q$ lattices may have a *non-collinear* intersection. But there is an escape: the good news is that any set of 3 non-collinear points of a $q\times q$ parallelogram lattice "nearly" determines the lattice (3-determinedness). Indeed, there are at most $\binom{q^2}{3}$ $q\times q$ parallelogram lattices containing a fixed set of 3 non-collinear points of the $N\times N$ board. In Theorem 8.2 $\binom{q^2}{3}$ is "very small" (a polylogarithmic function of N); thus we can (roughly) say that "$\binom{q^2}{3}$ is (nearly) as good as 1", where "1" represents the ideal case of planes (since planes intersect in lines or points).

We are aware that this argument is far too vague (though this *is* the basic idea), but at this point we just wanted to convince the reader that Theorem 40.1 is the critical new step, and the rest is just a long but routine adaptation. We conclude this section with an:

4. Outline of the proof of Theorem 40.1. First we explain why Theorem 37.5 (second Ugly Theorem) cannot directly prove Theorem 40.1 (i.e. why we need a new idea). In Theorem 37.5 there is an integral parameter k ($k\ge 8\log_2 n$) that can be freely chosen, but the hypothesis

$$2^{5k/2}D^{1+\frac{1}{k-2}} \le 2^n,$$

where $D = \text{MaxDeg}(\mathcal{F})$, puts a strong limitation on the Max Degree of the n-uniform hypergraph $\mathcal{F}$: D cannot go beyond $2^{n-\sqrt{n}}$ whatever way we choose parameter k (the optimal choice is around $k = \sqrt{n}$). The term $\frac{1}{k-2}$ in the power of D in the inequality above comes from the "centipedes" (see "Type 4" and formulas (39.3), (39.7)–(39.8)); we refer to this as the "centipede problem."

To replace the bound

$$D \le 2^{n-\sqrt{n}} \quad \text{with} \quad D \le 2^{n-cn^{1/2-\varepsilon}},$$

where $\varepsilon > 0$ is a fixed positive constant – i.e. "breaking the square-root barrier," which is the key point in Theorem 40.1 – requires a new idea. Formula (34.7) in Corollary 1 of Theorem 34.1 explains why Theorem 34.1 doesn't help either. We abandon Theorem 34.1, and try to modify the proof of Theorem 37.5.

A cure for the "centipede problem": hazardous sets. What is exactly the "centipede problem"? Well, the *secondary sets* may have very few marks of Maker, representing negligible "target value" – this explains why the proof technique developed in Section 39 is insufficient to break the "square-root barrier." We have to completely change the definition of *secondary sets* – to emphasize the change we introduce a new term: **hazardous sets**, which will play a *somewhat* similar role to that of the *secondary sets* in Section 39. The new notion of *hazardous sets* will lead to a new definition of the *small board*, and the *small game* in general.

There is only a minor change in the way how the *first emergency set* arises: when Maker (i.e. the first player) occupies the $(n-k)$th point of some *survivor* winning set $A \in \mathcal{F}$ for the first time (*survivor* means "Breaker-free" where Breaker is the second player); in that instant A becomes *dangerous*, and its k-element "blank" part is called an *emergency set*. Note in advance that parameter k will be specified as $k = n^{2/5}$. The k-element emergency set is removed from the *Big Board* (which was the whole board before) and added to the *small board* (which was empty before).

Note that we may have several dangerous sets arising at the same time; for each one the k-element "blank" part is a new emergency set, which is removed from the Big Board and added to the small board.

Since the Big Board is strictly smaller now than it was at the beginning, we may have an untouched survivor $A \in \mathcal{F}$ with $|A(Big; blank)| \le k$; in fact, we may have several of them: $A_i \in \mathcal{F}$, where $i \in I$. If $|A_i(small)| \le k$, then $A_i^* = A_i \setminus A_i(Big; Maker)$ is a new emergency set; if $|A_i(small)| \ge k+1$, then $A_i^* = A_i \setminus A_i(Big; Maker)$ is a *hazardous set*. In both cases A_i^* is removed from the Big Board and added to the small board; we do this for all $i \in I$.

At the end of this the Big Board becomes smaller, we may have an untouched survivor $A \in \mathcal{F}$ with $|A(Big; blank)| \le k$; in fact, we may have several of them:

$A_i \in \mathcal{F}$ where $i \in I$. If $|A_i(small)| \le k$ then $A_i^* = A_i \setminus A_i(Big; Maker)$ is a new emergency set; if $|A_i(small)| \ge k+1$, then $A_i^* = A_i \setminus A_i(Big; Maker)$ is a *hazardous set*. In both cases A_i^* is removed from the Big Board and added to the small board; we do this for all $i \in I$.

Repeating the previous argument, this *extension process* – which we call the "first extension process" – will eventually terminate, meaning that the inequality $|A(Big; blank)| \ge k+1$ holds for every untouched survivor $A \in \mathcal{F}$.

It may happen, however, that the inequality $k+1 \le |A(Big; blank)| \le |A(small)|$ holds for a family $A = A_i$, $i \in I$ of untouched survivors. Then again $A_i^* = A_i \setminus A_i(Big; Maker)$ is a new *hazardous set*, which is removed from the Big Board and added to the small board.

At the end of this the Big Board is decreased, so we may have an untouched survivor $A \in \mathcal{F}$ with $k+1 \le |A(Big; blank)| \le |A(small)|$. Then again $A^* = A \setminus A(Big; Maker)$ is a new *hazardous set* which is removed from the Big Board and added to the small board, and so on. Repeating this "second extension process," it will eventually terminate, meaning that the inequality $k+1 \le |A(Big; blank)| \le |A(small)|$ *fails* to hold for any untouched survivor $A \in \mathcal{F}$.

At the end of this there may exist an untouched survivor $A \in \mathcal{F}$ with $|A(Big; blank)| \le k$; in fact, we may have several of them: $A_i \in \mathcal{F}$, where $i \in I$. Applying again the "first extension process," and again the "second extension process," and again the "first extension process," and again the "second extension process," and so on, we eventually reach a stage where this process of alternating the "first extension" and the "second extension" terminates, meaning that *both* properties

$$|A(Big; blank)| \ge k+1; \tag{40.3}$$

$$|A(Big; blank)| > |A(small)| \tag{40.4}$$

hold for every untouched survivor $A \in \mathcal{F}$ – we refer to this as the *Closure Process*.

Up to this point the *small sets* are exactly the *emergency sets* and the *hazardous sets*; at a later stage, however, the small sets will *differ* from the emergency and hazardous sets as follows. Assume that, at a later stage the double requirement (40.3)–(40.4) is violated (due to the fact that the Big Board is shrinking and the small board is growing); it can be violated (in fact, "violated first") in two different ways:

$$\text{either} \quad k = |A(Big; blank)| > |A(small)| \quad \text{occurs}, \tag{40.5}$$

$$\text{or} \quad |A(Big; blank)| = |A(small)| \ge k+1 \quad \text{occurs}. \tag{40.6}$$

Case (40.5) will produce new emergency sets and case (40.6) will produce new hazardous sets as follows.

If (40.5) holds, then

$$S = S(A) = A(blank) = A(Big; blank) \cup A(small; blank) \qquad (40.7)$$

is a new *small set*, which is (as usual) removed from the Big Board and added to the small board. The (possibly) larger set

$$E = E(A) = A \setminus A(Big; Maker) = A(Big; blank) \cup A(small) \qquad (40.8)$$

is a new *emergency set*, and we refer to the small set $S = S(A)$ in (40.7) as a *small-em set* – meaning the "small-set part of emergency set $E = E(A)$ in (40.8)"; clearly $S \subseteq E$.

Next assume that (40.6) holds; then

$$S = S(A) = A(blank) = A(Big; blank) \cup A(small; blank) \qquad (40.9)$$

is a new *small set*, which is (as usual) removed from the Big Board and added to the small board. The (possibly) larger set

$$H = A \setminus A(Big; Maker) = A(Big; blank) \cup A(small) \qquad (40.10)$$

is a new *hazardous set*, and we refer to the small set $S = S(A)$ in (40.9) as a *small-haz set* – meaning the "small-set part of hazardous set $H = H(A)$ in (40.10)"; clearly $S \subseteq H$.

In other words, from now on the new *small sets* are either *small-em sets* ("blank parts of emergency sets") or *small-haz sets* ("blank parts of hazardous sets").

Notice that we returned to the *Same Board Principle* of Sections 35–36 – which was (possibly) violated in Section 39 – meaning that if Maker's last move was in the Big Board (resp. small board), then Breaker always replies in the Big Board (resp. small board).

Another new notion: the union of the *small-em sets* is called the **Emergency Room**, or simply the **E.R.**

Of course, we may have several violators of (40.3)–(40.4) arising at the first time: we repeat the previous argument for each one. Then again we may have several violators of (40.3)–(40.4): we repeat the previous argument for each; and so on. In other words, we apply a straightforward *Closure Process* – just like before – which terminates at a stage where again the double requirement (40.3)–(40.4) holds for every untouched survivor $A \in \mathcal{F}$. At a later stage the double requirement (40.3)–(40.4) may be violated; it can be violated (in fact, "violated first") in two different ways: (40.5) and (40.6), and so on.

This completes our description of the necessary changes in the definition of the small game.

5. How to solve the "centipede problem"? The main idea is the following lemma, which plays a key role in the proof of Theorem 40.1.

Hazardous Lemma: *Under the condition of Theorem 40.1, Breaker can force that, during the whole course of the play, every hazardous set H intersects the E.R. ("Emergency Room") in at least $|H|/4$ points.*

Remark. In view of (40.6), the Hazardous Lemma holds *trivially* for the *first* hazardous set (in fact holds with $|H|/2$ instead of $|H|/4$), but there is no obvious reason why the Hazardous Lemma should remain true at the later stages of the play (unless, of course, Breaker *forces* it by playing rationally).

Proof of the Hazardous Lemma (outline). To understand what Breaker has to prevent, assume that there is a "violator." Let H be the first violator of the Hazardous Lemma: H is the first hazardous set which intersects the E.R. in less than $|H|/4$ points. Since $|H| \geq 2k+2$ (see (40.6)), H must intersect the difference-set "smallboard minus E.R." in $\geq k/2$ points; this and Almost Disjointness together imply the existence of $k/2$ non-violator hazardous sets $H_1, H_2, \ldots, H_{k/2}$ such that $|H \cap H_j| = 1$, $1 \leq j \leq k/2$. Since each H_j is a non-violator of the Hazardous Lemma, again by Almost Disjointness each H_j must intersect at least $|H_j|/4 = l_j$ emergency sets $E_{j,1}, E_{j,2}, \ldots, E_{j,l_j}$ $(j = 1, 2, \ldots, k/2)$. Formally we have $1 + (k/2) + \sum_{j=1}^{k/2} l_j$ super-sets

$$A(H), A(H_1), \ldots, A(H_{k/2}), A(E_{1,1}), \ldots, A(E_{1,l_1}), A(E_{2,1}), \ldots, A(E_{2,l_2}),$$
$$A(E_{3,1}), \ldots, A(E_{3,l_3}), \ldots, A(E_{k/2,1}), \ldots, A(E_{k/2,l_{k/2}}), \qquad (40.11)$$

but there may occur some *coincidence* among these hyperedges of $\mathcal{F}$, or at least some "extra intersection" beyond the mandatory one-point intersections ("Almost Disjointness"). In (40.11) $A(\ldots)$ denotes the uniquely determined super-sets (uniquely determined because of Almost Disjointness).

The first case is when we have neither *coincidence* nor "extra intersection."

The Simplest Case: There is no coincidence among the $1 + (k/2) + \sum_{j=1}^{k/2} l_j$ super-sets in (40.11), and also there is no "extra intersection" beyond the mandatory 1-point intersections, except (possibly) the $k/2$ sets $A(H_1), \ldots, A(H_{k/2})$, which may intersect each other (this "minor overlapping" has a negligible effect anyway). Breaker wants to prevent the appearance of any "tree like configuration" of the *Simplest Case* (see figure below).

Breaker uses the Power-of-Two Scoring System ("Erdős–Selfridge technique"): the target value of a "tree like configuration" is at least

$$2^{\sum_{j=1}^{k/2}(n-|H_j|-j+(n-2k)|H_j|/4)}.$$

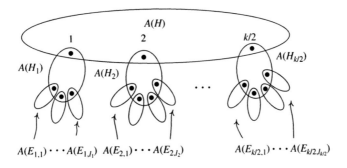

$$l_j = |H_j|/4, j = 1, \ldots, k/2$$

How many "tree like configurations" are there? Well, there are $|\mathcal{F}|$ ways to choose $A(H)$ ("the root"), at most $(nD)^{k/2}$ ways to choose $A(H_1), \ldots, A(H_{k/2})$ ("first neighborhood of the root"), and at most

$$(nD)^{(|H_1|+|H_2|+\ldots+|H_{k/2}|)/4}$$

ways to choose

$$A(E_{1,1}), \ldots, A(E_{1,l_1}), \ldots, A(E_{k/2,1}), \ldots, A(E_{k/2,l_{k/2}})$$

("second neighborhood of the root"), where $l_j = |H_j|/4$, $1 \le j \le k/2$. The Erdős–Selfridge technique works if

$$2^{\sum_{j=1}^{k/2}(n-|H_j|-j+(n-2k)|H_j|/4)} > |\mathcal{F}| \cdot (nD)^{(k/2)+(|H_1|+|H_2|+\ldots+|H_{k/2}|)/4}, \qquad (40.12)$$

where $D = \mathrm{MaxDeg}(\mathcal{F})$. Inequality (40.12) is equivalent to

$$\prod_{j=1}^{k/2} \left(\left(\frac{2^{n-2k}}{nD} \right)^{|H_j|/4} \cdot \frac{2^{n-|H_j|-j}}{nD} \right) > |\mathcal{F}|. \qquad (40.13)$$

(40.13) follows from the stronger inequality

$$\prod_{j=1}^{k/2} \left(\frac{2^{n-2k-4}}{nD} \right)^{1+|H_j|/4} > |\mathcal{F}|. \qquad (40.14)$$

By hypothesis

$$|\mathcal{F}| < 2^{n^{1.1}} \quad \text{and} \quad D < 2^{n-4n^{2/5}},$$

and also $|H_j|/4 \ge k/2 \approx n^{2/5}/2$, so (40.14) follows from the trivial numerical fact $3 \cdot 2/5 = 6/5 > 1.1$. This proves that Breaker can indeed prevent the appearance of a *first violator* of the Hazardous Lemma where the corresponding "tree like configuration" belongs to the *Simplest Case*.

Now it is easy to see that the *Simplest Case* is in fact the "worst case scenario," the rest of the cases are actually easier to handle. Indeed, if there is a (say) "extra intersection," then a *bond set* arises (see Section 39), which contains $(n-2k)$ marks of Maker in the Big Game: a huge "target value" of 2^{n-2k}.

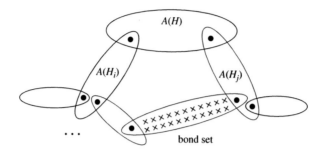

Every *bond set* represents, therefore, a "huge gain" for Breaker, explaining why the *Simplest Case* ("when there is no bond set") is the "worst case scenario." This completes our outline of the proof of the Hazardous Lemma.

But we didn't address the main question yet: "Why is the Hazardous Lemma so important?" It is important because it greatly improves the calculations in "Type 4" (see Section 39). Indeed, if H is a hazardous set, then by the Hazardous Lemma there exist $|H|/4 = l \geq k/2 \approx n^{2/5}/2$ emergency sets E_1, E_2, ..., E_l such that $A(H) \cap A(E_i)$, $i = 1, 2, \ldots, l$ form l distinct points (see figure).

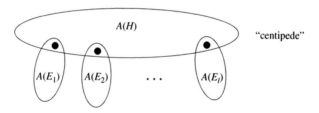

We can assume that $A(E_1)$, $A(E_2)$, ..., $A(E_l)$ are pairwise disjoint; indeed, otherwise a *bond set* arises, which means a "huge gain" for Breaker.

A *typical* appearance of "Type 4" is in "Case 4" (see Section 39); then the number of ways we can "grow" to $\mathcal{G}_{g(j+1)}$ from $\mathcal{G}_{g(j)}$ is bounded from above by

$$|\mathcal{U}_{g(j)}| \cdot D \cdot (nD)^l.$$

If in the "extraction procedure" (see Section 39) we **include** the "hazardous super-set" $A(H)$ and add it to the "emergency super-sets" $A(E_1)$, $A(E_2)$, ..., $A(E_l)$ – this *inclusion* is a key technical novelty of the proof! – then the "target value" of this configuration is ("Power-of-Two Scoring System")

$$2^{n-|H|+l(n-2k)} = 2^{n-|S|+(n-2k)|H|/4};$$

we assume the "worst case scenario": $A(E_1)$, $A(E_2)$, ..., $A(E_l)$ are pairwise disjoint. The following inequality certainly holds with $k = n^{2/5}$

$$\frac{|\mathcal{U}_{g(j)}| \cdot D \cdot (nD)^l}{2^{n-l+l(n-2k)}} \leq |\mathcal{U}_{g(j)}| \left(\frac{nD}{2^{n-2k-2}}\right)^{l+1} \leq 2^{-kl/2} \qquad (40.15)$$

$$= \text{extremely small if } |\mathcal{U}_{g(j)}| \leq 2^{\frac{n^{4/5}}{5}}.$$

Inequality (40.15) is more than enough to guarantee the truth of the "Big Set Lemma" (see Section 39) via the "Calculations" (see Section 39); this is how we take care of the "centipede problem" and manage to "break the square-root barrier."

   This completes the outline of the proof of Theorem 40.1 (heavily using of course the technique developed in Section 39).

41

Breaking the "square-root barrier" (II)

Recall Theorem 40.1: *If $\mathcal{F}$ is an n-uniform Almost Disjoint hypergraph such that*

$$|\mathcal{F}| \le 2^{n^{1.1}} \text{ and } \text{MaxDeg}(\mathcal{F}) \le 2^{n-4n^{2/5}},$$

then for $n \ge c_0$ ("n is sufficiently large") the second player can force a Strong Draw in the Positional Game on $\mathcal{F}$.

In Section 40 we looked at an outline of the proof; here we convert the outline into a precise argument. The reader needs to be warned that the details are rather tiresome. The key step is to complete the proof of the Hazardous Lemma. Recall the Hazardous Lemma: *Under the condition of Theorem 40.1, Breaker can force that, during the whole course of the play, every hazardous set H intersects the E.R. ("Emergency Room") in at least $|H|/4$ points.*

We assume that the reader is familiar with Sections 39–40.

1. How to reduce the proof of the Hazardous Lemma to the Simplest Case?
The Hazardous Lemma is already verified in the so-called *Simplest Case*, and the idea is to reduce the general case to the *Simplest Case*. We assume that Breaker is the second player. Again let H be the *first violator* of the Hazardous Lemma: H is the first hazardous set that intersects the E.R. in less than $|H|/4$ points. Since $|H| \ge 2k+2$ (see (40.6)), H must intersect the difference-set "smallboard minus E.R." in $\ge k/2$ points; this and Almost Disjointness together imply the existence of $k/2$ non-violator hazardous sets $H_1, H_2, \ldots, H_{k/2}$ such that $|H \cap H_j| = 1$, $1 \le j \le k/2$. Since each H_j is a non-violator of the Hazardous Lemma, again by Almost Disjointness each H_j must intersect at least $|H_j|/4 = l_j$ emergency sets $E_{j,1}$, $E_{j,2}, \ldots, E_{j,l_j}$ $(j = 1, 2, \ldots, k/2)$. Formally we have $1 + (k/2) + \sum_{j=1}^{k/2} l_j$ super-sets

$$A(H), A(H_1), \ldots, A(H_{k/2}), A(E_{1,1}), \ldots, A(E_{1,l_1}), A(E_{2,1}), \ldots, A(E_{2,l_2}),$$

$$A(E_{3,1}), \ldots, A(E_{3,l_3}), \ldots, A(E_{k/2,1}), \ldots, A(E_{k/2,l_{k/2}}), \tag{41.1}$$

but there may occur some *coincidence* among these hyperedges of $\mathcal{F}$, or at least some "extra intersection" beyond the mandatory one-point intersections

("Almost Disjointness"). In (41.1) $A(\ldots)$ denotes the uniquely determined super-sets (uniquely determined because of Almost Disjointness).

The set $A(H_j)$ together with the other sets $A(E_{j,1})$, $A(E_{j,2})$, ..., $A(E_{j,l_j})$ is called the jth "centipede"; $A(H_j)$ is the "head" and $A(E_{j,1})$, $A(E_{j,2})$, ..., $A(E_{j,l_j})$ are the "legs."

Statement 41.1 *For every i Breaker can prevent the appearance of $M > n^{4/5}$ pairwise disjoint sets $A(E_{i,j_1})$, $A(E_{i,j_2})$, ..., $A(E_{i,j_M})$ in the ith "centipede."*

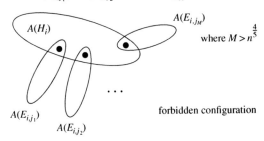

where $M > n^{\frac{4}{5}}$

forbidden configuration

Proof. Let M denote the maximum number of pairwise disjoint sets among the "legs." Breaker uses the Power-of-Two Scoring System; the target value of a "forbidden configuration" is at least $2^{(n-2k)M}$, and the number of ways to form a "forbidden configuration" is at most $|\mathcal{F}|(nD)^M$. Since $k = n^{2/5}$ and $M > n^{4/5}$, the "total danger"

$$|\mathcal{F}|(nD)^M 2^{-(n-2k)M} = |\mathcal{F}|\left(\frac{nD}{2^{n-2k}}\right)^M \leq 2^{n^{1.1}-n^{(2+4)/5}} = 2^{n^{1.1}-n^{1.2}} \qquad (41.2)$$

is "extremely small," proving that Breaker can indeed prevent the appearance of any "forbidden configuration" described in Statement 41.1. □

Statement 41.2 *Consider the ith "centipede" (i is arbitrary)*

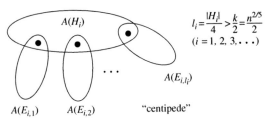

$$l_i = \frac{|H_i|}{4} > \frac{k}{2} = \frac{n^{2/5}}{2}$$
$$(i = 1, 2, 3, \ldots)$$

"centipede"

Breaker can force that, among the l_i super-sets $A(E_{i,1})$, $A(E_{i,2})$, ..., $A(E_{i,l_i})$ at least half (namely $\geq l_i/2$) are pairwise disjoint.

Proof. Let M denote the maximum number of pairwise disjoint sets among $A(E_{i,1})$, $A(E_{i,2})$, ..., $A(E_{i,l_i})$, and assume that $M < l_i/2$. For notational simplicity assume that $A(E_{i,j})$, $1 \leq j \leq M$ are pairwise disjoint, and each $A(E_{i,\nu})$ with $M < \nu \leq l_i$

intersects the union set $\bigcup_{1 \le j \le M} A(E_{i,j})$.

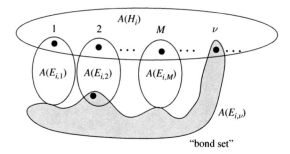

The $(l_i - M)$ sets $A(E_{i,\nu})$ with $M < \nu \le l_i$ are all *bond sets*. If Breaker uses the Power-of-Two Scoring System, the "total danger" is estimated from above by

$$|\mathcal{F}| \cdot (nD)^M 2^{-(n-2k)M} \cdot \prod_{\nu=1}^{l_i-M} \binom{(M+1)n}{2} 2^{-(n-2k-M-\nu)}.$$

Here the first part $|\mathcal{F}| \cdot (nD)^M 2^{-(n-2k)M}$ of the long product is the contribution of $A(H_i)$ and $A(E_{i,1}), A(E_{i,2}), \ldots, A(E_{i,M})$, and the rest of the long product is the contribution of $A(E_{i,M+\nu})$, $\nu = 1, 2, \ldots$.

Since $l_i - M \ge k/4$, the "danger" above is at most

$$\le |\mathcal{F}| 2^{-\frac{n}{2} \cdot \frac{k}{4}} \le 2^{n^{1.1} - n^{7/5}/8}, \tag{41.3}$$

which is "extremely small," proving that Breaker can indeed prevent the appearance of any "forbidden configuration" described in Statement 41.2. In the last step we used that $M = o(n)$, which follows from Statement 41.1. $\qquad\square$

Combining Statements 41.1-41.2 we obtain

$$\max_i |H_i| \le 4n^{4/5}. \tag{41.4}$$

The next statement excludes the possibility that too many hazardous super-sets $A(H_i)$ ("heads") share the same emergency super-set $A(E)$ ("leg").

Statement 41.3 *If Breaker plays rationally, then he can prevent that some $A(E)$ intersects $r \ge n^{1/5}$ distinct sets $A(H_i)$.*

Proof. The "total danger" of the "forbidden configurations" is at most

$$|\mathcal{F}|^2 \cdot \left(\binom{2n}{2} 2^{-(n-4n^{4/5}-n^{1/5})} \right)^{n^{1/5}} \le 2^{2n^{1.1} - n^{6/5}/2}, \tag{41.5}$$

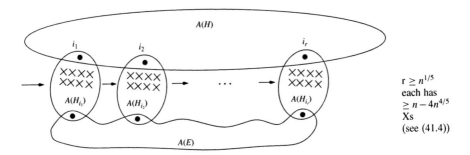

which is "extremely small." Note that the factor $|\mathcal{F}|^2$ on the left side of (41.5) bounds the number of ways to choose $A(H)$ and $A(E)$, and the rest of the left side is the contribution of the bond sets $A(H_{i_1})$, $A(H_{i_2})$, ..., $A(H_{i_{n^{1/5}}})$.

Notice that the term "$-n^{1/5}$" in $2^{-(n-4n^{4/5}-n^{1/5})}$ above comes from the fact that the sets $A(H_{i_1})$, $A(H_{i_2})$, ..., $A(H_{i_{n^{1/5}}})$ may intersect. □

Statement 41.4 *If Breaker plays rationally, then he can prevent the appearance of any forbidden configuration described by the picture*

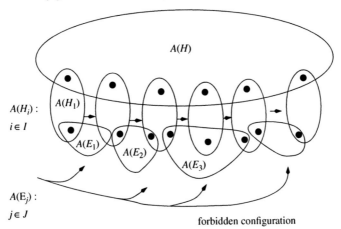

forbidden configuration

where the requirements are listed as follows:

(1) by dropping $A(H)$, the rest is still connected;
(2) $|I| \geq 2n^{1/5}$;
(3) every $A(E_j)$ intersects at least two $A(H_i)$, $i \in I$;
(4) every $A(E_j)$ intersects at least one new $A(H_i)$, $i \in I$.

Proof. In view of Statement 41.3 we can assume that $2n^{1/5} \leq |I| \leq 3n^{1/5}$, and want to show that the "danger" is "extremely small." For notational simplicity assume $J = \{1, 2, \ldots, J\}$; then the "danger" is less than the following very long product.

The first part is

$$|\mathcal{F}| \cdot (nD)2^{-(n-4n^{4/5})} \cdot (nD)2^{-(n-2k)} \cdot \left(\binom{2n}{2}2^{-(n-4n^{4/5}-|I|)}\right)^{a_1-1}$$

(explanation for the first part: here $|\mathcal{F}|$ is the number of ways to choose $A(H)$, the middle $(nD)2^{-(n-4n^{4/5})} \cdot (nD)2^{-(n-2k)}$ is the contribution of $A(H_1)$ and $A(E_1)$, a_1 is the number of neighbors of $A(E_1)$ among $A(H_i)$, $i \in I$, where "neighbor" of course means "intersecting," and the "-1" in $a_1 - 1$ is clear since $A(H_1)$ was already counted); the second part of the very long product is

$$\cdot (n^2 D)2^{-(n-2k-|I|)} \cdot \left(\binom{2n}{2}2^{-(n-4n^{4/5}-|I|)}\right)^{a_2}$$

(explanation: here $(n^2 D)$ bounds the number of ways to choose $A(E_2)$); and the rest of the very long product goes similarly as follows

$$\cdot (n^2 D)2^{-(n-2k-|I|)} \cdot \left(\binom{2n}{2}2^{-(n-4n^{4/5}-|I|)}\right)^{a_3} \cdots$$

where a_j $(j = 1, 2, \ldots, J)$ is the number of *new neighbors* of $A(E_j)$ among $A(H_i)$, $i \in I$ ("neighbor" of course means "intersecting"). Note that the term $-|I|$ in $2^{-(n-4n^{4/5}-|I|)}$ above comes from the fact that the sets $A(H_i)$, $i \in I$ may intersect.

Since $a_1 + a_2 + \cdots + a_J = |I| \geq 2n^{1/5}$, the very long product above is less than

$$\leq 2^{n^{1.1} - \frac{n}{2}(2n^{1/5}-1)}, \tag{41.6}$$

which is "extremely small." $\qquad\square$

Statement 41.4 was about the maximum size of a **component** of $A(H_i)$s (a **component** arises when $A(H)$ is dropped); the next result is about *all components*, each containing at least two $A(H_i)$s.

Statement 41.5 *If Breaker plays rationally, then he can prevent the appearance of any configuration described by the first picture on the next page,*
where there are j "components" with $t_1 + t_2 + \cdots + t_j \geq 4n^{1/5}$, where each $t_i \geq 2$ (i.e. each "component" has at least two $A(H_i)$s).

Proof. In view of Statement 41.4 we can assume that $4n^{1/5} \leq t_1 + t_2 + \cdots + t_j \leq 6n^{1/5}$, and want to show that the "danger" is "extremely small." From the proof of Statement 41.4 we can easily see that the "total danger" of the configurations described in Statement 41.5 is less than

$$|\mathcal{F}| \cdot 2^{-\frac{n}{2}((t_1-1)+(t_2-1)+\cdots+(t_j-1))}.$$

Since each $t_i \geq 2$, we have $(t_1 - 1) + (t_2 - 1) + \cdots + (t_j - 1) \geq (t_1 + t_2 + \cdots + t_j)/2 \geq 2n^{1/5}$, implying

$$|\mathcal{F}| \cdot 2^{-\frac{n}{2}((t_1-1)+(t_2-1)+\cdots+(t_j-1))} \leq 2^{n^{1.1} - \frac{n}{2} \cdot 2n^{1/5}}, \tag{41.7}$$

which is "extremely small," proving Statement 41.5. □

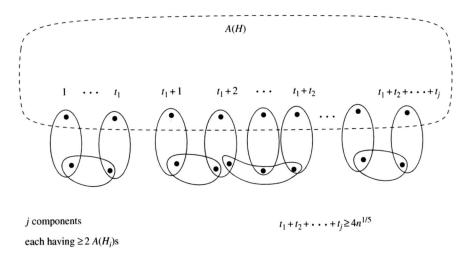

j components

each having $\geq 2 \, A(H_i)$s

$t_1 + t_2 + \cdots + t_j \geq 4n^{1/5}$

Combining Statements 41.2 and 41.5, and also inequality (41.4), we obtain the following:

Statement 41.6 *If Breaker plays rationally, then he can guarantee the following: if there is a violator of the Hazardous Lemma, then the* **first violator** *$A(H)$ generates a "tree like configuration"*

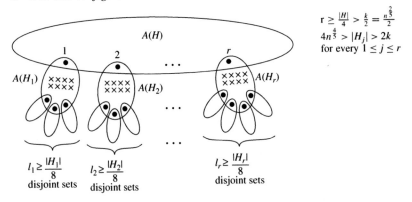

$r \geq \dfrac{|H|}{4} > \dfrac{k}{2} = \dfrac{n^{\frac{2}{5}}}{2}$

$4n^{\frac{4}{5}} > |H_j| > 2k$

for every $1 \leq j \leq r$

$l_1 \geq \dfrac{|H_1|}{8}$

disjoint sets

$l_2 \geq \dfrac{|H_2|}{8}$

disjoint sets

$l_r \geq \dfrac{|H_r|}{8}$

disjoint sets

where the $l_1 + l_2 + \cdots + l_r$ emergency super-sets are distinct (though they may have "extra intersections" among each other).

Next we study the possible "extra intersections" of these $l_1 + l_2 + \cdots + l_r$ emergency super-sets. *Question:* Given an emergency super-set, how many other emergency super-sets can intersect it? The following statement gives an upper bound:

Statement 41.7 *Consider the "tree like configuration" in Statement 41.6: let $A(E)$ be an arbitrary emergency super-set – the neighbor of (say) $A(H_{i_0})$ – and assume* that $A(E)$ has $t = y_1 + y_2 + \ldots + y_j$ "extra intersections" with other emergency super-sets. If Breaker plays rationally, then he can guarantee the upper bound $t = y_1 + y_2 + \ldots + y_j < n^{1/5}$ uniformly for all $A(E)$.

Proof. If $A(H)$, $A(H_{i_0})$, $A(H_{i_1})$, $\ldots A(H_{i_j})$, $A(E)$ are fixed, then the picture below shows $t = y_1 + y_2 + \ldots + y_j$ bond sets

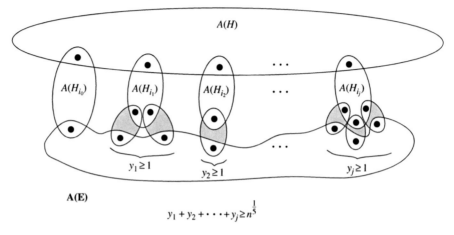

$$y_1 + y_2 + \cdots + y_j \geq n^{\frac{1}{5}}$$

where each $y_\nu \geq 1$, $1 \leq \nu \leq j$.

If the statement fails, then without loss of generality we can assume equality $y_1 + y_2 + \ldots + y_j = n^{1/5}$. We can estimate the "total danger" from above as follows

$$|\mathcal{F}| \cdot (nD)^{j+1} 2^{-(n - 4n^{4/5} - j - 1)(j+1)} \cdot (nD) 2^{-(n-2k)}$$

$$\cdot \left(\binom{(j+3)n}{2} 2^{-(n - 2k - n^{1/5})} \right)^{n^{1/5}} \leq |\mathcal{F}| \cdot 2^{-\frac{n}{2} \cdot n^{1/5}}, \qquad (41.8)$$

which is "extremely small." Explanation for (41.8): the first factor $|\mathcal{F}|$ comes from $A(H)$, the second factor $(nD)^{j+1} 2^{-(n - 4n^{4/5} - j - 1)(j+1)}$ comes from $A(H_{i_0})$, $A(H_{i_1})$, $A(H_{i_2})$, $\ldots$, $A(H_{i_j})$, the third factor $(nD) 2^{-(n-2k)}$ comes from $A(E)$, and finally

$$\cdot \left(\binom{(j+3)n}{2} 2^{-(n - 2k - n^{1/5})} \right)^{n^{1/5}}$$

is the contribution of the $y_1 + y_2 + \ldots + y_j = n^{1/5}$ bond sets.

We also used the trivial fact that $j \leq y_1 + y_2 + \ldots + y_j = n^{1/5}$. $\qquad \square$

Statement 41.8 *Consider the "tree like configuration" in Statement 41.6. By playing rationally Breaker can force that among these $l_1 + l_2 + \ldots + l_{k/2}$ emergency super-sets we can always select at least $(l_1 + l_2 + \ldots + l_{k/2}) - \frac{n^{4/5}}{16}$ pairwise disjoint sets.*

Proof. Let M denote the maximum number of pairwise disjoint sets among the $l_1 + l_2 + \ldots + l_{k/2}$ emergency super-sets, and assume $M < (l_1 + l_2 + \ldots + l_{k/2}) - \frac{n^{4/5}}{16}$.

This means, fixing the $k/2$ sets $A(H)$, $A(H_1)$, $A(H_2)$, ..., $A(H_{k/2})$, and also the M disjoint sets among the emergency super-sets, there are at least $\frac{n^{4/5}}{16}$ emergency super-sets which are bond sets. So the "total danger" is estimated from above as follows

$$|\mathcal{F}| \cdot (nD)^{k/2} 2^{-(n-4n^{4/5}-k/2)\frac{k}{2}} \cdot \left(n^2 D 2^{-(n-2k)}\right)^M \cdot \left(\binom{n^2}{2} 2^{-(n-2k-n^{1/5})}\right)^{\frac{n^{4/5}}{16}}. \qquad (41.9)$$

Explanation for (41.9): the factor $|\mathcal{F}| \cdot (nD)^{k/2} 2^{-(n-4n^{4/5}-k/2)\frac{k}{2}}$ is the contribution of $A(H)$, $A(H_1)$, ..., $A(H_{k/2})$, the factor $\left(n^2 D 2^{-(n-2k)}\right)^M$ is the contribution of the M disjoint emergency sets, and finally the "$n^{1/5}$" in the exponent of 2 comes from Statement 41.7. It follows that the "total danger" is less than

$$|\mathcal{F}| \cdot 2^{4n^{4/5} \cdot \frac{k}{2} - \frac{n}{2} \cdot \frac{n^{4/5}}{16}} \leq 2^{n^{1.1} + 2n^{1.2} - n^{1.8}/32}, \qquad (41.10)$$

which is "extremely small." $\qquad\qquad\square$

Notice that we basically succeeded to reduce the general case of the Hazardous Lemma to the *Simplest Case*: indeed, combining Statements 41.6 and 41.8 we obtain:

Statement 41.9 *By playing rationally Breaker can force that, if there is a first violator of the Hazardous Lemma, then its "tree like configuration" can be truncated (by throwing out sets if necessary) such that the $l_1 + l_2 + \ldots + l_{k/4}$ emergency super-sets are pairwise disjoint.* $\qquad\square$

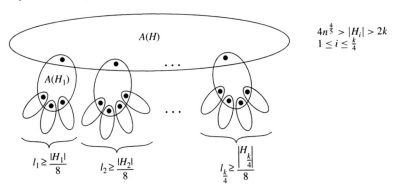

2. Conclusion. By using Statement 41.9 we are ready to complete the proof of the Hazardous Lemma. Indeed, by repeating the proof of the Simplest Case in Section 40 to the *truncated* "tree like configuration" in Statement 41.9, we obtain that Breaker can prevent the appearance of any *truncated* "tree like configuration," i.e. there is no first violator of the Hazardous Lemma. This completes the proof of the Hazardous Lemma. $\qquad\qquad\square$

The rest of the proof of Theorem 40.1 goes exactly the same way as was explained at the end of Section 40. The last technical problem is how to guarantee the "legs"

$A(E_1)$, $A(E_2)$, ..., $A(E_l)$ of the "centipede" to be pairwise disjoint. In view of Statement 41.2 the "disjointness" is guaranteed with "$l/2$" instead of "l," which is equally good in the calculations. This completes the proof of Theorem 40.1.　　□

Theorem 40.1 was about Almost Disjoint hypergraphs; in Theorems 8.2 and 12.6, however, the winning sets may have large intersections. This is a rather unpleasant new technical difficulty, which demands a change in the definition of the *small sets*, both the *emergency* and *hazardous sets*. We refer to this "change" as the RELARIN technique (named after RElatively LARge INtersections; the name will be justified later).

The simplest illustration of the RELARIN technique is the following proof of the Strong Draw part in Theorem 8.1 (Van der Waerden Game), see Section 42. (The Weak Win part was easy, see Section 13.) The good thing about Section 42 is that we can entirely focus on the RELARIN technique; we are not distracted by the (annoyingly complicated) technicalities of the method of "breaking the square-root barrier" (Sections 40–41).

42

Van der Waerden Game and the RELARIN technique

Recall the (N, n) Van der Waerden Game: the board is $[N] = \{1, 2, \ldots, N\}$ and the winning sets are the n-term A.P.s ("arithmetic progressions") in $[N]$. Note that the Weak Win part of Theorem 8.1 was a straightforward application of Theorem 1.2 (the "easy" Weak Win criterion); at the begining of Section 13 we already proved that, if

$$N > (1 + o(1)) \, n^3 \cdot 2^{n-3}, \tag{42.1}$$

then the first player can force a Weak Win in the (N, n) Van der Waerden Game.

Next we prove the *upper bound*: the Strong Draw part in Theorem 8.1. We employ the simplest BigGame–SmallGame Decomposition developed in Sections 35–36 (luckily we don't need the methods of Section 39), and combine it with a new idea called the **RELARIN technique** (named after RElatively LARge INtersections). First we have an:

1. Outline of the proof. Define an "auxiliary" m-uniform hypergraph $\mathcal{F}$ – note that $\mathcal{F}$ will be *different* from the family of all n-term A.P.s in $[N]$ – and as usual call an $A \in \mathcal{F}$ *dangerous* when Maker (the first player) occupies its $(m-k)$th point in the Big Game. The unoccupied (in the Big Game) k-element part of this dangerous $A \in \mathcal{F}$ *may* or *may not* become an emergency set as follows: let $E_1, E_2, E_3, \ldots$ be the complete list of the emergency sets arising in this order in the course of a play (if several emergency sets arise at the same time then we order them arbitrarily). Let $\widetilde{E}_1 = E_1$, $\widetilde{E}_2 = E_2 \setminus E_1$, $\widetilde{E}_3 = E_3 \setminus (E_1 \cup E_2)$, and, in general

$$\widetilde{E}_j = E_j \setminus \left(\bigcup_{i=1}^{j-1} E_i \right).$$

We call $\widetilde{E}_j$ the "on-line disjoint part" of E_j.

If $A \in \mathcal{F}$ is the next dangerous set, and A^* is the unoccupied (in the Big Game) k-element part of A, then we distinguish two cases: (1) if A^* intersects some "on-line disjoint part" $\widetilde{E}_j$ in $\geq k^\lambda$ points, then we *ignore* A saying that "A will be blocked inside $\widetilde{E}_j$," (i.e. A does *not* contribute a new emergency set); (2) if A^* intersects every "on-line disjoint part" $\widetilde{E}_j$ in less than k^λ points, then A^* is the next dangerous set.

Note that $0 < \lambda < 1$ is a fixed constant to be specified later; the intersection size k^λ justifies the name RElatively LARge INtersections (RELARIN).

In the proof of Theorem 34.1 (see Section 36) Breaker (the second player) used a trivial Pairing Strategy restricted to the "on-line disjoint parts" of the emergency sets (this was called the *small game* in Section 36). What we do here is more complicated. The similarity is that again in the small game Breaker plays many sub-games simultanously, where the disjoint sub-boards of the sub-games are the "on-line disjoint parts" $\widetilde{E}_j$: if Maker's last move was in a $\widetilde{E}_j$, then Breaker always replies in the same $\widetilde{E}_j$. The difference is that in every $\widetilde{E}_j$ Breaker replies by the Power-of-Two Scoring System (instead of making an "arbitrary reply," which was sufficient in Section 36).

After the outline, it is time now to discuss:

2. The details of the proof. We begin with the definition of "auxiliary" hypergraph $\mathcal{F}$: it is m-uniform where $m = n - \lfloor n^{1-\varepsilon} \rfloor$; the hyperedges in $\mathcal{F}$ are the m-term A.P.s in $[N]$ for which the lower integral part of the *Start/Gap* ratio is divisible by $\lfloor n^{1-\varepsilon} \rfloor$. More formally: the m-term A.P. $a, a+d, a+2d, \ldots, a+(m-1)d$ in $[N]$ is a hyperedge of $\mathcal{F}$ if and only if $\lfloor \frac{a}{d} \rfloor$ is divisible by $\lfloor n^{1-\varepsilon} \rfloor$; we refer to these A.P.s as *special m-term A.P.s*.

Notice that every ordinary n-term A.P. in $[N]$ contains a *special m-term A.P.s*. Indeed, if $s, s+d, s+2d, \ldots, s+(n-1)d$ is an arbitrary n-term A.P. in $[N]$, then among the first $r = \lfloor n^{1-\varepsilon} \rfloor$ members $x = s, s+d, s+2d, \ldots, s+(r-1)d$ there is exactly one for which $\lfloor \frac{x}{d} \rfloor$ is divisible by r; that member $s + jd = a$ is the starting point of a *special m-term A.P.* contained by $s, s+d, s+2d, \ldots, s+(n-1)d$. Breaker's goal is, therefore, to block the m-uniform hypergraph $\mathcal{F}$.

If two A.P.s of the same length have then same gap, we call them **translates**. Notice that in hypergraph $\mathcal{F}$ every $A \in \mathcal{F}$ has at most $2\lceil n^\varepsilon \rceil$ translates; this is a consequence of the definition of *special A.P.s*.

We need the following simple "selection lemma":

Lemma 42.1 *Let $A_1, A_2, \ldots, A_l$ be a family of m-term A.P.'s in the set of natural numbers such that no two have the same gap ("translates-free"). If $l \leq m$, then we can always find a sub-family $A_{i_1}, A_{i_2}, \ldots, A_{i_r}$ with $r = l^{1/5}$ such that*

$$\left| \bigcup_{j=1}^{r} A_{i_j} \right| \geq (r-1)m.$$

Proof. The proof uses the Pigeonhole Principle. The intersection $A_{i_1} \cap A_{i_2}$ of two A.P.s is always an A.P. itself; assume that the intersection has ≥ 2 elements. Let $A_{i_1} = \{a_{i_1} + jd_{i_1} : j = 0, 1, 2, \ldots, m-1\}$ and $A_{i_2} = \{a_{i_2} + jd_{i_2} : j = 0, 1, 2, \ldots, m-1\}$. Two-element intersection means that there exist b_{i_1} and b_{i_2} with $1 \leq b_{i_1} \leq m-1$, $1 \leq b_{i_2} \leq m-1$ such that $b_{i_1} \cdot d_{i_1} = b_{i_2} \cdot d_{i_2}$, i.e. $d_{i_2} = (b_{i_1}/b_{i_2}) \cdot d_{i_1}$.

Notice that we can estimate the intersection in terms of b_{i_1}, b_{i_2} as follows

$$|A_{i_1} \cap A_{i_2}| \leq \frac{2m}{\max\{b_{i_1}, b_{i_2}\}}. \tag{42.2}$$

Now we are ready to select a sub-family $A_{i_1}, A_{i_2}, \ldots, A_{i_r}$ with $r = l^{1/5}$ from $A_1, A_2, \ldots, A_l$ such that the union is "large." First pick A_1, and study those A_js for which $|A_1 \cap A_j| > \frac{2m}{M}$ (M will be specified later); we call these js "M-bad with respect to 1." By (42.2) we have $\max\{b_1, b_j\} \leq M$ where $d_j = (b_1/b_j) \cdot d_1$. If $\max\{b_1, b_j\} \leq M$, then the number of different ratios (b_1/b_j) is at most M^2, so the total number of indices j which are "M-bad with respect to 1" is $\leq M^2$. We throw out from $\{1, 2, \ldots, l\}$ all indices j which are "M-bad with respect to 1," and 1 itself, and pick an arbitrary index h_1 from the remaining set. We study those A_js for which $|A_{h_1} \cap A_j| > \frac{2m}{M}$; we call these js "M-bad with respect to h_1." By (42.2) we have $\max\{b_{h_1}, b_j\} \leq M$, where $d_j = (b_{h_1}/b_j) \cdot d_{h_1}$. If $\max\{b_{h_1}, b_j\} \leq M$, then the number of different ratios (b_{h_1}/b_j) is at most M^2, so the total number of indices j which are "M-bad with respect to h_1" is $\leq M^2$. We throw out all indices j which are "M-bad with respect to h_1," and h_1 itself, and pick an arbitrary index h_2 from the remaining set, and so on. Repeating this argument, we obtain a subsequence $h_0 = 1, h_1, h_2, h_3, \ldots$ of $\{1, 2, \ldots, l\}$ such that

$$|A_{h_i} \cap A_{h_j}| \leq \frac{2m}{M} \quad \text{holds for all } 0 \leq i < j < r = \frac{l}{M^2 + 1}. \tag{42.3}$$

By (42.3)

$$\left| \bigcup_{j=1}^{r} A_{h_j} \right| \geq r \cdot m - \binom{r}{2} \cdot \frac{2m}{M} \geq (r-1)m$$

if $M = l^{2/5}$ and $r = \sqrt{M}$. The completes the proof of Lemma 42.1. $\square$

A *survivor* $A \in \mathcal{F}$ is called a *secondary set* if (1) A intersects the *small board* in at least k points, (2) A intersects every $\widetilde{E}_j$ (i.e. "on-line disjoint part" of emergency set E_j) in less than k^λ points, (3) A is not dangerous.

The following lemma is the perfect analogue of Lemma 1 in Section 36; the concept of Big Game will be defined within the proof of this lemma.

Secondary Set Lemma: *Assume that Breaker has a winning strategy in the Big Game, and follows it in a play. Then there is no secondary set.*

Proof. Assume that A is a *secondary* set; from the existence of set A we want to derive a contradiction. Let $E_{i_1}, E_{i_2}, E_{i_3}, \ldots$ be the complete list of emergency sets intersecting A; assume that $1 \le i_1 < i_2 < i_3 < \cdots$. We claim that

$$A \cap (E_{i_1} \cup E_{i_2} \cup E_{i_3} \cup \ldots) = A \cap (\widetilde{E}_{i_1} \cup \widetilde{E}_{i_2} \cup \widetilde{E}_{i_3} \cup \ldots). \qquad (42.4)$$

To prove (42.4) note that, since E_{i_1} is the "first," its "on-line disjoint part" $\widetilde{E}_{i_1}$ must intersect A. For E_{i_2} we have two options: either (1) $A \cap E_{i_2} \subset \widetilde{E}_{i_1}$, or (2) $\widetilde{E}_{i_2}$ intersects A. The same two alternatives hold for $E_{i_3}, E_{i_4}, \ldots$; let E_{i_j} be the first one for which option (1) fails (i.e. option (2) holds); for notational convenience assume that $j = 2$, i.e. $\widetilde{E}_{i_2}$ intersects A. For E_{i_3} we have two options: either (1) $A \cap E_{i_3} \subset \widetilde{E}_{i_1} \cup \widetilde{E}_{i_2}$, or (2) $\widetilde{E}_{i_3}$ intersects A. The same two alternatives hold for E_{i_4}, $E_{i_5}, \ldots$; let E_{i_j} be the first one for which option (1) fails (i.e. option (2) holds); for notational convenience assume that $j = 3$, that is, $\widetilde{E}_{i_3}$ intersects A, and so on. This proves (42.4).

Since each intersection $A \cap \widetilde{E}_j$ has size less than k^λ and A intersects the small board in at least k points, in view of (42.4) there must exist at least $k^{1-\lambda}$ different "on-line disjoint parts" $\widetilde{E}_j$s which all intersect A.

Recall the following consequence of the definition of *special* A.P.s: in hypergraph $\mathcal{F}$ every $A \in \mathcal{F}$ has at most $2\lceil n^\varepsilon \rceil$ *translates*. Therefore, there exist an index-set J with $|J| \ge k^{1-\lambda}/2n^\varepsilon$ such that (1) the "on-line disjoint part" $\widetilde{E}_j$ intersects A for every $j \in J$, (2) the super-sets $A(E_j)$, $j \in J$ form a "translates-free" family (i.e. no two A.P.s have the same gap).

Now we apply Lemma 42.1: we can select a sub-family $A(E_i)$, $i \in I$ such that

$$|I| = r \ge \left(\frac{k^{1-\lambda}}{4n^\varepsilon}\right)^{1/5} \quad \text{and} \quad \left|\bigcup_{i \in I} A(E_i)\right| \ge (r-1)m. \qquad (42.5)$$

The reason why we wrote "4" instead of "2" in the denominator of the first inequality in (42.5) will become clear in the proof of the On-Line Disjoint Part Lemma below.

This family $A(E_i)$, $i \in I$ defines a Big Set $B = \bigcup_{i \in I} A(E_i)$. Big Set B has the following properties:

(1) $|B| \ge (r-1)m$;
(2) in the Big Game, Maker occupied at least $|B| - rk \ge (r-1)m - rk$ points from B;
(3) in the Big Game, Breaker couldn't put his mark in B.

What this means is that, in the Big Game, Maker managed to accomplish a Shutout of size $(r-1)m - rk$ in Big Set B. Breaker's goal in the Big Game is exactly to prevent every possible Shutout such as this!

Now we are ready to define the Big Game. Let

$$r = \left(\frac{k^{1-\lambda}}{4n^\varepsilon}\right)^{1/5}$$

(take the lower integral part); an r-element sub-family

$$\mathcal{G} = \{A_1, A_2, \ldots, A_r\} \subset \mathcal{F}$$

is called $\mathcal{F}$-*linked* if there is a set $A \in \mathcal{F}$ with $A \notin \mathcal{G}$ such that A intersects each element of family $\mathcal{G}$, i.e. $A \cap A_i \neq \emptyset$, $1 \leq i \leq r$. (Note that parameter r is an integer between 2 and $m/2$.)

The Big Game is played on the family $\mathcal{B}$ of *Big Sets*. What are the *Big Sets*? They are the *union sets* $\bigcup_{i=1}^r A_i$ of all possible $\mathcal{F}$-*linked* r-element sub-families $\mathcal{G} = \{A_1, A_2, \ldots, A_r\}$ of $\mathcal{F}$

$$\mathcal{B} = \left\{ B = \bigcup_{A \in \mathcal{G}} A : \mathcal{G} \subset \mathcal{F}, |\mathcal{G}| = r, \mathcal{G} \text{ is } \mathcal{F}-\text{linked} \right\}.$$

The total number of *Big Sets* is estimated from above by

$$|\mathcal{B}| \leq M \binom{m(D-1)}{k}. \tag{42.6}$$

Indeed, there are $|\mathcal{F}| = M$ ways to choose "linkage" A, there are at most $m(D-1)$ other sets intersecting A where $D = MaxDeg(\mathcal{F})$, and we have to choose r sets $A_1, A_2, \ldots, A_r$ among them.

We are going to show that Breaker can win the Big Game by using the Power-of-Two Scoring System. It suffices to check that

$$|\mathcal{B}| < 2^{(r-1)m-rk-1}. \tag{42.7}$$

Note that the board of the Big Game is shrinking during a play, but this does not cause any extra difficulty.

Now we are ready to complete the proof of the Secondary Set Lemma. If there exists a secondary set, then, in the Big Game, Maker can certainly achieve a Shutout of size $\geq (r-1)m - rk$ in some Big Set, which contradicts the assumption that Breaker has a winning strategy in the Big Game. $\qquad\square$

The next lemma is the perfect analogue of Lemma 2 in Section 36.

On-Line Disjoint Part Lemma: *Assume that Breaker has a winning strategy in the Big Game, and follows it in a play. Then every on-line disjoint part $\widetilde{E}_l$ has size $\geq k/2$.*

Proof. Assume $|\widetilde{E}_l| < k/2$. Let $E_{i_1}, E_{i_2}, E_{i_3}, \ldots$ be the complete list of emergency sets intersecting E_l; assume that $1 \leq i_1 < i_2 < i_3 < \cdots$. We have the analogue of (42.4)

$$E_l \cap (E_{i_1} \cup E_{i_2} \cup E_{i_3} \cup \ldots) = E_l \cap (\widetilde{E}_{i_1} \cup \widetilde{E}_{i_2} \cup \widetilde{E}_{i_3} \cup \ldots).$$

Since each intersection $E_l \cap \widetilde{E}_{i_j}$ has size less than k^λ and $|\widetilde{E}_l| < k/2$, there must exist at least $\frac{1}{2}k^{1-\lambda}$ different "on-line disjoint parts" $\widetilde{E}_i$s which all intersect E_l. This $\frac{1}{2}k^{1-\lambda}$ is half of what we had in the proof of the Secondary Set Lemma, explaining the factor of "4" (instead of "2") in (42.5). The rest follows from the previous proof.   □

3. Conclusion. The small game falls apart into disjoint sub-games, where each sub-game is played on some on-line disjoint part $\widetilde{E}_l$, and each *small set* has size $\geq k^\lambda$. Every *small set* is the intersection of $\widetilde{E}_l$ with an m-term arithmetic progression AP, implying that every small set in $\widetilde{E}_l$ is uniquely determined by a 4-tuple (F, L, i, j), where f and L are the first and last elements of the m-term arithmetic progression AP in $\widetilde{E}_l$, F is the ith and L is the jth elements of AP with $1 \leq i < j \leq m$. Thus the total number of small sets in $\widetilde{E}_l$ is

$$\leq \binom{|\widetilde{E}_l|}{2}\binom{m}{2} \leq \binom{k}{2}\binom{m}{2} \leq m^4.$$

By using the Power-of-Two Scoring System, Breaker can blocke every possible small set in every on-line disjoint part $\widetilde{E}_l$ if

$$m^4 < 2^{k^\lambda - 1}. \tag{42.8}$$

Moreover, by (42.6)–(42.7) Breaker can win the Big Game if

$$M\binom{m(D-1)}{k} < 2^{(r-1)m - rk - 1}, \tag{42.9}$$

where $|\mathcal{F}| = M$ and $D = MaxDeg(\mathcal{F})$.

Assume that both (42.8)–(42.9) hold; then Breaker can block every $A \in \mathcal{F}$, i.e. he can force a Strong Draw in $\mathcal{F}$. Assume that there is an $A_0 \in \mathcal{F}$ which is completely occupied by Maker; we want to derive a contradiction from this assumption.

If Maker is able to occupy $(m - k)$ points of A_0 in the Big Game, then A_0 becomes a dangerous set. We have two options: if (1) A_0 intersects some $\widetilde{E}_l$ in $\geq k^\lambda$ points, then A_0 will be blocked by Breaker inside $\widetilde{E}_l$; if (2) A_0 produces an emergency set E_0, then A_0 will be blocked inside $\widetilde{E}_0$.

If at the end of the Big Game Maker still couldn't occupy $(m - k)$ points of A_0, then again we have two options: either (i) at some stage of the play A_0 becomes a secondary set, which *contradicts* the Secondary Set Lemma, or (ii) A_0 intersects some $\widetilde{E}_l$ in $\geq k^\lambda$ points, which is case (1) above, i.e. A_0 will be blocked by Breaker inside $\widetilde{E}_l$. In each case we get a contradiction as we wanted.

The last step of the proof is to maximize the value of n under conditions (42.8)–(42.9). Recall that

$$m = n - n^{1-\varepsilon}, \quad r = \left(\frac{k^{1-\lambda}}{4n^{\varepsilon}} \right)^{1/5};$$

on the other hand, trivially $M = |\mathcal{F}| < N^2$ and $D = MaxDeg(\mathcal{F}) < N$. By choosing $\varepsilon = 1/8 = \lambda$ and $k = n^{7/8}$, inequalities (42.8)–(42.9) are satisfied with

$$n = \log_2 N + c_0 \cdot (\log_2 N)^{7/8} = \left(1 + \frac{c_0}{(\log_2 N)^{1/8}} \right) \log_2 N,$$

where c_0 is a large positive absolute constant. This completes the proof of the Strong Draw part of Theorem 8.1. $\square$

Chapter IX

Game-theoretic lattice-numbers

The missing Strong Draw parts of Theorems 8.2, 12.6, and 40.2 will be discussed here; we prove them in the reverse order. These are the most difficult proofs in the book. They demand a solid understanding of Chapter VIII. The main technical challenge is the lack of Almost Disjointness.

Chapters I–VI were about *Building* and Chapters VII–VIII were about *Blocking*. We separated these two tasks because undertaking them at the same time – ordinary win! – was hopelessly complicated. Now we have a fairly good understanding of *Building* (under the name of Weak Win), and have a fairly good understanding of *Blocking* (under the name of Strong Draw). We return to an old question one more time: "Even if ordinary win is hopeless, is there any other way to combine the two different techniques in a single strategy?" The answer is "yes," and some interesting examples will be discussed in Section 45. One of them is the proof of Theorem 12.7: "second player's moral-victory."

43

Winning planes: exact solution

The objective of this section is to prove the missing Strong Draw part of Theorems 12.6 and 40.2. The winning sets in these theorems are "planes"; two "planes" may be disjoint, or intersect in a point, or intersect in a "line." The third case – "line-intersection" – is a novelty which cannot happen in Almost Disjoint hypergraphs; "line-intersection" requires extra considerations.

1. A common generalization. Both Theorems 12.6 and 40.2 will be covered by the following generalization of Theorem 40.1:

Theorem 43.1 *Assume that $(\mathcal{F}, \mathcal{G})$ is a pair of hypergraphs which satisfies the following Dimension Condition (some kind of generalization of Almost Disjointness):*

(a) *$\mathcal{F}$ is m-uniform, and the hyperedges $A \in \mathcal{F}$ are called "planes";*
(b) *the hyperedges $B \in \mathcal{G}$ are called "lines," and each line has at most $2\sqrt{m}$ points;*
(c) *the intersection of two "planes" is (1) either empty, (2) or a point, (3) or a "line";*
(d) *$\mathcal{G}$ is Almost Disjoint, and $\mathcal{F}$ has the weaker property that 3 non-collinear (meaning "not on a line") points uniquely determine a "plane";*
(e) *the Max Pair-Degree $\Delta_2 = \Delta_2(\mathcal{F})$ of hypergraph $\mathcal{F}$ satisfies an upper bound*

$$\Delta_2 \le 2^{(1-\delta)m} \text{ with some positive constant } \delta > 0.$$

Then there exists a finite threshold constant $c_0 = c_0(\delta) < \infty$ depending on δ only such that, if

$$|\mathcal{F}| \le 2^{m^{1.1}} \text{ and } D = MaxDeg(\mathcal{F}) \le 2^{m-4m^{2/5}}$$

(which means: breaking the square-root barrier), and $m \ge c_0$, then the second player can force a Strong Draw in the positional game on hypergraph $\mathcal{F}$.

Before talking about the proof, let me explain how Theorems 12.6 and 40.2 follow from Theorem 43.1. Note that m is either $q^2 + q + 1$ ("projective planes"), or q^2 ("affine planes"), or n^2 ("comb-planes in the n^d torus"). Conditions (a)–(d) are trivially satisfied; but how about the Max Degree and the Max Pair-Degree? We have

$$D \approx q^{2d-4} \text{ and } \Delta_2 \approx q^{d-2} \approx D^{.5} \quad \text{(projective geometry)};$$

$$D \approx q^{2d} \text{ and } \Delta_2 \approx q^d \approx D^{.5} \quad \text{(affine geometry)};$$

$$D \approx 3^d \text{ and } \Delta_2 \approx 2^d \approx D^{\log 2/\log 3} = D^{.63} \quad \text{(planes in the } n^d \text{ torus)},$$

which settles condition (e).

2. The proof of Theorem 43.1. It is a combination of the proof of Theorem 40.1 (see Sections 39–41) and the RELARIN technique (see Section 42). The possibility of relatively large "line-intersections" demands a change in the definition of the *small sets*, both the *emergency* and *hazardous sets*. Let $k = m^{2/5}$.

There is no change in the way how the *first emergency set* arises: when Maker (the first player) occupies the $(m-k)$th point of some *survivor* (meaning "Breaker-free," where Breaker is the second player) winning set $A \in \mathcal{F}$ for the first time; in that instant A becomes *dangerous*. Note that we may have several dangerous sets arising at the same time: they are $A_i \in \mathcal{F}$ where $i \in I$. For simplicity we may assume that $I = \{1, 2, \ldots, s\}$, and proceed by induction. The first emergency set is the "blank" k-element part $A_1^* = A_1 \setminus A_1(Big; Maker)$ of A_1; A_1^* is removed from the Big Board and added to the small board. Next consider A_2; the first novelty is that set A_2 does *not* necessarily produce a new emergency set, i.e. A_2 may be "ignored." Indeed, if the "blank" k-element part $A_2^* = A_2 \setminus A_2(Big; Maker)$ of A_2 intersects an emergency set (at this stage this means the first emergency set A_1^*) in $\geq k^\lambda$ points, then we *ignore* A_2 by saying that "A_2 will be blocked inside of A_1," but if A_2^* intersects every emergency set (at this stage this means the first emergency set A_1^* only) in less than k^λ points, then A_2^* is the second emergency set; A_2^* is removed from the Big Board and added to the small board. (λ will be a positive absolute constant less than 1; note in advance that $\lambda = 1/2$ is a good choice.)

Similarly, let $1 < j \leq s$; if the "blank" k-element part $A_j^* = A_j \setminus A_j(Big; Maker)$ of A_j intersects an emergency set in $\geq k^\lambda$ points, then we *ignore* A_j by saying that "A_j will be blocked inside of some A_i with $i < j$," but if A_j^* intersects every emergency set in less than k^λ points, then A_j^* is a new emergency set; A_j^* is removed from the Big Board and added to the small board.

At the end of this the Big Board becomes smaller, so we may have an untouched survivor $A \in \mathcal{F}$ with $|A(Big; blank)| \leq k$; in fact, we may have several of them: $A_i \in \mathcal{F}$ where $i \in I$. For simplicity write $I = \{1, 2, \ldots, s\}$, and proceed by induction. Each A_i *may* or *may not* contribute a new emergency or hazardous set as follows. First assume that $|A_1(small)| \leq k$; let $A_1^* = A_1 \setminus A_1(Big; Maker)$; note that $|A_1^*| \geq k$. Consider the largest intersection of A_1^* with an emergency set; if the largest intersection is $\geq k^\lambda$, then (as usual) we ignore A_1. If the largest intersection is less than k^λ, then A_1^* is a new emergency set; A_1^* is removed from the Big Board and added to the small board.

Next assume that $|A_1(small)| \geq k+1$, and write $A_1^* = A_1 \setminus A_1(Big; Maker)$; note that $|A_1^*| \geq k$. Consider the largest intersection of A_1^* with an emergency set; if the largest intersection is $\geq k^\lambda$, then we ignore A_1. If the largest intersection is less than k^λ, then A_1^* is the *first* hazardous set; A_1^* is removed from the Big Board and added to the small board.

Similarly, let $1 < j \leq s$; write $A_j^* = A_j \setminus A_j(Big; Maker)$; if A_j intersects an emergency or hazardous set in $\geq k^\lambda$ points, then we *ignore* A_j by saying that "A_j will be blocked inside of some A_i with $i < j$," but if A_j^* intersects every emergency and hazardous set in less than k^λ points, then (1) A_j^* is a new emergency set if $|A(small)| \leq k$, and (2) A_j^* is a new hazardous set if $|A(small)| \geq k+1$; in both cases A_j^* is removed from the Big Board and added to the small board.

Repeating the previous argument, this *extension process* – that we call the "first extension process" – will eventually terminate, meaning that the inequality $|A(Big; blank)| \geq k+1$ holds for every untouched survivor $A \in \mathcal{F}$.

It may happen, however, that the inequality $k+1 \leq |A(Big; blank)| \leq |A(small)|$ holds for a family $A = A_i$, $i \in I$ of untouched survivors. For simplicity write $I = \{1, 2, \ldots, s\}$; we proceed by induction. Write $A_i^* = A_i \setminus A_i(Big; Maker)$; if A_i intersects an emergency or hazardous set in $\geq k^\lambda$ points, then we *ignore* A_i, but if A_i^* intersects every emergency and hazardous set in less than k^λ points, then A_i^* is a new hazardous set; A_i^* is removed from the Big Board and added to the small board.

At the end of this the Big Board is decreased, so we may have untouched survivor(s) $A = A_i$, $i \in I$ with $k+1 \leq |A(Big; blank)| \leq |A(small)|$, and so on. Repeating this "second extension process," it will eventually terminate, meaning that the inequality $k+1 \leq |A(Big; blank)| \leq |A(small)|$ *fails* to hold for any untouched survivor $A \in \mathcal{F}$.

At the end of this we may have an untouched survivor $A \in \mathcal{F}$ with $|A(Big; blank)| \leq k$; in fact, we may have several of them: $A_i \in \mathcal{F}$ where $i \in I$. Applying again the "first extension process," and again the "second extension process," and again the "first extension process," and again the "second extension process," and so on, we eventually reach a stage where this "alternating" terminates, meaning that *both* properties

$$|A(Big; blank)| \geq k+1; \tag{43.1}$$

$$|A(Big; blank)| > |A(small)| \tag{43.2}$$

hold for each untouched survivor $A \in \mathcal{F}$ – we refer to this as the *Closure Process*.

Up to this point the *small sets* are exactly the *emergency sets* and the *hazardous sets*; at a later stage, however, the small sets will *differ* from the emergency and hazardous sets as follows. Assume that, at a later stage the double requirement (43.1)–(43.2) is violated; it can be violated (in fact, "violated first") in two different ways

$$\text{either } k = |A(Big; blank)| > |A(small)| \quad \text{occurs,} \qquad (43.3)$$

$$\text{or } |A(Big; blank)| = |A(small)| \geq k+1 \quad \text{occurs.} \qquad (43.4)$$

Case (43.3) *will* or *will not* produce new emergency sets and case (43.4) *will* or *will not* produce new hazardous sets explained as follows.

If (43.3) holds, then write

$$S = S(A) = A(blank) = A(Big; blank) \cup A(small; blank), \qquad (43.5)$$

and consider the largest intersection of $S = S(A)$ with a small set. If the largest intersection is $\geq k^\lambda$, then we ignore A. If the largest intersection is less than k^λ, then $S = S(A)$ in (43.5) is a new small set, which is (as usual) removed from the Big Board and added to the small board; the (possibly) larger set

$$E = E(A) = A \setminus A(Big; Maker) = A(Big; blank) \cup A(small) \qquad (43.6)$$

is a new *emergency set*, and we refer to the small set $S = S(A)$ in (43.5) as a *small-em set* – meaning the "small set part of emergency set $E = E(A)$ in (43.6)"; clearly $S \subseteq E$.

Next assume that (43.4) holds. Again write

$$S = S(A) = A(blank) = A(Big; blank) \cup A(small; blank), \qquad (43.7)$$

and consider the largest intersection of $S = S(A)$ with a small set. If the largest intersection is $\geq k^\lambda$, then we ignore A. If the largest intersection is less than k^λ, then $S = S(A)$ in (43.7) is a new small set, which is (as usual) removed from the Big Board and added to the small board; the (possibly) larger set

$$H = H(A) = A \setminus A(Big; Maker) = A(Big; blank) \cup A(small) \qquad (43.8)$$

is a new *hazardous set*, and we refer to the small set $S = S(A)$ in (43.7) as a *small-haz set* – meaning the "small set part of hazardous set $H = H(A)$ in (43.8)"; clearly $S \subseteq H$.

In other words, from now on the new *small sets* are either *small-em sets* ("blank parts of emergency sets") or *small-haz sets* ("blank parts of hazardous sets").

The union of the *small-em sets* is called the *Emergency Room*, or simply the E.R.

Of course, we may have several violators of (43.1)–(43.2) arising at the first time: we repeat the previous argument for each one of them; then again we may have several violators of (43.1)–(43.2): we repeat the previous argument for each; and so on. In other words, we apply a straightforward *Closure Process* – just as before – which terminates at a stage where again the double requirement (43.1)–(43.2) holds for every untouched survivor $A \in \mathcal{F}$. At a later stage the double requirement (43.1)–(43.2) may be violated; it can be violated (in fact, "violated first") in two different ways: (43.3) and (43.4), and so on.

Summarizing, the new feature in the definition of the small game is to involve RElatively LARge INtersections of size $\geq k^\lambda$ (where λ is a positive absolute constant less than one) – we refer to this as the **RELARIN technique**. Roughly speaking, the proof of Theorem 43.1 is nothing else other than the proof of Theorem 40.1 modified by the RELARIN technique. The RELARIN technique forces us to modify the *Component Condition* (see Section 39): we have to guarantee that every component of the small game is less than $2^{\text{const}\cdot k^\lambda}$ (instead of the more generous upper bound $2^{\text{const}\cdot k}$ in Sections 39–41).

Again the main difficulty is to prove the Hazardous Lemma.

Hazardous Lemma (II): *Under the condition of Theorem 43.1, Breaker can force that, during the whole course of the play, every hazardous set H intersects the E.R. in at least $|H|/4$ points.*

Remark. Again Hazardous Lemma (II) holds *trivially* for the *first* hazardous set, but there is no obvious reason why it should remain true at the later stages of the play – unless, of course, Breaker forces it by playing rationally.

3. Proof of the Hazardous Lemma (II). To understand what Breaker has to prevent, assume that there is a "violator." Let H be the first violator of Hazardous Lemma (II): H is the first hazardous set which intersects the E.R. in less than $|H|/4$ points. Since $|H| \geq 2k+2$ (see (43.4)), H must intersect the difference-set "smallboard minus E.R." in $\geq k/2$ points. These $\geq k/2$ points are covered by *non-violator* hazardous sets $H_1, H_2, \ldots, H_q$, where $k^{1-\lambda}/2 \leq q \leq k/2$; note that an intersection of size $\geq k^\lambda$ would prevent the H to be a new hazardous set – this is a consequence of the RELARIN technique. Since each H_j, $1 \leq j \leq q$ is a non-violator of Hazardous Lemma (II), each H_j must intersect the E.R. in at least $|H_j|/4$ points, which are covered by emergency sets $E_{j,1}, E_{j,2}, \ldots, E_{j,l_j}$. Note that $k^{-\lambda}|H_j|/4 \leq l_j \leq |H_j|/4$ for every $j = 1, 2, \ldots, q$. Formally we have $1 + q + \sum_{j=1}^{q} l_j$ super-sets

$$A(H), A(H_1), \ldots, A(H_q), A(E_{1,1}), \ldots, A(E_{1,l_1}), A(E_{2,1}), \ldots, A(E_{2,l_2}),$$

$$A(E_{3,1}), \ldots, A(E_{3,l_3}), \ldots, A(E_{q,1}), \ldots, A(E_{q,l_q}), \tag{43.9}$$

but there may occur some *coincidence* among these hyperedges of $\mathcal{F}$, and also there may occur some "extra intersection" among these sets. As usual $A(\ldots)$ denotes a super-set (but the super-set is not uniquely determined anymore because we gave up on Almost Disjointness; the good news is that this ambiguity does not cause any problem).

In Section 41 we reduced the proof of the Hazardous Lemma to the *Simplest Case*; the reduction had 9 steps: Statements 41.1–41.9. Here we do something very similar: we reduce the proof of Hazardous Lemma (II) to an analogue of the *Simplest Case* via Statements 43.1–43.9.

Again we call $A(H_j)$ together with $A(E_{j,1})$, $A(E_{j,2})$, ..., $A(E_{j,l_j})$ the jth "centipede"; $A(H_j)$ is the "head" and $A(E_{j,1})$, $A(E_{j,2})$, ..., $A(E_{j,l_j})$ are the "legs." Also recall that $k = m^{2/5}$.

The first statement is the perfect analogue of Statement 41.1.

Statement 43.1 *For every possible i Breaker can prevent the appearance of $M > m^{4/5}$ pairwise disjoint sets $A(E_{i,j_1})$, $A(E_{i,j_2})$, ..., $A(E_{i,j_M})$ in the ith "centipede."*

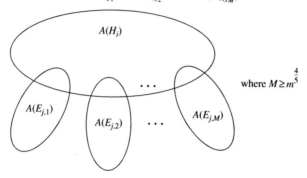

where $M \geq m^{\frac{4}{5}}$

Proof. Let M denote the maximum number of pairwise disjoint sets among the "legs." Breaker uses the Power-of-Two Scoring System; the target value of a "forbidden configuration" is at least $2^{(m-2k)M}$, and the number of ways to form a "forbidden configuration" is at most $|\mathcal{F}|(mD)^M$. Since $k = m^{2/5}$ and $M > m^{4/5}$, the "total danger"

$$|\mathcal{F}|(mD)^M 2^{-(m-2k)M} = |\mathcal{F}|\left(\frac{mD}{2^{m-2k}}\right)^M \leq 2^{m^{1.1}-2M \cdot m^{2/5}} \tag{43.10}$$

is "extremely small" if $M = m^{4/5}$, proving that Breaker can indeed prevent the appearance of any "forbidden configuration" described in Statement 43.1. $\quad\square$

Next we prove the analogue of Statement 41.2. The extra difficulty is that 2 points do *not* determine a winning set $A \in \mathcal{F}$ any more: we need 3 non-collinear points to determine a winning set $A \in \mathcal{F}$ (see condition (d) in Theorem 43.1).

Statement 43.2 *Consider the ith "centipede" (i is arbitrary)*

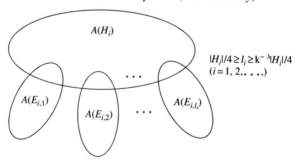

$|H_i|/4 \geq l_i \geq k^{-\lambda}|H_i|/4$
$(i = 1, 2, \ldots,)$

Breaker can force that, among the l_i super-sets $A(E_{i,1})$, $A(E_{i,2})$, ..., $A(E_{i,l_i})$ at least half, namely $\geq l_i/2$, are pairwise disjoint.

Proof. Let M_0 denote the maximum number of pairwise disjoint sets among $A(E_{i,1})$, $A(E_{i,2})$, ..., $A(E_{i,l_i})$, and assume that $M_0 < l_i/2$. For notational simplicity assume that $A(E_{i,j})$, $1 \leq j \leq M_0$ are pairwise disjoint, and each $A(E_{i,\nu})$ with $M_0 < \nu \leq l_i$ intersects the union set $\bigcup_{1 \leq j \leq M_0} A(E_{i,j})$.

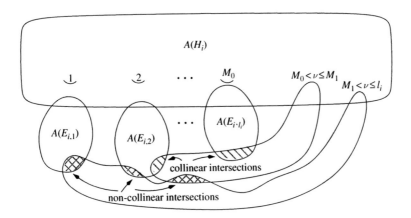

By rearranging the $(l_i - M_0)$ sets $A(E_{i,\nu})$ with $M_0 < \nu \leq l_i$ if necessary, we can find a threshold M_1 such that for every $M_0 < \nu \leq M_1$ the set $A(E_{i,\nu})$ has a collinear intersection with the union $\bigcup_{1 \leq j \leq \nu-1} A(E_{i,j})$ of the previous ones, and for every $M_1 < \nu \leq l_i$ the set $A(E_{i,\nu})$ has a non-collinear intersection with $\bigcup_{1 \leq j \leq M_1} A(E_{i,j})$.

First we show that $(M_1 - M_0) = o(l_i)$ ("small"). Indeed, the "total danger" is estimated from above by

$$|\mathcal{F}|(mD)^{M_0}2^{-(m-2k)M_0} \cdot \prod_{\nu=1}^{M_1-M_0} \left(\binom{(M_0+\nu+1)m}{2} \Delta_2 2^{-(m-2k-2\sqrt{m})} \right). \qquad (43.11)$$

Since

$$|\mathcal{F}| \leq 2^{m^{1.1}} \quad \text{and} \quad \Delta_2 \leq 2^{(1-\delta)m},$$

(43.11) becomes "extremely small" if $(M_1 - M_0) \geq m^{1/5}/\delta$.

Next we show that $(l_i - M_1) = o(l_i)$ ("small"). Note that every non-collinear set contains 3 points which are already non-collinear; it follows that for every ν with $M_1 < \nu \leq l_i$ there are (at most) 3 indices $\nu(1)$, $\nu(2)$, $\nu(3)$ such that $1 \leq \nu(1) \leq \nu(2) \leq \nu(3) \leq M_1$ and $A(E_{i,\nu})$ has a non-collinear intersection with $A(E_{i,\nu(1)}) \cup A(E_{i,\nu(2)}) \cup A(E_{i,\nu(3)})$. Since a non-collinear set uniquely determines an $A \in \mathcal{F}$ (see condition (d) in Theorem 43.1), we can estimate the "total danger" from above by

$$|\mathcal{F}|(mD)^R 2^{-(m-2k)R} \cdot \left(\binom{(M_1+1)m}{2} \Delta_2 2^{-(m-2k-4\sqrt{m})} \right)^S$$

$$\cdot \left(\binom{(M_1+1)m}{3} 2^{-(m-2k-2(R+S)\sqrt{m})} \right)^T , \tag{43.12}$$

where $R + S \le 3T$, $0 \le R$, $0 \le S$, and $1 \le T \le (l_i - M_1)$. By choosing $T \approx m^{1/5}$ in (43.12) we get an "extremely small" upper bound. $\square$

Combining Statements 43.1–43.2 we obtain

$$\max_i |H_i| \le m^{4/5}. \tag{43.13}$$

The next statement is the analogue of Statement 41.3.

Statement 43.3 *If Breaker plays rationally, then he can prevent that some $A(E)$ intersects $r \ge m^{1/5}$ distinct sets $A(H_i)$.*

Proof. The "total danger" of the "forbidden configurations" is at most

$$|\mathcal{F}|^2 \cdot \left(\binom{2m}{2} \cdot \Delta_2 \cdot 2^{-(m-m^{4/5}-2r\sqrt{m})} \right)^r \le$$

which is "extremely small." $\square$

The next statement is the perfect analogue of Statement 41.4.

Statement 43.4 *If Breaker plays rationally, then he can prevent the appearance of any forbidden configuration described by the picture*

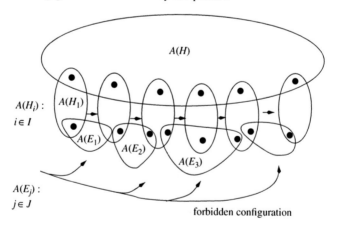

forbidden configuration

where the requirements are listed as follows:

(1) by dropping $A(H)$, the rest is still connected;
(2) $|I| \ge 2m^{1/5}$;
(3) every $A(E_j)$ intersects at least two $A(H_i)$, $i \in I$;
(4) every $A(E_j)$ intersects at least one new $A(H_i)$, $i \in I$.

Proof. In view of Statement 43.3 we can assume that $2m^{1/5} \le |I| \le 3m^{1/5}$, and want to show that the "danger" is "extremely small." For notational simplicity assume $J = \{1, 2, \ldots, J\}$; then the "danger" is less than a very long product which begins with

$$|\mathcal{F}| \cdot (mD)2^{-(m-m^{4/5})} \cdot (mD)2^{-(m-2k)} \cdot \left(\binom{2m}{2} \cdot \Delta_2 \cdot 2^{-(m-m^{4/5}-2\sqrt{m}|I|)} \right)^{a_1 - 1} .$$

(explanation for this part: $|\mathcal{F}|$ comes from choosing $A(H)$, $(mD)2^{-(m-m^{4/5})} \cdot (mD)2^{-(m-2k)}$ comes from choosing $A(H_1)$ and $A(E_1)$, a_1 is the number of neighbors of $A(E_1)$ among $A(H_i)$, $i \in I$, where "neighbor" of course means "intersecting," and the "-1" in $a_1 - 1$ is clear since $A(H_1)$ was already counted)

$$\cdot (m^2 D)2^{-(m-2k-2\sqrt{m}|I|)} \cdot \left(\binom{2m}{2} \cdot \Delta_2 \cdot 2^{-(m-m^{4/5}-2\sqrt{m}|I|)} \right)^{a_2} .$$

(explanation: the factor $(m^2 D)$ comes from choosing $A(E_2)$)

$$\cdot (m^2 D)2^{-(m-2k-2\sqrt{m}|I|)} \cdot \left(\binom{2m}{2} \cdot \Delta_2 \cdot 2^{-(m-m^{4/5}-2\sqrt{m}|I|)} \right)^{a_3} \cdots$$

where a_j $(j = 1, 2, \ldots, J)$ is the number of *new neighbors* of $A(E_j)$ among $A(H_i)$, $i \in I$ ("neighbor" of course means "intersecting"). Note that the term $-2\sqrt{m}|I|$ in the exponent of 2 above comes from the fact that the sets $A(H_i)$, $i \in I$ may intersect.

Since $a_1 + a_2 + \cdots + a_J = |I| \ge 2m^{1/5}$, the product above is "extremely small." $\qquad\square$

Statement 43.4 was about the maximum size of a **component** of $A(H_i)$s (a *component* arises when $A(H)$ is dropped); the next result – the perfect analogue of Statement 41.5 – is about *all components*, each containing at least two $A(H_i)$s.

Statement 43.5 *If Breaker plays rationally, then he can prevent the appearance of any configuration described by the figure*

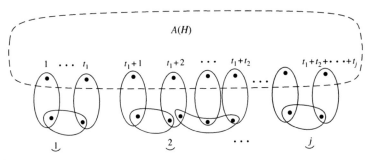

j components
each having ≥ 2 $A(H_i)$s

where there are j "components" with $t_1 + t_2 + \cdots + t_j \geq 4m^{1/5}$, where each $t_i \geq 2$ (i.e. each "component" has at least two $A(H_i)s$).

Proof. In view of Statement 43.4 we can assume that $4m^{1/5} \leq t_1 + t_2 + \cdots + t_j \leq 6m^{1/5}$, and want to show that the "danger" is "extremely small." From the proof of Statement 43.4 we can easily see that the "total danger" of the configurations described in Statement 43.5 is less than

$$|\mathcal{F}| \cdot 2^{-\delta m((t_1-1)+(t_2-1)+\cdots+(t_j-1))}.$$

Since each $t_i \geq 2$, we have $(t_1 - 1) + (t_2 - 1) + \cdots + (t_j - 1) \geq (t_1 + t_2 + \cdots + t_j)/2 \geq 2m^{1/5}$, implying that

$$|\mathcal{F}| \cdot 2^{-\delta m((t_1-1)+(t_2-1)+\cdots+(t_j-1))}$$

is "extremely small," proving Statement 43.5. □

Combining Statements 43.2 and 43.5, and also inequality (43.13), we obtain the following

Statement 43.6 *If Breaker plays rationally, then he can guarantee the following: if there is a violator of the Hazardous Lemma (II), then the* **first violator** $A(H)$ *generates a "tree like configuration"*

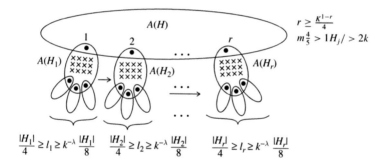

where the $l_1 + l_2 + \cdots + l_r$ emergency super-sets are distinct (though they may have "extra intersections" among each other). □

Next we study the possible "extra intersections" of these $l_1 + l_2 + \cdots + l_r$ emergency super-sets. *Question:* Given an emergency super-set, how many other emergency super-sets can intersect it? The following statement – the analogue of Statement 41.7 – gives an upper bound.

Statement 43.7 *Consider the "tree like configuration" in Statement 43.6: let $A(E)$ be an arbitrary emergency super-set – the neighbor of (say) $A(H_{i_0})$ – and assume that $A(E)$ has $t = y_1 + y_2 + \ldots + y_j$ "extra intersections" with other emergency* super-sets. *If Breaker plays rationally, then he can guarantee the upper bound* $t = y_1 + y_2 + \ldots + y_j < m^{1/5}$ *uniformly for all* $A(E)$.

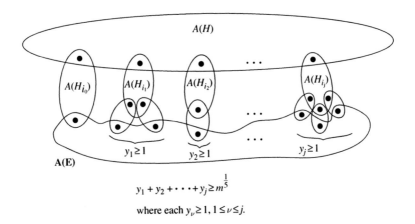

$$y_1 + y_2 + \cdots + y_j \geq m^{\frac{1}{5}}$$

where each $y_\nu \geq 1, 1 \leq \nu \leq j$.

Proof. If $A(H), A(H_{i_0}), A(H_{i_1}), \ldots A(H_{i_j}), A(E)$ are fixed, then the picture shows $t = y_1 + y_2 + \ldots + y_j$ 2-bond sets. If the statement fails, then without loss of generality we can assume equality $y_1 + y_2 + \ldots + y_j = m^{1/5}$. We can estimate the "total danger" from above as follows

$$|\mathcal{F}| \cdot \left((mD)2^{-(m - m^{4/5} - 2\sqrt{m}(j+1))} \right)^{j+1} \cdot (mD)2^{-(m-2k)}$$

$$\cdot \left(\binom{(j+3)m}{2} \cdot \Delta_2 \cdot 2^{-(m-2k-m^{1/5} \cdot 2\sqrt{m})} \right)^{m^{1/5}}$$

which is "extremely small." Here the first factor $|\mathcal{F}|$ comes from $A(H)$, the next factor

$$\left((mD)2^{-(m - m^{4/5} - 2\sqrt{m}(j+1))} \right)^{j+1}$$

comes from $A(H_{i_0}), A(H_{i_1}), A(H_{i_2}), \ldots, A(H_{i_j})$, the factor $(mD)2^{-(m-2k)}$ comes from $A(E)$, and finally

$$\left(\binom{(j+3)m}{2} \cdot \Delta_2 \cdot 2^{-(m-2k-m^{1/5} \cdot 2\sqrt{m})} \right)^{m^{1/5}}$$

is the contribution of the $y_1 + y_2 + \ldots + y_j = m^{1/5}$ 2-bond sets (see Δ_2).
We also used the trivial fact that $j \leq y_1 + y_2 + \ldots + y_j = m^{1/5}$. $\qquad \square$

Next we prove the analogue of Statement 41.8.

Statement 43.8 *Consider the "tree like configuration" in Statement 43.6:*

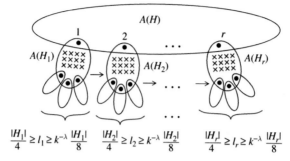

$$\frac{|H_1|}{4} \geq l_1 \geq k^{-\lambda}\frac{|H_1|}{8} \qquad \frac{|H_2|}{4} \geq l_2 \geq k^{-\lambda}\frac{|H_2|}{8} \qquad \frac{|H_r|}{4} \geq l_r \geq k^{-\lambda}\frac{|H_r|}{8}$$

By playing rationally Breaker can force that among these $l_1 + l_2 + \ldots + l_r$ emergency super-sets, where $r \geq k^{1-\lambda}/4$, we can always select at least $(l_1 + l_2 + \ldots + l_r) - m^{4/5-\lambda}$ pairwise disjoint sets.

Proof. Let M denote the maximum number of pairwise disjoint sets among the $l_1 + l_2 + \ldots + l_r$ emergency super-sets, and assume $M < (l_1 + l_2 + \ldots + l_r) - m^{4/5-\lambda}$. This means, fixing the $r+1$ sets $A(H)$, $A(H_1)$, $A(H_2)$, ..., $A(H_r)$, and also the M disjoint sets among the emergency super-sets, there are at least $m^{4/5-\lambda}$ emergency super-sets which are 2-bond sets. So the "total danger" is estimated from above as follows

$$|\mathcal{F}| \cdot (mD)^r 2^{-(m-m^{4/5}-r)r} \cdot \left(m^2 D 2^{-(m-2k)}\right)^M \cdot \left(\binom{m^3}{2} \cdot \Delta_2 \cdot 2^{-(m-2k-m^{1/5}\cdot 2\sqrt{m})}\right)^{m^{4/5-\lambda}}.$$

Here $|\mathcal{F}| \cdot (mD)^r 2^{-(m-m^{4/5}-r)r}$ is the contribution of $A(H)$, $A(H_1)$, ..., $A(H_r)$, the factor

$$\left(m^2 D 2^{-(m-2k)}\right)^M$$

comes from the M disjoint emergency sets

$$\left(\binom{m^3}{2} \cdot \Delta_2 \cdot 2^{-(m-2k-m^{1/5}\cdot 2\sqrt{m})}\right)^{m^{4/5-\lambda}}$$

comes from the 2-bond sets (see Δ_2). Finally, the "$m^{1/5}$" in the exponent of 2 comes from Statement 43.7. It follows that the "total danger" is "extremely small." □

Combining Statements 43.6 and 43.8 we obtain:

Statement 43.9 *By playing rationally Breaker can force that, if there is a first violator of Hazardous Lemma (II), then its "tree like configuration" can be truncated (by throwing out sets if necessary) such that the $l_1 + l_2 + \ldots + l_{r/2}$ emergency super-sets are pairwise disjoint, where $r \geq k^{1-\lambda}/4$.* □

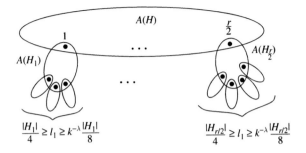

$$\frac{|H_1|}{4} \geq l_1 \geq k^{-\lambda}\frac{|H_1|}{8} \qquad\qquad \frac{|H_{r/2}|}{4} \geq l_1 \geq k^{-\lambda}\frac{|H_{r/2}|}{8}$$

Statement 43.9 can be interpreted as an analogue of *Simplest Case (II)*: there is no coincidence, and also there is no "extra intersection" beyond the mandatory 1-point intersections.

To complete the proof of Hazardous Lemma (II) we show that Breaker can prevent the appearance of any truncated "tree like configuration" in Statement 43.9. Let M be the maximum number of pairwise disjoint sets among $A(H_1), \ldots, A(H_{r/2})$. We distinguish two cases:

Case A: $M \geq r/4$
Then the originial argument at the end of Section 40 works without modification.

Case B: $M < r/4$
Then $(r/2 - M) > r/4 > M$. Breaker employs the usual Power-of-Two Scoring System; the "total danger" is estimated from above by

$$|\mathcal{F}| \cdot 2^{-\sum_{j=1}^{r/2}(m-|H_j|-2\sqrt{m}\cdot j+(m-2k)l_j)} \cdot (mD)^M \cdot \Delta_2^{\frac{r}{2}-M} \cdot (mD)^{l_1+\ldots+l_{r/2}}$$

$$= |\mathcal{F}| \cdot \prod_{j=1}^{r/2}\left(\left(\frac{mD}{2^{m-2k}}\right)^{l_j} \cdot 2^{-(m-|H_j|-2j\sqrt{m})} \cdot \sqrt{mD \cdot \Delta_2}\right). \qquad (43.14)$$

Note that $l_j \geq k^{-\lambda}|H_j|/8$, so returning to (43.14)

$$(43.14) \leq |\mathcal{F}| \cdot \prod_{j=1}^{r/2}\left(\left(\frac{mD}{2^{m-2k}}\right)^{k^{-\lambda}|H_j|/8} \cdot 2^{-\delta m/2+|H_j|+2j\sqrt{m}}\right),$$

which is "exponentially small," i.e. Breaker can indeed prevent it. This completes the proof of Hazardous Lemma (II). $\qquad\square$

Since hypergraph $\mathcal{F}$ is not Almost Disjoint, we have to slightly modify **Cases 1–8** in Section 39.

4. How to extract a Big Set from a large connected family of small sets?
Assume we are in the middle of a play; let $\mathcal{S}$ denote the family of all small sets (up to this point of the play); a small set is either a small-em set ("blank part of

an emergency set") or a small-haz set ("blank part of a hazardous set"). The union of the small sets is the small board; the union of the small-em sets is the E.R. (Emergency Room).

Let C be a component of S with $|C| \geq 2^{k^\lambda} \cdot m^{-2}$. Of course, a component is a connected family. It is trivial from Hazardous Lemma (II) that C must contain an emergency set.

We describe a **sequential extraction process** finding $T = 2^{k^\lambda/9} \cdot m^{-2}$ emergency sets in C such that the union of their *super-sets* in $\mathcal{F}$ form a Big Set. More precisely, we define a growing family $\mathcal{G}_j = \{E_1, E_2, \ldots, E_j\}$ of emergency sets, and the union $\mathcal{U}_j = \cup_{i=1}^{j} A(E_i)$ of their super-sets ($E_i \subset A(E_i) \in \mathcal{F}$) for $j = 1, 2, 3, \ldots, T$ where $T = 2^{k^\lambda/9} \cdot m^{-2}$ ("integral part").

The beginning is trivial: C must contain an emergency set E_1. Let $\mathcal{G}_1 = \{E_1\}$, and let $\mathcal{U}_1 = A(E_1)$ be the super-set of E_1.

To explain the general step, assume that we just completed the jth step: we have already selected a family $\mathcal{G}_{g(j)}$ of $g(j)$ distinct emergency sets, and the union of their emergency sets from $\mathcal{F}$ is denoted by $\mathcal{U}_{g(j)}$. Note that $g(j) \geq j$: indeed, in each step we shall find either 1 or $l = k^{-\lambda}|H|/8$ new emergency sets (the case of the "centipede"), i.e. either $g(j+1) = g(j)+1$ or $g(j+1) = g(j)+l$. By definition $|\mathcal{U}_{g(j)}| \leq g(j) \cdot m$.

Next we describe the $(j+1)$st step. Unfortunately we have to distinguish several cases. The case study is going to be very "geometric" in nature, so it is absolutely necessary to fully understand the corresponding figures.

Note in advance that the family $\mathcal{B}$ of Big Sets is defined as follows

$$\mathcal{B} = \left\{ B = \mathcal{U}_T : \text{ for all possible ways one can } \textbf{grow } \mathcal{U}_T \text{ of length } T = 2^{k^\lambda/9} \cdot m^{-2} \right.$$

$$\left. \textbf{in terms of } \mathcal{F} \text{ by using the Case Study below} \right\}$$

(we shall clarify "grow" and "in terms of $\mathcal{F}$" later).

Case 1A* is the same as Case 1 in Section 39; it is just repeated.

Case 1A*: The first $A(C)$-neighborhood of $\mathcal{U}_{g(j)}$ contains a super-set $A(E)$ of an emergency set E such that $|\mathcal{U}_{g(j)} \cap A(E)| = 1$. Then $\mathcal{G}_{g(j+1)} = \mathcal{G}_{g(j)} \cup \{E\}$, and so $\mathcal{U}_{g(j+1)} = \mathcal{U}_{g(j)} \cup A(E)$ satisfies $|\mathcal{U}_{g(j+1)}| = |\mathcal{U}_{g(j)}| + (m-1)$.

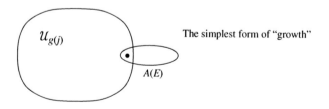

The simplest form of "growth"

We use the notation $A(\ldots)$ for a super-set in $\mathcal{F}$, and so "$A(\mathcal{C})$-neighborhood" means a family of super-sets of the members of component $\mathcal{C}$; observe that $A(\mathcal{C})$ is a connected sub-family of $\mathcal{F}$. Here the super-sets are not necessarily uniquely determined.

Since the hypergraph is not Almost Disjoint, we have a new case which turns out to be a huge "bonus" in the Calculations later.

Case 1B*: The first $A(\mathcal{C})$-neighborhood of $\mathcal{U}_{g(j)}$ contains a super-set A of a small set S such that $\mathcal{U}_{g(j)} \cap A$ is collinear. Then $\mathcal{G}_{g(j+1)} = \mathcal{G}_{g(j)} \cup \{S\}$, and so $\mathcal{U}_{g(j+1)} = \mathcal{U}_{g(j)} \cup A$ satisfies $|\mathcal{U}_{g(j+1)}| \geq |\mathcal{U}_{g(j)}| + (m - 2\sqrt{m})$.

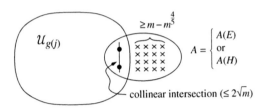

Why is Case 1B* a "bonus case"? Well, there are $\binom{|\mathcal{U}_{g(j)}|}{2}$ ways to choose 2 points of the collinear intersection $\mathcal{U}_{g(j)} \cap A$, and the Max Pair-Degree $\Delta_2 \leq 2^{(1-\delta)m}$; the product

$$\binom{|\mathcal{U}_{g(j)}|}{2} 2^{(1-\delta)m} \text{ is negligible compared to the target value } 2^{m - m^{4/5} - 2\sqrt{m}},$$

which comes from the fact that in the Big Game Maker has a Shutout of size $\geq m - m^{4/5} - 2\sqrt{m}$ in set A. Note that "$-m^{4/5}$" comes from (43.13) and "$-2\sqrt{m}$" comes from condition (b) in Theorem 43.1.

Next comes Case 2A* which is the analogue of Case 2 in Section 39.

Case 2A*: The second $A(\mathcal{C})$-neighborhood of $\mathcal{U}_{g(j)}$ contains a super-set $A(E)$ of an emergency set E such that there is a neighbor A_1 of $A(E)$ from the first $A(\mathcal{C})$-neighborhood of $\mathcal{U}_{g(j)}$ which intersects $\mathcal{U}_{g(j)}$ in at least three non-collinear points.

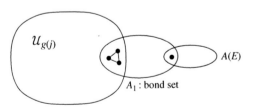

Then $\mathcal{G}_{g(j+1)} = \mathcal{G}_{g(j)} \cup \{E\}$, A_1 is an auxiliary **bond set**, and $\mathcal{U}_{g(j+1)} = \mathcal{U}_{g(j)} \cup A(E)$ satisfies $|\mathcal{U}_{g(j+1)}| = |\mathcal{U}_{g(j)}| + m$.

The name *bond set* comes from the fact that by condition (d) in Theorem 43.1 three non-collinear points determine a plane. *Bond sets* have little effect in the calculation of the *total number* of Big Sets, see the **Calculations** later. To illustrate what we mean, let's calculate the number of different ways we can "grow" to $\mathcal{G}_{g(j+1)}$ from $\mathcal{G}_{g(j)}$ in Case 2A*. A trivial upper bound on the number of possibilities is

$$\binom{|\mathcal{U}_{g(j)}|}{3}(m-3)\cdot D \le \frac{g^3(j)\cdot m^4\cdot D}{6},$$

where $D = D(\mathcal{F})$ is the Maximum Degree of $\mathcal{F}$.

Next comes another "bonus case."

Case 2B*: The second $A(\mathcal{C})$-neighborhood of $\mathcal{U}_{g(j)}$ contains the super-set $A(S)$ of a small set S such that there is a neighbor A_1 of $A(S)$ from the first $A(\mathcal{C})$-neighborhood of $\mathcal{U}_{g(j)}$ which (1) intersects $\mathcal{U}_{g(j)}$ in a non-collinear set, and (2) $A(S) \cap A_1$ has at least 2 elements.

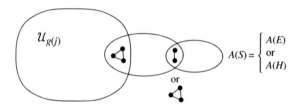

Since A_1 is a bond set, Case 2B* is a bonus case for the same reason as Case 1B*.

As in Section 39 we interrupt our case study for a moment, and make an observation about the *hazardous sets* of $\mathcal{C}$. Let $H \in S$ be an arbitrary small-haz set. We show that H can have a very limited number of "types" only. The types listed below form the analogue of Types 1–4 in Section 39.

Type 1*: $A(H)$ intersects $\mathcal{U}_{g(j)}$ in a non-collinear set.

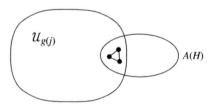

Type 2A*: $A(H)$ intersects $\mathcal{U}_{g(j)}$ in at least 2 elements.

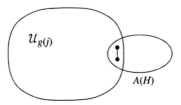

Observe that Type 2A* is covered by Case 1B*, so without loss of generality we can assume that there is *no* Type 2A* hazardous set in $\mathcal{C}$. Also Type 1* is excluded, unless, of course, $A(H)$ is in the first $A(\mathcal{C})$-neighborhood of $\mathcal{U}_{g(j)}$.

Type 2B*: There exist an emergency set E such that $|A(E) \cap A(H)| = 1$ and $|A(E) \cap \mathcal{U}_{g(j)}| = 1$.

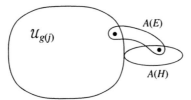

Observe that Type 2B* is covered by Case 1A*, so without loss of generality we can assume that there is *no* Type 2B* hazardous set in $\mathcal{C}$.

Type 2C*: There exist an emergency set E such that $|A(E) \cap A(H)| = 1$ and $A(E) \cap \mathcal{U}_{g(j)}$ is a collinear set of size ≥ 2.

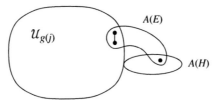

Observe that Type 2C* is covered by Case 1B*, so without loss of generality we can assume that there is *no* Type 2* hazardous set in $\mathcal{C}$.

Type 3*: There exist three emergency sets E_1, E_2, and E_3 such that each intersection $A(E_i) \cap \mathcal{U}_{g(j)}$ is non-collinear, $|A(E_i) \cap A(H)| = 1$ for $i = 1, 2, 3$, and the 3 points $A(E_i) \cap A(H)$ form a non-collinear set.

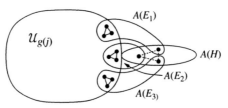

Type 4*: There exist $l = k^{-\lambda}|H|/8$ emergency sets $E_1, E_2, \ldots, E_l$ such that $|A(E_i) \cap \mathcal{U}_{g(j)}| = 0$ and $|A(E_i) \cap A(H)| = 1$ for $i = 1, 2, \ldots, l$; it is like a "centipede."

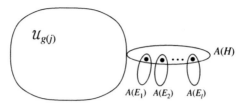

Note that for Types 2B*, 2C*, 3*, 4* the intersection size $|A(H) \cap \mathcal{U}_{g(j)}|$ is 0 or 1. In view of Statement 43.2 we can assume that the "legs" of the "centipede" are pairwise disjoint. The last type is another bonus.

Type 5*: $|A(H) \cap \mathcal{U}_{g(j)}|$ is 0 or 1, and there exist two emergency sets E_1 and E_2 such that $|A(E_i) \cap \mathcal{U}_{g(j)}| \geq 2$ and $|A(E_i) \cap A(H)| = 1$ for $i = 1, 2$.

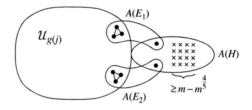

We have no more types, because the seemingly missing case "$A(H) \cap \mathcal{U}_{g(j)}$ is non-collinear" is already covered by Type 1*.

After the classification of "types," we can return to our case study. Assume that the second $A(\mathcal{C})$-neighborhood of $\mathcal{U}_{g(j)}$ contains a super-set $A(E)$ of an emergency set E. Let A_1 be a neighbor of $A(E)$ from the first $A(\mathcal{C})$-neighborhood of $\mathcal{U}_{g(j)}$. In view of Cases 1B*, 2A*, 2B* we can assume that A_1 intersects $\mathcal{U}_{g(j)}$ in exactly one point. By Case 1A* we know that $A_1 = A(H)$, i.e. A_1 has to be the super-set of a hazardous set H.

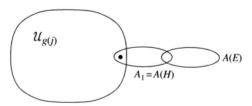

We have two possibilities: H is of Type 3* or Type 4*.

Case 3*: H is of Type 3*. Then $\mathcal{G}_{g(j+1)} = \mathcal{G}_{g(j)} \cup \{E\}$, $A(E_1), A(E_2), A(E_3), A(H)$ are auxiliary bond sets, and clearly $|\mathcal{U}_{g(j+1)}| = |\mathcal{U}_{g(j)}| + m$.

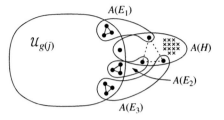

Case 4*: H is of Type 4*. Then $\mathcal{G}_{g(j+1)} = \mathcal{G}_{g(j)} \cup \{H, E_1, E_2, \ldots, E_l\}$, and $A(H)$ is an auxiliary set, but *not* a bond set. We have

$$|\mathcal{U}_{g(j+1)}| \geq |\mathcal{U}_{g(j)}| + lm + (m - l - 1).$$

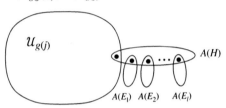

The number of ways we can "grow" to $\mathcal{G}_{g(j+1)}$ from $\mathcal{G}_{g(j)}$ in Case 4* is bounded from above by

$$|\mathcal{U}_{g(j)}| \cdot D \cdot (mD)^l \quad \text{where } l = k^{-\lambda}|H|/8$$

and $D = D(\mathcal{F})$ is the Maximum Degree. Similarly to the proof of Theorem 40.1, here in Case 4* we also include set $A(H)$, so the target value of this configuration is at least (note that $|H| = 8lk^\lambda$)

$$2^{(m-|H|)+l(m-2k)} = 2^{(l+1)m-2kl-8k^\lambda l} \geq 2^{(n-2k-8k^\lambda)(l+1)}.$$

It follows that the ratio

$$\frac{|\mathcal{U}_{g(j)}| \cdot D \cdot (mD)^l}{2^{(m-|H|)+l(m-2k)}} \leq |\mathcal{U}_{g(j)}| \cdot \left(\frac{mD}{2^{m-2k-8k^\lambda}}\right)^{l+1} \leq |\mathcal{U}_{g(j)}| \cdot 2^{-kl} \qquad (43.15)$$

is extremely small if $|\mathcal{U}_{g(j)}| \leq 2^{kl/2}$. Notice that (43.15) is the perfect analogue of (40.15).

In the rest of the case study we can assume that the second $A(\mathcal{C})$-neighborhood of $\mathcal{U}_{g(j)}$ consists of hazardous sets only.

Let H be a hazardous set in the second $A(\mathcal{C})$-neighborhood of $\mathcal{U}_{g(j)}$. Let A_1 be a neighbor of $A(H)$ which is in the first $A(\mathcal{C})$-neighborhood of $\mathcal{U}_{g(j)}$. Then:

(a) either $A_1 \cap \mathcal{U}_{g(j)}$ is non-collinear;
(b) or $A_1 \cap \mathcal{U}_{g(j)}$ is a collinear set of size ≥ 2;
(c) or $A_1 \cap \mathcal{U}_{g(j)}$ is a one-element set.

Observe that here the second option (b) is covered by Case 1B*.

Case 5A*: $A_1 \cap \mathcal{U}_{g(j)}$ is non-collinear and $A_1 \cap A(H)$ is a set of size ≥ 2.

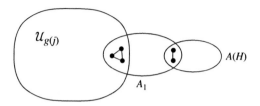

Then $\mathcal{G}_{g(j+1)} = \mathcal{G}_{g(j)} \cup \{H\}$. Since A_1 is a bond set, this case is basically the same as Case 1B*.

Case 5B*: $|A_1 \cap \mathcal{U}_{g(j)}| \geq 2$, $|A_1 \cap A(H)| = 1$, and H is a Type 4* hazardous set.

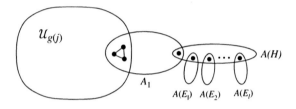

Then $\mathcal{G}_{g(j+1)} = \mathcal{G}_{g(j)} \cup \{E_1, E_2, \ldots, E_l\}$ with $l = k^{-\lambda}|H|/8$. Since A_1 is a bond set, this case is basically the same as Case 4*.

Note that the case "$A_1 \cap \mathcal{U}_{g(j)}$ is non-collinear, $|A_1 \cap A(H)| = 1$, and H is a Type 3* hazardous set" is covered by Case 3*.

Finally, consider the case $|A_1 \cap \mathcal{U}_{g(j)}| = 1$. Then $A_1 = A(H_1)$, where H_1 is either Type 4*, which is covered by Case 4*, or of Type 3*, which is covered by Case 3*. This completes the case study!

5. Calculations: We go through the cases discussed above; we give an upper bound to the number of ways one can "grow" to $\mathcal{G}_{g(j+1)}$ from $\mathcal{G}_{g(j)}$, and also we renormalize by dividing each with its own target value

$$|\mathcal{U}_{g(j)}| \cdot D \cdot 2^{-(m-2k)} \leq g(j) \cdot m \cdot D \cdot 2^{-(m-2k)} \quad \text{in Case 1A}^*;$$

$$\binom{|\mathcal{U}_{g(j)}|}{2} \Delta_2 \cdot 2^{-(m-m^{4/5})} \leq \frac{g^2(j) \cdot m^2 \cdot 2^{(1-\delta)m}}{2} \cdot 2^{-(m-m^{4/5})} \quad \text{in Case 1B}^*;$$

$$\binom{|\mathcal{U}_{g(j)}|}{3} \cdot mD \cdot 2^{-(m-2k)} \leq \frac{g^3(j) \cdot m^4 \cdot D}{6} \cdot 2^{-(m-2k)} \quad \text{in Case 2A}^*;$$

$$\binom{|\mathcal{U}_{g(j)}|}{3} \cdot \binom{m}{2} \Delta_2 \cdot 2^{-(m-m^{4/5})} \leq \frac{g^3(j) \cdot m^5 \cdot 2^{(1-\delta)m}}{12} \cdot 2^{-(m-m^{4/5})} \quad \text{in Case 2B}^*;$$

$$\left(\frac{|\mathcal{U}_{g(j)}|}{3}\right)^3 \cdot m^3 \cdot 2^{-(m-m^{4/5})} \leq \frac{g(j)^9 m^{12}}{216} \cdot 2^{-(m-m^{4/5})} \quad \text{in Case } 3^*;$$

$$|\mathcal{U}_{g(j)}| \cdot D \cdot (mD)^l 2^{-(m-|H|)-l(m-2k)} \leq g(j) \cdot 2^{-kl} \quad \text{in Case } 4^*,$$

where $l = k^{-\lambda}|H|/8$, and we used (43.15)

$$\left(\frac{|\mathcal{U}_{g(j)}|}{3}\right)\binom{m}{2}\Delta_2 \leq \frac{g^3(j) \cdot m^5 \cdot 2^{(1-\delta)m} \cdot 2^{-(m-m^{4/5})}}{12} \quad \text{in Case } 5A^*;$$

$$\left(\frac{|\mathcal{U}_{g(j)}|}{3}\right)(mD)^{l+1} 2^{-(m-|H|)-l(m-2k)} \leq g^3(j) \cdot 2^{-kl} \quad \text{in Case } 5B^*,$$

where $l = k^{-\lambda}|H|/8$, and we used (43.15).

Note that the "legs" $A(E_1), A(E_2), \ldots, A(E_l)$ of the "centipede" are pairwise disjoint. Indeed, in view of Statement 43.2 the "disjointness" is guaranteed with "$l/2$" instead of "l," which is irrelevant in the calculations.

6. Conclusion. In view of the RELARIN technique, the smallest winning set size in the small game is $\geq k^\lambda$. If two planes have an intersection of size ≥ 2, then the intersection is a line. Since 2 points uniquely determine a line (Almost Disjointness, see condition (d) in Theorem 43.1), every small set contains at most $\binom{m}{2}$ winning sets of size $\geq k^\lambda$.

It follows that, if every component of the family of small sets in the small game has size $\leq 2^{k^\lambda} \cdot m^{-2}$, then Breaker can block every winning set in the small game. This statement is the perfect analogue of the Growing Erdős–Selfridge Lemma in Section 39, and the proof is the same.

The next statement is the perfect analogue of the Big Set Lemma in Section 39: If at some stage of the play there is a component $\mathcal{C}$ of the family $\mathcal{S}$ of all small sets such that

$$|\mathcal{C}| \geq \frac{2^{k^\lambda}}{m^2},$$

then there is a $\mathcal{G}_{g(j)} \subset \mathcal{C}$ such that

$$g(j) \geq \frac{2^{k^\lambda/9}}{m^2}.$$

Here $\mathcal{G}_{g(j)}$ is a family of small sets selected from $\mathcal{C}$ such that in each one of the j steps the "growth" comes from one of the cases described above.

Indeed, we can clearly repeat the proof of the Big Set Lemma if the inequality

$$g^9(j) \cdot m^{12} < \frac{2^{k^\lambda}}{m^2}$$

holds, which is certainly the case with $g(j) = 2^{k^\lambda/9} \cdot m^{-2}$. The final step in the proof is to prevent the appearance of the corresponding huge Shutout in a Big Set $\mathcal{U}_T$ with

$$T = \frac{2^{k^\lambda/9}}{m^2}.$$

Similarly to Section 39, here again Breaker applies Lemma 5 from Section 36 to the family of all "Big Sets in the broad sense." By choosing the parameters as

$$\lambda = \frac{1}{2}, \quad k = m^{2/5}, \quad g(j) = T = \frac{2^{m^{1/5}/9}}{m^2},$$

and $m \geq c_0 = c_0(\delta)$, every term in the Calculations above becomes less than $1/2$ or $2^{-(l+1)}$, respectively, where $2^{-(l+1)}$ corresponds to the "centipedes." Therefore, Lemma 5 from Section 36 applies if

$$|\mathcal{F}| \left(\frac{1}{2}\right)^{T-1} < \frac{1}{2},$$

which is trivially true with

$$T = \frac{2^{m^{1/5}/9}}{m^2} \quad \text{and} \quad |\mathcal{F}| \leq 2^{m^{1.1}}$$

if m is large enough. This completes the proof of Theorem 43.1. $\qquad\square$

44

Winning lattices: exact solution

1. Taking care of the Translates. In the proof of Theorem 8.2 (we ignore case (h): complete bipartite graphs, which is basically covered by Theorem 6.4) we face a new technical difficulty: two $q \times q$ lattices may have *very large* non-collinear intersections, far beyond the square-root size intersection of "planes" in Section 43 – *translates* are particularly bad. Two $q \times q$ lattices are called *translates* of each other if one can be moved to the other by a translation, where the translation is given by a vector whose 2 endpoints belong to the $q \times q$ lattice.

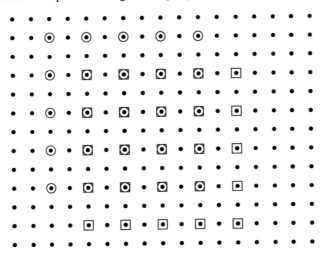

This means we have to modify the proof of Theorem 43.1; the RELARIN technique has to be combined with some new, *ad hoc* arguments – see the two **Selection Lemmas** below. The first Selection Lemma says that "a family of emergency sets arising in a play, where any two are translates of each other, cannot be too crowded." The second one, a 2-dimensional generalization of Lemma 42.1, says that, "if a family of $q \times q$ lattices doesn't contain translates, then we can always select a relatively large sub-family with a large union set."

First fix a type from (a) to (g) in Theorem 8.2, and let $\mathcal{F}$ denote the family of all $q \times q$ lattices of the fixed type in the $N \times N$ board. Let $m = q^2$, and assume that

$$|\mathcal{F}| \leq 2^{m^{1+\frac{\varepsilon}{2}}} = 2^{q^{2+\varepsilon}}. \tag{44.1}$$

Note in advance that $\varepsilon = 10^{-4}$ will be a good choice – see Theorem 44.1 at the end of the section – but at this stage we work with ε as an unspecified small positive absolute constant.

The precise form of the *First Selection Lemma* is the following; note that we use the notation of the RELARIN technique introduced in Section 43; here each winning set is a $q \times q$ lattice of fixed type (see (a)–(g) in Theorem 8.2).

First Selection Lemma: *Let $0 < \varepsilon \leq \mu \leq 1/100$ and $k = q^{1-\varepsilon}$. Consider a play in a Lattice Game in the $N \times N$ board where Breaker uses the RELARIN technique described in the proof of Theorem 43.1 up to Hazardous Lemma (II) in Section 43. (The value of the critical exponent $0 < \lambda < 1$ in the RELARIN technique is unspecified yet.) Let $A(E_1), A(E_2), \ldots, A(E_l)$ be l emergency super-sets such that:*

(1) any two are translates of each other, and
(2) every $A(E_i), i = 2, 3, \ldots, l$ intersects the first super-set $A(E_1)$. If q is sufficiently large depending on the value of the RELARIN constant λ, then $l < q^{1-\mu}$.

Remark. The proof employs an iterated *density growth* argument.

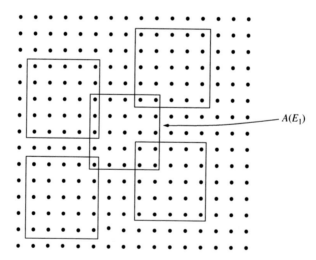

$A(E_1)$

Proof. Let $k = q^{1-\varepsilon}$ with $0 < \varepsilon \leq \mu \leq 1/100$. The hypothesis of the lemma implies that at the end of the play each $A(E_i)$, $1 \leq i \leq l$ has at least $(q^2 - 2k)$ marks of Maker in the Big Game. Here E_i denotes the ($\leq 2k$)-element emergency set. Let $\widetilde{E}_i$ denote the ($\geq k$)-element *small-em* subset $\widetilde{E}_i \subset E_i$; note that each E_i intersects every other small set $\widetilde{E}_j$ ($j \neq i, 1 \leq j \leq l$) in less than k^λ points – this is a consequence of the RELARIN technique.

Assume that every $A(E_i)$ with $i = 2, 3, 4, \ldots, l$ intersects $A(E_1)$. If $l \geq q^{1-\mu}$, then from this assumption we are going to derive a contradiction. Assumptions (1)–(2) above guarantee that the l translates $A(E_i)$ with $i = 1, 2, 3, \ldots, l$ can be all embedded in an appropriate "underlying lattice" of size $3q \times 3q$, see the figure above.

We divide this $3q \times 3q$ underlying lattice into $100q^{1+\mu}$ "boxes," where each "box" has the size $3q^{(1-\mu)/2}/10 \times 3q^{(1-\mu)/2}/10$. Every super-set $A(E_i)$ is a $q \times q$ lattice itself; we identify it with its "lower left corner"; we denote the "lower left corner" by LLC_i ($1 \leq i \leq l$). We distinguish two cases:

Case A: Some "box" contains at least 20 LLC_is

Let $\{LLC_i : i \in I\}$ denote the set of $|I| \geq 20$ "lower left corners" in the same "box." Project the $|I| \geq 20$ LLC_is on to the horizontal and vertical sides of the "box"; we get at most $2|I|$ points (some of the projections may coincide).

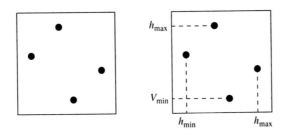

Let h_{max} and h_{min} denote the horizontal maximum and minimum among the projections, and, similarly, let v_{max} and v_{min} denote the vertical maximum and minimum among the projections. The four projections come from (at most) 4 LLC_is, which determine (at most) 4 supersets: for notational convenience we denote them by $A(E_1)$, $A(E_2)$, $A(E_3)$, $A(E_4)$ (there may occur some coincidence among them).

We study the following geometric question: Let $j \in I \setminus \{1, 2, 3, 4\}$; how large is the difference-set $\left(\bigcup_{i=1}^4 A(E_i)\right) \setminus A(E_j)$? Consider, for example, the lower left corner of $A(E_j)$, and assume that $A(E_1)$ corresponds to h_{min} and $A(E_2)$ corresponds to v_{min}.

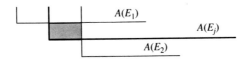

Then $A(E_1) \cup A(E_2)$ "almost" covers the *lower left corner neighborhood* of $A(E_j)$: the possibly uncovered part (the shaded region on the picture) has area less than that of the "box," i.e. less than $9q^{1-\mu}/100$. The same applies for the other 3 corners of $A(E_j)$. It follows that

$$\text{Area}\left(\left(\bigcup_{i=1}^{4}A(E_i)\right)\setminus A(E_j)\right)\le 4\cdot\frac{9}{100}q^{1-\mu}=\frac{9}{25}q^{1-\mu}. \tag{44.2}$$

Let $V(end; Big; Maker)$ denote the part of board $V = N \times N$, which is occupied at the end of the play by Maker in the Big Game. Clearly

$$A(E_i)\setminus V(end; Big; Maker) = E_i \supset \tilde{E}_i. \tag{44.3}$$

Since any 2 distinct $\tilde{E}_i$s have intersection $\le k^\lambda$, with $J = I\setminus\{1,2,3,4\}$ we have

$$\left|\bigcup_{j\in J}\tilde{E}_j\right|\ge\sum_{j\in J}\left|\tilde{E}_j\right|-\sum_{j_1<j_2}\left|\tilde{E}_{j_1}\cap\tilde{E}_{j_2}\right|$$

$$\ge|J|k-\binom{|J|}{2}k^\lambda\ge(|J|-1)k, \tag{44.4}$$

assuming q is sufficiently large depending on the value of (the unspecified yet) constant $0 < \lambda < 1$.

By (44.2) and (44.3)

$$\left|\bigcup_{i=1}^{4}E_i\right|\ge\left|\bigcup_{j\in J}\tilde{E}_j\right|-|J|\cdot\frac{9}{25}q^{1-\mu}. \tag{44.5}$$

But (44.5) contradicts (44.4): indeed

$$8k\ge\left|\bigcup_{i=1}^{4}E_i\right|\ge\left|\bigcup_{j\in J}\tilde{E}_j\right|-|J|\cdot\frac{9}{25}q^{1-\mu}\ge(|J|-1)k-|J|\cdot\frac{9}{25}q^{1-\mu}\ge 9k,$$

since $k = q^{1-\varepsilon}\ge q^{1-\mu}$. This proves that Case A is *impossible*.

If Case A *fails*, then at least $l/20$ "boxes" are non-empty; since $l \ge q^{1-\mu}$ and the total number of "boxes" is $100q^{1+\mu}$, we have the following:

Start: The density of the non-empty "boxes" among all "boxes" is at least

$$\frac{q^{1-\mu}/20}{100q^{1+\mu}}=\frac{q^{-2\mu}}{2000}.$$

Let us divide the $3q \times 3q$ underlying lattice into $6 \times 6 = 36$ equal parts, and for each part apply the 3×3 "Tic-Tac-Toe partition"

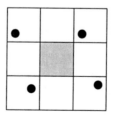

Assume first that there is a 3×3 "Tic-Tac-Toe partition" such that each one of the 4 corner squares contains at least one LLC_i – let $i = 1, 2, 3, 4$ – and also the middle square contains at least 20 LLC_js – let $j \in J$ where $|J| \geq 20$. Then we obtain a contradiction the same way as in Case A. Indeed

$$\bigcup_{i=1}^{4} A(E_i) \supset \bigcup_{j \in J} A(E_j), \tag{44.6}$$

and subtracting $V(end; Big; Maker)$ from both sides of (44.6), we have

$$\bigcup_{i=1}^{4} E_i \supset \bigcup_{j \in J} \widetilde{E}_j, \tag{44.7}$$

which is even stronger than (44.5). Now applying the argument in (44.4) we end up with a contradiction.

We can assume, therefore, that in each one of the 36 $q/2 \times q/2$ squares either one of the 4 small corner squares is empty, or the small middle square has less than 20 LLC_js ("almost empty"). Then in one of the $9 \cdot 36 = 324$ $q/6 \times q/6$ squares the density of the non-empty "boxes" goes up by at least 10%, i.e. we arrive at the:

First Step: There is a particular $q/6 \times q/6$ square such that the density of the non-empty "boxes" among its all "boxes" is at least

$$\left(1 + \frac{1}{10}\right) \frac{q^{-2\mu}}{2000}.$$

Let's divide this $q/6 \times q/6$ square into $3 \times 3 = 9$ smaller squares ("Tic-Tac-Toe partition").

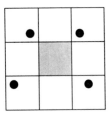

Assume first that each one of the 4 corner squares contains at least one LLC_i – let $i = 1, 2, 3, 4$ – and also the middle square contains at least 20 LLC_js – let $j \in J$ where $|J| \geq 20$. Then we obtain a contradiction the same way as in the Start.

We can assume, therefore, that either one of the 4 corner squares is empty, or the middle square has less than 20 LLC_js ("almost empty"). Then in one of the 9 $q/18 \times q/18$ squares the density of the non-empty "boxes" goes up by at least 10%, i.e. we arrive at the:

Second Step: There is a particular $q/18 \times q/18$ square such that the density of the non-empty "boxes" among its all "boxes" is at least

$$\left(1 + \frac{1}{10}\right)^2 \frac{q^{-2\mu}}{2000}.$$

Iterating this argument, in the rth step we get the:

General Step: There is a particular $q3^{-r}/2 \times q3^{-r}/2$ square such that the density of the non-empty "boxes" among its all "boxes" is at least

$$\left(1 + \frac{1}{10}\right)^r \frac{q^{-2\mu}}{2000}.$$

By choosing $r = r_0$ where $3^{r_0} \approx q^{1/3}$ we get a contradiction. Indeed, a density is always ≤ 1, so

$$\left(1 + \frac{1}{10}\right)^{r_0} \cdot \frac{1}{2000q^{2\mu}} \leq 1,$$

which clearly contradicts the choice $3^{r_0} \approx q^{1/3}$ if $\mu \leq 1/100$ and q is large enough. This contradiction proves the First Selection Lemma.   □

Second Selection Lemma: *Let $A_1, A_2, \ldots, A_l$ be l $q \times q$ parallelogram lattices in $\mathbb{Z}^2$. Assume that this is a translates-free family, i.e. there are no two which are translates of each other. If $l \leq q^4$, then we can always select a sub-family $A_{j_1}, A_{j_2}, \ldots, A_{j_\nu}$ such that $\nu \geq l^{1/33}$ and*

$$\left| \bigcup_{i=1}^{\nu} A_{j_i} \right| \geq (\nu - 1)q^2.$$

Proof. This lemma is a 2-dimensional generalization of Lemma 42.1; not surprisingly the proof is also a generalization of the proof of Lemma 42.1. Again the idea is to employ the Pigeonhole Principle. What can we say about the intersection $A_{i_1} \cap A_{i_2}$ of two $q \times q$ parallelogram lattices? The intersection is either *collinear* ("small" like $O(q)$) or *non-collinear* ("can be very large"). Let

$$A_{i_1} = \{\mathbf{u}_{i_1} + k\mathbf{v}_{i_1} + l\mathbf{w}_{i_1} : k = 0, 1, \ldots, q-1, l = 0, 1, \ldots, q-1\}$$

and

$$A_{i_2} = \{\mathbf{u}_{i_2} + k\mathbf{v}_{i_2} + l\mathbf{w}_{i_2} : k = 0, 1, \ldots, q-1, l = 0, 1, \ldots, q-1\}.$$

If the intersection is non-collinear, then there are integers $a_1, b_1, c_1, d_1, a_2, b_2, c_2, d_2$ such that

$$a_1\mathbf{v}_{i_1} + b_1\mathbf{w}_{i_1} = c_1\mathbf{v}_{i_2} + d_1\mathbf{w}_{i_2} \tag{44.8}$$

$$a_2\mathbf{v}_{i_1} + b_2\mathbf{w}_{i_1} = c_2\mathbf{v}_{i_2} + d_2\mathbf{w}_{i_2} \tag{44.9}$$

and the 2 vectors in (44.8) and (44.9) are non-parallel.

It is easy to see that

$$|A_{i_1} \cap A_{i_2}| \le \frac{2q^2}{\max\{|a_1|, |b_1|, |c_1|, |d_1|, |a_2|, |b_2|, |c_2|, |d_2|\}}. \qquad (44.10)$$

If $\mathbf{v} = (v^{(1)}, v^{(2)})$ and $\mathbf{w} = (w^{(1)}, w^{(2)})$, then (44.8) and (44.9) together can be expressed in terms of 2-by-2 matrices as follows

$$\begin{pmatrix} a_1 & b_1 \\ a_2 & b_2 \end{pmatrix} \begin{pmatrix} v_{i_1}^{(1)} & v_{i_1}^{(2)} \\ w_{i_1}^{(1)} & w_{i_1}^{(2)} \end{pmatrix} = \begin{pmatrix} c_1 & d_1 \\ c_2 & d_2 \end{pmatrix} \begin{pmatrix} v_{i_2}^{(1)} & v_{i_2}^{(2)} \\ w_{i_2}^{(1)} & w_{i_2}^{(2)} \end{pmatrix}, \qquad (44.11)$$

where (44.11) is a non-singular matrix.

By (44.11)

$$\begin{pmatrix} v_{i_2}^{(1)} & v_{i_2}^{(2)} \\ w_{i_2}^{(1)} & w_{i_2}^{(2)} \end{pmatrix} = \begin{pmatrix} c_1 & d_1 \\ c_2 & d_2 \end{pmatrix}^{-1} \begin{pmatrix} a_1 & b_1 \\ a_2 & b_2 \end{pmatrix} \begin{pmatrix} v_{i_1}^{(1)} & v_{i_1}^{(2)} \\ w_{i_1}^{(1)} & w_{i_1}^{(2)} \end{pmatrix}$$

$$= (c_1 d_2 - d_1 c_2)^{-1} \begin{pmatrix} d_2 & -d_1 \\ -c_2 & c_1 \end{pmatrix} \begin{pmatrix} a_1 & b_1 \\ a_2 & b_2 \end{pmatrix} \begin{pmatrix} v_{i_1}^{(1)} & v_{i_1}^{(2)} \\ w_{i_1}^{(1)} & w_{i_1}^{(2)} \end{pmatrix}. \qquad (44.12)$$

So by (44.12)

$$\begin{pmatrix} v_{i_2}^{(1)} & v_{i_2}^{(2)} \\ w_{i_2}^{(1)} & w_{i_2}^{(2)} \end{pmatrix} = \begin{pmatrix} \alpha & \beta \\ \gamma & \delta \end{pmatrix} \begin{pmatrix} v_{i_1}^{(1)} & v_{i_1}^{(2)} \\ w_{i_1}^{(1)} & w_{i_1}^{(2)} \end{pmatrix}, \qquad (44.13)$$

where $\alpha, \beta, \gamma, \delta$ are four rational numbers such that each numerator and each denumerator has absolute value $\le 2M^2$, where $M = \max\{|a_1|, |b_1|, |c_1|, |d_1|, |a_2|, |b_2|, |c_2|, |d_2|\}$.

Now we are ready to select a "relatively large sub-family of $A_1, A_2, \ldots, A_l$ which forms a large union set." First pick A_1, and study those A_js for which $|A_1 \cap A_j| > \frac{2q^2}{M}$ (M will be specified later as an integral parameter much less than q); we call these js "M-bad with respect to 1." By (44.10) we have $\max\{|a_1|, |b_1|, |c_1|, |d_1|, |a_j|, |b_j|, |c_j|, |d_j|\} \le M$ (see (44.8)–(44.9)).

If $\max\{|a_1|, |b_1|, |c_1|, |d_1|, |a_j|, |b_j|, |c_j|, |d_j|\} \le M$, then the number of different matrices $\begin{pmatrix} \alpha & \beta \\ \gamma & \delta \end{pmatrix}$ in (44.13) is at most $\left((4M^2+1)^2\right)^4$, so the total number of indices j which are "M-bad with respect to 1" is $(4M^2+1)^8$.

From $\{1, 2, \ldots, l\}$ we throw out all indices j which are "M-bad with respect to 1," and 1 itself, and pick an arbitrary index h_1 from the remaining set. We study those A_js for which $|A_{h_1} \cap A_j| > \frac{2q^2}{M}$; we call these js "M-bad with respect to h_1." Just as before we throw out all indices j which are "M-bad with respect to h_1," and h_1 itself, and pick an arbitrary index h_2 from the remaining set, and so on. Repeating this argument, we obtain a sub-sequence $h_0 = 1, h_1, h_2, h_3, \ldots$ of $\{1, 2, \ldots, l\}$ such that

$$|A_{h_i} \cap A_{h_j}| \le \frac{2q^2}{M} \text{ holds for all } 0 \le i < j < r = \frac{l}{(4M^2+1)^8+1}. \qquad (44.14)$$

By (44.14)

$$\left| \bigcup_{j=1}^{r} A_{h_j} \right| \ge r \cdot q^2 - \binom{r}{2} \cdot \frac{2q^2}{M} \ge (r-1)q^2$$

if $M = l^{2/33}$ and $r = \sqrt{M}$. This completes the proof of the Second Selection Lemma. $\qquad \square$

2. How to prove the Hazardous Lemma?

Again the main difficulty is how to prove the analogous Hazardous Lemma.

Hazardous Lemma (III): *Consider a play in a Lattice Game in the $N \times N$ board where Breaker uses the RELARIN technique with $k = q^{1-\varepsilon}$ (where $0 < \varepsilon \le 1/100$ will be specified later) developed at the beginning of Section 43 up to Hazardous Lemma (II). If Breaker extends the Big Game with some additional Forbidden Configurations, he can force that, during the whole course of the play, every hazardous set H intersects the E.R. ("Emergency Room") in at least $|H|/4$ points.*

Proof of the Hazardous Lemma (III). Recall that $\mathcal{F}$ denotes the family of all $q \times q$ sub-lattices of a fixed type in $N \times N$ (see types (a)–(g) in Theorem 8.2); $\mathcal{F}$ is an m-uniform hypergraph with $m = q^2$. (If the reader wants a concrete example, then it is a good idea to pick the most general lattice: the parallelogram lattice.)

The situation here is somewhat similar to Theorem 43.1: we have the following weaker version of condition (d):

(1) 3 non-collinear points in the $N \times N$ board nearly determine a winning lattice $A \in \mathcal{F}$ in the sense that there are at most $\binom{q^2}{3}$ winning lattices $A \in \mathcal{F}$ containing an arbitrary fixed non-collinear triplet;

(2) the Max Pair-Degree $\Delta_2 = \Delta_2(\mathcal{F})$ of hypergraph $\mathcal{F}$ satisfies an upper bound

$$\Delta_2 \le 2^{(1-\delta)m} = 2^{(1-\delta)q^2}$$

with some positive absolute constant $\delta > 0$.

To understand what Breaker has to prevent, assume that there is a "violator." Let H be the first violator of Hazardous Lemma (III): H is the first hazardous set which intersects the E.R. in less than $|H|/4$ points. Since $|H| \ge 2k+2$, H must intersect the difference-set "smallboard minus E.R." in $\ge k/2$ points. These $\ge k/2$ points are covered by *non-violator* hazardous sets $H_1, H_2, \ldots, H_q$ where $k^{1-\lambda}/2 \le q \le k/2$; note that an intersection of size $\ge k^\lambda$ would prevent H to be a new hazardous set – this is a consequence of the RELARIN technique. Since each H_j, $1 \le j \le q$ is a non-violator of the Hazardous Lemma (III), each H_j must intersect the E.R. in at least $|H_j|/4$ points, which are covered by emergency sets $E_{j,1}, E_{j,2}, \ldots, E_{j,l_j}$. Note that the inequality $k^{-\lambda}|H_j|/4 \leq l_j \leq |H_j|/4$ holds for every $j = 1, 2, \ldots, q$. Formally we have $1 + q + \sum_{j=1}^{q} l_j$ super-sets

$$A(H), A(H_1), \ldots, A(H_q), A(E_{1,1}), \ldots, A(E_{1,l_1}), A(E_{2,1}), \ldots, A(E_{2,l_2}),$$

$$A(E_{3,1}), \ldots, A(E_{3,l_3}), \ldots, A(E_{q,1}), \ldots, A(E_{q,l_q}), \tag{44.15}$$

but there may occur some *coincidence* among these hyperedges of $\mathcal{F}$, and also there may occur some "extra intersection" among these sets. As usual $A(\ldots)$ denotes a super-set (but the super-set is not uniquely determined anymore because we gave up on Almost Disjointness; this ambiguity doesn't cause any problem).

Again the idea is to reduce the proof of Hazardous Lemma (III) to an analogue of the *Simplest Case*. A novelty in the proof is that we need many more extra classes of Forbidden Configurations, and have to show that they are all "avoidable" in a rational play.

Let $k = m^{(1-\varepsilon)/2} = q^{1-\varepsilon}$ (the value of $0 < \varepsilon \leq 1/100$ is undefined yet; $\varepsilon = 10^{-4}$ will be a good choice).

Again we call $A(H_i)$ together with $A(E_{i,1})$, $A(E_{i,2}), \ldots$, $A(E_{i,l_i})$ the ith "centipede"; $A(H_i)$ is the "head" and $A(E_{i,1})$, $A(E_{i,2}), \ldots$, $A(E_{i,l_i})$ are the "legs."

A new technical difficulty is that the "legs" may have very large overlappings; the largest overlappings may come from "translates."

Step 1: Let $A(E_{i,j})$, $j \in J$ be a family of "legs" such that:

(1) $|A(H_i) \cap A(E_{i,j})| \geq 2$, $j \in J$, and
(2) any two $A(E_{i,j})$'s with $j \in J$ are translates of each other.

What is the largest possible size $|J|$ under conditions (1)–(2) above?

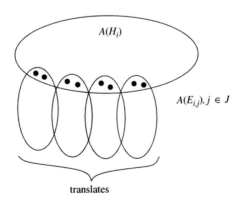

translates

It follows from the First Selection Lemma that there are $|J_0| \geq q^{-1+\mu}|J|$ pairwise disjoint "legs" $A(E_{i,j})$ with $j \in J_0 \subset J$. The "target value" of the system $\{A(H_i), A(E_{i,j}), j \in J_0\}$ is at most

$$|\mathcal{F}|\left(\binom{m}{2}\Delta_2 \cdot 2^{-(m-2k-2)}\right)^{|J_0|}$$

$$\leq |\mathcal{F}|\left(\binom{m}{2}\cdot 2^{(1-\delta)m-(m-2k-2)}\right)^{|J_0|},$$

which, in view of (44.1), is "extremely small" if $|J_0| \geq 2q^\varepsilon/\delta$. Thus the case

$$|J| \geq \frac{2}{\delta}q^{1-\mu+\varepsilon}$$

is avoidable if Breaker plays rationally in the Big Game.

***Step* 2:** Let $A(E_{i,p})$, $p \in P$ be the family of all "legs" such that $|A(H_i) \cap A(E_{i,p})| \geq 2$, $p \in P$. It follows from Step 1 that there is a subset $P_0 \subset P$ with

$$|P_0| \geq \frac{|P|}{\frac{2}{\delta}q^{1-\mu+\varepsilon}}$$

such that $A(E_{i,p})$, $p \in P_0$ is translates-free. By the Second Selection Lemma there is a subset $P_1 \subset P_0$ such that

$$|P_1| \geq |P_0|^{1/33} \quad \text{and} \quad |\bigcup_{p\in P_1} A(E_{i,p})| \geq (|P_1|-1)m.$$

The "target value" of the system $\{A(H_i), A(E_{i,p}), p \in P_1\}$ is at most

$$|\mathcal{F}| \cdot \left(\binom{m}{2}\Delta_2\right)^{|P_1|} \cdot 2^{-(|P_1|-1)(m-2k-2)}$$

$$\leq |\mathcal{F}| \cdot 2^m \cdot \left(\binom{m}{2}\cdot 2^{(1-\delta)m-(m-2k-2)}\right)^{|P_1|},$$

which is "extremely small" if

$$|P_1| \geq \frac{2}{\delta}q^\varepsilon \iff |P_0| \geq \left(\frac{2}{\delta}\right)^{33} q^{33\varepsilon}.$$

Therefore, the case

$$|P| \geq \left(\frac{2}{\delta}\right)^{33} q^{1-\mu+34\varepsilon}$$

is avoidable if Breaker plays rationally in the Big Game.

Recall that

$$l_i \geq k^{-\lambda}|H_i|/4 \geq \frac{1}{2}k^{1-\lambda} = \frac{1}{2}q^{(1-\varepsilon)(1-\lambda)}.$$

Thus if $(1-\varepsilon)(1-\lambda) > 1-\mu+34\varepsilon$, then the "legs" with (≥ 2)-element intersection in $A(H_i)$ form a small minority; we throw them all out. Therefore, in the rest of the proof we can assume that every "leg" $A(E_{i,j})$ intersects $A(H_i)$ in exactly 1 point.

Step 3: Let $A(E_{i,s})$, $s \in S$ be the family of all "legs" with the property that $A(E_{i,s})$ intersects another translate "leg." Let $A(E_{i,s})$, $s \in S_0$ be a maximum size sub-family of pairwise disjoint "legs"; of course, $S_0 \subset S$.

Case a: Among the "legs" $A(E_{i,s})$, $s \in S_0$ there is a sub-family of size $\geq \frac{2}{8}q^\varepsilon$ such that any 2 elements are translates of each other.

Let $S_1 \subset S_0$ with $|S_1| \geq \frac{2}{8}q^\varepsilon$ the index-set for this family of "pairwise translates." By definition each $A(E_{i,s})$ with $s \in S_1$ intersets another translate "leg"; these 2 intersecting $q \times q$ lattices together are contained in a $(2q) \times (2q)$ lattice, which intersets $A(H_i)$ in two points. This means we can repeat the argument of Step 1. Indeed, we can assume that the Max Pair-Degree for the family of all $(2q) \times (2q)$ lattices is $2^{(1-\delta)m}$; furthermore, there are at most $4q^2$ ways to localize a $q \times q$ lattice inside a $(2q) \times (2q)$ lattice. So the "target value" of the system $\{A(H_i), A(E_{i,s}), s \in S_1\}$ is at most

$$|\mathcal{F}| \cdot \left(\binom{m}{2} \cdot 2^{(1-\delta)m} \cdot 4m^2 \cdot 2^{-(m-2k-2)} \right)^{|S_1|},$$

which is "extremely small" if $|S_1| \geq \frac{2}{8}q^\varepsilon$. This proves that *Case a* is avoidable if Breaker plays rationally in the Big Game.

Case b: Among the "legs" $A(E_{i,s})$, $s \in S_0$ there is a sub-family of size

$$\geq \left(\frac{2}{\delta} \right)^{33} \cdot q^{33\varepsilon} \text{ which is translates} - \text{free.}$$

Now we can basically repeat the argument of Step 2 (applying the Second Selection Lemma!), and conclude that *Case b* is also avoidable by Breaker.

We can assume that

$$|S_0| < \left(\frac{2}{\delta} \right)^{34} \cdot q^{34\varepsilon}.$$

Indeed, otherwise either *Case a* or *Case b* holds, and both are avoidable. The First Selection Lemma gives that

$$|S| \leq |S_0| \cdot q^{1-\mu}, \text{ thus we have } |S| < \left(\frac{2}{\delta} \right)^{34} \cdot q^{1-\mu+34\varepsilon}.$$

If $(1 - \varepsilon)(1 - \lambda) > 1 - \mu + 34\varepsilon$, then the set $A(E_{i,s})$, $s \in S$ of "legs" forms a small minority among all "legs" of $A(H_i)$; we throw the "legs" $A(E_{i,s})$, $s \in S$ all out. Therefore, in the rest of the proof we can assume the following properties:

(A) any 2 "legs" of $A(H_i)$, which are translates of each other, are pairwise disjoint;
(B) the total number of "legs" of $A(H_i)$ is $\geq (1 - o(1))k^{-\lambda}|H_i|/4$; and from the argument at the end of Step 2:
(C) every "leg" intersects $A(H_i)$ in exactly 1 point (of course different "legs" intersect in different points).

Step 4: For notational convenience we can assume that the consecutive indices $A(E_{i,1})$, $A(E_{i,2})$, $A(E_{i,3}),\ldots,$ $A(E_{i,l_i})$ denote the set of "legs" of $A(H_i)$ satisfying properties (A)–(B)–(C) above. Consider the largest component of $\{A(E_{i,1}), A(E_{i,2}),\ldots, A(E_{i,l_i})\}$; again for simplicity we can assume that $\{A(E_{i,1}), A(E_{i,2}),\ldots, A(E_{i,t})\}$ is the largest component (the first t "legs"), and also we can assume that $A(E_{i,2})$ intersects $A(E_{i,1})$, $A(E_{i,3})$ intersects $A(E_{i,1})\cup A(E_{i,2})$, $A(E_{i,4})$ intersects $A(E_{i,1})\cup A(E_{i,2})\cup A(E_{i,3})$, and so on.

We compute the "target value" of $\{A(H_i), A(E_{i,1}), A(E_{i,2}), A(E_{i,3}),\ldots, A(E_{i,t})\}$. The first 2 sets $A(H_i), A(E_{i,1})$ represent "target value" at most

$$|\mathcal{F}|\cdot mD2^{-(m-2k-2)},$$

where $D = \text{MaxDegree}(\mathcal{F})$. The second "leg" $A(E_{i,2})$ intersects $A(H_i)$ in 1 point; moreover, $A(E_{i,2})$ intersects $A(E_{i,1})$ either (1) in 1 point, or (2) in a (≥ 2)-element collinear set, or (3) in a non-collinear set. Accordingly, the contribution of $A(E_{i,2})$ in the total "target value" is:

(1) either $\leq \binom{2m}{2}\cdot\Delta_2\cdot 2^{-(m-2k-4)}$;
(2) or $\leq \binom{m}{2}\cdot\Delta_2\cdot 2^{-(m-2k-3-q)}$;
(3) or $\leq \binom{m}{3}\cdot\Delta_3\cdot 2^{-M_2} \leq \binom{m}{3}\binom{m}{3}2^{-M_2}$,

    where M_2 is the number of Maker's marks in the difference set $A(E_{i,2})\setminus A(E_{i,1})$ in the course of the Big Game, and $\Delta_3 \leq \binom{m}{3}$ is a trivial upper bound on the Non-collinear Triplet-Degree Δ_3.

Similarly, the third "leg" $A(E_{i,3})$ intersects $A(H_i)$ in 1 point; moreover, $A(E_{i,3})$ intersects $A(E_{i,1})\cup A(E_{i,2})$ either (1) in 1 point, or (2) in a (≥ 2)-element collinear set, or (3) in a non-collinear set. Accordingly, the contribution of $A(E_{i,3})$ in the total "target value" is:

(1) either $\leq \binom{3m}{2}\cdot\Delta_2\cdot 2^{-(m-2k-5)}$;
(2) or $\leq \binom{2m}{2}\cdot\Delta_2\cdot 2^{-(m-2k-3-2q)}$;
(3) or $\leq \binom{2m}{3}\cdot\Delta_3\cdot 2^{-M_3} \leq \binom{2m}{3}\binom{m}{3}2^{-M_3}$,

    where M_3 is the number of Maker's marks in the difference set $A(E_{i,3})\setminus (A(E_{i,1})\cup A(E_{i,2}))$ in the course of the Big Game, and so on. Since

$$\Delta_2 \leq 2^{(1-\delta)m}, \; m = q^2, \; k = q^{1-\varepsilon},$$

both cases (1) and (2) contribute a very small factor $\leq 2^{-\delta m/2}$ for each new "leg," which obviously "kills" the big factor

$$|\mathcal{F}| \leq 2^{m^{1+\varepsilon/2}} = 2^{q^{2+\varepsilon}}$$

if say $t \geq q^{2\varepsilon}$ and cases (1)–(2) are the only options.

How about case (3)? We begin the discussion with the trivial lower bound: for any subset $I \subset \{1, 2, 3, \ldots, t\}$

$$M_2 + M_3 + \ldots + M_t \geq |\bigcup_{j \in I} A(E_{i,j})| - m - |I|(2k + 2), \qquad (44.16)$$

where the "$-m$" comes from $A(E_{i,1})$ and the "$-|I|(2k+2)$" comes from the condition "in the course of the Big Game."

Case a: Among the t "legs" $A(E_{i,j})$, $j = 1, 2, \ldots, t$ there is a subset of size $q^{2\varepsilon}$ with the property that any two are translates of each other.
Property (A) above implies that this set of $|I| = q^{2\varepsilon}$ "translates" consists of pairwise disjoint sets, which gives the trivial lower bound

$$|\bigcup_{j \in I} A(E_{i,j})| \geq q^{2\varepsilon} m.$$

Returning to (44.16) we have

$$M_2 + M_3 + \ldots + M_t \geq q^{2\varepsilon} m - 2m - q^{2\varepsilon}(2k + 2) \geq (q^{2\varepsilon} - 3)m.$$

This implies that the total contribution of case (3) "kills" the big factor

$$|\mathcal{F}| \leq 2^{q^{2+\varepsilon}},$$

which makes *Case a* "avoidable."

Case b: Among the t "legs" $A(E_{i,j})$, $j = 1, 2, \ldots, t$ there is a subset $A(E_{i,j})$, $j \in J$ with $|J| \geq q^{66\varepsilon}$ which is translates-free.
  Then we can apply the Second Selection Lemma: there is a subset $J_0 \subset J$ with $|J_0| = q^{2\varepsilon}$ such that

$$|\bigcup_{j \in J_0} A(E_{i,j})| \geq (|J_0| - 1)m = (q^{2\varepsilon} - 1)m.$$

Now we can argue similarly to *Case a*, and conclude that *Case b* is also "avoidable."

  We can assume that $t < q^{68\varepsilon}$; indeed, otherwise either *Case a* or *Case b* holds, and both are "avoidable." Keeping one "leg" from every component, we can modify property (B) above as follows:

(B*) the total number of "legs" of $A(H_i)$ is $\geq k^{-\lambda} |H_i| q^{-68\varepsilon}/5$.

  Of course, property (C) remains true (note that property (A) is covered by the new (B*)). This way obtain the following picture:

Step 5:

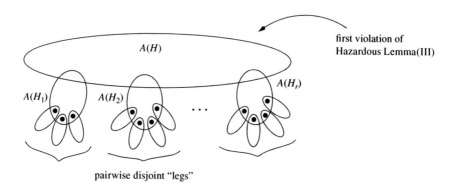

pairwise disjoint "legs"

The main point here is that, for every "head" $A(H_i)$, its "legs" $A(E_{i,j})$ satisfy properties (B∗) and (C); in particular, the "legs" $A(H_i)$ are pairwise disjoint.

We can assume that $|H_i| \leq m^{4/5}$ for every i. Indeed, $|H_i| > m^{4/5}$ means at least $k^{-\lambda}|H_i|q^{-68\varepsilon}/5 > m^{3/4}$ pairwise disjoint "legs" $A(E_{i,j})$. The "target value" of the system $\{A(H_i), A(E_{i,1}), A(E_{i,2}), \ldots, A(E_{i,m^{3/4}})\}$ is at most

$$|\mathcal{F}| \left(m \cdot D \cdot 2^{-(m-2k-2)} \right)^{m^{3/4}} \leq |\mathcal{F}| \cdot 2^{-k \cdot m^{3/4}},$$

which is extremely small, proving that $\max_i |H_i| > m^{4/5}$ is avoidable.

How about the "top centipede" where $A(H)$ is the "head" and $A(H_1)$, $A(H_2)$, $A(H_3), \ldots, A(H_r)$ with $r \geq k^{-\lambda}|H|/4$ are the "legs"? Can we apply Steps 1–5 to the "top centipede"? Well, an obvious technical difficulty is that the "legs" $A(H_i)$, $i = 1, 2, \ldots, r$ of the "top centipede" come from hazardous sets, and the First Selection Lemma is about emergency sets. Because H_i is a hazardous set, we cannot directly apply the First Selection Lemma, but we can still adapt its proof. This is how the adaptation goes.

Recall that at the end of a play each $A(H_i)$, $i = 1, 2, \ldots, r$ has $(m - |H_i|)$ marks of Maker made in the course of the Big Game. Let $\widetilde{H}_i$ denote the $(\geq |H_i|/2)$-element small-haz subset of hazardous set H_i. Note that each H_i intersects every other small set $\widetilde{H}_j$, $j \neq i$, $1 \leq j \leq r$ in less than k^{λ} points; this is a consequence of the RELARIN technique. Also $|H_i| \geq 2k+2$ for every i; this inequality (instead of the equality for the emergency sets) just *helps* in the adaptation of the proof technique of the First Selection Lemma. The only extra condition that we need in the proof is that the sizes are relatively close to each other (see condition (3) below).

Modified First Selection Lemma: *Let* $0 < \varepsilon \leq \mu \leq 1/100$ *and* $k = q^{1-\varepsilon}$. *Consider a play in a Lattice Game in the* $N \times N$ *board where Breaker uses the RELARIN technique described in the proof of Theorem 43.1 up to Hazardous Lemma (II) in Section 43. (The* *value of the critical exponent* $0 < \lambda < 1$ *in the RELARIN technique is unspecified yet.)* Let $A(H_1), A(H_2), \ldots, A(H_l)$ be l *hazardous super-sets such that:*

(1) any two are translates of each other, and
(2) every $A(H_i)$, $i = 2, 3, \ldots, l$ intersects the first super-set $A(H_1)$,
(3) the bounds $\frac{1}{2} \leq |H_i|/|H_j| \leq 2$ hold for any two sets.

If q is sufficiently large depending on the value of the RELARIN constant λ, then $l < q^{1-\mu}$.

By using the Modified First Selection Lemma (of course, condition (3) requires that first we put the sizes $|H_i|$ into power-of-two size boxes) instead of the original form, we can repeat Steps 1–5 for the "top centipede." The only irrelevant difference is that the factor

$$2^{-(m-|E_{i,j}|)} = 2^{-(m-2k-2)} \text{ is replaced by } 2^{-(m-|H_i|)} \leq 2^{-(m-m^{4/5})},$$

which is still small enough to "kill" the Max Pair-Degree $\Delta_2 \leq 2^{(1-\delta)m}$. This way we get:

Step 6:

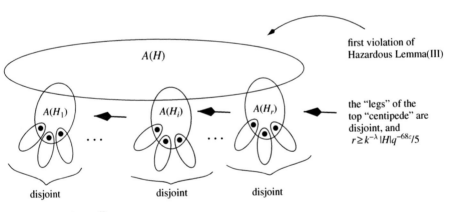

first violation of
Hazardous Lemma(III)

$A(H)$

the "legs" of the
top "centipede" are
disjoint, and
$r \geq k^{-\lambda} |H| q^{-68\varepsilon}/5$

$A(H_1)$ $A(H_i)$ $A(H_r)$

disjoint   disjoint   disjoint

$\geq k^{-\lambda} |H| q^{-68\varepsilon}/5$ legs

Of course, we still may have "extra overlappings" between "legs" with different "heads" like $A(E_{i_1,j_1})$ and $A(E_{i_2,j_2})$, $i_1 \neq i_2$; also between "legs" and "heads" like $A(E_{i_1,j_1})$ and $A(H_{i_2})$ where $i_1 \neq i_2$; or even a "coincidence" among these sets.

If there is no "extra overlapping" or "coincidence" among these sets, then we get the **Simplest Case**, which was already fully discussed at the end of Section 40. The standard idea that we always use is to reduce the General Case to the Simplest Case. This plan can be carried out by repeating the proof of Statements 43.3–43.9

(we already have the analogue of Statement 43.1, Statement 43.2, and inequality (43.13)).

3. Transference Principle. There is a good reason why can we repeat the proofs of Statements 43.3–43.9, and we call it a **Transference Principle**. In Section 43 two Planes intersect either in a Line with intersection size $\leq 2\sqrt{m}$, or in a point, or are disjoint. Here in Section 44 the situation is somewhat more complicated: two $q \times q$ lattices intersect either in a non-collinear set, or in a collinear set of size $\leq q = \sqrt{m}$, or in a point, or are disjoint. "Collinear intersection" in Section 44 is the perfect analogue of "line intersection" in Section 43; this part is obvious, but how about the "non-collinear intersection"? Well, for the Lattice Games the Non-collinear Triplet-Degree Δ_3 is clearly $\leq \binom{m}{3}$. Since $\binom{m}{3} = O((\log N)^{3/2})$ is a polylogarithmic function of N (which is far below the "square-root barrier"), "non-collinear intersection" in Section 44 corresponds to "coincidence of sets" in Section 43; at least the calculations give essentially the same result. This simple Transference Principle explains why we can "save" Statements 43.3–43.9 in Section 44, and can reduce the general case to the Simplest Case essentially the same way as we did in Section 43. This is how the proof of Hazardous Lemma (III) goes.

Next comes the BigGame–SmallGame Decomposition (combined with the RELARIN technique); in particular the step of *how to extract a Big Set from a large component of the family of small sets.* Again we can repeat the corresponding argument in Section 43 without any difficulty – of course, we always use the Transference Principle.

The last step is to specify the parameters. Let $\mu = 1/100$; in the arguments above we assumed the inequalities

$$68\varepsilon + (1-\varepsilon)\lambda < \frac{4}{5} - \frac{3}{4} = \frac{1}{20};$$

$$(1-\varepsilon)(1-\lambda) > 1 - \mu + 34\varepsilon;$$

$$3\left((1-\varepsilon)(1-\lambda) - 68\varepsilon\right) > 2 + \varepsilon,$$

where the last inequality is needed in the Simplest Case.

These 3 inequalities are clearly satisfied with the choice of $\lambda = 1/200$ and $\varepsilon = 10^{-4}$.

Summarizing, we have just proved the following result.

Theorem 44.1 *Fix an arbitrary Lattice Game among types (a)–(g) (see Section 8) in the $N \times N$ board. Let $\mathcal{F}$ denote the corresponding q^2-uniform hypergraph. Let $\varepsilon = 10^{-4}$; if the global and local sizes are bounded from above as*

$$|\mathcal{F}| \leq 2^{q^{2+\varepsilon}} \quad and \quad D = \mathrm{MaxDeg}(\mathcal{F}) \leq 2^{q^2 - 4q^{1-\varepsilon}},$$

and $q \geq q_0$, i.e. q is sufficiently large, then playing on $\mathcal{F}$ the second player can always force a Strong Draw. $\square$

The "largest" Lattice Game is the Parallelogram Lattice Game, and even then $|\mathcal{F}| \leq N^6$, so the global condition

$$|\mathcal{F}| \leq 2^{q^{2+\varepsilon}}$$

is trivially satisfied with $q = O(\sqrt{\log N})$, assuming q is large enough.

The local condition

$$D = \mathrm{MaxDeg}(\mathcal{F}) \leq 2^{q^2 - 4q^{1-\varepsilon}} \text{ with } \varepsilon = 10^{-4}$$

says that in the Lattice Games we succeeded to break the "square-root barrier"!

The application of Theorem 44.1 completes the proof of the Strong Draw part of Theorem 8.2 (a)–(g). (Note that the Strong Draw part of Theorem 8.2 (h): complete bipartite graph, is identical with the proof in Section 38.)

Finally, notice that the generous (super-exponential in $m = q^2$) upper bound

$$|\mathcal{F}| \leq 2^{q^{2+\varepsilon}} \text{ with } \varepsilon = 10^{-4}$$

on the global size proves the *Irrelevance of the Board Size*, an interesting feature mentioned in Section 9 after Open Problem 9.1.

45

I-Can-You-Can't Games – Second Player's Moral Victory

1. Building and Blocking in the same strategy. In Section 27 we introduced the class of Who-Scores-More Games; here another 2-hypergraph class is introduced: the class of *Who-Scores-First Games*. Let $\mathcal{F}$ and $\mathcal{G}$ be two hypergraphs on the same underlying set ("board") V. As usual, the players alternately occupy new points of the board: First Player's goal is to occupy an $A \in \mathcal{F}$ and Second Player's goal is to occupy a $B \in \mathcal{G}$; the player who achieves his goal *first* is declared the winner; otherwise the play ends in a draw. We refer to this game as the $(\mathcal{F}, \mathcal{G})$ **Who-Scores-First Game**. The symmetric case $\mathcal{F} = \mathcal{G}$ gives back the old concept of Positional Game. Just like the Positional Games, the Who-Scores-First Games in general are completely hopeless, but there is a sub-class for which some success can be reported: the I-Can-You-Can't Games, i.e. when one player's goal is just doable and the other player's goal is impossible.

Doable vs. impossible. Consider, for example, the Aligned Square Lattice vs. Aligned Rectangle Lattice Who-Scores-First Game on the $N \times N$ board where the square-lattice size is $q_1 \times q_1$ with

$$q_1 = \left\lfloor \sqrt{\log_2 N} + o(1) \right\rfloor \tag{45.1}$$

and the rectangle-lattice size is $q_2 \times q_2$ with

$$q_2 = \left\lceil \sqrt{2 \log_2 N} + o(1) \right\rceil. \tag{45.2}$$

In view of Theorem 8.2 (a) $q_1 = \left\lfloor \sqrt{\log_2 N} + o(1) \right\rfloor$ is the "largest achievable size," and in view of Theorem 8.2 (b) $q_2 = \left\lceil \sqrt{2 \log_2 N} + o(1) \right\rceil$ is the "smallest impossible size." This suggests that in this particular Aligned square-lattice vs. Aligned rectangle lattice Who-Scores-First Game the likely winner is First Player ("the Square Lattice guy"). It seems plausible that First Player's job is "merely" putting together a Weak Win Strategy and a Strong Draw Strategy. But this is exactly where the problem is: "How to put two very different strategies together?"

A simple arithmetic operation like "addition" or "multiplication" of the potential functions clearly doesn't work; we cannot expect such an elementary solution in general. Then what can we do? First of all:

What is the difficulty here? Because of the value of q_2 (see (45.2)), First Player has a Strong Draw strategy $Str\,(s.d.)$: he can block every $q_2 \times q_2$ aligned rectangle lattice in $N \times N$. This $Str\,(s.d.)$ is an advanced BigGame–SmallGame Decomposition where the *small game* falls apart into a huge number of disjoint sub-games with disjoint sub-boards ("components") in the course of a play.

Also First Player has a Weak Win Strategy $Str\,(w.w.)$ (an application of Theorem 1.2): he can occupy a whole $q_1 \times q_1$ aligned Square Lattice (see (45.1)). This $Str\,(w.w.)$ does not decompose; it remains a single coherent entity during the whole course of the play.

$Str\,(s.d.)$ decomposes into a huge number of components and First Player plays componentwise, on the other hand, $Str\,(w.w.)$ does *not* decompose! How to combine these two very different strategies into a single one? Even if there is no simple general recipe, in some special cases we can still succeed. Here one way to do it is described; call it:

2. The technique of super-polynomial multipliers. Assume that we are in the middle of a play, where First Player already occupied $x_1, x_2, \ldots, x_i$ and Second Player occupied $y_1, y_2, \ldots, y_i$ from the board $V = N \times N$. The question is how to choose First Player's next move x_{i+1}. Write $X(i) = \{x_1, x_2, \ldots, x_i\}$, $Y(i) = \{y_1, y_2, \ldots, y_i\}$, and

$$\mathcal{F}(i) = \{A \setminus X(i) : \; A \in \mathcal{F}, A \cap Y(i) = \emptyset\}, \tag{45.3}$$

where $\mathcal{F}$ is the family of all $q_1 \times q_1$ aligned Square Lattices in $N \times N$ (see (45.1)). Note that $\mathcal{F}(i)$ is the family of the unoccupied parts of the "survivors" in $\mathcal{F}$; the truncated $\mathcal{F}(i)$ can be a multi-hypergraph even if the original $\mathcal{F}$ is not. We use the Power-of-Two Scoring System

$$T(\mathcal{M}) = \sum_{M \in \mathcal{M}} 2^{-|M|}$$

where $\mathcal{M}$ is a multi-hypergraph, the sum $T(\mathcal{F}(i))$ represents the Opportunity Function: if First Player can guarantee that the Opportunity Function remains non-zero during the whole course of the play, then at the end First Player certainly owns a whole $A \in \mathcal{F}$, i.e. a $q_1 \times q_1$ aligned Square Lattice.

This is how $Str\,(w.w.)$ works: First Player tries to keep $T(\mathcal{F}(i))$ "large"; at the same time First Player also tries to block hypergraph $\mathcal{G}$ where $\mathcal{G}$ is the family of all $q_2 \times q_2$ aligned rectangle lattices in $N \times N$ (see (45.2)), i.e. he tries to employ strategy $Str\,(s.d.)$. $Str\,(s.d.)$ is rather complicated: it involves the concepts of Big

Game, small game, dangerous sets, emergency sets, hazardous sets, small-em sets, small-haz sets, the RELARIN technique, and the two Selection Lemmas.

Warning: The Big Game is a Shutout Game, not a Positional Game. This leads to a minor technical problem: "how to define Shutout Games in terms of hypergraphs." The notation

$$T(\mathcal{H}) = \sum_{A \in \mathcal{H}} 2^{-|A|}$$

was designed to handle Positional Games and Weak Win: if some winning set $A_0 \in \mathcal{F}$ is completely occupied by Maker, i.e. $A_0 \subset X(i)$ for some i, then $A_0 \setminus X(i) = \emptyset$ and

$$T(\mathcal{F}(i)) \geq 2^{-|A_0 \setminus X(i)|} = 1 \tag{45.4}$$

where $\mathcal{F}(i)$ is defined in (45.3).

In a Shutout Game Maker's goal is to put b marks in some $B \in \mathcal{B}$ before Breaker could put his first mark in it. Let $\mathcal{B}_m$ denote the multi-hypergraph where

every $B \in \mathcal{B}$ with $|B| \geq b$ has multiplicity $m(B) = 2^{|B|-b}$

and every $B \in \mathcal{B}$ with $|B| < b$ has multiplicity $m(B) = 0$. $\tag{45.5}$

If Maker wins the Shutout Game on $\mathcal{B}$ and B_0 is the "witness" of his win, then

$$T(\mathcal{B}_m(i)) \geq m(B_0) 2^{-|B_0 \setminus X(i)|} = 1, \tag{45.6}$$

which is the perfect analogue of (45.4) in terms of the auxiliary multi-hypergraph $\mathcal{B}_m$.

Let $\mathcal{G}_{BIG}$ denote the multi-hypergraph of the Big Sets with multiplicity (45.5); the Big Sets are defined in the "broad sense" (see the end of Section 39). Let V_{BIG} denote the Big Board, and, in particular, let $V_{BIG}(i)$ denote the Big Board at the stage $x_1, x_2, \ldots, x_i; y_1, y_2, \ldots, y_i$. Recall that the Big Board is "shrinking" in the course of a play.

In Section 44 we introduced several *Forbidden Configurations*; each one was a Shutout Game. In view of the *Warning* above each Shutout Game defines a hypergraph with multiplicity; the union of these multi-hypergraphs is denoted by $\mathcal{G}_{Forb}$ (again in the "broad sense"). Write

$$\mathcal{G}^*(i) = \{B \setminus (Y(i) \cap V_{BIG}(i)) : \ B \in \mathcal{G}, B \cap (X(i) \cap V_{BIG}(i)) = \emptyset\},$$

$$\mathcal{G}^*_{BIG}(i) = \{C \setminus (Y(i) \cap V_{BIG}(i)) : \ C \in \mathcal{G}_{BIG}, C \cap (X(i) \cap V_{BIG}(i)) = \emptyset\},$$

$$\mathcal{G}^*_{Forb}(i) = \{D \setminus (Y(i \cap V_{BIG}(i)) : \ D \in \mathcal{G}_{Forb}, D \cap (X(i) \cap V_{BIG}(i)) = \emptyset\},$$

and

$$\mathcal{F}^*(i) = \{A \setminus (X(i) \cap V_{BIG}(i)) : \ A \in \mathcal{F}, A \cap (Y(i) \cap V_{BIG}(i)) = \emptyset\}, \tag{45.7}$$

where the mark $\star$ is an abbreviation for the restriction to the actual Big Board $V_{BIG}(i)$.

An inspection of the calculations in Sections 39 and 44 immediately gives the following *super-polynomially small* upper bound

$$T(\mathcal{G}_{BIG}) + T(\mathcal{G}_{Forb}) \leq N^{-(\log N)^{\delta}} \tag{45.8}$$

with some positive absolute constant $\delta > 0$.

Now we are ready to choose First Player's next move x_{i+1}. We distinguish two cases:

Case 1: $y_i \notin V_{BIG}(i)$

In other words, Case 1 means that Second Player's last move was in the small board $V_{small}(i) = V \setminus V_{BIG}(i)$. Then First Player replies to y_i simply following the Strong Draw strategy $Str(s.d.)$: First Player replies in the same component of the hypergraph of all small sets where y_i was, and to find the optimal x_{i+1} he uses the *overcharged* version of the Power-of-Two Scoring System to enforce the Component Condition (see the proof of the *Growing Erdős–Selfridge Lemma* in Section 39).

Case 2: $y_i \in V_{BIG}(i)$

The main idea is the following: we work with the potential function

$$L_i = T(\mathcal{F}^*(i)) - \lambda_1 \cdot T(\mathcal{F}_{Forb}^*(i)) - \lambda_2 \cdot T(\mathcal{G}_{BIG}^*(i))$$
$$- \lambda_3 \cdot T(\mathcal{G}_{Forb}^*(i)) - \lambda_4 \cdot T(\mathcal{G}^*(i)); \tag{45.9}$$

in Case 2 First Player's choice for x_{i+1} is that unoccupied point $z \in V_{BIG}(i) \setminus (X(i) \cup Y(i))$ of the Big Board for which the function

$$L_i(z) = T(\mathcal{F}^*(i); z) - \lambda_1 \cdot T(\mathcal{F}_{Forb}^*(i); z) - \lambda_2 \cdot T(\mathcal{G}_{BIG}^*(i); z)$$
$$- \lambda_3 \cdot T(\mathcal{G}_{Forb}^*(i); z) - \lambda_4 \cdot T(\mathcal{G}^*(i); z) \tag{45.10}$$

attains its *maximum*. To understand what (45.9)–(45.10) means we note that

$$T(\mathcal{M}; z) = \sum_{z \in M \in \mathcal{M}} 2^{-|M|},$$

the families $\mathcal{G}_{BIG}^*(i)$, $\mathcal{G}_{Forb}^*(i)$, $\mathcal{G}^*(i)$ were defined in (45.7), $\mathcal{F}$ is the family of all $q_1 \times q_1$ aligned Square Lattices in $N \times N$ (see (45.1))

$$\mathcal{F}^*(i) = \{A \setminus (X(i) \cap V_{BIG}(i)) : A \in \mathcal{F}, A \cap Y(i) \cap V_{BIG}(i) = \emptyset\},$$

the multi-hypergraph $\mathcal{F}_{Forb}$ will be defined later. Even if we don't know $\mathcal{F}_{Forb}$ yet, we can nevertheless write

$$\mathcal{F}_{Forb}^*(i) = \{E \setminus (X(i) \cap V_{BIG}(i)) : E \in \mathcal{F}_{Forb}, A \cap Y(i) \cap V_{BIG}(i) = \emptyset\}.$$

It remains to specify the positive constants $\lambda_1, \lambda_2, \lambda_3, \lambda_4$. They are called *multipliers*, and are defined by the side condition

$$\lambda_1 \cdot T(\mathcal{F}_{Forb}) = \lambda_2 \cdot T(\mathcal{G}_{BIG}) = \lambda_3 \cdot T(\mathcal{G}_{Forb})$$

$$= \lambda_4 \cdot T(\mathcal{G}) = \frac{1}{8} T(\mathcal{F}). \tag{45.11}$$

It is clear from (45.8) that λ_2 and λ_3 are super-polynomially large. We are going to see later that λ_1 is also super-polynomially large; this explains the name "the technique of super-polynomial multipliers."

The auxiliary hypergraph $\mathcal{F}_{Forb}$ will be defined later as a "technical requirement"; right now we deal with $\mathcal{F}_{Forb}$ as an undefined parameter.

In Case 2 we have $y_i \in V_{BIG}(i)$ and $x_{i+1} \in V_{BIG}(i)$, where x_{i+1} is defined by the *maximum* property above. Let $y_{i+\nu}$ be the first move of Second Player in $V_{BIG}(i)$ with $\nu = \nu(i) \geq 1$. Since the moves after x_{i+1} and before $y_{i+\nu}$ are not in $V_{BIG}(i)$, we have the following inequality (see (45.9)–(45.10))

$$L_{i+\nu} \geq L_i + L_i(x_{i+1}) - L_i(y_{i+\nu}) - T(\mathcal{F}^*(i); x_{i+1}, y_{i+\nu}). \tag{45.12}$$

The maximum property of x_{i+1} implies

$$L_i(x_{i+1}) \geq L_i(y_{i+\nu}); \tag{45.13}$$

moreover, trivially

$$T(\mathcal{F}^*(i); x_{i+1}, y_{i+\nu}) \leq \Delta_2 \leq \binom{q_1^2}{2}, \tag{45.14}$$

where Δ_2 is the Max Pair-Degree of $\mathcal{F}$, where $\mathcal{F}$ denotes the family of all $q_1 \times q_1$ aligned Square Lattices in $N \times N$. Combining (45.12)–(45.14)

$$L_{i+\nu} \geq L_i - \binom{q_1^2}{2} \geq L_i - O\left((\log N)^2\right). \tag{45.15}$$

Side condition (45.11) implies that $L_0 = T(\mathcal{F})/2$, so by (45.15), (8.3) and (45.1)

$$L_{end} \geq L_0 - \frac{N^2}{2} O\left((\log N)^2\right)$$

$$= \frac{1}{2} T(\mathcal{F}) - O\left(N^2 (\log N)^2\right) \geq$$

$$= \frac{1}{4} T(\mathcal{F}) = \left(\frac{1}{12} + o(1)\right) \frac{N^3}{q_1} 2^{-q_1^2} > 0. \tag{45.16}$$

Thus by (45.10) and (45.16)

$$T(\mathcal{F}^*(end)) \geq \frac{1}{4} T(\mathcal{F}), \tag{45.17}$$

where hypergraph $\mathcal{F}^*(end)$ is defined by

$$V^* = V_{BIG}(end) = \bigcap_{i \geq 0} V_{BIG}(i)$$

and

$$\mathcal{F}^*(end) = \{A \setminus V^* : A \in \mathcal{F}, A \cap Y(end) \cap V^* = \emptyset\}.$$

Since

$$T(\mathcal{F}^*(end)) = \sum_{s=0}^{q_1^2} \sum_{A \in \mathcal{F}: \ A \cap Y(end) \cap V^* = \emptyset, |A \setminus V^*| = s} 2^{-s},$$

there must exist an integer s_0 in $0 \leq s_0 \leq q_1^2$ such that (see (45.16)–(45.17))

$$\sum_{A \in \mathcal{F}: \ A \cap Y(end) \cap V^* = \emptyset, |A \setminus V^*| = s_0} 2^{-s_0} \geq \frac{1}{q_1^2} T(\mathcal{F}^*(end))$$

$$\geq \left(\frac{1}{12} + o(1)\right) \frac{N^3}{q_1^3} 2^{-q_1^2}. \tag{45.18}$$

Note in advance that relation (45.18) will motivate our definition of the auxiliary hypergraph $\mathcal{F}_{Forb}$.

3. How to define $\mathcal{F}_{Forb}$? We start with the idea. First Player has two goals: (1) to completely occupy some $A_0 \in \mathcal{F}$, and to block every $B \in \mathcal{G}$ (where $B = q_2 \times q_2$ aligned rectangle lattice in $N \times N$). To own some $A_0 \in \mathcal{F}$ is equivalent to $A_0 \subseteq X(end)$. If the integer s_0 defined in (45.18) is $s_0 = 0$, then, of course, goal (1) is achieved. Thus in the rest we can assume $s_0 \geq 1$.

Here is the key idea about $\mathcal{F}_{Forb}$: $s_0 \geq 1$ means that in a certain sub-family of $\mathcal{F}$ First Player has a "very large" Shutout; on the other hand, working with a properly defined auxiliary hypergraph $\mathcal{F}_{Forb}$, First Player himself can prevent this "very large" Shutout. Thus the case $s_0 \geq 1$ will lead to a contradiction, proving that the "easy case" $s_0 = 0$ is the only possibility. Next we work out the details of this idea.

How to define $\mathcal{F}_{Forb}$: the details. Write (see (45.18))

$$\mathcal{F}_{s_0} = \{A \in \mathcal{F} : \ A \cap Y(end) \cap V^* = \emptyset, |A \setminus V^*| = s_0\}. \tag{45.19}$$

If there is no $A_0 \in \mathcal{F}$ with $A_0 \subseteq X(end)$, then every $A \in \mathcal{F}_{s_0}$ has a non-empty intersection with $Y(end) \cap V^{**}$, where $V^{**} = V \setminus V^*$.

How large is the small board V^{**}? Write

$$m = q_2^2 \text{ and } k = m^{\frac{1}{2} - \varepsilon} = q_2^{1-2\varepsilon}; \tag{45.20}$$

then, by definition the size of the E.R. (the Emergency Room) is estimated from above by

$$|\text{E.R.}| \leq |\mathcal{G}^*(end)| \cdot k = 2^k T(\mathcal{G}^*(end)) \mid \cdot k. \tag{45.21}$$

The Component Condition says that every component has size $\leq 2^{k^\lambda}$, thus by the Hazardous Lemma

$$|V^{**}| = |\text{small board}| \leq |\text{E.R.}| \cdot 2^{k^\lambda}. \tag{45.22}$$

Combining (45.20)–(45.22) we have

$$|V^{**}| = |\text{small board}| \leq T\left(\mathcal{G}^*(end)\right) | \cdot k \cdot 2^{k+k^\lambda} \leq T\left(\mathcal{G}^*(end)\right) | \cdot 2^{2m^{\frac{1}{2}-\varepsilon}}. \tag{45.23}$$

By (45.11)

$$\lambda_4 \cdot T(\mathcal{G}) = \frac{1}{8} T(\mathcal{F}),$$

so by (45.16), (45.9), and (45.23)

$$T(\mathcal{F}^*(end)) \geq \lambda_4 \cdot T(\mathcal{G}^*(end))$$

$$= \frac{T(\mathcal{F})}{8T(\mathcal{F})} \cdot T(\mathcal{G}^*(end)) \geq \frac{T(\mathcal{F})}{8T(\mathcal{F})} \cdot |V^{**}| \cdot 2^{-2m^{\frac{1}{2}-\varepsilon}}. \tag{45.24}$$

Combining (45.18), (45.19), and (45.24)

$$\frac{|\mathcal{F}_{s_0}|}{|V^{**}|} \geq \frac{2^{s_0}}{q_1^2} \cdot \frac{T(\mathcal{F})}{8T(\mathcal{G})} \cdot 2^{-2m^{\frac{1}{2}-\varepsilon}} = 2^{s_0-2m^{\frac{1}{2}-\varepsilon}-3} \frac{T(\mathcal{F})}{q_1^2 \cdot T(\mathcal{F})}. \tag{45.25}$$

We need to demonstrate that First Player owns a whole $A_0 \in \mathcal{F}$. If there is an $A_0 \in \mathcal{F}$ with $A_0 \subseteq X(end)$, then we are done. We can assume, therefore, that there is no $A_0 \in \mathcal{F}$ with $A_0 \subseteq X(end)$. Then every $A \in \mathcal{F}_{s_0}$ has a non-empty intersection with $Y(end) \cap V^{**}$ (where $V^{**} = V \setminus V^*$). By (45.19) and (45.25) there exists a $w_0 \in Y(end) \cap V^{**}$ such that

$$|\mathcal{F}_{s_0,w_0}| = |\{A \in \mathcal{F}_{s_0} : w_0 \in A\}| \geq N_0$$

$$\text{where } N_0 = N_0(s_0) = 2^{s_0-2m^{\frac{1}{2}-\varepsilon}-3} \frac{T(\mathcal{F})}{q_1^2 \cdot T(\mathcal{G})}. \tag{45.26}$$

4. Disjointness Argument: forcing a contradiction. Every $A \in \mathcal{F}_{s_0,w_0}$ contains point w_0; we call w_0 the *root* of hypergraph $\mathcal{F}_{s_0,w_0}$. Note that the Max Pair-Degree $\Delta_2(\mathcal{F})$ of the whole hypergraph $\mathcal{F}$ is $\leq \binom{q_1^2}{2}$, so by using a trivial greedy algorithm we can select a sub-family $\mathcal{F}_{s_0,w_0}(disj)$ of $\mathcal{F}_{s_0,w_0}$ satisfying (see (45.26))

(1) the hyperedges in $\mathcal{F}_{s_0,w_0}(disj)$ are pairwise disjoint apart from w_0, and

(2) $$|\mathcal{F}_{s_0,w_0}(disj)| \geq \frac{|\mathcal{F}_{s_0,w_0}|}{1+q_1^2 \Delta_2(\mathcal{F})} \geq \frac{N_0}{q_1^6} = N_1 = N_1(s_0). \tag{45.27}$$

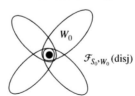

Notice that in the union set

$$\left(\bigcup_{A \in \mathcal{F}_{s_0, w_0}(disj)} A \right) \cap V^* \tag{45.28}$$

First Player has a Shutout of size

$$(q_1^2 - s_0)|\mathcal{F}_{s_0, w_0}(disj)| \ge (q_1^2 - s_0)N_1(s_0)$$

$$\ge (q_1^2 - s_0) \cdot 2^{s_0 - 2m^{\frac{1}{2} - \varepsilon} - 3} \frac{T(\mathcal{F})}{q_1^8 \cdot T(\mathcal{G})}, \tag{45.29}$$

where in the last step we used (45.26)–(45).

Recall that $m = q_2^2$. In order to get a contradiction First Player wants to prevent any kind of Shutout like (45.28)–(45.29).

What does it mean "any kind of Shutout like (45.28)–(45.29)"? Well, pick an arbitrary integer s in $1 \le s \le q_1^2$; let $\mathcal{H} \subset \mathcal{F}$ be an arbitrary sub-family satisfying the properties:

(α) $\bigcap_{A \in \mathcal{H}} A$ is a single point, the "root" of $\mathcal{H}$, and the sets $A \in \mathcal{H}$ are pairwise disjoint apart from the "root";

(β) $$|\mathcal{H}| = N_1(s) = 2^{s - 2m^{\frac{1}{2} - \varepsilon} - 3} \frac{T(\mathcal{F})}{q_1^8 \cdot T(\mathcal{G})}. \tag{45.30}$$

For every $\mathcal{H} \subset \mathcal{F}$ satisfying properties (α)–(β) above, consider the union set $\bigcup_{A \in \mathcal{H}} A$; the family of *all* union sets $\bigcup_{A \in \mathcal{H}} A$ is denoted by $\mathcal{F}^{(s)}$. The Shutout Game on hypergraph $\mathcal{F}^{(s)}$ with goal size $b = b(s) = (q_1^2 - s)N_1(s)$ defines a multi-hypergraph $\mathcal{F}_\mu^{(s)}$ with multiplicity function (45.5). Finally, let

$$\mathcal{F}_{Forb} = \bigcup_{s=1}^{q_1^2} \mathcal{F}_\mu^{(s)}. \tag{45.31}$$

If a Shutout (45.28)–(45.29) occurs in the ith round of the play, then $T(\mathcal{F}_{Forb}^*(i)) \ge 1$, so by (45.16), (45.9), and (45.11)

$$T(\mathcal{F}^*(i)) \ge \lambda_1 \cdot T(\mathcal{F}_{Forb}^*(i)) \ge \lambda_1 = \frac{T(\mathcal{F})}{8T(\mathcal{F}_{Forb})}. \tag{45.32}$$

We are going to see that (45.32) is a contradiction: the right-hand side of (45.32) is in fact much larger than the left-hand side. This contradiction will show that the case $s_0 \ge 1$ is impossible, so $s_0 = 0$, i.e. at the end of the play First Player owns a whole $A \in \mathcal{F}$.

5. Checking the contradiction. Estimate $T(\mathcal{F}_{Forb})$ from above: by (45.5), properties (α) and (β) above, and (45.32) with $\Delta_2 = \Delta_2(\mathcal{F}) \le \binom{q_1^2}{2}$ and $N_1 = N_1(s)$ (see

(45))

$$T(\mathcal{F}_{Forb}) \leq N^2 \sum_{s=1}^{q_1^2} \binom{N \cdot \Delta_2}{N_1} 2^{-N_1(q_1^2-1)} \cdot 2^{N_1(q_1^2-1)-N_1(q_1^2-s)}$$

$$= N^2 \sum_{s=1}^{q_1^2} \binom{N \cdot \Delta_2}{N_1} 2^{-N_1(q_1^2-s)} \leq N^2 \sum_{s=1}^{q_1^2} \left(\frac{e \cdot N \cdot q_1^4}{N_1 \cdot 2^{q_1^2-s}}\right)^{N_1}. \qquad (45.33)$$

We have

$$\frac{e \cdot N \cdot q_1^4}{N_1 \cdot 2^{q_1^2-s}} = \frac{e \cdot N \cdot q_1^4}{2^{s-2m^{\frac{1}{2}-\varepsilon}-3} \cdot \frac{T(\mathcal{F})}{q_1^8 \cdot T(\mathcal{G})} \cdot 2^{q_1^2-s}}$$

$$= \frac{e \cdot N \cdot q_1^{12} \cdot 2^{2m^{\frac{1}{2}-\varepsilon}-3} \cdot T(\mathcal{G})}{T(\mathcal{F}) \cdot 2^{q_1^2}};$$

and because $m = q_2^2$,

$$T(\mathcal{G}) = \left(\frac{1}{4} + o(1)\right) \frac{N^4 2^{-q_2^2}}{q_2^2},$$

and

$$T(\mathcal{F}) = \left(\frac{1}{3} + o(1)\right) \frac{N^3 2^{-q_1^2}}{q_1},$$

we obtain

$$\frac{e \cdot N \cdot q_1^4}{N_1 \cdot 2^{q_1^2-s}} \leq \frac{N^5 \cdot 2^{q_2^{1-\varepsilon}} \cdot 2^{-q_2^2}}{N^3} = N^2 \cdot 2^{q_2^{1-\varepsilon}-q_2^2}. \qquad (45.34)$$

We have a closer look at the last term

$$N^2 \cdot 2^{q_2^{1-\varepsilon}-q_2^2}$$

in (45.34). We need to involve the fractional parts

$$\left\{\sqrt{\log_2 N}\right\} = \theta_1 \quad \text{and} \quad \left\{\sqrt{2\log_2 N}\right\} = \theta_2;$$

by using the fractional parts we have

$$q_1 = \sqrt{\log_2 N} - \theta_1 \quad \text{and} \quad q_2 = \sqrt{2\log_2 N} + 1 - \theta_2. \qquad (45.35)$$

Since

$$q_2^2 = \left(\sqrt{2\log_2 N} + 1 - \theta_2\right)^2 = 2\log_2 N + 2(1-\theta_2)\sqrt{2\log_2 N} + O(1),$$

we have

$$N^2 2^{-q_2^2} = 2^{-2(1-\theta_2)\sqrt{2\log_2 N}+O(1)} = 2^{-2(1-\theta_2)q_2+O(1)}. \qquad (45.36)$$

If the inequality

$$1 - \theta_2 = 1 - \left\{\sqrt{2\log_2 N}\right\} \geq (\log_2 N)^{-\varepsilon/2} \qquad (45.37)$$

is satisfied, which happens for the overwhelming majority of Ns, then by (45.36) the last term in (45.34) can be estimated from above as follows

$$N^2 \cdot 2^{q_2^{1-\varepsilon}-q_2^2} = 2^{-2(1-\theta_2)q_2+q_2^{1-\varepsilon}+O(1)} \leq \frac{1}{2}. \tag{45.38}$$

Moreover, by (45.32)

$$N_1 = N_1(s) = 2^{s-2m^{\frac{1}{2}-\varepsilon}-3} \frac{T(\mathcal{F})}{q_1^8 \cdot T(\mathcal{G})}$$

$$\geq 2^{-q_2^{1-\varepsilon}} \cdot \frac{N^3 2^{-q_1^2}}{N^4 2^{-q_2^2}}$$

$$= 2^{-q_2^{1-\varepsilon}} \cdot \frac{N^3 2^{-\log_2 N + 2\theta_1 \sqrt{\log_2 N} + O(1)}}{N^4 2^{-2\log_2 N - 2(1-\theta_2)\sqrt{2\log_2 N} + O(1)}}$$

$$= 2^{-q_2^{1-\varepsilon}+\sqrt{2}(\theta_1+\sqrt{2}(1-\theta_2))q_2+O(1)}. \tag{45.39}$$

Assumption (45.37) already implies the inequality

$$-q_2^{1-\varepsilon} + \sqrt{2}(\theta_1 + \sqrt{2}(1-\theta_2))q_2 \geq q_2^{1-\varepsilon}, \tag{45.40}$$

so by (45.39) and (45.40)

$$N_1 \geq 2^{2q_2^{1-\varepsilon}} \geq 2^{q_2^{1/2}} \geq 2^{(\log_2 N)^{1/4}}. \tag{45.41}$$

Summarizing, by (45.33), (45.34), (45.38), and (45.41)

$$T(\mathcal{F}_{Forb}) \leq N^2 \cdot 2^{-2^{(\log_2 N)^{1/4}}}. \tag{45.42}$$

Inequality (45.42) shows that $T(\mathcal{F}_{Forb})$ is *super-polynomially* small in terms of N, so trivially

$$T(\mathcal{F}_{Forb}) \leq N^{-2}, \tag{45.43}$$

which suffices to get a contradiction.

The contradiction comes from the trivial bounds

$$T(\mathcal{F}^*(i)) \leq |\mathcal{F}| < N^3$$

and

$$T(\mathcal{F}) = \left(\frac{1}{3} + o(1)\right) \frac{N^3 2^{-q_1^2}}{q_1}$$

$$= N^{3+o(1)} 2^{-\log_2 N} = N^{2-o(1)},$$

combined with (45.32) and (45.43)

$$N^3 > T(\mathcal{F}^*(i)) \geq \frac{T(\mathcal{F})}{8T(\mathcal{F}_{Forb})} \geq \frac{N^{2-o(1)}}{8N^{-2}} = N^{4-o(1)}, \tag{45.44}$$

which is a contradiction if N is sufficiently large.

As said before, First Player has two goals that he simultaneously takes care of : the first goal is: (a) to completely occupy some $A \in \mathcal{F}$, and the second goal is (b) to block every $B \in \mathcal{G}$. The contradiction in (45.44) settles goal (a); it remains to settle goal (b).

6. How to block $\mathcal{G}$? Let's go back to Case 1 above; recall that in the *small game* First Player uses the $\mathcal{G}$-blocking Strong Draw strategy *Str (s.d.)*. This strategy enables First Player to block $\mathcal{G}$ as long as (1) he can win the Big Game, and (2) he can prevent the appearance of any Forbidden Configuration described in Section 44. A "failure" implies

$$\max \{ T(\mathcal{G}_{BIG}^*(end)), T(\mathcal{G}_{Forb}^*(end)) \} \geq 1,$$

thus by (45.9) and (45.16)

$$T(\mathcal{F}^*(end)) \geq \min \{ \lambda_2, \lambda_3 \} . \tag{45.45}$$

By (45.8) and (45.11)

$$\min \{ \lambda_2, \lambda_3 \} \geq \frac{T(\mathcal{F})}{8 N^{-(\log N)^\delta}} > N^{(\log N)^\delta} \tag{45.46}$$

with some positive absolute constant $\delta > 0$. By (45.45)–(45.46)

$$T(\mathcal{F}^*(end)) \geq N^{(\log N)^\delta}. \tag{45.47}$$

Inequality (45.47) is clearly false, since trivially

$$T(\mathcal{F}^*(end)) \leq |\mathcal{F}| < N^3.$$

This contradiction proves that a "failure" cannot occur, i.e. First Player can block hypergraph $\mathcal{G}$. This completes the proof of:

Theorem 45.1 *Consider the $q_1 \times q_1$ Aligned Square Lattice vs. $q_2 \times q_2$ Aligned Rectangle Lattice Who-Scores-First Game on an $N \times N$ board where*

$$q_1 = \left\lfloor \sqrt{\log_2 N} + o(1) \right\rfloor$$

(the "largest achievable size") and

$$q_2 = \left\lceil \sqrt{2 \log_2 N} + o(1) \right\rceil$$

(the "smallest impossible size"). If N is sufficiently large, First Player has a winning strategy. □

Notice that the term "$o(1)$" in $q_2 = \left\lceil \sqrt{2 \log_2 N} + o(1) \right\rceil$ takes care of condition (45.37) about the fractional part of $\sqrt{2 \log_2 N}$.

In the symmetric case, when First Player and Second Player have the same goal, namely to build a large Aligned Square Lattice, the proof of Theorem 45.1 gives the following:

Proposition: *Consider the game on the $N \times N$ board where both players want a large Aligned Square Lattice. For the overwhelming majority of Ns, Second Player can achieve two goals at the same time: (1) he can occupy a $q_1 \times q_1$ Aligned Square Lattice with*

$$q_1 = \left\lfloor \sqrt{\log_2 N} \right\rfloor .$$

and (2) he can prevent First Player from occupying a $q_2 \times q_2$ Aligned Square Lattice with $q_2 = q_1 + 1$.   □

Observe that the Proposition is exactly Theorem 12.7 about Second Player's Moral-Victory.

7. Switching goals. Let's return to Theorem 45.1. What happens if First Player and Second Player switch their goals: First Player wants an Aligned Rectangle Lattice and Second Player wants an Aligned Square Lattice? Again assume that First Player wants the "largest doable size" $q_1 \times q_1$ with

$$q_1 = \left\lfloor \sqrt{2 \log_2 N} + o(1) \right\rfloor \tag{45.48}$$

and Second Player wants the "smallest impossible size" $q_2 \times q_2$ with

$$q_2 = \left\lceil \sqrt{\log_2 N} + o(1) \right\rceil . \tag{45.49}$$

Who scores first?

Again it is most natural to expect First Player to have a winning strategy; and, indeed, we are going to supply a proof (with a minor weakness). Again the proof uses the technique of "super-polynomial multipliers."

Let $\mathcal{F}$ be the family of all $q_1 \times q_1$ aligned rectangle lattices in $N \times N$ (see (45.8)), and let $\mathcal{G}$ be the family of all $q_2 \times q_2$ aligned Square Lattices in $N \times N$ (see (45.9)). Assume that we are in the middle of a play, where First Player already occupied $x_1, x_2, \ldots, x_i$ and Second Player occupied $y_1, y_2, \ldots, y_i$ from the board $V = N \times N$. The question is how to choose First Player's next move x_{i+1}. Write $X(i) = \{x_1, x_2, \ldots, x_i\}$, $Y(i) = \{y_1, y_2, \ldots, y_i\}$, and for any hypergraph $\mathcal{H}$

$$\mathcal{H}(i) = \{A \setminus X(i) : A \in \mathcal{H}, A \cap Y(i) = \emptyset\}.$$

The first novelty is that the Max Pair-Degree $\Delta_2(\mathcal{F})$ of hypergraph $\mathcal{F}$ is *not* a polylogarithmic function of N. This is why we have to involve the Advanced Weak Win Criterion (Theorem 24.2) instead of the much simpler Theorem 1.2. This is why in "Case 2: $y_i \in V_{BIG}(i)$" First Player maximizes a more complicated function (compare (45.10) with (45.51) below). The new potential function is

$$L_i = \left(T(\mathcal{F}^*(i)) - \lambda_0 \cdot T(\mathcal{F}_2^{p*}(i)) \right) - \lambda_1 \cdot T(\mathcal{F}_{Forb}^*(i)) - \lambda_2 \cdot T(\mathcal{G}_{BIG}^*(i))$$

$$- \lambda_3 \cdot T(\mathcal{G}_{Forb}^*(i)) - \lambda_4 \cdot T(\mathcal{G}^*(i)), \tag{45.50}$$

where the new term "$-\lambda_0 \cdot T(\mathcal{F}_2^{p*}(i))$" is justified by the technique of Section 24.

First Player's choice for x_{i+1} is that unoccupied point $z \in V_{BIG}(i) \setminus (X(i) \cup Y(i))$ of the Big Board for which the function

$$L_i(z) = \left(T(\mathcal{F}^*(i); z) - \lambda_0 \cdot T(\mathcal{F}_2^{p*}(i); z) \right) - \lambda_1 \cdot T(\mathcal{F}_{Forb}^*(i); z)$$

$$- \lambda_2 \cdot T(\mathcal{G}_{BIG}^*(i); z) - \lambda_3 \cdot T(\mathcal{G}_{Forb}^*(i); z) - \lambda_4 \cdot T(\mathcal{G}^*(i); z) \qquad (45.51)$$

attains its *maximum*. Again the mark $\star$ indicates "restricted to the actual Big Board $V_{BIG}(i)$," the auxiliary hypergraph $\mathcal{F}_{Forb}$ will be defined later, and the positive constant factors ("multipliers") $\lambda_0, \lambda_1, \lambda_2, \lambda_3, \lambda_4$ are defined by the side condition

$$\lambda_0 \cdot T(\mathcal{F}_2^{p*}) = \lambda_1 \cdot T(\mathcal{F}_{Forb}) = \lambda_2 \cdot T(\mathcal{G}_{BIG})$$

$$= \lambda_3 \cdot T(\mathcal{G}_{Forb}) = \lambda_4 \cdot T(\mathcal{G}) = \frac{1}{10} T(\mathcal{F}). \qquad (45.52)$$

It is clear from (45.8) that the extra parenthesis $(\ldots)$ in (45.50)–(45.51) joining two terms together indicate that we apply the method of self-improving potentials (see Section 24). What it means is that we divide the course of the play in the actual Big Board $V_{BIG}(i)$ into several phases; in each phase we switch to a new potential function where multiplier λ_0 is replaced by $\lambda_0/2$, $\lambda_0/4$, $\lambda_0/8$, and so on, and each new phase in fact turns out to be a bonus (explaining the term *self-improving*).

The self-improving part works perfectly well; the first real challenge comes in the part of the Disjointness Argument (see the proof of Theorem 45.1), due to the fact that the Max Pair-Degree of $\mathcal{F}$ is not "small." Instead we have the following weaker property of $\mathcal{F}$: if two lattice points in $N \times N$ are not on the same horizontal or vertical line, then the pair-degree of this point pair is $\leq \binom{q_1^2}{2}$. This leads to the following:

8. Modified Disjointness Argument. The starting point is the same (see (45.26)): there is a $w_0 \in Y(end) \cap V^{**}$ such that

$$|\mathcal{F}_{s_0, w_0}| = |\{A \in \mathcal{F}_{s_0} : w_0 \in A\}| \geq N_0$$

$$\text{where } N_0 = N_0(s_0) = 2^{s_0 - 2m^{\frac{1}{2} - \varepsilon} - 3} \cdot \frac{T(\mathcal{F})}{q_1^2 \cdot T(\mathcal{G})} \qquad (45.53)$$

and $m = q_2^2$. Let N_1 be the maximum number of elements of $\mathcal{F}_{s_0, w_0}$ which are pairwise disjoint apart from the "root" w_0; clearly $N_1 \leq N$; let $\mathcal{F}_{s_0, w_0}(1) \subset \mathcal{F}_{s_0, w_0}$ be the sub-family of these N_1 sets.

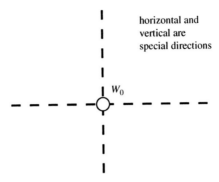

We extend $\mathcal{F}_{s_0, w_0}(1)$ by adding new sets

$$A_1, A_2, A_3, \ldots \in \mathcal{F}_{s_0, w_0} \setminus \mathcal{F}_{s_0, w_0}(1)$$

with the property that either $A_j \setminus \{$ horizontal w_0-line$\}$ or $A_j \setminus \{$ vertical w_0-line$\}$ is disjoint from

$$\left(\bigcup_{A \in \mathcal{F}_{s_0, w_0}(1)} A \right) \cup A_1 \cup A_2 \cup \cdots \cup A_{j-1}.$$

We keep doing this extension of $\mathcal{F}_{s_0, w_0}(1)$ as long as we can; let

$$\mathcal{F}_{s_0, w_0}(1) \cup \mathcal{F}_{s_0, w_0}(2)$$

denote the maximum extension; write $\mathcal{F}_{s_0, w_0}(2) = N_2$. The maximum property implies the analogue of (45)

$$N_3 = N_1 + N_2 \geq \frac{N_0}{1 + q_1^2(q_1^2)} \geq 2^{s_0 - 2m^{\frac{1}{2} - \varepsilon} - 3} \cdot \frac{T(\mathcal{F})}{q_1^8 \cdot T(\mathcal{G})}. \tag{45.54}$$

Write

$$\mathcal{F}_{s_0, w_0}(3) = \mathcal{F}_{s_0, w_0}(1) \cup \mathcal{F}_{s_0, w_0}(2);$$

in the union set restricted to the Big Board

$$\left(\bigcup_{A \in \mathcal{F}_{s_0, w_0}(1)} A \right) \cap V^* \tag{45.55}$$

Maker has a Shutout of size $\geq (q_1^2 - s_0)N_1 + (q_1^2 - q_1 - s_0)N_2$. \qquad (45.56)

The "loss" q_1 in the factor $(q_1^2 - q_1 - s_0)$ comes from the horizontal or vertical w_0-line.

Notice that Shutout (45.55)–(45.56) is the analogue of (45.28)–(45.29). Again Maker wants to avoid Shutout (45.55)–(45.56); the auxiliary hypergraph $\mathcal{F}_{Forb}$ (undefined yet) will be designed exactly for this purpose.

Fix an arbitrary integer s in $1 \le s \le q_1^2$; let $\mathcal{H} \subset \mathcal{F}$ be an arbitrary sub-family satisfying the properties:

(α) $\mathcal{H} = \mathcal{H}(1) \cup \mathcal{H}(2)$, where $\bigcap_{A \in \mathcal{H}(1)} A$ is a single point $w_0 = w_0(\mathcal{H})$, and the sets $A \in \mathcal{H}(1)$ are pairwise disjoint apart from w_0;

(β) for every $A' \in \mathcal{H}(2)$ either $A' \setminus \{\text{horizontal } w_0\text{-line}\}$ or $A' \setminus \{\text{vertical } w_0\text{-line}\}$ is disjoint from $\bigcup_{A \in \mathcal{H} \setminus \{A'\}} A$;

(γ) $|\mathcal{H}| = N_1 + N_2 = N_3 = N_3(s) = 2^{s - 2m^{\frac{1}{2} - \varepsilon} - 3} \frac{T(\mathcal{F})}{q_1^8 \cdot T(\mathcal{G})}$.

Fix an integer N_1 in $N_1 \le \min\{N, N_3\}$. For every $\mathcal{H} \subset \mathcal{F}$ satisfying properties (α)–(β)–(γ) and $|\mathcal{H}(1)| = N_1$ above, consider the union set $\bigcup_{A \in \mathcal{H}} A$; the family of *all* union sets $\bigcup_{A \in \mathcal{H}} A$ is denoted by $\mathcal{F}^{(s, N_1)}$. The Shutout Game on hypergraph $\mathcal{F}^{(s, N_1)}$ with goal size

$$b = b(s, N_1) = (q_1^2 - s)N_1 + (q_1^2 - q_1 - s)N_2 = (q_1^2 - q_1 - s)N_3 + q_1 N_1 \quad (45.57)$$

defines a multi-hypergraph $\mathcal{F}_\mu^{(s, N_1)}$ with multiplicity function (45.5). Finally, let

$$\mathcal{F}_{Forb} = \bigcup_{s=1}^{q_1^2} \bigcup_{N_1=1}^{\min\{N, N_3\}} \mathcal{F}_\mu^{(s, N_1)}. \quad (45.58)$$

If a Shutout (45.55)–(45.56) occurs in the ith round of the play, then $T(\mathcal{F}_{Forb}^*(i)) \ge 1$, and we get the analogue of (45.32)

$$T(\mathcal{F}^*(i)) \ge \lambda_1 \cdot T(\mathcal{F}_{Forb}^*(i)) \ge \lambda_1 = \frac{T(\mathcal{F})}{10 T(\mathcal{F}_{Forb})}. \quad (45.59)$$

Again we show that (45.59) is a contradiction: the right-hand side of (45.59) is in fact much larger than the left-hand side.

Estimate $T(\mathcal{F}_{Forb})$ from above: by (45.5), properties (α)–(β)–(γ) above, and (45.58), we obtain the analogue (45.33)

$$T(\mathcal{F}_{Forb}) \le N^2 \sum_{s=1}^{q_1^2} \sum_{N_1=1}^{\min\{N, N_3(s)\}} \binom{N \cdot \binom{q_1^2}{2}}{N_1} \cdot$$

$$\binom{N_1 \cdot q_1 \cdot \Delta_2}{N_3 - N_1} 2^{-N_1(q_1^2 - 1)} \cdot 2^{N_1(q_1^2 - 1) - (N_3 - N_1)(q_1^2 - q_1 - s)} \quad (45.60)$$

Note that $\Delta_2 = \Delta_2(\mathcal{F}) \le N$, so returning to (45.60) and using the trivial inequality $\binom{M}{r} \le (eM/r)^r$, we have

$$T(\mathcal{F}_{Forb}) \le N^2 \sum_{s=1}^{q_1^2} \sum_{N_1=1}^{\min\{N, N_3(s)\}} \left(\frac{e \cdot N^2 \cdot q_1^4}{2 N_1 \cdot 2^{q_1^2 - s}} \right)^{N_1} \left(\frac{e \cdot N \cdot N_1 \cdot q_1}{(N_3 - N_1) \cdot 2^{q_1^2 - q_1 - s}} \right)^{N_3 - N_1} \quad (45.61)$$

where

$$N_3 = N_3(s) = 2^{s - 2q_2^1 - 2\varepsilon - 3} \cdot \frac{T(\mathcal{F})}{q_1^8 \cdot T(\mathcal{G})}, \tag{45.62}$$

$$T(\mathcal{F}) = \left(\frac{1}{4} + o(1)\right) \frac{N^4 2^{-q_1^2}}{q_1^2}, \quad q_1 = \left\lfloor \sqrt{2\log_2 N} + o(1) \right\rfloor, \tag{45.63}$$

and

$$T(\mathcal{G}) = \left(\frac{1}{3} + o(1)\right) \frac{N^3 2^{-q_2^2}}{q_2}, \quad q_2 = \left\lceil \sqrt{\log_2 N} + o(1) \right\rceil. \tag{45.64}$$

By (45.61), with $N_2 = N_3(s) - N_1$, and using the fact $N_1 \le N$

$$\left(\frac{e \cdot N^2 \cdot q_1^4}{2N_1 \cdot 2^{q_1^2 - s}}\right)^{N_1} \left(\frac{e \cdot N \cdot N_1 \cdot q_1}{(N_3 - N_1) \cdot 2^{q_1^2 - q_1 - s}}\right)^{N_3 - N_1}$$

$$\le \left(\frac{N^2 \cdot 2^s}{2^{q_1^2}} \cdot \left(\frac{e q_1^4/2}{N_1}\right)^{N_1/N_3} \cdot \left(\frac{e N_1 q_1 2^{q_1}}{N \cdot N_2}\right)^{N_2/N_3}\right)^{N_3}$$

$$\le \left(\frac{N^2 \cdot 2^s}{2^{q_1^2}} \cdot \left(\frac{e q_1^4/2}{N_1}\right)^{N_1/N_3} \cdot \left(\frac{e N_1 q_1 2^{q_1}}{N_2}\right)^{N_2/N_3}\right)^{N_3}$$

$$\le \left(\frac{N^2 \cdot 2^s}{2^{q_1^2}} \cdot \frac{e \cdot q_1^4 \cdot 2^{q_1}}{N_3}\right)^{N_3}, \tag{45.65}$$

where in the last step we used the simple fact $N_1^{N_1} \cdot N_2^{N_2} \ge \left(\frac{N_1 + N_2}{2}\right)^{N_1 + N_2}$. By using the definition of $N_3 = N_3(s)$ (see (45.62)) in (45.61) and (45.65)

$$T(\mathcal{F}_{Forb}) \le N^2 \sum_{s=1}^{q_1^2} \sum_{N_1=1}^{\min\{N, N_3(s)\}} \left(\frac{N^2 \cdot 2^s}{2^{q_1^2}} \cdot \frac{e \cdot q_1^4 \cdot 2^{q_1}}{N_3}\right)^{N_3}$$

$$= N^2 \sum_{s=1}^{q_1^2} \sum_{N_1=1}^{\min\{N, N_3(s)\}} \left(\frac{N \cdot e q_1^{13} \cdot 2^{q_1 + 2q_2^1 - 2\varepsilon}}{2^{q_2^2}}\right)^{N_3}. \tag{45.66}$$

Consider the fractional part

$$\left\{\sqrt{\log_2 N}\right\} = \theta_2; \quad \text{then} \quad q_2 = \sqrt{\log_2 N} + 1 - \theta_2,$$

and

$$q_2^2 = \log_2 N + 2(1 - \theta_2)\sqrt{\log_2 N} + O(1) = \log_2 N + \sqrt{2}(1 - \theta_2) \cdot q_1 + O(1).$$

Therefore, returning to (45.66) we have

$$T(\mathcal{F}_{Forb}) \le N^2 \sum_{s=1}^{q_1^2} \sum_{N_1=1}^{\min\{N, N_3(s)\}} \left(e \cdot q_1^{13} \cdot 2^{q_1(1 - \sqrt{2}(1 - \theta_2)) + 2q_2^1 - 2\varepsilon}\right)^{N_3}. \tag{45.67}$$

Assume that

$$1 - \sqrt{2}(1 - \theta_2) < 0 \Longleftrightarrow \{\sqrt{\log_2 N}\} < 1 - \frac{1}{\sqrt{2}} = .293; \tag{45.68}$$

then by (45.67)

$$T(\mathcal{F}_{Forb}) \leq N^2 \sum_{s=1}^{q_1^2} \sum_{N_1=1}^{\min\{N, N_3(s)\}} \left(\frac{1}{2}\right)^{N_3(s)}. \tag{45.69}$$

Moreover, by (45.62)

$$N_3 = N_3(s) = 2^{s - 2q_2^{1-2\varepsilon} - 3} \frac{T(\mathcal{F})}{q_1^8 \cdot T(\mathcal{G})}$$

$$\geq 2^{-q_2^{1-\varepsilon}} \cdot \frac{N^4 2^{-q_1^2}}{N^3 2^{-q_2^2}}$$

$$= 2^{-q_2^{1-\varepsilon}} \cdot \frac{N^4 2^{-2\log_2 N + 2\theta_1 \sqrt{2\log_2 N} + O(1)}}{N^3 2^{-\log_2 N - 2(1-\theta_2)\sqrt{\log_2 N} + O(1)}}$$

$$= 2^{-q_2^{1-\varepsilon} + (2\sqrt{2}\theta_1 + (1-\theta_2))q_2 + O(1)}, \tag{45.70}$$

where $\theta_1 = \{\sqrt{2\log_2 N}\}$ (fractional part).

Assumption (45.68) already implies the inequality

$$-q_2^{1-\varepsilon} + (2\sqrt{2}\theta_1 + 2(1 - \theta_2))q_2 \geq q_2^{1-\varepsilon}, \tag{45.71}$$

so by (45.70)–(45.71)

$$N_3 = N_3(s) = \geq 2^{q_2^{1-\varepsilon}} \geq 2^{(\log_2 N)^{1/2}}. \tag{45.72}$$

Summarizing, by (45.69) and (45.72)

$$T(\mathcal{F}_{Forb}) \leq N^2 \sum_{s=1}^{q_1^2} \sum_{N_1=1}^{\min\{N, N_3(s)\}} 2^{-2^{(\log_2 N)^{1/2}}} \leq N^{-3}. \tag{45.73}$$

The contradiction comes from the trivial bounds

$$T(\mathcal{F}^*(i)) \leq |\mathcal{F}| < N^4$$

and

$$T(\mathcal{F}) = \left(\frac{1}{4} + o(1)\right) \frac{N^4 2^{-q_1^2}}{q_1^2}$$

$$= N^{4+o(1)} 2^{-2\log_2 N} = N^{2-o(1)},$$

combined with (45.59) and (45.73)

$$N^4 > T(\mathcal{F}^*(i)) \geq \frac{T(\mathcal{F})}{10 T(\mathcal{F}_{Forb})} \geq \frac{N^{2-o(1)}}{10 N^{-3}} = N^{5-o(1)},$$

which is a contradiction if N is sufficiently large.

The rest of the proof is the same as that of Theorem 45.1. Thus we obtain:

Theorem 45.2 *Consider the $q_1 \times q_1$ Aligned Rectangle Lattice vs. $q_2 \times q_2$ Aligned Square Lattice Who-Scores-First Game on an $N \times N$ board where*

$$q_1 = \left\lfloor \sqrt{2\log_2 N + o(1)} \right\rfloor$$

(the "largest achievable size") and

$$q_2 = \left\lceil \sqrt{\log_2 N + o(1)} \right\rceil$$

(the "smallest impossible size"). If N is sufficiently large and the fractional part $\{\sqrt{\log_2 N}\} < 1 - \frac{1}{\sqrt{2}} = .293$, then First Player has a winning strategy.

If $\{\sqrt{\log_2 N}\} \geq 1 - \frac{1}{\sqrt{2}} = .293$, then First Player still has a winning strategy, assuming we switch the value of q_2 to the one larger $q_2 = \left\lceil \sqrt{\log_2 N + o(1)} \right\rceil + 1$.

Comparing Theorem 45.2 with Theorem 45.1 there is an obvious weakness: we have the extra condition $\{\sqrt{\log_2 N}\} < 1 - \frac{1}{\sqrt{2}} = .293$ about the fractional part, which holds only for a "positive density" sequence of Ns (with an appropriate density concept) instead of the usual "overwhelming majority of Ns."

Chapter X
Conclusion

The reader is owed a few missing details such as (1) how to modify the Achievement proofs to obtain the Avoidance proofs, (2) the Chooser–Picker game, (3) the best-known Pairing Strategy Draw in the n^d hypercube Tic-Tac-Toe (part (b) in Open Problem 34.1).

Also we discuss a few new results: generalizations and extensions, such as what happens if we extend the board from the complete graph K_N and the $N \times N$ lattice to a typical sub-board.

We discuss these generalizations, extensions, and missing details in the last four sections (Sections 46–49).

46

More exact solutions and more partial results

1. Extension: from the complete board to a typical sub-board. The book is basically about two results, Theorems 6.4 and 8.2, and their generalizations (discrepancy, biased, Picker–Chooser, Chooser–Picker, etc.). Here is another, perhaps the most interesting, way to generalize. In Theorem 6.4 (a) the board is K_N, that is, a very special graph; what happens if we replace K_N with a *typical* graph G_N on N vertices?

Playing the usual (1:1) game on an arbitrary finite graph G, we can define the Clique Achievement (Avoidance) Number of G in the usual way, namely answering the question: "What is the largest clique K_q that Maker can build (that Forcer can force Avoider to build)?"

A *typical* sub-graph $G_N \subset K_N$ has about half of the edges of K_N, i.e. $(1 + o(1))N^2/4$ edges, and contains $(1 + o(1))\binom{N}{q}2^{-\binom{q}{2}}$ copies of K_q. Of course, a *typical* sub-graph $G_N \subset K_N$ is just an alternative name for the Random Graph $\mathbf{R}(K_N; 1/2)$ with edge probability $p = 1/2$.

The Meta-Conjecture predicts that the Clique Achievement (Avoidance) Number of a *typical* sub-graph $G_N \subset K_N$ is the lower integral part of the real solution $q = q(N)$ of the equation

$$\frac{\binom{N}{q}2^{-\binom{q}{2}}}{N^2/4} = 2^{\binom{q}{2}},$$

which is equivalent to

$$q = \log_2 N - \log_2 \log_2 N + \log_2 e - 1 + o(1). \tag{46.1}$$

And indeed, (46.1) gives the truth. The proof technique of Theorem 6.4 can be trivially adapted due to the fact that the Random Graph is very homogeneous and very predictable.

The corresponding Majority-Play Clique Number is the lower integral part of

$$q = \log_2 N - \log_2 \log_2 N + \log_2 e + 1 + o(1). \tag{46.2}$$

Comparing (46.1) to (46.2) we obtain the remarkable equality

Majority Play Clique Number$(G) - 2 =$ Clique Achievement Number(G)

$$= \text{Clique Avoidance Number}(G) \quad (46.3)$$

which holds for the overwhelming majority of all finite graphs G (G is the board)!

By the way, to decide whether or not a graph G_N contains a clique of $\log_2 N$ vertices, and, if it does, to find one, takes about $N^{\log_2 N}$ steps; on the other hand, the number of positions in a graph G_N with about $N^2/4$ edges is roughly $3^{N^2/4}$. What a big difference! This rough calculation justifies that the game numbers are much more difficult concepts than the Majority-Play Number; this makes the (typical) equality (46.3) even more interesting.

Of course, equality (46.3) is not true for every single graph. For example, if G consists of a huge number of vertex disjoint copies of K_N (say M copies), then the Clique Achievement Number remains the usual $(2 + o(1)) \log_2 N$ for every M, but the Majority-Play Clique Number becomes N if M is much larger than $2^{\binom{N}{2}}$.

What happens if the symmetric Random Graph $\mathbf{R}(K_N; 1/2)$ (meaning the *typical* sub-graph $G_N \subset K_N$) is replaced by the general Random Graph $\mathbf{R}(K_N; p)$ with an arbitrary edge probability $0 < p < 1$?

The Meta-Conjecture predicts that the Clique Achievement (Avoidance) Number of $\mathbf{R}(K_N; p)$ is the lower integral part of the real solution $q = q(N)$ of the equation

$$\frac{\binom{N}{q} p^{\binom{q}{2}}}{pN^2/2} = 2^{\binom{q}{2}},$$

which is equivalent to

$$q = \frac{2}{\log_2(2/p)} \left(\log_2 N - \log_2 \log_2 N + \log_2 e + \log_2 \log_2(2/p) \right) - 1 + o(1). \quad (46.4)$$

And again (46.4) gives the truth; the proof technique of Theorem 6.4 can be trivially adapted.

The corresponding Majority-Play Clique Number is the lower integral part of

$$q = \frac{2}{\log_2(2/p)} \left(\log_2 N - \log_2 \log_2 N + \log_2 e + \log_2 \log_2(2/p) \right) + 1 + o(1), \quad (46.5)$$

that is, we have the usual "gap 2" (see (46.3)) independently of the value of probability $0 < p < 1$.

Switching from ordinary graphs to 3-graphs "gap 2" becomes "gap 3/2." Since 3/2 is not an integer, this means that for a typical 3-graph the Majority-Play Clique Number differs from the Clique Achievement (Avoidance) Number either by 1 or by 2, and the two cases have the same fifty–fifty chance (this is why the average is 3/2).

In general, for k-graphs the "gap" is $k/(k-1)$; since $k/(k-1)$ is not an integer, this means that for a typical k-graph the Majority-Play Clique Number differs from the Clique Achievement (Avoidance) Number either by 1 or by 2, and the odds are $(k-2):1$ (this is why the average is $k/(k-1)$).

Next consider the Lattice Games. What happens to Theorem 8.2 if the $N \times N$ grid is replaced by a "random sub-set"? Let $0 < p < 1$ be the probability of keeping an arbitrary grid point; we decide independently. This is how we get the Random Sub-set $\mathbf{R}(N \times N; p)$ of the grid $N \times N$; the case $p = 1/2$ gives what is meant by a "typical sub-board."

Let us choose a lattice type; for example, consider the Parallelogram Lattice Game. Playing the usual (1:1) game on a Random Sub-set $\mathbf{R}(N \times N; p)$ of the grid $N \times N$, what is the largest value of q such that Maker can always occupy a $q \times q$ parallelogram lattice inside the given sub-board?

The Meta-Conjecture predicts that the largest value of the lattice size is the lower integral part of the real solution $q = q(N)$ of the equation

$$N^4 = \left(\frac{2}{p}\right)^{q^2},$$

which is equivalent to

$$q = 2\sqrt{\log_{(2/p)} N}. \tag{46.6}$$

The corresponding Majority-Play Number is

$$N^6 = \left(\frac{2}{p}\right)^{q^2} \iff q = \sqrt{6}\sqrt{\log_{(2/p)} N}. \tag{46.7}$$

In this case the *ratio* $\sqrt{6}/2 = \sqrt{3/2}$ remains the same independently of the value of probability $0 < p \le 1$.

The "invariance of the gap" in the Clique Games (gap 2 for graphs, gap 3/2 for 3-graphs, and so on) and the "invariance of the ratio" in the Lattice Games (where invariance means: independent of the value of probability $0 < p \le 1$) is another striking property of these game numbers. In the proofs switching from $p = 1$ to an arbitrary probability p between $0 < p < 1$ makes very little difference (because random structures are very "predictable").

2. Strategy Stealing vs. explicit strategy.
Consider now the "who can build a larger Square Lattice" game. First assume that the board is the complete $N \times N$ grid; the two players alternate the usual (1:1) way; that player is declared the winner who, at the end of the play, owns a *larger* Aligned Square Lattice (as for the winner who owns a $q \times q$ lattice and the opponent's largest lattice has size $(q-1) \times (q-1)$); in case of equality, the play ends in a draw. If N is odd, then the first player has a simple (at least) drawing strategy: his opening move is the center of the board, and in the rest of the play he takes the reflection of the opponent's moves.

If the board is an arbitrary sub-set of the $N \times N$ grid, then the first player still can force (at least) a draw. The reflection strategy above obviously breaks down, but the Strategy Stealing argument still works! Of course, Strategy Stealing doesn't say a word about how to actually force (at least) a draw. Can we find an *explicit* drawing strategy here? Well, we can solve this problem at least for a board which is a *typical* sub-set of $N \times N$. Of course, *typical* means the Random Sub-set $\mathbf{R}(N \times N; 1/2)$ of the $N \times N$ grid with inclusion probability $1/2$. In view of the Meta-Conjecture the largest achievable size is

$$ q = \left\lfloor \sqrt{\log_4 N} + o(1) \right\rfloor = \left\lfloor \sqrt{\frac{\log_2 N}{2}} + o(1) \right\rfloor. \tag{46.8} $$

A straightforward adaptation of the proof technique of Section 45 (in particular Theorem 45.1) gives that the first player can always build an Aligned Square Lattice of size $q \times q$, and at the same time he can prevent the second player from occupying a lattice of size $(q+1) \times (q+1)$, where q is defined in (46.8).

The main point here is that this explicit strategy (using the potential technique of Section 45) is much faster than the "backward labeling algorithm" (the general recipe to find a drawing strategy guaranteed by the Strategy Stealing argument); in fact, the running time of the explicit strategy is a *logarithmic* function of the running time of the "backward labeling." This is a huge difference!

The second player can do the same thing: he can also build an Aligned Square Lattice of size $q \times q$, and at the same time he can prevent the first player from occupying a lattice of size $(q+1) \times (q+1)$, where q is defined in (46.8).

Note that for the second player the Strategy Stealing argument does not seem to work; for the second player the potential technique seems to be the only way to force a draw (playing on a typical sub-board).

The last result of this section is also related to Section 45.

3. Maker's building when he is the topdog: how to involve the Cheap Building Lemma? We switch to biased games. The $(m:b)$ achievement version of the Meta-Conjecture with $m > b$, i.e. when Maker is the topdog, requires a correction: the usual threshold

$$ n = \log_{\frac{m+b}{m}} \frac{|\mathcal{F}|}{|V|} $$

is replaced by the more complicated threshold

$$ n = \log_{\frac{m+b}{m}} \frac{|\mathcal{F}|}{|V|} + \log_{\frac{m}{m-b}} |V|, \tag{46.9} $$

due to the effect of the Cheap Building Lemma (see Section 30). In Section 30 we gave an intuitive explanation for (46.9), we called it the "Random Play plus Cheap Building" intuition and said "it is surprisingly easy to make this intuition precise in the $(m:1)$ play by involving an auxiliary hypergraph," and promised to discuss the details in Section 46. What we are going to do is Maker's part (Breaker's part, unfortunately, remains unsolved). The following argument is similar to that of in Section 45, but the details here are simpler.

The real question is how to enforce the *Disjointness Condition* in the "Random Play plus Cheap Building" intuition. We illustrate the idea with a "typical" example: the (2:1) Parallelogram Lattice Game. Assume that we are in the middle of the First Stage, where Maker already occupied $X(i) = \{x_1^{(1)}, x_1^{(2)}, , x_2^{(1)}, x_2^{(2)}, \ldots, x_i^{(1)}, x_i^{(2)}\}$ and Breaker occupied $Y(i) = \{y_1, y_2, \ldots, y_i\}$ from the board $V = N \times N$. Let $\mathcal{F}$ denote the family of all $q \times q$ Parallelogram Lattices in $N \times N$, and write

$$\mathcal{F}(i) = \{A \setminus X(i) : A \in \mathcal{F}, A \cap Y(i) = \emptyset\}.$$

Note that $\mathcal{F}(i)$ is the family of the unoccupied parts of the "survivors" in $\mathcal{F}$; the truncated $\mathcal{F}(i)$ can be a multi-hypergraph even if the original $\mathcal{F}$ is not. We use the Power-of-(3/2) Scoring System: for an arbitrary finite hypergraph $\mathcal{H}$ let

$$T(\mathcal{H}) = \sum_{B \in \mathcal{H}} \left(\frac{3}{2}\right)^{-|B|} \quad \text{and} \quad T(\mathcal{H}; u_1, \ldots, u_m) = \sum_{B \in \mathcal{H}: \{u_1, \ldots, u_m\} \subset B} \left(\frac{3}{2}\right)^{-|B|}.$$

The new Potential Function is

$$L_i = \left(T(\mathcal{F}(i)) - \lambda_0 \cdot T(\mathcal{F}_2^p(i))\right) - \lambda_1 \cdot T(\mathcal{G}_1(i)) - \lambda_2 \cdot T(\mathcal{G}_2(i)),$$

where, of course, $T(\cdots)$ refers to the Power-of-(3/2) Scoring System. Here the part $(T(\mathcal{F}(i)) - \lambda_0 \cdot T(\mathcal{F}_2^p(i)))$ is clearly justified by the technique of Section 24, and the auxiliary hypergraphs $\mathcal{G}_1$ and $\mathcal{G}_2$ (motivated by the Disjointness Condition in the Second Stage) will be defined later.

The positive constants λ_0, λ_1, λ_2 are defined by the side condition

$$\frac{1}{6} T(\mathcal{F}) = \lambda_0 \cdot T(\mathcal{F}_2^p) = \lambda_1 \cdot T(\mathcal{G}_1) = \lambda_2 \cdot T(\mathcal{G}_2). \tag{46.10}$$

The First Stage ends when $|\mathcal{F}(i)| \leq N^2$, that is, when the number of "survivors" among the winning sets becomes less than the board size $N \times N = N^2$.

For notational simplicity, let $|\mathcal{F}(i_0)| = N^2$, i.e. i_0 is the end of the First Stage. Note that in the First Stage we apply the method of self-improving potentials (see Section 24), which means that we divide the play into several phases, in each phase we switch to a new potential where the multiplier λ_0 is replaced by $\lambda_0/2$, $\lambda_0/4$, $\lambda_0/8$, and so on, and each new phase turns out to be a bonus. It follows from the method that we can assume the following two facts

$$T(\mathcal{F}(i_0)) \geq \frac{1}{4} T(\mathcal{F}) \tag{46.11}$$

and

$$T(\mathcal{F}(i_0)) > \lambda_1 \cdot T(\mathcal{G}_1(i_0)) + \lambda_2 \cdot T(\mathcal{G}_2(i_0)). \tag{46.12}$$

Inequality (46.11) implies that there must exist an integer s_0 in $0 \leq s_0 < q^2$ such that

$$\sum_{B \in \mathcal{F}(i_0):\ |B| = s_0} \left(\frac{3}{2}\right)^{-s_0} \geq \frac{1}{4q^2} T(\mathcal{F}) = \frac{1}{4q^2} |\mathcal{F}| 2^{-q^2}. \qquad (46.13)$$

Write $\mathcal{F}_{s_0}(i_0) = \{B \in \mathcal{F}(i_0) :\ |B| = s_0\}$; for every $B \in \mathcal{F}_{s_0}(i_0)$ let $A(B) \in \mathcal{F}$ denote its ancestor, and write

$$\mathcal{F}_{s_0} = \{A(B) \in \mathcal{F} :\ B \in \mathcal{F}_{s_0}(i_0)\}.$$

We distinguish two cases:

Case 1: Family $\mathcal{F}_{s_0}$ contains at least

$$|\mathcal{F}_{s_0}| \cdot e^{-(\log N)^{1/3}}$$

pairwise disjoint sets.

Then Maker applies the Cheap Building Lemma (see Section 30). If the inequality

$$|\mathcal{F}_{s_0}| \cdot e^{-(\log N)^{1/3}} \geq 2^{s_0 - 2} \qquad (46.14)$$

holds, then Maker can occupy a whole $A \in \mathcal{F}_{s_0} \subset \mathcal{F}$. Recall inequality (46.13) above

$$\left(\frac{3}{2}\right)^{q^2 - s_0} \geq \frac{1}{4q^2} \frac{|\mathcal{F}|}{|\mathcal{F}_{s_0}|},$$

which is equivalent to

$$q^2 - s_0 \geq \log_{\frac{3}{2}} |\mathcal{F}| - \log_{\frac{3}{2}} |\mathcal{F}_{s_0}| - O(\log \log N). \qquad (46.15)$$

On the other hand, if (46.14) fails, then

$$\log_2 |\mathcal{F}_{s_0}| \leq s_0 + O\left((\log N)^{1/3}\right). \qquad (46.16)$$

Adding up (46.15) and (46.16), we obtain

$$q^2 \geq \log_{\frac{3}{2}} |\mathcal{F}| - \log_{\frac{3}{2}} |\mathcal{F}_{s_0}| + \log_2 |\mathcal{F}_{s_0}| - O\left((\log N)^{1/3}\right). \qquad (46.17)$$

Since $|\mathcal{F}_{s_0}| \leq N^2 = |V|$, from (46.17) we conclude that

$$q^2 \geq \log_{\frac{3}{2}} \frac{|\mathcal{F}|}{|V|} + \log_2 |V| - O\left((\log N)^{1/3}\right),$$

which contradicts the hypothesis

$$q = \left\lfloor \sqrt{\log_{\frac{3}{2}} (|\mathcal{F}|/|V|) + \log_2 |V|} - o(1) \right\rfloor$$

if $o(1)$ tends to 0 sufficently slowly.

It remains to show that Case 1 is the only alternative (i.e. Case 2 below leads to a contradiction).

Case 2: There are at least

$$\frac{e^{(\log N)^{1/3}}}{q^2}$$

sets in family $\mathcal{F}_{s_0}$ which contain the same point.
Let w_1 denote the common point, and write

$$\mathcal{F}_{s_0,w_1} = \{A \setminus \{w_1\} : w_1 \in A \in \mathcal{F}_{s_0}\}.$$

Let $\mathcal{H} \subset \mathcal{F}_{s_0,w_1}$ be the maximum size sub-family of $\mathcal{F}_{s_0,w_1}$ which consists of pairwise disjoint sets.

This is the part where we define and use the auxiliary hypergraphs $\mathcal{G}_1$ and $\mathcal{G}_2$. Again we distinguish two cases:

Case 2a: We have

$$|\mathcal{H}| \geq e^{(\log N)^{1/3}/2}$$

For notational simplicity write $m = e^{(\log N)^{1/3}/2}$. We define the auxiliary hypergraph $\mathcal{G}_1$ as follows: $B \in \mathcal{G}_1$ if and only if

$$B = \bigcup_{i=1}^{m} A_i \quad \text{where} \quad \left| \bigcap_{i=1}^{m} A_i \right| = 1, \quad \text{and}$$

$A_1, \ldots, A_m$ are m different sets in $\mathcal{F}$ such that they are pairwise disjoint apart from the one-element common part. Then by (46.10)

$$\frac{1}{6\lambda_1} T(\mathcal{F}) = T(\mathcal{G}_1) \leq N^2 \cdot \binom{N^4}{m} \left(\frac{3}{2}\right)^{-m(q^2-1)}. \tag{46.18}$$

In (46.18) the factor N^2 is the number of ways to fix the one-element common part, $\binom{N^4}{m}$ is an upper bound on the number of ways to choose $B = \bigcup_{i=1}^{m} A_i$, and the exponent $m(q^2 - 1)$ is explained by the disjointness apart from the one-element common part.

It follows from the definition of $\mathcal{H}$ that

$$T(\mathcal{G}_1) \geq \left(\frac{3}{2}\right)^{-m(s_0-1)}. \tag{46.19}$$

On the other hand, by (46.18)

$$\lambda_1 \geq \frac{T(\mathcal{F})}{6 \cdot N^2} \cdot \left(\frac{m \cdot (3/2)^{q^2-1}}{e \cdot N^4}\right)^m. \tag{46.20}$$

By (46.19)–(46.20)

$$\lambda_1 \cdot T(\mathcal{G}_1) \geq \frac{T(\mathcal{F})}{6 \cdot N^2} \cdot \left(\frac{m \cdot (3/2)^{q^2-s_0}}{e \cdot N^4}\right)^m. \tag{46.21}$$

Recall (46.13)

$$\left(\frac{3}{2}\right)^{q^2-s_0} \geq \frac{1}{4q^2}\frac{|\mathcal{F}|}{|\mathcal{F}_{s_0}|},$$

and since $|\mathcal{F}_{s_0}| \leq N^2$, we have

$$\left(\frac{3}{2}\right)^{q^2-s_0} \geq \frac{1}{4q^2}\frac{|\mathcal{F}|}{N^2}. \tag{46.22}$$

Trivially

$$N^6 > |\mathcal{F}| > \left(\frac{N}{2q}\right)^6, \tag{46.23}$$

so by (46.22)

$$\left(\frac{3}{2}\right)^{q^2-s_0} \geq \frac{(N/2q)^6}{4q^2 \cdot N^2} = \frac{N^4}{4^4 \cdot q^8}. \tag{46.24}$$

By (46.21) and (46.24) we have

$$\lambda_1 \cdot T(\mathcal{G}_1) \geq \frac{T(\mathcal{F})}{6 \cdot N^2} \cdot \left(\frac{m}{e \cdot 4^4 \cdot q^8}\right)^m, \tag{46.25}$$

and since $m = e^{(\log N)^{1/3}/2}$, (46.25) is a huge super-polynomial lower bound in terms of N (note that $T(\mathcal{F}) \geq 1$). This huge super-polynomial lower bound clearly contradicts the polynomial upper bound (46.12)

$$T(\mathcal{F}(i_0)) > \lambda_1 \cdot T(\mathcal{G}_1(i_0)).$$

Indeed, $T(\mathcal{F}(i_0)) \leq |\mathcal{F}| < N^6$, see (46.23). This contradiction proves that Case 2a is impossible.

If Case 2a is impossible, then we have

Case 2b: There are at least

$$\frac{e^{(\log N)^{1/3}/2}}{q^4}$$

sets in family $\mathcal{F}_{s_0,w_1}$ which contain the same point.

Let w_2 denote the common point; it means that there is a sub-family $\mathcal{H}_1$ of $\mathcal{F}_{s_0}$ such that

(1) $|\mathcal{H}_1| \geq \frac{e^{(\log N)^{1/3}/2}}{q^4}$, and

(2) every element of $\mathcal{H}_1$ contains the point pair $\{w_1, w_2\}$.

Let

$$\mathcal{H}_2 = \{A \setminus \{w_1 w_2 - \text{line}\} : A \in \mathcal{H}_1\},$$

that is, we throw out the whole $w_1 w_2$-line, and let $\mathcal{H}_3 \subset \mathcal{H}_2$ be the maximum size sub-family of $\mathcal{H}_2$ which consists of pairwise disjoint sets.

Three non-collinear points in the $N \times N$ grid "nearly" determine a $q \times q$ parallelogram lattice: the multiplicity is $\leq \binom{q^2}{3}$. It follows that

$$|\mathcal{H}_3| \geq \frac{|\mathcal{H}_2|}{1 + \binom{q^2}{3}} \geq \frac{e^{(\log N)^{1/3}/2}}{q^{10}}. \tag{46.26}$$

For notational simplicity write

$$r = \frac{e^{(\log N)^{1/3}/2}}{q^{10}}.$$

We define the auxiliary hypergraph $\mathcal{G}_2$ as follows: $B \in \mathcal{G}_2$ if and only if

$$B = \bigcup_{i=1}^{r} A_i \quad \text{where} \quad \bigcap_{i=1}^{r} A_i \text{ is } (\geq 2)\text{-element collinear, and}$$

$A_1, \ldots, A_r$ are r different sets in $\mathcal{F}$ such that they are pairwise disjoint apart from the line spanned by the collinear common part. Then by (46.10)

$$\frac{1}{6\lambda_2} T(\mathcal{F}) = T(\mathcal{G}_2) \leq \binom{N^2}{2} \cdot \binom{N^2}{r} \left(\frac{3}{2}\right)^{-r(q^2-q)}. \tag{46.27}$$

In (46.27) the factor $\binom{N^2}{2}$ is the number of ways to fix two different points, $\binom{N^2}{r}$ is an upper bound on the number of ways to choose $B = \bigcup_{i=1}^{r} A_i$, and the exponent $r(q^2 - q)$ is explained by the disjointness apart from the collinear common part (here we use that a line intersects a $q \times q$ parallelogram lattice in at most q points).

It follows from the definition of $\mathcal{H}_3$ that

$$T(\mathcal{G}_2) \geq \left(\frac{3}{2}\right)^{-r \cdot s_0}. \tag{46.28}$$

On the other hand, by (46.27)

$$\lambda_2 \geq \frac{T(\mathcal{F})}{3 \cdot N^4} \cdot \left(\frac{r \cdot (3/2)^{q^2-q}}{e \cdot N^2}\right)^r. \tag{46.29}$$

Notice that (46.27) is similar to (46.28); in fact, Case 2b is simpler than Case 2a, because in (46.27) (and so in (46.29)) we have the much smaller Pair-Degree $\leq N^2$ instead of N^4, implying that the crucial factor $\binom{N^2}{r}$ in (46.29) is much smaller than the corresponding factor $\binom{N^4}{m}$ in (46.18). This explains why repeating the argument of Case 2a, we obtain the same kind of contradiction here in Case 2b (in fact, Case 2b is simpler!).

This shows that Case 2 is really impossible, completing the proof of the Parallelogram Lattice Game.

The argument above is simpler than that of in Section 45; it can be easily extended to any other lattice game, and also to the Clique Games. Of course, in the Clique Games we need calculations similar to Section 25, but the idea remains the same.

47

Miscellany (I)

1. A duality principle. So far we have focused on the Achievement Games, and it was often stated that "of course the same holds for the Avoidance version" without going into the details. Here an honest effort is made to explain the striking equality

$$\text{Achievement Number} = \text{Avoidance Number} \qquad (47.1)$$

which holds for our "Ramseyish" games with quadratic goals (see Theorems 6.4, 8.2, and 12.6). We begin the precise discussion of (47.1) by recalling the simplest Achievement building criterion.

Theorem 1.2 *If $\mathcal{F}$ is n-uniform and*

$$|\mathcal{F}| > 2^{n-3} \cdot \Delta_2 \cdot |V|,$$

where Δ_2 is the Max Pair-Degree and V is the board, then at the end of the play on $\mathcal{F}$ Maker (the first player) can always occupy a whole winning set $A \in \mathcal{F}$.

The Reverse version goes as follows:

Reverse Theorem 1.2 *If $\mathcal{F}$ is n-uniform and*

$$|\mathcal{F}| > 2^{n-3} \cdot \Delta_2 \cdot |V|,$$

then Forcer (the second player) can always force Avoider to occupy a whole winning set $A \in \mathcal{F}$.

The proof of "Reverse Theorem 1.2" is the same as that of Theorem 1.2, except that Forcer always chooses a point of *minimum* value (note that Maker always chooses a point of *maximum* value).

Is it true that *every* Achievement building result in the book can be converted into a Forcer's win result in the Avoidance version by simply switching the role of maximum and minimum? The answer is "no," and an example is:

Theorem 20.3 *If* $b = (\frac{\log 2}{27} - o(1))n/\log n = (.02567 - o(1))n/\log n$, *then playing the (1:b) Hamiltoninan Cycle Game on* K_n, *underdog Maker can build a Hamiltonian cycle.*

What is so special about the proof of Theorem 20.3? Well, Maker's Hamilton cycle building strategy consists of two phases. In Phase 1 he guarantees that his graph is an Expander Graph with factor 2 (see property (α) in Section 20), and in Phase 2 he keeps creating a *longer* path by choosing an unoccupied Closing Edge from sub-graph $Close(G, P)$. Phase 1 is exactly like the proof of Theorem 1.2: every Maker's move is a **Potential Move**, meaning that Maker optimizes an appropriate potential function by choosing an edge of maximum value. There is no problem with Phase 1: it can be trivially converted into the Reverse Game by switching the roles of maximum and minimum.

Maker can guarantee that Phase 1 ends at a relatively early stage of the play, when the "good" sub-graph $Close(G, P)$ has plenty of unoccupied edges. Of course, in Phase 2 Maker jumps on the opportunity and picks an unoccupied Closing Edge from $Close(G, P)$ (creating a longer path, and he keeps repeating this). Now this is exactly where the problem with Phase 2 is: there is no way to force the reluctant Avoider to pick an edge from "good" sub-graph $Close(G, P)$ at an early stage; Avoider can wait until the very end when he has no other moves left, and then the proof may collapse in the next step (i.e. we may not be able to create a longer path in the next step)! This is why there is no obvious way to convert Phase 2 into the Reverse Game, and this leads to Open Problem 20.1.

Theorem 20.3 is a warning: there is no automatic "transference principle" here. The good news about the exact results is that their building part is proved by either Theorem 1.2 or Theorem 24.2, and in both strategies every Maker's move is a Potential Move (i.e. Maker optimizes a potential function). This fact can be checked by inspecting the proofs of Theorems 6.4, 8.2, and 12.6. Of course, Potential Moves are "safe": they can be trivially converted into the Reverse Game by switching the roles of maximum and minimum. The same applies for the *ad hoc* arguments in Section 23: every Maker's move is a Potential Move.

How about the blocking part of the exact solutions? The simplest blocking criterion is the Erdős–Selfridge Theorem (Theorem 1.4). It is obviously safe: every Breaker's move is a Potential Move. To prove the exact solutions we combined Theorem 1.4 (in fact its Shutout version) with the BigGame–SmallGame Decomposition technique (see Chapters VII–IX). The Big Game remains a single entity, and every Breaker's move in the Big Game is a Potential Move. The *small game*, on the other hand, falls apart into a large number of non-interacting components, and this needs a little bit of extra analysis.

Let us begin with Theorem 34.1 (the first Ugly Theorem). An inspection of Section 36 shows that the *small game* is a disjoint game: it is played on the on-line disjoint parts $\widetilde{S}_j$ of the emergency sets S_j, $j = 1, 2, 3, \ldots,$ and $|\widetilde{S}_j| \geq 2$ holds for every j. In the small game Breaker follows the trivial Same Set Rule: if Maker's last move was in an $\widetilde{S}_j$, then Breaker always takes another unoccupied point from the same $\widetilde{S}_j$ as long as he can. In the Reverse version Avoider follows the Same Set Rule: if Forcer's last move was in an $\widetilde{S}_j$, then Avoider always takes another unoccupied point from the same $\widetilde{S}_j$ as long as he can do it. Avoider can clearly avoid occupying a whole $\widetilde{S}_j$ if we have the slightly stronger condition: $|\widetilde{S}_j| \geq 3$ for every j. The original condition $|\widetilde{S}_j| \geq 2$ is not enough as the following simple counter-example shows.

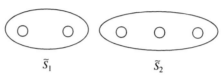

$$\widetilde{S}_1 \qquad\qquad\qquad \widetilde{S}_2$$

If Avoider starts in $\widetilde{S}_1$, then Forcer takes 2 points from $\widetilde{S}_2$, and even if Avoider follows the Same Set Rule, he will end up with both points of $\widetilde{S}_1$. Of course, this kind of "cheap" parity problem cannot occur if every $|\widetilde{S}_j| \geq 3$.

The new requirement "$|\widetilde{S}_j| \geq 3$ for every j" forces a slight change in the proof of Theorem 34.1: a survivor $A \in \mathcal{F}$ becomes dangerous when Maker occupies its $(|A| - k - 2)$th point, and the condition changes as follows: the term $k(k+1)$ is replaced by $k(k+2)$, which is, of course, irrelevant in the asymptotic behavior (compare (34.6) with (47.2) below).

Reverse Theorem 34.1 *Let $\mathcal{F}$ be an m-uniform Almost Disjoint hypergraph. The Maximum Degree of $\mathcal{F}$ is denoted by D, and the total number of winning sets is $|\mathcal{F}| = M$. If there is an integer k with $2 \leq k \leq m/2$ such that*

$$M \binom{m(D-1)}{k} < 2^{km - k(k+2) - \binom{k}{2} - 1}, \tag{47.2}$$

then Avoider can always avoid occupying a whole winning set $A \in \mathcal{F}$.

Next consider Theorem 37.5 (the second Ugly Theorem). Here the small game falls apart into many-many "sub-exponentially" large components (for the details of the proof, see Section 39). In each component Breaker follows the Same Component Rule (if Maker moves to a component of the small game, then Breaker replies in the same component), and chooses an unoccupied point of maximum value. In the Reverse version Avoider follows the Same Component Rule, and chooses a point of minimum value. Identical criterions give identical thresholds!

The same applies for the rest of Part D. This completes our explanation for the equality

$$\text{Achievement Number} = \text{Avoidance Number}$$

in Theorems 6.4, 8.2, and 12.6.

2. An extention of the duality. The equality above can be extended to a longer chain of equalities:

Achievement Number = Avoidance Number

= Chooser's Achievement Number = Picker's Avoidance Number. (47.3)

(47.3) involves two more games: the Chooser–Picker Game and the Picker–Chooser Game. These games are very different from the Maker–Breaker and Avoider–Forcer games: in each turn of the play Picker picks two previously unselected points of the board V, Chooser chooses one of them, and the other point goes back to Picker. The "first name" indicates the "builder": in the Chooser–Picker version Chooser is the "builder," he wins if he can occupy a whole winning set $A \in \mathcal{F}$; otherwise Picker wins. In the Reverse Picker–Chooser Game Picker is the "anti-builder": he loses if he occupies a whole winning set $A \in \mathcal{F}$; otherwise Picker wins. In the Chooser–Picker Game we have the following analogue of Theorem 1.2.

Chooser's building criterion ("linear"). *If $\mathcal{F}$ is n-uniform and*

$$|\mathcal{F}| > 2^{n-3} \cdot \Delta_2 \cdot |V|,$$

then at the end of the play on $\mathcal{F}$ Chooser can always occupy a whole winning set $A \in \mathcal{F}$.

The proof of this criterion is exactly the same as that of Theorem 1.2, except that, from the two points offered to him by Picker, Chooser always chooses a point of *larger* value.

In the Reverse Picker–Chooser Game we have:

Chooser's Picker-is-forced-to-build criterion ("linear"). *If $\mathcal{F}$ is n-uniform and*

$$|\mathcal{F}| > 2^{n-3} \cdot \Delta_2 \cdot |V|,$$

then Chooser can always force Picker to occupy a whole winning set $A \in \mathcal{F}$.

The proof of this criterion is the same, except that, from the two points offered to him by Picker, Chooser always chooses a point of *smaller* value.

Similarly, the Advanced Maker's Win Criterion (Theorem 24.2) can be converted into an Advanced Chooser's building criterion by simply replacing *maximum* with *larger*, and can be converted into an Advanced Chooser's Picker-is-forced-to-build criterion by switching to *smaller*. The same applies for the *ad hoc* arguments of Section 23.

In the Chooser–Picker Game we have the following version of the Erdős–Selfridge theorem (see Section 38).

Theorem 38.1 ("Picker's blocking") *If*

$$\sum_{A \in \mathcal{F}} 2^{-|A|} \leq \frac{1}{8(\|\mathcal{F}\| + 1)},$$

where $\|\mathcal{F}\| = \max_{A \in \mathcal{F}} |A|$ *is the rank of hypergraph* $\mathcal{F}$*, then playing the Chooser–Picker game on hypergraph* $\mathcal{F}$*, Picker can always block every winning set* $A \in \mathcal{F}$.

This criterion is not exactly the same as Theorem 1.4, but it is a *similar* Power-of-Two criterion; consequently, in the exact solutions – where the Erdős–Selfridge Theorem is combined with the BigGame–SmallGame Decomposition – the difference gives a negligible additive term $o(1)$, which tends to 0 as the board size N tends to infinity. Note that Picker can easily enforce the Same Component Rule in the small game by always picking his point pair from the same component (by using Theorem 38.1). This explains the equality

$$\text{Achievement Number} = \text{Chooser's Achievement Number.} \qquad (47.4)$$

In the Reverse Picker–Chooser Game we have the perfect analogue of Theorem 38.1.

Picker's anti-building criterion. *If*

$$\sum_{A \in \mathcal{F}} 2^{-|A|} \leq \frac{1}{8(\|\mathcal{F}\| + 1)},$$

where $\|\mathcal{F}\| = \max_{A \in \mathcal{F}} |A|$ *is the rank of* $\mathcal{F}$*, then, playing the Reverse Picker–Chooser game on* $\mathcal{F}$*, Picker can always avoid occupying a whole winning set* $A \in \mathcal{F}$.

The proof of this criterion is *exactly* the same as that of Theorem 38.1. This explains the equality

$$\text{Chooser's Achievement Number} = \text{Picker's Avoidance Number.} \qquad (47.5)$$

Combining (47.4)–(47.5) we obtain (47.3).

The Picker–Chooser and the Reverse Chooser–Picker Games have not been mentioned yet. From Section 22 we know the equality

$$\text{Majority Play Number} = \text{Picker's Achievement Number.} \qquad (47.6)$$

In the Reverse Chooser–Picker Game, Chooser loses if he occupies a whole winning set $A \in \mathcal{F}$; otherwise Chooser wins.

In the Picker–Chooser Game, we have (for the notation see Section 22)

Picker's building criterion: Theorem 22.1 *Consider the Picker–Chooser Game on hypergraph* $(V, \mathcal{F})$*. Assume that*

$$T(\mathcal{F}) \geq 10^{14} \|\mathcal{F}\|^{14} \left(\sqrt{T(\mathcal{F}_1^2)} + 1 \right).$$

Then Picker can always occupy a whole winning set $A \in \mathcal{F}$.

Exactly the same proof gives the following result in the Reverse Chooser–Picker Game:

Picker's Chooser-is-forced-to-build criterion. *Consider the Reversed Chooser–Picker Game on hypergraph* $(V, \mathcal{F})$. *Assume that*

$$T(\mathcal{F}) \geq 10^{14} \|\mathcal{F}\|^{14} \left(\sqrt{T(\mathcal{F}_1^2)} + 1 \right).$$

Then Picker can always force Chooser to occupy a whole winning set $A \in \mathcal{F}$.

Next we switch from building to blocking; in the Picker–Chooser Game we have:

Chooser's blocking criterion. *If* $\mathcal{F}$ *is n-uniform and* $|\mathcal{F}| < 2^n$, *then Chooser can always prevent Picker from occupying a whole* $A \in \mathcal{F}$ *in the Picker–Chooser Game.*

The proof of this criterion is totally routine: Chooser uses the Power-of-Two Scoring System, and in each turn among the two points offered to him by Picker he chooses a point of *larger* value.

A similar proof gives the following result in the Reverse Chooser–Picker Game.

Chooser's avoidance criterion. *If* $\mathcal{F}$ *is n-uniform and* $|\mathcal{F}| < 2^n$, *then Chooser can always avoid occupying a whole* $A \in \mathcal{F}$ *in the Reverse Chooser–Picker Game.*

The only change in the proof is that Chooser chooses the point of *smaller* value.

Identical criterions give identical thresholds; this explains why (47.6) can be extended by the extra equality

$$\text{Picker's Achievement Number} = \text{Chooser's Avoidance Number.} \qquad (47.7)$$

This concludes our discussion on the "duality" of the game numbers.

3. Balancing in discrepancy games. Let $1/2 < \alpha \leq 1$; in the α-Discrepancy Game Maker's goal is to occupy $\geq \alpha$ part of some $A \in \mathcal{F}$ (instead of occupying the whole $A \in \mathcal{F}$). The main result was formulated in Theorem 9.1, and the "lead-building" part was formulated in Sections 28–29. The balancing part is an adaptation of the BigGame–SmallGame Decomposition, developed for the special case $\alpha = 1$. In the Big Game, and also in every component of the small game, Breaker ("balancer") applies the Corollary of Theorem 16.1 (see below). We begin by recalling the:

Corollary of Theorem 16.1 *Let* $\mathcal{F}$ *be an n-uniform hypergraph, and consider the Balancer–Unbalancer game (introduced in Theorem 16.1) played on hypergraph* $\mathcal{F}$ *where Unbalancer's goal is to own at least* $\frac{n+\Delta}{2}$ *points from some* $A \in \mathcal{F}$. *If*

$$\Delta = \left(1 + O\left(\sqrt{\frac{\log |\mathcal{F}|}{n}} \right) \right) \sqrt{2n \log |\mathcal{F}|},$$

then Balancer has a winning strategy.

The key question is how to define the *emergency sets* in the α-Discrepancy Game. Assume that hypergraph $\mathcal{F}$ is *n*-uniform (i.e. $n = q \times q$; we cannot handle the Clique Games!). Let $0 < \delta < 1/2$ be a fixed constant; when an $A \in \mathcal{F}$ has the property for the first time that Maker's lead in A equals

$$(2\alpha - 1)n - \max \left\{ |A(blank)|^{\frac{1}{2}+\delta}, n^{\frac{1}{2}-\delta} \right\}, \tag{47.8}$$

set A becomes *dangerous*, and its blank part $A(blank)$ is the *first emergency set*. The emergency set $A(blank)$ is removed from the Big Board (which was the whole $N \times N$ board before) and added to the *small board* (which was empty before).

Note that we may have several dangerous sets arising at the same time: each one is removed from the Big Board and added to the small board.

The intuitive justification of (47.8) goes as follows. If Maker achieves an α-Discrepancy, then at the end of the play his lead in some $A \in \mathcal{F}$ is $\geq \alpha n - (1 - \alpha)n = (2\alpha - 1)n$. This is what Breaker (as balancer) wants to prevent, so a Maker's lead close to $(2\alpha - 1)n$ is obviously dangerous. The second term $|A(blank)|^{\frac{1}{2}+\delta}$ in (47.8) is motivated by the idea that, in each component of the small game, Breaker uses the Corollary of Theorem 16.1 to prevent a discrepancy of size $|A(blank)|^{\frac{1}{2}+\delta}$. Breaker succeeds if each component of the small game is "not too large," which component condition is enforced by the Big Game. This is exactly the basic idea of the BigGame–SmallGame Decomposition technique. Finally, the term $n^{\frac{1}{2}-\delta}$ shows up in (47.8) because it is much less than $\sqrt{n}$; this is what Breaker needs to break the "square-root barrier."

The rest of the adaptation is straightforward, see Sections 40–44.

4. Blocking in biased games. In the $(m : b)$ Achievement play the BigGame-SmallGame Decomposition can work only if $m \leq b$ (otherwise Breaker cannot keep up with Maker). If $m \leq b$ holds, then Breaker must "keep the ratio" in the Big Game and "keep it fair" in the components of the small game. For example, consider the (4:5) play; if Maker's next move is (say) $2 + 1 + 1$ in the sense that 2 points in the Big Game and the remaining 2 points split in 2 different components of the small game, then Breaker should reply the $3 + 1 + 1$ way, that is, 3 points in the Big Game and the remaining 2 points split in the same 2 components of the small game.

By the way, in both *exact* biased game results, Theorem 9.2 (the $(a:1)$ Avoidance game) and Theorem 32.1 (the $(1:b)$ Achievement game), number "1" cannot be divided into parts, which makes it even simpler! The decomposition techniques developed for the $(1:1)$ play in Part D can be trivially adapted for the $(1:b)$ Achievement games and the $(a:1)$ Avoidance games.

5. Blocking in biased Chooser–Picker games. Theorem 33.4 was about the biased $(1:s)$ Chooser–Picker game, where $s \geq 2$, i.e. Picker is the topdog. It proves what is conjectured to be the best possible Chooser's building criterion. Here we discuss the missing Picker's blocking part. The fair (1:1) case is solved in Theorem 38.1. This is particularly simple; unfortunately its proof does not extend to the biased (1:s) game, or at least we cannot see any straightforward generalization. The following, somewhat strange and technical, result is my best effort in the (1:s) case:

Theorem 47.1 *Assume that there is an integer t in* $1 \leq t \leq \|\mathcal{F}\|$ *such that*

$$(1) \qquad \sum_{A \in \mathcal{F}} (s+1)^{-|A|+t} \leq 1; \quad and$$

(2) *the restriction of* $\mathcal{F}$ *to any* $(s^{24}\|\mathcal{F}\|^{12})$-*element point-set spans* $\leq \frac{2^t}{8(t+1)}$ *different sets — we refer to this second condition as the Induced Sub-hypergraph Size Property.*

Then, playing the (1:s) *Chooser–Picker game on* $\mathcal{F}$, *topdog Picker can always block every* $A \in \mathcal{F}$.

Proof of Theorem 47.1. We are going to apply the Power-of-$(s+1)$ Scoring System: for an arbitrary hypergraph $(V, \mathcal{H})$, where V is the union set ("board"), write

$$T(\mathcal{H}) = \sum_{A \in \mathcal{H}} (s+1)^{-|A|},$$

and for any m-element sub-set of points $\{u_1, \ldots, u_m\} \subseteq V$ $(m \geq 1)$ write

$$T(\mathcal{H}; u_1, \ldots, u_m) = \sum_{A \in \mathcal{H}: \{u_1, \ldots, u_m\} \subseteq A} (s+1)^{-|A|},$$

of course, counted with multiplicity.

Assume that we are in the middle of a (1:s) Chooser–Picker play, where Chooser already occupied $X(i) = \{x_1, x_2, \ldots, x_i\}$, and Picker occupied

$$Y(i) = \left\{ y_1^{(1)}, \ldots, y_1^{(s)}, y_2^{(1)}, \ldots, y_2^{(s)}, \ldots, y_i^{(1)}, \ldots, y_i^{(s)} \right\}.$$

The question is how to pick Picker's next move $\{u_1, u_2, \ldots, u_{s+1}\}$. Of course, this $(s+1)$-element set equals $\{x_{i+1}, y_{i+1}^{(1)}, \ldots, y_{i+1}^{(s)}\}$, but Picker doesn't know in advance which one will be $x_{i+1} =$ Chooser's choice.

Let

$$\mathcal{F}(i) = \{A \setminus X(i) : \ A \in \mathcal{F}, \ A \cap Y(i) = \emptyset\};$$

$\mathcal{F}(i)$ is a multi-hypergraph.

We can describe the effect of the $(i+1)$st moves $x_{i+1}, y_{i+1}^{(1)}, \ldots, y_{i+1}^{(s)}$ as follows:

$$T(\mathcal{F}(i+1)) = T(\mathcal{F}(i)) + s \cdot T(\mathcal{F}(i); x_{i+1}) - \sum_{j=1}^{s} T(\mathcal{F}(i); y_{i+1}^{(j)})$$

$$- s \sum_{j=1}^{s} T(\mathcal{F}(i); x_{i+1}, y_{i+1}^{(j)}) + \sum_{1 \leq j_1 < j_2 \leq s} T(\mathcal{F}(i); y_{i+1}^{(j_1)}, y_{i+1}^{(j_2)})$$

$$+ s \sum_{1 \leq j_1 < j_2 \leq s} T(\mathcal{F}(i); x_{i+1}, y_{i+1}^{(j_1)}, y_{i+1}^{(j_2)}) - \sum_{1 \leq j_1 < j_2 < j_3 \leq s} T(\mathcal{F}(i); y_{i+1}^{(j_1)}, y_{i+1}^{(j_2)}, y_{i+1}^{(j_3)}) \mp \cdots$$

$$(47.9)$$

Identity (47.9) may seem rather complicated at first sight, but the underlying pattern is very simple: it is described by the expansion of the product

$$(1 + s \cdot x)(1 - y^{(1)})(1 - y^{(2)}) \cdots (1 - y^{(s)}) - 1. \qquad (47.10)$$

It follows from (47.9)–(47.10) that

$$T(\mathcal{F}(i+1)) \leq T(\mathcal{F}(i)) + \sum_{j=1}^{s} \left| T(\mathcal{F}(i); x_{i+1}) - T(\mathcal{F}(i); y_{i+1}^{(j)}) \right|$$

$$+ s \sum_{j=1}^{s} T(\mathcal{F}(i); x_{i+1}, y_{i+1}^{(j)}) + \sum_{1 \leq j_1 < j_2 \leq s} T(\mathcal{F}(i); y_{i+1}^{(j_1)}, y_{i+1}^{(j_2)})$$

$$+ s \sum_{1 \leq j_1 < j_2 \leq s} T(\mathcal{F}(i); x_{i+1}, y_{i+1}^{(j_1)}, y_{i+1}^{(j_2)}) + \sum_{1 \leq j_1 < j_2 < j_3 \leq s} T(\mathcal{F}(i); y_{i+1}^{(j_1)}, y_{i+1}^{(j_2)}, y_{i+1}^{(j_3)}) + \cdots$$

$$(47.11)$$

As said before, the key question is how to pick Picker's next move $\{u_1, u_2, \ldots, u_{s+1}\}$ $(= \{x_{i+1}, y_{i+1}^{(1)}, \ldots, y_{i+1}^{(s)}\})$ from the unoccupied part $V_i = V \setminus (X(i) \cup Y(i))$ of the board. To answer this question, list the following trivial inequalities

$$\sum_{u \in V_i} T(\mathcal{F}(i); u) \leq \|\mathcal{F}\| T(\mathcal{F}(i)), \qquad (47.12)$$

$$\sum_{\{u_1, u_2\} \in \binom{V_i}{2}} T(\mathcal{F}(i); u_1, u_2) \leq \binom{\|\mathcal{F}\|}{2} T(\mathcal{F}(i)), \qquad (47.13)$$

$$\sum_{\{u_1, u_2, u_3\} \in \binom{V_i}{3}} T(\mathcal{F}(i); u_1, u_2, u_3) \leq \binom{\|\mathcal{F}\|}{3} T(\mathcal{F}(i)), \qquad (47.14)$$

and so on, where the last one is

$$\sum_{\{u_1, \ldots, u_s\} \in \binom{V_i}{s}} T(\mathcal{F}(i); u_1, \ldots, u_s) \leq \binom{\|\mathcal{F}\|}{s} T(\mathcal{F}(i)). \qquad (47.15)$$

We need the following:

Deletion Lemma: *Assume that* $|V_i| \geq s\|\mathcal{F}\|^{12}$, *and choose*

$$m = \frac{|V_i|^{3/4}}{\|\mathcal{F}\|}.$$

Then there exists a sub-set $U_0 \subset V_i$ *such that*

$$(1) \qquad |U_0| \geq \frac{m}{2} = \frac{|V_i|^{3/4}}{2\|\mathcal{F}\|};$$

$$(2) \quad T(\mathcal{F}(i); u_1, \ldots, u_k) \leq \frac{s\|\mathcal{F}\|}{|V_i|^{1+(k-1)/4}} \cdot T(\mathcal{F}(i))$$

holds for every $\{u_1, \ldots, u_k\} \in \binom{U_0}{k}$ *and every* $2 \leq k \leq s$.

The **proof** of the Deletion Lemma is a routine application of Erdős's deletion technique, a standard idea in the Probabilistic Method.

A k-set $\{u_1, \ldots, u_k\} \in \binom{V_i}{k}$ with $2 \leq k \leq s$ is called a *bad* k-set if requirement (2) above is violated, i.e. if

$$T(\mathcal{F}(i); u_1, \ldots, u_k) > \frac{s\|\mathcal{F}\|}{|V_i|^{1+(k-1)/4}} \cdot T(\mathcal{F}(i)). \qquad (47.16)$$

Recall

$$m = \frac{|V_i|^{3/4}}{\|\mathcal{F}\|} \quad \text{and} \quad |V_i| \geq s\|\mathcal{F}\|^{12}.$$

Let $\mathbf{R}$ be a randomly chosen m-set in V_i, that is, all $\binom{|V_i|}{m}$ m-sets are equally likely. For every $2 \leq k \leq s$ let $B_k(\mathbf{R})$ denote the expected number of *bad* k-sets in $\mathbf{R}$. In view of (47.13)–(47.15) we have the trivial inequality

$$\binom{|V_i|}{m} \cdot B_k(\mathbf{R}) \cdot \frac{s\|\mathcal{F}\|}{|V_i|^{1+(k-1)/4}} \cdot T(\mathcal{F}(i)) \leq \binom{\|\mathcal{F}\|}{k} T(\mathcal{F}(i)) \binom{|V_i| - k}{m - k},$$

which is equivalent to

$$B_k(\mathbf{R}) \leq \frac{|V_i|^{1+(k-1)/4}}{s\|\mathcal{F}\|} \cdot \binom{\|\mathcal{F}\|}{k} \frac{\binom{|V_i|-k}{m-k}}{\binom{|V_i|}{m}}. \qquad (47.17)$$

Since

$$\frac{\binom{|V_i|-k}{m-k}}{\binom{|V_i|}{m}} = \frac{\binom{m}{k}}{\binom{|V_i|}{k}},$$

by (47.17) we have

$$B_k(\mathbf{R}) \leq \frac{|V_i|^{1+(k-1)/4}}{s\|\mathcal{F}\|} \cdot \frac{\binom{\|\mathcal{F}\|}{k}\binom{m}{k}}{\binom{|V_i|}{k}} \leq \frac{|V_i|^{1+(k-1)/4}}{s\|\mathcal{F}\|} \cdot \frac{1}{2} \left(\frac{\|\mathcal{F}\|m}{|V_i|}\right)^k$$

$$= \frac{|V_i|^{1+(k-1)/4}}{s\|\mathcal{F}\|} \cdot \frac{1}{2}|V_i|^{-k/4} = \frac{|V_i|^{3/4}}{2s\|\mathcal{F}\|} = \frac{m}{2s}. \qquad (47.18)$$

(47.18) gives that

$$\sum_{k=2}^{s} B_k(\mathbf{R}) \leq (s-1)\frac{m}{2s} < \frac{m}{2}. \tag{47.19}$$

It follows from inequality (47.19) about the expected value that the poorest m-set R_0 in V_i has the property

$$\sum_{k=2}^{s} B_k(R_0) < \frac{m}{2},$$

that is, the total number of bad sets in R_0 is less than $m/2$. Deleting 1 point from each bad set in R_0 we obtain a sub-set $U_0 \subset R_0$ such that $|U_0| > m/2$ and U_0 does not contain any bad set. This completes the proof of the Deletion Lemma.   □

It follows from (47.12) that

$$\sum_{u \in U_0} T(\mathcal{F}(i); u) \leq \|\mathcal{F}\| T(\mathcal{F}(i)), \tag{47.20}$$

where $U_0 \subset V_i$ is a sub-set satisfying the Deletion Lemma. By (47.20) at least half of $u \in U_0$ satisfy

$$T(\mathcal{F}(i); u) \leq \frac{2\|\mathcal{F}\|}{|U_0|} T(\mathcal{F}(i)),$$

that is, there is a sub-set $\{u_1, u_2, \ldots, u_l\} \subset U_0$ with $l \geq |U_0|/2 > m/4 = \frac{1}{4} V^{3/4}/\|\mathcal{F}\|$ such that

$$T(\mathcal{F}(i); u_j) \leq \frac{4\|\mathcal{F}\|^2}{|V_i|^{3/4}} T(\mathcal{F}(i)), \quad 1 \leq j \leq l. \tag{47.21}$$

Divide the interval

$$I = \left[0, \frac{4\|\mathcal{F}\|^2}{|V_i|^{3/4}} T(\mathcal{F}(i))\right] \quad \text{into} \quad \frac{|V_i|^{3/4}}{4s\|\mathcal{F}\|}$$

equal sub-intervals. By the Pigeonhole Principle one of the sub-intervals contains $\geq s+1$ elements of $\{u_1, u_2, \ldots, u_l\}$; for notational convenience denote them simply by $\{u_1, u_2, \ldots, u_{s+1}\}$. It follows from (47.21) that the inequality

$$|T(\mathcal{F}(i); u_p) - T(\mathcal{F}(i); u_q)| \leq \frac{\frac{4\|\mathcal{F}\|^2}{|V_i|^{3/4}} T(\mathcal{F}(i))}{\frac{|V_i|^{3/4}}{4s\|\mathcal{F}\|}}$$

$$= \frac{16s\|\mathcal{F}\|^3}{|V_i|^{3/2}} T(\mathcal{F}(i)) \tag{47.22}$$

holds for any $1 \leq p < q \leq s+1$. Summarizing, there is an $(s+1)$-element sub-set $\{u_1, u_2, \ldots, u_{s+1}\} \subset V_i$ such that (47.22) holds, and also, by the Deletion Lemma

$$T(\mathcal{F}(i); u_{j_1}, \ldots, u_{j_k}) \leq \frac{s\|\mathcal{F}\|}{|V_i|^{1+(k-1)/4}} T(\mathcal{F}(i)) \tag{47.23}$$

holds for any $1 \leq j_1 < j_2 < \cdots < j_k \leq s+1$ and any $2 \leq k \leq s$.

Let us return to (47.11): by (47.22) and (47.23) we have

$$T(\mathcal{F}(i+1)) \le T(\mathcal{F}(i))\left(1 + \frac{16s^2\|\mathcal{F}\|^3}{|V_i|^{3/2}} + \frac{2s^3\|\mathcal{F}\|}{|V_i|^{5/4}}\right.$$
$$\left. + \frac{s^4\|\mathcal{F}\|}{|V_i|^{6/4}} + \frac{s^5\|\mathcal{F}\|}{|V_i|^{7/4}} + \cdots\right). \qquad (47.24)$$

It follows from (47.24) that, if $|V_i| \ge s^{24}\|\mathcal{F}\|^{12}$, then

$$T(\mathcal{F}(i+1)) \le T(\mathcal{F}(i))\left(1 + \frac{1}{|V_i|^{9/8}}\right). \qquad (47.25)$$

Clearly

$$\sum_{i:\ |V_i| \ge M} \frac{1}{|V_i|^{9/8}} \le \int_M^\infty x^{-9/8}\, dx = \frac{8}{M^{1/8}}; \qquad (47.26)$$

moreover, by using the trivial inequality $1 + x \le e^x$, from (47.25) and (47.26) we obtain that

$$T(\mathcal{F}(i)) \le e \cdot T(\mathcal{F}(0)) = e \cdot T(\mathcal{F}) \qquad (47.27)$$

holds for all i with $|V_i| > M = s^{24}\|\mathcal{F}\|^{12}$.

Now we are ready to complete the proof of Theorem 46.1. Let $|V_{i_0}| = s^{24}\|\mathcal{F}\|^{12}$. It follows from (47.27) that every set in the truncated hypergraph $\mathcal{F}(i_0)$ has size $\ge t$. Indeed, otherwise $T(\mathcal{F}(i_0)) \ge (s+1)^{-t+1}$, and we get a contradiction from the first hypothesis

$$T(\mathcal{F}) \le (s+1)^{-t}$$

of Theorem 46.1 and inequality (47.27).

It remains to block $\mathcal{F}(i_0)$; call it the endplay. Since $\mathcal{F}(i_0)$ has less than $2^t/8(t+1)$ sets, and each set has size $\ge t$, Picker can block $\mathcal{F}(i_0)$ in a most trivial way by applying Theorem 38.1. This means Picker can ignore $(s-1)$ of his marks per move: the endplay is so simple that a $(1{:}1)$ play suffices to block $\mathcal{F}(i_0)$. $\qquad \square$

Application to the Clique Game. Consider the biased $(1{:}s)$ Chooser–Picker (K_N, K_q) game, i.e. K_N is the board, K_q is Chooser's goal (Chooser is the underdog). How to choose parameter t in Theorem 47.1? An arbitrary set of $s^{24}\binom{q}{2}^{12}$ edges in K_N covers $\le 2s^{24}\binom{q}{2}^{12}$ vertices in K_N. This gives the following trivial upper bound on the Induced Sub-hypergraph Size (see the second condition in Theorem 47.1)

$$\text{Induced Sub-hypergraph Size} \le \sum_{j=1}^q \binom{2s^{24}\binom{q}{2}^{12}}{j}.$$

Thus we have the following 2 requirements for parameter t

$$\sum_{j=1}^q \binom{2s^{24}\binom{q}{2}^{12}}{j} < \frac{2^t}{8(t+1)} \quad \text{and} \quad \binom{N}{q} \le (s+1)^{\binom{q}{2}-t}.$$

The first requirement is satisfied with the choice $t = 24q \log_2 q$; then the second requirement becomes

$$\frac{eN}{q} \leq (s+1)^{\frac{q}{2} - O(\log q)}.$$

It follows that, if $q = 2 \log_{s+1} N + O(\log \log N)$, then Theorem 47.1 applies and implies Picker's win. The value $q = 2 \log_{s+1} N + O(\log \log N)$ comes very close to the conjectured truth, which is the lower integral part of

$$2 \log_{s+1} N - 2 \log_{s+1} \log_{s+1} N + 2 \log_{s+1} e - 2 \log_{s+1} 2 - 1 - o(1). \qquad (47.28)$$

The discrepancy is $O(\log \log N)$, which is a logarithmic additive error (logarithmic in terms of the main term $2 \log_{s+1} N$).

Incidently, Theorem 33.4 proves exactly that, in the $(1{:}s)$ play in K_N, Chooser *can* build a clique K_q where q is defined in (47.28).

Application to the Lattice Games (Theorem 8.2). We can use an adaptation of Section 42, i.e. the simplest form of the RELARIN technique. This leads to an error term $O((\log N)^{\frac{1}{2} - \varepsilon})$ – note that the main term in Theorem 8.2 is constant times $\sqrt{\log N}$. The RELARIN technique in Section 42 involves Big Sets $\bigcup_{i=1}^{r} A_i$, where each A_i is a $q \times q$ lattice of the same type in the $N \times N$ board. The most general lattice in Theorem 8.2 (a)–(g) is the Parallelogram Lattice. Three non-collinear points of a $q \times q$ parallelogram lattice A nearly determine A: there are at most $\binom{q^2}{3}$ A's containing the same non-collinear triplet.

A Key Property of the family of Big Sets: the Size of the Induced Sub-hypergraph. Let X be an arbitrary $(s^{24}(q^2)^{12})$-element sub-set of the $N \times N$ board. Let $\mathcal{B}$ denote the family of all Big Sets $B = \bigcup_{i=1}^{r} A_i$ defined by the RELARIN technique in Section 42. We need an upper bound on the number of different sets in the induced sub-hypergraph $\mathcal{B}_X$, where $\mathcal{B}_X$ means the restriction of $\mathcal{B}$ to X; let $|\mathcal{B}_X|$ denote its size. Let $B = \bigcup_{i=1}^{r} A_i$ be arbitrary. If $|A_i \cap X| \geq 3$ and $A_i \cap X$ is non-collinear, then let A_i^* be an arbitrary non-collinear triplet in $A_i \cap X$; if $|A_i \cap X| \geq 3$ and $A_i \cap X$ is collinear, then let A_i^* be an arbitrary point-pair in $A_i \cap X$; and, finally, if $|A_i \cap X| \leq 2$, then let $A_i^* = A_i \cap X$. Let $B^* = \bigcup_{i=1}^{r} A_i^*$. The total number of sets B^* is trivially less than $|X|^{3r}$. Therefore, the size of the induced hypergraph

$$|\mathcal{B}_X| \leq |X|^{3r} \cdot \binom{q^2}{3}^r = (s^{24}q^{24})^{3r} \cdot \binom{q^2}{3}^r \leq (s^{24} \cdot q^{74})^r. \qquad (47.29)$$

By using inequality (47.29) it is easy now to apply Theorem 47.1 to the Lattice Games. The calculations go very similarly to the Clique Game. We stop here, and the details are left to the reader as an exercise.

Open Problem 47.1 *Formulate and prove a stronger version of Theorem 47.1 (Picker's blocking) which perfectly complements Theorem 33.4 (Chooser's building), i.e. which gives exact solutions in the biased (1:s) Chooser–Picker versions of the Clique and Lattice Games.*

48

Miscellany (II)

1. Fractional Pairing. Recall the stronger form of the Hales–Jewett Conjecture (part (b) in Open Problem 34.1): if there are at least twice as many points as winning lines in the n^d Tic-Tac-Toe board, then the Tic-Tac-Toe game is a Pairing Strategy Draw.

The "Degree Reduction for the n^d-hypergraph" (see Theorem 12.2) immediately gives the following: the n^d game is a Pairing Strategy Draw if $n \geq 4d$ (see Theorem 34.2 (iii)). Indeed, applying the "Degree Reduction" Theorem to the n^d-hypergraph with $\alpha = 1/4$, we obtain a $2\lfloor n/4 \rfloor$-uniform hypergraph with maximum degree $\leq d$. Then "Degree Criterion" Theorem 11.2 applies, and implies a Pairing Strategy Draw. Here we show how to improve the bound "$4d$" to "$3d$" by using the concept of Fractional Pairing; the result is due to Richard Schroeppel (Arizona).

What is a Fractional Pairing? Where did it come from? As far as we know, the very first "fractional" concept was van Neumann's idea of a *mixed strategy*. By extending the concept of ordinary strategy to *mixed strategy*, he could prove that every 2-player zero-sum game (i.e. pure conflict situation) has an "equilibrium," meaning the best compromise for both players. *Mixed strategy* means to randomly play a mixture of strategies according to a certain fixed probability distribution. The fixed probabilities are the "fractional weights," so a *mixed strategy* is nothing else other than a Fractional Strategy!

The concept of Fractional Matching is widely used with great success in Matching Theory. The Bigamy Version of Fractional Matching is exactly the concept of Fractional Pairing. It makes it much easier to check the Pairing Criterion (Theorem 11.1); it gives a lot of extra flexibility. The following result is from Schroeppel's unpublished (yet) manuscript (we include it here with his kind permission).

A Fractional Pairing formally means to fill out the Point-Line Incidence Matrix of the n^d board with real entries $a_{P,L}$ such that:

(1) $0 \leq a_{P,L} \leq 1$;

(2) $\sum_L a_{P,L} \leq 1$ for every point P;

(3) $\sum_P a_{P,L} \geq 2$ for every line L.

Schroeppel's Fractional Pairing Theorem.

(a) *If there is a Fractional Pairing, i.e. (1)–(3) are satisfied, then the n^d Tic-Tac-Toe has an ordinary Draw-Forcing Pairing ("0 − 1 pairing");*

(b) *If $n = 3d$, d is even, or if $n = 3d − 1$, d is odd, then the n^d game has a Pairing Strategy Draw.*

Remark. By using Lemma 1 below we obtain the following extension of (b): the n^d game has a Pairing Strategy Draw if $n \geq 3d$, d even, and $n \geq 3d − 1$, d odd. This is how close we can get to Open Problem 34.1 (b).

Proof. (a) The Pairing Criterion (Theorem 11.1) applies, since for any sub-family $\mathcal{L}$ of lines

$$2|\mathcal{L}| \leq \sum_{L \in \mathcal{L}} \sum_P a_{P,L} = \sum_P \sum_{L \in \mathcal{L}} a_{P,L} \leq \sum_{P \in \cup_{L \in \mathcal{L}} L} 1 = |\cup_{L \in \mathcal{L}} L|.$$

By choosing the common value $a_{P,L} = 2/n$, this argument gives the exponentially weak bound (34.1). The obvious advantage here is the extra flexibility that different entries $a_{P,L}$ may have different values (always between 0 and 1); in (b) below we will take advantage of this flexibility.

Next we prove (b). To find the appropriate weights $a_{P,L}$ we define the concepts of *point-type* and *line-type*. It is based on the concept of "coordinate-repetition" just as in the proof of the "Degree Reduction." We use the same notation: Let $\mathbf{P} = (a_1, a_2, a_3, \ldots, a_d)$, $a_i \in \{1, 2, \ldots, n\}$, $1 \leq i \leq d$ be an arbitrary point of the board of the n^d-game, let $b \in \{1, 2, \ldots, \lfloor (n+1)/2 \rfloor\}$ be arbitrary, and consider the *multiplicity of b and $(n+1−b)$ in $\mathbf{P}$*

$$M(\mathbf{P}, b) = |\{1 \leq i \leq d : a_i = b \text{ or } (n+1−b)\}| = M(\mathbf{P}, n+1−b).$$

(In the definition of multiplicity we identify b and $(n+1−b)$.) Let

$$M(\mathbf{P}, b_1) \geq M(\mathbf{P}, b_2) \geq M(\mathbf{P}, b_3) \geq \cdots \geq M(\mathbf{P}, b_\ell) \quad \text{where} \quad \ell = \lfloor (n+1)/2 \rfloor,$$

i.e. pair $(b_1, n+1−b_1)$ has the largest multiplicity in $\mathbf{P}$, pair $(b_2, n+1−b_2)$ has the second largest multiplicity in $\mathbf{P}$, pair $(b_3, n+1−b_3)$ has the third largest multiplicity in $\mathbf{P}$, and so on. For notational simplicity, write

$$M_1(\mathbf{P}) = M(\mathbf{P}, b_1), \ M_2(\mathbf{P}) = M(\mathbf{P}, b_2), \ M_3(\mathbf{P}) = M(\mathbf{P}, b_3), \ \cdots, \ M_\ell(\mathbf{P}) = M(\mathbf{P}, b_\ell).$$

We call the multiplicity vector $(M_1(\mathbf{P}), M_2(\mathbf{P}), M_3(\mathbf{P}), \ldots, M_\ell(\mathbf{P}))$ the *type* of point $\mathbf{P}$.

For example, the *type* of point

$$\mathbf{P} = (3, 7, 3, 5, 1, 3, 5, 4, 3, 1, 5, 3, 3, 5, 5, 1, 2, 6, 5, 2, 7) \in [7]^{21}$$

is $type(\mathbf{P}) = (12, 5, 3, 1)$.

Similarly, an *n-line* L of the n^d-game can be described by an x-vector $\mathbf{v} = (v_1, v_2, v_3, \ldots, v_d)$ where the ith coordinate v_i is either a constant c_i, or variable x, or variable $(n+1-x)$, $1 \le i \le d$, and for at least one index i, v_i is x or $(n+1-x)$.

For example, in the ordinary 3^2 Tic-Tac-Toe

$(1, 3)$	$(2, 3)$	$(3, 3)$
$(1, 2)$	$(2, 2)$	$(3, 2)$
$(1, 1)$	$(2, 1)$	$(3, 1)$

$\{(1, 1), (2, 2), (3, 3)\}$ is a winning line defined by the x-vector xx, $\{(1, 2), (2, 2), (3, 2)\}$ is another winning line defined by the x-vector $x2$, and finally $\{(1, 3), (2, 2), (3, 1)\}$ is a winning line defined by the x-vector xx', where $x' = (n+1-x)$.

The kth point $\mathbf{P}_k$ of x-vector $\mathbf{v}$ is obtained by specifying $x = k$, $1 \le k \le n$. The sequence $(\mathbf{P}_1, \mathbf{P}_2, \ldots, \mathbf{P}_n)$ gives an *orientation* of line L. Every n-line L has exactly two orientations: $(\mathbf{P}_1, \mathbf{P}_2, \ldots, \mathbf{P}_n)$ and $(\mathbf{P}_n, \mathbf{P}_{n-1}, \ldots, \mathbf{P}_1)$. The second orientation comes from x-vector $\mathbf{v}^*$ which is obtained from $\mathbf{v}$ by switching coordinates x and $(n+1-x)$.

Next we define the *type* of x-vector $\mathbf{v}$, i.e. of line L. Write $\ell = \lfloor (n+1)/2 \rfloor$, and let $b \in \{1, 2, \ldots, \ell\}$ be arbitrary, and consider the multiplicity of b and $(n+1-b)$ in x-vector $\mathbf{v}$

$$M(\mathbf{v}, b) = |\{1 \le i \le d : v_i = b \text{ or } (n+1-b)\}| = M(\mathbf{v}, n+1-b).$$

Similarly, consider the multiplicity of x and $(n+1-x)$ in x-vector $\mathbf{v}$

$$M(\mathbf{v}, x) = |\{1 \le i \le d : v_i = x \text{ or } (n+1-x)\}| = M(\mathbf{v}, n+1-x).$$

It follows that $M(\mathbf{P}_k, k) = M(\mathbf{v}, k) + M(\mathbf{v}, x)$, and $M(\mathbf{P}_k, b) = M(\mathbf{v}, b)$ if $k \notin \{b, n+1-b\}$, where $\mathbf{P}_k$ is the kth point of x-vector $\mathbf{v}$ (i.e. $\mathbf{P}_k$ is the kth point of line L in one of the two orientations), and $b \in \{1, 2, \ldots, \ell\}$.

Let

$$M(\mathbf{v}, b_1) \ge M(\mathbf{v}, b_2) \ge M(\mathbf{v}, b_3) \ge \cdots \ge M(\mathbf{v}, b_{\ell+1}) \text{ where } \ell = \lfloor (n+1)/2 \rfloor,$$

i.e. pair $(b_1, n+1-b_1)$ has the largest multiplicity in $\mathbf{v}$ ($b_1 \in \{x, 1, 2, \ldots, \ell\}$), pair $(b_2, n+1-b_2)$ has the second largest multiplicity in $\mathbf{v}$ ($b_2 \in \{x, 1, 2, \ldots, \ell\}$), pair $(b_3, n+1-b_3)$ has the third largest multiplicity in $\mathbf{v}$ ($b_3 \in \{x, 1, 2, \ldots, \ell\}$), and so on. For notational simplicity, write

$$M_1(\mathbf{v}) = M(\mathbf{v}, b_1), \ M_2(\mathbf{v}) = M(\mathbf{v}, b_2), \ M_3(\mathbf{v}) = M(\mathbf{v}, b_3), \ \cdots \ M_{\ell+1}(\mathbf{v}) = M(\mathbf{v}, b_{\ell+1}).$$

We call the multiplicity vector $(M_1(\mathbf{v}), M_2(\mathbf{v}), M_3(\mathbf{v}), \ldots, M_{\ell+1}(\mathbf{v}))$ the *type* of x-vector $\mathbf{v}$. Observe that it has $(\ell+1)$ coordinates instead of $\ell = \lfloor (n+1)/2 \rfloor$.

Let $M(\mathbf{v}, x) = M_q(\mathbf{v})$, that is, pair $(x, n+1-x)$ has the qth largest multiplicity in x-vector $\mathbf{v}$; q is between 1 and $(\ell+1)$.

The proof of **(b)** is somewhat technical. To illustrate the idea in a simpler case, first we prove a *weaker* statement:

(c) If $n \geq 4d - 2$, then the n^d-game has a Pairing Strategy Draw.

We begin the proof of **(c)** by recalling a variant of the *lexicographic order*. Let $\mathbf{u} = (k_1, k_2, k_3, \ldots)$ and $\mathbf{v} = (l_1, l_2, l_3, \ldots)$ be two real vectors such that the coordinates are decreasing, i.e. $k_1 \geq k_2 \geq k_3 \geq \cdots$ and $l_1 \geq l_2 \geq l_3 \geq \cdots$. We introduce a partial order: $\mathbf{u} < \mathbf{v}$ if there is a $j \geq 1$ such that $u_i = v_i$ for all $1 \leq i < j$ and $u_j < v_j$.

Let $\mathbf{P}$ be a point and L be a winning line of the n^d-game. If $\mathbf{P} \in L$, then clearly $type(\mathbf{P}) \geq type(L)$. Let $type(L) = (m_1, m_2, \ldots, m_q, \ldots, m_r)$; then rearranging the coordinates we can represent line L as follows

$$L: c_1 \ldots c_1 \ (m_1 \text{ times}) \ c_2 \ldots c_2 \ (m_2 \text{ times}) \ldots x \ldots x \ (m_q \text{ times}) \ldots c_r \ldots c_r \ (m_r \text{ times}), \tag{48.1}$$

where $m_1 \geq m_2 \geq \cdots \geq m_r \geq 1$, and c_i actually means "c_i or $(n+1-c_i)$" and x actually means "x or $(n+1-x)$".

Now we are ready to define the Fractional Pairing: let

$$a_{\mathbf{P},L} = \begin{cases} \frac{2}{n-2(r-1)}, & \text{if } \mathbf{P} \in L \text{ with } type(\mathbf{P}) = type(L), \text{ where } r \text{ is defined in (48.1);} \\ 0, & \text{otherwise.} \end{cases} \tag{48.2}$$

Then clearly

$$\sum_{\mathbf{P}: \ \mathbf{P} \in L} a_{\mathbf{P},L} = (n - 2(r-1)) \cdot \frac{2}{n - 2(r-1)} = 2$$

for every winning line, since in (48.1) x can be specified as any element of the set

$$[n] \setminus \{c_1, (n+1-c_1), \ldots, c_{q-1}, (n+1-c_{q-1}), c_{q+1}, (n+1-c_{q+1}), \ldots, c_r,$$
$$(n+1-c_r)\}.$$

It remains to show that for every point $\mathbf{P}$ of the n^d-board

$$\sum_{L: \ \mathbf{P} \in L} a_{\mathbf{P},L} \leq 1.$$

First assume that $n = 4d - 2$, so n is *even*.

By rearranging the coordinates we can represent point $\mathbf{P}$ as follows (the analogue of (48.1))

$$\mathbf{P}: c_1 \ldots c_1 \ (m_1 \text{ times}) \ c_2 \ldots c_2 \ (m_2 \text{ times}) \ldots c_k \ldots c_k \ (m_k \text{ times}), \tag{48.3}$$

where $m_1 \geq m_2 \geq \cdots \geq m_k \geq 1$, and c_i actually means "c_i or $(n+1-c_i)$."

There are exactly k winning lines L such that $\mathbf{P} \in L$ with $type(\mathbf{P}) = type(L)$. Indeed, replacing a *whole* block $c_i \ldots c_i$ (of length m_i) in (48.3) by $x \ldots x$ is the only way to get a line L satisfying $\mathbf{P} \in L$ with $type(\mathbf{P}) = type(L)$. Thus we have (see (48.2))

$$\sum_{L:\ \mathbf{P} \in L} a_{\mathbf{P},L} = k \cdot \frac{2}{n - 2(k-1)} \leq 1$$

if $4k - 2 \leq 4d - 2 = n$. This shows that (48.2) defines a Fractional Pairing if $n = 4d - 2$, so the $(4d-2)^d$-game has a Pairing Draw.

The case "$n > 4d - 2$" follows from "$n = 4d - 2$" and Lemma 1. This completes the proof of the weaker statement (**c**).

It is not difficult to improve on the previous argument; this is how we can prove the stronger statement (**b**); the details are left to the reader as an exercise.

Exercise 48.1 *By modifying the proof of (c) above, prove the stronger (b): If $n = 3d$, d even, or if $n = 3d - 1$, d odd, then the n^d Tic-Tac-Toe has a Pairing Strategy Draw.*

The following result shows how to extend a Pairing Draw to a larger cube:

Lemma 1 ("Extending Pairing Draws"):
(1) If there is a Pairing Strategy Draw (PSD) for n^d, and n is even, then there is a PSD for $(n+1)^d$.
(2) If there is a PSD for n^d, then there is a PSD for $(n+2)^d$.

Remarks. (1) and (2) imply that, if n^d has a PSD, and n is even, then m^d has a PSD for every $m \geq n$. Lemma 1 is due to R. Schroeppel.

Proof. We start with (2). First note that:

(3) If there is a Pairing Strategy Draw (PSD) for n^d, then there is a PSD for n^{d-1}.

Observe that (3) is completely trivial. Indeed, choose any section n^{d-1} from the n^d pairing. Dropping any cell which is paired *outside* of the section, the remaining pairs form a PSD for the section.

To prove (2) we assemble a pairing for $(n+2)^d$ out of various n^ks for dimensions $k \leq d$.

The *interior* n^d of $(n+2)^d$ is covered with the n^d pairing. For the *surface* of $(n+2)^d$ we apply (3): the surface of $(n+2)^d$ is covered with $2d$ n^{d-1} pairings, $2^2 \binom{d}{2} n^{d-2}$ pairings, $2^3 \binom{d}{3} n^{d-3}$ pairings, ..., and, finally, the 2^d unpaired corners (which are irrelevant – do not need to be paired). To prove that this pairing is a PSD, consider an arbitrary winning line in $(n+2)^d$. If the line penetrates the interior of $(n+2)^d$, then there are n interior cells, and the pairing of n^d blocks the line in a pair. If the line does not penetrate the interior, then it lies in a lower dimensional $(n+2)^k$ on the surface, which has an interior n^k. The pairing of n^k intersects the line in a pair.

To prove (1) we slice out all the middle sections $(n+1)^{d-1}$ from $(n+1)^d$. Then join the 2^d "octants" $(n/2)^d$ into an n^d, and apply the n^d pairing. For the middle sections $(n+1)^{d-1}$ we use (3) just like we did in the proof of (2). □

We conclude with a new game concept.

2. Coalition Games. One of the main results of the book is the "perfect solution" of the (1:1) Clique Game. We could determine the exact value of the Clique Achievement Number: playing on the complete graph K_N Maker can always build a K_q of his own if and only if

$$q \leq \lfloor 2\log_2 N - 2\log_2 \log_2 N + 2\log_2 e - 3 + o(1) \rfloor. \tag{48.4}$$

When Maker is the underdog, we know much less. For example, in the (1:2) version where Maker is the underdog, we conjecture, but *cannot* prove, that Maker can always build a K_q of his own if and only if

$$q \leq 2\log_3 N - 2\log_3 \log_3 N + O(1) \tag{48.5}$$

(the base of the logarithm is switched from 2 to 3).

Conjecture (48.5) is the best that we can hope for; this follows from the biased Erdős–Selfridge Blocking Criterion (see Theorem 20.1). Unfortunately we don't have a clue about Conjecture (48.5), *but* if two underdog players form a coalition, then it is possible to prove that at least one of them can always build a K_q of his own with

$$q = 2\log_3 N - 2\log_3 \log_3 N + O(1). \tag{48.6}$$

This justifies the vague intuition that coalition helps!

The notion of "coalition" is made precise in the following way (the author learned this concept from Wesley Pegden). We define the (2:1) **Coalition Game** on an arbitrary finite hypergraph $\mathcal{F}$ with vertex set V. In each turn of the (2:1) Coalition Game, Maker takes 2 new points from V, and colors them with 2 different colors, say red and white (the 2 points must have different colors!); Breaker takes 1 new point per move, and colors it blue. Maker's goal is to produce a red or white monochromatic $A \in \mathcal{F}$ (doesn't matter which color). If Maker succeeds to achieve his goal, he wins; if he fails to achieve his goal, Breaker wins (so draw is impossible by definition).

The name (2:1) *Coalition Game* is explained by the fact that in this game Maker represents a coalition of 2 players, Red and White, against a third player Blue ("Breaker").

In the rest we focus on a particular game: the (2:1) *Coalition Clique Game* played on K_N. At the end of a play each color class of Maker (red and white) has 1/3 of the edges of K_N; this indicates a similarity to the Random Graph $\mathbf{R}(K_N; 1/3)$ with edge probability 1/3. And, indeed, we will prove that Maker (coalition of Red and

White) can always build a red or white monochromatic K_q with

$$q = 2\log_3 N - 2\log_3 \log_3 N + O(1). \tag{48.7}$$

An easy calculation gives that (48.7) is the Clique Number of the Random Graph $\mathbf{R}(K_N; 1/3)$ apart from an additive constant $O(1)$.

This means that Maker, as the coalition of Red and White, can achieve the "Random Graph Clique Size". What we don't know is whether or not (48.7) is best possible. In other words, is it possible that the coalition beats the "Random Graph Clique Size"?

We are going to derive (48.7) from a (2:1) Coalition version of the Advanced Weak Win Criterion (Theorem 24.2); of course we rely heavily on the proof technique of Section 24.

Assume that there are two hypergraphs on the same board V: a "red hypergraph" $\mathcal{R}$ and a "white hypergraph" $\mathcal{W}$. Assume that we are in the middle of a play; Maker, the first player, owns ("r" indicates red and "w" indicates white)

$$X(i) = \{x_1^{(r)}, x_1^{(w)}, x_2^{(r)}, x_2^{(w)}, \ldots, x_i^{(r)}, x_i^{(w)}\},$$

and $Y(i) = \{y_1, y_2, \ldots, y_i\}$ is the set of Breaker's points. The question is how to choose Maker's $(i+1)$st move $x_{i+1}^{(r)}, x_{i+1}^{(w)}$. The following notation is now introduced

$$X^{(r)}(i) = \{x_1^{(r)}, x_2^{(r)}, \ldots, x_i^{(r)}\}, \ X^{(w)}(i) = \{x_1^{(w)}, x_2^{(w)}, \ldots, x_i^{(w)}\},$$
$$\mathcal{R}(i) = \{A \setminus X^{(r)}(i): \ A \in \mathcal{R}, A \cap (X^{(w)}(i) \cup Y(i)) = \emptyset\},$$
$$\mathcal{W}(i) = \{A \setminus X^{(w)}(i): \ A \in \mathcal{W}, A \cap (X^{(r)}(i) \cup Y(i)) = \emptyset\}.$$

We work with the usual potential function of Section 24 (see the proof of Theorem 24.2)

$$L_i(\mathcal{R}) = T_3(\mathcal{R}(i)) - \lambda \cdot T_3(\mathcal{R}_2^p(i)), \tag{48.8}$$
$$L_i(\mathcal{R}; u_1, \ldots, u_m) = T_3(\mathcal{R}(i); u_1, \ldots, u_m) - \lambda \cdot T_3(\mathcal{R}_2^p(i); u_1, \ldots, u_m),$$

and of course with $L_i(\mathcal{W})$ and $L_i(\mathcal{W}; u_1, \ldots, u_m)$. The index "3" in T_3 (see (48.8)) indicates that we use the Power-of-Three Scoring System

$$T_3(\mathcal{H}) = \sum_{H \in \mathcal{H}} 3^{-|H|} \ \text{ and } \ T_3(\mathcal{H}; u_1, \ldots, u_m) = \sum_{\{u_1, \ldots, u_m\} \subset H \in \mathcal{H}} 3^{-|H|};$$

the "3" comes from $2 + 1$ (in the (2:1) Coalition Game).

First let $\mathcal{H}(i) = \mathcal{R}(i)$ or $\mathcal{R}_2^p(i)$; the effect of the $(i+1)$st moves $x_{i+1}^{(r)}$, $x_{i+1}^{(w)}$ ("Maker") and $y(i+1)$ ("Breaker") is desribed by the following identity

$$T_3(\mathcal{H}(i+1)) = T_3(\mathcal{H}(i)) + 2T_3(\mathcal{H}(i); x_{i+1}^{(r)})$$
$$- T_3(\mathcal{H}(i); x_{i+1}^{(w)}) - T_3(\mathcal{H}(i); y_{i+1}) - 2T_3(\mathcal{H}(i); x_{i+1}^{(r)}, x_{i+1}^{(w)})$$
$$- 2T_3(\mathcal{H}(i); x_{i+1}^{(r)}, y_{i+1}) + T_3(\mathcal{H}(i); x_{i+1}^{(w)}, y_{i+1}) + 2T_3(\mathcal{H}(i); x_{i+1}^{(r)}, x_{i+1}^{(w)}, y_{i+1}). \tag{48.9}$$

The underlying pattern of the long (48.9) is rather simple: its terms come from the expansion of the product

$$(1 + 2x(r))(1 - x^{(w)})(1 - y) - 1 = 2x^{(r)} - x^{(w)} - y$$
$$- 2x^{(r)}x^{(w)} - 2x^{(r)}y + x^{(w)}y + 2x^{(r)}x^{(w)}y. \tag{48.10}$$

Identity (48.9) represents the "red case"; of course the "white case" is very similar: let $\mathcal{G}(i) = \mathcal{W}(i)$ or $\mathcal{W}_2^p(i)$; then

$$T_3(\mathcal{G}(i+1)) = T_3(\mathcal{G}(i)) + 2T_3(\mathcal{G}(i); x_{i+1}^{(w)})$$
$$- T_3(\mathcal{G}(i); x_{i+1}^{(r)}) - T_3(\mathcal{G}(i); y_{i+1}) - 2T_3(\mathcal{G}(i); x_{i+1}^{(w)}, x_{i+1}^{(r)})$$
$$- 2T_3(\mathcal{G}(i); x_{i+1}^{(w)}, y_{i+1}) + T_3(\mathcal{G}(i); x_{i+1}^{(r)}, y_{i+1}) + 2T_3(\mathcal{G}(i); x_{i+1}^{(w)}, x_{i+1}^{(r)}, y_{i+1}). \tag{48.11}$$

If u_1 and u_2 are two arbitrary unselected points of the common board V, then write

$$f_i(\mathcal{H}; \mathcal{G}; u_1; u_2) = T_3(\mathcal{H}(i); u_1) + T_3(\mathcal{G}(i); u_2) - T_3(\mathcal{H}(i); u_1, u_2) - T_3(\mathcal{G}(i); u_1, u_2). \tag{48.12}$$

By using notation (48.12) we can rewrite the sum of (48.9) and (48.11) as follows

$$T_3(\mathcal{H}(i+1)) + T_3(\mathcal{G}(i+1)) = T_3(\mathcal{H}(i)) + T_3(\mathcal{G}(i))$$
$$+ 2f_i(\mathcal{H}; \mathcal{G}; x_{i+1}^{(r)}; x_{i+1}^{(w)}) - f_i(\mathcal{H}; \mathcal{G}; x_{i+1}^{(w)}; y_{i+1}) - f_i(\mathcal{H}; \mathcal{G}; y_{i+1}; x_{i+1}^{(r)})$$
$$- 3T_3(\mathcal{H}(i); x_{i+1}^{(r)}, y_{i+1}) - 3T_3(\mathcal{G}(i); x_{i+1}^{(w)}, y_{i+1})$$
$$+ 2T_3(\mathcal{H}(i); x_{i+1}^{(r)}, x_{i+1}^{(w)}, y_{i+1}) + 2T_3(\mathcal{G}(i); x_{i+1}^{(w)}, x_{i+1}^{(r)}, y_{i+1}). \tag{48.13}$$

Combining (48.8) with (48.13), we have

$$L_{i+1}(\mathcal{R}) + L_{i+1}(\mathcal{W}) = L_i(\mathcal{R}) + L_i(\mathcal{W})$$

$$+ 2\left(f_i(\mathcal{R}; \mathcal{W}; x_{i+1}^{(r)}; x_{i+1}^{(w)}) - \lambda \cdot f_i(\mathcal{R}_2^p; \mathcal{W}_2^p; x_{i+1}^{(r)}; x_{i+1}^{(w)})\right)$$

$$- \left(f_i(\mathcal{R}; \mathcal{W}; x_{i+1}^{(w)}; y_{i+1}) - \lambda \cdot f_i(\mathcal{R}_2^p; \mathcal{W}_2^p; x_{i+1}^{(w)}; y_{i+1})\right)$$

$$- \left(f_i(\mathcal{R}; \mathcal{W}; y_{i+1}; x_{i+1}^{(r)}) - \lambda \cdot f_i(\mathcal{R}_2^p; \mathcal{W}_2^p; y_{i+1}; x_{i+1}^{(r)})\right)$$

$$- 3T_3(\mathcal{R}(i); x_{i+1}^{(r)}, y_{i+1}) - 3T_3(\mathcal{W}(i); x_{i+1}^{(w)}, y_{i+1})$$

$$+ 2T_3(\mathcal{R}(i); x_{i+1}^{(r)}, x_{i+1}^{(w)}, y_{i+1}) + 2T_3(\mathcal{W}(i); x_{i+1}^{(w)}, x_{i+1}^{(r)}, y_{i+1})$$

$$+ \lambda \left(3T_3(\mathcal{R}_2^p(i); x_{i+1}^{(r)}, y_{i+1}) + 3T_3(\mathcal{W}_2^p(i); x_{i+1}^{(w)}, y_{i+1}) \right.$$

$$\left. - 2T_3(\mathcal{R}_2^p(i); x_{i+1}^{(r)}, x_{i+1}^{(w)}, y_{i+1}) - 2T_3(\mathcal{W}_2^p(i); x_{i+1}^{(w)}, x_{i+1}^{(r)}, y_{i+1}) \right). \tag{48.14}$$

By using the notation

$$F_i(u_1; u_2) = f_i(\mathcal{R}; \mathcal{W}; u_1; u_2) - \lambda \cdot f_i(\mathcal{R}_2^p; \mathcal{W}_2^p; u_1; u_2), \tag{48.15}$$

and applying the trivial inequality

$$T_3(\mathcal{H}; u_1, u_2, u_3) \le T_3(\mathcal{H}; u_1, u_2),$$

identity (48.14) leads to the "Decreasing Property"

$$L_{i+1}(\mathcal{R}) + L_{i+1}(\mathcal{W}) \ge L_i(\mathcal{R}) + L_i(\mathcal{W}) +$$

$$2F_i(x_{i+1}^{(r)}; x_{i+1}^{(w)}) - F_i(x_{i+1}^{(w)}; y_{i+1}) - F_i(x_{i+1}^{(r)}; y_{i+1})$$

$$- 3T_3(\mathcal{R}(i); x_{i+1}^{(r)}, y_{i+1}) - 3T_3(\mathcal{W}(i); x_{i+1}^{(w)}, y_{i+1}). \tag{48.16}$$

Here is Maker's strategy: in his $(i+1)$st move Maker chooses that unoccupied ordered pair $u_1 = x_{i+1}^{(r)}$, $u_2 = x_{i+1}^{(w)}$ for which the function $F_i(u_1; u_2)$ (defined in (48.15)) attains its maximum. Combining this maximum property with (48.16), we obtain the *key inequality*

$$L_{i+1}(\mathcal{R}) + L_{i+1}(\mathcal{W}) \ge L_i(\mathcal{R}) + L_i(\mathcal{W})$$

$$- 3T_3(\mathcal{R}(i); x_{i+1}^{(r)}, y_{i+1}) - 3T_3(\mathcal{W}(i); x_{i+1}^{(w)}, y_{i+1}). \tag{48.17}$$

Inequality (48.17) guarantees that we can apply the technique of self-improving potentials (Section 24), and obtain the following (2:1) Coalition version of Theorem 24.2.

Advanced Weak Win Criterion in the (2:1) Coalition. *Let $\mathcal{R}$ and $\mathcal{W}$ be two hypergraphs with the same vertex set V ("board"). If there exists a positive integer $p \geq 2$ such that*

$$\frac{T_3(\mathcal{R}) + T_3(\mathcal{W})}{|V|} > p + 4p\left(T_3(\mathcal{R}_2^p)\right)^{1/p} + 4p\left(T_3(\mathcal{W}_2^p)\right)^{1/p},$$

then playing the (2:1) Coalition Game on $\mathcal{R} \cup \mathcal{W}$ Maker can always produce a red $A \in \mathcal{R}$ or a white $B \in \mathcal{W}$.

Combining this criterion with the sophisticated calculations in Section 25, (48.17) follows. The details are left to the reader.

To formulate a question, switch from the complete graph K_N to the complete 3-uniform hypergraph $K_N^3 = \binom{[N]}{3}$ (see Theorem 6.4 (b)). By playing the (2:1) Coalition Clique Game on K_N^3, Maker can always produce a red or a blue monochromatic sub-clique K_q^3 with $q = \sqrt{6 \log_3 N} + O(1)$ (see Theorem 6.4 (b)). This follows from an application of the Criterion above.

The question is about the ordinary (1:2) version where Maker is the underdog (no coalition!). Playing on K_N^3, can underdog Maker always occupy a sub-clique K_q^3 with $q = c \cdot \sqrt{\log N}$ where $c > 0$ is some positive constant? We don't know!

Note that for the ordinary graph case the Ramsey proof technique works, and guarantees a K_q for underdog Maker with $q = c \cdot \log N$ where $c > 0$ is some positive constant. (Of course we cannot prove the optimal constant, see Theorem 33.7.)

We have just discussed a single case, the (2:1) coalition version. How about the general coalition game? This is a most interesting open problem.

49

Concluding remarks

1. Is there an easy way to prove the Neighborhood Conjecture? The central issue of whole Part D was the Neighborhood Conjecture. Many pages were devoted to this single problem, proving several special cases. Still the general case remains wide open.

This is particularly embarrassing, because the Neighborhood Conjecture is simply a game-theoretic variant of the Erdős-Lovász 2-Coloring Theorem, and the latter has a 2-page proof (see Section 11). There is a fundamental difference, however, that needs to be emphasized: in Part D an *explicit* blocking strategy (using potential functions) was always supplied, but the 2-page proof is just an *existential* argument.

It is believed that there is no short proof of the Neighborhood Conjecture that would also supply an explicit blocking strategy, but perhaps there is a short existential proof that has been overlooked.

What kind of existential proof can we expect here? Here an example is given, an existential proof using a probabilistic result, which doesn't supply any explicit strategy. The example is due to Bednarska and Luczak [2000], whose result shows an exciting new analogy between the evolution of Random Graphs and biased Maker–Breaker graph games. The new idea is to use results from the theory of Random Graphs to show that the "random strategy" is optimal for Maker. Bednarska and Luczak investigated the following biased Maker–Breaker graph game $(K_N; 1, b; G)$. Here G is a given graph, the board is the complete graph K_N, in each round of the play Maker chooses one edge of K_N which has not been claimed before, and Breaker answers by choosing at most b new edges from K_N. The play ends when all the $\binom{N}{2}$ edges are claimed by either player. If Maker's graph contains a copy of G, then Maker wins; otherwise Breaker wins. Bednarska and Luczak considered only the case when G is *fixed*, i.e. it doesn't depend on N. This kind of game, when G is a clique, was proposed by Erdős and Chvátal [1978], who proved that the game $(K_N; 1, b; K_3)$ can be won by Maker if $b < (2N+2)^{1/2} - 5/2$, and by Breaker if $b \geq 2N^{1/2}$. In other words, the "threshold" for the $(K_N; 1, b; K_3)$ game is of the order $N^{1/2}$. It would be nice to know the exact constant factor!

What is the "threshold" for the $(K_N; 1, b; G)$ game in general? Bednarska and Luczak completely solved the problem apart from constant factors. For an arbitrary graph H let $v(H)$ and $e(H)$ denote the number of vertices and the number of edges, respectively. For a graph G with at least 3 vertices define

$$\alpha(G) = \max_{H \subseteq G: v(H) \geq 3} \frac{e(H) - 1}{v(H) - 2}.$$

Bednarska and Luczak proved that the "threshold" for the $(K_N; 1, b; G)$ game is of the order $N^{1/\alpha(G)}$.

Bednarska–Luczak Theorem. *For every graph G which contains at least 3 non-isolated vertices, there exist positive constants $c_1 = c_1(G)$, $c_2 = c_2(G)$, and $N_0 = N_0(G)$ such that for every $N \geq N_0$ the following holds:*

(i) If $b \leq c_1 \cdot N^{\frac{1}{\alpha(G)}}$, then Maker wins the $(K_N; 1, b; G)$ game.
(ii) If $b \geq c_2 \cdot N^{\frac{1}{\alpha(G)}}$, then Breaker wins the $(K_N; 1, b; G)$ game.

Note that the proof of part (ii) uses the biased version of the Erdős–Selfridge Theorem (Theorem 20.1).

The proof of part (i) is a novel *existential* argument, which goes like this. We can assume that G contains a cycle (otherwise the theorem is trivial). Let $(K_N; 1, b; G; *)$ denote the *modification* of game $(K_N; 1, b; G)$ in which Breaker has all the information about Maker's moves, but Maker cannot see the moves of Breaker. Thus, if Maker chooses an edge of K_N, it might happen that this edge has been previously claimed by Breaker. In such a case, this edge is marked as a *failure*, and Maker *loses* this move. We say that Maker plays according to the *random strategy* if in each move he selects an edge chosen uniformly at random from all previously unselected edges by him (Maker cannot see the edges of Breaker). Observe that if Breaker has a winning strategy for the $(K_N; 1, b; G)$ game, then the same strategy forces a win in the modified game $(K_N; 1, b; G; *)$ as well. Thus, in order to prove part (i), it is enough to show that if Maker plays according to the random strategy, there is a positive probability (in fact $> 1 - \varepsilon$) that he wins the modified game $(K_N; 1, b; G; *)$ against any strategy of Breaker. The proof of this statement is based on certain deep properties of Random Graphs. Let $b = \delta \cdot N^{\frac{1}{\alpha(G)}}/10$ where δ is a sufficiently small constant to be specified later, and consider the play of the modified game $(K_N; 1, b; G; *)$ in which Maker plays according to the *random strategy* and Breaker follows an arbitrary strategy *Str*. Let us consider the first

$$M = \frac{\delta}{2b} \binom{N}{2} < 2N^{2 - 1/\alpha(G)}$$

rounds of the play. Let $\mathbf{R}(K_N; M)$ denote the Random Graph chosen uniformly at random from the family of all sub-graphs of K_N with N vertices and M edges. Clearly

Maker's graph at the Mth round can be viewed as $\mathbf{R}(K_N; M)$, although some of these edges may be *failures*, i.e. they have been already claimed by Breaker (Maker cannot see Breaker's moves). Nevertheless, during the first M rounds of the play Breaker have selected less than $\delta/2$ part of the $\binom{N}{2}$ edges of K_N. So for every $i = 1, 2, \ldots, M$ the probability that Maker's ith move is a *failure* is bounded from above by $\delta/2$. Consequently, for large M, by the law of large numbers, with probability $> 1 - \varepsilon$ at most δM of Maker's first M moves are *failures*. Therefore, by the lemma below, with probability $> 1 - 2\varepsilon$ Maker's graph at his Mth move contains a copy of G.

Lemma A: *For every graph G containing at least one cycle there exist constants $0 < \delta = \delta(G) < 1$ and $N_1 = N_1(G)$ such that for every $N \geq N_1$ and $M = 2N^{2-1/\alpha(G)}$ the probability of the event "each sub-graph of $\mathbf{R}(K_N; M)$ with $(1 - \delta)M$ edges contains a copy of G" is greater than $1 - \varepsilon$.*

Lemma A easily follows from the following extremely good upper bound on the probability that a Random Graph contains no copies of a given graph G.

Lemma B (Janson, Luczak and Rucinski [2000]): *For every graph G containing at least one cycle there exist constants $c_3 = c_3(G)$ and $N_2 = N_2(G)$ such that for every $N \geq N_2$ and $M = N^{2-1/\alpha(G)}$, the probability of the event "$\mathbf{G}(N, M)$ contains no copy of G" is less than $\exp(-c_3 M)$.* □

This completes the outline of the proof of part (i) of the Bednarska–Luczak Theorem.

The proof shows that if Maker plays randomly, then he is able to build a copy of G with probabilty $> 1 - \varepsilon$. This ε is in fact exponentially small, so the random strategy succeeds with probability extremely close to 1. Still the proof doesn't give a clue of how to actually win!

2. Galvin's counter-example in the biased game. Theorem 6.1 says that if a finite hypergraph has chromatic number ≥ 3, then the (1:1) game on the hypergraph is a first player win, which of course implies Weak Win.

How about the (1:2) Maker–Breaker game where Maker is the underdog? Can large *chromatic number* help here? Is it true that, if the chromatic number of the hypergraph is sufficiently large, then playing the (1:2) game on the hypergraph, underdog Maker can achieve a Weak Win? Unfortunately, the answer is "no", even if the underdog is the first player. The following counter-example is due to Galvin. Since large chromatic number not necessarily enforces underdog's Weak Win, for biased games, Ramsey Theory gives very little help.

First we recall that the *rank* of a finite hypergraph $\mathcal{F}$ is $||\mathcal{F}|| = \max\{|A| : A \in \mathcal{F}\}$. A hypergraph consisting of r-element sets is an *r-uniform hypergraph*: a 2-uniform hypergraph is a *graph*, a 3-uniform hypergraph is a *triplet system*, 4-uniform hypergraph is a *quadruplet system*, and so on. The *complete r-uniform hypergraph* on a set V is $\binom{V}{k} = \{A \subset V : |A| = k\}$. A subset $S \subset V(\mathcal{F})$ of the pointset of hypergraph

$\mathcal{F}$ is *independent* if it does not contain a hyperedge of $\mathcal{F}$. A hypergraph $\mathcal{F}$ is *r-colorable* if its pointset $V(\mathcal{F})$ is the union of r independent sets. In other words, $\mathcal{F}$ is *r-colorable* if we can r-color the points such that there is no monochromatic hyperedge. The *chromatic number* of $\mathcal{F}$ is $\chi(\mathcal{F}) = \min\{r: \mathcal{F} \text{ is } r-colorable\}$; $\mathcal{F}$ is *r-chromatic* if $\chi(\mathcal{F}) = r$.

Fred Galvin showed that for every $r \geq 4$ there is an r-chromatic uniform hypergraph such that Breaker wins the biased (1:2) Maker–Breaker game (Maker takes 1 and Breaker takes 2 points per move).

We shall say that a hypergraph $\mathcal{F}$ is *2-fragile* if Breaker has a winning strategy in the biased (1:2) Maker–Breaker game on $\mathcal{F}$.

F. Galvin's counter-example: *There are 2-fragile uniform hypergraphs with arbitrarily high chromatic number.*

Proof. The first lemma almost proves the statement except that the constructed hypergraph is not uniform. □

Lemma 1: *For each $r \geq 3$, there is an r-chromatic hypergraph $\mathcal{H}$ such that the biased (1:2) Maker–Breaker game on $\mathcal{H}$ is a win for Breaker.*

Proof. We use induction on r. For $r = 3$, let $|V| = 6$ and $\mathcal{H} = \binom{V}{3}$. □

Now let $r \geq 3$, and suppose there is an r-chromatic hypergraph $\mathcal{G}$ such that the (1:2) game on $\mathcal{G}$ is a win for Breaker. Let $k = |V(\mathcal{G})|$, $m = 2^k + 1$, and let $W = V_1 \cup V_2 \cup \cdots \cup V_m$ where $V_1, \ldots, V_m$ are disjoint k-element sets. For each $i \in \{1, \ldots, m\}$, let $\mathcal{G}_i$ be an isomorphic copy of $\mathcal{G}$ with pointset $V(\mathcal{G}_i) = V_i$. Let $\mathcal{F} = \{V_1, V_2, \ldots, V_m\}$; and let $T(\mathcal{F}) = \{\{x_1, \ldots, x_m\}: x_i \in V_i, 1 \leq i \leq m\}$ be the set of all transversals of $V_1, \ldots, V_m$. Finally, let

$$\mathcal{H} = T(\mathcal{F}) \cup \mathcal{G}_1 \cup \mathcal{G}_2 \cup \cdots \cup \mathcal{G}_m.$$

We claim that the chromatic number of $\mathcal{H}$ is $r + 1$. First we prove that $\mathcal{H}$ is not r-colorable. Color the points of $\mathcal{H}$ with r colors. Since $\mathcal{G}_i$ is r-chromatic, the only way to avoid monochromatic hyperedges in $\mathcal{G}_i$ is to have each one of the r colors in every V_i, $1 \leq i \leq m$. Then $T(\mathcal{F})$ would have monochromatic edges, which proves that $\mathcal{H}$ is not r-colorable. On the other hand, $\mathcal{H}$ is $(r + 1)$-colorable. Indeed, we can color each $\mathcal{G}_i$ with r of the $(r + 1)$ colors in such a way that each one of the $(r + 1)$ colors is omitted from at least one of the $\mathcal{G}_i$s. This is clearly possible since $m \geq r + 1$: indeed, $r \leq k$ and $m = 2^k + 1$.

Next we need the following simple:

Lemma 2: *Let $k \geq 3$; let $V_1, V_2, \ldots, V_m$ be disjoint k-element sets; let $\mathcal{F} = \{V_1, V_2, \ldots, V_m\}$; and let $T(\mathcal{F}) = \{\{x_1, \ldots, x_m\}: x_i \in V_i, 1 \leq i \leq m\}$ be the set of all transversals of $V_1, \ldots, V_m$. If $m \geq 2^k + 1$, then the (1:2) game on $T(\mathcal{F})$ is a win for Breaker.*

Proof. Breaker's goal is to completely occupy some V_i. A play has k stages. In the 1st stage, at each of his turns Breaker chooses his 2 points from 2 different sets (say) V_j and V_ℓ such that none of them has at any point chosen by either player before. The first stage ends in 2^{k-1} turns, when there is an index-set $I_1 \subset \{1, 2, \ldots, m\}$ such that $|I_1| = 2^{k-1}$ and every V_i, $i \in I_1$ contains exactly 1 point of Breaker and none of Maker. Similarly, the second stage ends in 2^{k-2} turns, when there is an index-set $I_2 \subset I_1$ such that $|I_2| = 2^{k-2}$ and every V_i, $i \in I_2$ contains exactly 2 points of Breaker and none of Maker. Iterating this process, in the last stage there is an index-set $I_k \subset I_{k-1}$ such that $|I_k| = 1$ and V_i, $i \in I_k$ contains exactly k points of Breaker and none of Maker. □

Since $m = 2^k + 1$, By Lemma 2 the (1:2) game on $T(\mathcal{F})$ is a win for Breaker. We attempt to combine Breaker's winning strategy in the (1:2) game on $T(\mathcal{F})$ with his winning strategies in the (1:2) games on $\mathcal{G}_i$, $1 \leq i \leq m$ to produce a winning strategy in the (1:2) game on $\mathcal{H}$. As long as Maker chooses points from *different* V_is, Breaker responds according to his winning strategy in $T(\mathcal{F})$ (see Lemma 2). But whenever Maker returns to some V_i *second, third, fourth, ... time*, Breaker responds according his winning strategy in the (1:2) game on $\mathcal{G}_i$. This play doesn't hurt his chances of winning the (1:2) game on $T(\mathcal{F})$. Indeed, by the proof of Lemma 2, when Maker picks a point from some V_i, Breaker's winning strategy on $T(\mathcal{F})$ never requires Breaker to return to that V_i. But there is a minor technical problem! When Maker returns to some V_i for the *second time*, Breaker responds to 2 points of Maker, so Maker needs a winning strategy in the *modified* (1:2) game on $\mathcal{G}_i$ where Maker chooses 2 points at his first turn.

More precisely, the $(1^+:2)$ game on hypergraph $\mathcal{F}$ is the same as the (1:2) game on $\mathcal{F}$ except that Maker gets an extra point at the start. That is, Maker chooses 2 points at his first turn, but only 1 point at every sub-sequent turn, and Breaker chooses 2 points at every turn.

To be able to apply induction we have to work with the concept of the $(1^+:2)$ game instead of the (1:2) game. Therefore, in order to prove Lemma 2, we have to prove the slightly stronger $(1^+:2)$ versions of Lemmas 2 and 3.

Lemma 1′: *For each $r \geq 3$, there is an r-chromatic hypergraph $\mathcal{H}$ such that the biased $(1^+:2)$ game on $\mathcal{H}$ is a win for Breaker.*

Lemma 2′: *Let $k \geq 3$; let $V_1, V_2, \ldots, V_m$ be disjoint k-element sets; let $\mathcal{F} = \{V_1, V_2, \ldots, V_m\}$; and let $T(\mathcal{F}) = \{\{x_1, \ldots, x_m\} : x_i \in V_i, 1 \leq i \leq m\}$ be the set of all transversals of $V_1, \ldots, V_m$. If $m \geq 2^k + 2$. Then the $(1^+:2)$ game on $T(\mathcal{F})$ is a win for Breaker.*

The proof of Lemma 2′ is exactly the same as that of Lemma 2. □

Proof of Lemma 1'. We use induction on r. For $r = 3$, let $|V| = 7$ and $\mathcal{H} = \binom{V}{4}$.

The construction in the induction step is exactly the same as that of in Lemma 2. Let $r \geq 3$, and suppose there is an r-chromatic hypergraph $\mathcal{G}$ such that the $(1^+:2)$ game on $\mathcal{G}$ is a win for Breaker. Let $k = |V(\mathcal{G})|$, $m = 2^k + 2$, and let $W = V_1 \cup V_2 \cup \cdots \cup V_m$ where $V_1, \ldots, V_m$ are disjoint k-element sets. For each $i \in \{1, \ldots, m\}$, let $\mathcal{G}_i$ be an isomorphic copy of $\mathcal{G}$ with pointset $V(\mathcal{G}_i) = V_i$. Let $\mathcal{F} = \{V_1, V_2, \ldots, V_m\}$; and let $T(\mathcal{F}) = \{\{x_1, \ldots, x_m\} : x_i \in V_i, 1 \leq i \leq m\}$ be the set of all transversals of $V_1, \ldots, V_m$. Finally, let

$$\mathcal{H} = T(\mathcal{F}) \cup \mathcal{G}_1 \cup \mathcal{G}_2 \cup \cdots \cup \mathcal{G}_m.$$

Again the chromatic number of $\mathcal{H}$ is $r + 1$.

Since $m = 2^k + 2$, By Lemma 2' the $(1^+:2)$ game on $T(\mathcal{F})$ is a win for Breaker. Now Breaker's winning strategy in the $(1^+:2)$ game on $T(\mathcal{F})$ can be combined with his winning strategies in the $(1^+:2)$ games on $\mathcal{G}_i$, $1 \leq i \leq m$ to produce a winning strategy in the $(1^+:2)$ game on $\mathcal{H}$. As long as Maker chooses points from *different* V_is, Breaker responds according to his winning strategy in $T(\mathcal{F})$. But whenever Maker returns to some V_i *second, third, fourth, ... time*, Breaker responds according his winning strategy in the $(1^+:2)$ game on $\mathcal{G}_i$. This is how Breaker wins the $(1^+:2)$ game on $\mathcal{H}$, and the proof of Lemma 1' is complete.

Finally, Lemma 1' implies Lemma 2. □

Next we show that the distinction between uniform and non-uniform hypergraphs is more or less irrelevant to the question of finding 2-fragile hypergraphs. Indeed, a non-uniform 2-fragile hypergraph can be replaced by a somewhat larger uniform 2-fragile hypergraph with the same rank and chromatic number.

Lemma 3: *Suppose there is a 2-fragile hypergraph $\mathcal{H}_0$ of rank $||\mathcal{H}_0|| \leq k$ which has chromatic number $\chi(\mathcal{H}_0) \geq r$. Then there is a 2-fragile k-uniform hypergraph $\mathcal{H}$ with $\chi(\mathcal{H}) = r$.*

Proof. We use induction on r. If $r = 3$, then we can take $\mathcal{H} = \binom{V}{k}$ where $|V| = 2(k-1) + 1$. Now let $\mathcal{H}_0$ be a 2-fragile hypergraph of rank $||\mathcal{H}_0|| \leq k$ which has chromatic number $\chi(\mathcal{H}_0) \geq r \geq 4$. By the induction hypothesis, there is a 2-fragile k-uniform hypergraph $\mathcal{H}'$ with $\chi(\mathcal{H}') = r - 1$. Let $t = \max\{k - |H| : H \in \mathcal{H}_0\}$ be the maximum discrepancy from uniformity. Let $V = V_0 \cup V_1 \cup V_2 \cup \ldots \cup V_t$ where $V_0, V_1, V_2, \ldots, V_t$ are disjoint sets, $V_0 = V(\mathcal{H}_0)$, and $|V_1| = |V_2| = \ldots = |V_t| = |V(\mathcal{H}')|$. For each $i \in \{1, 2, \ldots, t\}$, let $\mathcal{H}_i$ be an isomorphic copy of $\mathcal{H}'$ with $V(\mathcal{H}_i) = V_i$. Finally let

$$\mathcal{H}^* = \{A \in \binom{V}{k} : A \cap V_0 \in \mathcal{H}_0\} \cup \mathcal{H}_1 \cup \mathcal{H}_2 \cup \cdots \cup \mathcal{H}_t.$$

The k-uniform hypergraph $\mathcal{H}^*$ is 2-fragile since Breaker can combine his winning strategies on $\mathcal{H}_0$ and on $\mathcal{H}_i$, $1 \leq i \leq t$.

The chromatic number of $\mathcal{H}^*$ is $\geq r$. Indeed, consider a coloring of the points of $\mathcal{H}^*$ with $r-1$ colors. Since the chromatic number $\chi(\mathcal{H}_i) = \chi(\mathcal{H}') = r-1$, the only way to avoid monochromatic hyperedges in $\mathcal{H}_i$ is to have each one of the $r-1$ colors in every V_i, $1 \leq i \leq t$. Since the chromatic number of $\chi(\mathcal{H}_0)$ is r, there is a monochromatic hyperedge, say, $H \in \mathcal{H}_0$ is red. Let $y_i \in V_i$, $1 \leq i \leq t$ be red points. Since $k - |H| = s \leq t$, the red set $A = H \cup \{y_1, \ldots, y_s\}$ is a hyperedge of $\mathcal{H}^*$, which proves that the chromatic number of $\mathcal{H}^*$ is $\geq r$. Since deleting a hyperedge can lower the chromatic number by no more than 1, there is a sub-hypergraph $\mathcal{H} \subseteq \mathcal{H}^*$ with chromatic number $\chi(\mathcal{H}) = r$, and Lemma 3 follows.

Lemmas 2 and 3 complete the proof.   □

3. Compactness–Keisler's Theorem. In this book we have been focusing on finite games, but it is impossible to entirely avoid infinity. Here is a nice example how infinity enters the story in a most natural way. Consider (say) the (K_N, K_5) Clique Game with $N \geq 49$. Since the Ramsey Number of K_5 is ≤ 49, by Theorem 6.1 the (K_N, K_5) Clique Game is a first player win for every $N \geq 49$. How about (K_∞, K_5)? (The board is the infinite complete graph.) Can the first player always win (i.e. build a K_5 first) in a finite number of moves? Well, we don't know; it is possible that in the ω-play the infinite Clique Game (K_∞, K_5) is a draw.

There is one thing, however, that we know: compactness holds, i.e. *if* the first player has a finite winning strategy in (K_∞, K_5), then he has a time-bounded winning strategy. In other words, finite is upgraded to bounded in time: if the first player can always win, then there is a natural number n_0 such that the first player can always win before his n_0th move. The proof is based on the following simple observation: let G be an arbitrary finite sub-graph of K_∞ and let $e \in K_\infty$ be a new edge; then there are only three possibilities: (1) e is vertex-disjoint from G, (2) e has exactly one common endpoint with the vertex-set of G, (3) both endpoints of e are in the vertex-set of G. This means a finite number of options, so the standard compactness argument works.

Next consider the 5-in-a-row on the infinite chessboard (the whole plane). It is widely believed ("folklore") that the first player has a winning strategy, but we don't know any strict proof. It is not even known whether a winning strategy for the first player, if any, can be bounded in time. In other words, is there a natural number n such that the first player can always win the 5-in-a-row on the infinite chessboard before his nth move? In this case compactness is not clear because the distance between two little squares is an arbitrary natural number (an infinite set!). The best compactness result is due to H. J. Keisler [1969] and goes as follows. If 5-in-a-row is played on *countably many* infinite chessboards, i.e. at every step the player chooses one of the boards and occupies a square on it, and if the first player has a winning strategy, then the first player has a winning strategy bounded in time.

The proof of this amusing theorem is relatively elementary. All that we need from set theory is the concept of *ordinals*.

Let G be a game between players I (the first player) and II (the second player) moving alternately on infinitely many infinite chessboards. By a position of length t we mean a sequence of moves $P = (p_1, \ldots, p_t)$ starting with all boards vacant. G is defined by declaring certain positions to be wins for player I. Say that two positions P and Q at time t are 0-equivalent if for each $i < j \leq t$ and direction d, p_j is adjacent to p_i in direction d if and only if q_j is adjacent to q_i in direction d. Assume that G has the following property: if P is a win for I and Q is 0-equivalent to P, then Q is a win for I; of course, the 5-in-a-row game has this property.

Notice that if t is even, player I has the next move, and otherwise player II has the next move.

Let us say that the Move Number of a (finite) positional game is n if the first player can win in n moves but not sooner.

Next consider a semi-infinite positional game, and assume that the first player has a winning strategy. The Move Number of this game is an *ordinal number* and can be defined as follows. We say that a position is *nice* if:

(1) it is an even position;
(2) no player has won yet;
(3) the first player still has a winning strategy.

An odd position has Move Number 0 if it is a win for the first player, i.e. if the first player already owns a whole winning set but the second player does not own any winning set yet. We assign ordinal Move Numbers $b > 0$ to the nice positions by using the following transfinite recursion. The nice predecessors of every position with Move Number 0 have Move Number 1; these are the maximal elements of the set $\mathcal{S}$ of all nice positions. We emphasize that the notion "maximal" is well-defined because there is no infinite increasing chain in $\mathcal{S}$ (due to the fact that the winning sets are all finite!). Let $\mathcal{S}_1$ denote the set of all nice positions with Move Number 1.

A nice position has Move Number 2 if and only if the first player can make such a move that, for the second player's any move, the new position (with two more marks) is nice and has Move Number 1; these are the maximal elements of the set $\mathcal{S} \setminus \mathcal{S}_1$. Let $\mathcal{S}_2$ denote the set of all nice positions with Move Number 2.

In general, a nice position P has Move Number $b > 1$ if and only if b is the least ordinal such that starting from position P the first player has such a move that, for the second player's any move, the new position (with two more marks) is nice and has Move Number less than b; these are the maximal elements of the set $\mathcal{S} \setminus \bigcup_{a < b} \mathcal{S}_a$. Let $\mathcal{S}_b$ denote the set of all nice positions with Move Number b.

The sets S_b are obviously disjoint, so eventually we decompose S as $S = \bigcup \{P_b : b < c\}$ for some ordinal c; this defines a Move Number function $m(P) = b$ if and only if $P \in S_b$.

Note that the Move Number is unique. Indeed, assume that m_1 and m_2 are two different Move Number functions; then $\{P \in S : m_1(P) \neq m_2(P)\}$ is non-empty; let Q be a maximal element of this set. The contradiction comes from the facts that:

(1) $\{m_1(P): Q \not\subseteq P, P \in S\} = \{m_2(P): Q \not\subseteq P, P \in S\}$;
(2) $m_1(Q) = l.s.u.b.\{m_1(P) : Q \not\subseteq P, P \in S\}$;
(3) $m_2(Q) = l.s.u.b.\{m_2(P) : Q \not\subseteq P, P \in S\}$;

where *l.s.u.b.* means the *least strict upper bound*.

The Move Number of the game is, of course, the Move Number of the starting position ("empty board"). Every semi-infinite positional game, which is a first player win, has a uniquely determined Move Number.

Lemma 2: *A position P has a finite Move Number at most n if and only if player I has a strategy which wins in at most n moves starting from P.*

Proof. An easy induction on n.   □

Lemma 3: *Player I has a winning strategy starting from position P if and only if the Move Number of P exists (and is an ordinal number).*

Proof. Suppose the Move Number of P is an ordinal b. Then player I has a strategy starting from P which guarantees that the Move Number decreases with each move. Since any decreasing sequence of ordinals is finite, the play must stop at Move Number 0 after a finite number of moves, and I wins.

If P does not have a Move Number, then II has a strategy which guarantees that starting from P no future position has a Move Number, and hence we cannot have a winning strategy at P.   □

Lemma 4: *Suppose there is a position with infinite Move Number. Then there is a position with Move Number ω.*

Proof. Let b be the least infinite ordinal such that there is a position P of Move Number b. Since there are no positions of infinite Move Number $< b$, the winning strategy of I guarantees that the next position has finite Move Number. So P has Move Number ω.   □

Definition. The distance between two points on the same board is the minimum path length connecting the points by a sequence of king moves. The distance between two points on different boards is infinite. Two positions P and Q of length t are n-equivalent if for each $i < j \leq t$, either the distances from p_i to p_j and from

q_i to q_j are both $> 2^n$, or the distances and the directions from p_i to p_j and from q_i to q_j are the same. (This agrees with the previous definition of 0-equivalent).

Lemma 5: *If P and Q are $(n+1)$-equivalent, then for every move r there is a move s such that (Q, s) is n-equivalent to (P, r).*

Proof. If r is within 2^n of some p_i, take s at the same distance and direction from q_i. Otherwise take s at distance $> 2^n$ from each q_i. $\square$

Lemma 6: *If P has Move Number n and Q is n-equivalent to P, then Q has Move Number n.*

Proof. By induction on n. The case $n = 0$ holds by hypothesis, and the step from n to $n+1$ follows from Lemma 5. $\square$

Keisler's Theorem. *If player I has a winning strategy in G, then there is a finite n such that player I can always win G in at most n moves.*

Proof. By Lemma 2, we must show that the starting position has a finite Move Number. Suppose not. By Lemmas 3 and 4, there is a position P of Move Number ω, with II to move.

Let r be a move on a vacant board. Then (P, r) has a finite Move Number n. If s is any move whose distance from P is $> 2^n$, then (P, s) is n-equivalent to (P, r), so (P, s) has Move Number n by Lemma 6. There are only finitely many moves s at distance at most 2^n from P. Thus there is a finite m such that (P, s) has Move Number $< m$ for every move s. Hence P has Move Number at most m, contrary to the hypothesis. $\square$

Corollary 1 *There are no positions Q of infinite Move Number in G.*

The proof is the same as that of the theorem but starting from position Q.

Now let G_k be the game G restricted to k boards.

Corollary 2 *If player I has a winning strategy for G in n moves, then he has a winning strategy for G_k in n moves.*

Proof. Lemmas 5 and 6 hold with the same proofs when P is a position for G and Q is a position for G_k.

The next corollary answers the question of why infinitely many boards forces a finite bound on the Move Number. $\square$

Corollary 3 *Suppose P is a position of infinite Move Number in the game G_k, and r is a move for G on a new board. Then player I cannot have a winning strategy in G from the position (P, r).*

Proof. Use Corollary 1 and the end of the proof of Keisler's Theorem. $\square$

This completes our short detour about Compactness.

4. Why games such as Tic-Tac-Toe? Why should the reader be interested in the "fake probabilistic method," and in games such as Tic-Tac-Toe in general? Besides the obvious reason that games are great fun and everybody plays Tic-Tac-Toe, or some grown-up version of it, at some point in his/her life, we are going to present 4 more reasons.

One extremely exciting aspect of studying these games is that it might help us to understand a little bit better how human intelligence works. It might have some impact on fundamental questions such as whether human understanding is a computational or a non-computational process. For example, in Japan there are several "perfect" players who, as the first player, can consistently win Go-Moku (the 5-in-a-row on the 19×19 Go board) but they are unable to *explain* what they are actually doing. Similarly, the unrestricted 6-in-a-row game (played on an infinite chessboard) between two reasonably good players always turns out to be a long, boring draw-game, but we don't know any *proof.* Or what is the reason that we cannot write good Go-playing computer programs? These are just three of the many examples where the human mind "knows" something that we are unable to convert into rigorous mathematical proofs.

It is interesting to know that the human approach to Chess is completely different from the way computer programs play. Humans search only a small set of positions in the game-tree, between 10 and 100. Surprisingly grand-masters do not search the game-tree deeper than less successful players, but consistently select only good moves for further study. Performance on *blitz* games, where the players do not have time to explore the game-tree to any depth, supports this idea. The play is still at a very high level, suggesting that in human Chess *pattern recognition* plays a much more important role than brute force search. What makes a master is the accumulation of problem-specific knowledge. A player needs at least five years of very intense chess study to become a grand-master. During this learning period a grand-master builds up an internal library of patterns, which is estimated to be around 50,000. Computer programs, on the other hand, typically examine millions of positions before deciding what to do next. We humans employ knowledge to compensate for our inherent lack of searching ability. How to "teach" this human knowledge to a computer is a puzzle that no one has solved yet. Maybe it is impossible to solve – this is what Roger Penrose, the noted mathematical physicist, believes. His controversial theory is that human intelligence is a *non-computational* process because it exploits physics at the *quantum-mechanical* level (see Penrose's two books, *The Emperor's New Mind* and the more recent *Shadows of the Mind*).

The second reason why the reader might find our subject interesting is that the "Tic-Tac-Toe Theory" forms a natural bridge between two well-established combinatorial theories: *Random Graphs* and *Ramsey Theory.*

Third, games are ideal models for all kinds of research problems. Expressing a mathematical or social problem in terms of games is "half of the battle."

Fourth, this theory already has some very interesting applications in combinatorics, algorithms, and complexity theory.

5. The 7 most humiliating open problems. A central issue of the book is the Neighborhood Conjecture (Open Problem 9.1). We were able to prove many partial results, but the general case remains wide open. The most embarassing special case is the following. Consider the usual (1:1) Maker–Breaker Degree Game played on an arbitrary n-regular graph (every degree is n). Breaker's goal is to prevent Maker from occupying a large star. At the end of Section 16, it was explained how Breaker can always prevent Maker from occupying a star of degree $> 3n/4$ (Breaker uses a simple Pairing Strategy). The problem is to "beat" the Pairing Strategy (it was asked by Tibor Szabó).

Open Problem 16.1 *Can the reader replace the upper bound $3n/4$ in the Degree Game above with some $c \cdot n$ where the constant $c < 3/4$? Is it possible to get $c = \frac{1}{2} + o(1)$?*

Another major open problem is to extend Theorem 6.4 ("Clique Games") and Theorem 8.2 ("lattice games") to the biased cases (and also the coalition versions). It is embarrassing that we cannot do it even for the (2:2) play, which seems to be the easiest case: it is the perfect analogue of the well-understood (1:1) play. The problem is just formulated in the special case of the Clique Game.

Open Problem 31.1

(a) *Is it true that playing the (2:2) Achievement Game on K_N, Maker can always build a clique K_q with $q = 2\log_2 N - 2\log_2 \log_2 N + O(1)$?*
(b) *Is it true that playing the (2:2) Achievement Game on K_N, Breaker can always prevent Maker from building a clique K_q with $q = 2\log_2 N - 2\log_2 \log_2 N + O(1)$?*
(c) *Is it true that the (2:2) Clique Achievement and Clique Avoidance Numbers are equal, or at least differ by at most $O(1)$?*

We know the exact solution of the (1:1) Clique Game, but the Tournament version remains wide open; this is the third problem in the list (see the *Tournament problem* below). First, recall that a *tournament* means a "directed complete graph" such that every edge of a complete graph is directed by one of the two possible orientations; it represents a tennis tournament where any two players play with each other, and an arrow points from the winner to the loser. The Tournament Game begins with fixing an arbitrary goal tournament T_q on q vertices. The two players are Red and Blue, who alternately take new edges of a complete graph K_N, and for each new edge choose one of the two possible orientations ("arrow"). Either player colors his arrow with his own color. At the end of a play, the players create a 2-colored tournament on N vertices. Red wins if there is a red copy of the goal tournament T_q; otherwise Blue wins.

Let $K_N(\uparrow\downarrow)$ denote the "complete tournament": it has $2\binom{N}{2}$ arrows, meaning that every edge of the complete graph K_N shows up with both orientations. At the end of a play in the Tournament Game Red owns 1/4 of the arrows in $K_N(\uparrow\downarrow)$. In the *random play* (where both Red and Blue play randomly) Red wins the Tournament Game with $q = (2 - o(1))\log_4 N$; Red wins with probability $\to 1$ as $N \to \infty$.

This motivates the following *tournament problem*: Is it true that Red can always win the Tournament Game with $q = (2 - o(1))\log_4 N$?

First, note that Theorem 33.8 proves only half of the conjectured truth.

Second, observe that the *Tournament problem* is solved for the Chooser–Picker version: it follows from Theorem 33.4. The Chooser–Picker version means that in each turn Picker picks 2 new edges of K_N, Chooser (Red) chooses one of them, and also chooses one of the two orientations; Chooser colors his arrow red. Chooser wins if there is a red copy of the goal tournament T_q; otherwise Picker wins. The Chooser–Picker version means a (1:3) play on the "complete tournament" $K_N(\uparrow\downarrow)$ where Chooser is the underdog. It follows that Theorem 33.4 applies with $s = 3$, and gives an affirmative answer to the corresponding *tournament problem*.

The fourth problem is about connectivity, the simplest graph-theoretic property. In spite of its simplicity, the game-theoretic problem is still unsolved. The Random Graph intuition makes the following conjecture (in fact, 3 conjectures) very plausible.

Open Problem 20.1 *Consider the (1:b) Connectivity Game on the complete graph K_n. Is it true that if $b = (1 - o(1))n/\log n$ and n is large enough, then underdog Maker can always build a spanning tree?*

Moreover, is it true that under the same condition Maker can always build a Hamiltonian cycle?

Finally, is it true that under the same condition Forcer can always force Avoider to build a Hamiltonian cycle?

The fifth one is the "pentagon problem." In Sections 1–2 we proved that the regular pentagon (of any fixed size) is a Weak Winner, and in the (1:1) play a player can build it in less than 10^{500} moves. Of course, the upper bound 10^{500} on the Move Number seems totally ridiculous. Can we prove a reasonable upper bound like ≤ 1000? Another natural question is whether or not the regular pentagon can be upgraded from Weak Winner to ordinary Winner, i.e. can the first player build it first?

The sixth problem is again about ordinary win. Ordinary win is such a difficult concept that it is unfair to call the following problem "humiliating," but we still list it here. The author tackled this question in 1980, more than 25 years ago, and does not believe it will be solved in the next 25 years.

Open Problem 4.6 (c) *Consider the Clique Game* (K_n, K_q) *where n is huge compared to q: is there a uniform upper bound for the Move Number? More precisely, is there an absolute constant* $C_4 < \infty$ *such that the first player can always win in less than* C_4 *moves in every* (K_n, K_4) *Clique Game with* $n \geq 18$?

Is there an absolute constant $C_5 < \infty$ *such that the first player can always win in less than* C_5 *moves in every* (K_n, K_5) *Clique Game with* $n \geq 49$?

In general, is there an absolute constant $C_q < \infty$ *such that the first player can always win in less than* C_q *moves in every* (K_n, K_q) *Clique Game with* $n \geq R(q)$ *where* $R(q)$ *is the Ramsey Number?*

Probably C_4 is finite, but it is perfectly possible that C_5 or C_6 are not finite (a curse of the Extra Set Paradox, see Section 5)!

Finally, the seventh humiliating problem is how to extend the Erdős–Selfridge Theorem in the biased $(a:f)$ Avoider-Forcer game. The special case "$f = 1$ and $a \geq 1$ is arbitrary" has a satisfying solution – see Theorem 20.4 and (30.15) – but the general case $f \geq 2$ remains a mystery. (It was Tibor Szabó who pointed out the difficulty of the general case.) Why is this innocent-looking problem so hard?

This was the author's list of the 7 most humiliating open problems. For convenience, in pages 716–23 the complete set of the open problems in the book is given. The reader is encouraged to make his/her own list of the most humiliating open problems.

Appendix A
Ramsey Numbers

For the sake of completeness we include a brief summary of *quantitative* Ramsey Theory.

A well-known puzzle states that in a party of six people there is always a group of three who either all know each other or are all strangers to each other. This puzzle illustrates a general theorem proved by Ramsey in 1929. The Ramsey Theorem is about partitions (colorings) of the edges of complete graphs and partitions of k-edges of complete k-uniform hypergraphs ($k \geq 2$). First consider the case of edges, i.e. $k = 2$.

Let $R_2(s, t)$, called *Ramsey Number*, be the minimum of n for which every 2-coloring (say, red and blue) of the edges of the complete graph K_n on n vertices yields either a red K_s or a blue K_t. A priori it is not clear that $R_2(s, t)$ is finite for every s and t.

The following result shows that $R_2(s, t)$ is finite for every s and t, and at the same time it gives a good upper bound on $R_2(s, t)$.

Theorem A1 (*Erdős and Szekeres*) *If $s \geq 3$ and $t \geq 3$, then*

$$R_2(s, t) \leq R_2(s - 1, t) + R_2(s, t - 1), \tag{A1}$$

and

$$R_2(s, t) \leq \binom{s + t - 2}{s - 1}. \tag{A2}$$

Proof. Let $n = R_2(s - 1, t) + R_2(s, t - 1)$, and consider an arbitrary 2-coloring (red and blue) of the edges of K_n. Assume that some vertex v has red degree at least $R_2(s - 1, t)$. Let X denote the set of other endpoints of these red edges. Since $|X| \geq R_2(s - 1, t)$, in the complete graph on vertex set X, either there is a blue K_t, and we are done, or there is a red K_{s-1}, which extends to a red K_s by adding the extra vertex v.

Similarly, if a vertex v has blue degree at least $R_2(s, t - 1)$, then either there is a red K_s, and we are done, or there is a blue K_{t-1}, and this extends to a blue K_t by adding the extra vertex v. The trivial equality red-degree(v)+blue-degree(v) $= n - 1$ guarantees that one of these possibilities occurs. This proves (A1). □

We prove (A2) by induction on $s + t$. Inequality (A2) holds if $s = 2$ or $t = 2$; in fact we have equality since $R_2(s, 2) = R_2(2, s) = s$. Then by (A1) we have

$$R_2(s, t) \le R_2(s - 1, t) + R_2(s, t - 1) \le \binom{s+t-3}{s-2} + \binom{s+t-3}{s-1}.$$

Since $\binom{s+t-3}{s-2} + \binom{s+t-3}{s-1} = \binom{s+t-2}{s-1}$, (A2) follows. □

If $s = t = 3$, then we get the bound

$$R_2(3, 3) \le \binom{3+3-2}{3-1} = \binom{4}{2} = 6,$$

which is in fact an equality. If $s = t = 4$, then we get

$$R_2(4, 4) \le \binom{4+4-2}{4-1} = \binom{6}{3} = 20,$$

which is a strict inequality, but it is very close to the truth since we know that $R_2(4, 4) = 18$. If $s = t = 5$, then we get

$$R_2(5, 5) \le \binom{5+5-2}{5-1} = \binom{8}{4} = 70,$$

which is rather far from the truth since we know that $43 \le R_2(5, 5) \le 49$. The exact value of $R_2(5, 5)$ is not known. Erdős once joked that if an alien being threatens to destroy the Earth unless we provide it the exact value of $R_2(5, 5)$, then we should set all the computers on the Earth to work on this problem. If the alien asks for $R_2(6, 6)$, then Erdős's advise is to try to destroy the alien. (It is known that $102 \le R_2(6, 6) \le 165$, and the Erdős–Szekeres bound $\binom{6+6-2}{6-1} = \binom{10}{5}$ is 252.)

The Erdős–Szekeres upper bound $\binom{2s-2}{s-1}$ to the *diagonal Ramsey Number* $R_2(s, s)$ is asymptotically $(1 + o(1))4^{s-1}/\sqrt{\pi s}$. An important result of Erdős from 1947 gives an exponential *lower* bound by using a *counting argument*, the simplest form of the *probabilistic method*.

Theorem A2 (Erdős [1947])

$$R_2(s, s) > 2^{s/2}$$

Proof. Notice that this is a special case of Theorem 11.3; for the sake of completeness the "counting argument" is repeated. Let $n = 2^{s/2}$; we are going to show that there is a simple graph G_n on n vertices (i.e. a sub-graph of K_n) such that neither G_n nor its complement $\overline{G_n}$ contains a K_s. Coloring the edges of G_n red and the edges of its complement blue, we get a 2-coloring of the edges of K_n such that there is *no* monochromatic K_s. This will prove the inequality $R_2(s, s) > 2^{s/2}$.

Consider the graphs with n *distinguished* (labeled) vertices. The total number of these labeled graphs is easy to compute: it is $2^{\binom{n}{2}}$. Each particular s-clique K_s

occurs in $2^{\binom{n}{2}-\binom{s}{2}}$ of these $2^{\binom{n}{2}}$ labeled graphs. There are $\binom{n}{s}$ ways to choose the set of s vertices for a K_s. So if for each of the $2^{\binom{n}{2}}$ labeled simple graphs G_n, either G_n or its complement $\overline{G}_n$ contains a K_s, then we have the inequality

$$\binom{n}{s}2^{\binom{n}{2}-\binom{s}{2}} \geq \frac{1}{2}2^{\binom{n}{2}}.$$

Therefore, the reversed inequality

$$\binom{n}{s}2^{\binom{n}{2}-\binom{s}{2}} < \frac{1}{2}2^{\binom{n}{2}} \tag{A3}$$

guarantees that there is a labeled simple graph G_n such that neither G_n nor its complement contains a K_s. Rough approximations yield that (A3) holds whenever $n = 2^{s/2}$. More careful approximation using Stirling's formula leads to the asymptotically stronger lower bound $R_2(s, s) > const \cdot s2^{s/2}$. $\qquad\square$

Theorems A1–2 imply that

$$\sqrt{2} \leq \liminf\left(R_2(s, s)\right)^{1/s} \leq \limsup\left(R_2(s, s)\right)^{1/s} \leq 4.$$

Determination, of this limit (including whether or not the limit exists) is a famous open problem in Combinatorics.

Theorem A1 extends to k-uniform complete hypergraphs for every $k \geq 2$. Let $R_k(s, t)$ be the minimum of n for which every 2-coloring (say, red and blue) of the k-edges of the complete k-uniform hypergraph $\binom{n}{k}$ (i.e. all the k-element subsets of an n-element set) yields either a red $\binom{s}{k}$ or a blue $\binom{t}{k}$. For $k \geq 3$ the thresholds $R_k(s, t)$ are called *higher Ramsey Numbers*. A priori it is not clear that $R_k(s, t)$ is finite for every k, s and t.

The following result shows that $R_k(s, t)$ is finite for every k, s and t. The proof is an almost exact replica of that of Theorem A1.

Theorem A3 *Let* $2 \leq k < \min\{s, t\}$; *then*

$$R_k(s, t) \leq R_{k-1}\left(R_2(s-1, t), R_k(s, t-1)\right) + 1.$$

Proof. Let

$$n = R_{k-1}\left(R_k(s-1, t), R_k(s, t-1)\right) + 1,$$

and let X be an n-element set. Given any red-blue 2-coloring $\mathcal{C}$ of the complete k-uniform hypergraph $\binom{X}{k}$, pick a point $x \in X$ and define an induced 2-coloring of the $(k-1)$-element subsets of $Y = X \setminus \{x\}$ by coloring $A \in \binom{Y}{k-1}$ with the $\mathcal{C}$-color of $A \cup \{x\} \in \binom{X}{k}$. By the definition of $n = |X|$, either there is an $R_k(s-1, t)$-element subset W of Y such that $\binom{W}{k-1}$ is monochromatically red, or there is an $R_k(s, t-1)$-element subset Z of Y such that $\binom{Z}{k-1}$ is monochromatically blue. By symmetry we may assume that the first case holds, i.e. there is an $R_k(s-1, t)$-element subset W of Y such that $\binom{W}{k-1}$ is monochromatically red in the induced 2-coloring.

Now let us look at the C-colors of the k-element subsets of $\binom{W}{k}$. Since $|W| = R_k(s-1,t)$, there are two alternatives. If W has a t-element subset T such that $\binom{T}{k}$ is monochromatically blue by C, then we are done. If the second alternative holds, i.e. if W has an $(s-1)$-element subset S' such that $\binom{S'}{k}$ is monochromatically red by C, then we are done again. Indeed, then every k-element subset of $S = S' \cup \{x\}$ is monochromatically red by 2-coloring C. $\qquad\square$

The proof of Theorem A3 easily extends to *infinite* graphs.

Theorem A4 *Let* $N = \{1, 2, 3, \ldots\}$ *denote the set of natural numbers. Let* $k \geq 1$ *be an integer, and let* $\binom{N}{k}$ *denote the set of all k-tuples of the set of the natural numbers* N. *Let* $C : \binom{N}{k} \to \{1, 2\}$ *be an arbitrary 2-coloring of the k-tuples of* N. *Then there is an infinite subset of* N *such that its k-tuples have the same color (we call it an infinite monochromatic subset of* N).

Proof. It goes by induction on k. The result is trivial for $k = 1$ so assume that $k \geq 2$ and the theorem holds for $k - 1$.

Pick $1 \in N$, and write $N_1 = N \setminus \{1\}$. As in the proof of Theorem A3, define an induced 2-coloring $C_1 : \binom{N_1}{(k-1)} \to \{1, 2\}$ of the $(k-1)$-tuples of N_1 by putting $C_1(A) = C(A \cup \{1\})$, where A is a $(k-1)$-tuple of N_1. By the induction hypothesis, N_1 contains an infinite set S_1 where all the $(k-1)$-tuples have the same color, say, $c_1 \in \{1, 2\}$. Let $n_2 \in S_1$ be arbitrary, and write $N_2 = S_1 \setminus \{n_2\}$. Define an induced 2-coloring $C_2 : \binom{N_2}{(k-1)} \to \{1, 2\}$ of the $(k-1)$-tuples of N_2 by putting $C_2(A) = C(A \cup \{n_2\})$ where A is a $(k-1)$-tuple of N_2. By the induction hypothesis, N_2 contains an infinite set S_2 where all the $(k-1)$-tuples have the same color, say, $c_2 \in \{1, 2\}$. Let $n_3 \in S_2$ be arbitrary, and write $N_3 = S_2 \setminus \{n_3\}$. Define an induced 2-coloring $C_3 : \binom{N_3}{(k-1)} \to \{1, 2\}$ of the $(k-1)$-tuples of N_3 by putting $C_3(A) = C(A \cup \{n_3\})$, where A is a $(k-1)$-tuple of N_3. By the induction hypothesis, N_3 contains an infinite set S_3 where all the $(k-1)$-tuples have the same color, say, $c_3 \in \{1, 2\}$, and so on. Repeating this argument, we obtain an infinite sequence of numbers $n_1 = 1, n_2, n_3, \cdots$, an infinite nested sequence of infinite sets $N = S_0 \supset S_1 \supset S_2 \supset S_3 \supset \cdots$, and an infinite sequence of colors $c_1, c_2, c_3, \cdots$, such that $n_i \in S_{i-1}$ and $n_i \notin S_i$, and all k-tuples where the only element outside of S_i is n_i have the same C-color $c_i \in \{1, 2\}$. The infinite sequence c_i must take at least one of the two values $1, 2$ infinitely often, say $c_{i_1} = 1$ for $1 \leq i_1 < i_2 < i_3 < \cdots$. Then by the construction, each k-tuple of the infinite set $\{n_{i_1}, n_{i_2}, n_{i_3}, \ldots\}$ has color 1, i.e. $\{n_{i_1}, n_{i_2}, n_{i_3}, \cdots\}$ is an *infinite monochromatic subset* of N. This completes the proof of Theorem A4. $\qquad\square$

Theorems A1, A3, and A4 easily extend to colorings with arbitrary finite number of colors. For example, if we have 3 colors, (say) red, green, and yellow, then we have the easy upper bound

$$R_k(s, s, s) \le R_k\big(s, R_k(s, s)\big). \tag{A4}$$

To prove (A4), we replace the last two colors green and yellow by a new color (say) blue. Then either there is a red $\binom{s}{k}$, and we are done, or there is a blue $\binom{R_k(s,s)}{k}$. But *blue* means *green or yellow*, so the green-blue 2-colored $\binom{R_k(s,s)}{k}$ contains either a green $\binom{s}{k}$ or a yellow $\binom{s}{k}$, which proves (A4).

Unfortunately, Theorem A3 provides very poor bounds, even for $k = 3$. Indeed, by Theorems A1–A2, $R_2(x, y)$ is exponential, so by repeated applications of Theorem A3 we obtain a tower function-like upper bound

$$R_3(s, s) < 2^{2^{2^{\cdot^{\cdot^2}}}}, \tag{A5}$$

where the height of the tower is $const \cdot s$.

Much better upper bounds come from an inequality of Erdős and Rado. For example, if $k = 3$, then the Erdős–Rado argument gives a double-exponential upper bound like 2^{2^s} instead of the tower function (A5). The proof works for arbitrary number of colors without any modification, so first we define the multi-color version of the Ramsey Number $R_k(s, t)$.

Let $R_k(s_1, s_2, \cdots, s_c)$ ($c \ge 2$ integer) be the minimum of n such that given any c-coloring of the k-edges of the complete k-uniform hypergraph $\binom{n}{k}$ (i.e. all the k-element subsets of an n-element set), there will be an i with $1 \le i \le c$ and a complete k-uniform hypergraph $\binom{s_i}{k}$ where all the k-edges are colored with the ith color.

In the diagonal case $s_1 = s_2 = \cdots = s_c = s$ we use the short notation $R_k(s|c) = R_k(s, s, \cdots, s)$. If $c = 2$, we simply write $R_k(s) = R_k(s|2)$.

Theorem A5 (Erdős and Rado)

$$R_{k+1}(s|c) < c^{\big(R_k(s|c)\big)^k}$$

Remark. If $k = 1$, then trivially $R_1(s|c) = sc + 1$. So for $c = 2$ we get $R_2(s) < 2^{2s+1} = 2 \cdot 4^s$, which is asymptotically just a little bit weaker than the Erdős–Szekeres bound $\binom{2s-2}{s-1} \approx const \cdot 4^s / \sqrt{s}$.

Proof. Let S denote the interval

$$\left\{1, 2, \cdots, c^{\big(R_k(s|c)\big)^k}\right\},$$

and let $\mathcal{C}$ be an arbitrary c-coloring of the $(k + 1)$-tuples of S. Unlike in the proof of Theorem 3, where we fixed a point and considered the induced coloring, here we fix a k-tuple and consider the induced coloring. Consider first the k-tuple $\{1, 2, \cdots, k\}$, and for any $m \in S$ with $k < m$, define the induced c-coloring $\mathcal{C}^*(m) = \mathcal{C}(\{1, 2, \cdots, k, m\})$. c-coloring $\mathcal{C}^*$ defines a partition of $S \setminus \{1, 2, \cdots, k\}$ into c color classes. Let S_1 denote the biggest color class, and let $\mathcal{C}^{**} = \mathcal{C}^{**}(\{1, 2, \cdots, k\})$ denote the corresponding color. We clearly have the inequality

$$|S_1| \geq \frac{|S| - k}{c}.$$

Next let i_{k+1} be the smallest number in S_1, and let A be an arbitrary k-tuple of $\{1, 2, \cdots, k, i_{k+1}\}$, which contains i_{k+1}. For this A and for any $m \in S_1$ with $i_{k+1} < m$, define the induced c-coloring $C_A^*(m) = C(A \cup \{m\})$. Let $A_1, A_2, \cdots, A_k$ be all k-tuples of $\{1, 2, \cdots, k, i_{k+1}\}$ which contain i_{k+1}, and consider the c^k-coloring C^*

$$C^*(m) = \left(C_{A_1}^*(m), C_{A_2}^*(m), \cdots, C_{A_k}^*(m) \right).$$

c^k-coloring C^* defines a partition of $S_1 \setminus \{i_{k+1}\}$ into c^k color classes. Let S_2 denote the biggest color class, and let $C^{**} = C^{**}(A)$ denote the corresponding color for each k-tuple A of $\{1, 2, \cdots, k, i_{k+1}\}$ with $i_{k+1} \in A$. We clearly have the inequality

$$|S_2| \geq \frac{|S_1| - 1}{c^k}.$$

Next let i_{k+2} be the smallest number in S_2, and let A be an arbitrary k-tuple of $\{1, 2, \cdots, k, i_{k+1}, i_{k+2}\}$ which contains i_{k+2}. For this A and for any $m \in S_2$ with $i_{k+2} < m$, define the induced c-coloring $C_A^*(m) = C(A \cup \{m\})$. Let $A_1, A_2, \cdots, A_r$, $r = \binom{k+1}{2} = \binom{k+1}{k-1}$ be all k-tuples of $\{1, 2, \cdots, k, i_{k+1}, i_{k+2}\}$ which contains i_{k+2}, and consider the c^r-coloring C^*

$$C^*(m) = \left(C_{A_1}^*(m), C_{A_2}^*(m), \cdots, C_{A_r}^*(m) \right).$$

Induced c^r-coloring C^* defines a partition of $S_2 \setminus \{i_{k+2}\}$ into c^r color classes. Let S_3 denote the biggest color class, and let $C^{**} = C^{**}(A)$ denote the corresponding color for each k-tuple A of $\{1, 2, \cdots, k, i_{k+1}, i_{k+2}\}$ with $i_{k+2} \in A$. We clearly have the inequality

$$|S_3| \geq \frac{|S_2| - 1}{c^{\binom{k+1}{k-1}}}.$$

Repeating this argument, we obtain a sequence of integers

$$i_1 = 1, i_2 = 2, \cdots, i_k = k < i_{k+1} < i_{k+2} < \cdots < i_q,$$

and a nested sequence of sets

$$S \supset S_1 \supset S_2 \supset S_3 \supset \cdots \supset S_{q-k}$$

such that $i_j \in S_{j-k}$, $k + 1 \leq j \leq q$

$$|S_{j+1}| \geq \frac{|S_j| - 1}{c^{\binom{k+j-1}{k-1}}},$$

and for any fixed k-tuple $(i_{j_1}, i_{j_2}, \cdots, i_{j_k})$ where $1 \leq j_1 < j_2 < \cdots < j_k < q$, the C-color of the $(k+1)$-tuple $(i_{j_1}, i_{j_2}, \cdots, i_{j_k}, m)$ is the *same* for any $i_{j_k} < m \in S_{j_k - k + 1}$.

This common color depends on the k-tuple $(i_{j_1}, i_{j_2}, \cdots, i_{j_k})$ only, and it is denoted by $C^{**}(i_{j_1}, i_{j_2}, \cdots, i_{j_k})$. Since

$$|S| = c^{\left(R_k(s|c)\right)^k},$$

and

$$|S_{j+1}| \geq \frac{|S_j| - 1}{c^{\binom{k+j-1}{k-1}}},$$

an easy calculation shows that the length q of the longest sequence $i_1 = 1, i_2 = 2, \cdots, i_k = k < i_{k+1} < i_{k+2} < \cdots < i_q$ can be as big as the Ramsey Number $R_k(s|c)$. So we write $q = R_k(s|c)$.

Now we are ready to complete the proof. Indeed, induced coloring C^{**} defines a c-coloring on the set of all k-tuples $(i_{j_1}, i_{j_2}, \cdots, i_{j_k})$ (where $1 \leq j_1 < j_2 < \cdots < j_k < q = R_k(s|c)$) of $i_1 = 1 < \cdots < i_q$. Since $q = R_k(s|c)$, by hypothesis there is a C^{**}-monochromatic s-set, i.e. there is an s-element subset Y such that its k-tuples have the same C^{**}-color. Finally, from the definition of induced coloring C^{**} it follows that this Y is a C-monochromatic set, too. In other words, all the $(k+1)$-tuples of Y have the same C-color. This completes the proof of Theorem A5.   □

We introduce the following *arrow-notation*: if G and H are k-uniform hypergraphs (graphs if $k = 2$), then $G \to H$ means that given any 2-coloring of the k-edges of G, there is always a monochromatic copy of H. Of course, $G \nrightarrow H$ means the opposite statement: there is a 2-coloring of the k-edges of G such that there is no monochromatic copy of H. In view of this, Theorems A1–A2 can be reformulated as follows

$$\binom{\binom{2s-2}{s-1}}{2} \to \binom{s}{2}$$

and

$$\binom{2^{s/2}}{2} \nrightarrow \binom{s}{2}.$$

The following result is usually referred as the "Stepping-Up Lemma":

Theorem A6 (*Erdős, Hajnal and Rado*) *If* $\binom{n}{k} \nrightarrow \binom{s}{k}$ *and* $k \geq 3$, *then*

$$\binom{2^n}{k+1} \nrightarrow \binom{2s+k}{k+1}.$$

Proof. We are going to transform an anti-Ramsey 2-coloring of $\binom{n}{k}$ into an anti-Ramsey 2-coloring of $\binom{2^n}{k+1}$. Let $C_1 \cup C_2 = \binom{n}{k}$ be a decomposition of the set of all k-tuples of $[n] = \{0, 1, \cdots, n-1\}$ into two classes such that no C_i ($i = 1$ or 2) contains an $\binom{s}{k}$, i.e. there is no *monochromatic* $\binom{s}{k}$ in $\binom{n}{k}$. The set of all 0–1 sequences of length n

$$S = \{\mathbf{a} = (a_1, a_2, \cdots, a_n) : a_i = 0 \text{ or } 1 \text{ for } 1 \leq i \leq n\}$$

is a natural choice for a set of cardinality 2^n. A string $\mathbf{a} = (a_1, a_2, \cdots, a_n)$, $a_i = 0$ or 1, corresponds to the binary form of an integer

$$f(\mathbf{a}) = a_1 2^{n-1} + a_2 2^{n-2} + \cdots + a_{n-1} 2 + a_n.$$

We say that $\mathbf{a} < \mathbf{b}$ if and only if $f(\mathbf{a}) < f(\mathbf{b})$. If $\mathbf{a} < \mathbf{b}$, then there is an index i, $1 \leq i \leq n$, such that $a_1 = b_1, \cdots, a_{i-1} = b_{i-1}$, but $a_i = 0$ and $b_i = 1$. We call this index i the *separating index* of the pair $(\mathbf{a}, \mathbf{b})$, and denote it by $d(\mathbf{a}, \mathbf{b})$. Note that:

(i) if $\mathbf{a} < \mathbf{b} < \mathbf{c}$, then $d(\mathbf{a}, \mathbf{b}) \neq d(\mathbf{b}, \mathbf{c})$, and both cases $d(\mathbf{a}, \mathbf{b}) < d(\mathbf{b}, \mathbf{c})$ and $d(\mathbf{a}, \mathbf{b}) > d(\mathbf{b}, \mathbf{c})$ are possible;

(ii) if $\mathbf{a}_1 < \mathbf{a}_2 < \cdots < \mathbf{a}_r$, then $d(\mathbf{a}_1, \mathbf{a}_r) = \min_{1 \leq i < r} d(\mathbf{a}_i, \mathbf{a}_{i+1})$.

We are going to define a 2-coloring $C_1 \cup C_2 = \binom{S}{k+1}$ of all the $(k+1)$-tuples of set S which doesn't contain a *monochromatic* $(2s+k)$-element subset. Let $\{\mathbf{a}_1, \mathbf{a}_2, \cdots, \mathbf{a}_{k+1}\}$ be a $(k+1)$-tuple of S. We can assume that $\mathbf{a}_1 < \mathbf{a}_2 < \cdots < \mathbf{a}_{k+1}$. Let $d_i = d(\mathbf{a}_i, \mathbf{a}_{i+1})$, $1 \leq i \leq k$. The sequence $d_1, d_2, \cdots, d_k$ does not necessarily means k different numbers. So first assume that the sequence d_i is monotonically decreasing or increasing, which automatically guarantees that we have k different numbers. If $d_1 > d_2 > \cdots > d_k$ or $d_1 < d_2 < \cdots < d_k$, then color the $(k+1)$-tuple $\{\mathbf{a}_1, \mathbf{a}_2, \cdots, \mathbf{a}_{k+1}\}$ by the color of the k-tuple $\{d_1, d_2, \cdots, d_k\}$, that is

$$\{\mathbf{a}_1, \mathbf{a}_2, \cdots, \mathbf{a}_{k+1}\} \in C_i$$

if and only if $\{d_1, d_2, \cdots, d_k\} \in C_i$, $i = 1$ or 2.

If the sequence d_i is *not* monotonically decreasing or increasing, then the color of the $(k+1)$-tuple $\{\mathbf{a}_1, \mathbf{a}_2, \cdots, \mathbf{a}_{k+1}\}$ will be determined later.

So far we defined only a *partial 2-coloring* of the $(k+1)$-tuples of S, but this is enough to make the following crucial observation:

Lemma 1: *Assume that* $\{\mathbf{a}_1, \mathbf{a}_2, \cdots, \mathbf{a}_m\}$ *is a partially monochromatic subset of* S, *i.e. all* $(k+1)$-*tuples of this m-element set which are colored under the partial 2-coloring have the same color. Without loss of generality, we can assume that* $\mathbf{a}_1 < \mathbf{a}_2 < \cdots < \mathbf{a}_m$. *Let* $d_i = d(\mathbf{a}_i, \mathbf{a}_{i+1})$, $1 \leq i < m$. *Then there is no index j such that the "subinterval"* $d_j, d_{j+1}, d_{j+2}, \cdots, d_{j+s-1}$ *is monotonically increasing or decreasing.*

Proof. Assume that the color class containing all partially colored $(k+1)$-tuples of $\{\mathbf{a}_1, \mathbf{a}_2, \cdots, \mathbf{a}_m\}$ is C_i ($i = 1$ or 2). We distinguish two cases.

Case 1: $d_j < d_{j+1} < d_{j+2} < \cdots < d_{j+s-1}$
Since $C_1 \cup C_2 = \binom{n}{k}$ was a 2-coloring satisfying the property $\binom{n}{k} \not\rightarrow \binom{s}{k}$, the s-element set $\{d_j, d_{j+1}, \cdots, d_{j+s-1}\}$ cannot have all its k-tuples in the same color class C_i. Therefore, there exists a k-tuple $\{i_1, i_2, \cdots, i_k\}$ with $j \leq i_1 < i_2 < \cdots < i_k \leq j+s-1$

such that $\{d_{i_1}, d_{i_2}, \cdots, d_{i_k}\}$ is in the other color class: $\{d_{i_1}, d_{i_2}, \cdots, d_{i_k}\} \in C_{3-i}$. Consider now the $(k+1)$-tuple

$$A = \{\mathbf{a}_{i_1}, \mathbf{a}_{i_2}, \cdots, \mathbf{a}_{i_k}, \mathbf{a}_{i_k+1}\}.$$

For $1 \le h < k$ we have

$$d(\mathbf{a}_{i_h}, \mathbf{a}_{i_{h+1}}) = \min_{i_h \le \ell < i_{h+1}} d_\ell,$$

and because in Case 1 we have $d_j < d_{j+1} < d_{j+2} < \cdots < d_{j+s-1}$, it follows that

$$d(\mathbf{a}_{i_h}, \mathbf{a}_{i_{h+1}}) = \min_{i_h \le \ell < i_{h+1}} d_\ell = d_{i_h}.$$

Therefore

$$d(\mathbf{a}_{i_1}, \mathbf{a}_{i_2}) = d_{i_1} < d(\mathbf{a}_{i_2}, \mathbf{a}_{i_3}) = d_{i_2} < \cdots < d(\mathbf{a}_{i_{k-1}}, \mathbf{a}_{i_k}) = d_{i_{k-1}} < d(\mathbf{a}_{i_k}, \mathbf{a}_{i_k+1}) = d_{i_k}.$$

So the partial 2-coloring applies, and the $(k+1)$-tuple $A = \{\mathbf{a}_{i_1}, \mathbf{a}_{i_2}, \cdots, \mathbf{a}_{i_k}, \mathbf{a}_{i_k+1}\}$ is colored by the color of $\{d_{i_1}, d_{i_2}, \cdots, d_{i_k}\}$. Since $\{d_{i_1}, d_{i_2}, \cdots, d_{i_k}\} \in C_{3-i}$, we conclude that $A \in C_{3-i}$, which contradicts the assumption that the color class containing all $(k+1)$-tuples of $\{\mathbf{a}_1, \mathbf{a}_2, \cdots, \mathbf{a}_m\}$ is C_i.

Case 2: $d_j > d_{j+1} > d_{j+2} > \cdots > d_{j+s-1}$
Just as in Case 1 we can assume that there exists a k-tuple $\{i_1, i_2, \cdots, i_k\}$ with $j \le i_1 < i_2 < \cdots < i_k \le j+s-1$ such that $\{d_{i_1}, d_{i_2}, \cdots, d_{i_k}\} \in C_2$. In this case we consider the slightly different $(k+1)$-tuple

$$A = \{\mathbf{a}_{i_1}, \mathbf{a}_{i_1+1}, \mathbf{a}_{i_2+1}, \cdots, \mathbf{a}_{i_k+1}\}.$$

Just as in Case 1 we obtain

$$d(\mathbf{a}_{i_1}, \mathbf{a}_{i_1+1}) = d_{i_1} > d(\mathbf{a}_{i_1+1}, \mathbf{a}_{i_2+1}) = d_{i_2} > \cdots > d(\mathbf{a}_{i_{k-1}+1}, \mathbf{a}_{i_k+1}) = d_{i_k}.$$

Now we get the contradiction exactly the same way as in Case 1. This completes the proof of Lemma 1. □

Next we extend the partial 2-coloring of the $(k+1)$-tuples of S to a proper 2-coloring. To do that we need the following elementary:

Lemma 2: Let $s \ge k \ge 3$, and let $d_1, d_2, d_3, \cdots, d_{2s+k-1}$ be an arbitrary sequence of $(2s+k-1)$ integers such that any two consecutive members of the sequence are different, i.e. $d_i \ne d_{i+1}$ for $1 \le i < 2s+k-1$. Then

(i) either there is subinterval $j, j+1, \cdots, j+s-1$ of length s such that the sequence $d_j, d_{j+1}, \cdots, d_{j+s-1}$ is monotonically increasing or decreasing;

(ii) or there are $j \le 2s$ and $\ell \le 2s$ such that $d_{j-1} < d_j$ and $d_j > d_{j+1}$, and $d_{\ell-1} > d_\ell$ and $d_\ell < d_{\ell+1}$.

Proof. We call a j a *local max* if $d_{j-1} < d_j$ and $d_j > d_{j+1}$, and a *local min* if $d_{j-1} > d_j$ and $d_j < d_{j+1}$.

If the sequence $d_1, \cdots, d_{2s+k-1}$ has a local max $j \le 2s$ and a local min $\ell \le 2s$, then we are done.

If the sequence $d_1, \cdots, d_{2s+k-1}$ has two local max's $j_1 \le 2s$ and $j_2 \le 2s$, then between the two local maxs there is always a local min, so this is the previous case. Of course, the same argument holds if there are two local mins.

If the sequence $d_1, \cdots, d_{2s+k-1}$ has *one* local max $j \le 2s$ and *no* local min $\ell \le 2s$, then:

(i) if $j < s$, the sequence $d_j, d_{j+1}, \cdots, d_{j+s-1}$ is monotonically decreasing;
(ii) if $j \ge s$, the sequence $d_{j+1-s}, d_{j+2-s}, \cdots, d_j$ is monotonically increasing.

The same argument holds if there is one local min and no local max. $\qquad\square$

Now we are ready to extend the partial 2-coloring to a complete 2-coloring $\mathbf{C}_1 \cup \mathbf{C}_2 = \binom{S}{k+1}$ of *all* the $(k+1)$-tuples of set S. Let $\mathbf{A} = \{\mathbf{a}_1, \mathbf{a}_2, \cdots, \mathbf{a}_{k+1}\}$ be an arbitrary $(k+1)$-tuple of S. We can assume that $\mathbf{a}_1 < \mathbf{a}_2 < \cdots < \mathbf{a}_{k+1}$. Let $d_i = d(\mathbf{a}_i, \mathbf{a}_{i+1})$, $1 \le i \le k$. The partial 2-coloring was defined for those $\mathbf{A}$s for which the sequence d_i is monotonically decreasing or increasing. Then the color of $\mathbf{A}$ was defined by "lifting up" the color of $\{d_1, d_2, \cdots, d_k\}$.

If the sequence d_i is *not* monotonically decreasing or increasing, then there is a local max or a local min (or both). There are three cases. Either 2 is a local max: $d_1 < d_2 > d_3$, or 2 is a local min: $d_1 > d_2 < d_3$, or the first local max is ≥ 3 and the first local min is ≥ 3. Then 2-color $(k+1)$-tuple $\mathbf{A}$ like this: if 2 is a local max, then $\mathbf{A} = \{\mathbf{a}_1, \mathbf{a}_2, \cdots, \mathbf{a}_{k+1}\} \in \mathbf{C}_1$; if 2 is a local min, then $\mathbf{A} = \{\mathbf{a}_1, \mathbf{a}_2, \cdots, \mathbf{a}_{k+1}\} \in \mathbf{C}_2$. (This is where we use the assumption that $k \ge 3$.) Finally, if the first local max is ≥ 3 and the first local min is ≥ 3, then color $\mathbf{A}$ *arbitrarily*.

Assume that $\{\mathbf{a}_1, \mathbf{a}_2, \cdots, \mathbf{a}_{2s+k}\}$ is a *monochromatic subset* of S, i.e. all the $(k+1)$-tuples of this $(2s+k)$-element set have the same color under this 2-coloring. Without loss of generality we can assume that $\mathbf{a}_1 < \mathbf{a}_2 < \cdots < \mathbf{a}_{2s+k}$. Let $d_i = d(\mathbf{a}_i, \mathbf{a}_{i+1})$, $1 \le i \le 2s+k-1$. In view of Lemma A8, there are two possibilities:

(i) There is an index j such that the "sub-interval" $d_j, d_{j+1}, d_{j+2}, \cdots, d_{j+s-1}$ is monotonically increasing or decreasing. But this is impossible by Lemma A7.
(ii) There are $j \le 2s$ and $\ell \le 2s$ such that j is a local max: $d_{j-1} < d_j > d_{j+1}$, and ℓ is a local min: $d_{\ell-1} > d_\ell < d_{\ell+1}$. Then by definition, $\{\mathbf{a}_{j-1}, \mathbf{a}_j, \mathbf{a}_{j+1}, \cdots, \mathbf{a}_{j+k-1}\} \in \mathbf{C}_1$ and $\{\mathbf{a}_{\ell-1}, \mathbf{a}_\ell, \mathbf{a}_{\ell+1}, \cdots, \mathbf{a}_{\ell+k-1}\} \in \mathbf{C}_2$, which is a contradiction. This completes the proof of Theorem A6.

The best lower bound for $R_3(s)$ is

$$R_3(s) > 2^{s^2/6},$$

proved by a simple adaptation of the proof of Theorem A2. Indeed, inequality

$$\binom{n}{s} 2^{\binom{n}{3} - \binom{s}{3}} < \frac{1}{2} 2^{\binom{n}{3}} \tag{A3$'$}$$

guarantees that there is a labeled simple 3-graph $G_{3,n}$ on n points such that neither $G_{3,n}$ nor its complement contains an $\binom{s}{3}$. Rough approximations yield that $(A3')$ holds whenever $n = 2^{s^2/6}$. It is a long-standing open problem to improve this lower bound to a doubly exponential bound.

Theorems A1–A2 and A5–A6 imply the following upper and lower bounds for $R_k(n)$:

Theorem A7 *For $k \geq 2$ we have*

$$2^{n/2} < R_2(n) < 4^n,$$

$$2^{n^2/6} < R_3(n) < 2^{2^{4n}},$$

$$2^{2^{n^2/24}} < R_4(n) < 2^{2^{2^{4n}}},$$

$$2^{2^{2^{n^2/96}}} < R_5(n) < 2^{2^{2^{2^{4n}}}},$$

and in general, for arbitrary $k \geq 3$

$$tower_{4^{3-k}n^2/6}(k-2) < R_k(n) < tower_{4n}(k-1),$$

where $tower_x(1) = 2^x$, and for $j \geq 2$, $tower_x(j) = 2^{tower_x(j-1)}$.

For more about the results of Erdős, Rado, and Hajnal, see Erdős and Rado [1952] and Erdős et al. [1965], and also the monograph Graham et al. [1980].

Appendix B
Hales–Jewett Theorem: Shelah's proof

If the chromatic number of the n^d-hypergraph is ≥ 3, i.e. if $d \geq HJ(n)$, then First Player has a winning strategy in the n^d Tic-Tac-Toe game. This is the only known "win criterion" – unfortunately it does not give a clue how to actually win.

For the sake of completeness we include a proof of Shelah's *primitive recursive* upper bound for the Hales–Jewett Number $HJ(n)$.

Note that the original van der Waerden–Hales–Jewett proof led to the notorious Ackermann function. It was a big step forward, therefore, when in 1988 Shelah was able to prove that the Hales–Jewett threshold $HJ(n, k)$ (to be defined below) is *primitive recursive* ($HJ(n)$ is the special case of $k = 2$).

We briefly recall the so-called Grzegorczyk hierarchy of recursive functions. In fact, we define the *representative* function for each class. For a more detailed treatment of primitive recursive functions we refer the reader to Mathematical Logic.

Let $g_1(n) = 2n$, and for $i > 1$, let $g_i(n) = g_{i-1}\big(g_{i-1}(\cdots g_{i-1}(1)\cdots)\big)$, where g_{i-1} is taken n times. An equivalent definition is $g_i(n+1) = g_{i-1}\big(g_i(n)\big)$. For example, $g_2(n) = 2^n$ is the exponential function

$$g_3(n) = 2^{2^{2^{\cdot^{\cdot^{\cdot^2}}}}}$$

is the "tower function" of height n. The next function $g_4(n+1) = g_3\big(g_4(n)\big)$ is what we call the "Shelah's super-tower function" because this is exactly what shows up in Shelah's proof. Note that $g_k(x)$ is the *representative* function of the $(k+1)$st Grzegorczyk class.

The Hales–Jewett Theorem is about monochromatic n-in-a-line's of the n^d hypercube. The proof actually gives more: it guarantees the existence of a monochromatic *combinatorial line*. Let $\Lambda = \{0, 1, 2, \ldots, n-1\}$. An *x-string* is a finite word $a_1 a_2 a_3 \cdots a_d$ of the symbols $a_i \in \Lambda \cup \{x\}$ where at least one symbol a_i is x. An x-string is denoted by $\mathbf{w}(x)$. For every integer $i \in \Lambda$ and x-string $\mathbf{w}(x)$, let $\mathbf{w}(x; i)$

denote the string obtained from $\mathbf{w}(x)$ by replacing each x by i. A *combinatorial line* is a set of n strings $\{\mathbf{w}(x; i) : i \in \Lambda\}$ where $\mathbf{w}(x)$ is an x-string.

Every combinatorial line is a geometric line, i.e. n-in-a-line, but the converse is not true: *not* every geometric line is a combinatorial line. A *geometric lines* can be described as an *xx'-string* $a_1 a_2 a_3 \cdots a_d$ of the symbols $a_i \in \Lambda \cup \{x\} \cup \{x'\}$ where at least one symbol a_i is x or x'. An xx'-string is denoted by $\mathbf{w}(xx')$. For every integer $i \in \Lambda$ and xx'-string $\mathbf{w}(xx')$, let $\mathbf{w}(xx'; i)$ denote the string obtained from $\mathbf{w}(xx')$ by replacing each x by i and each x' by $(n-1-i)$. A *directed geometric line* is a *sequence* $\mathbf{w}(xx'; 0)$, $\mathbf{w}(xx'; 1)$, $\mathbf{w}(xx'; 2)$, ..., $\mathbf{w}(xx'; n-1)$ of n strings where $\mathbf{w}(xx')$ is an xx'-string. Note that every geometric line has two orientations.

It is clear from the definition that there are sub-stantially more geometric lines than combinatorial lines. For example, in ordinary Tic-Tac-Toe

$(0,2)$	$(1,2)$	$(2,2)$
$(0,1)$	$(1,1)$	$(2,1)$
$(0,0)$	$(1,0)$	$(2,0)$

$\{(0,0), (1,1), (2,2)\}$ is a combinatorial line defined by the x-string xx, $\{(0,1), (1,1), (2,1)\}$ is another combinatorial line defined by the x-string $x1$, but the other diagonal

$$\{(0,2), (1,1), (2,0)\}$$

is a geometric line defined by the xx'-string xx'. So the other diagonal is a geometric line which is *not* a combinatorial line.

The Hales–Jewett Number $HJ(n, k)$ is the smallest integer d such that in each k-coloring of Λ^d there is a monochromatic *geometric* line. The modified Hales–Jewett Number $HJ^c(n, k)$ is the smallest integer d such that in each k-coloring of Λ^d there is a monochromatic *combinatorial* line. Clearly $HJ(n, k) \leq HJ^c(n, k)$.

In 1961 Hales and Jewett proved that $HJ(n, k) < \infty$ for all positive integers n and k; in fact, they proved the stronger statement that $HJ^c(n, k) < \infty$. Shelah's new proof gives the following explicit bound:

Theorem B1 (Shelah's upper bound) *For every $n \geq 1$ and $k \geq 1$*

$$HJ^c(n, k) \leq \frac{1}{(n+1)k} g_4(n+k+2).$$

That is, given any k-coloring of the hypercube $\{0, 1, \ldots, n-1\}^d$ where the dimension $d \geq \frac{1}{(n+1)k} g_4(n+k+2)$, there is always a monochromatic **combinatorial** *line.*

Proof. For each *fixed* value of $k \geq 2$, we are going to argue by induction on n that the threshold number $HJ^c(n-1, k)$ exists. However, when trying the induction step to show that $HJ^c(n, k)$ exists, we are only allowed to use the existence of

$HJ^c(n-1,k)$. If we also used the existence of $HJ^k(n-1,k')$ for very large values of k', then the argument would become a *double induction*, and this would lead to Ackermann-like bounds (see Section 7).

For $n=1$, the theorem is trivial. Assuming it holds for $n-1 \geq 1$ and $k \geq 2$, we prove it for n and for the *same* k. Let $h = HJ^c(n-1,k)$. Define a very rapidly increasing sequence m_i, $i = 1, 2, \ldots, h$ as follows. Let

$$m_1 = k^{n^h} = k^{n^{HJ^c(n-1,k)}},$$

and for $i \geq 2$

$$m_i = k^{n^{h+m_1+\ldots+m_{i-1}}}.$$

Let $d = m_1 + m_2 + \ldots + m_h$. We are going to show that $HJ^c(n,k) \leq d$. Let C be an arbitrary coloring of $\Lambda^d = \{0, 1, 2, \ldots, n-1\}^d$ by k colors. The idea of the proof is that in any k-coloring C of the d-dimensional cube Λ^d there is an $HJ^c(n-1,k)$-dimensional *sub-cube* in which "symbols 0 and 1 cannot be distinguished by k-coloring C." By hypothesis, this $HJ^c(n-1,k)$-dimensional *sub-cube* contains a *monochromatic* $(n-1)$-in-a-row. The $(n-1)$ points are those where the dynamic coordinates of the line simultaneously run through the values $1, 2, \ldots, n-1$. This *monochromatic* $(n-1)$-in-a-row extends to a combinatorial line of length n in Λ^d. The *new* point is that where the dynamic coordinates of the line are all 0. By using the key property that in the $HJ^c(n-1,k)$-dimensional *sub-cube* "symbols 0 and 1 cannot be distinguished by k-coloring C," it follows that the C-color of this *new* point is the same as the C-color of that point on the *monochromatic* $(n-1)$-in-a-row where the dynamic coordinates of the line are all 1. Therefore, the whole combinatorial line of length n is *monochromatic* as well. This will complete the proof. □

It is very important to clearly understand what a *sub-cube* means, so we show an example of a 3-dimensional sub-cube of a 11-dimensional cube $\{0, 1, 2\}^{11}$

0210, 100, 2002	1211, 100, 2002	2212, 100, 2002
0210, 111, 2002	1211, 111, 2002	2212, 111, 2002
0210, 122, 2002	1211, 122, 2002	2212, 122, 2002

0210, 100, 2112	1211, 100, 2112	2212, 100, 2112
0210, 111, 2112	1211, 111, 2112	2212, 111, 2112
0210, 122, 2112	1211, 122, 2112	2212, 122, 2112

0210, 100, 2222	1211, 100, 2222	2212, 100, 2222
0210, 111, 2222	1211, 111, 2222	2212, 111, 2222
0210, 122, 2222	1211, 122, 2222	2212, 122, 2222

This particular sub-cube can be described as a 3-parameter string x21x1yy2zz2 where the dynamic symbols x, y, z run through the values 0, 1, 2 independently of each other. The 1st, 4th, 6th, 7th, 9th, and 10th coordinates are the *dynamic* coordinates of this sub-cube. The rest of the coordinates, namely the 2nd, 3rd, 5th, 8th, and 11th coordinates remain fixed for all the $3^3 = 27$ points of this sub-cube. In this example we can separate symbols x, y, z from each other by using the comma-notation: x21x,1yy,2zz2. In other words, the string x21x1yy2zz2 can be decomposed into three sub-intervals x21x, 1yy, and 2zz2. This is why we can call the sub-cube x21x1yy2zz2 an *interval sub-cube*.

But this is not necessarily the case for a general sub-cube. For example, the 3-parameter string x21z1yx2zx2, where the symbols x, y, z run through the values 0, 1, 2 independently of each other, defines another 3-dimensional sub-cube of $\{0, 1, 2\}^{11}$, where the separation of symbols x, y, z by using comma is impossible. So the sub-cube x21z1yx2zx2 is *not* an interval sub-cube. But both sub-cubes x21x1yy2zz2 and x21z1yx2zx2 are *isomorphic* to the cube xyz (and so to each other) in a natural way.

As we said before, we are going to find an $HJ^c(n-1, k)$-dimensional sub-cube of Λ^d in which "symbols 0 and 1 cannot be distinguished by k-coloring C", and this sub-cube will be an *interval sub-cube*. In fact, its string of length d will decompose into $h = HJ^c(n-1, k)$ sub-intervals, where the ℓth sub-interval will have length m_ℓ, and it will look like $0\ldots0x_\ell\ldots x_\ell 1\ldots 1$ ($\ell = 1, 2, \ldots, h$). This means, each sub-interval will have the same form: it begins with a string of 0s, the middle part consists of a string of x_ℓs where x_ℓ is the ℓth independent parameter ("dynamic coordinates"), and the last part is a string of 1s. The independent parameters $x_1, x_2, \ldots, x_h$ run through the values $0, 1, \ldots, n-1$ independently of each other; this is how we get the n^h points of the sub-cube.

How to find this special $HJ^c(n-1, k)$-dimensional *interval sub-cube* of Λ^d? Well, we are going to construct the string of this interval sub-cube by descending induction on ℓ, i.e. $\ell = h, h-1, h-2, \ldots, 1$. This means first we find the string $0\ldots0x_h\ldots x_h 1\ldots 1$ of the last sub-interval of length m_h. The crucial fact is that m_h is much, much larger than $d - m_h = m_1 + m_2 + \ldots + m_{h-1}$ (see the definition of m_i), so we have plenty of room for an easy application of the *pigeonhole principle*. We claim that there are two strings ("closing strings") $\mathbf{t}_i = 0\ldots01\ldots1$ (where the first i symbols are 0) and $\mathbf{t}_j = 0\ldots01\ldots1$ (where the first j symbols are 0) with $i < j$, each of length m_h, such that given *any* string $\mathbf{s} \in \Lambda^{(d-m_h)}$ of length $(d - m_h)$ ("opening string"), the C-color of the union string $\mathbf{st}_i \in \Lambda^d$ is the *same* as the C-color of the other union string $\mathbf{st}_j \in \Lambda^d$. (Of course, $\mathbf{st}_i$ means that union which begins with $\mathbf{s}$ and ends with $\mathbf{t}_i$.) But this statement is a direct application of the pigeonhole principle if

$$m_h \geq k^{n^{d-m_h}}.$$

Indeed, there are $k^{n^{d-m_h}}$ c-colorings of the set of *all possible* "opening strings" $\mathbf{s} \in \Lambda^{(d-m_h)}$. On the other hand, there are $m_h + 1$ special strings

$$1, 1, 1, 1, 1, 1, 1, \ldots, 1$$

$$0, 1, 1, 1, 1, 1, 1, \ldots, 1$$

$$0, 0, 1, 1, 1, 1, 1, \ldots, 1$$

$$0, 0, 0, 1, 1, 1, 1, \ldots, 1$$

$$0, 0, 0, 0, 1, 1, 1, \ldots, 1$$

$$\cdots\cdots\cdots\cdots$$

$$0, 0, 0, 0, 0, 0, 0, \ldots, 0$$

of length m_h ("closing strings"). Each special string $\mathbf{t} = 0, \ldots, 0, 1, \ldots, 1$ on this list defines a k-coloring of the set of *all possible* "opening strings" $\mathbf{s} \in \Lambda^{(d-m_h)}$ as follows: the color of "opening string" $\mathbf{s}$ is the C-color of the union string $\mathbf{st}$. Therefore, if

$$m_h \geq c^{n^{d-m_h}},$$

then by the pigeonhole principle there exist two special strings $\mathbf{t}_i = 0 \ldots 01 \ldots 1$ (where the first i symbols are 0) and $\mathbf{t}_j = 0 \ldots 01 \ldots 1$ (where the first j symbols are 0) with $i < j$, each of length m_h, such that the C-color of $\mathbf{st}_i \in \Lambda^d$ is the *same* as the C-color of $\mathbf{st}_j \in \Lambda^d$ for *any* "opening string" $\mathbf{s} \in \Lambda^{(d-m_h)}$.

Note that inequality

$$m_h \geq k^{n^{d-m_h}} = k^{n^{m_1 + \ldots + m_{h-1}}}$$

holds by the definition of the m_is.

Now the string $\mathbf{w}(x_h) = 0 \ldots 0 x_h \ldots x_h 1 \ldots 1$ of the last sub-interval (of length m_h) of the $HJ^c(n-1, k)$-dimensional *interval sub-cube* of Λ^d is the following: the first part is a string of i 0s, the middle part is a string of $(j-i)$ x_hs ("dynamic coordinates"), and the last part is a string of $(m_h - j)$ 1s.

Clearly $\mathbf{w}(x_h; 0) = \mathbf{t}_i$ and $\mathbf{w}(x_h; 1) = \mathbf{t}_j$, so changing the dynamic coordinates of the last sub-interval simultaneously from 0 to 1 cannot be distinguished by c-coloring C.

The general case goes exactly the same way by using the pigeonhole principle. Suppose we have already defined the strings $\mathbf{w}(x_{\ell+1}), \mathbf{w}(x_{\ell+2}), \ldots, \mathbf{w}(x_h)$ of the last $(h - \ell)$ sub-intervals of the desired $HJ^c(n-1, k)$-dimensional *interval sub-cube* of Λ^d.

Again we claim that there are two strings ("middle strings" this time) $\mathbf{t}_i = 0 \ldots 01 \ldots 1$ (where the first i symbols are 0) and $\mathbf{t}_j = 0 \ldots 01 \ldots 1$ (where the first j symbols are 0) with $i < j$, each of length m_ℓ, such that given *any* "opening string" $\mathbf{s} \in \Lambda^{(m_1 + \ldots + m_{\ell-1})}$ of length $(m_1 + \ldots + m_{\ell-1})$ and given any "closing string"

$\mathbf{r} = r_1 r_2 \cdots r_{h-\ell} \in \Lambda^{(h-\ell)}$ of length $(h - \ell)$, the $\mathcal{C}$-color of the union string

$$\mathbf{st}_i \mathbf{w}(x_{\ell+1}; r_1) \mathbf{w}(x_{\ell+2}; r_2) \cdots \mathbf{w}(x_h; r_{h-\ell}) \in \Lambda^d$$

is the *same* as the $\mathcal{C}$-color of the other union string

$$\mathbf{st}_j \mathbf{w}(x_{\ell+1}; r_1) \mathbf{w}(x_{\ell+2}; r_2) \cdots \mathbf{w}(x_h; r_{h-\ell}) \in \Lambda^d.$$

(We recall that $\mathbf{w}(x; i)$ denotes the string obtained from $\mathbf{w}(x)$ by replacing each x by i.) Again the statement is a trivial application of the pigeonhole principle if

$$m_1 \geq k^{n^{m_1 + \cdots + m_{l-1} + (h-1)}}.$$

Indeed, there are $k^{n^{m_1 + \cdots + m_{\ell-1} + (h-\ell)}}$ k-colorings of the set of *all possible* union strings $\mathbf{sr}$ where $\mathbf{s} \in \Lambda^{(d-m_h)}$ is an arbitrary "opening string" and $\mathbf{r} = r_1 r_2 \cdots r_{h-\ell} \in \Lambda^{(h-\ell)}$ is an arbitrary "closing string." On the other hand, there are $m_1 + 1$ special strings

$$1, 1, 1, 1, 1, 1, 1, \ldots, 1$$

$$0, 1, 1, 1, 1, 1, 1, \ldots, 1$$

$$0, 0, 1, 1, 1, 1, 1, \ldots, 1$$

$$0, 0, 0, 1, 1, 1, 1, \ldots, 1$$

$$0, 0, 0, 0, 1, 1, 1, \ldots, 1$$

$$\cdots \cdots \cdots \cdots \cdots$$

$$0, 0, 0, 0, 0, 0, 0, \ldots, 0$$

of length m_ℓ ("middle strings"). Each special string $\mathbf{t} = 0, \ldots, 0, 1, \ldots, 1$ on this list defines a c-coloring of the set of *all possible* union strings $\mathbf{sr}$ where $\mathbf{s} \in \Lambda^{(d-m_h)}$ is an arbitrary "opening string" and $\mathbf{r} = r_1 r_2 \cdots r_{h-\ell} \in \Lambda^{(h-\ell)}$ is an arbitrary "closing string": the color of $\mathbf{sr}$ is the $\mathcal{C}$-color of

$$\mathbf{stw}(x_{\ell+1}; r_1) \mathbf{w}(x_{\ell+2}; r_2) \cdots \mathbf{w}(x_h; r_{h-\ell}) \in \Lambda^d.$$

Therefore, if

$$m_\ell \geq k^{n^{m_1 + \cdots + m_{\ell-1} + (h-\ell)}},$$

then there are two strings (" middle strings") $\mathbf{t}_i = 0 \ldots 01 \ldots 1$ (where the first i symbols are 0) and $\mathbf{t}_j = 0 \ldots 01 \ldots 1$ (where the first j symbols are 0) with $i < j$, each of length m_1, such that the $\mathcal{C}$-color of the union string

$$\mathbf{st}_i \mathbf{w}(x_{\ell+1}; r_1) \mathbf{w}(x_{\ell+2}; r_2) \cdots \mathbf{w}(x_h; r_{h-\ell}) \in \Lambda^d$$

is the *same* as the $\mathcal{C}$-color of the other union string

$$\mathbf{st}_j \mathbf{w}(x_{\ell+1}; r_1) \mathbf{w}(x_{\ell+2}; r_2) \cdots \mathbf{w}(x_h; r_{h-\ell}) \in \Lambda^d$$

for *any* "opening string" $\mathbf{s} \in \Lambda^{(m_1+\ldots+m_{\ell-1})}$ of length $(m_1 + \ldots + m_{\ell-1})$ and *any* "closing string" $\mathbf{r} = r_1 r_2 \cdots r_{h-\ell} \in \Lambda^{(h-\ell)}$ of length $(h - \ell)$.

Note that inequality

$$m_\ell \geq k^{n^{m_1+\ldots+m_{\ell-1}+(h-\ell)}}$$

holds by the definition of the m_is.

The string $\mathbf{w}(x_\ell) = 0 \ldots 0 x_\ell \ldots x_\ell 1 \ldots 1$ of the ℓth sub-interval (of length m_ℓ) of the $HJ^c(n-1, k)$-dimensional *interval sub-cube* of Λ^d is the following: the first part is a string of i 0s, the middle part is a string of $(j - i)$ x_ℓs ("dynamic coordinates"), and the last part is a string of $(m_h - j)$ 1s.

Clearly $\mathbf{w}(x_\ell; 0) = \mathbf{t}_i$ and $\mathbf{w}(x_\ell; 1) = \mathbf{t}_j$, so changing the dynamic coordinates of the ℓth sub-interval simultaneously from 0 to 1 cannot be distinguished by k-coloring C.

At the end of this procedure we obtain an h-parameter string $\mathbf{w}(x_1)\mathbf{w}(x_2) \cdots \mathbf{w}(x_h)$ of length d which defines an h-dimensional interval sub-cube of Λ^d having the following property. Let $\mathcal{I}$ be an arbitrary non-empty subset of $\{1, 2, \ldots, h\}$; then simultaneously changing the dynamic coordinates of all the ith sub-intervals, where $i \in \mathcal{I}$, from 0 to 1 cannot be distinguished by k-coloring C. This is the property referred to when we say that "in this sub-cube symbols 0 and 1 cannot be distinguished by k-coloring C."

The h-dimensional interval sub-cube defined by string $\mathbf{w}(x_1)\mathbf{w}(x_2) \cdots \mathbf{w}(x_h)$ is a $\{0, 1, 2, \ldots, n-1\}^h$ sub-cube. Consider that sub-cube of it where $\Lambda = \{0, 1, 2, \ldots, n-1\}$ is restricted to its $(n-1)$-element subset $\Lambda \setminus \{0\} = \{1, 2, \ldots, n-1\}$. This is a $\{1, 2, \ldots, n-1\}^h$ sub-cube. Since $h = HJ_*(n-1, c)$, this $\{1, 2, \ldots, n-1\}^h$ sub-cube contains a monochromatic $(n-1)$-in-a-row where the dynamic coordinates of this line simultaneously run through $\{1, 2, \ldots, n-1\} = \Lambda \setminus \{0\}$. This monochromatic $(n-1)$-in-a-row extends to a combinatorial line of length n in Λ^d, where the *new* point is that where the dynamic coordinates of the *line* are all 0.

By using the property that "in this sub-cube symbols 0 and 1 cannot be distinguished by k-coloring C," it follows that the C-color of this *new* point (i.e. where the dynamic coordinates of the line are all 0) is the same as the C-color of that point on the monochromatic $(n-1)$-in-a-row where the dynamic coordinates of the line are all 1. Therefore, the whole combinatorial line of length n is *monochromatic* as well.

This proves that

$$HJ^c(n, k) \leq d = m_1 + m_2 + \ldots + m_h.$$

It is trivial from the definition of the m_is that

$$m_1 + m_2 + \ldots + m_h \leq \frac{1}{(n+1)k} g_3(knh),$$

where $g_3(x)$ is the tower function. Since $h = HJ^c(n-1, k)$, by the induction hypothesis $HJ^k(n-1, k) \le \frac{1}{nk} g_4(n+k+1)$ we conclude that

$$HJ^c(n, k) \le \frac{1}{(n+1)k} g_3\left(nk\frac{1}{kn}g_4(n+k+1)\right) = \frac{1}{(n+1)k} g_3(g_4(n+k+1)).$$

By definition $g_3(g_4(n+k+1)) = g_4(n+k+2)$, so

$$HJ^c(n, k) \le \frac{1}{(n+1)k} g_3(g_4(n+k+1)) = \frac{1}{(n+1)k} g_4(n+k+2).$$

This completes Shelah's proof. $\qquad\square$

Appendix C
A formal treatment of Positional Games

Everything that we know about ordinary win in a positional game comes from Strategy Stealing. We owe the reader a truly precise treatment of this remarkable existence argument. Also we make the vague term "exhaustive search" precise by introducing a backtracking algorithm called "backward labeling". We start the formal treatment with a definite terminology (which is common sense anyway).

Terminology of Positional Games. There are some fundamental notions of games which are used in a rather confusing way in everyday language. First, we must distinguish between the abstract concept of a *game*, and the individual *plays* of that game.

In everyday usage, *game* and *play* are often synonyms. *Tennis* is a good example for another kind of confusion. To *play* a *game* of tennis, we have to win two or three sets, and to win a set, we must win six (or seven) *games*; i.e., certain components of the game are again called "games." If the score in a set is 6:6 – a "tie" – then, by a relatively new rule in tennis, the players have to play a "tie-break." We will avoid "tie," and use "draw" instead; "drawing strategy" sounds better than "tie, or tying, strategy."

In our terminology a **game** is simply the set of the rules that describe it. Every particular instance at which the game is played in a particular way from the beginning to the end is a **play**.

The same distinction should be made for the **available moves** and for the **personal moves**. At each stage of a play either player has a set of **available moves** defined by the rules of the game. An *available move* turns the given **position** of the play into one of its possible **options** defined by the rules. A **personal move** is a choice among the *available moves*.

A game can be visualized as a *tree of all possible plays*, where a particular play is represented by a *full branch* of the tree. This is the well-known concept of "game-tree"; We will return to it later in the section.

677

In a Positional Game:

1. there are **two players**: the first player, who starts, and the second player;
2. there are finitely many **positions**, and a particular **starting position**;
3. there are clearly defined rules that specify the **available moves** that either player can make from a given position to its **options**;
4. the players choose their **personal moves** alternately;
5. the rules are such that a sequence of alternating personal moves will always come to an end in a **finite** number of moves, and the "ends" are called **terminal positions**;
6. a complete sequence of alternating personal moves ending with a terminal position is called a **play**;
7. the **outcome** of a play is specified by the rules: it may be a win for the first player, or a win for the second player, or a draw;
8. both players know what is going on ("complete information");
9. there are no chance moves.

A **move** in a Positional Game simply means to claim a previously unselected point of the board.

What is a **position** in a Positional Game? Well, it is not completely obvious, because there are two natural interpretations of the concept. A position in the *broad sense* is called a **partial play**. It is the *sequence* of the alternating moves $x_1, y_1, x_2, y_2, \ldots$ made by the two players up to that point of the play. In other words, a linear ordering, namely the "history" of the play, is involved.

The notion used for **position** throughout this book is the "memoryless" position: the *ordered pair of sets* of the moves $\{x_1, x_2, \ldots\} \cup \{y_1, y_2, \ldots\}$ made by the first and second player up that point of the play ("position in a narrow sense"). This means the present configuration only is consided, and we ignore the way the "present" was developed from the "past" in the course of the play. A *position* in a Positional Game can be interpreted as a 3-coloring of the points of the board: the "colors" are X, O, and ?. X means the point was occupied by the first player, O means the point was occupied by the second player, and ? means "unoccupied yet."

In a Positional Game (i.e. such as the "Tic-Tac-Toe game") the *history* of a position is irrelevant; this is why we can *afford* working with the simpler "memoryless" concept of position. Chess is different: it is a game where the history of a play may become relevant. For example, a well-known "stop rule" says that the play is a draw if the same position ("in the narrow sense") occurs for the third time. This means in Chess we must work with "position in the broader sense."

Given a *partial play* $x_1, y_1, x_2, y_2, \ldots, x_i$ (respectively $x_1, y_1, x_2, y_2, \ldots, x_i, y_i$), we call the position $\{x_1, x_2, \ldots, x_i\}, \{y_1, y_2, \ldots, y_{i-1}\}$ (resp. $\{x_1, x_2, \ldots, x_i\}, \{y_1, y_2, \ldots, y_i\}$) the **end-position** of the partial play. In other words, the end-position is the "memoryless" partial play. Note that different partial plays may have the same *end-position*.

A **terminal position** in a Positional Game means a "halving" 2-coloring of the points of the whole board by X and O (naturally X means the first player and O means the second player). "Halving" means that the two classes either have the same size, or the number of X's is greater by one (depending on the parity of the size of the board). (A terminal position is the end-position of a *complete* play.)

A **drawing (terminal) position** means a position such that every winning set has at least one X and at least one O, i.e. every winning set is blocked by either player.

Let $N = |V|$ denote the size of the board V. What is the total number of *plays*? Well, the answer is simply $N!$ if we use the *Full Play Convention*: the players do not quit and keep playing until the whole board is completely occupied. $N!$ is obvious: every play (i.e. "full play") is a permutation of the board. What is the total number of *partial plays*? The answer is

$$N + N(N-1) + N(N-1)(N-2) + \ldots + N! = \lfloor e \cdot N! \rfloor,$$

that is, the lower integral part of $e \cdot N!$. What is the total number of *positions*? The exact answer is

$$\sum_{m=0}^{N} \binom{N}{m} \binom{m}{\lfloor m/2 \rfloor},$$

but we don't really need the exact answer, the easy upper bound 3^N is a good approximation. Notice that 3^N is the total number of 3-colorings of the board.

Of course, a hypergraph may have a lot of *ad hoc* symmetries. Identifying "isomorphic positions" may lead to reductions in the numbers $N!$, $\lfloor e \cdot N! \rfloor$, 3^N computed above.

Strategy. The main objective of Game Theory is to solve the **strategy problem**: *which* player has a winning (or drawing) strategy, and *how* does an optimal strategy actually look like? So the first question is: "What is a **strategy**?" Well, strategy is a rather sophisticated concept. It is based on the assumption that either player, instead of making each choice of his next personal move as it arises in the course of a play, makes up his mind *in advance* for all possible circumstances, meaning that, if there are more than one legitimate moves which the player can make, then a *strategy* prescribes unambiguously the next move. That is, a *strategy* is a complete plan which specifies what personal moves the player will make in every possible situation.

In the real world, when we are playing a concrete game, we are hardly ever prepared with a complete plan of behavior for all possible situations, i.e. we usually do not have a particular strategy (what we usually have instead is some kind of a vague "tactic"). All that we want is to win a *single* play, and the concept of *strategy* seems unnecessarily "grandiose." In Game Theory, however, *strategy* is the agreed-on primary concept – this idea goes back to J. von Neumann. This approach has some great advantages. It dramatically simplifies the mathematical model of game playing. In the von Neumann model, at the beginning of a play either player makes a single "ultimate move" by choosing his own strategy. If a definite strategy has been adapted by either player, then the two strategies together uniquely determine the entire course of the play, and, consequently, determine the *outcome* of the play. In other words, *strategy* is the primary concept, and the rest, like *play* and the *outcome of a play*, are "derivative concepts." This explains why Game Theory is often called the *Theory of Strategies*.

One more thing: it is important to see that the use of a strategy by no means restricts either player's freedom of action. A strategy does not force the player to make decisions on the basis of less information than there would be available for him in any instance during an actual play.

In a Positional Game, a *strategy* for player P (the first or second player) is a function that associates with every possible play of the opponent a unique counter-play of player P. This procedure is non-anticipative: at a certain stage of a play a strategy assigns the "next move" in such a way that it depends on the previously selected moves only. The precise formal definition is easy, and goes as follows:

Strategy: formal definition. Consider a Positional Game played on a finite hyper-graph $(V, \mathcal{F})$. A **strategy** for the first (resp. second) player formally means a function Str such that the domain of Str is a set of even (resp. odd) length sub-sequences of different elements of the board V, and the range is V. If the moves of the first player are denoted by $x_1, x_2, x_3, \ldots$, and the moves of the second player are $y_1, y_2, y_3, \ldots$, then the ith move x_i (resp. y_i) is determined from the "past" by Str as follows

$$x_i = Str(x_1, y_1, x_2, y_2, \ldots, y_{i-1}) \in V \setminus \{x_1, y_1, x_2, y_2, \ldots, y_{i-1}\}$$

$$\left(\text{resp. } y_i = Str(x_1, y_1, x_2, y_2, \ldots, y_{i-1}, x_i) \in V \setminus \{x_1, y_1, x_2, y_2, \ldots, y_{i-1}, x_i\} \right)$$

defines the ith move of the first (second) player.

In other words, a *strategy* for the *first* (*second*) player is a function that assigns a legal next move to all partial plays of *even* (*odd*) length.

A **winning** (or **drawing**) **strategy** Str for the first player means that in all possible plays where the first player follows Str to find his next move is a win for him (a win or a draw). Formally, each play

$$x_1 = Str(\emptyset), \forall y_1, x_2 = Str(x_1, y_1), \forall y_2, x_3 = Str(x_1, y_1, x_2, y_2), \forall y_3, \ldots, \forall y_{N/2} \tag{C.1}$$

if $N = |V|$ is *even*, and

$$x_1 = Str(\emptyset), \forall y_1, \ldots, \forall y_{(N-1)/2}, x_{(N+1)/2} = Str(x_1, y_1, x_2, y_2, \ldots, y_{(N-1)/2}) \tag{C.2}$$

if N is *odd*, is a win for the first player (a win or a draw).

Similarly, a winning (or drawing) strategy *Str* for the second player means that in all possible plays where the second player uses *Str* to find his next move is a win for him (a win or a draw). Formally, each play

$$\forall x_1, y_1 = Str(x_1), \ldots, \forall x_{N/2}, y_{N/2} = Str(x_1, y_1, x_2, y_2, \ldots, x_{N/2}) \qquad \text{(C.3)}$$

if N is *even*, and

$$\forall x_1, y_1 = Str(x_1), \forall x_2, y_2 = Str(x_1, y_1, x_2), \forall x_3, \ldots, \forall x_{(N+1)/2} \qquad \text{(C.4)}$$

if N is *odd*, is a win for the second player (a win or a draw). In both cases

$$x_i \in V \setminus \{x_1, y_1, x_2, y_2, \ldots, y_{i-1}\} \quad \text{and} \quad y_i \in V \setminus \{x_1, y_1, x_2, y_2, \ldots, y_{i-1}, x_i\} \quad \text{(C.5)}$$

hold for all $i \geq 1$.

The ultimate questions of Game Theory are about strategies, more precisely about **optimal strategies**. *Optimal strategies* are the (1) *winning strategies*, and the (2) *drawing strategies* when winning strategy does not exist. Unfortunately, counting the number of optimal strategies is hopelessly complicated. To count all possible strategies is a much easier task; we can do it as follows.

Again we use the Full Play Convention (the players do not quit and keep playing until the whole board is completely covered, even if the winner is known well before that). This assumption ensures that the structure of the hypergraph becomes irrelevant, and the answer depends on the size $N=|V|$ of the board only. The total number of strategies turns out to be a *doubly exponential* function of N. This quantitative result demonstrates the complexity of the concept of *strategy*; it justifies the common sense intuition that "*strategy* is a deep concept."

We begin the counting with an obvious analogy: *play* ↔ *function* and *strategy* ↔ *operator*. What this means is that a *play* is basically a *function*, and a *strategy* is basically an *operator*. Indeed, like an operator associates a function to a function, a strategy associates with every possible *play* of the opponent a *counter-play* of the player owning the strategy.

This analogy yields the trivial upper bound $N^{eN!}$ on the number of all strategies. Indeed, a strategy associates with every *accessible* partial play a unique next move. With a little bit more effort we can even find the exact answer.

Exact number of all strategies. Consider a concrete play $x_1, y_1, x_2, y_2, x_3, y_3, \ldots$; that is, $x_1, x_2, x_3, \ldots$ are the points claimed by the first player, and $y_1, y_2, y_3, \ldots$ are the points claimed by the second player in this order.

Let *Str* be a strategy of (say) the second player. A strategy for the second player is a function *Str* such that the ith move y_i is determined from the "past" by

$$y_i = Str(x_1, y_1, x_2, y_2, \ldots, y_{i-1}, x_i) \in V \setminus \{x_1, y_1, x_2, y_2, \ldots, y_{i-1}, x_i\}.$$

Therefore, given x_1 (the first move of the first player), *Str* uniquely determines y_1 ($\neq x_1$), the first move of the second player. Similarly, given x_1, x_2, *Str* uniquely determines y_2 ($\notin \{x_1, y_1, x_2\}$). Given x_1, x_2, x_3, *Str* uniquely determines y_3 ($\notin \{x_1, y_1, x_2, y_2, x_3\}$), and so on. This shows that we can write

$$y_i = Str(x_1, y_1, x_2, y_2, \dots, y_{i-1}, x_i) = Str_i(x_1, x_2, \dots, x_i)$$

since $y_1, y_2, \dots, y_{i-1}$ are already determined by $x_1, x_2, \dots, x_{i-1}$. Therefore, *Str* can be considered as a vector

$$Str = (Str_1, Str_2, Str_3, \dots),$$

and call Str_i the ith component of strategy *Str*. By definition, the total number of first components Str_1 is precisely $(N-1)^N$. Similarly, the total number of Str_2s is $(N-3)^{N(N-2)}$, the total number of Str_3s is $(N-5)^{N(N-2)(N-4)}$, and so on.

Two strategies, *str* and *STR*, are different if and only if there is an integer j, $1 \le j \le (N+1)/2$ and a sequence $x_1, x_2, \dots, x_j$ of length j such that

$$str_j(x_1, x_2, \dots, x_j) \ne STR_j(x_1, x_2, \dots, x_j).$$

It follows that the total number of strategies is the product

$$(N-1)^N \cdot (N-3)^{N(N-2)} \cdot (N-5)^{N(N-2)(N-4)} \dots$$

$$= \prod_{i=0}^{\lfloor N/2 \rfloor - 1} (N-1-2i)^{\prod_{j=0}^{i}(N-2j)} = e^{e^{N \log N/2 + O(N)}}. \tag{C.6}$$

This proves that *strategy* is a genuine doubly exponential concept.

In particular, ordinary 3^2 Tic-Tac-Toe has $9! \approx 3.6 \cdot 10^5$ possible *plays* ("Full Play Convention"), the total number of *partial plays* is

$$e \cdot 9! \approx 10^6,$$

the total number of *positions* is

$$\sum_{i=0}^{9} \binom{9}{i} \binom{i}{\lfloor i/2 \rfloor} \approx 7 \cdot 10^3,$$

and finally, the total number of *strategies* is

$$8^9 \cdot 6^{9 \cdot 7} \cdot 4^{9 \cdot 7 \cdot 5} \cdot 2^{9 \cdot 7 \cdot 5 \cdot 3} \approx 10^{500}.$$

To try out all possible strategies in order to find an *optimal* one is extremely impractical.

The next theorem says that every Positional Game is "determined." This is a special case of a more general theorem due to Zermelo [1912].

Theorem C.1 ("Strategy Theorem") *Let $(V, \mathcal{F})$ be an arbitrary finite hypergraph, and consider a Positional Game (normal or Reverse) on this hypergraph. Then there are three alternatives only: either the first player has a winning strategy, or the second player has a winning strategy, or both of them have a drawing strategy.*

Proof. We repeat the argument at the beginning of Section 5 with more rigor. It is a repeated application of De Morgan's law. Indeed, we have the following three possibilities:

(a) *either* the first player (**I**) has a winning strategy;
(b) *or* the second player (**II**) has a winning strategy;
(c) *or* the negation of (a)∨(b).

First assume that $N = |V|$ is *even*. In view of (C.1)–(C.5) case (a) formally means that

$$\exists x_1 \forall y_1 \exists x_2 \forall y_2 \cdots \exists x_{N/2} \forall y_{N/2} \qquad \text{(C.7)}$$

such that **I** wins (the sequence in (C.7) has to satisfy (C.5)).
    Indeed

$$Str(x_1, y_1, x_2, y_2, ..., y_{i-1}) = x_i \in V \setminus \{x_1, y_1, x_2, y_2, ..., y_{i-1}\}$$

defines a winning strategy *Str* for **I**.
    By the De Morgan's law, ¬(a) is equivalent to

$$\forall x_1 \exists y_1 \forall x_2 \exists y_2 \cdots \forall x_{N/2} \exists y_{N/2} \qquad \text{(C.8)}$$

such that **I** loses or it is a draw (the sequence in (C.8) has to satisfy (C.5)). Therefore, ¬(a) means that **II** has a drawing strategy.
    Similarly, ¬(b) means that **I** has a drawing strategy.
    Case (c) is equivalent to ¬(a)∧¬(b), which means that both players have a drawing strategy.
    Finally, we leave the case "N is odd" to the reader. □

The three alternatives of Theorem C.1 are the three possible **outcomes of a game**. Note that basically the same proof works for all finite *combinatorial Games* (to be defined later). This fact is traditionally expressed in the form that "every finite combinatorial game is **determined**"; this is often called *Zermelo's Theorem*. For a *constructive* proof of Theorem C.1, which gives an algorithm to actually *find* an explicit winning or drawing strategy, see Theorem C.3 later in the section (unfortunately the algorithm is impractical).
    In a Positional Game a particular play may end with Second Player's win, but only if First Player "made a mistake." If First Player plays "rationally" (i.e. he uses an optimal strategy), then he cannot lose a Positional Game (see also Theorem 5.1).

Theorem C.2 ("Strategy Stealing") *Let $(V, \mathcal{F})$ be an arbitrary finite hypergraph. Then playing the Positional Game on $(V, \mathcal{F})$, First Player can force at least a draw, i.e. a draw or possibly a win.*

Proof. Assume that Second Player (**II**) has a winning strategy *STR*, and we want to obtain a contradiction. The idea is to see what happens if First Player (**I**) steals and uses *STR*. A winning strategy for a player is a list of instructions telling the player that if the opponent does this, then he does that, so if the player follows the instructions, he will always win. Now **I** can use **II**'s winning strategy *STR* to win as follows. **I** takes an arbitrary first move, and *pretends* to be the second player (he ignores his first move). After **II**'s move, **I**, as a fake second player, reads the instruction in *STR* to take action. If **I** is told to take a move that is still available, he takes it. If this move was taken by him before as his ignored "arbitrary" first move, then he takes another "arbitrary move." The crucial point here is that an extra move, namely the last "arbitrary move," only *benefits* **I** in a Positional Game.

The precise formal execution of this idea is very simple and goes as follows. We use the notation $x_1, x_2, x_3, \ldots$ for the moves of **I**, and $y_1, y_2, y_3, \ldots$ for the moves of **II**. By using **II**'s moves $y_1, y_2, y_3, \ldots$ and **II**'s winning strategy *STR*, we are going to define **I**'s moves $x_1, x_2, x_3, \ldots$ (satisfying (4.5)), and also two auxiliary sequences $z_1, z_2, z_3, \ldots$ and $w_1, w_2, w_3, \ldots$. Let x_1 be an "arbitrary" first move of **I**. Let $w_1 = x_1$ and $z_1 = STR(y_1)$. We distinguish two cases. If $z_1 \neq w_1$, then let $x_2 = z_1$ and $w_2 = w_1$. If $z_1 = w_1$, then let x_2 be another "arbitrary" move, and let $w_2 = x_2$. Next let $z_2 = STR(y_1, z_1, y_2)$. Again we distinguish two cases. If $z_2 \neq w_2$, then let $x_3 = z_2$ and $w_3 = w_2$. If $z_2 = w_2$, then let x_3 be another "arbitrary" move, and let $w_3 = x_3$, and so on. In general, let $z_i = STR(y_1, z_1, y_2, z_2, \ldots, y_i)$. We distinguish two cases: if $z_i \neq w_i$, then let $x_{i+1} = z_i$ and $w_{i+1} = w_i$; and if $z_i = w_i$, then let x_{i+1} be another "arbitrary" move, and let $w_{i+1} = x_{i+1}$.

It follows from the construction that

$$\{x_1, x_2, \ldots, x_i, x_{i+1}\} = \{z_1, z_2, \ldots, z_i\} \cup \{w_{i+1}\} \tag{C.9}$$

for each $i \geq 1$. In view of (C.9) the "virtual play" $y_1, z_1, y_2, z_2, y_3, z_3, \ldots$ is a *legal* one, i.e. it satisfies (C.1)–(C.5). We call the two players of this "virtual play" Mr. Y (who starts) and Mr. Z (of course, Mr. Y is **II**, and Mr. Z is "almost" **I**). The only minor technical difficulty is to see what happens at the end. We consider two cases according to the parity of the board size. The complete "virtual play" between Mr. Y and Mr. Z is

$$y_1, z_1, y_2, z_2, y_3, z_3, \ldots, y_m, z_m, w_{m+1} \tag{C.10}$$

if the board size $|V| = 2m + 1$ is *odd*, and

$$y_1, z_1, y_2, z_2, y_3, z_3, \ldots, y_{m-1}, z_{m-1}, y_m, w_m \tag{C.11}$$

if the board size $|V| = 2m$ is *even*.

We recall that $z_i = STR(y_1, z_1, y_2, z_2, \ldots, z_{i-1}, y_i)$ for each $i \geq 1$. Since *STR* is a winning strategy for the second player, it follows that Mr. Z wins the virtual play (C.10) (i.e. when $|V|$ is odd) even if the last move w_{m+1} belongs to Mr. Y. In view of (C.9) this implies a **I**'s win in the "real play"

$$x_1, y_1, x_2, y_2, x_3, y_3, \ldots, x_m, y_m, x_{m+1}.$$

Similarly, Mr. Z wins the virtual play (C.11) (i.e. when $|V|$ is even) if the last move w_m belongs to him. This implies a **I**'s win in the "real play"

$$x_1, y_1, x_2, y_2, x_3, y_3, \ldots, x_m, y_m.$$

We used the fact that in a positional game an extra point cannot possibly harm **I**. This is how **I** "steals" **II**'s winning strategy *STR*.

The conclusion is that if **II** had a winning strategy, so would **I**. But it is clearly impossible that both players have a winning strategy. Indeed, if either player follows his own winning strategy, then the play has two winners, which is a contradiction. So the supposed winning strategy for **II** cannot exist. This implies that **I** can always force at least a draw. □

The main interest of Game Theory is to *solve* games, i.e. to determine which player has a winning or drawing strategy. In terms of *plays*, it means to predict the outcome of a play between *perfect* players.

In Game Theory we always *assume* that the players are perfect (gods like Pallas Athena and Apollo); what it means is that either player knows his/her *optimal* strategy (or at least one of them if there are several optimal strategies) even if this knowledge requires "supernatural powers."

Real world game playing is of course totally different. The players are anything but perfect, and make "mistakes" all the time. For example, when real world players are playing a Positional Game, it often happens that Second Player ends up winning a play by taking advantage of First Player's mistake(s). But, of course, an incident like this does not contradict Theorem C.2.

There is an important point that we have to clarify here. When we talk about "perfect players," it does not mean that the concept of winning (or drawing) strategy *presupposes* "rational" behavior of the opponent. Not at all. A winning (drawing) strategy for player P simply means that player P can always force a win (a draw or possibly a win), no matter what his opponent is doing. If player P has a winning strategy and follows the instructions of his winning strategy, it does not make any difference that the opponent is perfect or foolish, player P will win anyway. On the other hand, if player P has a drawing strategy, then the opponent can make a difference: by "making mistakes" the opponent may lose a potential draw-game, and player P may have a win after all.

Strategy Stealing may reveal who wins, but it does not give the slightest clue of how to actually *find* the existing drawing or winning strategy in a Positional Game. To *find* an explicit drawing or winning strategy we often have no choice but need to make an *exhaustive search*. For example, consider the $(K_{18}, K_4, -)$ Reverse Clique Game. In the normal version, First Player has a winning strategy (but we do not know how he wins); in the Reverse version one of the players must have a winning strategy, but we don't even know which one. To determine which player wins we need to make an exhaustive search. What is an *exhaustive search*? A trivial approach is to try out all possible strategies. Since the total number of strategies is doubly exponential, this is extremely impractical. A much faster way to do it is to search through all positions (instead of all strategies). A systematic way to search through all positions is now described. We begin with the concept of:

Game-Tree. Every Positional Game can be visualized as a "tree of all possible plays"; we call it the **game-tree**. The *game-tree* is a *labeled rooted directed tree*. The vertices represent the *partial plays*, the root is the starting position, and a directed edge goes from a partial play to one of the options of the end-position. The full branches of the *game-tree* represent the plays. The terminal positions ("leaves") are labeled by **I**, **II**, and **D** according to the outcome of the concrete play ending at that terminal position. **I** if it is a first player win, **II** if it is a second player win, and **D** if it is a draw.

Figure C.1 shows the game-tree of the 2^2-game.

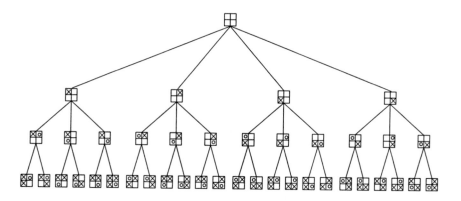

A good way to visualize a *strategy* is to look at it as a special sub-tree of the *game-tree*. Note that every vertex of the game-tree has a well-defined *distance* from the root. If this distance is *even*, then the first player moves next; if this distance is *odd*, then the second player moves next. The following is an alternative way to define *strategy*.

Strategy: an alternative definition. A *strategy* for the *first (second)* player is a sub-tree T' of the *game-tree* T such that:

(1) the root belongs to T';
(2) in T' each even-distance (odd-distance) vertex has out-degree 1;
(3) in T' each odd-distance (even-distance) vertex has the same out-degree as in the whole game-tree T.

A *drawing* or *winning* strategy for the *first (second)* player is a sub-tree T' of the game-tree T such that it satisfies (1)–(2)–(3) and each *leave* of T' is labeled by **D** or **I**, or **I** only (**D** or **II**, or **II** only). Figure C.2 shows a First Player's winning strategy in the 2^2-game.

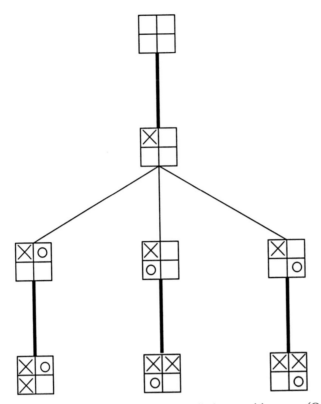

The next picture is the beginning of Qubic's distinct-position tree (Qubic is the $4 \times 4 \times 4$ version of Tic-Tac-Toe).

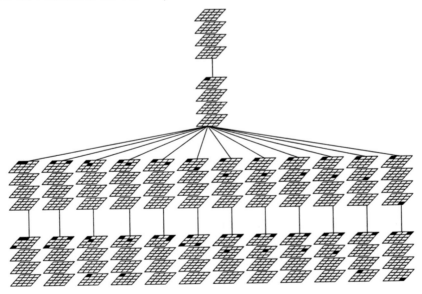

The picture represents levels 0,1,2, and 3. There are only 7 distinct 3-moves, reduced from 12 2-positions by choosing first player's optimal responses for these 12 positions.

Position-graph. Every Positional Game (normal or Reverse) has a **position-graph**. The *position-graph* is a *labeled rooted directed graph*. The vertices are the *positions*, the root is the starting position, and the directed edges go from a position to its options. The terminal positions ("leaves") are labeled by **I**, **II**, and **D** according to the outcome of the concrete play ending at that terminal position. **I** if it is a first player's win, **II** if it is a second player's win, and **D** if it is a draw.

Every vertex of the position-graph has a well-defined *distance* from the root: it is the number of moves made by the two players together. If this distance is *even* (*odd*), then the first (second) player moves next.

Memoryless strategy. A **memoryless strategy** for the *first* (*second*) player is a sub-graph **G′** of the *position-graph* **G** such that:

(i) the root belongs to **G′**;
(ii) in **G′** each even-distance (odd-distance) vertex has out-degree 1;
(iii) in **G′** each odd-distance (even-distance) vertex has the same out-degree as in the whole position-graph **G**.

A *drawing* or *winning* **memoryless strategy** for the *first* (*second*) player is a subgraph **G′** of the position-graph **G** such that it satisfies (i)-(ii)-(iii) and each *leave* (i.e. vertex with zero out-degree) of **G′** is labeled by **D** or **I**, or **I** only (**D** or **II**, or **II** only).

Note that a *strategy Str* is not necessarily a *memoryless strategy*. Indeed, let *pp* and *PP* be two distinct *partial plays* which have the same *end-position Pos*; then strategy *Str* may associate *different* next moves to *pp* and *PP*. If *Str* is an *optimal* strategy, then there is a simple way to reduce *Str* into an *optimal memoryless* strategy. Assume that *Str* is a (say) winning strategy for (say) the first player, but it is *not* a memoryless strategy. Then there exists a "violator," i.e. there exist two distinct partial plays *pp* and *PP* which:

(1) have the same end-position *Pos*;
(2) it is the first player's turn to move in *Pos*; and, finally,
(3) *Str* associates different next moves to *pp* and *PP*.

Let **T** be the game-tree. Note that *Str* is a special sub-tree of **T**. Let **T**(*pp*) and **T**(*PP*) denote, respectively, the labeled directed sub-trees of **T** which are rooted at *pp* and *PP*. By property (1) these two labeled sub-trees are identical (isomorphic). On the other hand, by property (3) *Str* ∩ **T**(*pp*) and *Str* ∩ **T**(*PP*) are *different* "sub-strategies." Replacing sub-tree *Str* ∩ **T**(*PP*) of *Str* with *Str* ∩

$T(pp)$, we obtain a winning strategy Str' of the first player with fewer number of "violators." Repeating this argument we have a sequence $Str, Str', Str'', \ldots$ of winning strategies with a strictly decreasing number of "violators," terminating (in a finite number of steps) in a *memoryless* winning strategy.

Note that the number of *memoryless strategies* is substantially less than the number of strategies (as always we use the Full Play Convention that the players do not quit and play until the whole board is completely covered). Indeed, by (C.6) the number of strategies is around

$$e^{e^{N \log N/2 + O(N)}} \tag{C.12}$$

where $N = |V|$ is the size of the board. On the other hand, since a memoryless strategy associates with every *accessible* position a unique next move, the number of memoryless strategies is less than

$$N^{3^N} = e^{3^N \log N} = e^{e^{O(N)}},$$

which is much less than (C.12).

A trivial *lower bound* on the number of memoryless strategies is

$$\left(\frac{N}{3} \right)^{2^{N/3}}. \tag{C.13}$$

Indeed, consider a concrete play $x_1, y_1, x_2, y_2, x_3, y_3, \ldots$, that is $x_1, x_2, x_3, \ldots$ are the points claimed by the first player, and $y_1, y_2, y_3, \ldots$ are the points claimed by the second player in this order. Let $V = V_1 \cup V_2$ be a decomposition of the board such that $|V_1| = N/3$ and $|V_2| = 2N/3$. Let Str be a *memoryless strategy* of (say) the second player. A memoryless strategy for the second player is a function which associates with an arbitrary accessible position $\{x_1, x_2, \ldots, x_i\} \cup \{y_1, y_2, \ldots, y_{i-1}\}$ ($i \geq 1$) a next move y_i such that $y_i \in V \setminus \{x_1, y_1, x_2, y_2, \ldots, y_{i-1}, x_i\}$. We follow the extra rule that if $\{x_1, x_2, \ldots, x_i\} \subseteq V_1$, then $y_i \in V_2$. Under this rule memoryless strategy Str associates with every subset $\{x_1, x_2, \ldots, x_i\} \subseteq V_1$ a next move $y_i \in V_2 \setminus \{y_1, y_2, \ldots, y_{i-1}\}$ where $|V_2 \setminus \{y_1, y_2, \ldots, y_{i-1}\}| \geq N/3$. This proves (C.13). (C.13) shows that the number of memoryless strategies is still *doubly* exponential.

The ultimate questions of game theory are "which player has a winning (or drawing) strategy" and "how to find one." From now on we can assume that *strategy* always means a *memoryless strategy*.

A simple linear time algorithm – the so-called "backward labeling" of the position-graph – settles both ultimate questions for any Positional Game (normal and Reverse). This algorithm gives an optimal memoryless strategy for either player in linear time: linear in size of the position-graph. What is more, the "backward labeling" provides the "complete analysis" of a Positional Game. Note that the position-graph is typically "exponentially large"; this is why the "backward labeling algorithm" is usually impractical. (To try out all possible strategies is much, much worse: the total number of strategies is a *doubly* exponential function of the size of the board.) Note that "backward labeling" applies for a larger class of games called *Combinatorial Games*.

Combinatorial Games. How to define *Combinatorial Games*? Here is a list of natural requirements:

1. There are two players: the first player (or **I**, or White, etc.) who starts the game, and the second player (or **II**, or Black, etc.).
2. There are finitely many **positions**, and a particular **starting position**.
3. There are clearly defined rules that specify the **moves** that either player can make from a given position to its **options**.
4. The players move alternately.
5. The rules are such that a sequence of alternating moves will always come to an end in a **finite** number of moves. The "ends" are called **terminal positions**.
6. A complete sequence of alternating moves ending with a terminal position is called a **play**.
7. The **outcome** of a play is specified by the rules. It may be a win for the first player, or a win for the second player, or a draw.
8. Both players know what is going on (complete information).
9. There are no chance moves.

Note that the 5th requirement prohibits perpetual draws, i.e. we exclude infinite "loopy" games.

Of course, *Combinatorial Games* must contain the class of Positional Games. With every Positional Game (normal and Reverse) we can associate its game-tree. The standard way to define an "abstract" combinatorial game is to reverse this process: an "abstract" combinatorial game is a "coin-pushing game" on a labeled rooted directed tree.

Definition of Combinatorial Games. A (finite) **combinatorial game** $\Gamma = (T, F)$ means a coin-pushing game on a finite rooted directed tree T. The root of T has in-degree zero, and any other vertex has in-degree exactly one. At the beginning of a play a coin is placed in the root ("starting position"). At each move the players alternately push the coin along a directed edge of tree T. F denotes the outcome-function ("labeling") which associates with each leaf (i.e. vertex with zero out-degree) of tree T a label. The three possible labels are **I**, **II**, and **D**.

Note that a vertex of the tree represents a "partial play," a leaf represents a "terminal position", and the unique full branch ending at that leaf represents a "play." Label **I** "means" that the particular play ending at that leaf is a first player win, **II** "means" a second player win, and **D** "means" a draw. We can identify a **combinatorial game** $\Gamma = (T, F)$ with its own game-tree.

We can easily define a *strategy* in a combinatorial game $\Gamma = (T, F)$ as a special sub-tree of T.

A Positional Game can be represented as a coin-pushing game on its own game-tree. A more "economic" representation is a coin-pushing game on its own *position-graph*. This representation belongs to the class of *Linear-Graph Games* (to be defined below). This representation implies a (rather irrelevant) formal restriction in the concept of *strategy*: it restricts strategies to *memoryless strategies*.

The position-graph of a Positional Game (without the labeling) belongs to the class of *rooted linear digraphs*. A *digraph* (i.e. *directed graph*) is a *linear digraph* if its vertex set has a partition $V_0, V_1, V_2, \ldots, V_k$ ($k \geq 1$) into non-empty sets such that:

(1) for each $v \in V_i$ with $1 \leq i \leq k$ there is a $u \in V_{i-1}$ such that the directed edge $u \rightarrow v$, which goes from vertex u to vertex v, belongs to the linear digraph,

(2) if $u \rightarrow v$ is a directed edge of the linear digraph, then there is an i with $1 \leq i \leq k$ such that $u \in V_{i-1}$ and $v \in V_i$.

A *rooted* linear digraph means that the first vertex-class V_0 consists of a single vertex called the root. Note that a rooted linear digraph is always *connected* as an ordinary graph, i.e. when we ignore the directions. In a *rooted linear digraph* every vertex has a well-defined *distance* from the root: if $u \in V_i$, then the distance of vertex u from the root is i.

Linear-graph game. A (finite) **linear-graph game** $\Gamma = (G, F)$ means a coin-pushing game on a finite rooted linear digraph G. At the beginning of a play a coin is placed in the root ("starting position"). At each move the players alternately push the coin along a directed edge of digraph G. F denotes the outcome-function ("labeling") which associates with each leaf (i.e. vertex with zero out-degree) of digraph G a label. The three possible labels are **I**, **II**, and **D**.

Note that a vertex of the digraph represents a "position," a leaf represents a "terminal position," label **I** "means" that the terminal position is a first player win, **II** "means" a second player win, and **D** "means" a draw.

By following We leave to the reader to define a *strategy* in a linear-graph game $\Gamma = (G, F)$ as a special sub-graph of G.

Since a rooted directed tree is a rooted linear digraph, the class of *Combinatorial Games* is a sub-class of the class of *linear-graph games*. "Backward labeling" (see below) will give a precise meaning to the vague expression "exhaustive search"; "backward labeling" applies for the whole class of *Linear-Graph Games*.

Complete Analysis. "Backward labeling" is an algorithm for extending the **I, II, D**-labeling F of the terminal positions in a linear-graph (or combinatorial) game Γ to *all* positions. The resulting label of the starting position answers the first ultimate question "who wins?".

Let $\Gamma = (G, F)$ be an arbitrary linear-graph game, where $G = (V, E)$ is a finite rooted linear digraph, V is the vertex-set, and E is the edge-set. Let $Z \subset V$ be the set of terminal positions (i.e. vertices with zero out-degree). The outcome-function F cuts Z into 3 parts ("I wins," "II wins," "draw")

$$Z_I = \{u \in Z : F(u) = I\},$$

$$Z_{II} = \{u \in Z : F(u) = II\},$$

$$Z_D = \{u \in Z : F(u) = D\}.$$

We are going to extend this partition $Z = Z_I \cup Z_{II} \cup Z_D$ into a partition of $V = V_I \cup V_{II} \cup V_D$ satisfying $V_I \supseteq Z_I$, $V_{II} \supseteq Z_{II}$ and $V_D \supseteq Z_D$.

Let $u \in V$ be an arbitrary position (i.e. vertex). Its options are

$$\text{Options}(u) = \{v \in V : u \to v \in E\}.$$

By definition, all terminal positions are labeled by **I**, **II**, and **D**. In a *rooted linear digraph* every vertex has a well-defined *distance* from the root. If the distance is *even* (*odd*), then the first (second) player moves next.

We use the natural rules (1)-(2)-(3) and (1')-(2')-(3') below to extend the **I,II,D**-labeling F *backward* from Z to the whole vertex-set V in $O(|E|)$ steps ("*iff*" stands for *if and only if*):

(1) $u \in V_I$ iff $\text{Options}(u) \cap V_I \neq \emptyset$,
(2) $u \in V_D$ iff $\text{Options}(u) \cap V_I = \emptyset$ and $\text{Options}(u) \cap V_D \neq \emptyset$,
(3) $u \in V_{II}$ iff $\text{Options}(u) \subseteq V_{II}$,

if vertex u has an *even* distance from the root; and

(1') $u \in V_{II}$ iff $\text{Options}(u) \cap V_{II} \neq \emptyset$,
(2') $u \in V_D$ iff $\text{Options}(u) \cap V_{II} = \emptyset$ and $\text{Options}(u) \cap V_D \neq \emptyset$,
(3') $u \in V_I$ iff $\text{Options}(u) \subseteq V_I$,

if vertex u has an *odd* distance from the root.

More precisely, let $s \in V$ denote the starting position. Let $V_0 = \{s\}$, $V_1 = \text{Options}(s)$, and in general

$$V_{i+1} = \bigcup_{u \in V_i} \text{Options}(u).$$

$V_0, V_1, V_2, \ldots$ is a partition of V. Let m be the *largest* index such that V_m is *not* contained by set Z. Then depending on the *parity* of m we use rules (1)-(2)-(3) if m is *even* and rules $(1') - (2') - (3')$ if m is *odd* to extend the **I,II,D**-labeling to V_m, then to V_{m-1}, then to V_{m-2}, and so on back to V_0.

Let $\overline{F}$ denote this extension of labeling F from the set of the terminal positions Z to the whole V. We call $\overline{\Gamma} = (G, \overline{F})$ the **complete analysis** of game $\Gamma = (G, F)$. The reason why we call $\overline{\Gamma}$ the **complete analysis** is that it answers the two ultimate questions of "who wins" and "how to win." Rules $(1),(2),(3),(1'), (2'), (3')$ guarantee that if a position is labeled **I** (**II**) then the first (second) player can force a win in the *sub-game* starting from this particular position. Label **D** means that both players can maintain a draw in this *sub-game*. If the position is the *starting position*, then the sub-game is the whole game Γ. In other words, the label of the starting position answers the question of "who wins."

The next question is "how to win." The *complete analysis* $\overline{\Gamma}$ provides an explicit way to *find* either player's optimal strategy in game Γ.

(i) If the starting position is labeled by **I** (i.e. the first player has a winning strategy), then the first player's explicit winning strategy is to keep moving to any option with label **I** (he can always do that).

(ii) If the starting position is labeled by **II**, then the second player's explicit winning strategy is to keep moving to any option with label **II**.

(iii) If the starting position is labeled by **D**, then the first (second) player's explicit drawing strategy is to avoid **II** (**I**).

Note that the complete analysis carries much more information than merely answering the two ultimate questions. Indeed, by using the complete analysis we can easily obtain the *complete list* of all possible optimal strategies in a linear-graph game (though this list is usually *absurdly* long).

Furthermore, the complete analysis $\overline{\Gamma}$ of Γ answers the two ultimate questions in every *sub-game* where the players start from an *arbitrary* position, not necessarily the starting position.

For example, in Section 3 we gave an explicit Second Player's Drawing Strategy in ordinary Tic-Tac-Toe. But an optimal strategy is far less than the Complete Analysis; an optimal strategy does *not* solve, among many others, the following Tic-Tac-Toe puzzle: "What is Second Player's (i.e. Os) optimal reply in the position on Picture 1?". Note that Second Player can force a draw here, but the five moves, marked by ?, all lead to Second Player's loss (why?). (That is, First Player gives the opponent several "chances" to lose; this is all what we need against a poor player.)

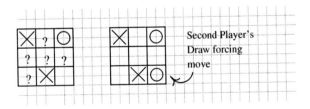

Note that the position on in the figure cannot show up if Second Player follows his Section 3 Drawing Strategy. Of course, the Complete Analysis of ordinary Tic-Tac-Toe solves this particular puzzle, and all possible puzzles like that.

Here is a game where it is easy to find a winning strategy but the Complete Analysis remains a mystery. In the following *Domino Game* the players take turns placing a domino everywhere on a rectangular board. Each domino must be put down flat, within the border of the rectangle and without moving a previously placed piece. There are more than enough dominoes to cover the whole rectangle. The player who puts down the last piece is the winner. A play cannot end in a draw, so which player has a winning strategy? Of course, First Player can force a win: his opening move is to place the first domino exactly at the center of the board, and thereafter to copycat his opponent's play by placing symmetrically opposite. It is obvious that whenever Second Player finds an open spot, there will always be an open spot to pair with.

Here comes the twist: if First Player's opening move is *not* the center, then we do *not* know whether or not he can still force a win.

A similar example is "The-longer-the-better Arithmetic Progression Game" played on the interval $\{1, 2, \ldots, 2N\}$ of even size. The players alternately select previously unselected integers from the interval. The winner is the player whose longest arithmetic progression at the end of the play is *longer* than the opponent's longest progression; in case of equality the play is a draw. The mirror image pairing $\{i, N+1-i\}$, $i = 1, 2, \ldots, N$ is obviously a drawing strategy for Second Player. However, if Second Player's first move is *not* the reflection of the opponent's opening move, then again we do *not* know whether or not he can still force a draw. In these two games it was very easy to find an optimal strategy, but the Complete Analysis is not known.

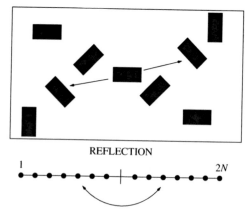

REFLECTION

We summarize the properties of the "backward labeling" in the following general statement.

Theorem C.3 ("Backward Labeling") *Let* $\Gamma = (G, F)$ *be a linear-graph game. The Backward Labeling algorithm determines the complete analysis of* Γ *– in particular, it finds the outcome of the game, and provides either player's optimal strategy – in computing time* $O(|G|)$*, i.e. linear in size of the graph G.*

In particular, for a Positional Game $(V, \mathcal{F})$ *the computing time is linear in size of the position-graph, i.e.* $O(N \cdot 3^N)$ *where* $N = |V|$*.*

In the special case of Combinatorial Games $\Gamma = (T, F)$ *the computing time is linear in size of the game-tree T.*

Theorem C.3 seems to be *folklore*; a variant was published in the classical book of von-Neumann and Morgenstern [1944].

The Main Problem of Positional Game Theory. Consider a Positional Game (normal or Reverse): if the size of the board V is N, then the size of the game-tree is clearly $O(N!)$. Note that $O(N \cdot 3^N)$ is substantially less than $O(N!)$; this saving – $O(N \cdot 3^N)$ instead of $O(N!)$ – was the reason to introduce the concept of Linear-Graph Games.

The Backward Labeling algorithm describes the winner and provides an explicit winning or drawing strategy in *linear time*. In particular, for Positional Games the running time is $O(N \cdot 3^N)$.

Backward Labeling answers the ultimate questions of Game Theory. This is great, but there are several reasons why we are still "unhappy." First, the Backward Labeling is hardly more than mindless computation. It lacks any kind of "understanding."

Second, a 3^N-step algorithm is impractical. To perform 3^N operations is far beyond the capacity of the fastest computers, even for relatively small board-size like (say) $N = 100$. The laws of physics suggest a universal speed limit which is much smaller than (say) 3^{100} operations per second (or minute, or hour, or year, it does not really matter). Right now 10^{10} operations per second is about what the fastest computers are capable of doing.

This means that, unless we find some substantial shortcut, for a game with board-size ≥ 100 Backward Labeling is intractable. Human brain can sometime diagnose shortcuts, but we cannot expect substantial shortcuts too often, certainly not for the *whole* class of Positional Games. In other words, even small Positional Games are so *complex* that it is possible that the existing winning or drawing strategies may never be found. (In general we cannot expect too many *symmetries*, and it is not clear how to make use of *sequences of forced moves*.)

Is there any escape from the "trap" of *exponential running time*? At first sight, *exponential running time* seems inevitable: a *complete analysis* has to describe a unique next move in every possible position, and in a Positional Game the total number of positions is around 3^N (i.e. exponential). But in a particular play the players face no more than $N = |V|$ positions, and the game is described by a hypergraph $\mathcal{F}$. What we are really interested in is *not* the complete analysis, but to achieve the optimal outcome *in every particular play*. The real question is: can we implement an optimal strategy in *polynomial* time – polynomial in terms of $O(|V|+|\mathcal{F}|)$ where $(V, \mathcal{F})$ is the given hypergraph – *in every single play?*

Now we understand why the Backward Labeling, although theoretically very important, cannot be the last word of the theory. The "catch" of Theorem C.3 is that Backward Labeling answers the two ultimate questions – "who wins" and "how to win" – in the fastest way in terms of the position-graph (or game-tree), but the position-graph (or game-tree) is an **inefficient** way to describe a particular play.

For example, if the Erdős–Selfridge Strong Draw criterion applies: $\mathcal{F}$ is n-uniform and $|\mathcal{F}| < 2^{n-1}$, then the power-of-two scoring system provides an explicit drawing strategy for both players. The computational complexity ("running time") of the power-of-two scoring system is **polynomial** in terms of $O(|V|+|\mathcal{F}|)$ *for every single play.*

Similarly, every other theorem in this book describes classes of Positional Games (i.e. hypergraphs) for which we know the outcome ("who wins"), and the proof supplies an *explicit* strategy for either player (with the exception of those using Strategy Stealing). The implementation of these explicit strategies is always *polynomial* in the sense that the necessary computation is polynomial in terms of $O(|V|+|\mathcal{F}|)$ *for every single play.*

Can we do it in general? This is a major open problem.

Open Problem C.1 *Consider the class of all Positional Games.*

(a) *Does there exist an algorithm that determines the outcome of the game on an arbitrary hypergraph $(V, \mathcal{F})$ in* **polynomial** *time,* **polynomial** *in terms of* $O(|V|+|\mathcal{F}|)$?

(b) *Does there exist an algorithm that determines an optimal strategy for either player, and* **for every single play** *the computational complexity of the implementation of this optimal strategy is* **polynomial** *in terms of* $O(|V|+|\mathcal{F}|)$?

In view of the famous conjecture P$\neq$NP (where P and NP are the well-known complexity classes), it seems to be hopeless to give a positive solution to Open Problem C.1. We conclude, therefore, that Theorem C.3 is not the end, but rather the starting point of the theory. We consider Backward Labeling the "worst case scenario," and the main problem of Positional Game Theory is to describe large classes of games (i.e., hypergraphs) for which there is a much more efficient way to answer "who wins" and "how to win."

A complementary "negative" approach is to try to prove that some classes of Positional Games are *hard* in the sense that every *algorithm*, which determines the "outcome" (or an "optimal strategy") for *all* games in the class, must necessarily take as much computations as the Backward Labeling, or at least a *very large* amount of computations. Unfortunately no one knows how to prove unconditional computational complexity lower bounds for "natural games," only for "artificial games," so the "negative" approach is quite unexplored yet. For a survey paper about games and computational complexity, see Fraenkel [2005].

Simulation. Positional Games form a very narrow sub-class of Combinatorial Games. Indeed, Combinatorial Games like Chess and Checkers are "dynamic" in the sense that the players can repeatedly *relocate* and also *remove* pieces from the board. On the other hand, Positional Games are "static" in the sense that the players are not allowed to relocate or remove their marks. It is rather surprising, therefore, that the sub-class of Positional Games is, in fact, *universal*: every combinatorial game can be *simulated* by a Positional Game, and the simulation can be described by a fairly simple general construction.

First, however, we have to overcome an obvious obstacle: in a Positional Game Second Player cannot have a winning strategy. To resolve this difficulty, we introduce the concept of **Snub** Games. Every combinatorial game can be played in the **Snub** way: it just means a single *premove*. First player decides whether he wants to keep his role (to be the first player) or he wants to switch role and become the second player.

Snub Games are in favor of the first player because the second player cannot have a winning strategy. This peculiar resemblance to Positional Games suggests that they are closely related. And so they are: every Snub Combinatorial Game is "equivalent" to a Positional Game. This surprising result is due to Ajtai, Csirmaz, and Nagy [1979].

What does "equivalent" mean? Well, two Combinatorial Games are *strictly equivalent* if the labeled game-trees are isomorphic in the standard graph-theoretic sense. This implies that finding the *complete analysis* in two strictly equivalent games are *identical* problems. But what we are really interested in is *not* the complete analysis: all what we want is a single Optimal (winning or drawing) Strategy. So it is reasonable to consider two games *equivalent* if any winning or drawing strategy in either game can be converted into a winning or drawing strategy of the other game in "linear time". We stop here and refer the reader to the paper of Ajtai, Csirmaz and Nagy [1979].

Undetermined Games Exist! Let us return to the Backward Labeling algorithm. An important byproduct is that every finite combinatorial game is *determined*. The proof of Theorem C.1 gave a pure *existence argument*, and Backward Labeling (Theorem C.3) supplied an *algorithm*.

Determined means that either player has a winning strategy, or both of them have a drawing strategy. These three alternatives are the possible **outcomes** of a combinatorial game. Formally, we distinguish this from the concept of the outcome of a *perfect play*, which means the outcome of a particular *play* where either player knows and follows one of his optimal strategies. There is, however, a natural connection between these two concepts. If the *outcome of a game* is that the first player (second player, neither player) has a winning strategy, then the *outcome of a perfect play* in the same game is a first player win (second player win, draw), and vice versa. So the loosely stated first ultimate question "who wins a game" more precisely means "who has a winning strategy", or equivalently, "who wins a perfect play." It is a completely different question whether real-world players know an optimal strategy or not. In advanced games like Chess and Go nobody knows an optimal strategy, and exactly this *ignorance* keeps the game alive for competitive play ("ignorance is fun")!

If we extend the concept of Combinatorial Games to Infinite Games, then the Backward Labeling argument obviously breaks down. If a proof breaks down, it doesn't mean that the theorem itself is necessarily false. But it *is* false: there exist infinite Combinatorial Games which are *undetermined!* This is a striking result, a "paradox" in the same league as the more well-known Banach-Tarski "paradox" of *doubling the ball*.

Banach-Mazur Game. The first example of an *undetermined* game was invented by the Polish mathematician S. Mazur around 1928. Let A be an arbitrary subset of the closed unit interval $[0,1]$, and let B be the complementary set $B = [0, 1] \setminus A$. The (A, B)-game is played by two players: player A and player B, who move alternately. Player A begins the play by choosing an arbitrary closed sub-interval $I_1 \subset [0, 1]$, next player B chooses a closed sub-interval $I_2 \subset I_1$, then again player A chooses a closed sub-interval $I_3 \subset I_2$, then player B chooses a closed sub-interval $I_4 \subset I_3$, and so on. At the end of a play the two players determine a nested infinite sequence of closed intervals I_n, $n = 1, 2, 3, \ldots$; player A chooses those with odd indices, and player B chooses those with even indices. If the intersection set $\bigcap_{n=1}^{\infty} I_n$ has at least one common point with set A, then player A wins; otherwise player B wins.

Banach and Mazur proved the following perfect characterization of the (A, B)-game.

Exercise C.1

(a) *Playing the (A, B)-game Player B has a winning strategy if and only if A is a first category set (i.e. a countable union of nowhere dense sets).*

(b) *On the other hand, player A has a winning strategy in the (A, B)-game if and only if $B \cap I_1$ is a first category set for some proper sub-interval $I_1 \subset [0, 1]$.*

Is it possible that neither of these two criterions hold? Yes! By using transfinite induction we can prove the *existence* of a subset S of the real numbers such that both S and its complement intersect all uncountable closed subsets of the reals. Such a set S is called a *Bernstein Set*.

Exercise C.2 *Prove that Bernstein Set exists.*

Now if A is the intersection of the unit interval $[0,1]$ with a Bernstein set S, then the Banach–Mazur (A, B)-game is clearly *undetermined*.

Gale–Stewart game. More than 20 years after Mazur, in 1953, D. Gale and F. M. Stewart discovered a much simpler undetermined game, the so-called *Infinite 0–1 Game*. The fact that the *Infinite 0–1 Game* is undetermined can be proved by simple "cardinality considerations" instead of the more delicate topological proof of the Banach–Mazur Theorem. Next we discuss the *Infinite 0–1 Game* in detail.

Let $A \subset \{0, 1\}^\omega$ be an arbitrary subset, and $B = \{0, 1\}^\omega \setminus A$. The game-tree of the Gale–Stewart Infinite 0-1 Game is the infinite binary tree of height ω (the first infinite ordinal number), and the labeling is determined by sets A and B. The coin-pushing game on the infinite labeled binary tree of height ω means the following. The two players, player A and player B, alternately say 0 or 1, which at the end gives an infinite 0–1 sequence. If this infinite 0–1 sequence belongs to set $A \subset \{0, 1\}^\omega$, then player A wins; otherwise player B wins.

Theorem C.4 (Gale and Stewart) *There exists a set $A \subset \{0, 1\}^\omega$ such that the corresponding Infinite 0–1 Game is undetermined.*

Proof. A strategy for player A or B is a rule that specifies the next move. It is a function F which associates with every finite 0–1 sequence $(a_1, a_2, a_3, \ldots, a_{2n+\delta})$, $a_i = 0$ or 1, $1 \le i \le 2n + \delta$, the next move

$$a_{2n+1+\delta} = F(a_1, a_2, a_3, \ldots, a_{2n+\delta}),$$

where $\delta = 0$ for player A and 1 for player B, and $n \ge 0$. Let STR_A and STR_B be, respectively, the set of all strategies for players A and B.

Let $F \in STR_A$ be a strategy of player A and $G \in STR_B$ be a strategy of player B. If both players follow their strategies, then we get a unique *play*. This means an *infinite 0–1 sequence* that we denote by $\langle F, G \rangle$.

Lemma 1:

(i) *Both sets STR_A and STR_B have cardinality $2^{\aleph_0}$.*

(ii) *For each $F \in STR_A$ the cardinality of the set of plays (i.e. infinite 0–1 sequences) $\mathcal{P}_F = \{\langle F, G \rangle : G \in STR_B\}$ is $2^{\aleph_0}$. Similarly, for each $G \in STR_B$ the cardinality of the set of plays $\mathcal{P}_G = \{\langle F, G \rangle : F \in STR_A\}$ is $2^{\aleph_0}$.*

Proof.

(i) Let $\{0, 1\}^{<\omega}$ denote the set of all finite 0-1 sequences. Set $\{0, 1\}^{<\omega}$ is clearly countable. Let $\mathcal{F}$ be the set of all functions $f : \{0, 1\}^{<\omega} \to \{0, 1\}$. The cardinality of $\mathcal{F}$ is $2^{\aleph_0}$. We can associate with every $f \in \mathcal{F}$ a strategy $F \in STR_A$ of player A as follows

$$a_{2n+1} = F(a_1, a_2, a_3, \ldots, a_{2n}) = f(a_2, a_4, \ldots, a_{2n}).$$

Since different functions $f \in \mathcal{F}$ define different strategies $F \in STR_A$, the cardinality of STR_A is the same as the cardinality of $\mathcal{F}$, namely $2^{\aleph_0}$. The same argument works for STR_B.

(ii) Let $\mathbf{b} = (b_1, b_2, b_3, \ldots)$, $b_i = 0$ or 1, be an arbitrary infinite 0–1 sequence. The cardinality of the set of these $\mathbf{b}$s is $2^{\aleph_0}$. We associate with every such $\mathbf{b}$ a strategy $G_{\mathbf{b}} \in STR_A$ of player B as follows: his ith move is b_i for every $i \geq 1$. Now let $F \in STR_A$ be an arbitrary strategy of player A. If $\mathbf{b} \neq \mathbf{c}$, then $\langle F, G_{\mathbf{b}} \rangle \neq \langle F, G_{\mathbf{c}} \rangle$ (indeed, the $(2i)$th coordinate of $\langle F, G_{\mathbf{b}} \rangle$ is b_i and the $(2i)$th coordinate of $\langle F, G_{\mathbf{c}} \rangle$ is c_i). This proves that the cardinality of $\mathcal{P}_F$ is $2^{\aleph_0}$ for every $F \in STR_A$. The same holds for $\mathcal{P}_G$. $\qquad \square$

Now we are ready to complete the proof of Theorem C.4. By using transfinite induction we define subset $A \subset \{0, 1\}^\omega$ such that:

(1) for each strategy $F \in STR_A$ of player A there is a "witness" play $\mathbf{t}(F) \in \mathcal{P}_F \cap B$, which proves that F is not a winning strategy;
(2) for each strategy $G \in STR_B$ of player B there is a "witness" play $\mathbf{s}(G) \in \mathcal{P}_G \cap A$, which proves that G is not a winning strategy;
(3) different strategies have different "witnesses."

We cannot give an explicit construction of set A. The proof uses the *Axiom of Choice* via the well ordering.

By the well-ordering principle, the class STR_A of strategies of player A can be indexed by the ordinal numbers α less than $2^{\aleph_0}$: $STR_A = \{F_\alpha : \alpha < 2^{\aleph_0}\}$. Similarly, we can index the set $STR_B = \{G_\alpha : \alpha < 2^{\aleph_0}\}$ of strategies of player B.

Choose $\mathbf{t}_0 \in \mathcal{P}_{F_0}$ arbitrarily. Choose $\mathbf{s}_0 \in \mathcal{P}_{G_0}$ such that $\mathbf{s}_0 \neq \mathbf{t}_0$.

Proceed inductively. If $0 < \alpha < 2^{\aleph_0}$ and if $\mathbf{t}_\beta$ and $\mathbf{s}_\beta$ have been defined for all $\beta < \alpha$, then the sets $\{\mathbf{s}_\beta : \beta < \alpha\}$ and $\{\mathbf{t}_\beta : \beta < \alpha\}$ have cardinality less than $2^{\aleph_0}$. So the set

$$\mathcal{P}_{F_\alpha} \setminus (\{\mathbf{s}_\beta : \beta < \alpha\} \cup \{\mathbf{t}_\beta : \beta < \alpha\})$$

is non-empty. Choose one of its elements and call it $\mathbf{t}_\alpha$. Similarly

$$\mathcal{P}_{G_\alpha} \setminus (\{\mathbf{s}_\beta : \beta < \alpha\} \cup \{\mathbf{t}_\beta : \beta \leq \alpha\})$$

is non-empty. Choose one of its elements and call it $\mathbf{s}_\alpha$. By definition the sets $\mathbf{S} = \{\mathbf{s}_\alpha : \alpha < 2^{\aleph_0}\}$ and $\mathbf{T} = \{\mathbf{t}_\alpha : \alpha < 2^{\aleph_0}\}$ are disjoint. Let $\mathbf{S} \subset A$ and $\mathbf{T} \subset B = \{0, 1\}^\omega \setminus A$.

Since every play has a winner, it is enough to check that neither player has a winning strategy. First we show that player A does not have a winning strategy.

Let $F = F_\alpha \in STR_A$ be an arbitrary strategy of player A. By the construction there exists a play $\mathbf{t}_\alpha \in \mathcal{P}_{F_\alpha}$ such that $\mathbf{t}_\alpha \in B$. By definition $\mathbf{t}_\alpha = \langle F_\alpha, G \rangle$ for some strategy $G \in STR_B$ of player B. This means that if player A uses strategy $F = F_\alpha$ and player B uses strategy G, then player A loses the play.

Symmetrically we show that player B does not have a winning strategy. Let $G = G_\alpha \in STR_B$ be an arbitrary strategy of player B. By the construction, there exists a play $\mathbf{s}_\alpha \in \mathcal{P}_{G_\alpha}$ such that $\mathbf{s}_\alpha \in A$. By definition $\mathbf{s}_\alpha = \langle F, G_\alpha \rangle$ for some strategy $F \in STR_A$ of player A. This means that if player B uses strategy $G = G_\alpha$ and player A uses strategy F, then player B loses the play. This completes the proof of Theorem C.4. $\qquad\square$

The Ultrafilter Game: an undetermined positional game. Theorem C.4 was an undetermined combinatorial game; next we show a *positional game* which is undetermined. The underlying set can be ω, but the *length* of a play may go beyond ω, the first infinite ordinal number. Let $\mathcal{F}$ be an arbitrary infinite hypergraph, and suppose Maker (the first player) and Breaker play a positional game on $\mathcal{F}$, and they play not only until they have made their nth move for every natural number, but they continue to make moves as long as there is any unoccupied vertex of $\mathcal{F}$. In this case a play is a transfinite sequence of moves, and for an ordinal number α, the αth move of that play is the αth element of the sequence. The players move alternately, but the limit moves have no immediate predecessor, so we have to decide separately about them. We admit the most natural possibility and offer these limit moves to Maker. This type of game will be called *infinite full games*.

If the hypergraph $\mathcal{F}$ has *finite edges* only, then the following compactness result holds (we mention it without proof): the infinite full game on $\mathcal{F}$ is a win for Maker if and only if, for some *finite* sub-family $\mathcal{G} \subset \mathcal{F}$, the finite game on $\mathcal{G}$ is a win for Maker. In other words, if Maker has a winning strategy in an infinite full game, then, for some natural number n, he can win within n moves. It follows that if $\mathcal{F}$ has finite edges only, then the weak infinite full game on $\mathcal{F}$ is determined.

To construct an undetermined positional game, the edges of the hypergraph are required to be infinite. We recall the concept of the *non-trivial ultrafilter*. A filter $\mathcal{H}$ on an infinte set S is a family of subsets of S such that:

(i) if $A, B \in \mathcal{H}$, then $A \cap B \in \mathcal{H}$,
(ii) if $A \in \mathcal{H}$ and $A \subset B$, then $B \in \mathcal{H}$, and
(iii) $\mathcal{H} \neq 2^S$, i.e. $\mathcal{H}$ doesn't contain the empty set.

By Zorn's lemma, every filter is contained in a maximal filter, called *ultrafilter*. If $\mathcal{U}$ is an ultafilter, then for every $A \subset S$ either $A \in \mathcal{H}$ or $(S \setminus A) \in \mathcal{H}$. This implies that every ultrafilter $\mathcal{U}$ defines a *finitely additive* 0–1 measure $\mathcal{M}$ on 2^S: $\mathcal{M} = 1$ if $A \in \mathcal{H}$ and $\mathcal{M} = 0$ if $(S \setminus A) \in \mathcal{H}$. Conversely, every finitely additive 0–1 measure on 2^S defines an ultrafilter. If there is a *finite* set of measure 1, then one of the elements, say, $x \in S$ has measure 1, too. So $\mathcal{U} = \{A \subset S : x \in A\}$. These ultrafilters are called *trivial*. Not every ultrafilter is trivial: any ultrafilter containing the filter $\mathcal{H} = \{A \subset S : S \setminus A \text{ is finite}\}$ is non-trivial.

Theorem C.5 (McKenzie and Paris) *Suppose that the edges of a hypergraph form a non-trivial ultrafilter of* $\omega = \{0, 1, 2, \cdots\}$. *Then the infinite full game played on it is undetermined.*

Proof. When an infinite full play ends, all the elements of the board are occupied, so either the set of the points of Maker, or the set of the points of Breaker is in the given ultrafilter $\mathcal{U}$, but not both. This means that every play has a winner. This proves that Breaker cannot have a winning strategy in the ultrafilter game. Indeed, a straigthforward adaptation of the finite strategy stealing argument shows that the second player (Breaker) cannot have a winning strategy in an infinite full positional game.

Now assume that Maker has a winning strategy: *Str*. We show that if Maker uses strategy *Str*, then Breaker can win the strong play in ω moves, which clearly contradicts the fact that *Str* is a winning strategy.

Let Breaker play 3 simultaneous plays of length ω against Maker's winning strategy *Str*: we call it the solitary 3ω game. The board of the game is 3 copies of ω

$$\omega_1 = \{0_1, 1_1, 2_1, 3_1, 4_1, \cdots\}$$
$$\omega_2 = \{0_2, 1_2, 2_2, 3_2, 4_2, \cdots\}$$
$$\omega_3 = \{0_3, 1_3, 2_3, 3_3, 4_3, \cdots\}.$$

Breaker is the only player in this game, since Maker's moves will be uniquely determined by winning strategy *Str*. The first move is special: Maker occupies 3 points of the board, but after that Breaker and Maker alternately occupy 1 new point per move only. Let $m^* \in \omega$ denote Maker's first move advised by winning strategy *Str*: $Str(\emptyset) = m^*$; then Maker's first move in the 3ω game is to occupy all the 3 elements of the set $\{m_1^*, m_2^*, m_3^*\}$. Then Breaker occupies an arbitrary new element from one of the 3 sub-boards ω_{i1} ($i1 = 1$ or 2 or 3): let $b_{i1} \in \omega_{i1}$ denote Breaker's first move. Then Maker occupies another element from the same sub-board ω_{i1} by using strategy *Str*: let $m_{i1} = Str(m_{i1}^*, b_{i1}) \in \omega_{i1}$ be Maker's second move. Then again Breaker occupies a new element from one of the 3 sub-boards $b_{i2}' \in \omega_{i2}$ ($i2 = 1$ or 2 or 3); then Maker occupies another element from the same sub-board ω_{i2} by using strategy *Str* as follows:

(i) if $i2 \neq i1$, then let $m_{i2}' = Str(m_{i2}^*, b_{i2}') \in \omega_{i2}$ be Maker's third move;
(ii) if $i2 = i1$, then let $m_{i1}' = Str(m_{i1}^*, b_{i1}, m_{i1}, b_{i1}') \in \omega_{i1}$ be Maker's third move.

Then again Breaker occupies a new element from one of the 3 sub-boards $b''_{i3} \in \omega_{i3}$ ($i3 = 1$ or 2 or 3); then Maker occupies another element from the same sub-board ω_{i3} by using strategy *Str* as follows:

(i) if $i3 \notin \{i1, i2\}$, then let $m''_{i3} = Str(m^*_{i3}, b''_{i3}) \in \omega_{i3}$ be Maker's 4th move;

(ii) if $i3 = i1$ and $i3 \neq i2$, then let $m''_{i1} = Str(m^*_{i1}, b_{i1}, m_{i1}, b''_{i1}) \in \omega_{i1}$ be Maker's 4th move;

(iii) if $i3 = i2$ and $i3 \neq i1$, then let $m''_{i2} = Str(m^*_{i2}, b'_{i2}, m'_{i2}, b''_{i2}) \in \omega_{i2}$ be Maker's 4th move;

(iv) if $i1 = i2 = i3$, then let $m''_{i1} = Str(m^*_{i1}, b_{i1}, m_{i1}, b'_{i1}, m'_{i1}, b''_{i1}) \in \omega_{i1}$ be Maker's 4th move, and so on.

The length of a play is ω, i.e. the players take turns for every natural number n. At the end of a play Breaker wins the solitary 3ω game if

(i) all the elements of the board $\omega_1 + \omega_2 + \omega_3$ are occupied, and

(ii) for every $j \in \omega$ with $j \neq m^*$, Breaker occupied at least one element of the 3-set $\{j_1, j_2, j_3\}$.

In other words, Breaker loses the solitary 3ω game if either there remains an unoccupied element of the board or if for some $j \neq m^*$, Maker can occupy all the 3 elements of the set $\{j_1, j_2, j_3\}$. (Note that Maker occupies $\{m^*_1, m^*_2, m^*_3\}$ in his opening move.) We show that by using a "forced blocking and filling up the holes" combination strategy, Breaker can win the solitary 3ω game. ☐

Lemma. *Breaker has a winning strategy in the solitary 3ω game.*

Proof. Breaker's winning strategy consists of two kinds of moves: "forced moves to block" and "filling up the holes" moves. To define these moves, we introduce some notation. Suppose that Maker has already made $n \geq 2$ moves, Breaker made $(n - 1)$ moves and it is his turn to make his nth move. Let $M(n)$ denote the set of Maker's moves; in particular, let $m(n) \in M(n)$ be Maker's nth move, and let $B(n - 1)$ denote the set of Breaker's moves at this stage of the play. Let (D stand for "danger" and H stands for "hole"):

$$D_n = \{i \in \omega \setminus \{m^*\} : M(n) \cap \{i_1, i_2, i_3\} \neq \emptyset\}$$

and

$$H_{n-1} = \{i \in \omega \setminus \{m^*\} : \{i_1, i_2, i_3\} \not\subset M(n) \cup B(n - 1)\}.$$

Now we are ready to define Breaker's nth move. If $D_n = D_{n-1}$, i.e. $m(n) \in \{j_1, j_2, j_3\}$ for some $j \in D_{n-1}$, then Breaker has no choice and have to make a "forced move to block": he occupies the third element of the set $\{j_1, j_2, j_3\}$. If $D_n \neq D_{n-1}$, i.e. $m(n) \in \{j_1, j_2, j_3\}$ for some $j \notin D_{n-1}$, then Breaker has time and doesn't rush to block the set $\{j_1, j_2, j_3\}$. Instead he makes a "filling up the holes" move: if i is the smallest element of H_{n-1}, then Breaker occupies an available element of the set $\{i_1, i_2, i_3\}$. By using this strategy, among his first n moves Breaker makes at least $n/2$ "filling up the holes" moves, so at the end of a play, Breaker can block every 3-set $\{j_1, j_2, j_3\}$ where $j \neq m^*$, and at the same time he can guarantee that every element of the board is occupied by either player. This completes the proof of the Lemma.   □

Finally observe that Breaker's winning strategy in the solitary 3ω game contradicts the fact that *Str* is a winning strategy for Maker in the ultrafilter game on $\mathcal{U}$. Indeed, consider a play in the 3ω game where Breaker uses his winning strategy. Since Maker's moves are determined by strategy *Str*, and at the end of the ω-play all the elements of the board are occupied, Maker wins the ultrafilter game in each "row" ω_i, $i = 1, 2, 3$. So the set of Maker's points in ω_i belongs to ultrafilter $\mathcal{U}$ (we identify ω_i with ω). Since the ultrafilter is non-trivial, the intersection of these 3 sets must be infinite, which contradicts the Lemma (indeed, by the Lemma, the intersection is one-element and contains m^* only). This proves that if Maker uses strategy *Str*, then Breaker wins at least one of the 3 strong plays of the ultrafilter game on $\mathcal{U}$ in ω moves.

In the previous argument we can switch the role of Maker and Breaker, and we obtain the following result. If Breaker uses any strategy, then Maker can win at least one of the 3 strong plays of the ultrafilter game on $\mathcal{U}$ in ω moves. This gives an alternative proof of the fact that Breaker cannot have a winning strategy, avoiding the infinite strategy stealing argument. This completes the proof of Theorem C.5.   □

Appendix D

An informal introduction to game theory

Every "Theory" of Games concentrates on one aspect only, and pretty much neglects the rest. For example:

(I) Traditional Game Theory (J. von Neumann, J. Nash, etc.) focuses on the *lack* of complete information (for example, card games like Poker). Its main result is a minimax theorem about *mixed* strategies ("random choice"), and it is basically Linear Algebra. Games of complete information (like Chess, Go, Checkers, Nim, Tic-Tac-Toe) are (almost) completely ignored by the traditional theory.

(II) One successful theory for games of complete information is the "Theory of Nim like compound games" (Bouton, Sprague, Grundy, Berlekamp, Conway, Guy, etc. – see volume one of the *Winning Ways*). It focuses on "sum-games", and it is basically Algebra ("addition theory").

(III) In this book we are tackling something completely different: the focus is on "winning configurations," in particular on "Tic-Tac-Toe like games," and develop a "fake probabilistic method." Note that "Tic-Tac-Toe like games" are often called *Positional Games*.

Here in Appendix D a very brief outline of (I) and (II) is given. The subject is *games*, so the very first question is: "What is a *game*?". Well, this is a hard one; an easier question is: "How can one classify games?" One natural classification is the following:

(a) games of pure chance;
(b) games of mixed chance and skill;
(c) games of pure skill.

Another reasonable way to classify games is:

(i) games of complete information;
(ii) games of incomplete information.

(I) Neumann's theory: understanding poker. Traditional game theory, initiated by John von Neumann in an early paper from 1928, and broadly extended in the monumental *Theory of Games and Economic Behavior* by von Neumann and Morgenstern in 1944, deals with an extremely wide concept of games, including all the $3 \times 2 = 6$ classes formed by pairing (a)–(b)–(c) with (i)–(ii), with two or more players, of mixed chance and skill, arbitrary payoff functions. In the traditional theory each player has a choice, called *strategy*, and his objective is to maximize a payoff which depends both on his own choice and on his opponent's. The crucial *minimax* theorem for 2-player zero-sum games (i.e. pure conflict situation) is that it is always possible for either player to find a *mixed* strategy forming an "equilibrium," which means the best compromise for *both* players. Mixed strategy means to randomly play a mixture of strategies according to a certain fixed probability distribution. The philosophically interesting consequence of Neumann's minimax theorem is that the best play (often) requires *random*, unpredictable moves.

The contrasting plan of playing one strategy with certainty is called a *pure strategy*.

Mixed strategies are necessary to make up for the lack of a saddle point in the payoff matrix, i.e. for the discrepancy between the row maximin and the column minimax (note that in general row-maximin$\leq$column-minimax). If the payoff matrix has a saddle point, then there is no need for mixed strategies: the optimal strategies are always (deterministic) pure strategies.

The payoff matrix of the so-called "coin-hiding" game is a particularly simple example for row-maximin$\neq$column-minimax: in fact row-maximin$=-1$ and column-minimax$=1$:

$$
\begin{array}{c c}
 & \begin{array}{c c} L & R \end{array} \\
\begin{array}{c} L \\ R \end{array} & \begin{pmatrix} 1 & -1 \\ -1 & 1 \end{pmatrix},
\end{array}
$$

where L is for *left* and R is for *right*. In the "coin-hiding" game the first player has a coin that he puts behind his back in his right or left fist. Then he shows his closed fists to the second player, who has to guess where the coin is. They do this a number of times, and in the end they count how many times the second player won or lost. If the first player puts the coin in the same hand or if he simply alternates, the second player will soon notice it and win. Similarly, a clever opponent will eventually see through any such "mechanical" rule. Does it mean that a clever second player must necessarily win in the long run? Of course not. If the first player put the coin *at random* with probability one-half in either hand, and if his successive choices are *independent*, then the second player, no matter how smart or foolish, will make a correct guess asymptotically half of the time. On average the second player will neither win nor lose – this is the optimal strategy.

Note that Poker was Neumann's main motivation to write his pioneering paper on games in 1928. He could prove the fundamental *minimax* theorem, but the "computational complexity" of the problem prevented him (and anybody else since) from finding an explicit optimal mixed strategy for Poker. The next example is a highly simplified version of *Poker* (Poker itself is far too complex); the example is due to von Neumann, Morgenstern, and A. W. Tucker.

Suppose we have a deck of n cards (n is a large even number), half of which are marked H (high) and the other half marked L (low). Two players A and B start with an initial "bid" of a (i.e. each of them puts a dollars into the "pot"), and are dealt one card each, which they look at. Now A can "See" (i.e. demand B to expose his card), or "Raise" by an amount of b (i.e. put an extra b dollars into the "pot"). If A chooses "See," then B has no choice but to expose his card. B gains a (A's bid) if his card is H, while A's card is L, and loses a (i.e. his own bid) if his card is L, while A's card is H. The "pot" is split (gain 0) if both have H or both have L. However, if A "Raises," then B has a choice either to "Pass" (i.e. he is willing to lose a dollars without further argument) or to "Call" (i.e. to put in the extra b dollars, forcing A to expose his card). Again the stronger card wins everything, and they split if the cards have the same value.

Player A has 4 pure strategies:

(1) (S,S), a "See–See": "See" regardless of whether he has been dealt H or L;
(2) (S,R), a "See–Raise": "See" if he has been dealt H and "Raise" if he has been dealt L;
(3) (R,S), a "Raise–See": "Raise" if he has been dealt H and "See" if he has been dealt L;
(4) (R,R), a "Raise–Raise": "Raise" regardless of whether he has been dealt H or L.

We recall that if A chooses "See," then B has no choice. If A chooses "Raise," then B has the options of "Pass" and "Call". Therefore, player B has four pure strategies: A chooses "Raise" then B has the four options of

(1) (P,P), a "Pass–Pass": "Pass" regardless of whether he has been dealt H or L;
(2) (P,C), a "Pass–Call": "Pass" if he has been dealt H and "Call" if he has been dealt L;
(3) (C,P), a "Call–Pass": "Call" if he has been dealt H and "Pass" if he has been dealt L;
(4) (C,C), a "Call–Call": "Call" regardless of whether he has been dealt H or L.

Note that (S,S) and (S,R) are not good strategies for A, because neither takes advantage of the good luck of having been dealt a high card. Similarly, (P,P) and (P,C) are not good for B, because they require him to "Pass" with a high card.

Therefore, if we assume that both A and B are intelligent players, then we can disregard outright (S,S) and (S,R) for A, and (P,P) and (P,C) for B.

It follows that the payoff matrix can be reduced to the following 2-by-2 matrix

$$
\begin{array}{cc}
 & (C,P) \quad (C,C) \\
\begin{array}{c} (R,S) \\ (R,R) \end{array} & \begin{pmatrix} 0 & b/4 \\ (a-b)/4 & 0 \end{pmatrix}
\end{array}
$$

The four entries of the payoff matrix come from the following consideration. Assume, for example, that A chooses (R,S) and B chooses (C,C). Then

(1) if A is dealt H and B is dealt H, then A gains 0;
(2) if A is dealt H and B is dealt L, then A gains $a+b$;
(3) if A is dealt L and B is dealt H, then A gains $-a$;
(4) if A is dealt L and B is dealt L, then A gains 0.

If n is very large, the four combinations (H,H), (H,L), (L,H), (L,L) appear with the same frequency 1/4. On the average A gains $((a+b)-a)/4 = b/4$ dollars per game. This is how we get the second entry in the first row of the payoff matrix. The other 3 entries can be obtained by similar considerations.

Assume first that $a < b$. Then the payoff matrix

$$
\begin{array}{cc}
 & (C,P) \quad (C,C) \\
\begin{array}{c} (R,S) \\ (R,R) \end{array} & \begin{pmatrix} 0 & b/4 \\ (a-b)/4 & 0 \end{pmatrix}
\end{array}
$$

has a saddle point: row-maximin=column-minimax=0. It means the optimal strategies are pure: A chooses (R,S) and B chooses (C,P). In other words, both players best choice is to play "conservative": "Raising" with H, "Seeing" with L, "Calling" a "Raise" with H, and "Passing" a "Raise" with L. This game is fair: each player has an average gain of 0.

Next assume $a > b$. Then the payoff matrix does *not* have a saddle point. A's optimal strategy is a *mixed strategy* where A chooses (R,S) with probability $(a-b)/a$ and (R,R) with probability b/a. Then A can achieve an average gain of at least $(a-b)b/4a$ per game independently of B's choice. B's optimal strategy is to choose (C,P) with probability b/a and (C,C) with probability $(a-b)/a$. This way B can prevent A from averaging more than $(a-b)b/4a$ per game. This game is not fair (due to the asymmetry that A is the only player who can "Raise").

We thus see that in his optimal strategy A must "bluff", i.e. "Raise" with a low card, at least part of the time (if $a > b$). Similarly, in order to minimize his loss, B must give up the "conservative" play. He has to play risky by "Calling" a "Raise" with a low card in part of the time.

In the years since 1944 traditional game theory has developed rapidly. It plays a fundamental role in mathematical fields like Linear Programming and Optimization. It has an important place in Economics, and it has contributed non-trivial insights to many areas of social science (Management, Military Strategy, etc.). Traditional game theory is very successful in games of *incomplete* information. It means games in which one player knows something that the other does not, for example a card that a player has just drawn from a pack. However, for Combinatorial Games like Chess, Go, Checkers, Tic-Tac-Toe, Hex, traditional theory does not give too much insight. Indeed, from the viewpoint of traditional game theory Combinatorial Games are "trivial." For games of complete information the minimax theorem becomes the rather simplistic statement that each player has a *pure* optimal strategy, i.e. there is no need for mixed strategies whatsoever. No one may have discovered what an optimal deterministic strategy actually looks like (since the exhaustive search through the complex "game-tree" requires enormous amount of time), but it definitely exists. (In fact, usually there are several optimal strategies.) Combinatorial Games are, therefore, the simplest kind of pure conflict ("zero-sum") situation with no room for coalition, where the problem is "merely" the computational *complexity* of the exhaustive analysis. The above-mentioned coin-hiding game and the simplified poker with their very small 2×2 payoff matrices

$$
\begin{array}{cc}
 & L \quad R \\
\begin{array}{c} L \\ R \end{array} & \begin{pmatrix} 1 & -1 \\ -1 & 1 \end{pmatrix}
\end{array}
\qquad
\begin{array}{cc}
 & (C,P) \quad (C,C) \\
\begin{array}{c} (R,S) \\ (R,R) \end{array} & \begin{pmatrix} 0 & b/4 \\ (a-b)/4 & 0 \end{pmatrix}
\end{array}
$$

are terribly misleading examples. Indeed, they are hiding the weak point of the payoff matrix approach, which is exactly the *size* of the matrix (i.e. the total number of strategies). In a typical board-game the number of strategies is usually a *doubly* exponential function of the size of the board, making the payoff matrix setup **extremely** *impractical*.

It is worth while quoting von Neumann; this is what he said about games of complete information. "Chess is not a game. Chess is a well-defined form of computation. You may not be able to work out the answers, but in theory there must be a solution, a right move in any position."

For complete-information games, including the sub-class of Combinatorial Games, *exhaustive search* is much more efficient than the payoff matrix setup (i.e. to try out all possible strategies). Unfortunately the running time of the *exhaustive search* is "exponential" in terms of the board-size; this is still *impractical*. Note that *exhaustive search* is officially called Backward Labeling.

We can say, therefore, that the basic challenge of the theory of Combinatorial Games is the *complexity problem*, or as it is often called, the *combinatorial chaos*. Even for games with relatively small boards, the Backward Labeling of the game-tree or any other brute force case study is far beyond the capacity of the fastest computers. This is the reason why computers could not make such a big impact on the theory of Combinatorial Games.

The original 1944 edition of the *Theory of Games and Economic Behavior* was written before modern computers became available (somewhat later von Neumann himself made a considerable contribution to the early developments of the electronic computer), and the "complexity" issue was not addressed. Since then the main trend in traditional game theory have been shifted from zero-sum 2-player games to study games where the conflict is mixed with opportunities for coalition like *more*-than-2-player games. These games often lead to absolute, unsolvable conflict between individual rationality and collective rationality. A well-known example is the so-called *prisoner's dilemma*. The new trend is in a far less pleasing state than the elegant von Neumann's theory for the pure conflict case.

The complexity problem, i.e. our inability to make a complete analysis of Combinatorial Games like Chess, Go, Checkers without impractically detailed computation, has one apparent compensation: it leaves the "hard" games alive for competition. Once a Combinatorial Game has been fully analyzed, like Nim or the $4 \times 4 \times 4$ version of Tic-Tac-Toe, then, of course, it is competitively dead.

Let us say a few words about game-playing computer programs: we compare the popular games of Chess, Go, and Checkers. First note that Go is not "computer-friendly." In spite of serious efforts, Go-playing programs are nowhere close to the level of good human players. On the other hand, computer programs can play Chess much better; they reached the level of the top human grand-masters.

A good illustration is the story of the Kasparov vs. Deep Blue (later Deep Junior) matches. In February of 1996 Garry Kasparov beat the supercomputer Deep Blue 4-2. In Deep Blue 32 separate computers operated in parallel. It could analyze more than 200 million moves per second and had "studied" thousands of Chess's most challenging matches, including hundreds involving Kasparov. To beat Deep Blue Kasparov had to play many moves in advance which extended beyond the computer's horizon.

Fifteen months later in the rematch the improved version of Deep Blue made history by beating Kasparov 3.5–2.5. This time Deep Blue was programmed to play "positional Chess" like a grand-master.

In January–February of 2003 there was a highly publicized third match between Kasparov and the further improved Deep Blue, called Deep Junior, which ended in a 3–3 draw.

"Positional Chess," as opposed to "tactical Chess," involves a situation in which there are no clear objectives on the board, no obvious threats to be made. It means that the two sides are maneuvering for a position from which to begin long-term plans. It is the kind of Chess-playing in which grand-masters used to do so much better than machines. The surprising loss of Kasparov in 1997, and the equally disappointing draw in 2003, are owe much to the fact that the new supercomputers evidently "understand" positional Chess far beyond the level of the old machines.

Deep Blue's victory in 1997 was a real breakthrough. It is safe to say that now the best Chess-playing programs are at the same level, or perhaps even better, than the top human players.

Finally, in the game of Checkers, a computer program is so good that it consistently beats everyone but the world champion.

What is the basic difference between Go, Chess, and Checkers? In Checkers the average number of options per move is not very large, say, about 4, which enables a computer to calculate to a considerable depth in the available time, up to about 20 moves, which means 4^{20} alternatives. On the other hand, in Go the number of options per move is much greater – something like 200 – so the same computer in the same time could analyze Go no more than 5 moves deep: 200^5 alternatives. Note that very roughly $4^{20} \approx 200^5$. The case of Chess is somewhat intermediate. Consequently, games for which the number of options per move is large, but can effectively be cut down by *understanding* and *judgment*, are relatively speaking to the advantage of the human player.

Let us return to the complexity problem. The brute force way to analyze a *position* in a game is to examine all of its options, and all the options of those options, and all the options of the options of those options, and so on. This exhaustive search through the game-tree usually takes a tremendous amount of time. The objective of Combinatorial Game Theory is to describe wide classes of games for which there is a substantial "shortcut." This means to find a fast way of answering the question of "who wins," and also, if possible, to find a tractable way of answering the other question of "how to win," avoiding the exhaustive search in full depth. A natural way to cut down the alternatives is to be able to judge the value of a position at a level of a few moves depth only. This requires human *intelligence*, an essential thing that contemporary computers lack.

A complementary "negative" approach is to try to prove that some other classes of games are *hard* in the sense that any *algorithm* which determines the "outcome" (i.e. which player has a winning or drawing strategy) for *all* of the games in the class must necessarily take as much, or nearly as much, computations as the "brute force" complete analysis. Unfortunately there are very few *unconditional* lower bound results in Complexity Theory, see pp. 218–9 in the *Winning Ways* and also Fraenkel [1991]. This is the reason why the "negative" approach, due to the lack of effective methods, turned out to be much less successful, at least so far.

(II) Theory of NIM like games. A well-known and very advanced branch in the "positive direction" is the beautiful theory of Nim like *compound* games, played with the *normal play convention*: a player unable to move loses. The most famous example of such sum-games is of course **Nim** itself.

The ancient game of Nim is played by two players with heaps of coins (or stones, or beans). Suppose that there are $k \geq 1$ heaps of coins that contain, respectively, $n_1, n_2, \ldots, n_k$ coins. The players alternate turns, and each player, when it is his turn, selects one of the heaps and removes at least one coin from the selected heap (the player may take all coins from the selected heap, which is now "out of play"). The play ends when all the heaps are empty, and the player who takes the last coin(s) is the winner. A century ago, in 1902 Bouton found the following surprisingly simple solution to Nim. Express each one of the heap-sizes n_i in binary form

$$n_i = a_0^{(i)} + a_1^{(i)} \cdot 2 + a_2^{(i)} \cdot 2^2 + \ldots + a_{s_i}^{(i)} \cdot 2^{s_i}.$$

By including 0s we can clearly assume that all of the heap sizes have the *same* number of binary digits, say, $s + 1$ where $s = \max_{1 \leq i \leq k} s_i$. We call a Nim game *balanced* if and only if the $s + 1$ (complete) sums

$$a_0^{(1)} + a_0^{(2)} + \ldots + a_0^{(k)}$$

$$a_1^{(1)} + a_1^{(2)} + \ldots + a_1^{(k)}$$

$$\ldots\ldots\ldots\ldots\ldots\ldots\ldots\ldots$$

$$a_s^{(1)} + a_s^{(2)} + \ldots + a_s^{(k)}$$

are *all even*, i.e. all sums are 0 (mod 2). A Nim game which is not balanced is called *unbalanced* (at least one sum is odd). We mention two simple facts (we leave them to the reader as an easy exercise):

(1) whatever move made in a balanced Nim game, the resulting game is always unbalanced;
(2) starting from an unbalanced Nim game, there is always a move which balances it.

Now it is obvious that first player has a winning strategy in an arbitrary *unbalanced* Nim game: first player keeps applying (2). On the other hand, second player has a winning strategy in *balanced* Nim games: indeed, first player's opening move unbalances the game, and second player keeps applying (2).

Reverse Nim: Suppose we change the objective of Nim so that the player who takes the last coin *loses*. Will this make much difference? Not if at least one heap has ≥ 2 coins. Then the ordinary and Reverse Nim behave alike, and the following is a winning strategy (we leave the proof to the reader): Play as in ordinary Nim until all but exactly one heap contains a single coin. Then remove either all or all but one of the coins of the exceptional heap so as to leave an *odd* number of heaps of size 1.

If all the heaps are of size 1, then of course the game becomes trivial ("mechanical"), and the two versions have opposite outcomes: a player wins in the ordinary (Reverse) Nim if and only if the number of singletons (heaps of size 1) is odd (even).

After Bouton's complete solution of Nim (and Reverse Nim), the theory of "Nim-like games" was greatly developed by Sprague and Grundy: they discovered large classes of games which are "Nim in disguise" (*impartial games*). More recently, Berlekamp, Conway, Guy and others discovered large classes of games for which the "Nim-addition," or a similar "addition theory" works and solves the game. Note that the "Nim-sum" of $n_1, n_2, \ldots, n_k$ is 0 if the Nim game is *balanced*, and non-zero if the game is *unbalanced*. In the unbalanced case the "Nim-sum" is a sum of distinct powers of 2; namely those powers of 2 which show up an *odd* number of times in the binary representations of $n_1, n_2, \ldots, n_k$. We refer to two remarkable books: (1) *On Games and Numbers* by Conway (which gives, among others, a striking new way to construct the set of real numbers); and (2) volume I of the *Winning Ways*. The theory developed in volume I can be employed in games in which the positions are composed of several non-interacting very simple games. (Note that Nim is a "sum" by definition: a "sum" of one-heap games.) Then the first thing to do is to associate values (numbers, *nimbers*, and other *"Conway numbers"*) with these components. Next comes the problem of finding ways of determining the outcome of a *sum* of games given information only about the values of the separate components. (For example, in Nim we apply the "Nim-addition.") This *addition theory* is where the shortcut comes from. This theory has been quite successful in analyzing endgame problems in a wide range of "real" games including Go – see e.g. Berlekamp [1991]. The reason why it works is that positions which tend to occur in the later stages of a game like Go *decompose* into separate, independent regions. Therefore, it is natural to apply the addition theory. On the other hand, positions which occur in the early and middle stages of a game like Go do *not* seem to decompose.

We give a thumbnail summary of this *addition theory* – we refer the reader to David Gale's excellent introductory article in *The Mathematical Intelligencer*, 16 (2).

The main theory treats only win–lose games with the normal play convention. Although this seems rather special, many games can be put into this form by making simple changes to the rules. The two players are called Left and Right, and the concrete rules of the game specify who moves first. Such games are represented by trees: the moves of Left are represented by leftward slanting edges, and those for Right by rightward slanting ones. We call them left–right trees, and they are kinds of game-trees.

Next comes Conway's arithmetic. There is a natural notion of *addition* on the set of these games: the sum $G + H$ of two games G, H is defined to be the two games played simultaneously. A player in his turn may play in either game. The negative $-G$ of a game G is to be G with the roles of the players interchanged. Conway defines an equivalence relation $\sim$ by writing $G \sim H$ if $G - H$ is a win for the second player. Observe that $G \sim G$; indeed, the second player can win $G - G$ by playing "copycat": whenever the first player moves in G $(-G)$, he makes the same move in $-G$ (G). It is easy to see that the set of equivalence classes forms an Abelian group whose 0 element is the set of all second player wins. This group is called Γ, and it is the object of study in the algebraic theory. Of course many different left–right trees correspond to the same group element, i.e. equivalence class. One of the important theorems of the theory states that every group element has a "canonical form," meaning a unique left–right tree with the fewest number of edges. This canonical tree is that Conway calls the *value* of the game. Given a concrete game, how do we find its *value*? Unfortunately, the general problem is NP-hard.

There is a natural order on group Γ. Call G positive, $G > 0$, if G is a win for Left, no matter who starts, and negative if G is a win for Right, no matter who starts. Note that $>$ is well defined. Every game is either positive, negative, zero, or none of the above, meaning a first player win. Defining $G > H$ if $G - H > 0$ gives a partial ordering of Γ.

What does it mean to *solve* a game in this theory? Well, it means to find the *value* of the game – instead of directly answering the natural questions of "who wins" and "how to win." Unfortunately, knowing the *value*, i.e. the canonical tree, does not mean that we can necessarily obtain a winning strategy without performing a huge amount of computation. But if "Combinatorial Game Theory" does not tell us how to win a game, what good is it? The theory turned out to be very useful in those games which can be expressed as a *sum* of simple games with explicitly known *values*. Indeed, there is often a fast way to determine the *value* of the sum-game from the *values* of the components, and that sum-value sometimes happens to be simple. An essential part of the theory is a wonderful "botanical garden" of simple games like Numbers, Nimbers, Star, Up, Down, Tiny, Miny, Double-Up-Star, etc. satisfying surprising identities.

If games like Chess and Go are seemingly far too difficult for "brute force" analysis then how can we write good game-playing computer programs? Well, we have to make a compromise. Instead of finding the *best possible* next move, the existing computer programs usually make use of some kind of *evaluation function* to efficiently judge the "danger" of a position without analyzing the game-tree in full depth. This approximation technique produces at least a "reasonably good" next move. As a result, computer-Chess (-Go, -Checkers) is "just" a bunch of good heuristic arguments rather than a "theorem-proof" type rigorous mathematical theory.

The considerable success of the addition theory of Nim-like sum-games might suggest the optimistic belief that we are at the edge of "understanding" Combinatorial Games in general: it is just a matter of being clever enough and working hard, and then sooner-or-later we are going to find a shortcut way to quickly predict the winner, and supplement it with an explicit winning strategy.

Well, let us disagree; this optimism is wishful thinking, and for the overwhelming majority of Combinatorial Games there are *no* such shortcuts. It seems inevitable that an efficient method has to *approximate* – just like all existing Chess-playing programs approximate the optimal strategy. In this book we use evaluation techniques which approximate the "danger" level of a single position without analyzing the whole game-tree in full depth.

Complete list of the Open Problems

1. Open Problem 3.1 *Consider the S-building game introduced in Section 1: is there a finite procedure to decide whether or not a given finite point set S in the plane is a Winner? In other words, is there a way to characterize those finite point sets S in the plane for which Maker, as the first player in the usual (1:1) play, can always build a congruent copy of S in the plane* **first** *(i.e. before the opponent could complete his own copy of S)?*

2. Open Problem 3.2 *Is it true that 5^3 Tic-Tac-Toe is a draw game? Is it true that 5^4 Tic-Tac-Toe is a first player win?*

3. Open Problem 4.1 *Is it true that unrestricted 5-in-a-row in the plane is a first player win?*

4. Open Problem 4.2 *Is it true that unrestricted n-in-a-row is a draw for $n = 6$ and $n = 7$?*

5. Open Problem 4.3 *Consider Harary's Animal Tic-Tac-Toe introduced in Section 4. Is it true that "Snaky" is a Winner? In particular, is it true that "Snaky" is a Winner on every $n \times n$ board with $n \geq 15$, and the first player can always build a congruent copy of "Snaky" first in at most 13 moves?*

6. Open Problem 4.4 *Is it true that Kaplansky's 4-in-a-line is a draw game? Is it true that Kaplansky's n-in-a-line is a draw game for every $n \geq 4$?*

7. Open Problem 4.5 *Consider Hex (see Section 4); find an explicit first player ("White") winning strategy in $n \times n$ Hex for every $n \geq 8$. In particular, find one for the standard size $n = 11$.*

8. Open Problem 4.6

(a) *Find an explicit first player winning strategy in the (K_{18}, K_4) Clique Game.*
(b) *Which player has a winning strategy in the Reverse Clique Game $(K_{18}, K_4, -)$? If you know who wins, find an explicit winning strategy.*

(c) *Consider the Clique Game* (K_n, K_q) *where n is huge compared to q; is there a uniform upper bound for the Move Number? More precisely, is there an absolute constant* $C_4 < \infty$ *such that the first player can always win in less than* C_4 *moves in every* (K_n, K_4) *Clique Game with* $n \geq 18$?

Is there an absolute constant $C_5 < \infty$ *such that the first player can always win in less than* C_5 *moves in every* (K_n, K_5) *Clique Game with* $n \geq 49$?

In general, is there an absolute constant $C_q < \infty$ *such that the first player can always win in less than* C_q *moves in every* (K_n, K_q) *Clique Game with* $n \geq R(q)$ *where* $R(q)$ *is the Ramsey Number?*

9. Open Problem 5.1 *Consider the unrestricted 5-in-a-row; can the first player always win in a bounded number of moves, say, in less than 1000 moves?*

Notice that Open Problems 4.1 and 5.1 are two different questions! It is possible (but not very likely) that the answer to Open Problem 4.1 is a "yes," and the answer to Open Problem 5.1 is a "no."

10. Open Problem 5.2 *Is it true that if the* n^d *Tic-Tac-Toe is a first player win, then the* n^D *game, where* $D > d$, *is also a win?*

The twin brother of Open Problem 5.2 is

11. Open Problem 5.3 *Is it true that if the* n^d *Tic-Tac-Toe is a draw game, then the* $(n+1)^d$ *game is also a draw?*

12. Open Problem 5.4 *Consider the Hypergraph Classification introduced at the end of Section 5. Is it true that each hypergraph class contains infinitely many* n^d *games? The unknown cases are Class 2 and Class 3.*

13. Open Problem 6.1

(a) *Find an explicit first player winning strategy in the* (K_{49}, K_5) *Clique Game.*
(b) *Find an explicit first player winning strategy in the* (K_{165}, K_6) *Clique Game.*

14. Open Problem 6.2 *Which player has a winning strategy in the Reverse Clique Game* $(K_{49}, K_5, -)$? *How about the* $(K_N, K_5, -)$ *game with* $N \geq 49$ *where* $N \to \infty$?
How about the Reverse Clique Game $(K_{165}, K_6, -)$? *How about the* $(K_N, K_6, -)$ *game with* $N \geq 165$ *where* $N \to \infty$? *In each case find an explicit winning strategy.*

The next few problems are about the Win Number.

15. Open Problem 6.3

(i) *Is it true that* $w(K_q) < R(q)$ *for all sufficiently large values of q? Is it true that*

$$\frac{w(K_q)}{R(q)} \longrightarrow 0 \text{ as } q \to \infty?$$

(ii) *Is it true that* $w(K_q; -) < R(q)$ *for all sufficiently large values of* q? *Is it true that*

$$\frac{w(K_q; -)}{R(q)} \longrightarrow 0 \ \text{as} \ q \to \infty?$$

16. Open Problem 6.4 *Is it true that* $w(K_4) < w(K_4; -) < 18 = R(4)$?

17. Open Problem 6.5 *Consider the* (K_N, K_q) *Clique Game, and assume that the Erdős–Szekeres bound applies:* $N \geq \binom{2q-2}{q-1}$. *Find an explicit first player winning strategy.*

18. Open Problem 6.6

(a) *What is the relation between the Weak Win and Reverse Weak Win Numbers* $ww(K_q)$ *and* $ww(K_q; -)$? *Is it true that* $ww(K_q) \leq ww(K_q; -)$ *holds for every* q?

(b) *Is it true that* $ww(K_q) < w(K_q)$ *for all sufficiently large values of* q? *Is it true that*

$$\frac{ww(K_q)}{w(K_q)} \longrightarrow 0 \ \text{as} \ q \to \infty?$$

(c) *Is it true that* $ww(K_q; -) < w(K_q; -)$ *for all sufficiently large values of* q? *Is it true that*

$$\frac{ww(K_q; -)}{w(K_q; -)} \longrightarrow 0 \ \text{as} \ q \to \infty?$$

(d) *Is it true that*

$$\frac{ww(K_q)}{R(q)} \longrightarrow 0 \ \text{and} \ \frac{ww(K_q; -)}{R(q)} \longrightarrow 0 \ \text{as} \ q \to \infty?$$

19. Open Problem 6.7 *Is it true that* $ww(K_q) = ww(K_q; -)$ *for every* q? *Is it true that* $ww(K_q) = ww(K_q; -)$ *for all but a finite number of* qs? *Is it true that they are equal for infinitely many* qs?

Recall that $HJ(n)$ is the Hales–Jewett number (see Sections 7–8 and Appendix B).

20. Open Problem 7.1 *Is it true that* $w(n-\text{line}) < HJ(n)$ *for all sufficiently large values of* n? *Is it true that*

$$\frac{w(n-\text{line})}{HJ(n)} \longrightarrow 0 \ \text{as} \ n \to \infty?$$

21. Open Problem 7.2 *Consider the* (N, n) *van der Waerden Game where* $N \geq W(n)$; *for example, let*

$$N \geq 2 \uparrow 2 \uparrow 2 \uparrow 2 \uparrow 2 \uparrow (n+9)$$

(Gowers's bound on the van der Waerden Number). Find an **explicit** *first player winning strategy.*

22. Open Problem 7.3 *Is it true that* $w(n-\text{term A.P.}) < W(n)$ *for all sufficiently large values of n? Is it true that*

$$\frac{w(n-\text{term A.P.})}{W(n)} \longrightarrow 0 \text{ as } n \to \infty?$$

23. Open Problem 8.1 *(a) Is it true that* $ww(n-\text{line}) < w(n-\text{line})$ *for all sufficiently large values of n? Is it true that*

$$\frac{ww(n-\text{line})}{w(n-\text{line})} \longrightarrow 0 \text{ as } n \to \infty?$$

24. Open Problem 8.2

(a) Is it true that $ww(n-\text{term A.P.}) < w(n-\text{term A.P.})$ *for all sufficiently large values of n? Is it true that*

$$\frac{ww(n-\text{term A.P.})}{w(n-\text{term A.P.})} \longrightarrow 0 \text{ as } n \to \infty?$$

(b) Is it true that

$$\frac{ww(n-\text{term A.P.})}{W(n)} \longrightarrow 0 \text{ as } n \to \infty?$$

25. Open Problem 9.1 ("Neighborhood Conjecture")

(a) Assume that $\mathcal{F}$ is an n-uniform hypergraph, and its Maximum Neighborhood Size is less than 2^{n-1}. Is it true that playing on $\mathcal{F}$ the second player has a Strong Draw?

Maybe the sharp upper bound $< 2^{n-1}$ is not quite right, and an "accidental" counter-example disproves it. The weaker version (b) below would be equally interesting.

Open Problem 9.1

(b) If (a) is too difficult (or false), then how about if the upper bound on the Maximum Neighborhood Size is replaced by an upper bound $2^{n-c}/n$ on the Maximum Degree, where c is a sufficiently large positive constant?

(c) If (b) is still too difficult, then how about a polynomially weaker version where the upper bound on the Maximum Degree is replaced by $n^{-c} \cdot 2^n$, where $c > 1$ is a positive absolute constant?

(d) If (c) is still too difficult, then how about an exponentially weaker version where the upper bound on the Maximum Degree is replaced by c^n, where $2 > c > 1$ is an absolute constant?

(e) How about if we make the extra assumption that the hypergraph is Almost Disjoint (which holds for the n^d Tic-Tac-Toe anyway)?

26. Open Problem 10.1 *Is it true that the "Maker's building" results in the book, proved by using explicit potentials, can be also achieved by a Random Strategy (of course, this means in the weaker sense that the strategy works with probability tending to 1 as the board size tends to infinity)?*

27. Open Problem 12.1

(a) *Which order is the right order of magnitude for* $\mathbf{w}(n\text{--line})$ *: (at least) exponential or polynomial?*

(b) *For every* n^d *Tic-Tac-Toe, where the dimension* $d = d(n)$ *falls into range (12.30), decide whether it belongs to Class 2 or Class 3.*

Theorem 12.5 gave a quadratic bound for the "phase transition" from Weak Win to Strong Draw in the n^d Tic-Tac-Toe game. The best that we could prove was an inequality where the upper and lower bounds for the dimension d differ by a factor of $\log n$.

28. Open Problem 12.3 *Phase transition for* n^d *hypercube Tic-Tac-Toe: Which order of magnitude is closer to the truth in Theorem 12.5,* $n^2/\log n$ *("the lower bound") or* n^2 *("the upper bound")?*

29. Open Problem 12.4 *Are there infinitely many pairs* (N, n) *for which the* (N, n) *Van der Waerden Game is a second player moral-victory?*

30. Open Problem 16.1 *Can we replace the pairing strategy upper bound* $3n/4$ *in the Degree Game on* K_n *with some* $c \cdot n$ *where* $c < 3/4$ *? Is it possible to get* $c = \frac{1}{2} + o(1)$?

The next two problems are justified by the Random Graph intuition.

31. Open Problem 20.1 *Consider the* $(1{:}b)$ *Connectivity Game on the complete graph* K_n. *Is it true that, if* $b = (1 - o(1))n/\log n$ *and* n *is large enough, then underdog Maker can build a spanning tree?*

32. Open Problem 20.2 *Consider the Reverse Hamiltonian Game, played on* K_n, *where Avoider takes 1 and Forcer takes* f *new edges per move; Forcer wins if at the end Avoider's graph contains a Hamiltonian cycle. Is it true that, if* $f = c_0 n/\log n$ *for some positive absolute constant and* n *is large enough, then Forcer can force Avoider to own a Hamiltonian cycle?*

33. Open Problem 25.1 *For simplicity assume that the board is the infinite complete graph* K_∞ *(or at least a "very large" finite* K_N*); playing the usual* $(1{:}1)$ *game, how long does it take to build a* K_q*?*

34. Open Problem 28.1 *Consider the discrepancy version of the Clique Game. Is it true that the* α*-Clique Achievement Number* $\mathbf{A}(K_N; \text{clique}; \alpha)$ *is the lower integral part of*

$$q = q(N, \alpha) = \frac{2}{1 - H(\alpha)} \left(\log_2 N - \log_2 \log_2 N + \log_2(1 - H(\alpha)) + \log_2 e - 1 \right) - 1 + o(1),$$

or at least the distance between the two quantities is $O(1)$? Here the function $H(\alpha) = -\alpha \log_2 \alpha - (1 - \alpha) \log_2(1 - \alpha)$ is the well-known Shannon's entropy.

35. Open Problem 30.1 *Playing the (2:1) game in an $N \times N$ board, what is the largest $q \times q$ aligned Square Lattice that topdog Maker can always build?*

36. Open Problem 30.2 ("Biased Clique Game") *Is it true that, in the (m:b) Biased Clique Achievement Game with $m > b$, played on K_N, the corresponding Clique Achievement Number is*

$$A(K_N; \text{clique}; m:b) = \lfloor 2\log_c N - 2\log_c \log_c N + 2\log_c e - 2\log_c 2 - 1 + \frac{2\log c}{\log c_0} + o(1)\rfloor,$$

where $c = \frac{m+b}{m}$ and $c_0 = \frac{m}{m-b}$?
Is it true that, in the (m : b) Biased Clique Achievement Game with $m \leq b$, played on K_N, the corresponding Clique Achievement Number is

$$A(K_N; \text{clique}; m:b) = \lfloor 2\log_c N - 2\log_c \log_c N + 2\log_c e - 2\log_c 2 - 1 + o(1)\rfloor$$

where $c = \frac{m+b}{m}$?
Is it true that the Avoidance Number

$$A(K_N; \text{clique}; a:f; -) = \lfloor 2\log_c N - 2\log_c \log_c N + 2\log_c e - 2\log_c 2 - 1 + o(1)\rfloor$$

where the base of logarithm is $c = \frac{a+f}{a}$?

37. Open Problem 30.3 ("Biased Lattice Games") *Is it true that, in the (m : b) Biased Lattice Achievement Game with $m > b$ played on an $N \times N$ board, the corresponding Lattice Achievement Number is*

(1a)
$$A(N \times N; \text{ Square Lattice}; m:b) = \left\lfloor \sqrt{\log_c N + 2\log_{c_0} N} + o(1) \right\rfloor,$$

(1b)
$$A(N \times N; \text{ rectangle lattice}; m:b) = \left\lfloor \sqrt{2\log_c N + 2\log_{c_0} N} + o(1) \right\rfloor,$$

(1c)
$$A(N \times N; \text{ tilted Square Lattice}; m:b) = \left\lfloor \sqrt{2\log_c N + 2\log_{c_0} N} + o(1) \right\rfloor,$$

(1d)
$$A(N \times N; \text{ tilted rectangle lattice}; m:b) = \left\lfloor \sqrt{2\log_c N + 2\log_{c_0} N} + o(1) \right\rfloor,$$

(1e)
$$A(N \times N; \text{ rhombus lattice}; m:b) = \left\lfloor \sqrt{2\log_c N + 2\log_{c_0} N} + o(1) \right\rfloor,$$

(1f)

$$A(N \times N; \text{ parallelogram lattice}; m : b) = \left\lfloor 2\sqrt{\log_c N + 2\log_{c_0} N} + o(1) \right\rfloor,$$

(1g)

$$A(N \times N; \text{ area one lattice}; m : b) = \left\lfloor \sqrt{2\log_c N + 2\log_{c_0} N} + o(1) \right\rfloor$$

where $c = \frac{m+b}{m}$ and $c_0 = \frac{m}{m-b}$?

Is it true that, in the $(m : b)$ Biased Lattice Achievement Game with $m \le b$ played on an $N \times N$ board, the corresponding Lattice Achievement Number is

(2a)

$$A(N \times N; \text{ Square Lattice}; m : b) = \left\lfloor \sqrt{\log_c N} + o(1) \right\rfloor,$$

(2b)

$$A(N \times N; \text{ rectangle lattice}; m : b) = \left\lfloor \sqrt{2\log_c N} + o(1) \right\rfloor,$$

(2c)

$$A(N \times N; \text{ tilted Square Lattice}; m : b) = \left\lfloor \sqrt{2\log_c N} + o(1) \right\rfloor,$$

(2d)

$$A(N \times N; \text{ tilted rectangle lattice}; m : b) = \left\lfloor \sqrt{2\log_c N} + o(1) \right\rfloor,$$

(2e)

$$A(N \times N; \text{ rhombus lattice}; m : b) = \left\lfloor \sqrt{2\log_c N} + o(1) \right\rfloor,$$

(2f)

$$A(N \times N; \text{ parallelogram lattice}; m : b) = \left\lfloor 2\sqrt{\log_c N} + o(1) \right\rfloor,$$

(2g)

$$A(N \times N; \text{ area one lattice}; m : b) = \left\lfloor \sqrt{2\log_c N} + o(1) \right\rfloor$$

where $c = \frac{m+b}{m}$?

Is it true that the Lattice Avoidance Number in the $(a : f)$ play is the same as in (a2)–(g2), except that $c = \frac{a+f}{a}$?

38. Open Problem 31.1

(a) *Is it true that, playing the $(2:2)$ Achievement Game on K_N, Maker can always build a clique K_q with $q = 2\log_2 N - 2\log_2 \log_2 N + O(1)$?*

(b) *Is it true that, playing the $(2:2)$ Achievement Game on K_N, Breaker can always prevent Maker from building a clique K_q with $q = 2\log_2 N - 2\log_2 \log_2 N + O(1)$?*

39. Open Problem 31.2 *Is it true that Theorem 31.1 is best possible? For example, is it true that, given any constant $c > 2$, playing the (2:2) game on an $N \times N$, Breaker can prevent Maker from building a $q \times q$ parallelogram lattice with $q = c\sqrt{\log_2 N}$ if N is large enough?*

Is it true that, given any constant $c > \sqrt{2}$, playing the (2:2) game on an $N \times N$, Breaker can prevent Maker from building a $q \times q$ aligned rectangle lattice with $q = c\sqrt{\log_2 N}$ if N is large enough?

40. Open Problem 32.1 *Is it true that Theorem 31.1' is best possible? For example, is it true that, given any constant $c > 2$, playing the (k:k) game on an $N \times N$, Breaker can prevent Maker from building a $q \times q$ parallelogram lattice with $q = c\sqrt{\log_2 N}$ if N is large enough?*

41. Open Problem 33.1 *Let K_q(red, blue) be an arbitrary fixed 2-colored goal graph, and let $q = (2 - o(1))\log_2 N$. Is it true that, playing on K_N, Red has a winning strategy in the K_q(red, blue)-building game?*

42. Open Problem 34.1 ("Hales–Jewett Conjecture")

(a) *If there are at least twice as many points (i.e., cells) as winning lines, then the n^d Tic-Tac-Toe game is always a draw.*

(b) *What is more, if there are at least twice as many points as winning lines, then the draw is actually a Pairing Strategy Draw.*

43. Open Problem 47.1 *Formulate and prove a stronger version of Theorem 47.1 (Picker's blocking) which perfectly complements Theorem 33.4 (Chooser's building), that is, which gives exact solutions in the biased (1:s) Chooser–Picker versions of the Clique and Lattice Games.*

What kinds of games? A dictionary

Throughout this list we will describing 2-player games played on an arbitrary (finite) hypergraph. The players take turns occupying points of the hypergraph. The play is over when all points have been occupied. We begin with the *straight complete occupation games*.

Positional Game: Such as Tic-Tac-Toe: Player 1 and Player 2 take turns occupying new points. The player who is the first to occupy all vertices of a hyperedge (called winning set) is the winner. If there is no such a hyperdge, the play is declared a draw. See p. 72.

Maker–Breaker Game: Maker and Breaker take turns occupying new points. Maker wins if at the end of the play he occupies all points of some winning set; otherwise Breaker wins (i.e. Breaker wins if he can occupy at least 1 point from each winning set). See pp. 27 and 83.

Picker–Chooser Game: In each turn Picker picks two unoccupied points, Chooser chooses one from the pair offered, the other goes to Picker. Picker wins if at the end of the play he occupies all points of some winning set; otherwise Chooser wins. See p. 320.

Chooser–Picker Game: In each turn Picker picks two unoccupied points, Chooser chooses one from the pair offered, the other goes to Picker. Chooser wins if at the end of the play he occupies all points of some winning set; otherwise Picker wins.

In the *reverse* versions described below, the objective of the game is to force your opponent to occupy a whole hyperedge, rather than occupying a hyperedge yourself. See p. 320.

Reverse Positional Game: Again Player 1 and Player 2 take turns occupying new points. The player who is the first to occupy all vertices of a hyperedge is the loser. If there is no such a hyperdge, the play is declared a draw. See p. 76.

Reverse Maker–Breaker Game=Avoider–Forcer Game: Avoider and Forcer take turns occupying new points. Forcer wins if at the end of the play Avoider occupies all points of some winning set; otherwise Avoider wins. See p. 93.

Reverse Picker–Chooser Game: In each turn Picker picks two unoccupied points, Chooser chooses one from the pair offered, the other goes to Picker. Picker loses if at the end of the play he occupies all points of some winning set; otherwise Picker wins. See p. 623.

Reverse Chooser–Picker Game: In each turn Picker picks two unoccupied points, Chooser chooses one from the pair offered, the other goes to Picker. Chooser loses if at the end of the play he occupies all points of some winning set; otherwise Chooser wins. See p. 624.

Here is a new concept.

Improper–Proper Game: Mr. Improper and Ms. Proper take turns occupying new points, Mr. Improper colors his points red and Ms. Proper colors her points blue. Ms. Proper wins if and only if at the end of the play they produce a proper 2-coloring of the hypergraph (i.e., no hyperedge is monochromatic). Mr. Improper wins if at the end they produce an improper 2-coloring of the hypergraph (there is a monochromatic hyperedge), so a draw is impossible by definition. See p. 134.

Notice that this game has a one-sided connection with both the Maker–Breaker and the Avoider–Forcer Games: Maker's winning strategy is automatically Mr. Improper's winning strategy (he can force a red hyperedge), and, similarly, Forcer's winning strategy is automatically Mr. Improper's winning strategy (he can force a blue hyperedge).

If the points of a hypergraph are 2-colored (say, red and blue), the *discrepancy* of a hyperedge is the absolute value of the difference of the number of red points and the number of blue points in that hyperedge.

Notice that every play defines a 2-coloring: one player colors his moves (i.e., points) red and the other player colors his moves (points) blue.

In a **Discrepancy Game** one player is trying to force the existence of a hyperedge with large discrepancy. See p. 231.

In a **One-Sided Discrepancy Game** one player is trying to force the existence of a hyperedge with large discrepancy where he owns the majority. See p. 231.

A **Shutout** of a hyperedge is the number of points that one player occupies in a hyperedge before the opponent places his first mark in that hyperedge.

In a **Shutout Game** a player is trying to force some hyperedge to have a large Shutout. See p. 85.

Finally, a bunch of concrete games.

Animal Tic-Tac-Toe: See Section 4, p. 60.

Bridge-it: A particular "connectivity game"—see Section 4, p. 66.

Clique Games: Graph games in which the goal is to build (or to avoid building) a large complete graph (often called clique)—see Section 6, p. 92.

Connectivity Games: See Sections 4 and 20, pp. 67 and 286.

Degree Game: Maker's goal is to build a large maximum degree in a complete graph – see Sections 16–17, pp. 231–59.

Hamilton Cycle Game: Maker's goal is to build a Hamilton cycle – see Section 20, p. 291.

Hex: A particular "connectivity game" – see Section 4, p. 65.

Kaplansky's k-in-a-line game: See Section 4, p. 64.

n^d**-game (multidimensional Tic-Tac-Toe):** The board is the d-dimensional hypercube, where each side is divided into n equal pieces. It consists of n^d little "cells." The winning sets are the n-in-a-row sets. See Section 3, p. 44.

Row-Column Game: See Sections 16–19, pp. 231–85.

Unrestricted n-in-a-row: The players play on the infinite chessboard, and the goal is to have n consecutive little squares horizontally, vertically, or diagonally (slopes ± 1). See the end of Section 10, p. 157.

Dictionary of the phrases and concepts

Almost Disjoint Hypergraphs: Any two different hyperedges intersect in at most one common point (like a collection of lines), p. 27.

Achievement Number: See Section 6, p. 104.

Avoidance Number: See Section 6, p. 104.

Backward Labeling: Extension of the labeling (**I** wins, or **II** wins, or **Draw**) to the whole Position-Graph (or Game-Tree), see Appendix C, p. 691.

Board of a Hypergraph Game: The underlying set of the hypergraph; it usually means the union set of all hyperedges (winning sets), p. 72.

Chooser Achievement Number: See Section 47, p. 623.

Chooser Avoidance Number: See Section 47, p. 625.

Chromatic Number: The chromatic number of a graph (hypergraph) is the minimum number of colors needed to color the vertices such that no edge (hyperedge) is monochromatic.

Classification of finite Hypergraphs: Six classes defined at the end of Section 5, p. 88.

Combinatorial Games: A very large class of games defined in Appendix C, p. 690.

Complete Analysis of a Game: Complete solution of a game—see Appendix C, p. 691.

Degree (Maximum, Average) of a Graph or Hypergraph: The degree of a vertex is the number of edges (hyperedges) containing the vertex. The Maximum (Average) Degree means to take the maximum (average) over all vertices.

Determined and Undetermined Games: Determined means that either the first player has a winning strategy, or the second player has a winning strategy, or both players have a drawing strategy. There exist infinite games which are undetermined – see Appendix C, p. 697.

Drawing Terminal Position: A Halving 2-Coloring of the board such that no winning set is monochromatic, p. 51.

Drawing Strategy: By using such a strategy a player cannot lose a play (i.e., either wins or forces a draw), p. 680.

End-Position of a Partial Play: The last position of a partial play, p. 678.

Game: A set of rules.

Game-Tree: See Appendix C, p. 686.

Hales-Jewett Conjecture(s): If the number of points is at least twice as large as the number of winning lines in an n^d Tic-Tac-Toe game, then the second player can force a draw (or even a pairing draw) – see Section 34, p. 463.

Hypergraph: A collection of sets; the sets are called hyperedges or winning sets.

Linear-Graph Game: See Appendix C, p. 691.

Pairing Criterions: Criterions which guarantee a draw-forcing pairing strategy – see Theorems 11.1–11.2, p. 164.

Pairing Draw (Draw-Forcing Pairing): A decomposition of the board (or a subset of the board) into pairs such that every winning set contains a pair, p. 163.

Partial Play: It means an initial segment of a play with "history" – see Appendix C, p. 678.

Picker Achievement Number: See Section 22, p. 322.

Picker Avoidance Number: See Section 47, p. 623.

Position: See Appendix C, p. 678.

Position-Graph: See Appendix C, p. 688.

Power-of-Two Scoring System: The basic idea of the proof of the Erdős–Selfridge Theorem (Theorem 1.4), pp. 28 and 152.

Proper 2-Coloring: A 2-coloring of the board with the property that there is no monochromatic winning set, p. 51.

Ramsey Criterions: See Theorems 6.1-2, pp. 98–9.

Resource Counting: See Section 10, p. 149.

Reverse Games: Avoidance versions, p. 76.

Strategy Stealing: See Section 5 and Appendix C, p. 74.

Strong Draw: When a player, usually the second player (or Breaker), can occupy at least one point from every winning set, p. 46.

Terminal Position: The "end" of a play; when the board is completely occupied.

Weak Win: A player can completely occupy a winning set, but not necessarily first, p. 46.

Weak Win Number: See Section 6, p. 97.

Winning Sets: An alternative name for the hyperedges of a hypergraph.

Winning Strategy: A player using such a strategy wins every possible play.

Win Number: See Section 6, p. 97.

References

Ajtai, M., L. Csirmaz, and Zs. Nagy (1979), On the generalization of the game Go-Moku, *Studia Sci. Math. Hungarica*, **14**, 209–226.

Alon, N., M. Krivelevich, J. Spencer and T. Szabó (2005), Discrepancy Games, *Electronic Journal of Combinatorics*, **12** (1), publ. R51.

Alon, N. and J. Spencer (1992), *The Probabilistic Method*, New York: Academic Press.

Baumgartner, J., F. Galvin, R. Laver and R. McKenzie (1973), Game theoretic versions of partition relations, *Colloq. Math. Soc. János Bolyai: "Infinite and Finite Sets,"* Hungary: Keszthely, pp. 131–135.

Beasley, J. (1992), *The Ins and Outs of Peg Solitaire: Recreations in Mathematics*, Vol. III, Oxford, New York: Oxford University Press.

Beck, J. (1981a), On positional games, *J. of Comb. Theory*, **30** (*ser. A*), 117–133.

Beck, J. (1981b), Van der Waerden and Ramsey type games, *Combinatorica*, **2**, 103–116.

Beck, J. (1982), Remarks on positional games – Part I, *Acta Math. Acad. Sci. Hungarica*, **40**, 65–71.

Beck, J. (1983), There is no fast method for finding monochromatic complete sub-graphs, *J. of Comb. Theory*, **34** (*ser. B*), 58–64.

Beck, J. (1985), Random Graphs and positional games on the complete graph, *Annals of Discrete Math.*, **28**, 7–13.

Beck, J. (1991), An algorithmic approach to the Lovász Local Lemma: I, *Random Structures and Algorithms*, **2**, 343–365.

Beck, J. (1993a), Parallel matching complexity of Ramsey's theorem, *DIMACS Series*, 13, AMS, 39–50.

Beck, J. (1993b), Achievement games and the probabilistic method, *Combinatorics, Paul Erdős is Eighty, Keszthely*, Vol. I, Bolyai Soc. Math. Studies, pp. 51–78.

Beck, J. (1993c), Deterministic graph games and a probabilistic intuition, *Combinat. Probab. Comput.*, **3**, 13–26.

Beck, J. (1996), Foundations of positional games, *Random Structures and Algorithms*, **9**, 15–47.

Beck, J. (2002a), Ramsey games, *Discrete Math.*, **249**, 3–30.

Beck, J. (2002b), The Erdős–Selfridge Theorem in positional game theory, *Bolyai Society Math. Studies, 11: Paul Erdős and His Mathematics, II*, Budapest, pp. 33–77.

Beck, J. (2002c), Positional games and the second moment method, *Combinatorica*, **22** (2), 169–216.

Beck, J. (2005), Positional games, *Combinatorics, Probability and Computing*, **14**, 649–696.

Beck, J. and L. Csirmaz (1982), Variations on a game, *J. of Comb. Theory*, **33** (*ser. A*), 297–315.

Bednarska, B. and T. Luczak (2000), Biased positional games for which random strategies are nearly optimal, *Combinatorica*, **20** (4), 477–488.

Berge, C. (1976), Sur les jeux positionelles, *Cahiers Centre Études Rech. Opér.*, **18**.

Berlekamp, E. R. (1968), A construction for partitions which avoid long arithmetic progressions, *Canadian Math. Bull.*, **11**, 409–414.

Berlekamp, E. R. (1991), Introductory overview of mathematical Go endgames, *Combinatorial Games*, Proceedings of Symposia in Applied Mathematics, Vol. 43, AMS, pp. 73–100.

Berlekamp, E. R., J. H. Conway and R. K. Guy (1982), *Winning Ways*, Vols 1–2, London: Academic Press.

Bollobás, B. (1982), Long paths in sparse Random Graphs, *Combinatorica*, **2**, 223–228.

Bollobás, B. (1985), *Random Graphs*, London: Academic Press.

Cameron, P. J. (1994), *Combinatorics: Topics, Techniques, Algorithms*, Cambridge University Press.

Chvátal, V. and P. Erdős (1978), Biased positional games, *Annals of Discrete Math.*, **2**, 221–228.

Conway, J. H. (1976), *On Numbers and Games*, London: Academic Press.

Dumitrescu, A. and R. Radoicic (2004), On a coloring problem for the integer grid, *Towards a Theory of Geometric Graphs*, edited by J. Pach, *Contemporary Mathematics*, Vol. 342, AMS, pp. 67–74.

Edmonds, J. (1965), Lehman's Switching Game and a theorem of Tutte and Nash-Williams, *J. Res. Nat. Bur. Standards*, **69B**, 73–77.

Erdős, P. (1947), Some remarks on the theory of graphs, *Bull. Amer. Math. Soc.*, **53**, 292–294.

Erdős, P. (1961), Graph Theory and probability, II, *Canadian Journ. of Math.*, **13**, 346–352.

Erdős, P. (1963), On a combinatorial problem, I, *Nordisk Mat. Tidskr.* **11**, 5–10.

Erdős, P. and L. Lovász (1975), Problems and results on 3-chromatic hypergraphs and some related questions, in *Infinite and Finite Sets*, edited by A. Hajnal *et al.*, *Colloq. Math. Soc. J. Bolyai*, **11**, Amsterdam: North-Holland, pp. 609–627.

Erdős, P., A. Hajnal, and R. Rado (1965), Partition relations for cardinal numbers, *Acta Math. Acad. Sci. Hungar.* **16**, 93–196.

Erdős, P. and R. Rado (1952), Combinatorial theorems on classifications of subsets of a given set, *Proc. London Math. Soc.* **3**, 417–493.

Erdős, P. and J. Selfridge (1973), On a Combinatorial Game, *Journal of Combinatorial Theory*, **14** (*Ser. A*), 298–301.

Fraenkel, A. S. (1991), Complexity of games, *Combinatorial Games*, Proc. *Symposia in Applied Mathematics*, Vol. 43, AMS, pp. 111–153.

Gale, D. (1979), The game of Hex and the Brower fixed-point theorem, *American Math. Monthly* (December), 818–828.

Gardner, M. (2001), *The Colossal Book of Mathematics*, New York: Norton & Company. London.

Golomb, S. W. and A. W. Hales (2002), Hypercube Tic-Tac-Toe, *More on Games of No Chance*, MSRI Publications **42**, 167–182.

Gowers, T. (2001), A new proof of Szemerédi's theorem, *Geometric and Functional Analysis*, **11**, 465–588.

Graham, R. L., B. L. Rothschild and J. H. Spencer (1980), *Ramsey Theory*, Wiley-Interscience Ser. in Discrete Math., New York: Wiley.

Hales, A. W. and R. I. Jewett (1963), On regularity and positional games, *Trans. Amer. Math. Soc.*, **106**, 222–229.

Harary, F. (1982), Achievement and avoidance games for graphs, in *Graph Theory (Cambridge 1981)* Annals of Discrete Math. **13**, Amsterdam–New York: North-Holland, pp. 111–119.

Harary, F. (1993), Is Snaky a winner? *Combinatorics*, 2 (April), 79–82.

Hardy, G. H. and E. M. Wright (1979), *An Introduction to the Theory of Numbers,* 5th edition, Oxford: Clarendon Press.

Hefetz D., M. Krivelevich, and T. Szabó (2007), Avoider Enforcer games, *Journal of Combinatorial Theory,* **114** (ser. A), 840–853.

Janson S., T. Luczak, and A. Rucinski (2000), *Random Graphs,* Wiley-Interscience.

Keisler, H. J. (1969), Infinite quantifiers and continuous games, in *Applications of Model Theory to Algebra, Analysis and Probability,* New york: Holt, Rinehart & Winston, pp. 228–264.

Kleitman, D. J., and B. L. Rothchild (1972), A generalization of Kaplansky's game, *Discrete Math.,* **2**, 173–178.

Lehman, A. (1964), A solution to the Shannon switching game, *SIAM J. Appl. Math.,* **12**, 687–725.

Lovász, L. (1979), *Combinatorial Problems and Exercises,* North-Holland and Akadémiai Kiadó.

Nash-Williams C. St. J. A. (1961), Edge-disjoint spanning trees of finite graphs, *Journal London Math. Soc.,* **36**, 445–450.

von Neumann, J. and O. Morgenstern (1944), *Theory of Games and Economic Behavior,* Princeton, NJ: Princeton University Press.

O'Brian, G. L. (1978–79), The graph of positions for the game of SIM, *J. Recreational Math.,* **11** (1), 3–8.

Patashnik, O. (1980), Qubic: $4 \times 4 \times 4$ Tic-Tac-Toe, *Mathematics Magazine,* **53**: 202–216.

Pegden, W. (2005), A finite goal set in the plane which is not a Winner, manuscript, submitted to *Discrete Mathematics.*

Pekec, A. (1996), A winning strategy for the Ramsey graph game, *Comb. Probab. Comput.,* **5**, 267–276.

Pluhár, A. (1997), Generalized Harary Games, *Acta Cybernetica,* **13**, 77–83.

Pósa, L. (1976), Hamilton circuits in Random Graphs, *Discrete Math.,* **14**, 359–364.

Schmidt, W. M. (1962), Two combinatorial theorems on arithmetic progressions, *Duke Math. Journal,* **29**, 129–140.

Shelah, S. (1988), Primitive recursive bounds for van der Waerden numbers, *Journal of the American Math. Soc.* 1 3, 683–697.

Simmons, G. J. (1969), The game of SIM, *J. Recreational Math.,* **2:6**.

Székely, L. A. (1981), On two concepts of discrepancy in a class of Combinatorial Games, *Colloq. Math. Soc. János Bolyai 37 "Finite and Infinite Sets,"* North-Holland, pp. 679–683.

Székely, L. A. (1997), Crossing numbers and hard Erdős problems in discrete geometry, *Combinatorics, Prob. Comput.,* **6**, 353–358.

Tutte, W. T. (1961), On the problem of decomposing a graph into n connected factors, *Journal London Math. Soc.,* **36**, 221–230.

van der Waerden, B. L. (1927), Beweis einer Baudetschen Vermutung, *Niew Archief voor Wiskunde,* **15**, 212–216.

Zermelo, E. (1912), Uber eine Anwendung der Mengenlehre und der Theorie des Schachspiels, *Proceedings of the Fifth International Congress of Mathematicians,* Cambridge, pp. 501–504.

LaVergne, TN USA
05 November 2009
163244LV00001B/1/P